Lecture Notes in Networks and Systems

Volume 100

The series "Lecture Notes in Networks and Systems" publishes the latest developments in Networks and Systems—quickly, informally and with high quality. Original research reported in proceedings and post-proceedings represents the core of LNNS.

Volumes published in LNNS embrace all aspects and subfields of, as well as new challenges in, Networks and Systems.

The series contains proceedings and edited volumes in systems and networks, spanning the areas of Cyber-Physical Systems, Autonomous Systems, Sensor Networks, Control Systems, Energy Systems, Automotive Systems, Biological Systems, Vehicular Networking and Connected Vehicles, Aerospace Systems, Automation, Manufacturing, Smart Grids, Nonlinear Systems, Power Systems, Robotics, Social Systems, Economic Systems and other. Of particular value to both the contributors and the readership are the short publication timeframe and the world-wide distribution and exposure which enable both a wide and rapid dissemination of research output.

The series covers the theory, applications, and perspectives on the state of the art and future developments relevant to systems and networks, decision making, control, complex processes and related areas, as embedded in the fields of interdisciplinary and applied sciences, engineering, computer science, physics, economics, social, and life sciences, as well as the paradigms and methodologies behind them.

**** Indexing: The books of this series are submitted to ISI Proceedings, SCOPUS, Google Scholar and Springerlink ****

More information about this series at http://www.springer.com/series/15179

Rajesh Kumar Shukla · Jitendra Agrawal ·
Sanjeev Sharma · Narendra S. Chaudhari ·
K. K. Shukla
Editors

Social Networking and Computational Intelligence

Proceedings of SCI-2018

Springer

Editors
Rajesh Kumar Shukla
Department of Computer Science
and Engineering
Sagar Institute of Research and Technology
Bhopal, Madhya Pradesh, India

Sanjeev Sharma
School of Information Technology
Rajiv Gandhi Technical University
(State Technological University)
Bhopal, Madhya Pradesh, India

K. K. Shukla
Department of Computer Science
and Engineering
Indian Institute of Technology BHU
Varanasi, Uttar Pradesh, India

Jitendra Agrawal
Department of Computer Science
and Engineering, University Teaching
Department
Rajiv Gandhi Technical University
(State Technological University)
Bhopal, Madhya Pradesh, India

Narendra S. Chaudhari
Department of Computer Science
and Engineering
Indian Institute of Technology Indore
Indore, Madhya Pradesh, India

Visvesvaraya National Institute
of Technology
Nagpur, Maharashtra, India

ISSN 2367-3370 ISSN 2367-3389 (electronic)
Lecture Notes in Networks and Systems
ISBN 978-981-15-2070-9 ISBN 978-981-15-2071-6 (eBook)
https://doi.org/10.1007/978-981-15-2071-6

This Springer imprint is published by the registered company Springer Nature Singapore Pte Ltd.
The registered company address is: 152 Beach Road, #21-01/04 Gateway East, Singapore 189721, Singapore

Contents

Social Networking

Editors and Contributors

About the Editors

Dr. Rajesh Kumar Shukla is the Dean (R&D) and Head of the Department of Computer Science and Engineering, Sagar Institute of Research and Technology, Bhopal. He holds B.E. (CSE), M.Tech. (CSE), and Ph.D. (CSE) degrees, and has served as the Head of Department, Dean, and Vice-Principal of various institutions. He is authored several books, including Analysis and Design of Algorithms (A beginners Approach), Data structure and Files, Basics of Computer Engineering, Data Structure using C and C++, Object Oriented Programming in C++ (all published with Wiley India) and Theory of Computation and Formal Languages and Automata Theory (Published with Cengage Learning). His research interests include recommendation systems, social networking, machine learning, computational intelligence, and data mining, and he has published over 40 papers in international journals and conferences. He has received several awards, including the Chapter Patron Award by the Computer Society of India in 2018, Significant Contribution Award under CSI Service Award by CSI India in 2017, ISTE U.P. Government National Award in 2015, and Bharat Excellence Award in 2015. He has also been active a number of professional societies, and is currently the Chairman of ACM and CSI Bhopal Chapter.

Dr. Jitendra Agrawal works at the Department of Computer Science & Engineering at the Rajiv Gandhi Proudyogiki Vishwavidyalaya, MP, India. He is a teacher, researcher, and consultant in the field of computer science and information technology. His research interests include databases, data structure, data mining, soft computing, and computational intelligence. He has published more than 60 papers in international journals and conferences along with two books entitled "Data Structures" and "Advanced Database Management System".

Dr. Sanjeev Sharma works as a Professor and the Head of the School of Information Technology at Rajiv Gandhi Proudyogiki Vishwavidyalaya, Bhopal, India, and has received the World Education Congress's Best Teacher Award in Information Technology. He graduated in Electrical & Electronics from Samrat Ashok Technical Institute, India, and holds a postgraduate qualification in Microwave and Millimeter from Maulana Azad College of Technology, India. He completed his Doctorate in Information Technology at Rajiv Gandhi Proudyogiki Vishwavidyalaya. He has over 29 years' teaching and research experience and his areas of interest include mobile computing, ad hoc networks, data mining, image processing, and information security. He has edited the proceedings of several national and international conferences and published more than 150 research papers in respected journals.

Dr. Narendra S. Chaudhari is an established researcher in Computer Science and Engineering, and has made significant contributions to engineering education as an institute developer and to professional societies. As a Dean of the Faculty of Engineering Sciences, Devi Ahilya Vishwavidyalaya (DAVV), Indore, from 1995 to 1998, he initiated the Institute of Engineering and Technology, which is now the leading engineering institute in central India. At VNIT Nagpur, he promoted institute-wide research with multi-disciplinary projects, student mentorship programs, and involvement of alumni in entrepreneurship among students. He also founded the innovation center at VNIT Nagpur and led product development that resulted in patents and technology transfer for engineering products. He has also been involved in technical education at national level as: (i) chairman of the Central Regional Committee, AICTE, MHRD, Government of India and (ii) co-convener and secretary of the standing council of NITs, MHRD, Government of India. His research contributions are in the areas of network security and mobile computing, game AI, novel neural network models like binary neural nets and bidirectional nets, context-free grammar parsing, optimization, and graph isomorphism problems.

Dr. Narendra S. Chaudhari was a member of the academic delegation for the Honorable President of India's state visits to Sweden and Belarus in 2015, and to the People's Republic of China in 2016. He was also part of FICCI's higher education delegation to Germany, France, and the Netherlands for in 2015. He represented VNIT, Nagpur, at the first BRICS-Network University (NU) Conference at Yekaterinburg, Russia.

He has published more than 340 papers in journals and conferences, and completed eight R&D projects, funded by DST, UGC, AICTE, and MHRD. He has been a reviewer for DST and UGC projects and has contributed collaborative research for other pilot projects on computing techniques and industry interaction funded by ST-Engg, DSTA, and A*STAR in Singapore.

Dr. K. K. Shukla is a Professor of Computer Science and Engineering and Dean (Faculty Affairs), Indian Institute of Technology, BHU, Varanasi. He has 35 years of research and teaching experience. Professor Shukla has published more than 160 research papers in leading journals and conferences. He has written 5 books, and contributed chapters to or edited many other books. He holds 24 intellectual property rights in the area of socially relevant computing.

Contributors

Jayashree Agarkhed P.D.A College of Engineering, Kalaburagi, Karnataka, India

Navanshu Agarwal Department of Computer Science and Engineering, National Institute of Technology Hamirpur, Hamirpur, Himachal Pradesh, India

Vidushi Agarwal Department of Computer Science and Engineering, National Institute of Technology Hamirpur, Hamirpur, Himachal Pradesh, India

Chetan Agrawal Computer Science and Engineering, RITS Bhopal, Bhopal, India

Lalita Agrawal Maulana Azad National Institute of Technology, Bhopal, India

Madhuri Agrawal UIT, RGPV, Bhopal, Madhya Pradesh, India

Shikha Agrawal UIT, RGPV, Bhopal, Madhya Pradesh, India

Faraz Ahmad Department of Computer Science, Jamia Millia Islamia, New Delhi, India

Sadare Alam Maulana Azad National Institute of Technology, Bhopal, India

Abdul Alim Department of Computer Science and Applications, Dr. Harisingh Gour Vishwavidyalaya, Sagar, Madhya Pradesh, India

Amisha National Institute of Technology Jamshedpur, Jamshedpur, India

Jayesh Amrutphale Malwa Institute of Technology, Indore, India

Yashasvee Amrutphale Malwa Institute of Technology, Indore, India;
Rajiv Gandhi Proudyogiki Vishwavidyalaya, Bhopal, India

Ridam Arora Department of Computer Science and Engineering, National Institute of Technology Hamirpur, Hamirpur, Himachal Pradesh, India

Goldi Bajaj Sardar Vallabhbhai Polytechnic College, Bhopal, India

Noopur Ballal Department of Computer Science and Engineering, Maulana Azad National Institute ofTechnology, Bhopal, India

Pavan Bansal National Institute of Technology Hamirpur, Hamirpur, Himachal Pradesh, India

Pratosh Bansal Department of Information Technology, IET DAVV, Indore, Madhya Pradesh, India

Jyoti Bharti Maulana Azad National Institute of Technology, Bhopal, India

Vishal Bhatnagar Ambedkar Institute of Advanced Communication Technologies and Research, New Delhi, India

Rajesh Kumar Chakrawarti Department of Computer Engineering, IET, DAVV, Indore, Madhya Pradesh, India

Satish Chand School of Computer and Systems Sciences, Jawaharlal Nehru University, New Delhi, India

Neeraj Chandnani Devi Ahilya University, Indore, Madhya Pradesh, India; Military College of Telecommunication Engineering, Mhow, Madhya Pradesh, India

Yash Chaudhary National Institute of Technology Hamirpur, Hamirpur, Himachal Pradesh, India

Sujata Chiwande Department of Electronics and Telecommunication Engineering, YCCE, Nagpur, Nagpur, India

Lokesh Chouhan Department of Computer Science and Engineering, National Institute of Technology Hamirpur, Hamirpur, Himachal Pradesh, India

Satyendra Singh Chouhan Shri Govindram Seksaria Institute of Technology and Science, Indore, India

Rattan K. Datta Mohyal Educational and Research Institute of Technology, New Delhi, India

Manjima De Sarkar Department of Electronics and Communication Engineering, National Institute of Technology Arunachal Pradesh, Yupia, Arunachal Pradesh, India

Rekha Deshlahre Maulana Azad National Institute of Technology, Bhopal, Madhya Pradesh, India

Apurva Deshpande P.D.A College of Engineering, Kalaburagi, Karnataka, India

Himanshu Dhiman Department of Computer Science and Engineering, National Institute of Technology Hamirpur, Hamirpur, Himachal Pradesh, India

Atrayee Dutta Department of Electronics and Communication Engineering, National Institute of Technology Arunachal Pradesh, Yupia, Arunachal Pradesh, India

Divya Gautam Amity University Madhya Pradesh, Gwalior, India

Richa Golash E&I Department, Samrat Ashok Technological Institute, Vidisha, Madhya Pradesh, India

Monika Rani Golla Department of Computer Science and Engineering, Visvesvaraya National Institute of Technology, Nagpur, India

Chetan Gupta Department of Computer Science and Engineering, SIRTS, Bhopal, India

Shikhin Gupta Department of Computer Science and Engineering, National Institute of Technology Hamirpur, Hamirpur, Himachal Pradesh, India

Puneet Himthani Department of Computer Science and Engineering, TIEIT (TRUBA), Bhopal, India

Manoj S. Ishi Department of Computer Engineering, R. C. Patel Institute of Technology, Shirpur, Maharashtra, India

Aditi Jain Shri Govindram Seksaria Institute of Technology and Science, Indore, India

Sweta Jain Maulana Azad National Institute of Technology, Bhopal, India

Yogendra Kumar Jain E&I Department, Samrat Ashok Technological Institute, Vidisha, Madhya Pradesh, India

Rekh Ram Janghel National Institute of Technology Raipur, Raipur, India

Nisheeth Joshi Department of Computer Science and Engineering, Banasthali Vidyapith, Vanasthali, Rajasthan, India

Dhruba Jyoti Kalita Gaya College of Engineering, Gaya, India

Kanishka National Institute of Technology Jamshedpur, Jamshedpur, India

Vandana Kate Institute of Engineering and Technology, Indore, India

Ashish K. Kaushal Department of Computer Science and Engineering, National Institute of Technology Hamirpur, Hamirpur, Himachal Pradesh, India

Chandrakant N. Khairnar Faculty of Communication Engineering, Military College of Telecommunication Engineering, Mhow, Madhya Pradesh, India

Mohammed Sameer Khan Department of Computer Science and Engineering, Oriental Group of Institutes, Bhopal, India

Vasima Khan SIRT, Bhopal, India

Yunus Khan Shri Govindram Seksaria Institute of Technology and Science, Indore, Madhya Pradesh, India

Ankit Khede Malwa Institute of Technology, Indore, India

Anand Kumar Department of Computer and Information Sciences, J. R. Handicapped University, Chitrakoot, Uttar Pradesh, India

Rakesh Kumar Computer Science and Engineering, Rabindranath Tagore University, Bhopal, India

Sanjay Kumar National Institute of Technology Jamshedpur, Jamshedpur, India

Shishir Kumar Computer Science and Engineering, Jaypee University of Engineering and Technology, Guna, India

Vinay Kumar National Institute of Technology Jamshedpur, Jamshedpur, India

Amit Kumar Mishra Department of Computer Science and Engineering, Banasthali Vidyapith, Vanasthali, Rajasthan, India

Sangeeta Kumari National Institute of Technology Raipur, Raipur, Chhattisgarh, India

Shailja Kumari National Institute of Technology Jamshedpur, Jamshedpur, India

Mohit Kushwaha Computer Science and Engineering, RITS Bhopal, Bhopal, India

Bimalkant Lauhny National Institute of Technology Hamirpur, Hamirpur, Himachal Pradesh, India

Jitendra Madarkar Department of Computer Science and Engineering, Visvesvaraya National Institute of Technology, Nagpur, India

Shruti Mahajani Malwa Institute of Technology, Indore, India;
Rajiv Gandhi Proudyogiki Vishwavidyalaya, Bhopal, India

Priti Maheshwary Computer Science and Engineering, Rabindranath Tagore University, Bhopal, India

Timothy Malche Computer Science and Engineering, Rabindranath Tagore University, Bhopal, India

Vijay Malviya Malwa Institute of Technology, Indore, India;
Rajiv Gandhi Proudyogiki Vishwavidyalaya, Bhopal, India

Mohan Rao Mamidkar National Institute of Technology Jamshedpur, Jamshedpur, India

Monica Manviya Maulana Azad National Institute of Technology, Bhopal, India

Iti Mathur Department of Computer Science and Engineering, Banasthali Vidyapith, Vanasthali, Rajasthan, India

Aiman Meenai UIT-RGPV, Bhopal, India

Ram Milan Department of Computer Science and Applications, Dr. Harisingh Gour Vishwavidyalaya, Sagar, Madhya Pradesh, India

Subhra Shriti Mishra Maulana Azad National Institute of Technology, Bhopal, Madhya Pradesh, India

Sarla More Dr. Harisingh Gour University, Sagar, Madhya Pradesh, India

Anand Motwani VIT Bhopal University, Sehore, India

Sushma Nagdeote Department of Electronics Engineering, Fr. CRCE, Mumbai, India

Sairaj Nemmaniwar National Institute of Technology Jamshedpur, Jamshedpur, India

Shivam Nilosey Malwa Institute of Technology, Indore, India

Shalini Zanzote Ninoria Department of Mathematics and Computer Science, RDVV, Jabalpur, Madhya Pradesh, India;
Department of Applied Mathematics, Jabalpur Engineering College, Jabalpur, Madhya Pradesh, India

Tayyaba Nousheen P.D.A College of Engineering, Kalaburagi, Karnataka, India

Kavita Pabreja Maharaja Surajmal Institute, GGSIPU, New Delhi, India

A. P. Padhy National Institute Technology, Raipur, India

Nikhil Paliwal Madhav Institute of Technology and Science, Gwalior, Madhya Pradesh, India

Kamlesh Kumar Pandey Department of Computer Science and Applications, Dr. Harisingh Gour Vishwavidyalaya, Sagar, Madhya Pradesh, India

Manjaree Pandit Madhav Institute of Technology and Science, Gwalior, Madhya Pradesh, India

Ishita Parmar Department of Computer Science and Engineering, National Institute of Technology Hamirpur, Hamirpur, Himachal Pradesh, India

Shadab Pasha Khan Department of Information Technology, Oriental Group of Institutes, Bhopal, India

Margi Patel IIST, Indore, India

J. B. Patil Department of Computer Engineering, R. C. Patel Institute of Technology, Shirpur, Maharashtra, India

Abhishek Pipliya Malwa Institute of Technology, Indore, India

K. Shreesha Prabhu Maulana Azad National Institute of Technology, Bhopal, India

Aditi Biswas Purba National Institute of Technology Jamshedpur, Jamshedpur, India

Lalit Purohit Shri Govindram Seksaria Institute of Technology and Science, Indore, India

Kapil Dev Raghuwanshi Department of Computer Science and Engineering, TIEIT (TRUBA), Bhopal, India

Prerana Rai Computer Science and Engineering, Jaypee University of Engineering and Technology, Guna, India

Nidhi Rajak Department of Computer Science and Applications, Dr. Harisingh Gour Vishwavidyalaya, Sagar, Madhya Pradesh, India

Anand Rajavat Department of Computer Science and Engineering, SVIIT, SVVV, Indore, Madhya Pradesh, India

Akhtar Rasool Maulana Azad National Institute of Technology, Bhopal, Madhya Pradesh, India

Chandra Prakash Rathore National Institute of Technology Raipur, Raipur, India

Nitin Rathore IIST, Indore, India

Pavan Rathore Malwa Institute of Technology, Indore, India

Amit Rege Medicaps University, Indore, India

S. A. M. Rizvi Department of Computer Science, Jamia Millia Islamia, New Delhi, India

Sahadev Roy Department of Electronics and Communication Engineering, National Institute of Technology Arunachal Pradesh, Yupia, Arunachal Pradesh, India

Nidhi Ruthia Department of Computer Science and Engineering, SIRTS, Sagar Group of Institute, Bhopal, India

Santosh Sahu School of Information Technology, Rajiv Gandhi Proudyogiki Vishwavidyalaya, Bhopal, Madhya Pradesh, India

Satya Prakash Sahu National Institute of Technology Raipur, Raipur, India

Harshit Saini National Institute of Technology Jamshedpur, Jamshedpur, India

Mukesh Sakle Shri Govindram Seksaria Institute of Technology and Science, Indore, India

Neeti Sangwan GGS Indraprastha University, Dwarka, India;
MSIT, New Delhi, India

Ankita Saraf P.D.A College of Engineering, Kalaburagi, Karnataka, India

Sri Khetwat Saritha Department of Computer Science and Engineering, Maulana Azad National Institute of Technology, Bhopal, India

Kanak Saxena Department of Computer Application, Samrat Ashok Technological Institute, Vidisha, India

Sheirsh Saxena Department of Computer Science and Engineering, National Institute of Technology Hamirpur, Hamirpur, Himachal Pradesh, India

Poonam Sharma Department of Computer Science and Engineering, Visvesvaraya National Institute of Technology, Nagpur, India

Pratyush Sharma Malwa Institute of Technology, Indore, India;
Rajiv Gandhi Proudyogiki Vishwavidyalaya, Bhopal, India

Priyamwada Sharma School of Information Technology, Rajiv Gandhi Proudyogiki Vishwavidyalaya, Bhopal, Madhya Pradesh, India

Sanjay Sharma Oriental Institute of Science & Technology, Bhopal, India

Urvashi Sharma IPS Academy, Indore, India

Vedant Sharma University Institute of Technology, Rajiv Gandhi Proudyogiki Vishwavidyalaya, Bhopal, Madhya Pradesh, India

Diwakar Shukla Department of Computer Science and Applications, Department of Mathematics and Statistics, Dr. Harisingh Gour Vishwavidyalaya, Sagar, Madhya Pradesh, India

Pragya Shukla Institute of Engineering and Technology, Indore, India

Summaiya Siddiqua P.D.A College of Engineering, Kalaburagi, Karnataka, India

Ravi Sindal IET Devi Ahilya University, Indore, India

Arun Pratap Singh The Right Click Services Pvt. Ltd., Bhopal, India

B. P. Singh Dayalbagh Educational Institute, Agra, Uttar Pradesh, India

Lokesh Singh National Institute of Technology Raipur, Raipur, India

Pardeep Singh School of Computer and Systems Sciences, Jawaharlal Nehru University, New Delhi, India

Shailendra Singh National Institute of Technical Teachers' Training & Research, Bhopal, Madhya Pradesh, India

Sonam Singh Parul Institute of Engineering and Technology, Vadodara, India

V. P. Singh National Institute Technology, Raipur, India

Vaishali Singh The Right Click Services Pvt. Ltd., Bhopal, India

Vibhav Prakash Singh Motilal Nehru National Institute of Technology, Allahabad, Prayagraj, India

Dharna Singhai Department of Computer Science and Engineering, SIRTS, Bhopal, India

Swarnangini Sinha Department of Computer Science and Engineering, Banasthali Vidyapith, Vanasthali, Rajasthan, India

Laxmi Srivastava Madhav Institute of Technology and Science, Gwalior, Madhya Pradesh, India

Lubna Tahreem P.D.A College of Engineering, Kalaburagi, Karnataka, India

Himani Thakur Department of Computer Science and Engineering, SVIIT, SVVV, Indore, Madhya Pradesh, India

S. S. Thakur Department of Mathematics and Computer Science, RDVV, Jabalpur, Madhya Pradesh, India;
Department of Applied Mathematics, Jabalpur Engineering College, Jabalpur, Madhya Pradesh, India

Usha D. Tikale PIET, Nagpur, India

Namita Tiwari Maulana Azad National Institute of Technology, Bhopal, Madhya Pradesh, India

Abhigyan Tiwary Department of Computer Science and Engineering, SIRTS , Sagar Group of Institute, Bhopal, India

Vrinda Tokekar IET, DAVV, Indore, India

Sunita Varma Shri Govindram Seksaria Institute of Technology and Science, Indore, Madhya Pradesh, India

Dinesh Kumar Verma Computer Science and Engineering, Jaypee University of Engineering and Technology, Guna, India

Nishant Vijayvargiya Malwa Institute of Technology, Indore, India;
Rajiv Gandhi Proudyogiki Vishwavidyalaya, Bhopal, India

Anshuma Yadav Department of Computer Science and Engineering, SVIIT, SVVV, Indore, Madhya Pradesh, India

Himanshu Yadav Computer Science and Engineering, RITS Bhopal, Bhopal, India

Snehlata Yadav Maulana Azad National Institute of Technology, Bhopal, India

S. D. Zade PIET, Nagpur, India

Cloud Computing

An Efficient Honey Bee Approach for Load Adjusting in Cloud Environment

Sangeeta Kumari and Shailendra Singh

Abstract Cloud computing is an Internet-based approach that delivers on-demand processing resources and information to the users in a shared mode. At the serving end, there is a prerequisite of proper scheduling and load adjusting to deal with the enormous measure of data. Our algorithm aims to distribute the equal load on each server in the cloud network and additionally enhances the asset usage. With the proposed approach, the honey bee inspired load adjusting (HBI-LA) method has been used for balancing the load of the virtual machine and schedule the task with respect of their priorities. Because of over-burdening of the task on a machine, there may be a chance of CPU crash. To overcome this problem, aging is applied to gradually enhance the priority of those jobs having longer waiting time as compared to the predefined time. At last, we compared our proposed work with the existing HBB-LB in terms of CPU time, execution time and waiting time. The examination of these three parameters demonstrates that the proposed algorithm requires less CPU time, less execution time and less waiting time than existing algorithm, hence it shows better performance and less energy consumption than the existing one.

Keywords Load balancing · Aging · Honey bee behavior · Cloud computing

1 Introduction

As the quantity of cloud clients is expanding in an exponential way, the duty of the cloud benefit provider is also increased to adjust the total workload among the different hubs in the cloud. Computing services are virtualized and delivered to the customer as a service. Weinman has given a term "*Cloudonomics*" [1], which define cloud computing from an economical perspective. Virtual machine (*VM*) gives a

S. Kumari (✉)
National Institute of Technology Raipur, Raipur, Chhattisgarh 492010, India
e-mail: sangeetak2606@gmail.com

S. Singh
National Institute of Technical Teachers' Training & Research, Bhopal, Madhya Pradesh 462002, India
e-mail: ssingh@nitttrbpl.ac.in

R. K. Shukla et al. (eds.), *Social Networking and Computational Intelligence*, Lecture Notes in Networks and Systems 100, https://doi.org/10.1007/978-981-15-2071-6_1

programmable framework [2]. *VMs* are allocated and deallocated to the cloud clients on interest. All the service models [3] provides the highest performance and load balancing, but the load balancing and performance is the serious parameter of cloud computing. The web is the prime prerequisite for using the cloud administrations, so an unavoidable issue is that framework bottleneck often happening when large information is exchanged over the network, it is essential to deal with every one of the assets like *CPU*, memory, in a server effectively [4]. For a proficient system, the aggregate exertion and the preparation time for all the client solicitations ought to be as low as could be expected under the circumstances, while being capable to deal with the different influencing requirements, for example, heterogeneity and high system delays [5, 6]. Specialized objectives of load adjusting for the most part manages issues identified with the registering system, i.e. all the specialized issues with respect to the figuring system.

A load balancer must ensure that the framework ought to be steady all through the calculation work in the system. To have the capacity to receive all the future adjustments in the framework, it might augment the assets in the system or it might help up the limit of effectively existing assets. A heap balancer must be fit for guaranteeing the accessibility of the administrations or assets at whatever point required by the client [7]. Commercial cloud solutions have boosted dramatically in the last few years and promoted organization reallocation from company-owned resources for peruse service-based models. Some of the most popular cloud projects are Amazon *EC2* [8], Amazon *S3* [9], Google App Engine [10], Map Reduce [11] etc. and some of the active projects include XtreemOS [12], OpenNebula [13], etc.

Load balancing is a standout amongst the most critical elements which influence the general execution of the framework. It can give client better nature of administrations and cloud specialist cooperative can have higher throughput with better asset use [14]. It is essential for each cloud specialist cooperative to make its load balancer working in the most ideal way to enhance the performance of cloud services and to reduce the load in cloud architecture. The system's focal point of consideration was on enhancing the basic execution parameters like *CPU* time, execution time, waiting time and performance of the system. In the proposed work, we consider load adjusting problem in which we applied honey bee inspired load adjusting (*HBI-LA*) method with concept of aging to balance the load between *VMs* and schedule the job with higher priorities.

The remaining work is illustrated as follows: Sect. 2, presents a brief discussion of the related work on load balancing in cloud computing. In Sect. 3, proposed work is presented. Result analysis has been discussed in Sect. 4. Section 5, conclude the proposed work. Section 6, presents future work.

2 Related Work

In this section, a brief overview of the load balancing and scheduling in cloud computing environment is discussed.

Kansal and Chana [15] focused on existing procedures went for diminishing related overhead, benefit reaction time and enhancing execution of the method. The paper gives insights about different parameters and every parameter play an essential part in overseeing the overall execution of the system. A more proficient load adjusting calculation *LB3M* [16], was proposed by Hung, et al. In [17], the thought is to locate the best cloud asset by utilizing co-operative power aware scheduled load balancing. In *PALB* [18] method, usage rates of every node are assessed. This calculation has three segments: Balancing segment decides on the premise of use rates where virtual machines will be instantiated. Upscale section power-on the extra compute nodes and downscale section shutdowns the idle calculated nodes.

Accessibility of resources in a cloud domain, and additionally different components like scaling of resources and power utilization, in a distributed computing environment are one of the vital worries that needs an awesome consideration. Load adjusting strategies ought to be with the end goal that to get quantifiable upgrades in resources usage and accessibility of a distributed computing environment [19]. There are some methodologies that utilization load as a parameter for the distribution of the cloud assets, fuzzy based technique [20], *CLBDM* [21], active monitoring load balancer [22], evolution of gang scheduling [23], throttled load balancing [24], dynamic request management algorithm [25, 26]. In [27] dissected the performance of distributed computing administrations for investigative registering loads. They performed probes genuine experimental registering workloads on many-task computing (*MTC*) clients. *MTC* clients utilize loosely coupled applications including numerous errands to accomplish their technical ideas.

In [28], Bayes and clustering based scheme is applied for load adjusting which enhanced the throughput and performance of the system. In [29], dynamic weighted scheme have been considered to migrate the workload among *VMs* and also analysis the energy efficiency of the system using linear regression technique. It shows higher accuracy and more stability as compared with existing work. Chen et al. [30] have been illustrated the idea of dynamic balancing strategy to resolve the issue of static balancing approach.

Sethi et al. [31] has been developed a load adjusting technique using the concept of fuzzy logic system as a distributed computing system. In [32], ant colony optimization (*ACO*) technique have been proposed to balance the workloads between datacenters in a cloud computing system. Babu and Krishna [33] have been considered the idea of honey bee for load balancing in cloud computing environment. In this system, task is scheduled on the basis of priorities from one *VM* to another *VM*. It shows less execution time and also less overloading as compared to the existing system. In [34], priority pre-emptive scheduling with aging technique is used to overcome the starvation problem while scheduling the job from one place to another.

3 Proposed HBI-LA

The proposed load adjusting calculation goes for appropriating the aggregate workload from the diverse cloud clients among different nodes in the data center. The *VM* which are instantiated by the client are mapped onto the physical servers in the data center. A node in view of its arrangement permits a *VM* to be distributed to it. A specific *VM* allocation strategy is required to distribute the nodes to the various *VM*. At the time of allocation there might be a *CPU* crash because of overburden of the tasks, so to mitigate this issue we use the idea of honey bee which are discussed in Sect. 3.1. Further, we collaborate the *HBI-LA* with aging technique which is used to gradually increase the priority of those jobs which waits in the system longer with respect to their waiting time.

3.1 Overview of Honey Bee Method

The honey bee method depends on the behavior of the honey bees it has two sorts that are: finders and reapers. Finder first goes outside of the honeycomb and locates the honey sources, after searching they return to the honeycomb and do the waggle move to show the quality and amount of the honey sources and afterwards reapers goes outside the honeycomb and harvest the honey from those assets, in the wake of gathering the nectar they return and again, do the waggle move to demonstrate the amount of sustenance is left. For this situation, the server is gathered as a *VM* and every *VM* have a procedure queue, subsequent to handle the solicitation it computes the benefit and furthermore augments the throughput. The current workload of the *VM* is ascertained, then it chooses the *VM* status, whether it is underloaded, overloaded or adjusted by the current workload of the *VM* they are grouped. The priority of the undertaking is thought about after expelling from the over-burden *VM* which are sitting tight for the *VM*, after then the work is transferred to the underloaded *VM*. *HBI-LA* is used to adjust the load and expand the execution of the system [33]. The algorithmic steps of our proposed work are described in Sect. 3.2.

3.2 Description of Proposed HBI-LA *Scheme with Concept of Aging*

In this section, we present steps of the proposed technique for scheduling the task according to their priority.

Step 1: Estimate the capacity of a *VM* (C_k)

Capacity of a single *VM* is based on the available information that is number of cores, *mips* and bandwidth to identify the overloaded and underloaded *VMs* which is calculated using Eq. (1)

$$C_k = numP_k \times mipsP_k + ComBd_k \tag{1}$$

where C_k is the capacity of kth *VM*, $numP_k$, $mipsP_k$ are the number of cores and million instruction per second of all cores in VM_k respectively, $ComBd_k$ is the communication bandwidth of *VM*.

Step 2: Calculate the workload of a *VM* (L_k)

Number of jobs allocated to the single *VM* is known as workload, which is calculated by dividing number of jobs in a queue to the service rate of *VM* at time *ti*.

$$L_k = \frac{numJ}{S\_VM_k} \tag{2}$$

where $numJ$, S_VM_k are the number of jobs and service rate of *VM* at time *ti*.

Step 3: Find the processing time of all *VM* ($PrTi$)

First, we calculated the processing time of a *VM* which is defined as the ratio of workload and capacity of single *VM*.

$$PrTi_k = \frac{L_k}{C_k} \tag{3}$$

Next, we evaluated the processing time of all *VMs* with the help of total workload and capacity. It is used to check the processing time of all *VMs* at time *ti* and identify the overloaded machine. Next, we calculate the standard deviation of a load using Eq. (5)

$$PrTi = \frac{L}{C} \tag{4}$$

$$SD = \frac{1}{Z}\sum_{i=1}^{z}(PrTi_k - PrTi)^2 \tag{5}$$

where L_k, C_k are the capacity and load of kth *VM* respectively. Z is the number of *VMs*.

Step 4: Find the overburden group

At the point when the present workload of the *VM* bunch goes beyond the extreme capacity of the *VM* then it is overburdened and load balancing is unstable and also we check the stability of the system using Eq. (6).

$$if\ SD \le thv\ then\ system\ is\ stable$$
$$else$$

$$system\ is\ unstable \tag{6}$$

where thv is the threshold value which is in the range of [0–1].

Step 5: Apply aging concept

While scheduling jobs, priority of the recently arrived job is more than the existing job than non-pre-emptive scheduling takes the job and put it at the front of the queue, if this occurs again and again than there may be a chance of starvation. To overcome this problem, aging is applied to gradually enhance the priority of those jobs which are having longer waiting time as compared to the predefined time. Thus by applying aging it tends to enhance the priority of jobs which are idle on the system for longer time.

4 Performance Analysis

The performance of the system has been illustrated by implementing the concept in Java [35] using CloudSim [36] simulator. The simulation has been carried out to analyze the performance of two algorithms based on *CPU* time, execution time and waiting time. At first, simulation is done for the existing load balancing algorithm, namely *HBB-LB* and secondly, the proposed algorithm was simulated by varying various parameters which are *CPU* time, execution time and waiting time. Then the results of both algorithms are compared and a graphical analysis is also done in order to have a clear vision of both the schemes. The graphical analysis of both the algorithm is mentioned in Figs. 1, 2, 3, 4 and 5 in details.

In Fig. 1, we have taken number of user is 1, number of *VM* are 20 and number of cloudlets or request (*CL*) are 40 then plotted the graph between *CPU* time and

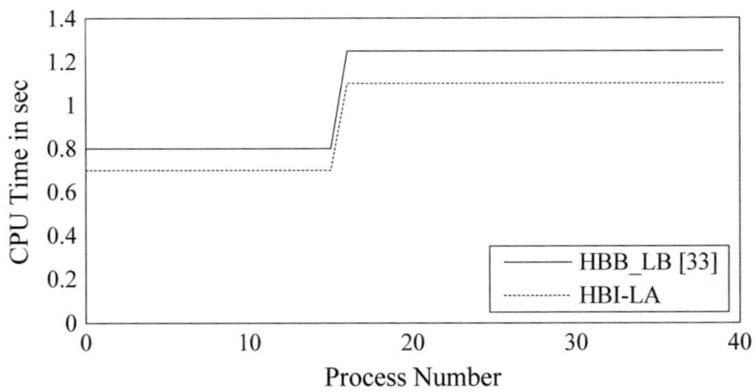

Fig. 1 *CPU* time versus process number

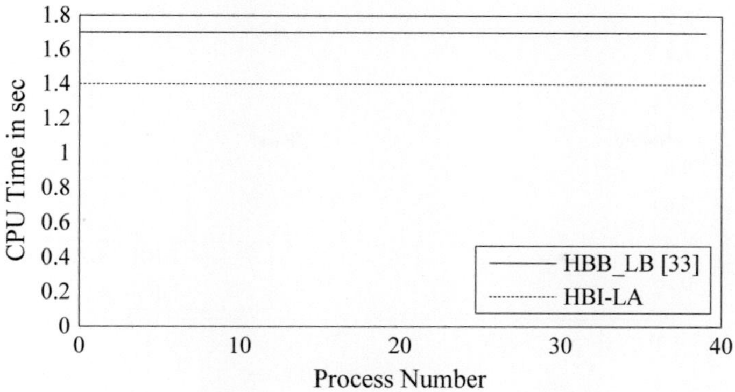

Fig. 2 *CPU* time versus process number

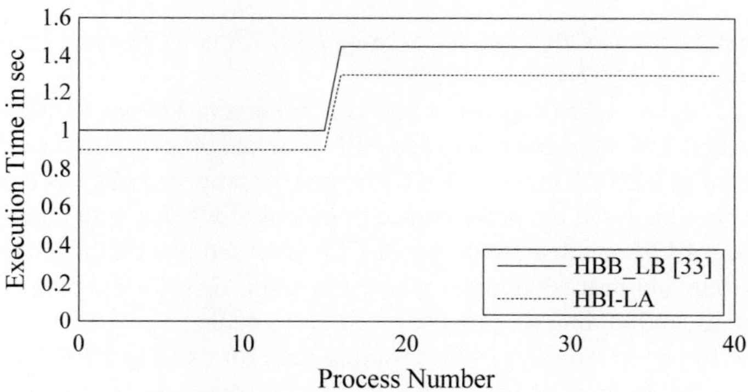

Fig. 3 Execution time versus process number

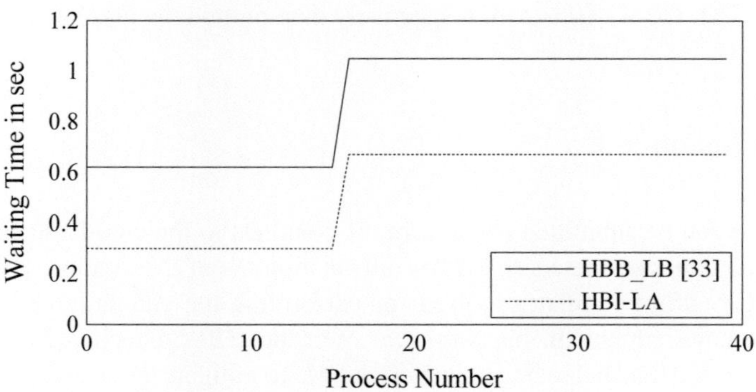

Fig. 4 Waiting time versus process number

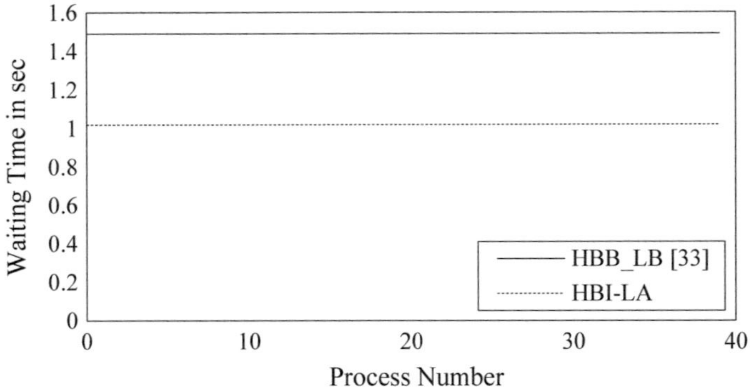

Fig. 5 Waiting time versus process number

process number where *CPU* time for existing *HBB-LB* is 43.28 s and *CPU* time for proposed *HBI-LA* is 37.04 s.

In Fig. 2, we have taken number of user is 1, number of *VM* are 10 and number of *CL* are 40 then plotted the graph between *CPU* time and process number where *CPU* time for existing *HBB-LB* is 66.8 s and *CPU* time for proposed *HBI-LA* is 57.19 s. In Fig. 3, we have analysed the performance of proposed scheme with number of user is 1, number of *VM* are 20 and number of *CL* are 40 and plotted the graph between execution time and process number where execution time for existing *HBB-LB* is 51.28 s and execution time for proposed *HBI-LA* is 45.04 s.

It can be observed from Fig. 4 that waiting time for existing *HBB-LB* is 35.28 s and waiting time for proposed *HBI-LA* is 21.04 s where number of user is 1, number of *VM* are 20 and number of *CL* are 40. In Fig. 5, waiting time for existing *HBB-LB* is 58.8 s and waiting time for proposed *HBI-LA* is 41.2 s with number of user is 1, number of *VM*, *CL* are 10 and 40 respectively then plotted the graph between them.

5 Conclusion

The request can be submitted in the form of cloudlets to the cloud datacenter. The cloudlets contain parameters that have information about the amount of resources required. CloudSim simulator is used for performing the simulation of the cloud computing environment. In our work we used the *HBI-LA* technique for energy consumption and load balancing between the *VM*. To mitigate the starvation problem, schedule the task from higher to lower priorities with the concept of aging. Also compared the proposed method with existing one in terms of three parameters that is *CPU*, execution and waiting time for 10 VMs and 20 VMs and plot the graph between these parameters and process number. After analysis of the results we found that the proposed algorithm shows better performance, while scheduling the task with priority.

6 Future Work

As the *IT* World is shifting towards the cloud computing, the demand of a new type of services is increasing day by day. The proposed approach can be further modified by setting an appropriate threshold to achieve power saving while making the overall system energy efficient.

References

1. Weinman J (2011) Cloudonomics: a rigorous approach to cloud benefit quantification. J Softw Technol Cloud Comput 14(4):44
2. Zeng W, Zhao Y, Ou K, Song W (2009) Research on cloud storage architecture and key technologies. In: Proceedings of the 2nd international conference on interaction sciences: information technology, culture and human ICIS'09, pp 1044–1048
3. Armbrust M, Fox A, Griffith R, Joseph A, Katz RH (2009) Above the clouds: a Berkeley view of cloud computing. Technical report UCB, 07–013. University of California, Berkeley
4. Armbrust M et al (2010) A view of cloud computing. Commun ACM 53(4):50
5. Randles M, Lamb D, Taleb-Bendiab A (2010) A comparative study into distributed load balancing algorithms for cloud computing. In: IEEE 24th international conference on advanced information networking and applications workshops, WAINA 2010, pp 551–556
6. Gopinath PPG, Vasudevan SK (2015) An in-depth analysis and study of load balancing techniques in the cloud computing environment. Procedia Comput Sci 50:427–432
7. Mesbahi M, Rahmani AM (2016) Load balancing in cloud computing: a state of the art survey. Int J Mod Educ Comput Sci 8(3):64–78
8. Amazon web services. https://aws.amazon.com/ec2/
9. Amazon simple storage service. https://aws.amazon.com/s3/
10. Google app engine. https://code.google.com/appengine/
11. Dean J, Ghemawat S (2008) MapReduce: simplified data processing on large clusters. Commun ACM 51(1):107
12. XtreemOS Linux based operating system. http://www.xtreemos.eu/
13. OpenNebula cloud management platform. https://dev.opennebula.org/
14. Ghomi EJ, Rahmani AM, Qader NN (2017) Load-balancing algorithms in cloud computing: a survey. J Netw Comput Appl 88:50–71
15. Kansal NJ, Chana I (2012) Existing load balancing techniques in cloud computing: a systematic review. J Inf Syst Commun 3(1):87–91
16. Hung C, Wang H, Hu Y (2012) Efficient load balancing algorithm for cloud computing network case study. In: International conference on information science and technology (IST 2012), Apr 2012, pp 28–30
17. Anandharajan T, Bhagyaveni M (2011) Co-operative scheduled energy aware load-balancing technique for an efficient computational cloud. Int J Comput Sci 8(2):571–576
18. Galloway JM, Smith KL, Vrbsky SS (2011) Power aware load balancing for cloud computing. In: Proceedings of the world congress on engineering and computer science, vol I, Oct 2011, pp 122–128
19. Zenon C, Venkatesh M, Shahrzad A (2011) Availability and load balancing in cloud computing. In: International conference on computer and software modeling, IPCSIT, vol 14. IACSIT Press, Singapore, pp 134–140
20. Leontiou N, Dechouniotis D, Denazis S, Papavassiliou S (2018) A hierarchical control framework of load balancing and resource allocation of cloud computing services. Comput Electr Eng 67:235–251

21. Radojevic B, Zagar M (2011) Analysis of issues with load balancing algorithms in hosted (cloud) environments. In: 2011 proceedings of the 34th international convention MIPRO, pp 416–420
22. Sharma M, Sharma P (2012) Efficient load balancing algorithm in VM cloud environment, vol 8491, pp 439–441
23. Moschakis IA, Karatza HD (2012) Evaluation of gang scheduling performance and cost in a cloud computing system. J Supercomput 59(2):975–992
24. Tyagi V, Kumar T (2015) ORT broker policy: reduce cost and response time using throttled load balancing algorithm. Procedia Comput Sci 48:217–221
25. Panwar R, Mallick B (2015) Load balancing in cloud computing using dynamic load management algorithm. In: International conference on green computing and internet of things (ICGCIoT). IEEE, pp 773–778
26. Ningning S, Chao G, Xingshuo A, Qiang Z (2016) Fog computing dynamic load balancing mechanism based on graph repartitioning. China Commun 13(3):156–164
27. Iosup A, Ostermann S, Yigitbasi N, Prodan R, Fahringer T, Epema D (2011) Performance analysis of cloud computing services for many-tasks scientific computing. IEEE Trans Parallel Distrib Syst 22(6):931–945
28. Zhao J, Yang K, Wei X, Ding Y, Hu L, Xu G (2016) A heuristic clustering-based task deployment approach for load balancing using Bayes theorem in cloud environment. IEEE Trans Parallel Distrib Syst 27(2):305–316
29. Zuo L, Shu L, Dong S, Zhu C, Zhou Z (2017) Dynamically weighted load evaluation method based on self-adaptive threshold in cloud computing. Mob Netw Appl 22(1):4–18
30. Chen S-L, Chen Y-Y, Kuo S-H (2017) CLB: a novel load balancing architecture and algorithm for cloud services. Comput Electr Eng 58:154–160
31. Sethi S, Sahu A, Jena SK (2012) Efficient load balancing in cloud computing using fuzzy logic. IOSR J Eng 2(7):2250–3021
32. Nishant K et al (2012) Load balancing of nodes in cloud using ant colony optimization. In: 2012 UKSim 14th international conference on computer modelling and simulation, pp 3–8
33. Babu LDD, Krishna PV (2013) Honey bee behavior inspired load balancing of tasks in cloud computing environments. Appl Soft Comput J 13(5):2292–2303
34. Satapathy SC et al (eds) (2014) ICT and critical infrastructure: proceedings of the 48th annual convention of computer society of India—volume I. Hosted by CSI Vishakapatnam chapter, vol 248
35. Varalakshmi P, Deventhiran H (2012) Integrity checking for cloud environment using encryption algorithm. In: 2012 international conference on recent trends in information technology, ICRTIT 2012, pp 228–232
36. Buyya R, Ranjan R, Calheiros RN (2009) Modeling and simulation of scalable cloud computing environments and the CloudSim toolkit: challenges and opportunities. In: Proceedings of the 2009 international conference on high performance computing & simulation, HPCS 2009, pp 1–11

A Novel Approach of Task Scheduling in Cloud Computing Environment

Nidhi Rajak and Diwakar Shukla

Abstract The recent era of technology is rapidly increasing with cloud environment, and it is just nothing but Internet computing via Internet technology. The demand of cloud computing is everywhere such as healthcare, physics, online marketing and so on. Scheduling problem is referred as NP-complete, and it is basically, allocation of the tasks in the available virtual machines. The primary objective of any task scheduling approach is to reduce the overall execution time. This paper presents an algorithm which is based on two steps process such as first step, to find the minimum value of all the tasks between entry task(s) and all nonentry tasks and to sort the tasks as per minimum value. This is basically used for priority of the tasks. Secondly, to allocate the tasks on available virtual machines. Proposed algorithm is tested on two different DAGs with ten tasks and fifteen tasks which gives better scheduling length as compared to heuristic algorithm such as HEFT algorithm.

Keywords Cloud computing · Task scheduling · DAG · Scheduling length · VM · EST

1 Introduction

Cloud computing is optimized the technology in the era of computing fields. It has usually linked with Internet technology due to its fast speed and reducing the cost in the present time. Cloud computing is also known as the Internet-based computing, and it is a unified computing resources based on a pay-per-use basis [1]. It has number of applications in various fields such astronomy, physics, bioinformatics, healthcare, DNA computing and various advanced scientific computing.

N. Rajak (✉) · D. Shukla
Department of Computer Science and Applications, Dr. Harisingh Gour Vishwavidyalaya, Sagar, Madhya Pradesh 470003, India
e-mail: nidhi.bathre@gmail.com

D. Shukla
e-mail: diwakarshukla@rediffmail.com

R. K. Shukla et al. (eds.), *Social Networking and Computational Intelligence*, Lecture Notes in Networks and Systems 100, https://doi.org/10.1007/978-981-15-2071-6_2

There has been three-stage evaluation for cloud computing development, which were distributed, parallel and grid computing [2]. It is the next generation of computing because it is based on two computing models such as *data sharing computing model and service sharing computing model* [1].

Cloud computing is again classified into two basic models such as deployment model and service model. Deployment model [3–5] is classified into four categories such as *community cloud, public cloud, private cloud and hybrid cloud*. The above categories of this models are based on three factors such as location, infrastructure and availability.

Service model is also classified into three categories such as Software as a Service (SaaS), Platform as a Service (PaaS) and Infrastructure as a Service (IaaS). These models provide resource services to user on the basis: *Pay-Use Computing resources*. Here the resources are memory, servers, network resources etc. The objective of cloud computing is to get maximum benefits by using available resources.

Task scheduling in cloud computing area is recent and burning topic of research, and it is also known as NP-complete problem. The process of allocating the tasks in the available virtual machines is known as task scheduling. Task scheduling having three stage process such as find priority of the tasks of given DAG using priority attribute method, sort the tasks as per priority value of the tasks and finally allocate these tasks in the available virtual machines. There are number pre-assumed which are number of tasks in DAG known in advance, allocation of tasks statically, no deadline for tasks, task priority known in advance and allocation in batch mode as per priority. Various types of different task scheduling such as preemptive, non preemptive, static, dynamic, distributed, centralized, etc. which has their owned merits and demerits.

Task scheduling of the given DAG is done by using task scheduler, and the main role of task scheduler is to schedule the meta task (number of tasks) is mapped into available resources, i.e., virtual machines. Here in Fig. 1 shown role of task scheduler [6]. To reduce overall execution time in task scheduling is the primary objective.

The paper presents new task scheduling method which solve the scheduling problem, and it is based on minimum value all tasks of given DAG use as priority of the tasks. The proposed method works on two stages. First stage, it will compute minimum value of tasks using communication time among the tasks. Second stage is to sort tasks in increasing order as per the minimum value. The performance of this method gives better result in respect of scheduling length as compared with well-known heuristics algorithm HEFT [7].

This paper is orderly arranged as follows: Sect. 1 discussed basic cloud computing, task scheduling and its objective. Section 2 will define problem statement, basic terminologies and priority attributes. Section 3 shows task scheduling method with illustrative examples and finally concludes the paper in Sect. 4.

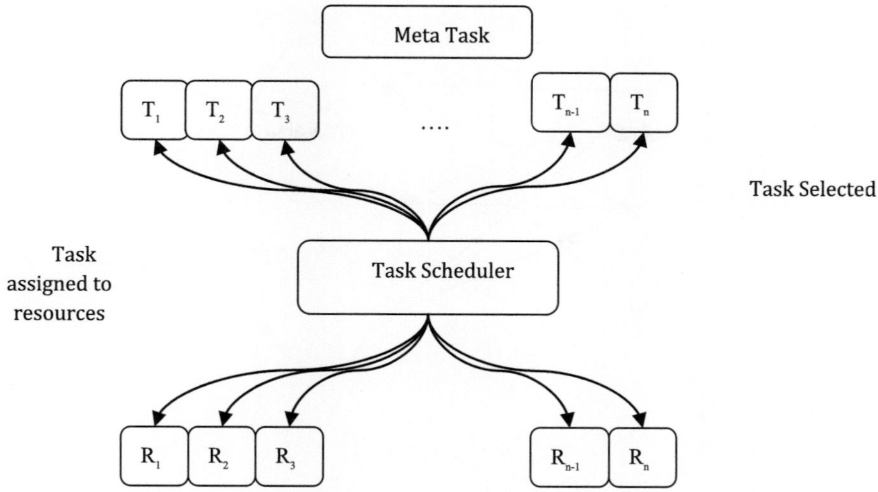

Fig. 1 Task scheduler

2 Scheduling Model

Scheduling model is discussed into four part such as application models, system resource model, attributes and objective function.

2.1 Application Model

Directed acyclic graph (DAG) is used to represent the given application model, and DAG can be defined as a graph G which consists three tuples such as finite number of tasks (T), communication time (CT) between the tasks and weight of communication link (W), i.e., $G = \{T, E, W\}$. Formally, $T = \{t_1, t_2, \ldots, t_n\}$ is a set of finite tasks, $E = \{e_{i,j} \in (t_i, t_j)\}$ is a set of dependency edges between any two task t_i and t_j. $W = \{\mathrm{CT}(t_i, t_j) | t_i, t_j \in T\}$ is the communication time between any two task t_i and t_j.

An entry task t_i of given DAG is defined as a task t_i, which does not consist any predecessor (pred) tasks, i.e., pred(t_i) = Φ. Similarly, an exit task t_i of given DAG is defined as a task t_i, which does not consist any successor (succ) tasks, i.e., succ(t_i) = Φ. DAG model should be stratified the precedence constraint which stated as task t_j cannot be started until execution of t_i has been finished.

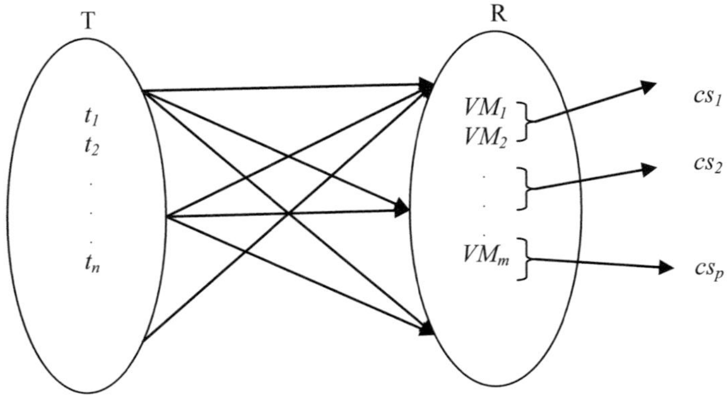

Fig. 2 Mapping T to R

2.2 System Resource Model

System resource computing model is basically cloud resource computing model. This model consists of p number of cloud servers, and it is represented by $S = \{cs_1, cs_2, \ldots, cs_p\}$, which again consists of m resources $R = \{VM_1, VM_2, \ldots, VM_m\}$. Formally, system resource model is mapping of given a set of tasks $T = \{t_1, t_2, \ldots, t_n\}$ of DAG to the available m resources $R = \{VM_1, VM_2, \ldots, VM_m\}$. This resources are considered as heterogeneous. The communication time between two virtual machine VM_i and VM_j is zero if they are in the same cloud server otherwise it will be taken.

A set task during processing in cloud model, it is assumed that there is no preemption between tasks, i.e., no any interrupt. The mapping between the tasks (T) and Resources (R) is shown in Fig. 2 [8].

2.3 Preliminaries of Task Scheduling

This section discusses some attributes and the basic notations which will be used in this paper.

i. ***Estimated Computation Time ECT*** [9]: It is defined as each task t_i is executed on a VM_j, and it is mathematically expressed as a ratio the size of task (MI) and the running speed of VM (MIPS). A matrix form representation of ECT [10] as follows:

$$\mathrm{ECT}_{ij} = \begin{bmatrix} \mathrm{ECT}_{11} & \mathrm{ECT}_{12} & \cdots & \mathrm{ECT}_{1n} \\ \mathrm{ECT}_{21} & \mathrm{ECT}_{22} & \ldots & \mathrm{ECT}_{2n} \\ \mathrm{ECT}_{m1} & \mathrm{ECT}_{m2} & \ldots & \mathrm{ECT}_{mn} \end{bmatrix} \quad (1)$$

where ECT_{ij} is estimated computation time of task t_i on resource R_j. The value of i is $1 \leq i \leq n$, and the value of j is $1 \leq j \leq m$.

ii. ***Earliest Start Time EST*** [9, 11]: It is defined as the time taken by each task on VM which should take minimum starting time. It is assumed that every entry tasks having EST will be zero because entry task did no any parents task. The EST of nonentry tasks can be expressed mathematically as follows:

$$\mathrm{EST}(t_i, \mathrm{VM}_j) = \left\{ \begin{array}{ll} 0 & \text{if } t_i \in t_{\text{entry}} \\ \max\limits_{t_j \in \mathrm{pred}(t_i)} \left\{ \mathrm{EFT}(t_j, \mathrm{VM}_j) + \mathrm{MET}(t_i) + \mathrm{CT}(t_i, t_j) \right\} & \text{otherwise} \end{array} \right\} \tag{2}$$

where $\mathrm{MET}(t_i)$ is defined as minimum execution time of task t_i on all available VM_m, i.e.,

$$\mathrm{MET}(t_i) = \min \cdot \{\mathrm{ECT}(t_i, \mathrm{VM}_m)\} \tag{3}$$

iii. ***Earliest Finished Time EFT*** [9]: It is computation of earliest finished time of a given task t_i on VM_m as follows (Table 1):

$$\mathrm{EFT}(t_i, \mathrm{VM}_j) = \mathrm{ECT}_{ij} + \mathrm{EST}(t_i, \mathrm{VM}_j) \tag{4}$$

Table 1 Notations used

Notation	Meaning
DAG	Directed acyclic graph
T	Set of finite tasks
n	Total n number of tasks
p	Total p number of cloud servers
m	Total m number of virtual machines
$\mathrm{CT}(t_i, t_j)$	Communication time between t_i and t_j
$\mathrm{Pred}(t_i)$	Predecessor of task t_i
VM_m	Virtual machine
cs	Cloud server
ECT	Estimated computation time
EST	Earliest start time
EFT	Earliest finished time
MET	Minimum execution time
$\mathrm{Sch}_{\mathrm{Len}}$	Scheduling length
MV	Minimum value of the tasks

2.4 *Task Scheduling Objective*

Task scheduling method is designed to mapping of n number of tasks of a given DAG into m number of resources in a such manner, so that it should be minimized overall execution time or makespan on cloud computing environment. Makespan can be defined as execution time by the exit task t_{exit} of a given DAG on available resource VM_j. It is also known as scheduling length. An objective function can be expressed as follows:

$$\text{Sch}_{\text{Len}} = \min\left\{\text{EFT}\left(t_{\text{exit}}, \text{VM}_j\right)\right\} \tag{5}$$

3 Proposed Algorithm

This section will be discussing on the proposed task scheduling algorithm in cloud computing environment, and algorithm gives optimal scheduling length, i.e., minimum scheduling length as compared to heuristic algorithm such as HEFT. It is designed for both single entry task and multiple entry tasks of given DAGs. Proposed algorithm works in two phases such as:
Phase-I: It will find the priority of the tasks by using communication time as minimum value between an entry task(s) and nonentry tasks. This value should be considered minimum while computing it. Let us consider Dis[1, 2, …, n] is a distance array of size n, which consists of minimum value of the task $T = \{t_1, t_2, \ldots, t_n\}$. Here, Dis [1] is for entry task and Dis[n] is for exit task in case of single entry task and Dis[1, 2, … m] is for m entry task and Dis[n] is for single exit task. MV is represented as minimum value of the tasks, and it is defined by summation of MV of marked task and communication time between marked task and its successor task. Initially, it is assumed that entry task(s) having distance zero because it has no any parents and all nonentry tasks are initialized with infinity. That is,

If DAG model having a single entry $\{t_1\}$, in that case MV of the tasks will be as follows:

$$\text{Dis}[1] = \text{MV}\left(t_{\text{entry}}\right) = 0, \text{Dis}[2] = \text{MV}(t_2),$$
$$\text{Dis}[3] = \text{MV}(t_3) = \infty, \ldots, \text{Dis}[n] = \text{MV}(t_n) = \infty$$

If DAG model having multiple m entry tasks $\{t_1, t_2, \ldots t_m\}$ in that case MV of the tasks will be as follows

$$\text{Dis}[1] = \text{MV}(t_1) = 0, \text{Dis}[2] = \text{MV}(t_2), \ldots \text{Dis}[m] = \text{MV}(t_m) = 0,$$
$$\text{Dis}[m+1] = \text{MV}(t_{m+1}), \ldots, \text{Dis}[n] = \text{MV}(t_n) = \infty$$

There are two cases which will be explained how to compute MV of the tasks, and it allocate in Dis[] array for entry task(s) and nonentry tasks.

Case 1: If task is entry task t_{entry} then find MV between t_{entry} and its successor task(s) t_i by using following

$$MV(t_i) = MV\left(t_{entry}\right) + CT\left(t_{entry}, t_i\right)$$
$$If\ Dis[i] > MV(t_i)$$
$$Dis[i] = MV(t_i)$$

Case 2: If task is nonentry task then find the MV of smallest task t_s in Dis[] which is unmarked.

$$Find\ MV\ between\ task\ t_s\ and\ its\ successor\ tasks\ t_i.$$
$$MV(t_i) = MV(t_s) + CT(t_s, t_i)$$
$$If\ Dis[i] > MV(t_i)$$
$$Dis[i] = MV(t_i)$$
$$Marked\ the\ smallest\ task.$$
$$Above\ steps\ will\ be\ repeated\ until\ all\ tasks\ of\ DAG\ will\ be\ marked.$$

Now, Dis[1, 2, …, *n*] have minimum value of all tasks and sort this tasks in increasing order as per Dis[] value. Allocate these sorted tasks in priority queue PQ.

Phase-II: This phase is basically mapping of tasks onto available Virtual Machine VM. Now, remove the tasks one by one from PQ and check its precedence constraint PC before allocate to the available Virtual Machine VM. The allocation of each task onto VM is done by computing EST and EFT as per proposed algorithm.

Detail of proposed algorithm is mentioned in Table 2.

3.1 Illustrative Examples

To understand proposed algorithm, we have taken two different examples of DAG models. DAG_1 model [9] consists ten tasks with fifteen communication links, i.e., dependency edges which is shown in Fig. 4 [9].

This model has to schedule in three, VM_1, VM_2 and VM_3, and these VMs are in two cloud servers CS_1 and CS_2. This model having single entry task t_1. The Estimated Computation Time (ECT) is also taken into account for corresponding VMs which is shown in Table 3 [10].

Now, we have to find distance array Dis[] for DAG_1 model which is shown in Table 4. This value is found as per first phase in proposed algorithm. Here, coloured value means that the task having the smallest MV from entry task and marked it.

Final MV of the tasks t_1, t_2, t_3, t_4, t_5, t_6, t_7, t_8, t_9, t_{10} are 0, 18, 12, 9, 11, 14, 35, 29, 24, 37. Sorted the tasks as per *MV* in increasing order as follows: t_1, t_4, t_5, t_3, t_6, t_2, t_9, t_8, t_7, t_{10}. These are called priority of the given tasks, and it should be

Table 2 Proposed Algorithm

Step 1. *Read DAG with n^{th} tasks*
Step 2. *Check whether given DAG having single entry task or multiple entry tasks.*
Step 3. *If Given DAG have single entry task then*
3.1. *If DAG model having a single entry{t_1}, in that case MV of the tasks will be as follows:* *Initialization entry task with zero and all non entry task with infinity.* $Dis[1] = MV(t_{entry}) = 0, Dis[2] = MV(t_2), Dis[3] = MV(t_3) = \infty, \ldots, Dis[n] = MV(t_n) = \infty$
3.2. a. *if task is entry then find MV between t_{entry} and its successor task(s) t_i by using following* $MV(t_i) = MV(t_{entry}) + CT(t_{entry}, t_i)$ *If Dis[i]>MV(t_i)* $Dis[i] = MV(t_i)$ ***b.*** *If task is nonentry task then Find the MV of smallest task t_s in Dis[] which is unmarked.* *Find MV between task t_s and its successor tasks t_i.* $MV(t_i) = MV(t_s) + CT(t_s, t_i)$ *If Dis[i]>MV(t_i)* *Dis[i]=MV(t_i)* *Marked the smallest task and this step b will be repeated until all tasks of DAG are marked*
Step 4. *Otherwise more than entry tasks then*
4.1. *If DAG model having multiple m entry tasks{$t_1, t_2, \ldots t_m$} in that MV of the tasks will be as follows* *Initialization all entry tasks with zero and all non entry task with infinity.* $Dis[1] = MV(t_1) = 0, Dis[2] = MV(t_2), \ldots Dis[m] = MV(t_m) = 0, Dis[m+1] = MV(t_{m+1}), \ldots, Dis[n] = MV(t_n) = \infty$
4.2. a. *if task is entry then find MV between t_{entry} and its successor task(s) t_i by using following* $MV(t_i) = MV(t_{entry}) + CT(t_{entry}, t_i)$ *If Dis[i]>MV(t_i)* $Dis[i] = MV(t_i)$ ***b.*** *If task is nonentry task then Find the MV of smallest task t_s in Dis[] which is unmarked.* *Find MV between task t_s and its successor tasks t_i.* $MV(t_i) = MV(t_s) + CT(t_s, t_i)$ *If Dis[i]>MV(t_i)* *Dis[i]=MV(t_i)* *Marked the smallest task and this step b will be repeated until all tasks of DAG are marked*
Step 5. *Repeat step. 3 to 4 until all tasks minimum distance will be found.*
Step 6. *Now we have minimum distance between entry task and all non entry tasks.*
Step 7. *Sort the tasks in increasing order as per minimum distance value of the tasks.*
Step 8. *Keep sorted tasks in priority queue(PQ).*
Step 9. *While PQ is not empty then*
Remove task from PQ
If task is not satisfied Precedence Constraint then
Removed task and insert at end of PQ
Else
Allocate to Available Virtual Machine with EST and EFT.
End while
Step 10. *Compute Scheduling length .i.e.* $Sch_{Len} = \min\{EFT(t_{exit}, VM)\}$
Step 11. *Stop.*

mapped into VMs as per proposed algorithm. So, the complete mapping of task-VM is shown in Fig. 3. The proposed algorithm gives 64 units of scheduling length which is minimum as compared to HEFT algorithm [7] gives 73 units (Fig. 4).

DAG_2 model [12] consists fifteen tasks with twenty four communication links, i.e., dependency edges which is shown in Fig. 5 [12]. This model has to schedule in four, VM_1, VM_2, VM_3, and VM_4, and these VMs are in two cloud servers CS_1, CS_2. Estimated Computation Time (ECT) is also taken into account for corresponding VMs are shown in Table 5 [12].

Table 3 ECT matrix for DAG_1 model

	t_1	t_2	t_3	t_4	t_5	t_6	t_7	t_8	t_9	t_{10}
CS_1 VM_1	14	13	11	13	12	13	7	5	18	21
CS_1 VM_2	16	19	13	8	13	16	15	11	12	7
CS_2 VM_3	9	18	19	17	10	9	11	14	20	16

Table 4 Dis[1, 2, 3 … 10] for MV of tasks

t_1	t_2	t_3	t_4	t_5	t_6	t_7	t_8	t_9	t_{10}
0	∞	∞	∞	∞	∞	∞	∞	∞	∞
0	18	12	9	11	14	∞	∞	∞	∞
0	18	12	9	11	14	∞	36	32	∞
0	18	12	9	11	14	∞	36	24	∞
0	18	12	9	11	14	35	36	24	∞
0	18	12	9	11	14	35	29	24	∞
0	18	12	9	11	14	35	29	24	∞
0	18	12	9	11	14	35	29	24	37
0	18	12	9	11	14	35	29	24	37
0	18	12	9	11	14	35	29	24	37
0	18	12	9	11	14	35	29	24	37

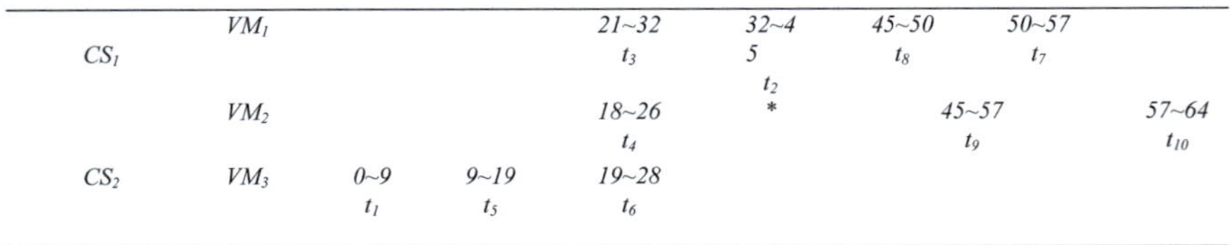

Fig. 3 Gantt chart for proposed algorithm for DAG$_1$ model

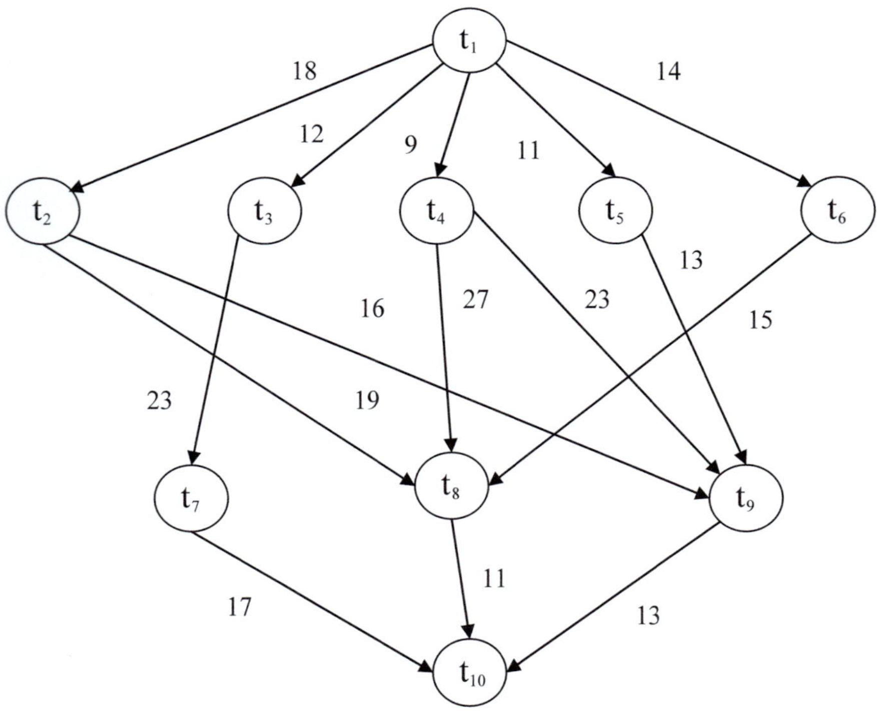

Fig. 4 DAG$_1$ model with ten tasks

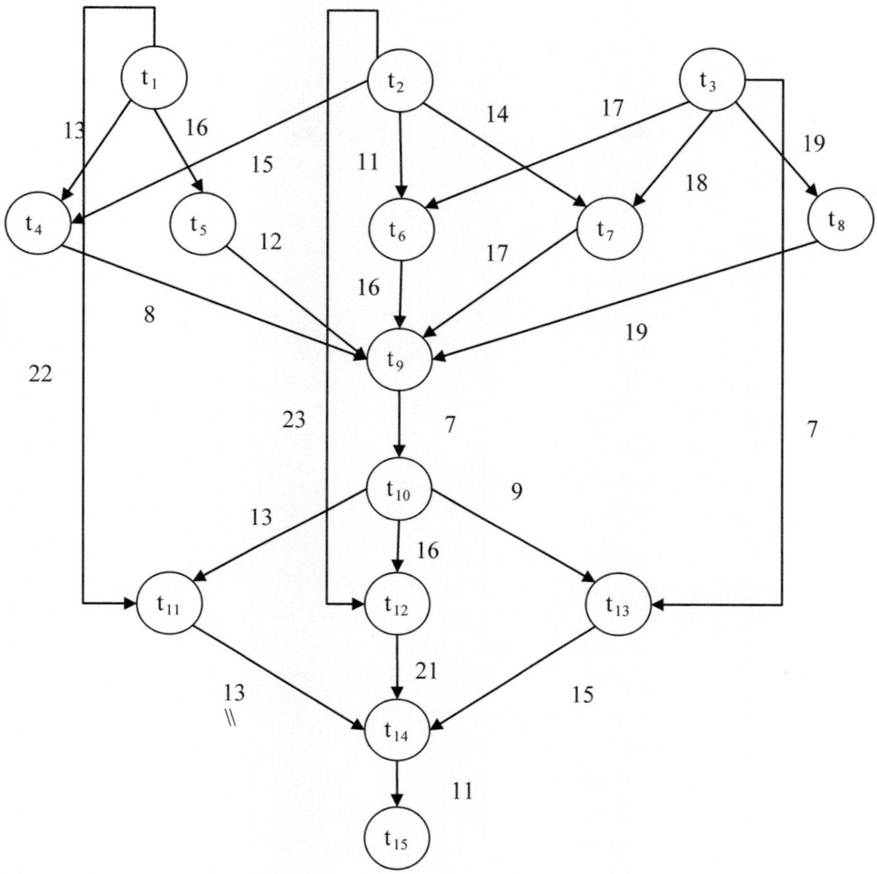

Fig. 5 DAG_2 model with fifteen tasks

Similarly, we have to find distance array Dis[] for DAG_2 model which is shown in Table 6. This value is found as per first phase in proposed algorithm. Here, coloured value means task is the smallest and marked it.

Final MV value of the tasks t_1, t_2, t_3, t_4, t_5, t_6, t_7, t_8, t_9, t_{10}, t_{11}, t_{12}, t_{13}, t_{14}, t_{15} are corresponding to 0, 0, 0, 13, 16, 11, 14, 19, 21, 28, 22, 23, 7, 22, 33. Sorted the tasks as per their MV in increasing order as follows: t_1, t_2, t_3, t_{13}, t_6, t_4, t_7, t_5, t_8, t_9, t_{11}, t_{14}, t_{12}, t_{10}, t_{15}. These are called priority of the given tasks, and it should be mapped into VMs as per proposed algorithm. So, the complete mapping of task-VM is shown in Fig. 6. Scheduling length of new method is 145 units which is minimum as compared to HEFT algorithm [12] gives 152 units (Fig. 7).

Table 5 ECT matrix for DAG_2 model

	t_1	t_2	t_3	t_4	t_5	t_6	t_7	t_8	t_9	t_{10}	t_{11}	t_{12}	t_{13}	t_{14}	t_{15}
CS_1 VM_1	17	14	19	13	19	13	15	19	13	19	13	15	18	20	11
CS_1 VM_2	14	17	17	20	20	18	15	20	17	15	22	21	17	18	18
CS_2 VM_3	13	14	16	13	21	13	13	13	13	16	14	22	16	13	21
CS_2 VM_4	22	16	12	14	15	18	14	18	19	13	12	14	14	16	17

Table 6 Dis[1, 2, 3 …, 15] for MV of tasks

t_1	t_2	t_3	t_4	t_5	t_6	t_7	t_8	t_9	t_{10}	t_{11}	t_{12}	t_{13}	t_{14}	t_{15}
0	0	0	∞	∞	∞	∞	∞	∞	∞	∞	∞	∞	∞	∞
0	0	0	13	16	∞	∞	∞	∞	∞	22	∞	∞	∞	∞
0	0	0	13	16	11	14	∞	∞	∞	22	23	∞	∞	∞
0	0	0	13	16	11	14	19	∞	∞	22	23	7	∞	∞
0	0	0	13	16	11	14	19	∞	∞	22	23	7	22	∞
0	0	0	13	16	11	14	19	27	∞	22	23	7	22	∞
0	0	0	13	16	11	14	19	21	∞	22	23	7	22	∞
0	0	0	13	16	11	14	19	21	∞	22	23	7	22	∞
0	0	0	13	16	11	14	19	21	∞	22	23	7	22	∞
0	0	0	13	16	11	14	19	21	∞	22	23	7	22	∞
0	0	0	13	16	11	14	19	21	28	22	23	7	22	∞
0	0	0	13	16	11	14	19	21	28	22	23	7	22	∞
0	0	0	13	16	11	14	19	21	28	22	23	7	22	33
0	0	0	13	16	11	14	19	21	28	22	23	7	22	33
0	0	0	13	16	11	14	19	21	28	22	23	7	22	33
0	0	0	13	16	11	14	19	21	28	22	23	7	22	33

CS_1	VM_1	0~14 t_2	14~26 *	26~39 t_4							
	VM_2	0~29 *	29~49 t_5								
CS_2	VM_3	0~13 t_1	13~25 *	25~38 t_6	38~51 t_8	51~61 *	61~74 t_9	74~87 *	87~101 t_{11}	101~115 *	115~128 t_{14}
	VM_4	0~12 t_3	12~28 *	28~42 t_7	42~74 *	74~87 t_{10}	87~101 t_{13}	101~115 t_{12}	115~128 *	128~145 t_{15}	

Fig. 6 Gantt chart for proposed algorithm for DAG_2 model

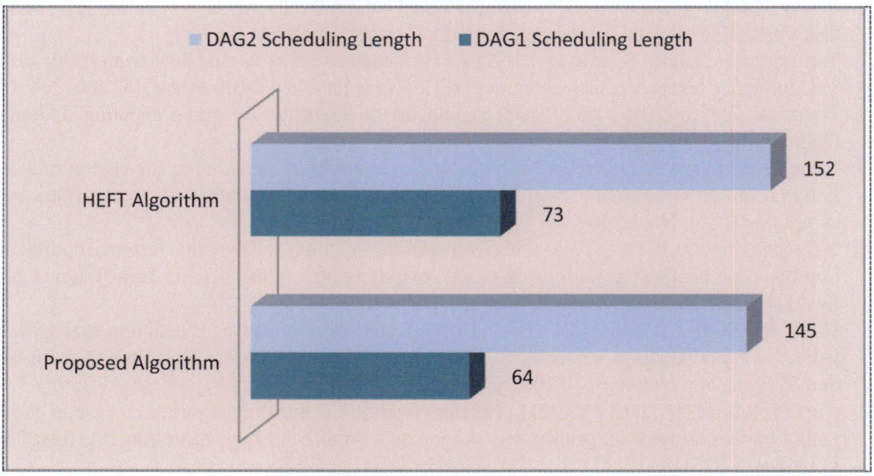

Fig. 7 Scheduling length for both different DAGs

4 Conclusion

This paper presented new task scheduling algorithm based on minimum value attribute. The performance evaluation of the proposed algorithm has done on two different DAG models such as DAG_1 model with 10 tasks, and it has 15 dependent edges. This model has analysis on two cloud servers which consist three virtual machines and proposed algorithm on this model gives 64 units of scheduling length. Proposed algorithm gives less scheduling length as compared to HEFT algorithm which has 73 units scheduling length. Similarly, second DAG_2 model also gives better results as compared HEFT. This model is evaluated on two cloud servers but having four virtual machines. Overall, the proposed model gives better result as compared to HEFT algorithm.

References

1. Xue S, Shi W, Xu X (2016) A heuristic scheduling algorithm based on PSO in the cloud computing environment. Int J u- e-Serv Sci Technol 9(1):349–362
2. Buyya R, Yeo CS, Venugopal S, Broberet J, Brandic I (2009) Cloud computing and emerging IT platforms: vision, hype and reality for delivering computing as the 5th utility. Future Gener Comput Syst 25(6):599–616
3. Sosinsky B (2011) Cloud computing bible. Wiley Publishing Inc., Canada
4. Pachghare VK (2016) Cloud computing. PHI
5. Badger L, Grance T, Corner RP, Voas J (2011) Draft cloud computing synopsis and recommendations, May. NIST, Department of Commerce, U.S.
6. Patel G, Mehta R, Bhoi U (2015) Enhanced load balanced min-min algorithm for static meta task scheduling in cloud computing. Procedia Comput Sci 57:545–553
7. Topcuoglu H, Hariri S, Wu M-Y (2002) Performance-effective and low-complexity task scheduling for heterogeneous computing. IEEE Trans Parallel Distrib Syst 13(3):260–274
8. Pooranian Z et al (2015) An efficient meta-heuristic algorithm for grid computing. J Comb Optim 30(3):413–423
9. Kumar MS, Gupta I, Jana PK (2017) Delay-based workflow scheduling for cost optimization in heterogeneous cloud system. In: 2017 tenth international conference on contemporary computing (IC3), Noida, pp 1–6
10. NZanywayingoma F, Yang Y (2017) Effective task scheduling and dynamic resource optimization based on heuristic algorithms in cloud computing environment. KSII Trans Internet Inf Syst 11(12):5780–5802
11. Haidri RA, Katti CP, Saxena PC (2017) Cost effective deadline aware scheduling strategy for workflow applications on virtual machines in cloud computing. J King Saud Univ Comput Inf Sci. In Press. https://doi.org/10.1016/j.jksuci.2017.10.009
12. Gupta I, Kumar MS, Jana PK (2018) Efficient workflow scheduling algorithm for cloud computing system: a dynamic priority-based approach. Arab J Sci Eng. https://doi.org/10.1007/s13369-018-3261-8

Development and Design Strategies of Evidence Collection Framework in Cloud Environment

Yunus Khan and Sunita Varma

Abstract Nowadays, cloud computing is one of the popular and widely used concepts in information technology paradigm. It is committed to improving the IT business technically and economically. On the other hand, digital forensic is the process of collection, identification, preservation, examination, and analysis of data or information for the proof in the court of law as an evidence. It is very difficult and challenging to apply digital forensic operation in a cloud environment because CSPs are dependent on each other either they provide IaaS, PaaS, or SaaS. So the cloud forensic, one of the applications of digital forensic in a cloud environment, is just a subset of network forensic. It is a cross-field of digital forensic and cloud computing. In this paper, we investigate all the research issues, problems, and implementation ethics of cloud forensic from the initial level. We found that lots of issues and challenges are remaining to address in this domain. Some major research domains are architectures, data collection and analysis, anti-forensic, incident first responders, roles and responsibilities, legal, standards, and some learning issues. In our research work, we mainly focus on the data collection and cloud forensic architectures and also implement a cloud forensic framework in the context of cloud service models. This research work is tested using different private cloud solutions such as eucalyptus, OpenNebula, VMware, vCloud, and Hadoop platform. In our research work, we implement pattern search facility using the proposed approach in open-source software called digital forensic framework. We also implement in near future digital forensic triage using Amazon Elastic MapReduce. In this research, we also implement designed and development of forensic method for the PaaS and SaaS delivery models of cloud computing, also apply machine learning principles to design and develop new digital forensic methods, and improve the efficiency of investigation using machine learning algorithms for feature extraction and priority of evidence classification of evidence in virtual machines.

Y. Khan · S. Varma (✉)
Shri Govindram Seksaria Institute of Technology and Science, Indore, Madhya Pradesh, India
e-mail: sunita.varma19@gmail.com

Y. Khan
e-mail: callyunuskhan@gmail.com

R. K. Shukla et al. (eds.), *Social Networking and Computational Intelligence*, Lecture Notes in Networks and Systems 100, https://doi.org/10.1007/978-981-15-2071-6_3

Keywords Digital forensic · Data collection · Evidence segregation · Dependency chains · IDS · Multiple jurisdictions and tenancy · IaaS · SaaS · PaaS · SLA · Virtual environment · VMware

1 Introduction

Cloud computing is an evolving technology which offers opportunities for the development of the IT organizations and their business using high available computing resources, pay-per-use model, on-demand broad network access, and software and hardware resource pooling. Characteristics of cloud computing have drastically minimized information technology cost and motivated business and government for adopting the cloud environment. The digital forensic plays an important role in the field of cloud computing and have following characteristics like rapid elasticity, resource pooling, metered and measured service, on-demand self-service, and broad network access. Cloud forensics is a combination of cloud computing and digital forensics. Digital forensic is the new similar subfield of computer forensic science to perform the forensic investigation functions and tasks. Cloud forensic involves after attack analysis of the attack information and evidence collection. Evidence collection is difficult in cloud forensic because of the multiple geographical locations of the stored data and privacy issues. Cloud forensic uses the features and properties of digital forensic and cloud computing to achieve some forensic investigation objectives. Cloud Service Providers (CSP) maintains the data centers all over the world for gaining the high availability of service and effectiveness of the cost. To avoid the data loss risk from natural disasters, the stored data are replicated around the multiple data centers. Data acquisition and collection are the methods of recognizing, naming, gathering, recording, and collecting forensic data shown in Fig. 1. The forensic data contain the information, knowledge, and parameters of the customer or user premises and the service provider artifacts reside on the CSP's infrastructure. Lots of tools are available to collect the forensic data on the basis of the roles and responsibility of

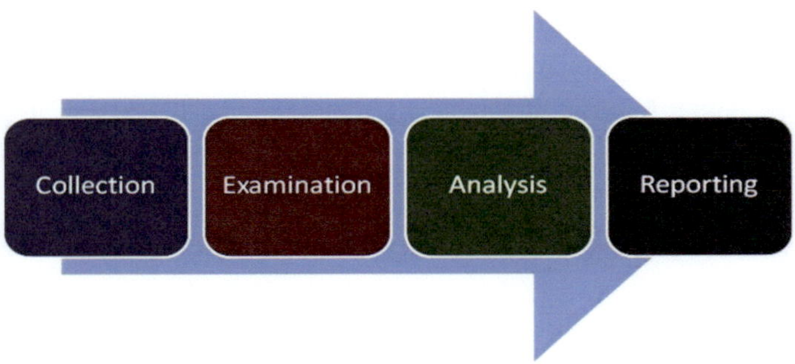

Fig. 1 Process of forensic evidence collection

the model used. During the data collection process, user must be careful about the integrity of the data and separation of duties between client and service provider.

Investigation procedure based on the hypervisor is limited. Loss of data control is another challenge which we found in the cloud forensic science. The time of capturing and collecting the evidence is another issue while taking into consideration the court and laws [1, 2]. So the development tools and procedure have the feature to timestamp the collected data in a cloud environment.

Why improved mechanism of evidence collection in cloud environment is required?

In traditional digital forensic, forensic experts have complete control over the data and evidence like network, mobile, and disk-related data. In cloud computing, cloud customer has limited control on software as a service and better control over the infrastructure as a service. It makes difficult to evidence collection process in the cloud. So the cloud customer only accesses the application log and depends on CSPs for the operating system, database disk logs, and network logs as shown in Fig. 2. Because of the feature of multi-tenancy, it is very difficult to prove and justify to the court that the collected evidence actually belongs to the suspect (privacy of tenants). Chain of custody: Traditional evidence collections methods are depends on CSP's and not provide grantee about the integrity of the evidence and is a challenge is cloud forensic [3].

Fig. 2 Data collection scenario

2 Comparative Research of Existing Data and Evidence Collection Tools, Techniques, Frameworks, Algorithm, and Approaches in Cloud Forensics

Limited access to forensic data is a big challenge for the cloud customer because CSPs hide the physical locations of the customer's data. Limited access means cloud customers have zero knowledge about the physical locations of their data. They know only the high level of data abstraction regarding their data in the form of object or container. Data movement and duplication details are hidden from the customers by service providers. Even they are not providing the links or interaction to access or collect the forensic data to the customers. For example, SaaS providers do not allow the users to see the logs, disk, and virtual machine information. Cloud customers have very less access to file logs and data dictionary among all levels, as well as the limited ability to audit and conduct monitoring on the data stored [4]. For the analysis purpose, cloud forensic is a cross-discipline of cloud computing and digital forensics study all the existing frameworks of cloud forensic data collection and challenges in cloud forensic, the existing systems and frameworks are:

Alqahtany develops a novel architecture to support forensic evidence collection and analysis of infrastructure as a service (IaaS) in cloud environment formally known as cloud forensic acquisition and analysis system without depending on cloud service provider and third party. This approach also provides the access of deleted data and overwritten data files which is not provided in existing forensic investigation techniques [5]. Piwari develops a model for the evidence forensic named development and implementation of evidence collection strategy in cloud environment within cloud client and service providers using Microsoft server 2012 database and hyper-V [6]. An implementation of cloud forensic done by Katz and Ryan at LCDI in November 2013 which is based on skydrive, Google drive, and dropbox on some CSV file artifacts.

Almarzooqi develops a model which provide roadmap and proper guideline to the cloud forensic area researchers that how to build digital forensic framework and apply in cybercrime actual scene [7, 8]. The author aimed and provided a solution to the isolation and segregation of evidence in cloud environment [9, 10]. The researcher shows relation between digital forensic investigation and cloud computing shown by different steps and phases comparison in terms of I-SIRIDE process model for risk assessment, mitigation, and investigation cycle based on eucalyptus [11].

Dykstra and Sherman developed and investigated digital forensic for infrastructure as a service (IaaS) is cloud computing in which they investigated how it is more difficult to handle malicious activities in the cloud, trustworthiness of cloud forensic tools, and saving, collection, and modification of digital evidence in the cloud 2013 [12]. Daryabar develops a mechanics for the mobile Smartphone named Digital forensic Framework for investigation client cloud storage application on smartphone which deal with how forensic is applied in smartphone-based cloud computing [13]. Hewling designed a novel approach for the computer-based cybercrime malicious activities which manage computer security using digital forensics [14]. Chaurasia addresses the issue of evidence collection in cloud environment like identifying the target, jurisdiction problem, collection of evidence, chain of custody, third party issues, and privacy and security of data [15]. The researcher implemented AUDIT tool which is designed to configure and integrate open-source tools for investigation of disk and ranking of forensic tools (Linux-based) [16]. Kebande implements a novel cloud forensic readiness service model, and researcher focuses the forensic application in cloud environment and develops a model, cloud forensic readiness as a service software prototype [17].

Mustafa discovered and investigated the evidence recovered from the Zen Cloud Platform using available tools. Basically, the research focuses on the three areas like apply existing tools in cloud environment, collect artifacts and evidences from cloud, and analyze the value in collected evidence. As a part future direction, we can implement in near future existing tools with platform as a services (PaaS) and software as a service (SaaS) (or all service type models in one framework). Finally, this research focused on recovery XCP with file system-based and LVM-based storage repositories (SRS) [18]. Clark contributed to their work mainly on exif metadata contained in JPEG image files. In near future, all testings are performed in various other file formats like pdf, word, excel, ppt, and others [19]. Krishnan discusses about the major current trend issues of security and privacy in the cloud computing and also categorizes security and privacy issues into security only issues, privacy only issues, and intertwined security privacy and security issues [20]. Sibiya in their research describes the requirements for a cloud forensics system and what standard procedures followed during the cloud forensic process and how a cloud forensics system can be designed and also cloud forensic as a service CFAAS architecture [21].

De Marco implements in their research digital forensic readiness capability in the cloud using natural language based on service level agreement (SLA's) clauses, cloud logs and utilize tuple, set theory and functions; this research is totally based on service level agreements [22]. Povar covers the challenges like explore the challenges and requirements of applying forensic in the cloud, design a digital forensic framework for the cloud computing environment architecture, and also address the issues of dead live forensic analysis within and outside the virtual machine which runs in cloud environment, and examination and analysis of evidence [23].

Manoj and Bhaskari present secure cloud framework for the cybercrime with the help of team support of cloud user, cloud service providers, trusted third party, and forensic investigators [24]. Alex and Kishore develop a system in case of denial of service (DDoS) attack whether the forensic management plane (FMP) collects the data about the fraudulent activities for forensic analysis. In near future, we can implement the whole attack scenario in cloud platform [25, 26]. Pichan et al. in their research provide a systematic solution of analysis of cloud forensic challenges, possible solution of each phase, and summary of forensic as a service model [27]. Roussev et al. in their investigation applied analysis and acquisition on SaaS and tested the result in their case studies: Kumodd: tools used for cloud drive acquisition, Kumodocs: a tool for Google Docs acquisition and analysis, and Kumofs: a tool for remote preview and screening of cloud drive data [28].

3 Noteworthy Contributions in the Field of Proposed Work

Primary goal of computer forensic is to provide structured phenomena for investigation and fact collection. It creates bridge between live evidences and security measurement process. Study of conventional cloud forensic explores that several security challenges and loopholes are required to be resolved with high priority. It expects that enhanced security policy should overcome the security overhead and increase trust of user by improving accuracy in security measurement process. Although there are several solutions developed and suggested by researchers, still a gap has been observed between traditional security policies and security requirement. Table 1 shows existing work done in the field cloud forensic by various researchers.

Cloud forensic may not only add a new dimension into data collection but will also create enhanced toughness for security threats and simplify chain of custody. The complete work observes that proposed solution will help to enhance data collection policy in cloud forensic.

We found that very less work has been done in Data Collection [38, 39], Evidence Segregation, and Dependency Chains. In our proposed work, we work on the following research objectives:

- Investigation and analysis of issues, challenges, and future directions of cloud forensics.
- Comparative research of existing data and evidence collection tools, techniques, frameworks, algorithm, and approaches.
- Develop, design, and validate a modified data and evidence collection framework in cloud environment (without compromising the evidence integrity).

After the complete review of area, so many methods suggest by author a novel digital forensic framework for cloud computing is tested using the private cloud test bed setup using OpenStack cloud solution. This can also be tested using different private cloud solution such as eucalyptus, OpenNebula, VMware, vCloud, and Hadoop in our proposed research work.

In our proposed research work, we implement pattern search facility using the proposed approach in open-source software called digital forensic framework. We also implement in near future digital forensic triage using Amazon Elastic MapReduce. In this work, researcher targeted delivery model as a part of future work, and we can also implement design and development of forensic method for the PaaS and SaaS delivery models of cloud computing, also apply machine learning principles to design and develop new digital forensic methods, and improve the efficiency of investigation using machine learning algorithms for future extraction priority of evidence classification of evidence in virtual machines.

Table 1 Study and analysis of recent existing work done in cloud forensics

S. no.	Researcher/author name	Year	Investigated cloud forensic issues	Proposed mechanism
1	Kebande [17]	2018	A novel cloud forensic readiness service model	Develop a model cloud forensic readiness as a service software prototype
2	Alqahtany [5]	2017	Architecture of cloud computing based on IaaS	Forensic evidence collection and analysis of infrastructure as a service in cloud environment
3	Choo et al. [29]	2017	Privacy and accuracy of forensic data and its issues	Future directions of cloud forensic
4	Alex and Kishore [30]	2017	Discuss issues forensic framework for cloud computing	Forensic data collection using FMP and FTK analyzer with DDoS attack
5	Roussev et al. [28]	2016	Cloud forensic tools development	Result analysis on Kumodd, Kumodocs, and Kumofs case studies
6	Alex and Kishore [26]	2016	Data collection issue, trust issues	Forensic management plane (FMP)
7	Manoj and Bhaskari [24]	2016	Investigation of cyberattacks in cloud environment	Trusted third party (TTP) and cloud forensic investigation team (CFIT)
8	Morioka and Sharbaf [31]	2016	Need of tools, algorithm and procedure for cloud forensics	Acquiring evidence using amazon web services
9	Mustafa [18]	2016	Assessing the evidential value of artifacts recovered from the cloud	Investigated the evidence recovered from the (Zen Cloud Platform) using available tools
10	Povar and Geethakumari [23]	2016	Cloud-based digital forensic architecture	Tested using the private cloud test bed setup using OpenStack cloud
11	De Marco [22]	2015	Forensic readiness capability for cloud computing	Service level agreement (SLA's) clauses, cloud logs and utilize tuple, set theory and functions
12	Sibiya [21]	2015	Digital forensic model for a cloud environment	Design cloud forensic as a service architecture (CFAAS)

(continued)

Table 1 (continued)

S. no.	Researcher/author name	Year	Investigated cloud forensic issues	Proposed mechanism
13	Pichan et al. [27]	2015	Comparative analysis of cloud forensics research	Discuss about problems associated in cloud-based forensic environment
14	Reichert et al. [36]	2014	Automatic forensics data acquisition in the cloud	Forensic evidence collection using snapshots Google rapid response (GRR)
15	Patrascu and Patriciu [4]	2014	Implementation of cloud computing framework for cloud forensics	Various factors of security and architecture of data center with reason
16	Shah and Malik [33]	2013	Multiple challenges during forensic steps	Mining-based cloud forensic architecture
17	Martini and Choo [34]	2013	Challenges during forensic steps and client–server system	Design of client and server forensic setup
18	Sharevski [35]	2013	Challenges during evidence collection, preservation and acquisition phase	Analysis of various aspects in network, storage and virtualization
19	Zargari and Benford [37]	2012	Map of multiple cloud forensic concepts, issues, and challenges	Comparison between computer forensics and cloud forensics
20	Martini and Choo [32]	2012	Compare issues of NIST and McKemmish framework	Refinement of forensic framework

4 Expected Outcome of the Proposed Work

The expected outcome of these research is that will achieve all the research objectives and this research work is able to investigation and analysis of issues, challenges and future directions of cloud forensics, comparative research of existing data and evidence collection tools, techniques, frameworks, algorithm, and approaches, develop, design and validate a modified data and evidence collection framework in cloud environment (without compromising the evidence integrity) and attribute deleted data in the cloud, recovery of deleted data before overwrites, single point of failures, no single point of failure for criminals, multiple geo-location of data, problems during the auditing like accuracy and correctness of data, separatism of evidence data in a multi-tenancy, system boundaries need to be defined, limited access and control of customers on cloud data, dependency of chains around the globe, locating evidence in a large and changing system, backup and capture the activity of data transfer, limited time data access of newly added VMs and recognition of data storage media. At the end of this research, we implement a modified evidence collection framework in cloud environment based on PaaS and SaaS service models and improve the efficiency of evidence investigation process using machine learning algorithms for future extraction priority of evidence classification of evidence in virtual machines.

5 Conclusion

So the conclusion of this paper is that so many issues are remaining to work like integrity and data confidentiality of the collected data during the forensic process and also analyze how to implement the forensic data collection in a cloud environment for the cloud service models, for that, we also check the implementation feasibility with open-source tools for the framework. In near future, we implement a cloud forensic data collection framework based on PaaS and SaaS, improve the efficiency of existing evidence investigation process using machine learning process, and overcome all the challenges in more accurately and efficiently. So the objective of our research work is investigation and analysis of issues, challenges, and future directions of cloud forensics, comparative analysis of existing data and evidence collection tools, techniques, frameworks, algorithms, and approaches and finally develop design and is to validate a modified data and evidence collection framework in cloud environment (without compromising the evidence integrity).

References

1. Kaur M, Kaur N, Khurana S (2016) A literature review on cyber forensic and its analysis tools. Int J Adv Res Comput Commun Eng 5(1). ISSN (Online) 2278-1021. ISSN (Print) 2319 5940
2. Zhou G, Cao Q, Mai Y (2011) Forensic analysis using migration in cloud computing environment. In: Information and management engineering, pp 417–423
3. Zawoad S, Hasan R (2013) Digital forensics in the cloud. In: Securing the cloud. Crosstalk, Sept/Oct. University of Alabama, Birmingham
4. Patrascu A, Patriciu VV (2014) Implementation of a cloud computing framework for cloud forensics. In: Proceedings of the 18th international conference on system theory, control and computing, Sinaia, Romania, 17–19 Oct. ISBN 978-1-4799-4601-3/14/$31.00 ©2014 IEEE
5. Alqahtany SS (2017) A forensically-enabled IaaS cloud computing architecture. Thesis, University of Plymouth, Jan 2017. http://hdl.handle.net/10026.1/9508
6. Piwari MTM (2016) Digital forensics in the cloud: the reliability and integrity of the evidence gathering process. Thesis, Auckland University of Technology, New Zealand
7. Almarzooqi A, Jones A (2016) A framework for assessing the core capabilities of a digital forensic organization. In: IFIP international conference on digital forensics, Jan 2016. Springer International Publishing, pp 47–65
8. Almarzooqi A, Jones A, Howley R (2016) Applying grounded theory methods to digital forensics research. In: The 11th annual ADFSL conference on digital forensics, security and law, May 2016
9. Delport W, Olivier MS, Kohn M (2011) Isolating a cloud instance for a digital forensic investigation. In: Information security South Africa conference (ISSA)

10. Delport W, Olivier MS (2012) Isolating instances in cloud forensics. In: Advances in digital forensic VIII IFIP, vol 383. Springer, Berlin, pp 187–200
11. James JI, Shosha AF, Gladyshev P (2012) Digital forensic investigation and cloud computing. ResearchGate, Dec 2012
12. Dykstra JABS (2013) Digital forensics for infrastructure-as-a-service cloud computing. Dissertation, Faculty of the Graduate School of the University of Maryland, Baltimore County
13. Daryabar F (2015) Digital forensics framework for investigating client cloud storage applications on smartphones. Thesis, University Putra Malaysia, May 2015
14. Hewling MO (2013) Digital forensics: an integrated approach for the investigation of cyber computer related crime. Thesis, University of Bedfordshire
15. Chaurasia G (2015) Issues in acquiring digital evidence from cloud. J Forensic Res S3. https://doi.org/10.4172/2157-7145.1000s3-001
16. Karabiyik U (2015) Building an intelligent assistant for digital forensic. Thesis, Florida State University
17. Kebande VR, Venter HS (2018) Novel digital forensic readiness techniques in the cloud environment. Aust J Forensic Sci
18. Mustafa ZS (2016) Assessing the evidential value of artifacts recovered from the cloud. Cranfield University
19. Clark P (2011) Digital forensics tool testing image metadata in the cloud. Gjovik University College, Norway
20. Krishnan R (2017) Security and privacy in the cloud computing. Western Michigan University
21. Sibiya MG (2015) Digital forensic model for a cloud environment. University of Pretoria, Feb 2015
22. De Marco L (2015) Forensic readiness capability for cloud computing. Università Degli Studi Di Salerno
23. Povar D, Geethakumari G (2016) Digital forensic architecture for cloud computing systems: methods of evidence identification, segregation, collection and partial analysis. In: The third international conference on information systems design and intelligent applications-India-2016. Advances in intelligent systems and computing (AISC) series
24. Manoj SK, Bhaskari DL (2016) Cloud forensics—a framework for investigating cyber attacks in cloud environment. Procedia Comput Sci 85:149–154
25. Dykstra J, Sherman AT (2012) Acquiring forensic evidence from infrastructure-as-a-service cloud computing: exploring and evaluating tools, trust, and techniques. Digit Investig 9(Supplement):S90–S98
26. Alex ME, Kishore R (2016) Forensic model for cloud computing. In: IEEE WiSPNET conference
27. Pichan A, Lazarescu M, Soh ST (2015) Cloud forensics: technical challenges, solutions and comparative analysis. Digit Investig
28. Roussev V, Ahmed I, Barreto A, McCulley S, Shanmughan V (2016) Cloud forensics—tool development studies and future outlook. Digit Investig
29. Choo KKR, Esposito C, Castiglione A (2017) Evidence and forensics in the cloud: challenges and future research directions. IEEE Cloud Comput
30. Alex ME, Kishore R (2017) Forensics framework for cloud computing. Comput Electr Eng
31. Morioka E, Sharbaf MS (2016) Digital forensics research on cloud computing: an investigation of cloud forensics solutions. IEEE. ISBN 978-1-5090-0770-7
32. Martini B, Choo KKR (2012) An integrated conceptual digital forensic framework for cloud computing. Digit Investig 9:71–80. Journal homepage: www.elsevier.com/locate/diin
33. Shah JJ, Malik LG (2013) Cloud forensics: issues and challenges. In: 2013 sixth international conference on emerging trends in engineering and technology. ISBN 978-1-4799-2560-5/13 © 2013 IEEE 2013. IEEE Computer Society. https://doi.org/10.1109/icetet.2013.44
34. Martini B, Choo KKR (2013) Cloud storage forensics: own cloud as a case study. Digit Investig 17–36
35. Sharevski F (2013) Digital forensic investigation in cloud computing environment: impact on privacy. In: International conference IEEE Louisville chapter 2013, pp 1–6

36. Reichert Z, Richards K, Yoshigoe K (2014) Automated forensic data acquisition in the cloud. In: IEEE international conference computer society
37. Zargari S, Benford D (2012) Cloud forensics: concepts, issues, and challenges. In: 2012 third international conference on emerging intelligent data and web technologies. IEEE Computer Society. https://doi.org/10.1109/eidwt.2012.44. ISBN 978-0-7695-4734-3/12 © 2012
38. NIST Cloud Computing Forensic Science Working Group (2014) NIST cloud computing forensic science challenges. Draft NISTIR 8006. Information Technology Laboratory, 23 June 2014
39. U.S. Department of Justice (2015) Research and development in forensic science for criminal justice purposes. OMB No. 1121-0329. Office of Justice Programs. Approval expires 31 July 2016

A Systematic Analysis of Task Scheduling Algorithms in Cloud Computing

Nidhi Rajak and Diwakar Shukla

Abstract Today is an era of the fastest technology which is growing in every field such as medical, marketing, aerospace and high-level computing. Cloud computing is new area of research which is used in every IT industry. It is basically on demand of resources via Internet. Here, resource can be storage, server, networks, etc. Task scheduling is NP-complete problem, and it is a mechanism to allocate the tasks on available resources. So that it can be minimized the following parameters such as execution time, cost and maximized the utilization of resources. In this paper, we have surveyed various task scheduling algorithms with their brief description, scheduling parameter and tools used. Also, we have discussed various basic tasks scheduling models and scheduling attributes.

Keywords Cloud computing · DAG · Task scheduling · Scheduling length · Virtual machine · Cost

1 Introduction

Cloud computing is a recent area of research in computer science, and it is growing at the rapid speed in the technology world. It is an extension of conventional computing such as parallel, distributed and grid [1]. It can be formally defined as a model which provides service to the user on demand via Internet for which user needs to pay. There are three basic characteristics [2] of cloud computing such as *dynamic extendibility, virtualization and distribution*. Cloud computing can be classified based on deployment model and service model. *Deployment model* provides three types of cloud such as public cloud, private cloud and hybrid cloud, whereas *service model* [3] provides three types of services such as *Platform as Service* (*PaaS*), *Software as*

N. Rajak (✉) · D. Shukla
Department of Computer Science and Applications, Dr. Harisingh Gour Vishwavidyalaya, Sagar, Madhya Pradesh 470003, India
e-mail: nidhi.bathre@gmail.com

D. Shukla
e-mail: diwakarshukla@rediffmail.com

R. K. Shukla et al. (eds.), *Social Networking and Computational Intelligence*, Lecture Notes in Networks and Systems 100, https://doi.org/10.1007/978-981-15-2071-6_4

Service (*SaaS*) and *Infrastructure as Service* (*IaaS*). Here, virtualization method is used for mapping of resources and virtual machines.

Cloud computing is also based on large-scale distributed computing model [4], and it depends on abstract, virtualized and dynamic. Following are the major concerned with cloud computing platform such as how to optimized the computing power, storage management, and different types of service etc. These can be allocated to user as per their demands via Internet.

Task scheduling is known as NP-complete problem [5]. The major objectives of any scheduling algorithms are *to minimize scheduling length, to minimize load balancing, to maximize resource utilization and to improve the Quality of Service (QoS).*

There are three basic generalized steps [6] of any task scheduling methods in cloud computing environment which are as following: *resource discovering and filtering, resource selection, task submission* as shown in Fig. 1 [7].

Task scheduling is a process to allocate the task to the available resource that is mapping of each task to the available virtual machine. There are many scheduling algorithms has been developed which are based on various parameters such as Scheduling Length, Cost, Load Balancing, QoS, Resource Utilization and Energy Consumption etc. These algorithms are proposed based on some scheduling parameters such as makespan, cost, Quality of Service (QoS), resources utilization, processing time, execution time, performance, energy consumption, and load balancing.

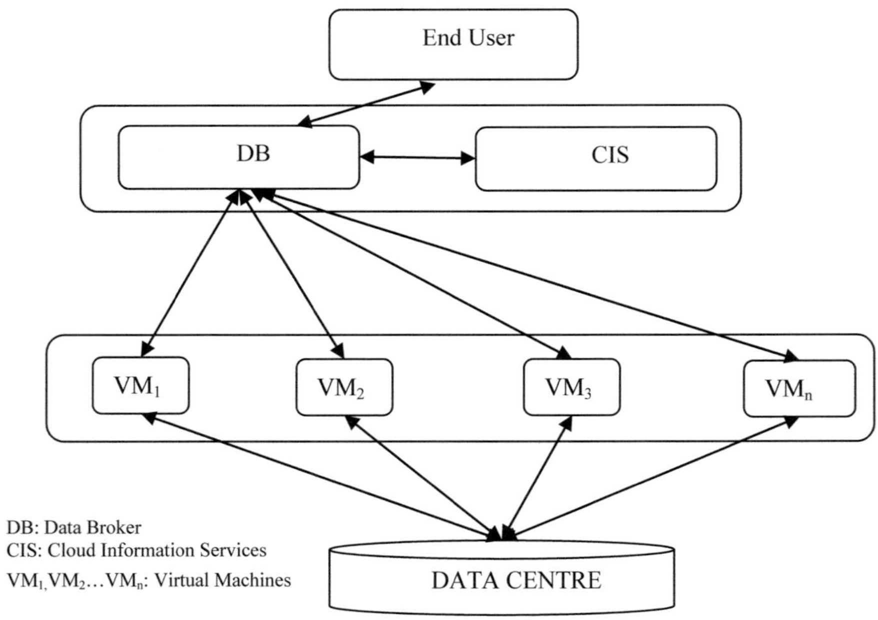

Fig. 1 Steps of scheduling

Review of scheduling algorithms has been done based on scheduling parameters and simulation tools used in tabulation form.

This paper is organized as follows: Sect. 1 discussed basic cloud computing, task scheduling and its objective. Various scheduling models in Sects. 2 and 3 will define basic task scheduling problem definition, basic terminologies and attributes in Sect. 4. Task scheduling review and analysis of exiting task scheduling algorithms in Sect. 5. Finally concludes the paper in Sect. 6.

2 Task Scheduling Models Classification

Task scheduling models [8] are classified into eleven categories which are briefly discussed in Table 1.

3 Basic Task Scheduling Problem Definition

The system model for the scheduling in the cloud computing is divided into three parts: *Task Graph, Resource Graph and Cloud Infrastructure.*

Task Graph is represented by directed acyclic graph (DAG), and it is defined by $G1 = (T, E, C)$. Where T is a set of finite n tasks of DAG which can be defined as $T = \{T_1, T_2, \ldots, T_n\}$ and E is an edge between the tasks T_i and T_j. The precedence constraint always maintained among the tasks during the execution of the tasks that is if there is an edge between T_i and T_j, then T_j will start execution only after the completion of the task T_i. C is the communication time between the tasks. Every DAG has entry and exit task where entry task can be defined as a task which has no parent, and exit task is a task having no children.

Resource Graph represents the virtual machines or virtual resources which can be defined as the total m number of virtual machines or resources which are available in the cloud, and it can be represented by $R = \{R_1, R_2, \ldots R_m\}$. The mapping function $F: T \rightarrow R$ is the function which maps the nth tasks onto the mth resources which is shown in Fig. 2 [9].

Cloud Infrastructure is the collection of interconnected physical computer shown in Fig. 3 [10].

4 Scheduling Attributes

This section discusses various scheduling attributes which are used in task scheduling methods for allocating tasks to available virtual machine. These are following attributes:

Table 1 Classifications: task scheduling models

S. no.	Scheduling models	Brief descriptions
1	Static scheduling model	• Simplest model • Also known as compile-time scheduling • All information related to tasks and resources are known in advanced
2	Dynamic scheduling model	• Flexible model • Also known as execution time scheduling • All information related to tasks and resources are known at execution time, i.e., dynamically
3	Hybrid scheduling model	• Combination of static and dynamic scheduling models • Follows as per static and dynamic model
4	Distributed scheduling model	• More convenient and realistic model • Unreliable as compared to centralized model
5	Centralized scheduling model	• Simple implementation • Better manageable method • Does not support for huge collections of clusters due to lack of scalability
6	Cooperative scheduling model	• Two or more different schedulers perform simultaneously in a synchronized manners • Worked as common system scheduler • Follows set defined of rules • Every scheduler performs predefined tasks in the system and performed their duty
7	Non-cooperative scheduling model	• No different schedules work as group • Not allowed to execute tasks by two or more schedulers simultaneously
8	Batch mode heuristic algorithm (BMHA)/off-line mode	• Collection of task in batches • Each batch executes the tasks • Also called off-line mode scheduling model
9	Prompt/on-line mode	• Task received and executed in scheduler • Not wait for any batches • Also called on-line mode
10	Pre-emptive scheduling model	• Suspended low priority task by high priority task • Execute high priority task first • Resumed low priority task for execution
11	Non-pre-emptive scheduling model	• No preemption is allowed for any tasks during execution • Not allowed interrupt by high priority task

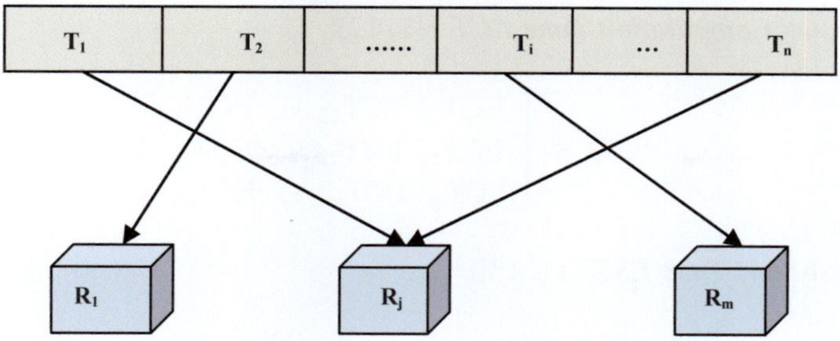

Fig. 2 Mapping of *n* tasks onto *m* resources

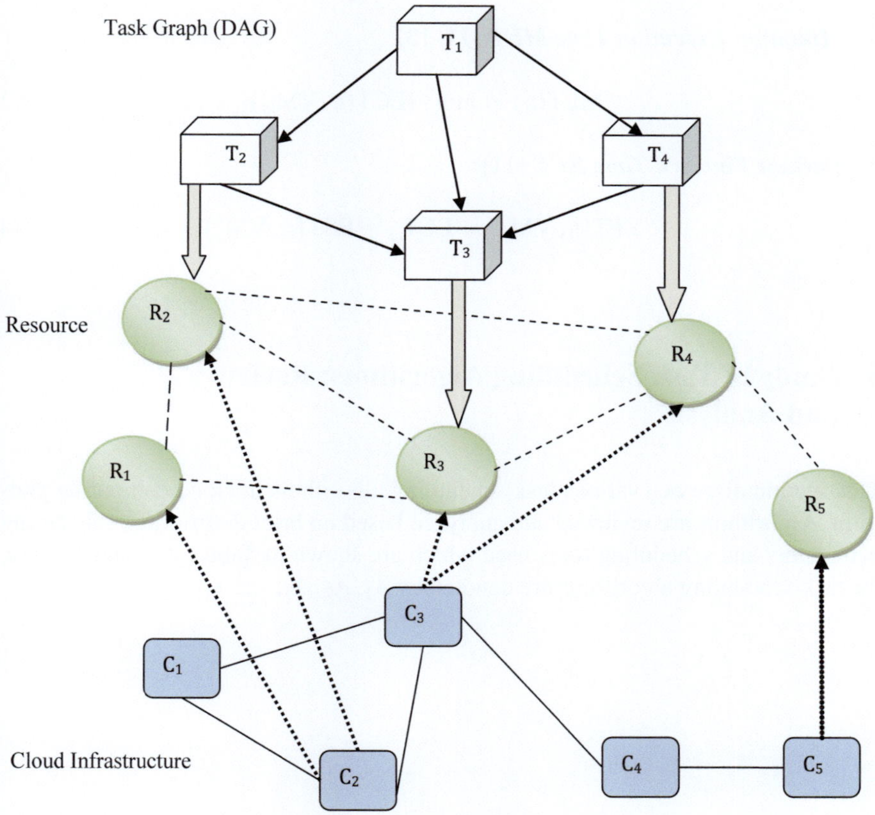

Fig. 3 System model of DAG scheduling service on clouds [10]

a. ***Estimated Computation Time ECT*** [11, 12]:

$$\mathrm{ECT}_{ij} = \begin{bmatrix} \mathrm{ECT}_{11} & \mathrm{ECT}_{12} & \cdots & \mathrm{ECT}_{1n} \\ \mathrm{ECT}_{21} & \mathrm{ECT}_{22} & \ldots & \mathrm{ECT}_{2n} \\ \mathrm{ECT}_{m1} & \mathrm{ECT}_{m2} & \ldots & \mathrm{ECT}_{mn} \end{bmatrix} \tag{1}$$

b. ***Earliest Start Time EST*** [11, 13]:

$$\mathrm{EST}(t_i, \mathrm{VM}_j) = \begin{Bmatrix} 0 & \text{if } t_i \in t_{\text{entry}} \\ \max\limits_{t_j \in \mathrm{pred}(t_i)} \{\mathrm{EFT}(t_j, \mathrm{VM}_j) + \mathrm{MET}(t_i) + \mathrm{CT}(t_i, t_j)\} & \text{otherwise} \end{Bmatrix} \tag{2}$$

c. ***Minimum Execution Time MET*** [11, 13]:

$$\mathrm{MET}(t_i) = \min \cdot \{\mathrm{ECT}(t_i, \mathrm{VM}_m)\} \tag{3}$$

d. ***Earliest Finished Time EFT*** [11]:

$$\mathrm{EFT}(t_i, \mathrm{VM}_j) = \mathrm{ECT}_{ij} + \mathrm{EST}(t_i, \mathrm{VM}_j) \tag{4}$$

5 Study of Task Scheduling Algorithms: Review and Analysis

This section discusses various task scheduling algorithms in cloud computing platform. Algorithms are reviewed and analyzed based on brief description, scheduling parameters and scheduling tools used which are shown in Tables 2, 3 and 4. Here, the task scheduling algorithms are denoted by $A_1, A_2, A_3, \ldots, A_{11}$.

Table 2 Study of task scheduling algorithms

Algorithms	Task scheduling name	Brief descriptions
A_1	Improved cost-based algorithm for task scheduling in cloud computing [2]	• Better mapping between the tasks and virtual machines • Assigned priority to the tasks • Better resource cost • Better computation performance • Better computation and communication ratio (CCR)
A_2	Scheduling scientific workflows elastically for cloud computing [14]	• Set of resources are divided into clusters • Different resources having same computing capability such as network communication are assigned into same cluster • Data rate transfer same for resources within same cluster • Support both homogenous and heterogeneous computing capability • More scalability • Better execution time
A_3	A multiple QoS constrained scheduling strategy of multiple workflows for cloud computing [15]	• Support multiple workflows and multiple QoS • Increased scheduling access rate • Given minimum makespan and cost
A_4	A compromised-time-cost scheduling algorithm in SwinDeW-C for instance-intensive cost-constrained workflows on cloud computing platform [16]	• Support pay per uses principle of cloud computing characteristic • Follow time-cost relationship during execution of workflow • Main attributes are time and cost
A_5	A particle swarm optimization-based heuristic for scheduling workflow applications in cloud computing environments [17]	• Considered both costs such as computation and data transmission • Used in workflow with varying of both communication cost and computation cost
A_6	An efficient multiqueue job scheduling for cloud computing [18]	• Reduced memory space and cost due to fragmentation in previous methods such as FCFS and round robin • Sort the task in increasing order • Divided into three queues such as small queue, medium queue and large queue • Tasks are allocated to virtual machine using meta scheduler • Give good performance

(continued)

Table 2 (continued)

Algorithms	Task scheduling name	Brief descriptions
A_7	HEFT-based workflow scheduling algorithm for cost optimization within deadline in hybrid clouds [19]	• Major concerned how to process of given resources by the tasks • It is based on public cloud • Reduced cost during selection of resources
A_8	Deadline and budget distribution-based cost-time optimization algorithm [20]	• Based on two conditions: deadline and budget • Complete the execution of all tasks within deadline time • Also reduced computation cost while executing the tasks
A_9	Independent tasks scheduling based on genetic algorithm in cloud computing [21]	• Use genetic algorithm to reduce execution time • Based on centralized scheduling model • The preconditioned for the tasks such a periodic, non-pre-emptive • Process unit access by each task by using two methods: shared and exclusive • Optimized the resource and time utilization
A_{10}	Context-aware scheduling [22]	• Based on two applications such as voice assistant and shopping assistant • Both assistants follow user context • Reduced the wastage of resources • Worked better for normal workload
A_{11}	A hyper-heuristic scheduling algorithm for cloud [23]	• Based on two operators: diversity detection operator and improvement detection operator • Decision taken based on above operators • Reduced computational time

6 Conclusion

Task scheduling in cloud computing platform plays important role and critical part for execution of the tasks. There are eleven scheduling algorithms are investigated which have own merits and demerits. The target of any scheduling algorithm in cloud environment is to maximize resources utilization, efficient resources management, to reduce completion time, less consumption of computing power, etc. We have studied exiting task scheduling algorithms based on the scheduling parameters and what scheduling tools used in simulation process. We have found that some algorithms have better makespan, less cost of computation and better scalability as previous developed

Table 3 Comparisons based on scheduling parameters

Parameters algorithms	Performance	Execution time	Cost	Makespan	Scalability	Processing time	Quality of services (QoS)	Time	Resource utilization
A_1	✓	x	✓	x	x	x	x	x	x
A_2	x	✓	x	x	✓	x	x	x	x
A_3	x	x	✓	✓	x	x	x	x	x
A_4	x	x	✓	x	x	x	x	✓	x
A_5	x	x	x	x	x	x	x	✓	✓
A_6	x	x	x	x	x	✓	x	x	x
A_7	✓	x	✓	x	x	x	x	x	x
A_8	x	x	✓	x	x	x	x	✓	x
A_9	x	✓	x	x	x	x	x	x	x
A_{10}	x	x	x	x	x	x	✓	x	✓
A_{11}	x	x	x	✓	x	x	x	x	x

Table 4 Comparisons based on simulation tools used

Tools used algorithms	SwinDeW	Amazon EC2	CloudSim	Java environment	Event-driven simulator
A_1	x	x	✓	x	x
A_2	x	x	✓	x	x
A_3	x	x	✓	x	x
A_4	✓	x	x	x	x
A_5	x	✓	x	x	x
A_6	x	x	✓	x	x
A_7	x	x	✓	x	x
A_8	x	x	x	✓	x
A_9	x	x	✓	x	x
A_{10}	x	x	x	x	✓
A_{11}	x	x	✓	x	x

methods. Some researchers solved scheduling problem using priority attributes and genetic algorithms. All the eleven algorithms have improved some parameters of scheduling which shows excellent results.

References

1. Tilak S, Patil D (2012) A survey of various scheduling algorithm in cloud environment. Int J Eng Invent 1(2):36–39
2. Selvarani S, Sadhasivam GS (2010) Improved cost-based algorithm for task scheduling in cloud computing. In: IEEE international conference on computational intelligence and computing research, Coimbatore, pp 1–5
3. Parikh SM (2013) A survey on cloud computing resource allocation techniques. In: IEEE international conference on engineering (NUiCONE), Ahmedabad, pp 1–5
4. Singh RM, Paul S, Kumar A (2014) Task scheduling in cloud computing: review. Int J Comput Sci Inf Technol 5(6):7940–7944
5. Pinedo ML (2008) Scheduling: theory, algorithm and system, 3rd edn. Springer, Berlin
6. Salot P (2013) A survey of various scheduling algorithm in cloud computing environment. Int J Res Eng Technol 2(2):131–135
7. Thakur P, Mahajan M (2017) Different scheduling algorithm in cloud computing: a survey. Int J Mod Comput Sci (IJMCS) 5(1):44–50
8. Awan M, Shah MA (2015) A survey on task scheduling algorithms in cloud computing environment. Int J Comput Inf Technol 4(2)
9. Zhan Z, Liu XF, Gong Y, Zhang J, Chung HS, Li Y (2015) Cloud computing resource scheduling and a survey of its evolutionary approaches. ACM Comput Surv 47:63:1–63:33
10. Wu CQ, Lin X, Yu D et al (2015) End-to-end delay minimization for scientific workflows in clouds under budget constraint. IEEE Trans Cloud Comput 3(2):169–181
11. Kumar MS, Gupta I, Jana PK (2017) Delay-based workflow scheduling for cost optimization in heterogeneous cloud system. In: Tenth international conference on contemporary computing (IC3), Noida, pp 1–6

12. NZanywayingoma F, Yang Y (2017) Effective task scheduling and dynamic resource optimization based on heuristic algorithms in cloud computing environment. KSII Trans Internet Inf Syst 11(12):5780–5802
13. Haidri RA, Katti CP, Saxena PC (2017) Cost effective deadline aware scheduling strategy for workflow applications on virtual machines in cloud computing. J King Saud Univ Comput Inf Sci. https://doi.org/10.1016/j.jksuci.2017.10.009 (in press)
14. Lin C, Lu S (2011) Scheduling scientific workflows elastically for cloud computing. In: IEEE 4th international conference on cloud computing, Washington, pp 746–747
15. Xu M, Cui L, Wang H, Bi Y (2009) A multiple QoS constrained scheduling strategy of multiple workflows for cloud computing. In: IEEE international conference on parallel and distributed processing with applications, pp 629–634
16. Liu K, Yang Y, Chen J, Liu X, Yuan D, Jin H (2010) A compromised-time-cost scheduling algorithm in SwinDeW-C for instance-intensive cost-constrained workflows on cloud computing platform. Int J High Perform Comput Appl 24(4):445–456
17. Pandey S, Wu L, Guru SM, Buyya R (2010) A particle swarm optimization-based heuristic for scheduling workflow applications in cloud computing environments. In: Proceedings of the 24th IEEE international conference on advanced information networking and applications, 20–23 Apr 2010, pp 400–407
18. Karthick AV, Ramaraj E, Subramanian RG (2014) An efficient multi queue job scheduling for cloud computing. In: IEEE conference world congress on computing and communication technologies, Tiruchirappalli, pp 164–166
19. Chopra N, Singh S (2013) HEFT based workflow scheduling algorithm for cost optimization within deadline in hybrid clouds. In: IEEE fourth international conference on computing, communications and networking technologies (ICCCNT), Tiruchengode, pp 1–6
20. Verma A, Kaushal S (2014) Deadline constraint heuristic based genetic algorithm for workflow scheduling in cloud. J Grid Util Comput 5(2):96–106
21. Zhao C, Zhang S, Liu Q, Xie J, Hu J (2009) Independent tasks scheduling based on genetic algorithm in cloud computing. In: IEEE international conference on wireless communications, networking and mobile computing, pp 1–4
22. Assuncao MD, Netto MAS, Koch F, Bianchi S (2012) Context-aware job scheduling for cloud computing environments. In: 5th international IEEE conference on utility and cloud computing (UCC), pp 255–262
23. Tsai C-W, Huang W-C, Chiang M-H, Chiang M-C, Yang C-S (2014) A hyper-heuristic scheduling algorithm for cloud. IEEE Trans Cloud Comput 2(2):236–250

A Survey on Cloud Federation Architecture and Challenges

Lokesh Chouhan, Pavan Bansal, Bimalkant Lauhny and Yash Chaudhary

Abstract Conventional Cloud Computing system possess some limitations. It is difficult for users to switch between various cloud providers due to lack of a standard architecture. There is no standard metering system for cloud computing services. For cloud providers, it is difficult to maintain performance transparency due to changing user requirements, varying loads and limited resources, especially for small and medium providers. Federation of Clouds is the possible solution for these problems. In this paradigm, various cloud providers can federate with each other to expand their business opportunities while achieving optimal resource utilization and cost effectiveness. This collaboration of clouds is fruitful to users in terms of better pricing options, availability of services and overall better Quality of Service. Unlike current Cloud Computing, Federation of Clouds requires a standard architecture to which every participating cloud provider must comply. In nutshell, Federation of Clouds opens a domain of infinite possibilities to reshape the existing world of Cloud Computing and Information Technology, in general. It will provide a level playing field for emerging small and medium level cloud providers to compete among big players. It also has the promise to provide users with better ways to access great computational power and resources.

Keywords Cloud computing · Federation of clouds · Hybrid clouds · Federation of hybrid clouds

1 Introduction

Cloud computing is a paradigm which deals with providing pay-per-use, metered services and computational resources to the users on internet. Cloud computing has gained lot of popularity and has grown up to be fifth largest utility after water, electricity, gas and telephony with a total market cap of $210 billion in 2016 [1]. This has attracted interest of lots of players across the industry. Cloud computing has gained the investment of mega-providers like Amazon, Google, Microsoft and

L. Chouhan · P. Bansal (✉) · B. Lauhny · Y. Chaudhary
National Institute of Technology Hamirpur, Hamirpur, Himachal Pradesh 177005, India
e-mail: bansal.pavan92@gmail.com

R. K. Shukla et al. (eds.), *Social Networking and Computational Intelligence*, Lecture Notes in Networks and Systems 100, https://doi.org/10.1007/978-981-15-2071-6_5

many other small and medium scale companies. Cloud computing opens innumerable doors for innovation across various domains. There is an evergrowing necessity of improving customer satisfaction and quality of services. A logical step in this direction is federation of cloud. Federation of cloud is a system in which more than one cloud providers with variations in their architectures, services provided, technologies policies and pricing, federate together with each other to provide user with single point access of services. The collaboration between different cloud providers leads to proper utilization of resources, costeffectiveness, proper pricing policies and increase in business opportunities. The Cloud Federation Architecture [2] explained in this work, focuses on standards for provisioning of services to users. This becomes difficult since it requires various Cloud Providers to come together and follow a common standard. Also, analysis of various types of services from Cloud Providers and users perspective is required. The system tries to define a simple framework which is used for searching services. It also analyzes the scenarios for interfacing service from different Cloud vendors. For efficient communication between multiple vendors in a Federated Cloud, resource brokering is a must. In order to achieve this resource brokering in a transparent and effective manner, the system discusses the Cloud Federation platform in cases where it acts as a normalization layer. This is an important scenario for interoperability among traditional Cloud Computing platforms. The architecture deals with a logical framework, which enhances the resource allocating capabilities of a Cloud provider forming a federation with other vendor. The other provides can offer them their unused resources and capabilities. Thus each participant in a Cloud Federation ensures overall profits. The system is described as "Cloud Orchestration and Federation" platform in the paper. It is based on a centralized Orchestrator. Orchestrator can be explained as a mediator between the users requests and the resources available. In this way, the platform ensures transparency in federation of resources among multiple Cloud providers, and, allows dynamic allocation of resources. It acts as a balance between user requests and the platform constraints. From users point of view, they gain the ability to choose among a wide plethora of resources from a common single interface. This provides various pricing facilities and greater extent of availability for them. The system suffers from various challenges like Inter-domain Communication, Resource and Energy Utilization, Decentralized Resource Finding and Allocation, End-to-End Security etc. Remaining of this paper is organized into four sections. Literature Review is given in Sect. 2. Section 3 presents the basic architecture of Federated Clouds. In Sect. 4, Broker based Framework is discussed. Further, Sect. 5 gives an overview of Federation of Hybrid Clouds. Section 6 puts some light on the open challenges of this architecture. Finally, Sect. 7 conclude this paper.

2 Literature Review

Zangara et al. [2] have described a simple prototype of Cloud Federation Architecture. It describes the implementation of the complete platform by proposing the implementation methods for basic modules involved. There are proposed implementation of metering and billing modules, which are quite robust and have been presented as an effective solution. There is a discussion on some important aspects that must be dealt with in production scenarios, primarily, the communication gap between the Cloud providers and the Cloud Federation platform, dynamic allocation of resources and the associated changes in cost, management of multiple customers trying to access a single instance, and, security issues and policies related to membership of a Cloud provider in a federation. Experiments have been performed on a prototype developed on the basis of the proposed architecture. Experimental results show that the prototype successfully federates the providers used i.e. OpenStack and CloudStack. Customers have a choice to select the best service on the basis of their technical requirements and the available pricing options. Customers do not need register separately to each participating Cloud Providers in the federation. Assis and Bittencourt [3] have discussed some inherent limitations to cloud computing and have given a reasoning on why Cloud Federation of Providers is a solution to these. There is a discussion on the characteristics of a simple Inter-Cloud arrangement. The focus of the paper is on voluntary aspects of the participating Cloud providers in a Cloud Federation. A Voluntary Index has been defined on the basis of Volunteer Characteristics, which describes three categories of Cloud Federation—proto, middle and full. Finally, there is a discussion on voluntary characteristics and voluntary index of some existing Cloud Federation technologies like Fogbow, Resource Market, RESERVOIR, Broker Multicloud and EGI Cloud Federation. Patel and Dahiya [1] discuss about the definitions and architectures of existing cloud computing infrastructure with the limitations associated with single cloud providers (existing trend) like scalability, availability and describe the concept of aggregating cloud providers in a collaborative manner. They discuss about a multi-cloud environment where multiple cloud providers collaborate with each other and enlist various motivations that are driving the researchers towards this goal like resource constraints, vender lock problem, portability etc. They discuss about various challenges associated with federating cloud providers like varying architectures, virtualizations, service level agreements etc. They talk about various models and approaches proposed by various researchers in wake of the challenges discussed and also try to analyse the gaps between existing between solutions and problems that they need to address. Messina et al. [4] discuss about the reliance of the cloud architectures on a centralised control which exposes them to bottlenecks and adversely affect the Quality of Service. They talk about the resource finding and allocation problems that arises in cloud computing in general and limitation which are posed by centralized architecture. They propose a decentralized resource finding and allocation architecture named DEVIRA (Decentralized Virtual Resource finding and Allocation) where the virtual resources of various cloud platforms are organised in a structure of an overlay network. Basic model for DEVIRA is discussed with algorithm. The performed experiments show insensitivity towards the increasing network status and loading conditions thus warranting the proposed approach. Biran et al. [5] discussed about the proper utilization of energy and resources which are the major factors in the efficiency of federated cloud. There is some discussion on the consequences of cloud on ecosystem like the percentage of carbon emission increasing year by year which puts bad impact on environment. There was a discussion about multi-platform computing which tends to use multiple-clouds for service. There is another counter trend which is the formation of mega cloud organizations as single source providers of infrastructure and services. It has some limitations. The system of system model was introduced in order to minimize energy and utilize resources by having the option of shared resources rather than making out of computing resources within each cloud service provider. There are certain solutions to minimize the energy and resource utilization like we can provide some kind of scheduling to the resources and by lowering the data-centers deployment to provider ratio. Demchenko et al. [6] discussed about the Intercloud Architecture Framework that addresses the problem of multi-domain heterogeneity to allow on-demand provisioning. The subsequent part of paper described general use cases to provide cloud based infrastructure like IT infrastructure disaster recovery that not only require backing up the data but also the whole supporting infrastructure restoration on entirely new cloud platform which may be software or hardware platform. The further part of paper described the requirements and four main components of proposed architecture like CSM, ICCMP, ICFF and ICOF. It then described the multi-layer Cloud Services Model that combines commonly adopted cloud service models in one multilayer model.

3 The Basic Architecture

The proposed architecture [2] is based on three basic components:

- Front-end—the point of user entry to access to the Cloud Federation platform
- Resource Broker—where the Orchestrator Engine takes the customer requirements and matches it with the available computing resources, dynamically. It is supported by Metering and Billing modules
- Cloud Interface—the connection to the federated resources at the end-point (Fig. 1).

3.1 Frontend Module

It is the point of entry into the Cloud Federation Platform for the customer. It basically allows user to navigate through the available resources, choose a suitable service and go through the billing process, leading to the activation of the service. It contains three basic submodules:

- Web Frontend, enabled with an intuitive interface to handle the individual users.
- API Frontend, with interface to provide access to automated operations. It has restricted access, only available to authorized software. It is also an interface

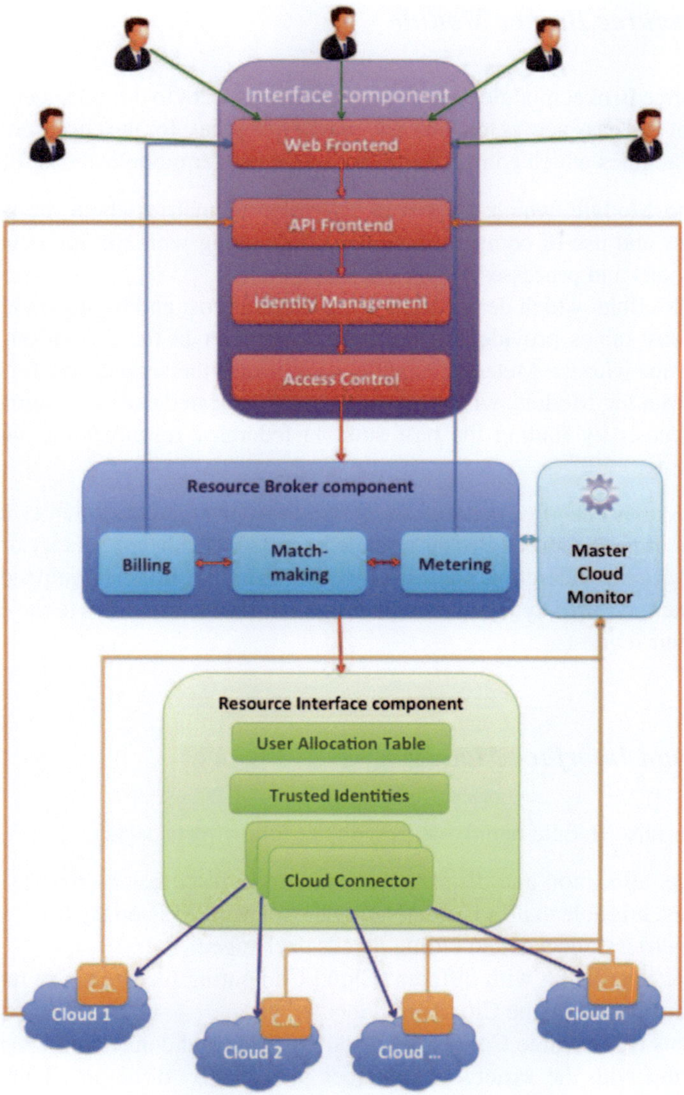

Fig. 1 Cloud Federation—a basic architecture

to access the major functioning like searching a service, selecting one and then activating it for customer, and, then notifying the platform.

- Identity Management module, which takes care of the credentials of each registered customer.

Any Cloud platform that is willing to merge its services with the Federation platform, may easily use the API Frontend to interface with, and, start using the other resources available within the federation.

3.2 Resource Broker Module

The Resource Broker module ensures high quality services to the customer. It ensures transparent and easy access to cloud computing resources, for the end-user. The three basic submodules which constitute the Resource Broker module are as follows:

- Metering Module, which deals with collection of metrics about the users level of access and use of computational resources, along with specific details of the applications and processes run by the user.
- Billing Module, which deals with management of cost and billing data. It makes use of cost-tables provided by each cloud provider in the federation. Thus, in compliance with the Metering module, it calculates the service cost for the users
- Match-Making Module, which makes use of a dedicated decision engine to serve user requests by finding the best suitable federated resources on the basis of availability.

The user may specify a generic configuration for the type of resources it needs. It is the duty of the Resource Broker module to abstract the heterogeneity of resource availability in the federation platform. An Orchestration Engine equipped with the expert system ensures dynamic provisioning of available resources to the users, that best fits their requests.

3.3 Cloud Interface Module

Cloud Interface Module mainly deals with the following aspects:

- Purchase, allocation and dismissal of computing resources by the customers or end users. It is able to do so by activating the relevant APIs on the basis of specific usage metrics collected and billing details exchanged.
- Dynamic interfacing with different Cloud Computing platforms, by making use of APIs provided by the Cloud providers.
- It consists of Resource Connector, made up of several connectors which help in interfacing with the variety of resources provided by different Cloud Provider platform, which are now part of the Federated Platform.
- To ensure smooth Cloud Interoperability, it has a Master Cloud Monitor (MCM). MCM collects Cloud Agent data and analyzes it to perform its tasks.

Thus, this module performs the overall monitoring of resources and their interfacing among heterogenous platforms. It also manages the task of interoperability among Cloud Providers.

4 A Broker Based Framework

According to Wikipedia, A broker is an individual entity who charges as the commission for every transaction between buyer and seller for managing transactions between them. Now, Cloud broker is an individual who is having mutual agreements with cloud vendors and claim the commission on the service provided to the customers of the cloud. They also help the customers to choose their cloud vendors according to their requirements. In Cloud environment, a broker mediates, aggregates, and arbitrates the services among Cloud Providers and Cloud Consumers. It provides users with the ability to interact using a single interface which brings together numerous Cloud Service providers. More precisely, CSB (Cloud Service Brokerage) is the service partner that negotiates relationship and provides interoperability between CSPs (cloud service providers) and CSCs (cloud service customers). Sustainable models for broker business must ensure that CSCs and CSPs have real interest in making use of broker services. For CSCs, a broker must help by providing more economical options to buy resources and solve the problem of vendor lock-in. For CSPs, brokers should create better growth opportunities resulting better sales and higher profits. CSBs must accomplish this by ensuring interoperability among distinct CSPs and provide scalability and availability of cloud resources. The proposed framework [7] is a centralized broker based framework which manages the allocation of various resources in a Federated Cloud environment, considering the QoS requirements of the user applications. It has following main functions (Fig. 2).

4.1 Resource Discovery

The first job of a CSB is to identify appropriate resources whenever it receives a request from a CSC to run a particular application. This task is performed by Resource Discovery Component. The component is provided with a resource information service directory, which contains the information about the resources and their availability. It searches the directory to find and select the appropriate resources based on the request of the user or demand of the application. The results are then sent to the Resource Provisioning Component.

4.2 Resource Provisioning

Once the Resource Discovery Component has determined pool of cloud resources that a particular application or job would use, it would provide that pool of resources to the end user. Based on the users request the component can allocate resources either statically or dynamically (on-demand provisioning). The SLA (Service Level

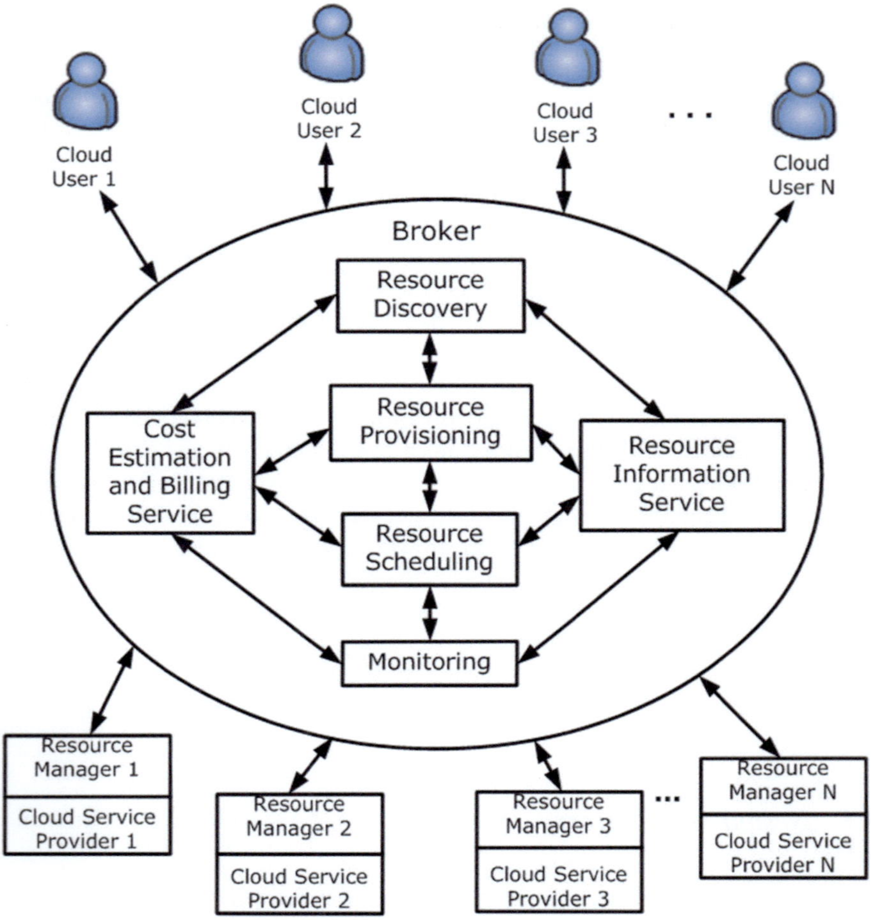

Fig. 2 Cloud Federation—a broker based architecture

Agreement) between CSPs and CSC is taken care of by the component while carrying out the process of resource provisioning. It also takes help of Cost Estimation and Billing component to calculates the cost. The component makes sure that the provisioned resources are reserved for the user, and, they are always ready to be scheduled whenever the user requests to use them. It also ensures that the resources are not under-provisioned or over-provisioned. Another important aspect of resource provisioning with which this component deals is consideration of the QoS parameters like response time, availability, performance, trust, security, reliability etc. without any violation of the SLA.

4.3 Resource Scheduling

The Resource Scheduling Component is responsible for one to one mapping of the job or application to the given VM (Virtual Machine). This is a challenging task due to the heterogeneous nature of the resources in a federated cloud environment. Also, the problem of resource scheduling in a Cloud environment is well known to be NP complete. The component plays the role of scheduling the tasks to the resources. It has to consider the user demands, QoS requirements and maximize the utilization of resources at the same time.

4.4 Monitoring

After the scheduling or provisioning of resources, they need to be monitored. Tracking the resources, their usage details, deployment details, their allocation and deallocation status etc. are some of the tasks performed by the Monitoring Component. Cost Estimation and Billing component uses the information provided by this component to perform its functioning and generate billing information for the Cloud user.

4.5 Cost Estimation and Billing

The heterogeneity of resources is an inherent problem in a Federated Cloud environment. Therefore, there are different costs of usage for the resources. The Cost and Billing Component receives the requests by users to generate the billing details. It gathers the resources usage information from the Monitoring Component and estimates the cost. It then sends the generated bill to the cloud user who requested for the same.

4.6 Resource Information Service

As the name suggests, the Resource Information Service component maintains information related to resources, more specifically, their availability status. This information keeping about cloud resources is necessary in a federated cloud environment. It helps to keep the resources leased by various CSPs at one place, so that there easy access is ensured. The component uses a Resource Information Service Directory to store the required information about resources, like location, cost of usage, SLAs, etc.

5 Federation of Hybrid Clouds

5.1 Hybrid Clouds

Hybrid Clouds have became an active area of research in the recent times. Basically a hybrid cloud is mixture of private and third party on demand public clouds to provide user with smooth integration of private and public cloud (remote). They were basically designed to handle load peaks when additional work is transferred to remote cloud.

5.2 Hierarchical Structure of Cloud

Hierarchical structure has been implemented in many contemporary systems of clouds. OpenStack (supported by IBM, Dell, HP) uses the concept of cells. Here, the whole structure of cloud has a tree like formation comprising of parent and child clouds. Child clouds are basically the sub clouds which contain the subset of the resources of their parent clouds. Eucalyptus is open-source implementation of clouds with Amazon EC2 APIs. Eucalyptus is controlled by a cloud controller which is master of all operation along with being access point for user interaction to whole infrastructure. Under the command of cloud controller, there are cluster controllers which manage the operation of a subset of clouds. Node controllers are the point of disembarkation which are responsible for the operation of final nodes. As we can see, there are many implementations of cloud which are based on hierarchical structure and this structure can be utilized for the federation of clouds.

5.3 Architecture

Proposed by Sitaram et al. [8] (Simple Cloud Federation), a way that can utilize the hierarchical structure of the already implemented clouds for federation. The basic idea comprises of making a cloud a child of cloud who is asking for its resources. Let cloud X request the resources of cloud Y, X makes cloud Y its child. X can use the services provided by Y according to the agreements but Y can not. For Y, X is just a client in need of resources. Similarly if Y wants to utilize the resources of X, it can just make X its child node. In this way, the vendor-lock in problem does not occur, because, an abstraction layer is created between the clouds. This system overcomes the problems of implementation that other designs have, like in broker based systems, a broker has to control things from top and make decisions. This development is costly and leads to redundancy. Deployment of this design is also easy as the required design is already the part of existing ones which also result in easy usage.

6 Challenges

6.1 Energy and Resource Utilization

To provide any computing services the provider (of these services) need to invest in the computational resources like storage and processors (and related infrastructure) and the energy required to operate the former. So, it is of paramount importance that the underlying architecture and protocols guarantee the their optimal utilization. The Cloud Federation provides it own unique challenges to achieve these goals. With more than one clouds independent of each others, participating as federation pose a serious challenge of resource allocation. Proper policies are needed to be formed to avoid failure for Cloud Service Providers to meet its projected goals and over-utilization of their resources, both resulting in sub-optimal performance. Task distribution among resources of different Cloud Service Providers is also big issue to be considered. Balanced distribution among pool of resources among every participating cloud improves the performance of the task giving user the best experience and providers with the fair share of their revenue. This pose as a difficult challenge because of varying regularity sovereignties of various providers, some of them very big and powerful players. Energy efficiency during operation is directly related to cost-saving and sustainability. Complementing the resource utilization problem, the energy consumption highly depends upon the load placement policies of federation. Load balancing policies across the federation needs to be made in consent with optimal energy utilization. Federation of clouds also leads to high interoperability between different clouds which in itself consume energy, so needs to be implemented with care of energy efficiency. Need for a sustainable, scalable solution for energy efficiency is of great importance for the vision of Cloud Federation. The efficiency of the architecture can be improved by efficiently handle the requests from the user [9]. This approach emphasize multitenancy as a crucial factor.

6.2 Distributed Resource Finding and Allocation

The conventional Cloud Federation architecture explained in this paper is based on a centralized system in which a central system acts as a Orchestrator. This kind of implementation of Cloud Federation introduces a bottleneck to the Quality of Service (QoS) of the whole system. So this is a major challenge for the cloud providers to implement a decentralized approach where the resource finding and allocation is done in decentralized manner. In decentralized manner, multiple systems provide fault tolerance and increase the efficiency of the system. But the synchronization among various system is the serious challenge in the federated cloud system. Distributed cloud architectures can provide greater performance in terms of communication overheads, overall durability of system, cost, latencies etc. With distributed architectures the need for centralized system in removed so protocols need to be designed to provide fairness in resource allocation keeping in consideration the peculiarities of distributed systems like latencies in message passing and other distributed algorithms required for its proper operations. Distributed resource finding provide a better way for resource allocation, more durability, more fairness and is worth all the investments that can come its way.

6.3 Resource Provisioning

In federated cloud there are multiple cloud providers providing service together to their users. Users have their specific requirements which they expect to fulfill from the cloud service provider. They choose their cloud service provider wisely according to the rank of cloud service provider. But due to multiple cloud service provider in the federated clouds, it is difficult for the users to choose cloud service provider of their specific need. The reason is some cloud service provider provides best performance at their best and some are not able to achieve that level of good performance in some parameters. The user should not be bothered with the inner selection of clouds in federation. User should be provided with a single point access to the Cloud Federation environment and all the resource provisioning should be done in optimal manner. The participating cloud providers should agree upon different protocols to provide best service to the users while keeping fairness for cloud providers in check. Load balancing plays an important role in the performance of the system. By implementing efficient load balancing, the throughput of the system can be increased and the reliability will also increase [10].

6.4 Network Monitoring in Federated Clouds

Seamless networking in cloud is somewhat difficult to achieve and it becomes more tedious in case of federated clouds. This complexity is introduced due the heterogeneity of the cloud resources in federated clouds. Federated networks not only deals with the heterogeneity of the resources but also with the different types or multi-layered virtualization. The network monitoring system also need to function with various architectures of Cloud Federation like broker based, hybrid or peer to peer. With federation of clouds comes also the issue of fairness of network resource distribution which plays essential role in overall performance of a cloud provider and thus play a key role in resource distribution across them. Full aggregated of federated network is needed to deals with the monitoring and analysis of such complex heterogeneous environment.

6.5 Security and Privacy in Federated Clouds

Security and Privacy is the primary concern in any system so is in the federated clouds. In a federated cloud, there is a big cloud provider at higher level which comprises of several small cloud providers at the lower level. So there are high chances of malicious attacks on the system and the confidentiality of data may be compromised. Federated cloud systems works on corporation with each other so any cloud provider can become source for attack on all systems so a collaborated effort is needed to provide overall security i.e. Global security policies are needed to be implemented. Integration of security features in existing infrastructure is also a challenge so that efficiency is not hampered. So the security of federated clouds is a major challenge and is a current active research topic. If we can implement end-to-end encryption within the system then the end-to-end security can be achieved.

6.6 Inter-domain Communication

There are lot of big cloud players in the market and it is very difficult for small cloud players to compete with them. So, there is a possible solution that those small cloud players can get together and can make a federated cloud. But this approach suffers from the challenge which is inter-domain communication which means how different small cloud players will communicate to act as a single big cloud player for the outside users. Global policies on matters like optimal resource allocation, fairness, security, optimal resource utilization, energy efficiency etc. need to be maintained, providing user with best experience without hassle of managing large number of small clouds and achieving good profit across the federation.

7 Conclusion

Cloud Computing has emerged as a paradigm that provides computational resources and processing power to users on a pay-per-use basis. Cloud Providers are focusing on schemes to provide best services to their customers while earning profits themselves. Forming a Cloud Federation platform appears to be a promising step in this direction. Basically, a Cloud Federation is a platform where multiple Cloud vendors come together to provide better services to their users. Such a platform is beneficial for users as it solves vendor-lock-in problem, results in better resource utilization and provide users with better pricing options. At the same time, it is profitable for participating Cloud providers because of improved resource utilization and enhanced user satisfaction, resulting in more users.

A basic architecture for a Cloud Federation contains three main modules viz Fronted module, Resource Broker module and Cloud Interface module. The front-end module provide users with a single interface to choose their preferred service and access Cloud resources. Broker module deals with pricing and billing estimation. Cloud Interface module works at core of the architecture to dynamically allocate resources to users, whenever they demand. A resource broker plays major role in a Cloud Federation platform. Broker based architecture focuses on the resource broker as a centralized entity. The steps take place in-order as resource discovery, provisioning, scheduling, monitoring and cost estimation. Federation of hybrid clouds tries to utilize the hierarchical structure of hybrid clouds to form a federation platform. This approach is a development over resource broker based architecture, because, it does not need to rely on brokers for all sorts of functions in a federated platform. Any, Cloud Federation platform has to deal with certain challenges in order to be useful to customers and Cloud providers. Resource finding, allocation and provisioning in a distributed environment is one of the major challenges faced by Cloud Federation architectures. From users point of view, network monitoring, security and privacy are important issues. Other challenges are inter-domain communication among provided in a federated platform, and, schemes for energy conservation.

Acknowledgements This research was supported by National Institute of Technology Hamirpur. I would also like to show my gratitude to the Dr. Lokesh Chouhan, Assistant Professor, National Institute of Technology Hamirpur for sharing his pearls of wisdom with me during the course of this research, and I thank my friends for their so-called insights. We thank our colleagues from National Institute of Technology Hamirpur who provided insight and expertise that greatly assisted the research, although they may not agree with all of the interpretations/conclusions of this paper.

References

1. Patel R, Dahiya D (2015) Aggregation of cloud providers: a review of opportunities and challenges. In: International conference on computing, communication & automation
2. Zangara G, Terrana D, Corso PP, Ughetti M, Montalbano G (2015) A cloud federation architecture. In: 10th international conference on P2P, parallel, grid, cloud and internet computing
3. Assis MRM, Bittencourt LF (2015) An analysis of the voluntary aspect in cloud federations. In: 8th international conference on utility and cloud computing, pp 500–505
4. Messina F, Pappalardo G, Santoro C (2014) Decentralised resource finding and allocation in cloud federations. In: International conference on intelligent networking and collaborative systems
5. Biran Y, Collins G, Azam S, Dubow J (2017) Federated cloud computing as system of systems. In: Workshop on computing, networking and communications (CNC)
6. Demchenko Y, Ngo C, de Laat C, Rodriguez J, Contreras LM (2013) Intercloud architecture framework for heterogeneous cloud based infrastructure services provisioning on-demand. In: 27th international conference on advanced information networking and applications workshops
7. Chauhan SS, Pilli ES, Joshi RC (2016) A broker based framework for federated cloud environment. In: 2016 international conference on emerging trends in communication technologies (ETCT)

8. Sitaram D, Phalachandra HL, Harwalkar S, Murugesan S, Sudheendra P, Ananth R, Vidhisha B, Kanji AH, Bhat SC, Kruti B (2014) Simple cloud federation
9. Habibi M, Fazli M, Movaghar A (2018) Efficient distribution of requests in federated cloud computing environments utilizing statistical multiplexing. Future Gener Comput Syst
10. Levin A, Lorenz D, Merlino G, Panarello A, Puliafito A, Tricomi G (2018) Hierarchical load balancing as a service for federated cloud networks. Comput Commun 129:125–137

Multi-tier Authentication for Cloud Security

Kapil Dev Raghuwanshi and Puneet Himthani

Abstract Cloud computing is one of the latest fields of Computer Science Engineering that deals with providing services to the users of the system/network as per their requirements and on pay-per-use basis. It provides a simple, flexible, heterogeneous, and architecture-neutral platform from where the user can access the desired services with ease. It can be considered as a modified form of distributed computing. The entire operation of cloud computing is based on Internet; hence, we can consider it as Internet-based computing. Security for cloud systems is an important factor that ensures trust between the cloud service provider and the users. As cloud provides a centralized repository where all services and resources reside, it is very important to ensure the authorized access to them. As cloud is based on Pay-per-Use model, it is necessary that the user should access only those services and resources for which they had subscribed. Any type of unauthorized access will lead to some type of loss. In this paper, a technique to ensure authorized access by users to cloud resources and services has been proposed, so as to overcome the above-stated issues.

Keywords Cloud Computing · Cloud Security · SPI Model · Security Threats · Authentication

1 Introduction

One of the most recent fields of Computer Science Engineering in which a huge amount of research is going on is cloud computing. In general, cloud computing can be defined as "A model for providing resources and services to its Users through the underlying Networks with ease and minimal Service Provider's Interaction. This model is Architecture Neutral and based on the concept of Pay per Use basis" [1].

In general, cloud computing provides a flexible and heterogeneous platform from where a user/a number of users can get the desired services as per their requirements.

K. D. Raghuwanshi (✉) · P. Himthani
Department of Computer Science and Engineering, TIEIT (TRUBA), Bhopal, India
e-mail: dev_2988@yahoo.co.in

P. Himthani
e-mail: puneethimthani@gmail.com

R. K. Shukla et al. (eds.), *Social Networking and Computational Intelligence*, Lecture Notes in Networks and Systems 100, https://doi.org/10.1007/978-981-15-2071-6_6

It provides a type of distributed network from where a number of users can access the services directly without intervention of others.

2 Characteristics of Cloud Computing

The essential characteristics of cloud computing [1] include:

2.1 On Demand Self Service

The user can access the services and resources provided by the cloud as and when required. There is no time boundation; only requirement is the availability of Internet [1].

2.2 Broad Network Access

The underlying network architectures are sufficient enough to enable users to access the cloud systems. There is no need to deploy new infrastructures for this [1].

2.3 Resource Pooling

All services and resources can be provided to the users by the cloud in the form of a centralized repository or pool. The user can access the desired services or resources but the user is unaware about their location (Location independence). Lack of Service Provider Interaction and Multi Tenant Model are salient features of cloud computing [1].

2.4 Rapid Elasticity

Resources and services can be added or removed from the system as and when required. Similarly, number of users can be increased or reduced. These issues will not degrade the performance of the cloud systems. Cloud systems will work effectively and efficiently, even in heavy loads with minimal resources [1].

2.5 *Measured Service*

The most striking feature of cloud systems is that it provides the services on pay-per-use basis, i.e., the customer will have to pay the amount only for the services that it had utilized [1]. Another important feature is that it is totally based on Internet. By this, sharing of information, data, and resources has also been possible between users in a network, thus utilization/efficiency of the network/systems has also been improved.

Also, it provides a heterogeneous environment, i.e., all the systems that are present in a network can be of same type or different type. We can consider cloud computing as "XAAS" where "X" is a service provided by the cloud system and "AAS" means "As a Service." From the available pool of resources and services, the user can select the resources and the services that he/she want to access. Then, the user should specify the time duration, as it is a measured service and while subscribing for the resources and services, the user will need to pay the charges applicable. The users can access the services, resources, and data provided by the cloud through the various Cloud Models.

3 Cloud Delivery Models

The services of cloud systems are provided to its users in the form of models. The Cloud Models are classified into two categories, as "SPI Model" and "Cloud Deployment Models" [1].

3.1 *SPI Model*

It is the basic Cloud Model in which we classify the services provided by the cloud systems into three broad categories, as software (SaaS), platform (PaaS), and infrastructure (IaaS) [1].

SaaS: "Software as a Service" or the SaaS Model deals with providing various software/applications to the users of the cloud systems. These applications are already deployed on the cloud systems [1].
PaaS: "Platform as a Service" or the PaaS Model enables the cloud users to access the Cloud Infrastructure for Deployment of the Software/Applications developed by them. By this, users can get the domain where they can register their applications, which can be used by others in future. These services can be accessed by using the Internet through the Web Browsers [1].

IaaS: "Infrastructure as a Service" or the IaaS Model allows the users to use the cloud system resources like servers to carry out the desired operation. The cloud infrastructure is accessible to the user and the user can use it to store its data/information [1].

3.2 Cloud Deployment Models

The Cloud Deployment Models are mainly classified into four categories, as private cloud, community cloud, public cloud, and hybrid cloud [1].

Private Cloud: The cloud infrastructure can be used solely by an Individual or a Single Organization. It can be handled by the Person/Member of the Organization to which it belongs [1].
Community Cloud: The cloud infrastructure can be shared by a specific group of people or by a specific group of organizations. It can be managed by these organizations or by any third party [1].
Public Cloud: The cloud infrastructure and its services are made available to a large group of people or a number of different organizations. It can be managed by the cloud service providers [1].
Hybrid Cloud: This type of cloud infrastructure is a combination of two or more Cloud Models (private, community, or public clouds) [1]. Here, some services and resources are only accessible to internal users of the system, while other services and resources are accessible to internal users as well as external users of the system, hence called as hybrid cloud infrastructure.

4 Security for Cloud Systems

Security is the branch of computer science that deals with keeping the data/information of the user inside the system or in the network during transmission intact. It also makes sure that the sensitive data of the users can not be accessed by others or by those who are not authorized for it.

The set of all the tools and technologies employed to keep the data inside the system intact and hidden from unauthorized access is called system security. The set of all the tools and technologies that are used to keep the data to be transmitted across the network from source to destination intact, so that it can be kept secure from unauthorized access by any third party during transmission is called network security.

4.1 Security Parameters

Authentication: Authentication checks whether the user is correct or some fake user is pretending to be an authorized user. It is basically used to validate the identity of a user [2].
Authorization: The user is accessing the service for which it has proper privileges [2].
Integrity: It refers to the consistency and correctness of the user data [2].
Confidentiality: The communication between the two parties cannot be penetrated by any third party [2].

4.2 Security Threats

Interruption: The communication between the two parties can be monitored by any third party.
Interception: The third party can act as a receiver (unauthorized access) and hijack the Message.
Modification: The third party can access the message, modifies/alters it and then sends it to the intended recipient.
Fabrication: The third party can create a message by itself and sends it to the receiver, pretending it to be the sender (unauthorized access). For cloud systems, it is important to maintain the proper secrecy of user data and the users of the system must be properly authenticated before granting access to the systems and also they can be checked about the services to be provided to them. This is because of the pay-per-use concept.

If a third party (unauthorized user) can access the system by pretending himself as an authorized user (fake identity) then it can get access over all the resources and services for which that user has proper rights. This will create issues for the user who had subscribed for the resources and services. Gaining access to someone's static ID's and passwords is very easy in today's world. So, there has to be a mechanism by which one can ensure that no one can pretend himself to be that user and gain access to the privileges to the actual user.

5 Security Measures

In normal situations, passwords, artifacts, and biometrics are used to keep the systems/computers/workstations/servers secure so that data/information stored in them should remain intact. These techniques can also be employed in cloud environments to keep the information stored in a system secure and free from unauthorized access of outsiders. To keep data to be transmitted across the network secure, we can use cryptography (Encryption and Decryption Techniques). For cloud systems, it is important that only valid users can access the system and then through some mechanism, prove that they have access to specified resources and services. Then only, the user will be allowed to access the cloud resources and services.

6 Literature Review

Gunjan et al. [3] suggested that for proper authentication of users that access the cloud systems and for proper utilization of resources and services provided to these users, identity management should be performed. The identity of the users can be verified, based on passwords, artifacts, and biometrics or by any combination of these. For this, there has to be an identity provider that generates new identities, an identity verifies that verifies the user based on the identities provided by it, an entity or the user for which the identity has been generated and the service provider that provides the services to the users on proper verification of its identity by the identity verifier. The problem is that there is no proper framework specified by them.

Govinda and Ravitheja [4] suggested a group digital signature based on which the identities are managed in the cloud systems. In his approach, there is a group manager that generates a Public Key, based on RSA (Rivest, Shamir, and Adleman) encryption algorithm. Then, it generates a shared key based on Diffie Hellman Algorithm. Then, these keys are provided to the cloud users and service providers. These keys are used to send the data between users. There is a cloud provider that stores the user data on the cloud and the data to be stored is encrypted by using the DES (data encryption standard) algorithm. The drawback of this technique is that they had not specified about the working of the system in case some nodes are need to be added or deleted.

Eludiora et al. [5] proposed an algorithm for verification of users based on the identities provided to them in which a number of entities are there to take part in the entire process which includes cloud service providers, cloud registry, bank for financial operations, Cloud Service Metering Unit, Cloud Service Billing Unit, Internet service providers, and cloud costumers. Each of them has to play a specific role in the process. But this technique is not good as its results show. Also, the billing scheme is not properly specified by them.

Karunanithi et al. [6] suggested that a One Click Log In Portal can play a sufficient role in providing Identity Management in Cloud Systems. This portal can provide authentication, password management, and security policies for proper management of cloud system.

Angin et al. [7] proposed a zero-knowledge proof authentication scheme commonly called IDM wallet, based on active bundles.

They suggested that the identities to the users are provided in such a way that the identity cannot depict the actual information of the user. For this, they suggested the use of digital identities or dynamic identities. This scheme is independent of any third party, keeps the personal information of the user secret and can unambiguously identify the users.

Horrow and Sardana [8] proposed an identity management technique, which is divided into two main modules, viz. authentication module and the authorization module. Sensors, receivers, network, services, and environment are the entities to be considered in the system. Authentication module authenticates/verifies the sensors, receivers, and services. Authorization module defines the accessibility of a service to a sensor and the accessibility of the receiver to the information provided by a particular service. They do not specify any kind of implementation or protocols regarding their approach.

7 Proposed Algorithm

For proper user identification, we are proposing a multi-tier authentication technique. This approach works as follows:

Step I: In this step, the user has to provide his credentials (LOG IN ID and password) to the system. The system then validates the credentials. If the credentials are correct, then the user successfully completes the first level of authentication.
Step II: In this step, the user has to select a combination of 3 images. This is basically a Graphical Based Authentication Level. Here, the user has to select 3 images that he/she had already specified during the registration phase.

After the second step is completed successfully, the identity of the user has been successfully verified. If the user is verified through both the levels, the user is valid. The user can now access the system and can look for the resources and services that he/she wants to access and then subscribe for them, so as to finally access them and carry out his required tasks.

Step III: Here, the user will select the resources and the services that he/she wants to access and subscribe for them by paying the specific tariffs according to the time duration for which they are going to be utilized. Once the payment has been made by the user, a "Unique Service Number" will be generated for the user. Through this unique service number, the user can get access to the specified resources and services.
Step IV: If the unique service number provided by the user is incorrect or does not recognized by the cloud system, the user will not be allowed to access the resources and services (Fig. 1).

In step 2, we had used recognition-based graphical passwords. Another option is to use recall-based graphical passwords (pattern lock) for this.

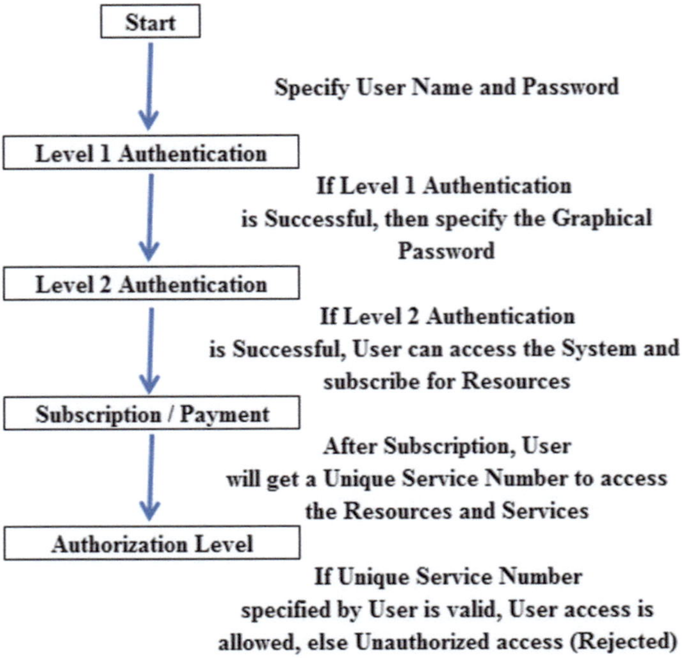

Fig. 1 Flow chart of the proposed algorithm

7.1 Modules of the Proposed System

The basic modules in the proposed system are as follows:

Registration Module: The user has to specify his credentials (USER ID and password) and graphical password for accessing the selected services/resources in later phases.

Change Password Module: In this, the user has the flexibility to change his password, as well as graphical password as per his/her needs.

LOG IN Module: The user accesses his system with the user ID and password. If credentials are correct, then the user will have to provide the graphical password. If it is also correct, then the user authentication is successful.

Authentication Module: In the first step, it validates the credentials provided by the user. If correct, user goes to second level of authentication. In the second step, it validates the graphical password provided by the user. If correct, the user successfully completes the second level of authentication. It means user is valid.

Subscription Module: Here, the verified user will look for the resources and the services that he/she want to access. Then, after specifying the time duration and resources to access, the user will make payment as per the tariffs. If payment is successful, the user will get a unique service number through which the user will access the subscribed resources and services.

Authorization Module: To access the subscribed service or resource, the user will provide the unique service number to the system. The system will validate it. If the unique service number is valid, the user will be allowed to access that service or resource, else the request will be rejected, as it will be unauthorized access if unique service number is invalid.

8 Conclusion

Cloud computing is a very interesting and efficient platform from where a number of users can get the service that they require. Also, it is very easy to access as almost everyone in this world is using Internet now a day. But everything thing in this world has some positives as well as some negatives. The major advantage is that it is flexible, scalable, heterogeneous, distributed, and architecture neutral.

But we have to make sure that it can be accessed only by the authorized/authenticated users. Otherwise, it can create serious troubles for its users as they had already paid the subscription amount for the utilization of system and its resources by unauthorized people pretending as authorized ones. There has to be a proper verification of identities of all the users who want to access the cloud and also their privileges should also be verified before providing any service to them.

By proposed algorithm, the chances of unauthorized access by the users to the cloud resources and services are nearly reduced to 0%. Yes, the complexity of the proposed algorithm is high, but if we can ignore this, then it is a good solution for implementing trust and achieving identity management in cloud systems.

References

1. Mell P, Grance T (2011) The NIST definition of cloud computing, NIST, Sept
2. Hashizume K, Rosado DG, Fernández-Medina E, Fernandez EB (2013) An analysis of security issues for cloud computing. J Internet Serv Appl
3. Gunjan K, Sahoo G, Tiwari RK (2012) Identity management in cloud computing—a review. Int J Eng Res Technol
4. Govinda K, Ravitheja P (2012) Identity anonymization and secure data storage using group signature in private cloud. In: ICACCI, Chennai
5. Eludiora S, Abiona O, Oluwatope A, Oluwaranti A, Onime C, Kehinde L (2011) A user identity management protocol for cloud computing paradigm. Sci Res Int J Commun Netw Syst Sci
6. Karunanithi D, Kiruthika B, Sajeer K (2011) Different patterns of identity management implemented in cloud computing. In: ICAIT 2011, IPCSIT
7. Angin P, Bhargava B, Ranchal R, Singh N, Linderman M, Othmane LB, Lilien L (2010) An entity centric approach for privacy and identity management in cloud computing
8. Horrow S, Sardana A (2012) Identity management framework for cloud based internet of things. In: Secure IT conference, India, ACM Journal

Investigations of Microservices Architecture in Edge Computing Environment

Nitin Rathore, Anand Rajavat and Margi Patel

Abstract Purpose of Internet of Thing (IoT) is to carry each entity on the web, consequently creating tremendous measure of information that can crush network data transfer capacity. To provide cloud services in boundary of end user edge computing turned into promising fashion, to conquer such problem, from centralized computation to decentralized computation. Over the last one decade the advancement of Internet services has constrained outlook changes in the course of the monolithic architectures to Service Oriented Architecture (SOA) and thusly from SOA to microservices. So in this paper we will try to investigate the suitability of microservices architecture style in edge computing environment and determine some similarities in the goals of microservices architecture style and edge computing environment.

Keywords Cloud computing · Edge computing · Microservices · IoT

1 Introduction

Cloud computing is Internet based computing that offers pooled computer processing resources/data/information to computers and other gadgets on request, over the World Wide Web (WWW). Everyone on the globe wish to be interconnected and access programs/data at any site, from anyplace, whenever, has to utilize application of cloud computing. Over the last one decade business applications are moving to the cloud, a shift from traditional software models to the Internet based computing. To save and process data/information produced by computer system, mobile devices such as cell phones, tablets and smart cameras; about all cloud based applications are making use of a data center as a centralized server. Mobile devices such as smartphones, tablets, and smart cameras, called as edge devices. However, usage of cloud as a centralized

N. Rathore (✉) · M. Patel
IIST, Indore, India
e-mail: nitin.rathore08@gmail.com

A. Rajavat
Department of Computer Science and Engineering, SVIIT, SVVV, Indore, Madhya Pradesh 453111, India
e-mail: anandrajavat@yahoo.co.in

R. K. Shukla et al. (eds.), *Social Networking and Computational Intelligence*, Lecture Notes in Networks and Systems 100, https://doi.org/10.1007/978-981-15-2071-6_7

server essentially expands the recurrence of correspondence between client devices and unified server [1], as they may be situated away from client devices, this is restricting for the application that require real time response. Consequently there has been a need for processing and storing data towards the edge of network. This practice is known as edge computing.

Edge Computing is characterized as moving some of the computational load from centralized server towards boundary of network (edge nodes such as switches, routers) to cater computational capabilities and reduce network traffic that may lead to congestion. The aim of edge computing is to explore possibilities of performing computations on edge nodes through which network traffic is directed, for example routers, switches and base stations, refer to as edge nodes. As evaluated by Cisco Global Cloud Index, by 2019, information created by individuals, things and devices will achieve 5 ZB; however, the worldwide IP activity will merely arrive at 10.4 ZB by 2019 [2]. About 45% of IoT generated information will be saved, investigated, filtered and acted upon near to, or at the edge of the network, by 2019 [3]. Goal of Internet of Things (IoT) is to carry each object (e.g. smart cameras, wearable, ecological sensors, home equipment's, and means of transportation) online, consequently creating tremendous measure of information that can crush network data transfer capacity [4]. So there is a required to process data at the boundary of network.

Advantages of edge computing over cloud computing:

1. **Minimizing network congestion**—Because of enormous volumes of information and constrained network bandwidth; in edge computing, edge nodes will channel and lesson information earlier then being dispatched to cloud server.
2. **Analyzing the data at the edge of network**—Devices located at the edge of network (edge nodes) will analyze data streams generated by end devices. The key advantage of performing this, if the consumer of the data is local to the data generating device, there is no need to send the data to centralized server. Analysis of the data is done perfectly, if performed local to source of data where it is generated.
3. **Response in real time**—In edge computing, it is possible to perform real-time data analytics near the data generating nodes. For applications, like self-driving cars, visual guiding service, decisions based on sensor data often must be made in real-time.
4. **Privacy**—Edge computing ensures that privacy and information security are provided at the edge of the network [5].

While there has been significant research has been done in edge computing, most of it has focus on system aspect of edge computing that comprises of network system, middleware and provide support for edge computing on cloud [6]. However, creating applications utilizing edge computing resources is problematic on the ground that it includes organizing extremely dynamic, heterogeneous resources at various stages of network hierarchy to support low latency and scalability requirements of applications [7]. Till date has been not as quite a bit of work done, starting at application development, that still remains a sophisticated procedure because of intrinsic nature of the IoT as IoT condition made of heterogeneous, widely distributed devices and computing resource.

Over the last one decade the advancement of Internet services has constrained outlook changes in the course of the monolithic architectures to Service Oriented Architecture (SOA) and thusly from SOA to microservices [8]. Definition and advantages of microservices architecture described below. Lewis and Fowler defined the microservices architectural style as "an approach for building a single application as a suite of tiny services, each running in its own process and communicating with lightweight components, often an HTTP resource API" [7].

Advantages of microservices architecture:

- The main advantage of using microservices is that they are build around business capabilities.
- In monolithic approach, a change made to a small part of the application, requires the entire system to be rebuilt and deployed. In contrast to microservices, applications are composed of small, independently deployable processes.
- By utilizing microservices, a system can be developed and deployed in a way that helps to ensure that each microservice is built, operated, extended, maintained, and eventually retired or replaced by the same team [9].
- A microservice is relatively small, and therefore a developer can understand it easily and build it quickly.
- Implementing new microservices, new capabilities can be added to a system relatively easily. An existing microservice can be updated and deployed independently, enabling development teams to practice continuous delivery of features and functions.
- There is at least centralized administration for these services also microservices may be written in diverse programming languages utilize diverse information storage technology.
- Microservices eliminates long term commitment to a single technology stack, since individual services are decoupled; moving to a new technology stack on an individual service is now possible.

Rest of this paper is organized as follows. Section 2 presents literature review of programming model for edge computing environment and microservices architecture. Section 3 describes objectives of proposed work and lastly Sect. 4 concludes our work.

2 Review of Literature

2.1 Programming Model for Edge Computing Environment

In this section we analyze some existing programming model that leverages the edge computing. Hong et al. [10] proposed a programming model for large-scale situation awareness applications named as Mobile Fog. Mobile Fog requires location and hierarchy conscious handling to address the data streams from broadly dispersed edge devices. Dispersed mobile fog processes in this programming model are delineated onto scattered computing resources in the fog and cloud, and in addition to numerous edge devices. Application must implement logic that contains a number of occurrence handlers and it also call a set of functions. In response to certain occasion's Mobile fog run time system calls the occasion handlers and runs the identical program on different devices, along with cell phones, savvy cameras and computing nodes placed in the fog and cloud environment. With a purpose to permit software code to carry out its assignment concerning to the location, accessible system resources and network hierarchy level unique information about the underlying physical device is also furnished by Mobile Fog.

Giang et al. [6] proposed Distributed Dataflow (DDF) programming model for the IoT that utilizes computing infrastructures from edge to Fog and Fog to Cloud. Fog based IoT platform must be able to maintain four desirable characteristic as proposed by Giang in his work—have to maintain different PA cycles, scalability, device distinctness, and motility. DDF programming model is based on Dataflow programming model where program dialectic is derived as a directed graph (flow based), in which every node be capable of accepting input, producing output. DDF deploys flow on numerous physical devices with the goal that each physical device may be in charge for the execution of at least one node in the flow. It is also possible that a few inner segmentation information exchanges be permitted among devices so to support this DDF requires mechanism that provides correspondence connecting two nodes on various gadgets.

As the edge nodes are most likely heterogeneous devices; developer faces tremendous challenges to compose a service that might be set up in the edge computing model. Shi et al. [5] proposed the idea of building stream that is portrayed as a sequential of processing connected on the information along the information transmission way, to address the programmability of edge computing paradigm. In edge computing environment, processing can happen anyplace on the way as long as the application characterizes where the computing should be performed and the functions/computing could be whole or part functionalities of an application. The computing stream is software defined computing flow. Using the logic of this, data/application can be handled in scattered and productive way on information creating devices, edge nodes and the centralized server. A great deal of processing should be possible at the edge rather than the centralized cloud, as defined in edge computing. In this case, the computing stream flow can help the application developer to figure out what functions/processing should be performed and how information is engendered after the processing occurred at the edge node.

Gupta et al. [4] also presented a simulator named iFogSim: that defines three main services named Monitoring components, Resource management and Power monitoring that are described below.

Resource consumption and accessibility of sensors, actuators, Fog machine and network elements is monitored by Monitoring component. In order to meet application level QoS limitation and to minimize resource wastage, iFogSim provides resource management as central part of architecture that compromise of segments that logically administrates resources. Power monitoring component is essential in IoT because Fog computing includes large number of devices with heterogeneous power utilization, making vitality administration hard to accomplish. Power monitoring component is in-charge for observing and revealing the energy utilization of Fog devices in the mock-up. Programming model of iFogSim is based on distributed data flow model which is described in previous review paper.

2.2 Microservices Architecture

In this section we explore review of architecting microservices, design pattern of microservicse and their application in the cloud environment. Pahl and Jamshidi [11] presented a systematic mapping study of microservices in which they presented a microservices architecture framework, architecting principles of microservices and their design patterns. Based on [7], Pahl and Jamshidi characterized microservices through a number of principles:

1. Association around business ability.
2. Developmental outline.
3. Deployment/infrastructure automation.
4. Knowledge in the endpoints.
5. Heterogeneity and decentralized control.
6. Decentralized control of data.
7. Design for failure.

From the above mentioned point, characteristic number three to seven directly narrates to the platform on which architecture is deployed (cloud in author context). Gupta [12] in his blog "microservices design patterns", proposed design pattern on how to make up microservices together.

- **Aggregator Microservices Design Pattern**—Aggregator might be a straightforward web page or application program that summons various services to accomplish the usefulness required by the program. In this a service invokes others to retrieve/process data.
- **Proxy Microservices Design Pattern**—This is a variation of Aggregator with no aggregator, in this case a proxy may be a just dumb that delegates the request/job to one of the service.
- **Chained Microservices Design Pattern**—As the name suggests in this case, a single consolidated response to the request is produced and client is blocked until the complete chain of request/response is completed.
- **Branch Microservices Design Pattern**—It extends the Aggregator design pattern and permits synchronous reaction handling from probably totally unrelated chains of microservices. Based upon the business needs, this pattern can also be used to call different chains, or a single chain.

- **Shared Data Microservices Design Pattern**—It provides autonomy through full-stack services having control of all components like UI, middleware and transaction. Some microservices, possibly in a chain, may share cache and database storage.
- **Asynchronous Messaging Microservices Design Pattern**—REST has the limitation of being synchronous and thus blocking. So this design pattern uses message queues instead of REST request/response to remove blocking.

Sill [13] covered areas related to bundling and distribution of microservices in containers, data exchange, and their data formats, messaging and networking. Alan Sill suggested, employing container for the delivery of microservices because containers segregate execution condition from each other and also they arrive themselves to be versatile. Despite these benefits of using containers, there are also some issues regarding network communication between microservices and complication of their utilization on various physical hosts or on hosts situated in different data-centers needs to be considered. To deal with these complications requires the use of standards. Docker and CoreOS are two well known standards that provide application to packetized into containers.

In order to make microservices work in practice information exchange and control passing feature must take place at microservices boundaries. So software engineer must manage problems handling data/information exchange and put in force these services with appropriate orchestration and control. For such data exchange standards like JSON (JavaScript Object Notation) and XML exist. In cloud computing JSON and XML are most popular standard. Sensor Network Object Notation (SNON) is a depiction based on JSON, for IoT and sensor-oriented setting. To manage wide range of formats for datasets, without being bolted into a specific format, general data standard are available. For getting to microservices effective API requirements are the RESTful API Markup Language (RAML) and Swagger which has formed bit by bit into the open API inventiveness. Next step is to move toward messaging and application control. For messaging in microservices a number of messaging standard are available, among which HTTP and its variant secure HTTPs are familiar. Another example is Constrained Application Protocol (CoAP) which provides manufacturing-relevant specialized transfer. Another popular middleware messaging standard is Advanced Message Queuing Protocol (AMQP).

IoT consist of a number of heterogeneous devices and processing is distributed throughout the system so Butzin et al. [14] presented microservices approach for IoT as many of the requirements in microservices are similar to those of IoT. Assessment of microservices and IoT approach is shown in Table 1.

Bjorn Butzin concluded that Application building objectives of both microservices and IoT are very comparable anyway practice once in a while is diverse as appeared in Table 1.

Table 1 Analogy between microservices and IoT approach

Feature	Microservices	IoT
Self containment—Services ought to contain all that they have to satisfy their activity all alone, incorporates its business rationale as well as its front- and backend, just as required libraries	Build in the order to satisfy business goal and eliminates reliance, supporting APIs composed with service	Build around device capabilities and libraries not packed with application
Choreography over orchestration—Choreography utilizes a decentralized approach for service composition and it describes the interactions between multiple services from a global perspective, where as orchestration represents control from one party's perspective	Microservices preferred choreography	Often orchestration (when using HTTP or CoAP)
Container virtualization	Yes (Docker and CoreOS)	No (but similar, OSGi—Open Service Gateway Initiative)
Continuous delivery	Yes (short release cycle)	No (rare updates)
Continuous integration	Yes (test whole application)	Partly (in single vendor context)

3 Objectives

Objectives of writing this paper are as follows:

1. To develop framework for microservices architecture in edge computing environment.
2. To identify efficient parameters for quality improvement.
3. To develop fundamental design decisions and mechanism to measure performance impact.
4. To develop scalable edge computing environment.

4 Conclusion

In this paper we have presented some similarities in the goals of microservices architecture style and edge computing environment, and also suggested how microservices architecture is suitable for building edge computing environment. We discussed the core requirements that edge computing environment need to meet heterogeneous resources at different levels of network hierarchy to support low latency and scalability requirements of applications. In future we will try to explore possibilities of developing framework for microservices architecture in edge computing environment.

References

1. Varghese B, Wang N, Barbhuiya S, Kilpatrick P, Nikolopoulos DS (2016) Challenges and opportunities in edge computing. URL: https://arxiv.org/pdf/1609.01967.pdf, 7 Sept 2016
2. Cisco global cloud index: forecast and methodology, 2014–2019 white paper (2014)
3. Evans D (2011) The internet of things: how the next evolution of the internet is changing everything. CISCO White Pap 1:1–11
4. Gupta H, Vahid Dastjerdi A, Ghosh SK, Buyya R (2016) iFogSim: a toolkit for modeling and simulation of resource management techniques in internet of things, edge and fog computing environments. URL: arXiv:1606.02007v1 [cs.DC], 7 June 2016
5. Shi W, Cao J, Zhang Q, Li Y, Xu L (2016) Edge computing: vision and challenges. IEEE Internet Things J 3(5)
6. Giang NK, Blackstock M, Lea R, Leung V (2015) Developing IoT applications in the fog: a distributed dataflow approach. In: 2015 5th international conference on the internet of things (IOT). IEEE, pp 155–162
7. Fowler M, Lewis J (2014) Microservices a definition of this new architectural term. URL: http://martinfowler.com/articles/microservices.html. Accessed 2014
8. Maresca P (2015) From monolithic three-tiers architectures to SOA vs microservices. URL: https://thetechsolo.wordpress.com/2015/07/05/from-monolith-three-tiers-architectures-to-soa-vs-microservices/
9. Ahmed M (2015) Microservices architecture: redefining enterprise systems. URL: https://canvas.harvard.edu/courses/4437/files/785336/download?verifier…wrap
10. Hong K, Lillethun D, Ramachandran U, Ottenwälder B, Koldehofe B (2013) Mobile fog: a programming model for large-scale applications on the internet of things, Aug 2013. ACM. ISBN 978-1-4503-2180-8
11. Pahl C, Jamshidi P (2016) Microservices: a systematic mapping study. In: Proceedings of the 6th international conference on cloud computing and services science, Rome, Italy, pp 137–146
12. Gupta A (2015) Microservice design patterns. URL: http://blog.arungupta.me/microservice-design-patterns/
13. Sill A (2016) The design and architecture of microservices. IEEE Cloud Comput
14. Butzin B, Golatowski F, Timmermann D (2016) Microservices approach for the internet of things. In: IEEE 21st international conference on emerging technologies and factory automation (ETFA)

Improving Reliability of Mobile Social Cloud Computing using Machine Learning in Content Addressable Network

Goldi Bajaj and Anand Motwani

Abstract Mobile social cloud computing (MSCC) is a paradigm that focuses on sharing data and services between end-users over a scalable network of cloud servers, mobile, computers, and web services. Quality of Service (QoS) based task provisioning in MSCC is one of the most eminent optimization problems, also used in improving the performance of system and efficient service delivery. Cloud based social networking service (SNS) is an application platform where individuals with like interests, family, and friends communicate with each other and share the data with less or no authentication. In MSCC, the user mobility is supported by infrastructure like access points (APs) and networking protocols. Content Addressable Network (CAN) is used to provide logical structure to resources (mobile devices and servers) and look up any resource on cloud servers. MSCC performance essentially includes QoS requirement that evaluates the quality of MSCC. Apart from basic QoS like time and cost, extended QoS is crucial for evaluating these networks. In this work, a machine learning-based framework is proposed for improving QoS of MSCC through reliability. This framework not only optimizes QoS but also restrains the malicious nodes by taking feedback from ML method.

Keywords Cloud computing · Social Cloud · Mobile Social Cloud · Content Addressable Network (CAN) · Malicious node · Fault tolerance · Machine learning · Quality of Service (QoS)

G. Bajaj (✉)
Sardar Vallabhbhai Polytechnic College, Bhopal, India
e-mail: mahak.motwani27@gmail.com

A. Motwani
VIT Bhopal University, Sehore, India
e-mail: motwani.personal@gmail.com

R. K. Shukla et al. (eds.), *Social Networking and Computational Intelligence*, Lecture Notes in Networks and Systems 100, https://doi.org/10.1007/978-981-15-2071-6_8

1 Introduction

In the modern era of computing, cloud computing (CC) frameworks have become progressively more popular among IT developers and organizational clients. Simultaneously, we have seen an exceptional boost in the usage and deployment of smartphone platforms and social networking applications worldwide. Mobile devices with time are becoming essential part of one's life as these are most convenient and effective communication tools and not bounded by time and place. The rapid progress of mobile computing (MC) [1] becomes a powerful trend in the development of IT technology as well as commerce and industry fields.

Cloud computing is a way for enabling ubiquitous, convenient, and on-demand network access to a shared pool of configurable resources that can be provisioned and released with minimum management effort or service provider interaction [2]. Cloud computing (CC) [3] has been extensively accepted as the next-generation computing framework. CC offers some use of infrastructure, platforms, and software provided by cloud providers at low cost. In addition, CC enables users to access on-demand, pay-as-you-use, and elastically utilizable resources. As a result, mobile applications can be quickly provisioned and released with just effortless management by service provider. Three types of mobile clouds are as follows: offloading to a remote server, to a local cloudlet, and sharing work in a mobile p2p network [4]. At the same time, users' demands of services and storage are tremendously increasing.

Social networking (SN) or computing is related with social behavior and computing systems. In SN, individuals build relationships online to communicate with each other. In SN, an inherent level of trust is build among users and they share files including media among themselves with almost no authentication. Also, members of same social network are willing to provide their mobile service and their data to other members.

Mobile social cloud computing (MSCC) is a radical rise in social networking applications, and the hold of CC infrastructure for diverse mobile users, MSCC is evolved as an assimilation of mobile social networking and cloud computing. Authors [5] discussed the current state of the art in the merger of these two popular technologies that we refer to as mobile cloud computing (MCC). MSCC is a paradigm that converges cloud computing, mobile application and social networking. The dominating influence of mobile computing, and growth of social networks for business, and continuing rise of cloud computing has led to great use of MSCC. In MSCC, when a mobile user request service from cloud server, it is being informed about the closest mobile device of a user who belongs to the same social network and able to provide required service. The user request is seamlessly fulfilled without further authentication; MSCC is a magnificent use of cloud computing, social networking, and mobile devices to provide service.

The major issues with MSCC are related with mobile device issues as battery availability, user mobility, inherent software, hardware problems of mobile device, and malicious behavior. Malicious nodes are one that only utilizes cloud services provided from other users and refrains from providing such services to other devices.

Machine learning (ML) techniques in cloud have been in use to design and develop the models to improve security and quality of services [6–8]. The effects of malicious nodes on quality of service (QoS) parameters are presented which is rectified through proposed framework.

This paper proposes a ML-based framework that adapts the proposed service delivery (SD) algorithm to share cloud services directly among mobile users in MSCC with higher reliability. The proposed service delivery algorithm takes feedback from ML method to test whether a node is malicious or not. ML algorithm enables server to make decision for choosing genuine and nearest service provider or node. Using proposed algorithm, this work offers improvement in basic QoS such as average response time and execution time, and addresses extended QoS parameters like reliability.

The rest of the paper is structured as follows: Sect. 2 introduces studies related to this work, QoS parameter for MSCC, and Content Addressable Network (CAN) in MSCC environment; Sect. 3 describe MSCC architecture; Sect. 4 provides the working environment and proposed framework; Sect. 5 details the experiment setup, scenario, and simulation configuration and presents results; and Sect. 6 summarizes the results and concludes.

2 Related Work

2.1 Fault Tolerance in Computing Environment

See Table 1.

2.2 Quality of Service in Computing Environment

QoS is necessary parameter to evaluate its quality. Depending on the research areas, QoS is defined in different ways by different researchers. Qian et al. [9] defined basic and extended QoS for evaluating scheduling algorithms. Time and cost are under basic QoS while reliability, availability, security/privacy, and reputation enclosed in extended QoS. Besides basic QoS, extended QoS such as reputation, reliability, availability, and security can also be addressed. Cloud service replication aids in enhancing availability and prevails faults. Replication also diminishes waiting time for service requests thus enhancing performance. Proficient resource scheduling also aids in improving QoS. Reputation determines whether a mobile device is in a MSCC network is malicious. The higher value of reputation represent more reliability while lower value specifies maliciousness. A malicious node is one that uses cloud services from other users and avoids granting cloud services to others for one of the other reason.

Table 1 Related work

Authors	Proposed work
Hu et al. [6]	Multi-step-ahead load forecasting method by using support vector regression algorithm and Kalman smoother. An efficient strategy is proposed which is based on predicted results. The strategy is reducing resource requirements while maintaining service-level agreements (SLAs)
Choi et al. [7]	QoS scheduling using CAN in MSCC and fault tolerance
Marinelli [8]	Addressed major challenges and faults including frequent disconnections and scarcity
Qian et al. [9]	Discussed several issues including: privacy, security, trust, data management, and operational and end-user-level issues
Rahimi et al. [10]	Discussed the application of MCC in different fields including SN, health/wellness, learning, and commerce
Dinh et al. [11]	Presents a survey of MSCC, architecture, issues, solutions, and applications
Goettelmann et al. [12]	Proposed a framework with security emphasis for business processes deployment
Bankole and Ajila [13]	Analyzed the resource provisioning from client's perspective to take scaling decisions for hosted applications. The prediction model proposed is based on ML techniques: linear regression, neural network, and support vector machine
Varghese and Buyya [14]	Stated that ML libraries that use less memory would benefit data mining tasks for edge nodes, i.e., mobile base stations and nodes, routers etc.

2.3 Content Addressable Network

CANs are self-organizing, fault-tolerant, completely scalable, and peer-to-peer networks. CAN being a distributed infrastructure is suitable for MSCC. It offers hash table-like functionality, and request messages are forwarded by CAN node using CAN route mechanism for every key. Messages include insert, delete, and lookup. For routing, intermediate nodes forward the messages to the zone whose CAN node contains that key for routing. The IP addresses of the nodes available in routing table are used. The node checks for the nearest neighboring zone to the target point and looks up the IP address of a node. In this paper, CAN structure is employed to administer and represent network space having cloud servers and mobile devices.

CAN is a distributed network in which keys are mapped onto values. CAN is represented using Fig. 1. Keys hashed into *N*-dimensional space with 2 interfaces: Insert (key, value) and Retrieve (key).

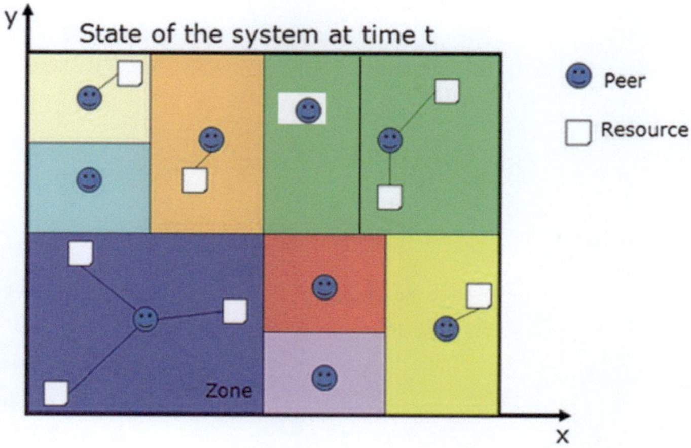

Fig. 1 Example CAN

3 Mobile Social Cloud Computing (MSCC)

Mobile social cloud paradigm is developed in view of mobility. Practically, a mobile device may fit into several SNs. The cloud services are shared among SN members with almost no authentication. The mobile devices in network are frequently requesting services and providing computing resources for serving the requests.

3.1 MSCC Architecture

MSCC architecture is depicted in Fig. 2. The connections are enabled between APs, some servers, and mobile devices. The devices form a SN with cloud server (CS) and other devices. A CS can be a part of every SN to provide computing service over cloud. Each CS manages a CAN structure to manage mobile devices and mobility. To enable CAN routing, every mobile device is indexed with CAN along with CAN coordinates. The coordinates is also called a virtual logical address.

MSCC composed of mobile infrastructure and wired servers to support mobility of users and devices. The requests are generated from smartphones, PDAs, laptops etc. APs enable communication link between wired computers and mobile devices. CSs offer services to a number of mobile devices. CS manages all the mobile devices that are registered under it. The information from these devices conveyed periodically to the server. Without further authentication, the users share the data and services in any SN. The mobile device can act as resource now. Mobile devices locate their position using GPS in the network and determine other's position from CS. The basic information about members is retrieved from primary servers. Accordingly, a device

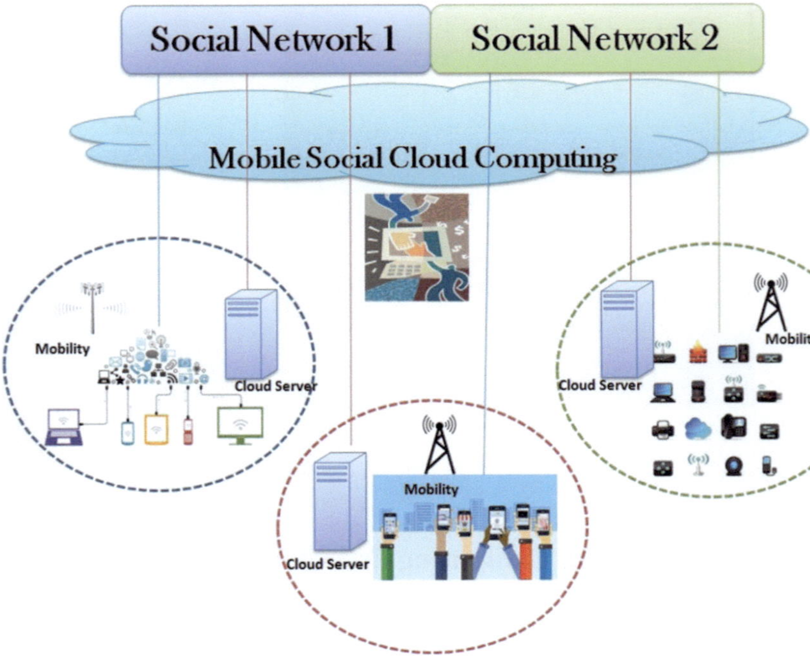

Fig. 2 Architecture of MSCC environment

requests services from the closest CS or from device which is a member of SN as well. It results in improved service response time.

3.2 Preamble

Figure 3 shows the computing environment of MSCC with a global view and communication format over MSCC. MSCC employs CAN. To understand, assume that User-A and User-B are on same SN. When service is requested by User-A from server, the server sends device information of User-B. Ultimately, both user connect and communicate to share the resources and/or services, with almost no authentication.

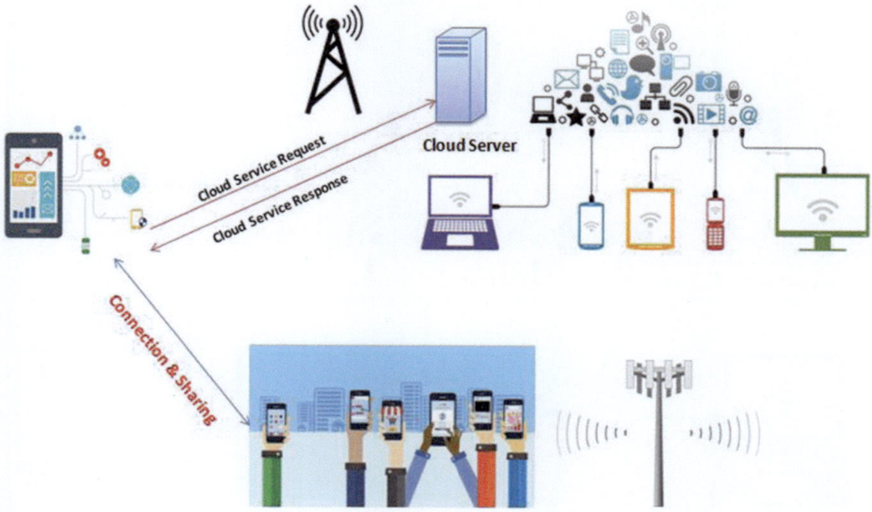

Fig. 3 Connection and sharing in MSCC environment

4 Proposed Work

4.1 Proposed Framework

In MSCC, mobile devices complete service requests with almost no authentication, but it is not mandatory that each device will share their resource information like performance, battery power etc. There are instances where devices cannot share their resource information with all requesting devices. In that case, reputation is major metric that is maintained for choosing the reliable device in terms of service delivery. Reputation factor is evaluated after each successful service delivery. Reputation factor and QoS provisioning are promising areas of research in cloud. QoS is determined by efficient provisioning of virtualized services and resources in cloud.

MSCC is a dynamic network where the malicious behavior of nodes is common and ever changing. The proposed framework is depicted in Fig. 4, where cloud server uses the ML method to detect the malicious nodes in the MSCC.

The information of same is provided to service delivery algorithm with each service provisioning. The algorithm, in addition to basic and extended QoS parameters, uses feedback from server regarding malicious nodes.

The proposed framework includes reliability centric QoS algorithm that secures the network against malicious nodes.

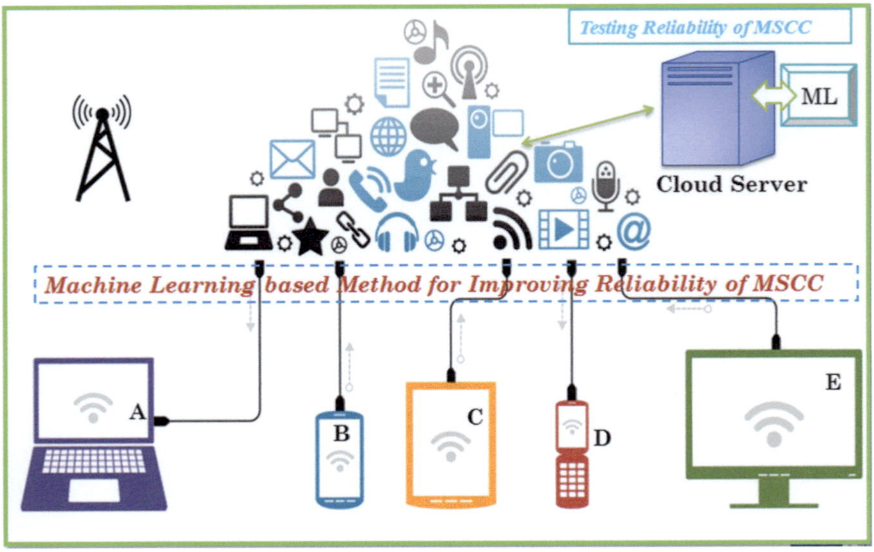

Fig. 4 Machine learning-based method for improving reliability of MSCC

4.2 Proposed Algorithm

In this work, reliability and error rate are considered as extended mobile QoS parameters. In addition, the reputation factor specifies the performance of mobile device since it computes the reputation of device on each successful delivery of service, thus providing input for ML framework.

The MSCC information managed by cloud server includes service id, provider's id, social network id, members of social network, co-ordinate of device, and reputation factor. Below, we mention service provisioning algorithm which calls the searchResource() function. Reputation of device is recomputed using a logarithmic function on each successful delivery of service using formula mentioned below.

$$\begin{aligned} Reputation_{k+1} &= Reputation_k (if\ k = 0) \\ Reputation_{k+1} &= Reputation_k + Reputation_k * log_{10}\, e^{(Reputation_k)} \end{aligned} \tag{1}$$

QoS Provisioning Algorithm

```
serviceDelivery(s_id) {
        if (s_id) {
        sf = sf + 1;
          locate mobile device that is requesting a service;
                  if (same AP) {
                                  pr_id = sr_id;
                                }
                  else {
                       pr_id = searchResource(m_id, s_id);
                       }
             return pr_id;
          else {
               sf = sf - 1;
              }
if (rs = 1) {
          Mobile device sends finish message to server;
        for (sdf == 1) {
                      computeReputationFactor(r_id );
                      }
          }
}

searchResource (m_id, s_id) {
save s_id with r_id managed by server into 1st candidate group.
for each (r_id in 1st candidate group)
  {
          if (same sn_id) {
                         Save resource searched in 2nd Candidate group
                         Apply extended QoS (Reputation, RAM, storage etc.)
                         Apply Machine learning approach for filtering malicious
nodes.
for each node if (satisfies Mobile QoS) {
                                      save the resources in 3rd candidate group
      for each (r_id in candidate group) {
                                   Compute Distance_XY between mobile devices.
                                      }
                                      }
                       }
Select the r_id with least distance (in candidate group)
return p_id
}
```

where: *s_id*: service id; *sf*: service frequency; *sr_id*: server id; *p_id*: provider id; *m_id*: mobile id; *rs*: receive service; *sdf*: service delivery flag; *r_id*: id of mobile device providing service; *sn_id*: social network id.

If coordinates of mobile device X is $X(\varphi_1, \varphi_2)$, and coordinates of Y is $Y(\gamma_1, \gamma_2)$, then Euclidian distance between A and B is represented using Eq. (2)

$$\text{Distance}_{\mathbf{XY}} = \sqrt{\sum_{i=1}^{n} (\varphi_i - \gamma_i)^2} \tag{2}$$

5 Simulation Scenario, Experimental Setup, and Result Analysis

We used CloudSim [11] to simulate the MSCC environment. CloudSim [11] is a popular simulation platform for defining and simulating CC infrastructure and services. The Java-based tool supports a large-scale cloud experiment with little or no simulation overhead. BRITE [12] (Boston University Representative Internet Topology gEnerator) is used to build the mixed environment with mobile devices and infrastructure-based network including data center and cloud servers. This tool generates realistic Internet topologies. For experimentation, mobile device in MSCC is represented as CloudSim entity to act as physical computing resource in a cloud. In the SNS, malicious users are those who have low reputation value as they take cloud services from other nodes and restrain from providing services to other nodes [1]. Malicious nodes affect the QoS at much greater extent. The experiment is done in the presence of such nodes in the network.

5.1 Simulation Configuration

To simulate the performance of proposed framework and algorithms, 100 mobile devices and 50 cloudlets are taken into consideration. The configurations and scenario are mentioned in Table 2.

5.2 Experiment Scenarios

For experiment purpose, 05 scenarios are configured considering existing and proposed work. The simulation cases are categorized in Table 3.

Table 2 Simulation configuration

Entities	Nos.
Data center (DC)	03
Operating system at DC	Linux
Broker	04
Virtual machine	30
Hosts (mobile devices)	100
Cloudlet (varying length)	50
VMM	Xen
VM configuration	RAM = 512; MIPS = 250; PEs = 1

Table 3 Simulation cases

	SNS with CAN	SNS with malicious nodes in MSCC with CAN	Reliable service delivery (SD) algorithm	Machine learning-based SD algorithm	Service replication
Case 1	N	N	N	N	N
Case 2	Y	N	N	N	Y
Case 3	Y	Y	N	N	Y
Case 4	Y	Y	Y	N	Y
Case 5	Y	Y	Y	Y	Y

Y = Yes; N = No

Exhaustive experiments have been conducted to test and evaluate the proposed work. In this work, result of 10 simulation runs for each case is presented. The average value of execution time and response time, for reliable SD algorithm and ML-based SD algorithm, are compared. The proposed work is also evaluated for reliability of algorithm and error rate.

Basically, the experiment consists of 05 cases. The experiment is first run for without SNS and without CAN (refer Case 1 in Table 3). Later on, to see the actual work, MSCC is implemented for SNS using CAN and simulated for remaining four cases.

Also, the impact of malicious nodes is shown in Case 3, where reliable SD algorithm and ML method are not applied.

Case 4 shows the application of proposed reliable SD algorithm in environment with malicious nodes.

Then, in Case 5, ML-based reliable SD algorithm is applied for achieving network reliability along with QoS. The proposed framework is providing optimal solution for efficient service delivery and better network reliability.

5.3 *Results and Analysis*

5.3.1 Average Execution Time

It is the time taken to carry out the service, which is demanded by mobile device. The results are presented in Table 4. Figure 5 shows the comparison on execution time.

Table 4 Average execution time

Instances	SNS with CAN	SNS with malicious nodes in MSCC with CAN	Reliable SD algorithm	Machine learning-based reliable SD algorithm	Service replication
1	288.2	301.75	317.96	307.8	310.6
2	288.2	301.75	321.6	305.8	306.9
3	288.2	301.75	318.25	304.4	307.3
4	288.2	301.75	315.24	307.97	310.7
5	288.2	301.75	316.23	306.4	314.2
6	288.2	301.75	318.3	302.74	315
7	288.2	301.75	319.26	302.85	309.1
8	288.2	301.75	316.28	302.74	300.77
9	288.2	301.75	315.81	304.5	309.74
10	288.2	301.75	317	301.87	312.12

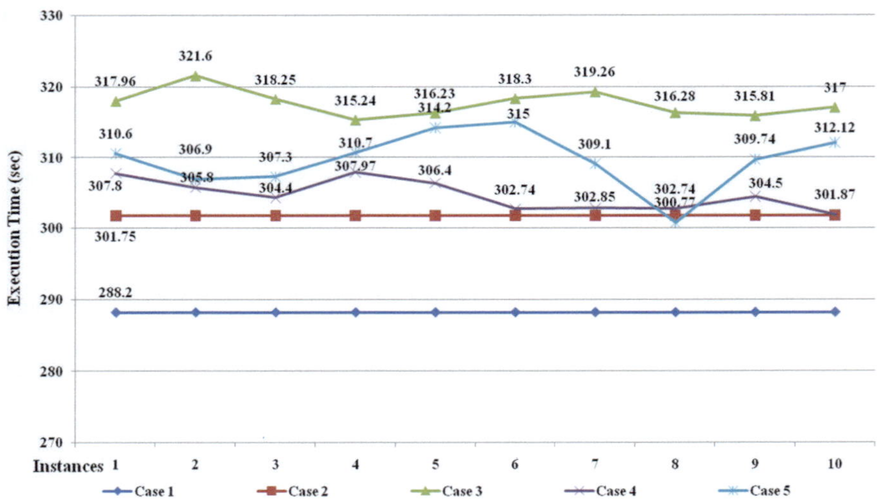

Fig. 5 Execution time result of all cases

Table 5 Average response time

Instances	SNS with CAN	SNS with malicious nodes in MSCC with CAN	Reliable SD algorithm	Machine learning-based reliable SD algorithm	Service replication
1	0.15	13.75	23.9	19.83	18.88
2	0.15	13.75	33.61	17.93	18.88
3	0.15	13.75	30.25	16.4	19.27
4	0.15	13.75	27.22	19.97	22.69
5	0.15	13.75	28.23	18.4	26.33
6	0.15	13.75	30.33	14.74	27
7	0.15	13.75	34.12	14.85	21.12
8	0.15	13.75	27.84	14.74	22.4
9	0.15	13.75	27.16	16.53	21.74
10	0.15	13.75	28.06	13.87	24.14

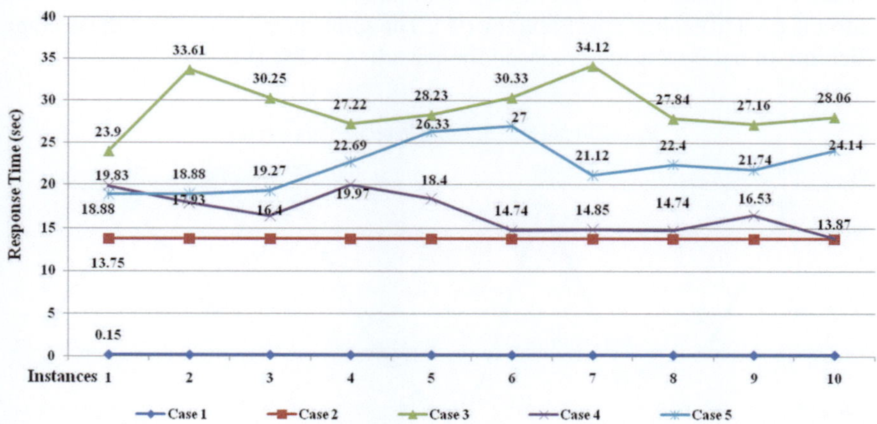

Fig. 6 Response time result of all cases

5.3.2 Average Response Time

It refers to the time an application server takes to respond a request before serving. For average response time results, refer Table 5 (Fig. 6).

5.3.3 Network Reliability

Network reliability metric is evaluated to demonstrate the efficacy of proposed method. A newer method of reliability calculation is chalked out in this research. The

Table 6 Reliability

Scenarios	Machine learning-based reliable SD algorithm (%)
1	99
2	94
3	93
4	99
5	91
6	94
7	99
8	95
9	99
10	95

proposed framework improves service delivery as well as security from malicious nodes, thus enhancing network reliability. Table 6 represents the reliability results.

For scoring the 'Reliability of algorithm' used within the proposed framework, a machine learning-based data mining method (classification) is used. Confusion matrix is used to evaluate the accuracy (%). The same metric shows that the average reliability of network using proposed framework is 95.8% (Fig. 7).

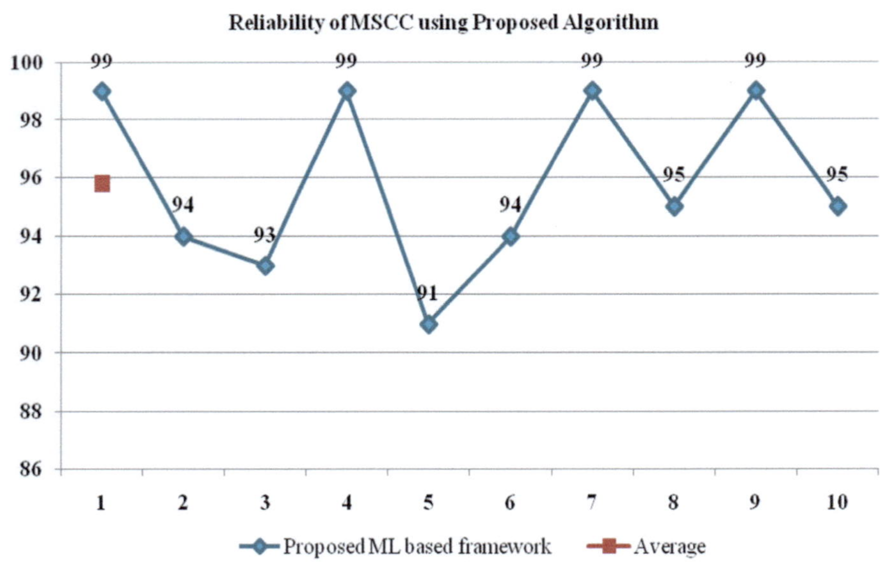

Fig. 7 Reliability of proposed algorithm

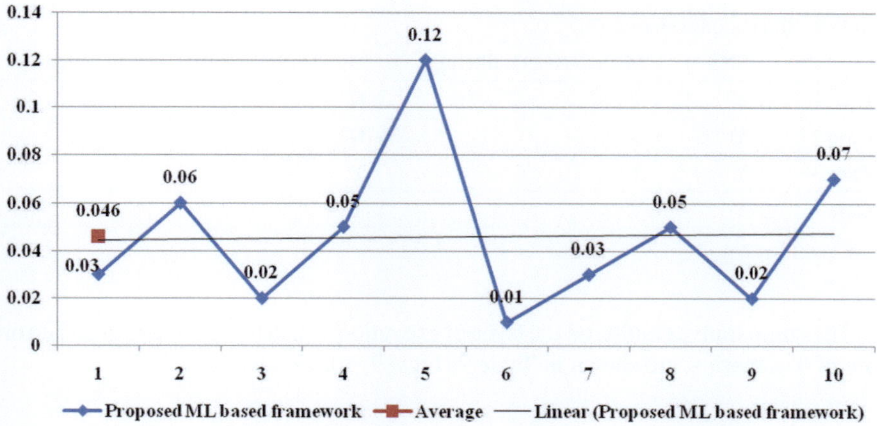

Fig. 8 Error rate of proposed algorithm

Table 7 The error (misclassification) rate is 0.046

Scenarios	ML-based reliable SD algorithm	Average
1	0.03	0.046
2	0.06	
3	0.02	
4	0.05	
5	0.12	
6	0.01	
7	0.03	
8	0.05	
9	0.02	
10	0.07	

5.3.4 Error Rate

The error rate shows the number of nodes that are not correctly identified as malicious or not in the MSCC network (Fig. 8; Table 7).

6 Comparative Study

The proposed work is also compared with the fault tolerance and QoS scheduling using CAN in MSCC by Choi et al. [7] in 4 different cases since the similar working configuration is used in [7] except ML framework. The comparative study with respect to various criteria is shown in tabular format and with the help of graphs (Table 8).

Table 8 Simulation scenarios [7]

	SNS	Malicious node filtering	User QoS	Service replication
Case 1	Y	Y	N	N
Case 2	Y	Y	N	Y
Case 3	Y	Y	Y	N
Case 4	Y	Y	Y	Y

Y = Yes; N = No

The comparative results, on the basis of execution, finish time, reliability and error rate of 4 scenarios, are shown in Table 9 (Figs. 9, 10, 11, and 12).

Table 9 Comparative study

Case	Execution time (s)		Finish time (s)		Reliability		Error rate	
	Base [7]	Proposed	Base [7]	Proposed	Base [7]	Proposed	Base [7]	Proposed
1	492.4	485.5	2121.6	2032.6	0.93	0.94	0.07	0.04
2	415.7	389.97	1642.8	1629.6	0.91	0.97	0.06	0.03
3	344.9	344.42	1501.6	1466.8	0.97	0.99	0	0.01
4	327.8	309.64	1340	1308.2	0.92	0.98	0	0.02

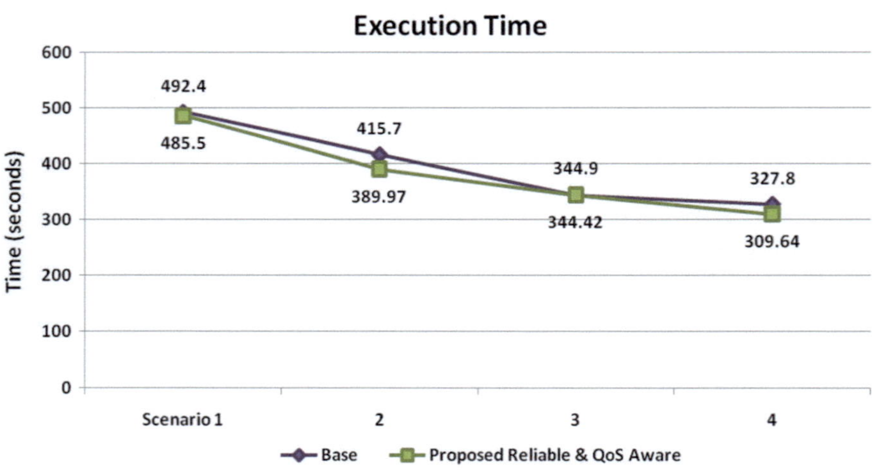

Fig. 9 Comparison of proposed and base work in terms of execution time

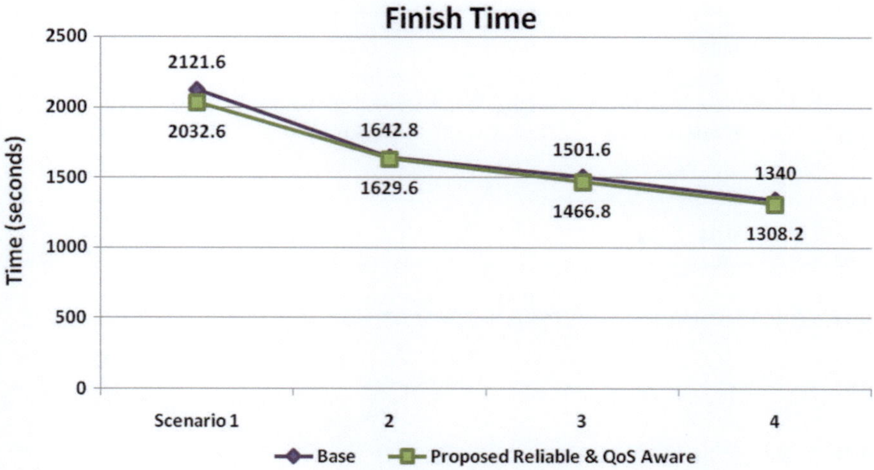

Fig. 10 Comparison of proposed and base work in terms of response time

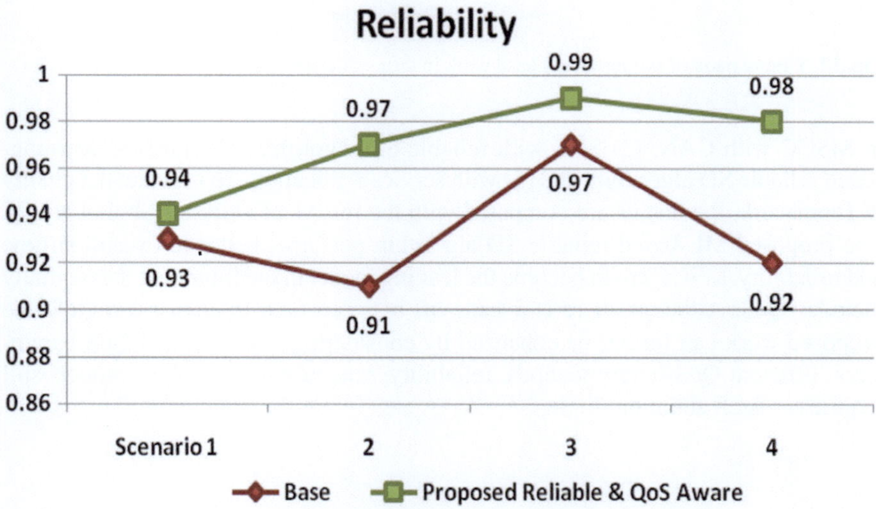

Fig. 11 Comparison of proposed and base work in terms of reliability

7 Conclusion and Prospects

MSCC is a SN-based cloud computing environment supporting mobility, QoS, and sharing of cloud services. Members of a social network share resource and services with other members with almost no authentication. A CAN structure is a distributed network structure to manage mobile devices in the computing environment. We proposed a machine learning-based service delivery algorithm in MSCC. The algorithm is implemented in five scenarios: (1) SNS with CAN, (2) SNS with malicious nodes

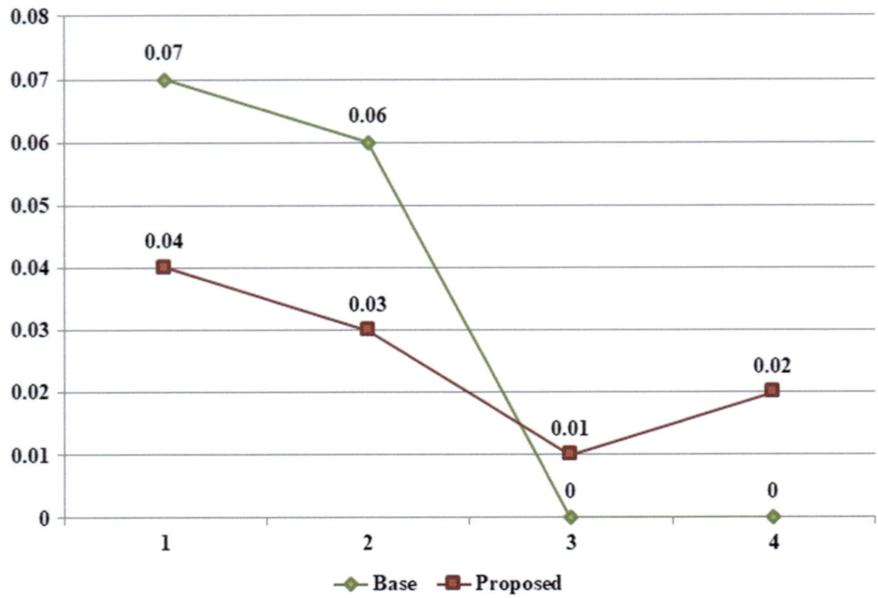

Fig. 12 Comparison of proposed and base work in terms of error rate

in MSCC with CAN, (3) SNS with reliable SD algorithm, (4) machine learning-based reliable SD algorithm, and (5) with service replication. To evaluate the quality of framework, the results are compared with the results of work by Choi et al. [7]. The proposed ML-based reliable SD algorithm performs better in execution time and reliability. MSCC could become the leading model in the future but it essentially includes QoS. Although there is a trade-off between fault tolerance and QoS, the proposed work can further be enhanced by considering few extended QoS parameters. Efficient QoS-based methods, reliability, scheduling, and robust models still require research attention.

References

1. Satyanarayanan M (2010) Proceedings of the 1st ACM workshop on mobile cloud computing & services: social networks and beyond (MCS)
2. Peter M, Timothy G (2011) The NIST definition of cloud computing. National Institute of Science and Technology, Special Publication 800-145
3. Mell P, Grance T (2010) The NIST definition of cloud computing, National Institute of Standards and Technology, ver. 15, 9 July 2010
4. Fernando N, Loke SW, Rahayu W (2013) Mobile cloud computing: a survey. Future Gener Comput Syst 29(1):84–106. ISSN 0167-739X
5. Rahimi MR, Ren J, Liu CH et al (2014) Mobile cloud computing: a survey, state of art and future directions. Mobile Netw Appl 19:133
6. Hu R, Jiang J, Liu G, Wang L (2014) Efficient resources provisioning based on load forecasting in cloud. Sci World J 2014:12 pp, Article ID 321231
7. Choi SK, Chung KS, Yu H (2013) Fault tolerance and QoS scheduling using CAN in mobile social cloud computing. Cluster Comput. https://doi.org/10.1007/s10586-013-0286-3
8. Marinelli EE (2009) Hyrax: cloud computing on mobile devices using MapReduce. Masters thesis, Carnegie Mellon University
9. Qian T, Huiyou C, Yang Y, Chunqin G (2010) A trustworthy management approach for cloud services QoS data. In: ICMLC, pp 1626–1631
10. Rahimi MR, Ren J, Liu CH, Vasilakos AV, Venkatasubramanian N (2013) Mobile cloud computing: a survey, state of art and future directions. Springer Science + Business Media, New York
11. Dinh HT, Lee C, Niyato D, Wang P (2011) A survey of mobile cloud computing: architecture, applications, and approaches. Wirel Commun Mob Comput
12. Goettelmann E, Fdhila W, Godart C (2013) Partitioning and cloud deployment of composite web services under security constraints. In: IEEE international conference on cloud engineering, pp 193–200
13. Bankole AA, Ajila SA (2013) Predicting cloud resource provisioning using machine learning techniques. In: 2013 26th IEEE Canadian conference on electrical and computer engineering (CCECE), Regina, SK, pp 1–4
14. Varghese B, Buyya R (2017) Next generation cloud computing: new trends and research directions. Future Gener Comput Syst. ISSN: 0167-739X. Elsevier Press, Amsterdam, The Netherlands

Data De-duplication Scheme for File Checksum in Cloud

Jayashree Agarkhed, Apurva Deshpande and Ankita Saraf

Abstract Data de-duplication is a significant source of security technique in the cloud environment. It accomplishes identical reproductions of the redundant records to save cloud storage. The de-duplication has been used to preserve bandwidth and confidentiality of the user. As de-duplication has many advantages, safety and protection of data is the primary concern as the data of the clients are susceptible to various types of attacks.

Keywords Cloud computing · Duplication · Checksum · Secure hashing algorithm

1 Introduction

Distributed computing gives lots of resources to the clients over the entire internet but concealing the platform and usage details. But nowadays, the cloud providers are providing the considerable processing assets at a very low expense. A huge amount of data is being stored in distributed computing that can be shared among various clients from the cloud. The privileges help in identifying the rights to access the stored data due to the feature of scalability and the information administration in the cloud. Data de-duplication has become an important technique which automatically attracts multiple users. The technique of information de-duplication is vital ideas for disposing of the same copy of data from the cloud pool, which helps in the practical usage of storage and applied to bandwidth also.

The remaining of this paper is given as follows. The overview of the work resented in Sect. 2. Section 3 presents the system architecture. Section 4 includes the results part and Sect. 5 highlights the conclusion of work.

J. Agarkhed (✉) · A. Deshpande · A. Saraf
P.D.A College of Engineering, Kalaburagi, Karnataka 585102, India
e-mail: jayashreeptl@yahoo.com

A. Deshpande
e-mail: apurvadeshpande857@gmail.com

A. Saraf
e-mail: ankitasaraf2106@gmail.com

R. K. Shukla et al. (eds.), *Social Networking and Computational Intelligence*, Lecture Notes in Networks and Systems 100, https://doi.org/10.1007/978-981-15-2071-6_9

2 Related Work

The following sections briefs some of the encryption algorithms for data de-duplication process method in the area of cloud computing.

Authors in [1] use distributed computing which gives a method for putting away voluminous information and can be effortlessly reached to anyplace. The work bargains about the avoidance of a copy of the record in the cloud. To be specific, the proprietor of the information who created will store data in the cloud. The client can be a legitimate individual who will download the record in the wake of giving the appropriate accreditations. The client provides the scrambled key which got through e-mail to download a document from the cloud. Here the third vital segment is the cloud. The record is given extraordinary esteem which can be utilized to recognize it as a Secure Hash Algorithm (SHA). For the protected encryption and unscrambling way, RSA calculation is used. The client needs to give key to decode and acquire document guarantee secrecy of information. A proprietor can transfer a document and moves a duplicate copy of the record to the cloud server. This copy avoidance in a cloud guarantees that the memory space is adequately used along with these lines decreasing the handling overhead.

Authors in [2] have used data information through de-duplication technique which gets rid of duplicate data copies to decrease storage space and exchange information transmission. Promising as it is by all accounts, a developing test is to perform secure information de-duplication in appropriated capacity. But blended encryption has grasped for secure information de-duplication. An essential issue of making simultaneous encryption conventional is to gainfully and continuously manage a significant number of joined keys. The attempt has been made to formally address the issue of finishing capable and crucial robust organization in secure information de-duplication.

Authors in [3] have given public and private clients who send the information to the distributed storage. The late information break makes encryption at end-to-end an undeniably apparent necessity. Semantically the secure encryption plans render different financially enhancement strategies, like information de-duplication and ineffectual. The information has been separated as per their fame. The information de-duplication helps for the success of public information and semantically secure encryption guarantee disagreeable substance. The author has demonstrated that the plan is safe under the Symmetric External Decisional Diffie-Hellman Assumption in the arbitrary prophet show.

Authors in [4] have provided private information de-duplication convention. In any secret information, the de-duplication convention permits a customer who holds private information. It demonstrates the server details which hold a rundown string of the information which is the proprietor of that information without uncovering additional data to the server. This idea can provide the supplement to the cutting edge open information de-duplication conventions. The security of private information de-duplication conventions unique in the recreation-based structure with regards to two-party calculations. Development of individual information de-duplication conventions in light of the standard cryptographic presumptions is then exhibited and examined.

Authors in [5] proposed information de-duplication method which is known to dispense with copies adequately. It presents a fracture that debases read execution. The de-duplication framework that enhances peruses to the most recent reinforcements of virtual machine pictures utilizing reverse information de-duplication. The regular information de-duplication expels copies from new information and ejects prints from old data, in this way moving fracture to old details while keeping the format of further information as successive as would be prudent.

Authors in [6] have provided data de-duplication as a technique for limiting the measure of capacity estimate for the association needs. Some document might show up in a few distinctive parts by various clients. Authors in [7] have given priority to the corporate revelation on PCs. The declaration of de-duplication work is without a doubt supportive in exceed figuring and has many parts for subsidence the capacity numerous a moon and watchful data communication. Examination of the security indicates for the last time that our program is light in limitation of the methodologies specified in the possible model. Here, we are patched to comprise of a portrayal of our coming approved copy seek to plan and burrow test-bed tests utilizing our sort.

Authors in [8] has found the quickly expanding measures of information delivered around the world, organized and multi-client record frameworks are winding up extremely prominent. The worries over data protection keep numerous clients moving data to remote storing. Furthermore accomplishing information de-duplication in an approved way be appeared in security conspire.

The authors in [9] have given the information data de-duplication necessary data weight frameworks for removing duplicate prints. It is used to confine the measure of storage space and extra exchange speed. For supporting the data de-duplication, the protection of needed data is ensured. The simultaneous encryption framework has been used to encrypt the information previously outsourcing. The advantage level of the customer checked with a particular objective to verify customer. Security examination demonstrates that the specified arrangement is secure. This endorsed duplicate check plot which has unnecessary overhead appeared differently about regular operations through a model of sanctioned duplicate check design.

Authors in [10] have proposed the privacy preservation scheme and cloud security issue is the most critical challenge that we are facing in today's world in cloud computing. The authors in [11] found the advanced cryptographic techniques proposed for performing security much stronger way. Many people from different field especially for business in companies are relying on cloud for storage purpose.

3 System Architecture

Three modules provide the detailed description about the proposed work. They are listed as follows.

- Data Users
- Private Cloud
- Cloud Storage Service Provider (CSSP)

A. *Data Users*

The responsibility of the user is to send data for storage to CSSP for future access. To protect the bandwidth, the user uploads the unique data elements. The individual files have their encryption key along with the privilege key to realize the authenticate de-duplication along with different privileges.

B. *Private Cloud*

For using secure cloud services this entity is being used. Here a private cloud manages private keys for all privileges and provides file token to users. A public cloud cannot be trusted, and therefore the resources at user or the data owner are restricted.

C. *CSSP Cloud Storage Service Provider*

For storing data in public cloud, CSSP entity form is used. This entity stores the data on behalf of users. It drops the duplicate data elements in aspect to reduce the storage cost. It can eradicate duplicate data by performing de-duplication, and only single data copy is being stored in the scheme. Here CSSP has abundant storage space along with computational power [12].

- **Security model**: In cloud computing, privacy is preserved in data de-duplication and develops another data de-duplication framework that may support. The proposed model is having the data security features which scrambles the data element before sending it to the cloud and also maintains the confidentiality of storing data along with data de-duplication check. The proposed technique enables the cloud to perform data de-duplication on the cipher text. The encryption of the data elements uses a key which is obtained by deriving hashing the data elements. An unauthorized user is unable to decrypt the cipher text using CSSP. In the system architecture, the data files are stored after encrypting the files in the public cloud [13].
- **De-duplication model**: The upload action is completed only when there are no duplicate files already stored in the cloud. If duplicate file is present, then that data will not be stored. Instead a pointer is sent to users so that users can access that file which helps in reducing the storage space by eliminating the duplicate copies. CSSP helps in eliminating the duplicates copies of the same files. This helps in addressing the problem of data de-duplication, along with data security [14].
- **Differential service model**: The proposed system considers different privileges of users in duplicate check beside data itself. The application will provide much security for the use of dynamic trustworthiness checking [15]. The proposed technique supports different privileged users which are also checked along with the data element itself. Differential Authorization carried out depending upon privileges and separate tokens are given to each user of the file to check for duplicity. Authorized duplicate check is done for producing a query for each block of the

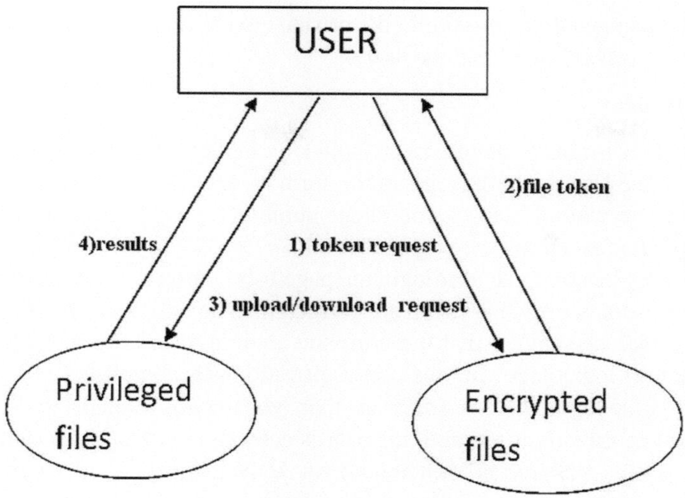

Fig. 1 Architecture for authorized de-duplication

files. A separate authorized user uses own private keys. Checking of duplicate files will be done by a public cloud [16]. Figure 1 shows the system architecture of the proposed technique.

The work describes a company by where the employee's details such as password, name, contact number, email id are registered through admin or owner of the company. Then using user id and password, the employees are able to carry out actions like file upload download and duplicate checks on the data based on their privileges. After data encryption is done for content, similar data copies are generated with the same keys with the cipher text. This helps in maintaining the confidentiality of data elements. It provides more security in comparison with existing techniques [17, 18].

4 Results and Discussion

In this section, the snapshots of the proposed system are presented. Further the results of the frame work are evaluated.

4.1 Snapshots

Snapshots show the pictorial representation of how the project design is viewed in a web view. This method refers to a private cloud page, and there is login option. By entering the private cloud details as individual cloud name, and password and the secret cloud name and password into the private cloud. There are four modules like home, admin, private cloud and the users.

A. *User module*

First, if the user is not registered, he will click on the registration option in the user module then the User login the registration form by entering the user details such as User Name, Password, Email-id, and Phone number. Figure 2 shows the home page of the work after executing the proposed work.

After the registration, the user login the page by entering the username and password. If the username and password are correct, then he will be logged in otherwise will have a message saying that the username and password are invalid. Figure 3 depicts the user login page after the user registration is accomplished.

Even after the login, there is a message that "you are not yet registered". Because login cannot be directly made until the user is activated by the private cloud. Then, it will display the message as "you are not yet Activated user".

B. *Cloud module*

Private cloud module login is made then after the authenticated user login. Figure 4 shows the private cloud server login page.

The user gets the E-mail in the mailbox and finds the token and right to update and download and update permission. Figure 5 shows the token details in the mail box received in the E-mail.

The token got in E-mail should be copied from E-mail id. By entering the correct token here one can login the page. Figure 6 specifies the login page by entering the token number.

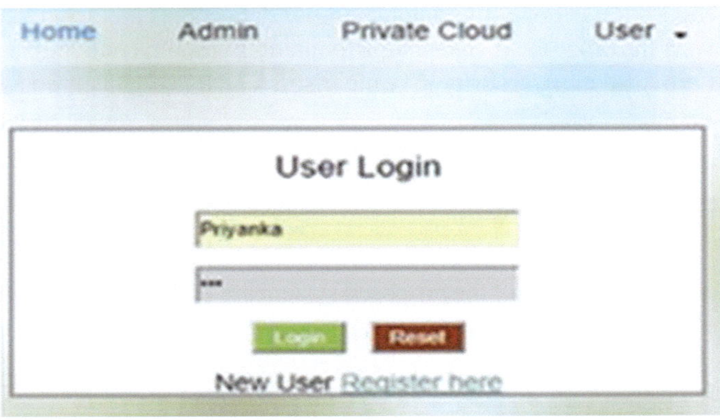

Fig. 2 User login page

Fig. 3 User login page after registration

Fig. 4 Private cloud login page

Fig. 5 The user can see their rights in mail-id and token

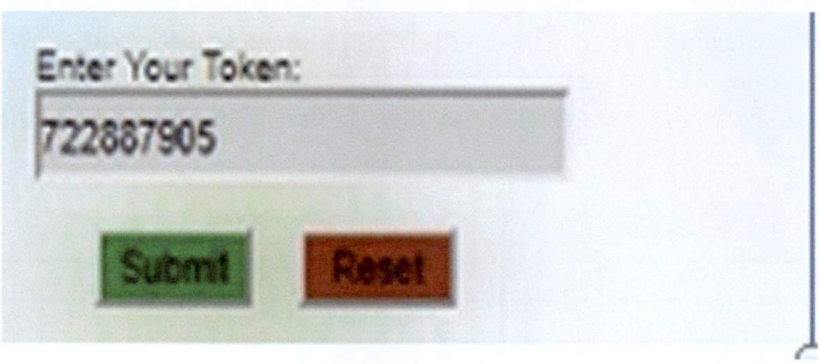

Fig. 6 User login page by entering token

Fig. 7 DriveHQ uploaded file open page

C. *Storage module*

If we want to open a file which we do not have permission, it cannot be done. This is because in DriveHQ, the data are stored in encrypted format. If anyone wants to open the uploaded file, then the content of the file will be displayed in the encrypted format. Figure 7 depicts the DriveHQ uploaded file page in encrypted form.

Figure 8 specifies the downloaded file page. In the download option, we can see all the uploaded files in the list and we can download any file. The details of the files such as file name, owner name, upload date time, size are viewed.

Here is updating a file. Previously the file contains only "hi" and now by adding "hello…" to the next line we are updating a file. Again by opening a file, the updated file can be viewed. Figure 9 depicts updating a file page.

Figure 10 specifies the administrator module. The admin login page is opened to enter the admin username and password.

Figure 11 specifies the registration page. The registration form to be downloaded for the user in the above page. The details of the files such as file name, owner name, upload date time, size are viewed.

Figure 12 shows the downloaded file page. It shows the details of the file that are

FILE NAME	OWNER NAME	UPLOAD TIME	SIZE	DOWNLOAD
iu.txt	zeba	2017/05/15 17:29:57	24bytes	Download
hello.txt	Priyanka	2017/06/09 18:31:07	24bytes	Download
hello.txt	Priyanka	2017/06/09 18:31:04	24bytes	Download
ab.txt	Priyanka	2017/06/09 22:51:35	24bytes	Download
b.txt	zeba	2017/06/30 15:33:12	24bytes	Download
abcd.txt	Priyanka	2017/06/30 17:26:31	24bytes	Download

Fig. 8 Download file page

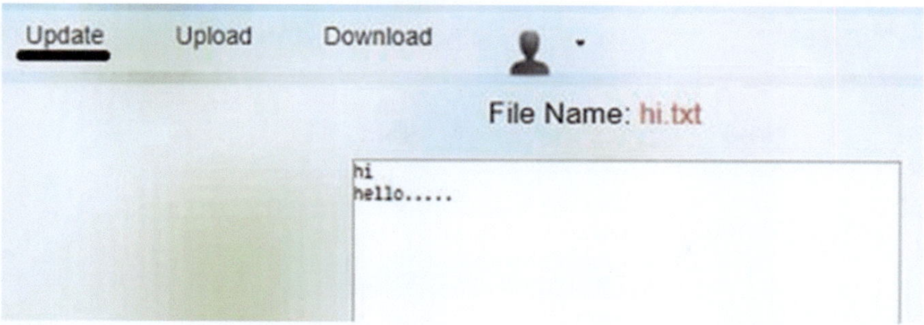

Fig. 9 Updating a file

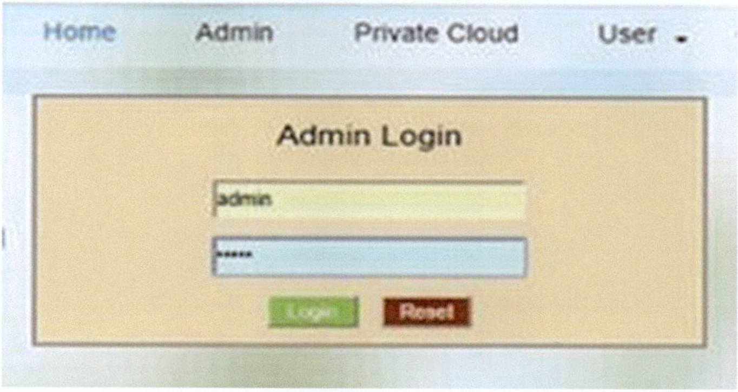

Fig. 10 Admin login page

downloaded from the cloud server. This can be done if it has permission to download otherwise it cannot.

In Fig. 13, by clicking the update option, select any one of file which one want to be updated. The update files are shown in the process at Fig. 13.

In Fig. 14, there are again options like user and user request by clicking on the user button. One can see all the details of the already registered users. Here, we have to change the status and user activated by the private cloud by clicking the action on *Activate*. The private cloud page after login is given in Fig. 14.

After the user is activated by the private cloud, then there is a token. It also gives the upload, download and update rights to the user after the submit button and the token is sent to the particular user email-id. Figure 15 gives the rights to the activated user page.

After entering the token there are options like upload, update and download. First we choose the option upload and then comes the button by pressing the button we can select the File upload. The file to be uploaded is shown in the uploaded page in Fig. 16.

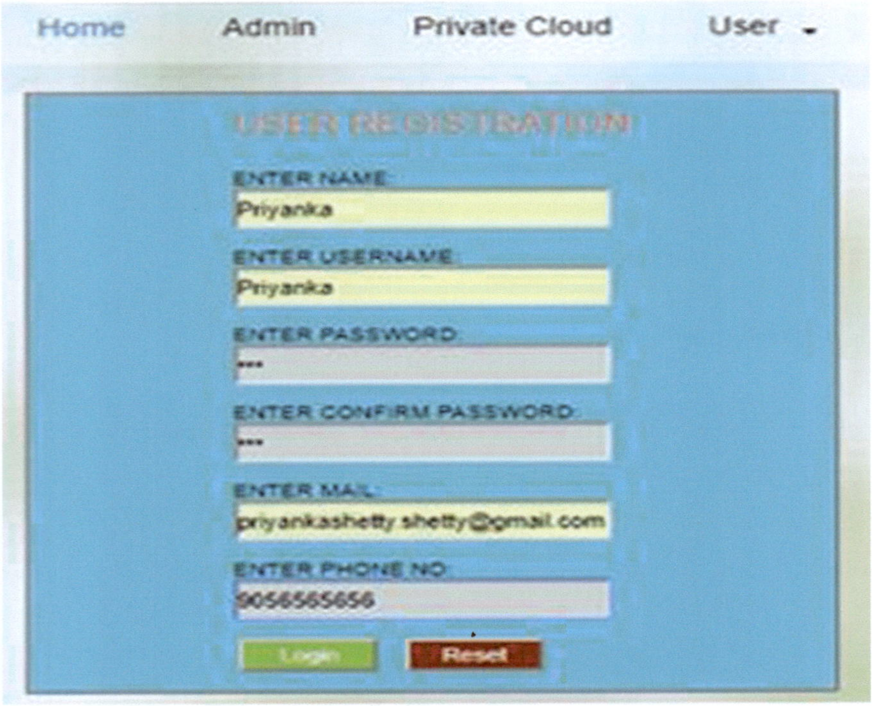

Fig. 11 Registration form

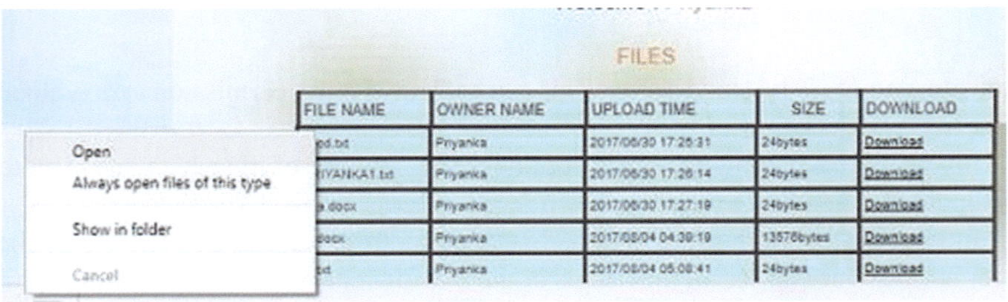

Fig. 12 Downloaded file page

If we want to upload the file which is already present in the cloud, we can upload the same file again then it will display the message as "file already exists". Hence the file duplication is restricted. The uploading the same file page is given in Fig. 17.

DriveHQ is the cloud service provider. All uploaded files are stored in DriveHQ in the drive, and gives the details of the data such as name, action, creates time modified time size etc. Next, in admin page, by entering admin details such as admin name and password as admin we can log in the page. The process is given in Fig. 18.

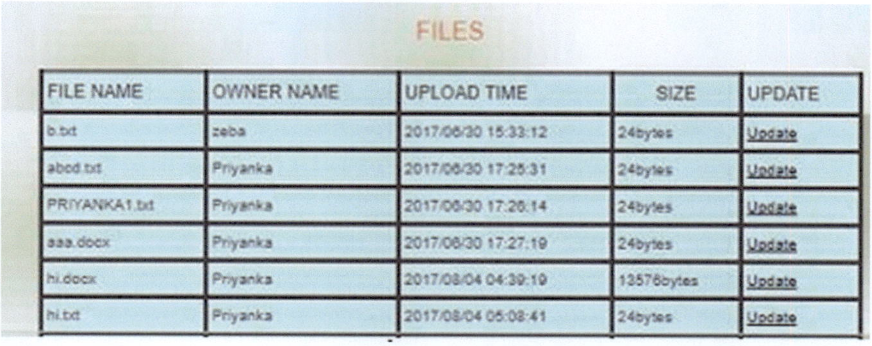

FILES

FILE NAME	OWNER NAME	UPLOAD TIME	SIZE	UPDATE
b.txt	zeba	2017/06/30 15:33:12	24bytes	Update
abcd.txt	Priyanka	2017/06/30 17:25:31	24bytes	Update
PRIYANKA1.txt	Priyanka	2017/06/30 17:26:14	24bytes	Update
aaa.docx	Priyanka	2017/06/30 17:27:19	24bytes	Update
hi.docx	Priyanka	2017/08/04 04:39:19	13576bytes	Update
hi.txt	Priyanka	2017/08/04 05:08:41	24bytes	Update

Fig. 13 Update files

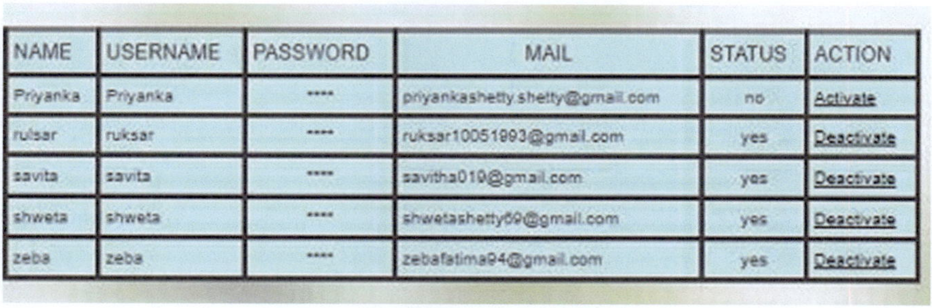

NAME	USERNAME	PASSWORD	MAIL	STATUS	ACTION
Priyanka	Priyanka	****	priyankashetty.shetty@gmail.com	no	Activate
rulsar	ruksar	****	ruksar10051993@gmail.com	yes	Deactivate
savita	savita	****	savitha019@gmail.com	yes	Deactivate
shweta	shweta	****	shwetashetty69@gmail.com	yes	Deactivate
zeba	zeba	****	zebafatima94@gmail.com	yes	Deactivate

Fig. 14 Private cloud page after login

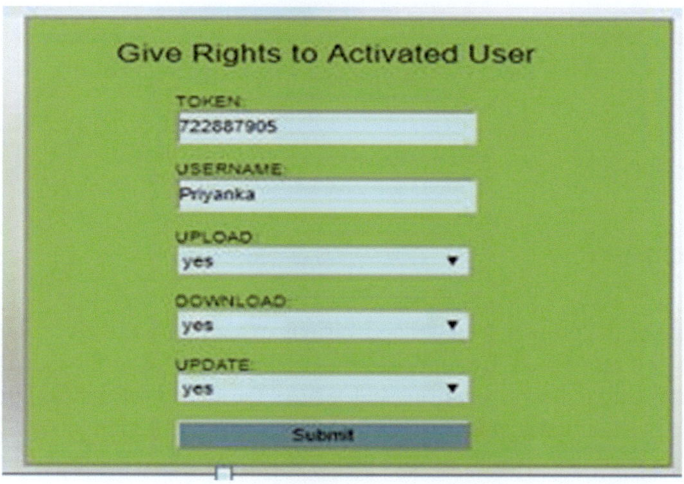

Fig. 15 Rights to activated user page

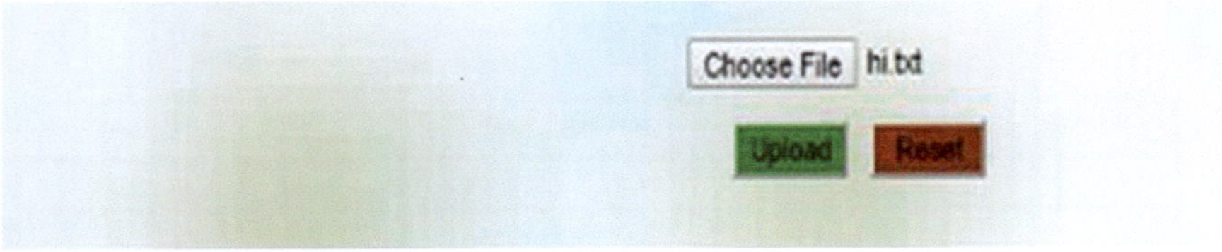

Fig. 16 File upload page

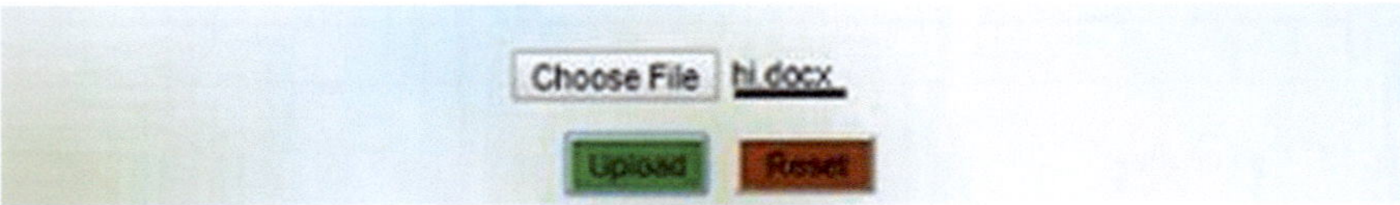

Fig. 17 Uploading same file page

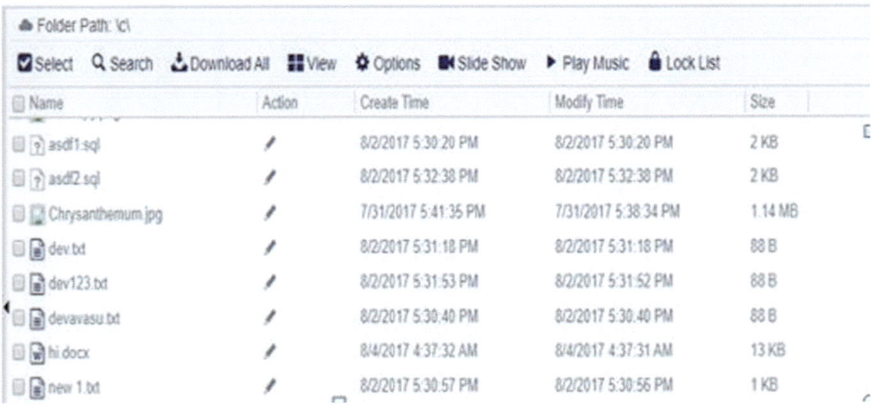

Fig. 18 DriveHQ page

5 Conclusion

A prototype of the proposed model and various experiments are given in the work. The developed models ensure minimal overhead as it relates to the existing one. It provides approval to the private firms and secures the secrecy of critical information. The existing techniques make use of different users for encryption in varying stages. By making use of their private keys, same data copies are stored by different users of the cloud so that the de-duplication is made impossible.

References

1. Ng CH, Lee PP (2013) Revdedup: a reverse deduplication storage system optimized for reads to latest backups. In: Proceedings of the 4th Asia-Pacific workshop on systems, July 2013. ACM, p 15
2. Huang CK, Chien LF, Oyang YJ (2003) Relevant term suggestion in interactive web search based on contextual information in query session logs. J Assoc Inf Sci Technol 54(7):638–649
3. Bugiel S, Nurnberger S, Sadeghi A, Schneider T (2011) Twin clouds: an architecture for secure cloud computing. In: Workshop on cryptography and security in clouds (WCSC 2011), vol 1217889, Mar 2011
4. Ng WK, Wen Y, Zhu H (2012) Private data deduplication protocols in cloud storage. In: Proceedings of the 27th annual ACM symposium on applied computing, Mar 2012. ACM, pp 441–446
5. Di Pietro R, Sorniotti A (2012) Boosting efficiency and security in proof of ownership for deduplication. In: Proceedings of the 7th ACM symposium on information, computer and communications security, May 2012. ACM, pp 81–82
6. Ducasse S, Rieger M, Demeyer S (1999) A language independent approach for detecting duplicated code. In: Proceedings IEEE international conference on software maintenance, 1999 (ICSM'99). IEEE, pp 109–118
7. Jin K, Miller EL (2009) The effectiveness of deduplication on virtual machine disk images. In: Proceedings of SYSTOR 2009: the Israeli experimental systems conference, May 2009. ACM, p 7
8. Tan Y, Jiang H, Feng D, Tian L, Yan Z, Zhou G (2010) SAM: a semantic-aware multi-tiered source de-duplication framework for cloud backup. In: 2010 39th international conference on parallel processing (ICPP), Sept 2010. IEEE, pp 614–623
9. Bellare M, Keelveedhi S, Ristenpart T (2013) DupLESS: server-aided encryption for deduplicated storage. IACR Cryptol ePrint Arch 429
10. Agarkhed J, Ashalatha R (2017) A privacy preservation scheme in cloud environment. In: 2017 third international conference on advances in electrical, electronics, information, communication and bio-informatics (AEEICB), Feb 2017. IEEE, pp 549–552
11. Ashalatha R, Agarkhed J, Patil S (2016) Data storage security algorithms for multi cloud environment. In 2016 2nd international conference on advances in electrical, electronics, information, communication and bio-informatics (AEEICB), Feb 2016. IEEE, pp 686–690
12. Bellare M, Keelveedhi S, Ristenpart T (2013) Message-locked encryption and secure deduplication. In: Annual international conference on the theory and applications of cryptographic techniques, May 2013. Springer, Berlin, Heidelberg, pp 296–312
13. Anderson P, Zhang L (2010) Fast and secure laptop backups with encrypted de-duplication. In: Proceedings of the 24th international conference on large installation system administration, pp 29–40
14. Halevi S, Harnik D, Pinkas B, Shulman-Peleg A (2011) Proofs of ownership in remote storage systems. In: Proceedings of the 18th ACM conference on computer and communications security, Oct 2011. ACM, pp 491–500
15. Li J, Chen X, Li M, Li J, Lee PP, Lou W (2014) Secure deduplication with efficient and reliable convergent key management. IEEE Trans Parallel Distrib Syst 25(6):1615–1625
16. Ni J, Zhang K, Yu Y, Lin X, Shen XS (2018) Providing task allocation and secure deduplication for mobile crowdsensing via fog computing. IEEE Trans Dependable Secure Comput
17. Jiang T, Chen X, Wu Q, Ma J, Susilo W, Lou W (2017) Secure and efficient cloud data deduplication with randomized tag. IEEE Trans Inf Forensics Secur 12(3):532–543
18. Youn TY, Chang KY, Rhee KH, Shin SU (2018) Efficient client-side deduplication of encrypted data with public auditing in cloud storage. IEEE Access 6:26578–26587

A Survey on Cloud Computing Security Issues and Cryptographic Techniques

Vidushi Agarwal, Ashish K. Kaushal and Lokesh Chouhan

Abstract Cloud computing is an Internet-based computing model, having various resources used by distinct users in a concurrent manner. Apart from all of its advantages, it faces a major setback due to various data security issues. To overcome these issues, various security mechanisms have been proposed, such as cryptography and authentication. Cryptography can be used to provide data integrity, authorization for data manipulation, and also making the data unreadable to an interceptor through encryption. There are various classifications of models in cloud computing. The service models are classified as Software as a Service (SaaS), Platform as a Service (PaaS), and Infrastructure as a Service (IaaS). There are several deployment models mainly distinguished by ownership which consists of public cloud, private cloud, and hybrid cloud. This survey mainly focuses on security issues in cloud service models and cloud deployment models along with various cryptographic mechanisms of data protection, such as symmetric key cryptography, asymmetric key cryptography, and their encryption algorithms.

Keywords Asymmetric algorithms · Cloud computing · Cryptography · Symmetric algorithms · Quantum cryptography

1 Introduction

With the advancement of technology, cloud computing is nowadays the most expanding and emerging technology in the field of computer science [1]. In today's world, everyone uses the cloud in one way or the other. Gmail, Facebook, Twitter, Dropbox,

V. Agarwal · A. K. Kaushal · L. Chouhan (✉)
Department of Computer Science and Engineering, National Institute of Technology Hamirpur, Hamirpur, Himachal Pradesh, India
e-mail: lokeshchouhan@gmail.com

V. Agarwal
e-mail: vidushiagarwal524@gmail.com

A. K. Kaushal
e-mail: akaushal4004@gmail.com

R. K. Shukla et al. (eds.), *Social Networking and Computational Intelligence*, Lecture Notes in Networks and Systems 100, https://doi.org/10.1007/978-981-15-2071-6_10

and most of the widely used applications generally use cloud in their basic functioning. Cloud provides various advantages [2] like access to all the information anytime and anywhere, low cost of infrastructure, less access cost, etc. But along with these advantages, we also face some challenges like securing user's data, absence of expertise and resources, etc. In this paper, security issues related to various cloud models and the techniques of cryptography to protect cloud data are described.

2 Security Issues of Deployment Models in Cloud

A brief description of cloud deployment models is given in Fig. 1. All the available types of cloud deployment models named as public cloud, private cloud, and hybrid cloud have certain issues when it comes to security among which a few are given below.

2.1 Security Concerns in Public Cloud

A public cloud can be accessed by many customers publicly enabling the use of shared computing resources. Service provider provides the infrastructure security required. Some issues faced in a public cloud are as follows [3]:

(a) Security mechanisms in a public cloud are controlled by the service provider which makes it difficult to secure data. Therefore, meeting the basic security requirements, namely integrity, availability, and confidentiality, becomes

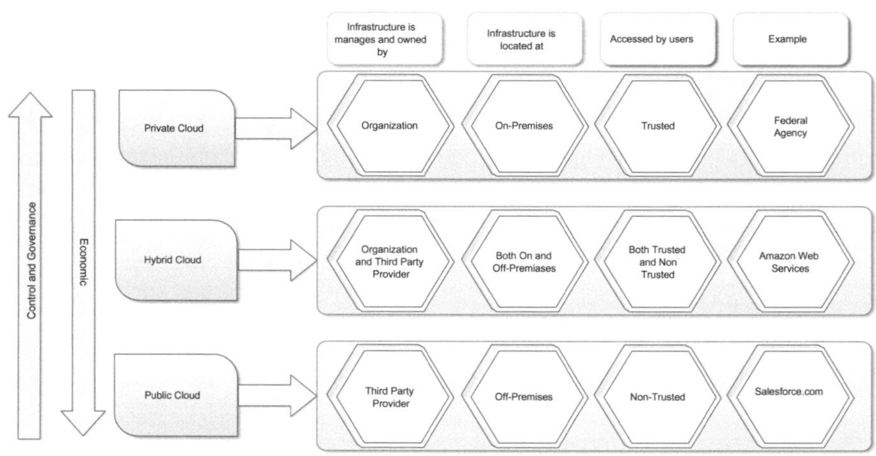

Fig. 1 Brief description of cloud deployment models

a complicated situation to protect data for its lifetime during various stages of operation.

(b) Due to the shared nature of public cloud, chances of a breach in data are very high. Hence, the service provider has to be chosen carefully to avoid such risks.
(c) If a third-party vendor is used by a cloud service provider, the customer should verify the service level agreements (SLAs) as well as contingency plans in case of any failures [2].
(d) To prevent insider attacks, SLAs should be verified along with the levels to which data will be encrypted and authenticated to protect it from malicious intruders.

2.2 *Security Concerns in Private Cloud*

A private cloud is exclusively controlled by a single customer or an organization giving greater control over the cloud resources and flexibility to implement any security practice. However, certain security issues need to be considered [4]:

(a) Due to virtualization techniques, it may be possible that virtual machines (VMs) communicate with the wrong VMs, leading to various security risks. To prevent this, proper encryption and authentication mechanisms should be implemented.
(b) To avoid any risk, the operating system acting as the host should be prevented from any malware threat. Moreover, communication between the host operating system and guest VMs should be via physical interfaces not directly.
(c) In private clouds, users can control a part of the cloud and use this infrastructure by an HTTP end point or a web interface. In such a case, the interfaces need to be developed properly and various HTTP requests need to be protected by the use of standard web-based application security techniques.
(d) Apart from providing standard security, a security policy should be used to prevent attacks originating from inside the organization [3].

2.3 *Security Concerns in Hybrid Cloud*

A hybrid cloud employs both public and private cloud services with orchestration between the two platforms. Various security issues found in a hybrid cloud are as follows:

(a) To maintain a uniform security policy across the entire network, a proper infrastructure policy, such as IPS signatures, firewall rules, and user authentication, should be applied.
(b) To ensure the compliance of public and private cloud provider and maintaining coordination between them is difficult to ensure in a hybrid cloud.

(c) When public and private clouds are integrated in a hybrid environment, security management is essential. Hence, existing policies, such as authentication, authorization, and identity management, need to be modified to address these complex integration issues.
(d) Hybrid clouds need to manage tasks across multiple domains and not many administrators have this kind of experience and knowledge exposing it to various risks.

3 Security Issues in Cloud Service Models

All the types of cloud service models have certain issues when it comes to security. A brief description of cloud service models is given in Fig. 2. Some of the security concerns associated with the service models are as follows.

3.1 Security Concerns in Software as a Service (SaaS)

The overwhelming benefits, desired convenience, and relaxed technological environment to the software suppliers and users caused many people and organizations to establish their software products in Software-as-a-Service model. With these increasing dependencies on SaaS model, the concerns of how much trustworthy, reliable,

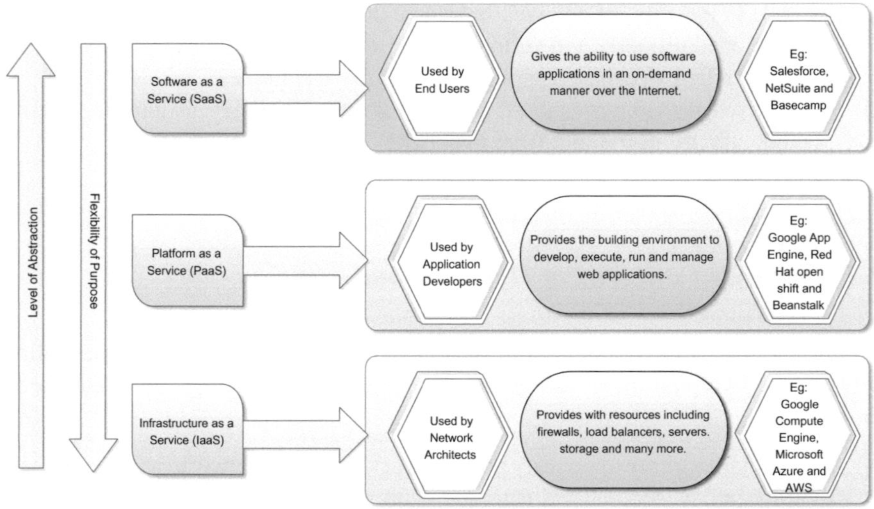

Fig. 2 Brief description of cloud service models

and safe these models are, are also rising. To define the security challenges in this model, they are categorized into many groups as defined below [5]:

(a) Multi-tenancy: Multi-tenancy being a fundamental requirement of SaaS model allows data of multiple users to be kept at the same site. If any malicious attack turned out to be successful even on a single instance, data of all the other instances located at the same site (server) will be at high risk. For stored data, to ensure high reliability and availability, data mirroring and redundancy is done at multiple locations across countries. During this data travel, there is more probability that sensitive information might be stolen by unauthorized intruders. To protect the data from such occasional leaks, SaaS providers might adopt data encryption methods to ensure data confidentiality. However, encryption techniques have their own challenges associated with them which increase the complexity of the entire system.
(b) Data Security: Data security refers to the protection of data belonging to the users and the owners. In SaaS, security can be compromised due to intentional or careless actions caused by an insider trusted partner. Hence, cloud provider must check that access to the database will be provided to the users only through authorization along with ensuring the compatibility of a user to the data it is allowed to access. Along with database access, data recovery and backup are important in data administration. If any hardware failures or data losses occur, SaaS users entirely dependent on SaaS provider might face valuable data loss. Hence, counteractions to restore the database should be present so that data always remain in a consistent state.
(c) Data Accessibility: Accessing SaaS applications through computer or mobile devices over the Internet exposes these services with the security risks associated with illegal accessing of information on Internet like data stealing, ID management, etc. [6].
(d) Application Security: It refers to the protection of application against the attacker from making malicious changes, getting administrative access, etc. Software design of conventional client-server model is different from SaaS model, and various APIs supported by SaaS model may get exploited due to the vulnerabilities present in those APIs. Systems interact with web services by exchanging structured information through SOAP messages. SOAP messages contain the vital information that should not be tampered with during the procedure of transfer. Hence, web services of this model are prone to many attacks such as XML injection, DoS Attack, DDoS Attack, and XML wrapping Attack. Moreover, backdoor, debug options, lack of unpredictable cross-site request forgery (CSRF) token, and hidden field manipulation are major threats to SaaS applications.

3.2 Security Concern in Platform as a Service (PaaS)

In PaaS, cloud service provider does not provide the users with entire application. Rather, they provide the users with functional control to build the application by providing their platform. However, all the security services below the application level are still the responsibility of the service provider. Some of the challenges faced by PaaS are [7]:

(a) Interoperability: When different clouds at three different levels (PaaS, SaaS, and IaaS) talk to each other, coding should be done such that it works with every cloud provider regardless of the differences between them. Common interfaces to objects can be provided for accessing resources to maintain interoperability. Code should be implemented in a way that avoids complexity and security flaws, such as attacks to hosts from objects.

(b) Host and Object Vulnerability: Hosts can be vulnerable to attacks in an environment where user objects are spread over multi-user interconnected hosts. If necessary security measures are not taken, an attacker will get the access to host's resources and also to the tenant objects. There are high chances of object security breach in PaaS in various ways. Service providers can have the access to user objects residing on its hosts. Cryptographic defense in such a case is computationally expensive so this attack can be avoided only through trust relations between provider and the user. Users that are tenants of the same host may attack each other's objects mutually because they share the same resources. Moreover, a user object may be attacked directly by a third party. Hence, the provider must protect privacy and integrity of a user object residing on a host [8].

(c) Access Control: Access control includes authentication, authorization, and traceability. Authentication means the parties need to prove that their identities are valid before an interaction. Authorization determines which user can access which object. It becomes difficult to maintain authorization in PaaS when objects migrate and uphold the policies during host reconfiguration. Traceability is employed by keeping history of every event occurring in a system to measure the service characteristics.

(d) Underlying infrastructure security: It is the responsibility of the provider to secure the underlying infrastructure as developers cannot approach basic layers regularly. Therefore, any security below the application level has to be assured by the provider because the developers only manage their applications on top of the platform. The development tools provided by the service provider to the clients should also be secure.

3.3 Security Concerns in Infrastructure as a Service (IaaS)

The cloud provider that provide IaaS to the users provide them with hardware (server, communication network and channel, storage, and processing), software (file system, operating system, and virtual environment) to run on that hardware, monitoring of system, overall maintenance and management, and administration for the operation of these installed hardware devices. Due to its high investment requirement and high administration cost, all the issues related to IaaS should be handled carefully. Some of the major issues are [9]:

(a) Leakage of data and monitoring: Data stored in the cloud both private and public should be confidential. Only authorized users should be allowed the privilege of accessing the data, and it should be well known how the information was accessed and from which location it was accessed. Policies should be created to restrict all the critical information usage, and it should be monitored continuously.
(b) Authentication and Authorization: For data loss prevention, robust authorization and authentication methods are needed. Authentication methods such as user name and password or multi-factor authentication can be used. Tiering of access policies can be done on the basis of level of trust or trust policy for any identity provider within the IaaS cloud.
(c) Logging end-to-end and data publishing: For effective and efficient deployment of IaaS, comprehensive logging and publishing should be in place. User's information might be at any place with time because of virtual machines, as they are moved dynamically in an array between servers over time and spun up automatically. Hence, robust reporting and logging are needed to keep track of the location of the information, the machines handling it, and the storage arrays keeping it.
(d) Virtualization: Rather than the numerous advantages virtualization provides, it becomes a bane when it comes to security, providing opportunities to intruders because of the additional layer which needs to be protected. Virtualized environments are prone to all kinds of attacks because of complexity in interconnections and more than one point of entry. Virtual machines residing on same site server can share computing resources, memory, I/O, and other distributed resources compromising with the security of the IaaS cloud.
(e) End-to-End Encryption: Encryption in IaaS cloud, both public and private, should be end-to-end. It should be such that not just data files but whole disks including all the data on them are encrypted. Moreover, all the communications to virtual machines (or between the virtual machines) or host operating systems should be encrypted. Homomorphic encryption can be used to keep communications between the end users safe and secure [10].

4 Cryptographic Techniques for Cloud

The use of Internet is increasing day by day, due to which it has become increasingly important to secure the information we are transmitting. This is where cryptography comes into play by encrypting the data we transmit over the web. Cryptography covers the process of altering data into imperceptible code and then transmitting it so that only those for whom it is intended can get the actual information. The right to digital privacy and cryptography thus go hand in hand. A brief description of various cryptographic techniques used in cloud is given in Table 1. These techniques are divided into three distinct types of algorithms: asymmetric key, symmetric key, and hash function cryptographic algorithms, which are elicited below [11].

4.1 Asymmetric Key Algorithms

In asymmetric key cryptography, two different types of keys are used for the purpose of encryption and decryption of data, namely public key which can be accessed by everyone and private key which is kept confidential. It has a better way to secure information transmitted via communication and provides authentication and confidentiality.

(a) RSA Cryptosystem: RSA (Rivest-Shamir-Adleman) [12] takes the advantage of difficulty we face in factoring the product of two large prime numbers. It consists of a public key used to encrypt messages which can be decrypted only by using the private key. The security of RSA depends on the encryption function and key generation which are both one-way functions.
(b) Diffie-Hellman Key Exchange Method: It is a method [13] given by two cryptographers Whitfield Diffie and Martin Hellman in 1976 in which exchange of cryptographic keys securely over a public channel is done with the help of a special mechanism. The security issues in this algorithm are based on the difficulty associated with discrete logarithm problem that uses specific numbers raised to specific powers to produce the associated decryption keys. The major disadvantage of Diffie-Hellman is that it is prone to man-in-the-middle-attack.

4.2 Symmetric Key Algorithms

In symmetric key cryptography, a single private key is maintained in the encryption and decryption algorithm at the sender and receiver, respectively. This is also known by the name of private key cryptography. This type is cost effective in terms of algorithm processing and key producing process. It also provides an acceptable amount of authentication because we need the same key for both the sides [14].

Table 1 Brief overview of various cryptographic techniques

S. No.	Cryptographic technique	Type of techniques				Pros	Cons	Future scope
		Asymmetric	Symmetric	Authentication scheme	Other(s)			
1.	RSA Algorithm [12]	✓				Hard to crack prime number factorization makes it safe and secure	Slow when encrypting large amount of data, third-party reliability on public key sharing	Processing can be made faster, and also, computational power can be reduced
2.	Diffie-Hellman Key Exchange [13]	✓				Communication can take place through an insecure connection, and also, sharing a secret key is safe	Highly susceptible to man-in-the-middle-attack and also cannot be used for asymmetric key exchange	Can be improved and made susceptible to man-in-the-middle-attack
3.	Blowfish Algorithm [15]		✓			Throughput is high, power consumption is less, hard to crack because of key expansion	Presence of considerable number of classes of weak keys	It can be made more efficient in terms of energy consumption and security

(continued)

Table 1 (continued)

S. No.	Cryptographic technique	Type of techniques				Pros	Cons	Future scope
		Asymmetric	Symmetric	Authentication scheme	Other(s)			
4.	Data Encryption Standard Algorithm [16]		✓			Harder to crack because of many rounds for encrypting message, and it is faster than many other methods	S-box can give same output for two chosen input, brute force is now possible on DES, and DES fails in front of linear cryptanalysis	Modification in functional implementation, S-box design, and replacing the XOR with another function can improve security
5.	Advanced Encryption Standard Algorithm [18]		✓			Implemented in both hardware and software; hence, it is robust security protocol	Hard to implement with software and also each block is encrypted in the same way	Selection of larger key will make it more secure, and selection of larger text will increase the throughput
6.	Remote User Authentication Scheme [21]			✓		Suitable for digital libraries and smart card applications	If using proxy servers, problem of autoconfiguration of proxy servers and multiple proxy servers is there	Channel security and user's anonymity can be improved by using some new mechanism of working

(continued)

Table 1 (continued)

S. No.	Cryptographic technique	Type of techniques				Pros	Cons	Future scope
		Asymmetric	Symmetric	Authentication scheme	Other(s)			
7.	ID-based password Authentication scheme [22]			✓		Simple to use and deployment of mechanism is easy	Security is completely dependent upon strength of password and confidentiality, no other strong means of identity check	The true user authentication by biometrics, user's procession can help it making more secure
8.	DNA Cryptography [24]				✓	High speed of processing, minimal requirement for storage and low power requirements	Stability depends on the environmental conditions, and time complexity is of around few hours	Still a fresh research issue with lots of scope in DNA steganography and triple-stage DNA cryptography
9.	Quantum Cryptography [28]				✓	Simple to use and virtually unhackable, require fewer resources for maintenance, and can also detect eavesdropping	Polarization can change during channel and lack many features like digital signature, and the need of dedicated channel results in high cost	Performance of this cryptosystem can be improved in a scalable manner

(a) Blowfish: This can be used as a significantly faster substitute to DES and IDEA and was devised by Bruce Schneier in 1993 [15]. This comes under the category of symmetric block ciphers that employ a variable-length key, from 32 bits to 448 bits with a block size of 64 bits. The only setback to this algorithm is the presence of considerable number of classes of weak keys.
(b) DES (Data Encryption Standard) Algorithm: DES [16] is the most popular block cipher technique in the world and employs 64-bit data blocks with 56-bit key. The disadvantage of DES is that encryption process is rather lagging and small key size cannot offer appropriate security. Further improvement on DES resulted in double DES and triple DES algorithms [17] which used multiple instances of DES in succession for encryption and decryption process.
(c) AES (Advanced Encryption Standard): This algorithm [18] provides an improvement over DES, was first published in the year 1998 by Belgian cryptographers Vincent Rijmen and Joan Daemen, and was initially established by US National Institute of Standards and Technology (NIST) in 2001. AES is also known as Rijndael cipher. This symmetric cipher works on 128-bit block size and three keys of different sizes: 128, 192, and 256 bits and varying number of rounds, i.e., 10, 12, and 14. Because of its structure, it has shown effective resistance to cryptanalytic attacks but bicyclic attacks with computational complexities of $2^{126.1}$ are known to break the cipher. This cipher is faster than DES because of its compactness and also invulnerable to collision attacks.

4.3 *Hash Function Algorithms*

A cryptographic hash function takes message or data as input and returns an alphanumeric string as output. An ideal hash function has 3 properties: It is fairly easy to compute hash on any given input, it is computationally challenging to calculate an alphanumeric text from a given hash, and no two different input values will have the same hash value as an output. Some of the methods are given below:

(a) MD5 (Message-Digest algorithm 5): This comes under the message-digest family which includes various hash functions named as MD2, MD4, MD5, and MD6. It was mentioned and specified as Standard RFC 1321 and was made by Ronald Rivest in 1991 [19]. MD5 is popularly used 128-bit hash function that provides an acceptable level of integrity to transmitted files. MD5 is vulnerable to collision attacks.
(b) Secure Hash Function (SHA): SHA [20] consists of 4 versions: SHA0, SHA1, SHA2, and SHA3. All these have different structures. The initial version is SHA0 which is a 160-bit hash function and was published by NIST in 1993. SHA21 was designed as advancement on SHA0 in 1995 and is employed in protocols like secure socket layer (SSL) security. SHA1 was found to be vulnerable to multiple collisions in 2005 and SHA2 was formulated with four versions: SHA-224, SHA-256, SHA-384, and SHA-512 depending upon the

number of bits in their hash value. SHA2 is a strong hash function but the basic structural design of SHA2 is based on SHA0. Hence, SHA3 or Keccak algorithm was chosen by NIST in 2012 which offers more efficient performance and certain level of invulnerability toward attacks.

4.4 *Authentication Schemes*

Authentication is a process to determine whether a user is the person who it is claiming to be. Authentication technique provides access for systems by checking a user's credentials via matching the credentials in a database of the users who are authorized.

(a) Remote User Authentication Scheme: Remote user authentication (RUA) [21] is a mechanism to authenticate remote users securely over a network which is not secure. User's IDs and passwords are maintained in a password table by the remote server. To verify whether a remote user is legitimate, a remote server is used before the user is provided the access to resources. First, the user uses a remote server for any service, and then, a secret communication path is established between the server and the user by using a session key.

(b) ID-based password Authentication scheme: Password verification [22] is a method by which a user provides his identification name and password value and he is given access only if the entered credentials match the stored value in the database. Many applications based on this scheme have been developed, such as ATM, remote login, database management systems, and private organizations. The major drawback is that the password can be known to server's administrator if stored in a normal text form. Moreover, any intruder can impersonate a user by stealing his name and password.

5 Recent Trends and Algorithms for Cloud Security

5.1 *Securing Cloud Computing Environment Based on DNA Cryptography*

DNA cryptography is a promising field in the future of information cryptography in which information is placed in a DNA molecule and hidden among other molecules of DNA. This concept was first introduced by Adleman [23] in 1994 which used DNA for carrying information. Then, DNA cryptography was proposed based on this idea which uses modern biological techniques as a tool for implementation, and the exceptional energy efficiency, huge amount of parallelism, and extra data depth present in DNA molecules can be used for cryptographic purposes [24]. This field is now being explored for providing security to cloud data with combinations of steganography and other hybrid algorithms [25, 26].

5.2 Quantum Cryptography for Secure Cloud Computing

Quantum cryptography is an approach which exploits quantum mechanical properties to provide cryptography by using photons to generate a cryptographic key and returning this key to the recipient using an appropriate relationship path [27]. Quantum cryptography can be used in cloud computing for protecting data. Since, it uses protons to generate encryption key, it becomes almost impossible to break [28]. It also helps in managing data easily and securely from anywhere with super-fast speed and efficiency.

5.3 Hybrid Cryptographic Algorithms

There are numerous limitations for both symmetric and asymmetric encryption algorithms, such as the problem of key maintenance and over consumption of computing resources, respectively. Therefore, to ensure security, a new hybrid cryptography protocol is needed. Using hybrid algorithms in cloud improves the security of other encryption algorithms because the data is encrypted by using more than just one algorithm which also minimizes the time taken by the algorithms altogether. Both symmetric and asymmetric cryptographic techniques can be combined to form a new hybrid combination which provides the required security [29–31].

6 Conclusion

Cloud computing provides users with on-demand delivery to computing resources for a friendly computational environment. Cloud technologies can help in cost reduction, reduction in management responsibilities, and increased efficiency of organization. Although cloud providers ensure to provide security and confidentiality, security attacks can still take place leading to loss of user's data. To protect data from these attacks, various techniques are implemented in different ways. This paper provided a survey on major security problems in cloud and challenges arising from unique characteristics of cloud models. Methods to overcome these issues using various cryptographic algorithms were also explored. Future research needs to be focused on mitigation techniques against security attacks emerging due to rapid increase in number of users each day and developing even more secure cryptographic algorithms using the latest available technologies.

References

1. Vasiljeva T, Shaikhulina S, Kreslins K (2017) Cloud computing: business perspectives, benefits and challenges for small and medium enterprises (Case of Latvia). Procedia Eng 178:443–451
2. Lee J (2013) A view of cloud computing. Int J Networked Distrib Comput 1(1):2
3. Poniszewska-Maranda Aneta (2014) Selected aspects of security mechanisms for cloud computing—current solutions and development perspectives. J Theor Appl Comput Sci 8(1):35–49
4. Bhadauria R, Sanyal S (2012) Survey on security issues in cloud computing and associated mitigation techniques. Int J Comput Appl 47(18):0975–888
5. Chouhan PK, Yao F, Sezer S (2015) Software as a service: understanding security issues. In: Science and information conference 2015, 28–30 July 2015, London, UK
6. Ochani A, Dongre N (2017) Security issues in cloud computing. In: International conference on I-SMAC, IoT in social, mobile, analytics and cloud (I-SMAC 2017)
7. Chowdhury RR (2014) Security in cloud computing. Int J Comput Appl 96(15):0975–8887
8. Devi T, Ganesan R (2015) Platform-as-a-Service (PaaS): model and security issues. TELKOMNIKA Indonesian J Electr Eng 15(1):151–161
9. Jaiswal PR, Rohankar AW (2014) Infrastructure as a service: security issues in cloud computing. Int J Comput Sci Mob Comput 3(3):707–711
10. Murray A, Begna G, Nwafor E, Blackstone J, Patterson W (2015) Cloud service security and application vulnerability. In: Proceedings of the IEEE Southeast Conference 2015, 9–12 Apr 2015, Fort Lauderdale, Florida
11. Chatterjee R, Roy S (2017) Cryptography in cloud computing: a basic approach to ensure security in cloud. IJESC 7(5)
12. Zhou X, Tang X (2011) Research and implementation of RSA algorithm for encryption and decryption. In: Proceedings of 2011 6th international forum on strategic technology, 2011, pp 1118–1121
13. Zhang C, Zhang Y (2008) Authenticated Diffie-Hellman key agreement protocol with forward secrecy. Wuhan Univ J Nat Sci 13(6):641–644
14. Geetha V, Laavanya N, Priyadharshiny S, Sofeiyakalaimathy C (2016) Survey on security mechanisms for public cloud data. In: 2016 international conference on emerging trends in engineering, Technology and Science (ICETETS)
15. Schneier B (1994) Description of a new variable-length key, 64-bit block cipher (Blowfish). Springer, Berlin, Heidelberg, pp 191–204
16. Han S-J, Oh H-S, Park J (1996) The improved data encryption standard (DES) algorithm. In: Proceedings of ISSSTA'95 international symposium on spread spectrum techniques and applications, vol 3, pp 1310–1314
17. Coppersmith D, Johnson DB, Matyas SM (1996) A proposed mode for triple-DES encryption. IBM J Res Dev 40(2):253–262
18. NIST, "FIPS PUB 197: specification for the advanced encryption standard (AES) (2001)
19. Rivest R (1992) The MD5 message-digest algorithm. RFC 1321, Apr 1992
20. "Secure Hash Standard", United States of American, National Institute of Science and Technology, Federal Information Processing Standard (FIPS) 180-1, Apr 1993
21. Shen J-J, Lin C-W, Hwang M-S (2003) A modified remote user authentication scheme using smart cards. IEEE Trans Consum Electron 49(2):414–416
22. Kim H-S, Lee S-W, Yoo K-Y (2003) ID-based password authentication scheme using smart cards and fingerprints. ACM SIGOPS Oper Syst Rev 37(4):32–41
23. Adleman L (1994) Molecular computation of solutions to combinatorial problems. Science 266:1021–1023
24. Xiao G, Lu M, Qin L, Lai X (2006) New field of cryptography: DNA cryptography. Sci Bull 51(12):1413–1420
25. Prajapati Ashishkumar B, Barkha P (2016) Implementation of DNA cryptography in cloud computing and using socket programming. In: 2016 international conference on computer communication and informatics (ICCCI), 2016, pp 1–6

26. Gugnani G, Ghrera SP, Gupta PK, Malekian R, Maharaj BTJ (2016) Implementing DNA encryption technique in web services to embed confidentiality in cloud. Springer, New Delhi, pp 407–415
27. Wiesner SJ (1983) Conjugate coding. SIGACT News 15(1):78–88
28. Olanrewaju RF, Islam T, Khalifa OO, Anwar F, Pampori BR (2017) Cryptography as a service (CaaS): quantum cryptography for secure cloud computing. Indian J Sci Technol 10(7):1–6
29. Abdelminaam DS (2018) Improving the security of cloud computing by building new hybrid cryptography algorithms. 8(1):40–48
30. Moghaddam FF, Alrashdan MT, Karimi O (2013) A hybrid encryption algorithm based on RSA Small-e and Efficient-RSA for cloud computing environments. J Adv Comput Netw 238–241
31. Arockiam L, Monikandan S (2013) Data security and privacy in cloud storage using hybrid symmetric encryption algorithm. Int J Adv Res Comput. Commun Eng 2(8):3064–3070

Machine Learning

Features Identification for Filtering Credible Content on Twitter Using Machine Learning Techniques

Faraz Ahmad and S. A. M. Rizvi

Abstract In the present era of Internet, Twitter is one of the pivotal platforms for sharing the views and opinions of an individual related to any topic by means of tweets. However, the credibility of such tweets is unidentified. Demonetization is one of the events in India when lots of chaos happened among public, and people posted tweets whose authenticity was trailed by the question mark. In this study, we have discovered the credibility of user content on Twitter network with the help of 26 different features. The experiments have been carried out on more than 1 k user tweets related to demonization. For classifying the tweets into four different credibility classes (acceptable, somewhat acceptable, somewhat unacceptable, and unacceptable) several machine learning techniques such as random forest, Naive Bayes, and support vector machine has been utilized. Out of all the selected classifiers, random forest has been observed as the best classifier with accuracy and f1 score as 0.977 and 0.9911, respectively. Furthermore, out of 26 identified features, we have recognized 10 most distinctive features to efficiently distinguish the user tweets in different credibility classes.

Keywords Classifiers · Credibility · Machine learning · Tweets · Twitter

1 Introduction

Twitter provides a platform for every individual to share text, media or information in the form of tweets, pictures, and videos on online social network. It is the most prevalent microblogging network, as over 330 million users actively participated from all over the world, and approximately 30.4 million Twitter users are from India itself, making it the most significant platform. Tweet using characters such as "@userhandle" is being used to refer a user name or company name by which user handle is made, or # (hashtag) symbol is used to mark tweet to appropriate topic so

F. Ahmad · S. A. M. Rizvi (✉)
Department of Computer Science, Jamia Millia Islamia, New Delhi, India
e-mail: sarizvi@jmi.ac.in

F. Ahmad
e-mail: faraz159020@st.jmi.ac.in

R. K. Shukla et al. (eds.), *Social Networking and Computational Intelligence*, Lecture Notes in Networks and Systems 100, https://doi.org/10.1007/978-981-15-2071-6_11

that it would easily display in Twitter search. Retweet functionality is an effective feature for faster propagation of the tweets posted by others.

There are several means of posting the tweets on Twitter, like using mobile phone, PC, e-mail, etc. In emergency conditions or high impact event when something foremost happened, most of the people turn toward these microblogging Web sites either for sharing news or information regarding the particular incident. Some people are also turn toward social media just for finding the insights related to the event. Twitter is providing the platform where one can effectively broadcast information/news. However, false information is also diversified with the genuine content. In critical condition, when everyone is fractious, information whose authenticity is unverified can create lots of chaos and agitation. If these rumored/false contents are allowed to spread through the network, the results will be unforeseen and can even lead to the riots.

The main purpose of this paper is to develop a model to find the credibility of posted content on Twitter. In this paper, approximately around 3250 tweets were crawled related to "Demonetization" which has taken place in the month of November 2016, it is possibly one of the major incidents happened in India in last three years. Afterward, these tweets are preprocessed and all the duplicate and irrelevant tweets are omitted. All those tweets which were written in some other language except English were also omitted. After preprocessing of the data, it has narrowed down to finest 1000 unique tweets which further outsourced to five different Ph.D. scholars from different departments for labeling them into four levels like credible, somewhat credible, somewhat incredible, and incredible.

Tweet consists of two main components, one is the content and its associated meta features and second is its author who composed the tweet. These two given features help to find the credibility of posted content on Twitter. This paper has used twenty-six such features, including three features in which we are finding the sentiments, emotions, and polarity of tweets with the help of IBM Watson for Natural Language Understanding and Meaning Cloud. Lastly, supervised learning methods for classifying the tweets in four different categories are applied. The authors have used random forest, support vector machine, and Naive Bayes classifiers. Random forest will give maximum accuracy of 0.9772, and f1 score as 0.9911 has been found

The structure of the rest of the paper is as follows: Sect. 2, the review of the literature concerning credibility of the posted contents is presented and discussed, Sect. 3 describes the definition of credibility of the posted content in social network given by other researchers, Sect. 4 discuss about the data crawling, preprocessing, and labeling the tweets. Section 5 discusses about the feature selection and further classification of tweets with different supervised learning approaches, Sect. 6 we discussed about best features selection, and finally in Sect. 7 presents the conclusion and future work.

2 Background Study

The existing research has attempted to resolve the problem of trust and integrity on online social networks (OSN) using different methods and techniques. Numerous researches in various domains of social media platforms have been done, especially in identifying rumored and unreliable content whose authenticity has not yet been acknowledged. This unauthenticated content must require to get filtered before it creates chaos or unevenness in OSN. In this section, we discuss researches which have been done to identify rumored and unauthenticated content on OSN.

Ratkiewicz et al. [1] developed Truthy system for the real-time analysis of meme dissemination on Twitter by mining, classifying, visualizing, mapping, and modeling immense content related to various microblogging events, which help in detecting the smear campaigns, astroturfing, and other deception related to political election in U.S. Further, they presented some cases of abusive behavior of users on Twitter. Gupta et al. [2] developed a ranking model which is used to rank the tweets using SVM rank based on their credibility scale of one (least credible) to seven (most credible). They have trained their model on six different disasters event happened in 2013 and used set of forty-five different features which is further sub-divided into 6 major categories for calculating the credibility of posted content on Twitter.

Xia et al. [3] proposed the method for calculating the credibility of content on Twitter in emergency conditions. They developed a model named Twitter Monitor for observing Twitter to detect the emergency situation. It is implemented using an unsupervised learning algorithm (K-means) which is used to detect the highest density cluster of tweets. Afterward, they have mined the collective components in the cluster and analyzed it by the experts for deciding the emergency situation arises or not. Finally, authors used a set of features which is broadly divided into four major categories for classifying the tweets into two classes, credible and incredible. Castillo et al. [4] proposed a method on automatic credibility assessment of tweets. Their primary focus was to analyze the tweets related to "trending topics." Based on the features extracted from the tweets, they classified them into two major categories as credible or non-credible. Authors have used several features such as user features, content-related features, topic-related features, and propagation features to classify the tweets. For classification purpose, first step is to label each tweet for news-related topics which is fulfilled by Mechanical Turk workers, and then applied several supervised learning algorithm and achieved best results from J48 decision tree with 86% accuracy, precision, and recall from 70 to 80%.

Lorek et al. [5] proposed a technique for finding the credibility of tweets in two different steps. Authors gathered the data (tweets) based on natural environment preservation using Twitter river plugin to get access to Twitter Stream API known as Elastic search. In first step, they select set of features related to tweets and its user, and then manually tagged each tweet by two different individual who gave their independent notes related to the tweets and also, assigning one of the following credibility levels (Highly Credible, Highly Non-Credible, Neutral, and Controversial). If content or link associated with the tweets is not retrievable, then it was assigned as an error.

Andin second step known as reconcile step, they reconciled the separate tags given by both the individuals, and each tweet acknowledged with one of the four credibility scores. For classification purpose, authors have used random forest algorithm and analyzed three separate results. The first is Twitter features-based classification, the second is reconcile feature-based classification, and the third is combination of both. They found 89% precision while combining both of the features.

Zhang et al. [6] proposed a classification problem for detecting rumor which consists of three main parts. Foremost is the data cleaning which involves the filtering of spam messages and those messages which contains only punctuations or URL. Second part is feature extraction which consists of extracting relevant features from the given microblog content, and the third is implementing a classification model. They crawled 3229 rumored and 12,534 non-rumored microblog from the Chinese microblogging Web site Sina Weibo. Authors suggested an automatic rumor detection technique which is based on shallow and implicit features. Shallow features consist of standard user or content features, whereas the proposed implicit features comprise profound information regarding user and its posted content. Further, they train support vector machine classifier with derived features and accompanied three types of experiments to identify which one should give better classification results. Lastly, to evaluate the efficacy of implicit features, they used features such as shallow user based, shallow content based and implicit user based, implicit content based for detecting rumor alone and concluded that implicit features outperformed shallow features in detecting rumors.

O'Donovan et al. [7] gathered Twitter data (tweets) from eight diverse crawls and extracted set of features for finding the credibility of the tweets. Majorly, they classified features into three main categories such as social, content, and behavioral. For annotation purpose, they hired 236 Amazon Mechanical Turk workers and annotated the tweets on Likert scale from 1 to 5 (least credible to most credible). Authors skip the tweets which had been assigned score 3 to reduce the ambiguity. Lastly, an analysis of the features which were distrusted through dyadic pairs of tweets and retweets of different length is defined. Their results showed the features who are the best indicator of credibility like mentions, URLs, tweet, and retweet length comes more predominantly in the tweets.

Resnick et al. [8] described a technique called RumorLens for systematically identifying the new rumors on Twitter. This tool is highly dependent on human annotators to label them as spreading, correcting, or not related to the rumor. The architecture of RumorLens system consists of three main stages. First one is Rumor Detector, which uses Twitter Garden Hose API for mining the rumors from the tweets and produces the cluster of the tweets that come across to be a rumor. An interface is provided by RumorLens Web site which allows other user to upgrade the output provided by Rumor Detector component. Second is ReQuery-ReClassify (ReQ-ReC), which retrieves and classifies the tweets associated with specific rumor. Judgment has been provided by different people, which leads to update the classifier (ReC part) and sometimes system also generates additional query (ReQ part). The third stage is Interactive Visualization which allows users to further explore the data provided by ReQ-ReC system. It facilitates the exact assessment of the impact of rumors on users. etc.

2.1 User Specified Features for Finding the Credibility of Tweets

Abbasi and Liu [9] proposed a CredRank algorithm for finding the credibility of user in OSN. They considered the condition in such a way that they do not need to assess the credibility of posted content or user's information displayed on his/her profile. For this research, authors' crawled the data of U.S. Senators' voting records from 1989 to 2012 and proposed algorithm are used to analyze correlation between the votes. This also detects users with coordinated behavior and cluster them using hierarchical clustering technique. They assigned lesser credibility score to the users which have higher coordinated behavior and higher credibility to lesser coordinated behavior. Lastly, algorithm helps in detecting those users who are using multiple accounts to diffuse irrelevant information into social media to prevent the dissemination of fake content. Westerman et al. [10] studied the behavior of users' perception related to source credibility. A total of 289 participants from two different universities were participated and were randomly assigned to one of the six mock Twitter webpage. Based on the heuristics given on the webpage, participants need to assign the credibility of the source. After analyzing the given mock pages, authors' find out that the source with too many and too few followers were perceived less credible as compared with the sources which have moderate number of followers. Similarly, one more observation is the ratio of number of followers the users have to the number of people they followed must be of two different kinds, narrow and wide. Narrow ratio gap perceived more credible as compared with wide ratio gap. Authors' applied several statistical test such as ANOVA and MANOVA for finding such results.

Cha et al. [11] proposed three measures to find how much an influence of particular user varies over a variety of topics. Three measures are in degree (number of followers), retweets, and mentions. In degree represents how much a user is popular on OSN, retweets measure the value of posted contents and mention represents the value of the user. Findings showed that huge number of followers would provide the popularity but it delivers lowers influence score such as involving audiences. Further, users' higher in degree could not necessarily provide higher retweets and mentions. News, media channels spawned higher level of retweets over variety of topics and influentially interacts with their audiences. Whereas, celebrities were better in inducing mentions, it provides better impact on influence scale because of their name value. Lastly, authors' mentioned that the influence of any user can not gain abruptly, and it could take lots of self-involvement to gain the influence.

3 Explaining Credibility

Researchers described several features that define the credibility of posted content in OSN. These features help in finding the true authentic value of the tweet and its author who posted the tweet.

These features are majorly classified into two categories; first consists of tweet content features. These features are extracted by deeply analyzing the content and associated data with the tweet, like sentiment, polarity, number of special symbol, tweet length, etc. Second category contains user features, like number of followers, number of friends; user is verified by twitter, user's signup duration on Twitter, etc.

3.1 Evaluating Credibility

If social media users want to acquire knowledge about any particular topic, the most important task is to subscribe or follow a user with high credibility and most substantial content. Although, evaluating the credibility of the tweets by other users is always influenced by heuristics or biased due to their trust relationships. It is difficult to gain all the details about any particular incident by individual tweets, but sometime people use very strong negative and positive emotion words like (#pappu, #kejru, #feku, etc.) or strong/abusive words against any religion, cast, or community, that may effects other users and will end into a disastrous condition. So, this is the task of primary importance to find these words and their users' irrespective of their heuristics who are responsible for spreading these types of content before it may create any chaos.

4 Data Crawling and Labeling Credibility

In this paper, authors have collected tweets related to an incident that is demonetization using Twitter rest API and gathered it into an external database. For data collection purpose, we used tending hashtags like #Demonetization, #Modi, #NoteBan, #cash, #Blackmoney, #BJP, #RBI, #cashless, #currency, #digitalpayments and handles like @NarendraModi, @PMOIndia, @OfficeOfRG, @RBI, and many more. Approximately 8,000 tweets were collected in the period of 4 months, all the tweets that were duplicate, less than of 15 words, whose author sign up with twitter for less than six months, which are in any other language except for English and Hindi and tweets that only contain special symbols or emoticons were omitted. After applying these filters, only the finest 1000 tweets were narrowed that are directly related to the topic.

For labeling credibility classes, highly qualified human annotators (Ph.D. students from 5 different departments) were selected who have a significant amount of

Table 1 Credibility class distribution in data set

Credibility classes	Distribution
Unacceptable	287
Slightly unacceptable	250
Slightly acceptable	306
Acceptable	157
Total	1000

knowledge related to the topic. After having a well-defined data set, the next task was to label the tweets into one of the four possible credibility levels such as acceptable, somewhat acceptable, somewhat unacceptable, and unacceptable.

Acceptable are those tweets which are informative related to the event, which are not an opinion or sarcastic comment given by user, and do not contain any abusive or negative words for any person, religion, cast, and creed. *Somewhat Acceptable* are those tweets which are not very much informative but still giving relevant information to other users. It does not contain any personal comment or words that bring adverse impact for the society. *Somewhat Unacceptable* are the unrelated tweets with the same hashtag #demonetization, not given any relevant information and possess personal comments. Whereas, *Unacceptable* are those tweets which consist of personal opinion, comments, negative sentiment words like #pappu, #kejru, #feku, or any abusive word. The distribution of these credibility classes are shown in (Table 1).

5 Feature Analysis and Classification

We proposed a set of content and user based features that are used as credibility indicators and abet in classifying tweets in one of the four above mentioned credibility classes. These features were either crawled directly by using Twitter API or captured by processing tweets. Feature set is shown in Table 2. In this paper, authors have used IBM Watson Natural Language Understanding & Meaning Cloud for finding the exact sentiments, emotions, and polarity related to the tweets. Sentiment measures on a scale of -1 to $+1$, 0 to -1 for negative sentiment and 0 to $+1$ for positive sentiment. However, emotion can be measured on 0 to $+1$ into five different categories like anger, fear, disgust, sadness, and joy. Whereas polarity of the tweets is measured on six pointer scale such as P+, P, Neutral, NONE, N, and N+. For the sake of simplicity, we merge NONE into Neutral category as we find out both of them belongs to the same labeled class.

We used 1000 tweets related to demonetization incident and split the number of tweets into 70/30 ratio for training and testing purpose. Further, we implemented machine learning using Naive Bayes, random forest, and support vector machine classifier and implement in R for developing a classification model. Random forest will give 97.72% accuracy which is best among other classifiers. Whereas, Naive

Table 2 Features used for classifying tweets in credibility classes

S. No.	Feature name	Feature description
1	User verified	Is user verified by Twitter?
2	User favourites count	Number of times user liked tweets of other users in their life time on Twitter
3	User followers count	User followed by the number of other users
4	User friends count	Number of other users the user is following on Twitter
5	User statuses count	Total number of status posted by user
6	Tweet favorite count	Number of times tweet is liked by other users
7	SENTIMENT_userdesc	Sentiment of user description on a scale of -1 to $+1$ given by IBM Watson Natural Language Understanding
8	ANGER_userdesc	Emotion of user description on a scale of 0–1 (angry)
9	DISGUST_userdesc	Emotion of user description on a scale of 0–1 (disgust)
10	FEAR_userdesc	Emotion of user description on a scale of 0–1 (fear)
11	JOY_userdesc	Emotion of user description on a scale of 0–1 (joy)
12	SADNESS_userdesc	Emotion of user description on a scale of 0–1 (sadness)
13	Sentiment_tweet	Sentiment of tweet description on a scale of -1 to $+1$ given by IBM Watson natural language understanding
14	ANGER_tweet	Emotion of tweet description on a scale of 0–1 (angry)
15	DISGUST_tweet	Emotion of tweet description on a scale of 0–1 (disgust)
16	FEAR_tweet	Emotion of tweet description on a scale of 0–1 (fear)
17	JOY_tweet	Emotion of tweet description on a scale of 0–1 (joy)
18	SADNESS_tweet	Emotion of tweet description on a scale of 0–1 (sadness)
19	Polarity	Polarity of the tweet calculated by Meaning Cloud on a scale of P+, P, None, Neutral, N, N+
20	Number of @ symbols	Total number of (mention(s)) @ in tweet
21	Number of hashtag# symbols	Total number of (hashtag) # in tweet

(continued)

Table 2 (continued)

S. No.	Feature name	Feature description
22	Number of question mark? symbols	Total number of (question mark)? in tweet
23	Number of exclamation mark! symbols	Total number of (exclamation mark)! in tweet
24	Percentage of uppercase letter	Percentage of text written in capital letter
25	Percentage of lowercase letter	Percentage of text written in small letter
26	Retweet count	Number of times the tweet is retweeted

Table 3 Overall statistics

	Random forest	SVM	Naive Bayes
Accuracy	0.9772	0.9064	0.8732
95% CI	(0.9536, 0.9908)	(0.8675, 0.9369)	(0.8289, 0.9096)
No information rate	0.2866	0.3043	0.331
P-value [Acc > NIR]	<2.2e−16	<2.2e−16	<2.2e−16
Kappa	0.9693	0.8729	0.8246

Bayes gives 87.32%, and SVM gives 90.64% accuracy. Summary of classifiers is displayed in Table 3.

Kappa statistics compares observed accuracy with expected accuracy; it is not only used to estimate a single classifier, but also used for evaluating classifiers among themselves. It tells the predictability of our classifier is better than a random predictor. The details of evaluation w.r.t. each classifier are shown in Table 4. The performance of all four classes for distinct classifiers is almost similar. F1 score is high for all the given classes, indicates a better stability between precision and recall.

Random forest demonstrates the multi-class receiver operation characteristics with 99.2986% area under the curve in Fig. 1.

6 Best Features Analysis

Random forest classifier is giving a best classification results like accuracy, precision, recall, and F1 score. To illustrate further, we are interested in finding the best features which is helpful in better classification. For this paper, we find mean decrease accuracy and mean decrease Gini for top 10 features and the results are shown in Fig. 2. Mean decrease accuracy graph test how worse the model perform if we remove a particular variable. Whereas, mean decrease Gini graph measures how pure the node

Table 4 Precision, recall, and F1 Score for credibility classes

	Acceptable	Slightly acceptable	Slightly unacceptable	Unacceptable
Random forest				
Precision	1	0.964705882	0.962025316	0.977272727
Recall	0.98245614	0.987951807	0.95	0.977272727
F1 score	0.991150442	0.976190476	0.955974843	0.977272727
SVM				
Precision	0.936170213	0.934065934	0.8	0.953488372
Recall	0.897959184	0.913978495	0.923076923	0.891304348
F1 score	0.916666667	0.923913043	0.857142857	0.921348315
Naive Bayes				
Precision	0.885714286	0.893617021	0.771428571	0.929411765
Recall	0.96875	0.884210526	0.794117647	0.887640449
F1 score	0.925373134	0.888888889	0.782608696	0.908045977

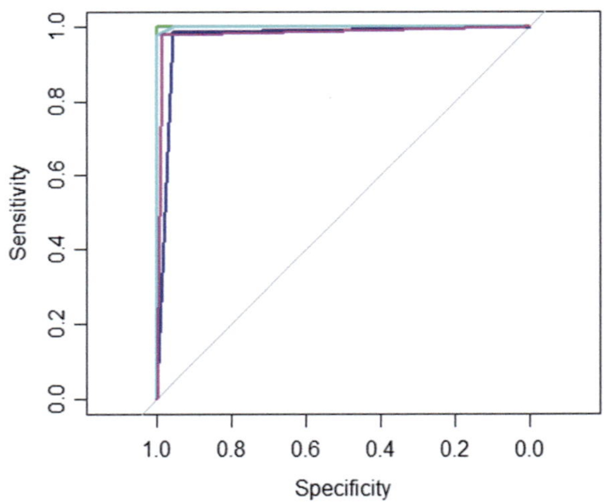

Fig. 1 Multi-class receiver operating characteristic (ROC) curve

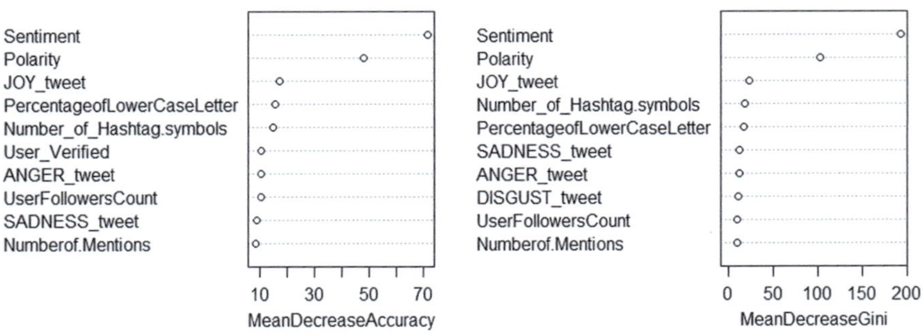

Fig. 2 Top 10 features for classifying tweets using random forest

are at the end of the tree without each variables, as a result we find that sentiment and polarity of the tweets comes out to be the most important variables in both the cases.

Further, in Fig. 3 shows box plot analysis of those features that demonstrate significant difference among all the classes. Tweets having higher positive sentiment score fall into the category of acceptable class; however, tweets with higher negative score will fall into unacceptable category. We also find that acceptable class tweets are having higher emotions of joy and lesser emotions of disgust, sadness, anger, and fear. Polarity of the tweet which was being calculated by the Meaning Cloud (a well-established tool given by IBM) is also very helpful measure for classifying the tweets into acceptable and unacceptable categories. Other features like user favorite count, retweet count, user friend count, status count, etc. are not putting any great effect on model accuracy.

7 Conclusion and Future Work

Twitter provides flexibility to the people throughout the world to post content and media of their choice. Nowadays, whenever a high impact event is happen, most of the people turn toward various social media platforms like Twitter, either to gather information, news, or for sharing their views related to the event. However, the trustworthiness of the tweets which has been shared on Twitter is always trailed by an interrogation mark. There is a dire need of time to develop a model which can filter out all the irrelevant and biased content from social media platform.

We have crawled 3250 tweets from twitter in the period of three months, after that we preprocessed it and narrowed it to 1000 tweets. These tweets are further analyzed by IBM Watson Natural Language Understanding tool for calculating sentiment and emotion (joy, anger, disgust, fear, and sadness). We also calculate the polarity of the tweets with the help of tool given by IBM named Meaning Cloud on a scale of (P+, P, Neutral, NONE, N, N+). For the sake of simplicity, we merge NONE into Neutral category as we find out both of them belongs to the same labeled class.

We have developed a machine learning model which can classify the tweets in four different classification categories with the help of 26 distinct features. Random forest, support vector machine, and Naive Bayes algorithm for classifying the tweets were used, and it has been found that the best result is being given by random forest 0.977% accuracy along with f1 score as 0.9911.

For future perspective, we would try to develop a model for calculating the sentiment, emotion, and polarity of the tweets which are written in both Hindi and English languages. We will also focus to work on larger set of tweets with various real world events.

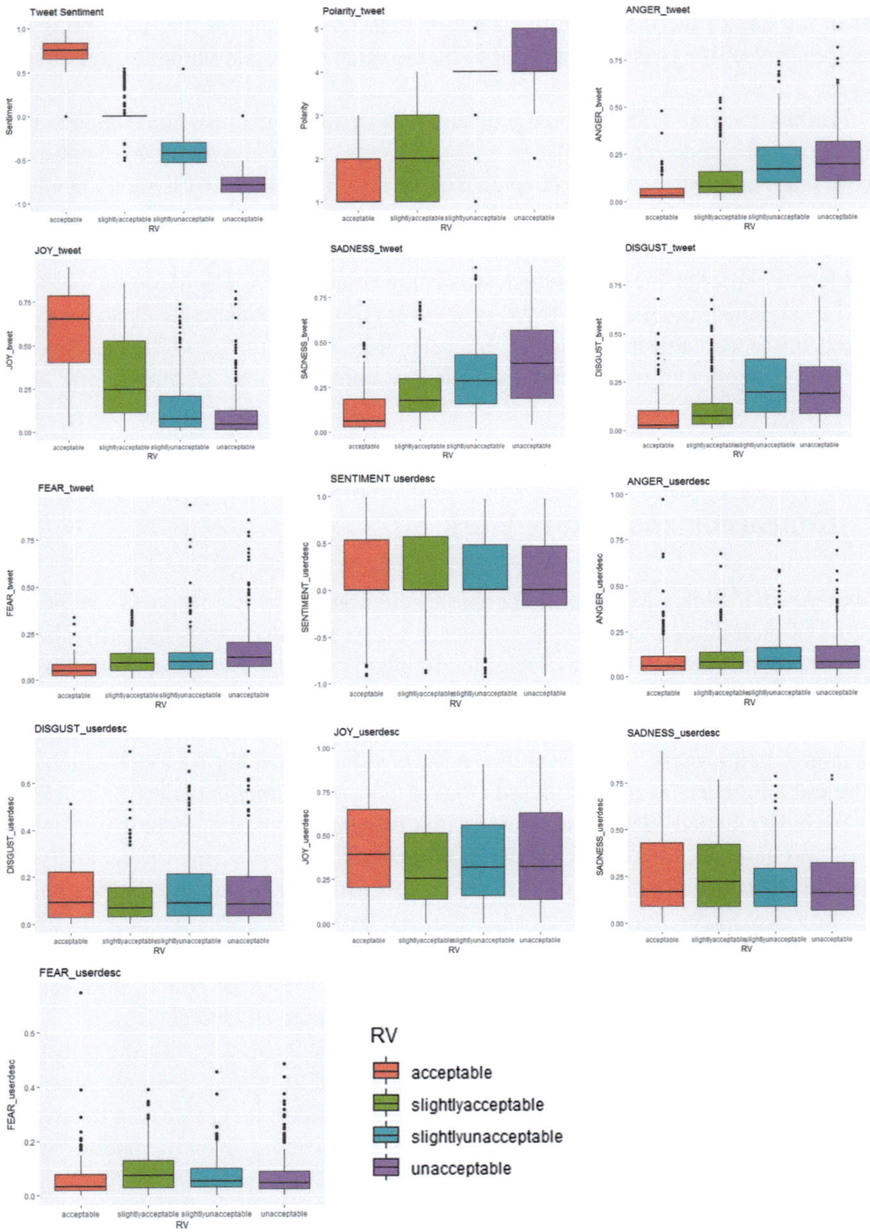

Fig. 3 Box plot analysis of sentiment, polarity, and emotions of tweets

References

1. Ratkiewicz J, Conover M, Meiss MR, Gonçalves B, Flammini A, Menczer F (2011) Detecting and tracking political abuse in social media. ICWSM 11:297–304
2. Gupta A., Kumaraguru P, Castillo C, Meier P (2014) Tweetcred: real-time credibility assessment of content on twitter. In: International conference on social informatics. Springer, Cham, pp 228–243
3. Xia X, Yang X, Wu C, Li S, Bao L (2012) Information credibility on twitter in emergency situation. In: Pacific-Asia workshop on intelligence and security informatics. Springer, Berlin, Heidelberg, pp 45–59
4. Castillo C, Mendoza M, Poblete B (2011) Information credibility on twitter. In: Proceedings of the 20th international conference on World wide web. ACM, pp 675–684
5. Lorek K, Suehiro-Wiciński J, Jankowski-Lorek M, Gupta A (2015) Automated credibility assessment on twitter. Comput Sci 16(2):157–168
6. Zhang Q, Zhang S, Dong J, Xiong J, Cheng X (2015) Automatic detection of rumor on social network. In: Natural language processing and chinese computing. Springer, Cham, pp 113–122
7. O'Donovan J, Kang B, Meyer G, Hollerer T, Adalii S (2012) Credibility in context: an analysis of feature distributions in twitter. In: Privacy, security, risk and trust (PASSAT), 2012 international conference on and 2012 international conference on social computing (SocialCom). IEEE, pp 293–301
8. Resnick P, Carton S, Park S, Shen Y, Zeffer N (2014) Rumorlens: a system for analyzing the impact of rumors and corrections in social media. In: Proceedings of computational journalism conference
9. Abbasi MA, Liu H (2013) Measuring user credibility in social media. In: International conference on social computing, behavioral-cultural modeling, and prediction. Springer, Berlin, Heidelberg, pp 441–448
10. Westerman D, Spence PR, Van Der Heide B (2012) A social network as information: the effect of system generated reports of connectedness on credibility on Twitter. Comput Hum Behav 28(1):199–206
11. Cha M, Haddadi H, Benevenuto F, Gummadi PK (2010) Measuring user influence in twitter: the million follower fallacy. ICWSM 10(10–17):30

Perspectives of Healthcare Sector with Artificial Intelligence

Mohammed Sameer Khan and Shadab Pasha Khan

Abstract Artificial Intelligence is the technology that processes data and makes the machine learn from the data, using a complex mathematical function. The technology has many applications in industries such as Management and Health care. To overcome the limitations of the traditional methods used in the present scenario, we can use AI-based technologies coupled with Internet-of-Things (IoT) to record electronic health record (EHR). Furthermore, the data is stored in the Cloud and is efficiently managed using Big Data. The data is processed through machine learning algorithms which find patterns in the data, to identify symptoms of diseases. This helps in better diagnosis of various diseases. AI will not only provide relief to the patients but also help medical professionals to improve their learnings about chronic diseases such as cancers and coronary heart diseases. This paper aims to address all the techniques involved in making the maximum usage of Artificial Intelligence to bring efficiency and effectiveness in the Healthcare Industry.

Keywords Computation with modeling · Artificial intelligence (AI) · Healthcare system and technologies · Machine learning (ML) · Neural networks

1 Introduction

1.1 What Is Artificial Intelligence?

When computers were still in their initial stage, scientists and researchers were wondering if they could make these inhumane machines that worked on cogs and mechanical inputs, to think like a human. The feat has not been achieved yet, but that idea has been conceptualized into a discipline popularly known as Artificial

M. S. Khan
Department of Computer Science and Engineering, Oriental Group of Institutes, Bhopal, India
e-mail: sameerkhanofficial@gmail.com

S. Pasha Khan (✉)
Department of Information Technology, Oriental Group of Institutes, Bhopal, India
e-mail: shadabpasha@gmail.com

R. K. Shukla et al. (eds.), *Social Networking and Computational Intelligence*, Lecture Notes in Networks and Systems 100, https://doi.org/10.1007/978-981-15-2071-6_12

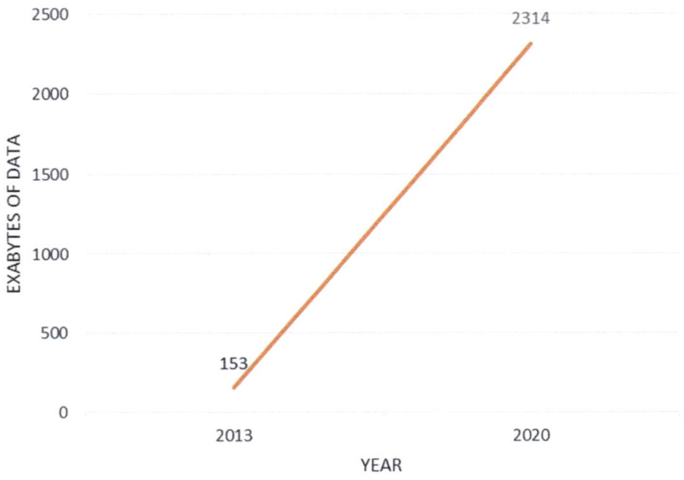

Fig. 1 Increase in the amount of data on health care

Intelligence. Based on the studies in the structure of the human mind, scientists coined the study of a similar phenomenon in the machine, dubbing it as "Neural Networks," in 1943 [1].

1.2 The Need for Artificial Intelligence

Artificial Intelligence shows promising results in the healthcare sector, where there was a dire need for improvement. AI can be easily leveraged to minimize medical errors made, which comes into several categories such as wrong prescription, wrong analysis of reports and missing out on the opportunities of leveraging the past data of the patient, and the data already present in the field (Fig. 1).

AI can prove beneficial in lowering down the mortality rates by avoiding such errors in the analysis and even provide better results by weighing in the factors in analysis which gets neglected even by the best physicians.

1.3 How Is Artificial Intelligence Helping the Patients?

Artificial Intelligence has shown promising results in the field of health care. AI technologies can cater to the need of the patients and simultaneously help the doctors and administrators. It has proved its potential of creating intelligent systems, by learning to identify patterns and deriving satisfactory conclusions BY just analyzing data. The technique of machine learning has proven to be most successful in deriving a sense from the electric health records (EHR) taken from umpteen number of patients

[2]. Learning algorithms are fed data that they process by making use of mathematical functions and techniques like classification and regression to identify patterns in data. Thus, an AI learns from the data, and when given any other data, it makes use of the previously learned information and processes the result based on that. It is quite efficient, exponentially fast, with minimized errors.

1.4 Using AI to Help Medical Professionals

The advancements in the field of AI can help medical professionals predict and deliver much more accurate diagnostics on any given diseases. Instances such as using collected data on the occurrence of kidney stones in patients, and making AI algorithms to learn and predict the possibility of having kidney stones in other patients prove that [3]. The ability of AI to learn from data without the need to prior programming has intrigued both researchers and professionals in the field of health care. Medical Professionals can make use of AI to better predict diseases and minimize any human errors, which is a frequent occurrence in traditional diagnostic methods. AI has even exceeded the experience of medical professionals in medical domains like detection of Tumor Mutation, Pneumonia, Coronary Heart Diseases and Autism [4–6].

1.5 Administrators Leverage AI to Plan Efficiently

The technology is also being used to bring efficiency into an already placed system for the healthcare organization. This proves to be beneficial to administrators which can use it to provide the best service and experience to their patients. Instances like the UK-based Harrow Council collaborating with the IBM Watson Care Manager team to reduce costs of operations, and ensure security and transparency in the process of providing services to the patient, only make the health care more optimistic about the future of more advanced use case of AI [7].

1.6 Limitations of Artificial Intelligence

AI falls short of providing viable results due to the lack of rich data in healthcare industry. From a technical perspective, sometimes the data is not interoperable between different digital management systems used by the hospitals or universities keeping the records. Furthermore, the algorithms used can act as a black box, because they change their method of learning with every data point to further understand and process the data. Creating algorithms fed with the same kind of data can introduce increased selectiveness and thus bring an increase in error. Bias introduced by the developers in the learning algorithm can make the algorithm faulty, both in design and ethics. Furthermore, the ethics of keeping data secure and private, and using it only for research purposes is difficult, when individual data gets involved. The consent of the patient and making the data anonymous to further ensure privacy also limit the process of learning [8].

This paper deals with the usage of Artificial Intelligence in facilitating better diagnosis and care to the patient, providing confirmation and new insights into the diagnosis for medical professionals and efficient planning for healthcare administrators.

2 Applications of AI in the Healthcare Sector

Artificial Intelligence analyses a humongous amount of data to learn the patterns and predict the expected outcome from the algorithm it built using mathematical functions. Healthcare industry with the advent of modern techniques of data collections has the datasets required for an AI to learn. The technology can be leveraged in fields that possess the data.

Researchers have been creating machine learning algorithms to identify patterns of diseases in the patient's electronic health records. Researchers made use of the support vector machines, a concept from supervised machine learning to process medical resonance imaging (MRI) images of the brain and predict whether a patient would show signs of hemorrhage. The algorithm showed better results than an ordinary CT scan to evaluate the signs of the disease [9] (Fig. 2).

Furthermore, scholars researching in medical imaging often make use of deep learning to make use of Artificial Intelligence on a complex dataset that requires more data processing and learning. Deep learning algorithms, such as the neural network,

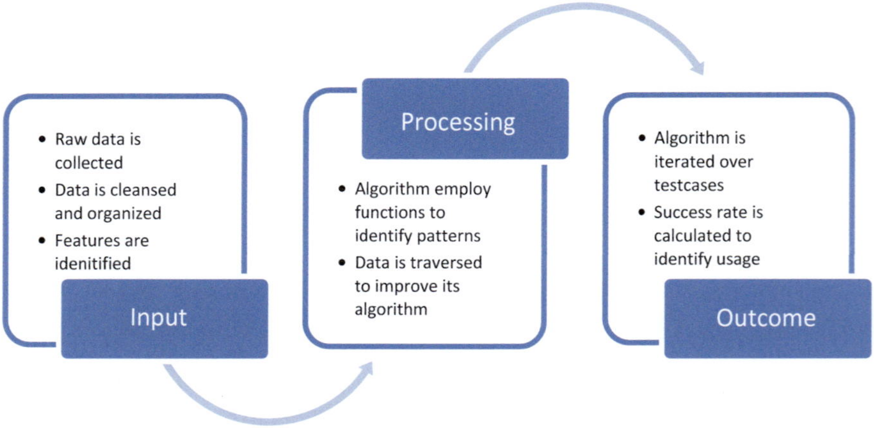

Fig. 2 How machine learning algorithms work over a dataset

are modeled after the human brain. Each node in the neural network represents synapses that grow as the information processed by the node validates to the correct output. As a large dataset gets processed on the network, the nodes are sufficiently strong enough to predict the desired/expected outcome from the data. The technique is often employed in image recognition and speech processing.

Ophthalmologists employ high-definition cameras to take images of the eyeball in order to detect diseases of the eyes and heart. These cameras would capture the eye from a different orientation and provide retinal images, which can be analyzed by any skilled specialist of the eye. Furthermore, a manual check for the disease gets the job done of confirming it and providing any further treatment available to the patient. But, the process becomes cumbersome when we weigh in pragmatic factors such as the lack of skilled ophthalmologists and the toll it takes on the expenses of the patient.

Due to the availability of the data on thousands of such retinal images, a deep learning model was trained with the inputs from few skilled eye specialists to improve and verify the results. The model was successful in providing satisfactorily accurate results and could even be used when cameras with relatively lower resolution were employed [10]. This could prove to be a boon by solving the problem of availability of the skill in the sector and proving to be economical once low-budget cameras work with the image processing deep learning model.

Researchers also employed several other learning algorithms for processing data present on other diseases. From using machine learning algorithms to analyze X-ray data to detect pneumonia to predicting kidney stone, using state-of-the-art IBM Watson to get an ideal 99% success rates in detecting cancers [11], Artificial Intelligence can help in detecting diseases efficiently and effectively.

The mortality rate in developing countries in India is declining with the advent of modern healthcare technologies and the accessibility of these technologies to the masses [12]. From witnessing a high of 8.88 in the year 2000 to a low of 7.3 in 2017, the mortality rate was reduced only with technologies such as modern cameras and machines. Using Artificial Intelligence technology to minimize the cost, and the time factor that plagues the healthcare sector, the mortality rate is predicted to fall below 5 and even expected to touch a 1 mark that is prevalent in developed countries such as the United States (Fig. 3).

The technology can be used for increasing the user experience in medical apps to assess patients and decreases the need for doctor visits. This has been tried in action by researchers, who made a tool for patients suffering from arthritis to provide them with optimal information and cutting down the unnecessary information. This could help users in getting the best course of action for the future [13].

3 Artificial Intelligence for Medical Professionals

Medical Professionals are expected to maintain the quality of their work when assessing a patient or when analyzing a report. Even something as simple as making a wrong

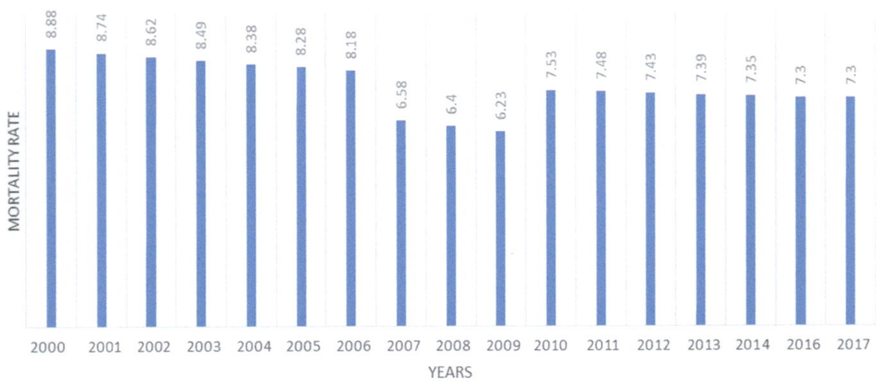

Fig. 3 Mortality rate of India over the years

prescription can take a toll on someone's life. A medical professional has to spend time in analyzing several reports and assess patients, which results in a lack of time, and becomes a factor in the reduction of quality of analysis, and increase the medical errors [14] (Fig. 4).

These errors can be minimized if a medical professional utilizes the advancements in machine learning to confirm their analysis or even use it as the sole tool for disease detection.

The World Health Organization predicts that if medical errors are minimized using learning algorithms to confirm the results or even replace traditional methods of analyzing diseases the mortality rate can be brought down exponentially [14] (Fig. 5).

It can be used to provide data even about rare diseases or epidemics. This could help professional specializing in such field, to avert any epidemic without carrying out much activity or cumbersome procedures. It can be used to find the optimal drug

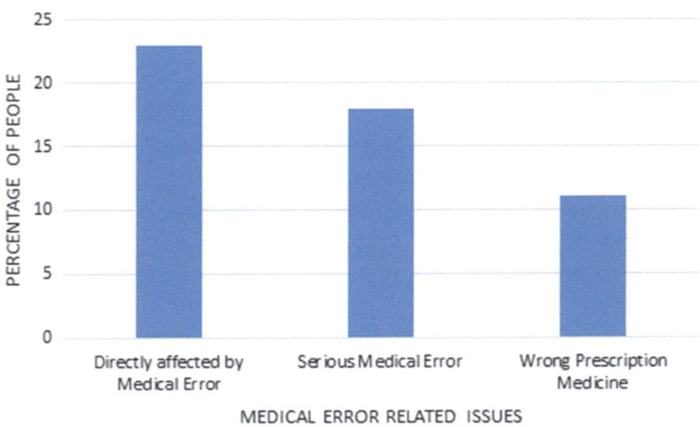

Fig. 4 Deaths related to medical errors

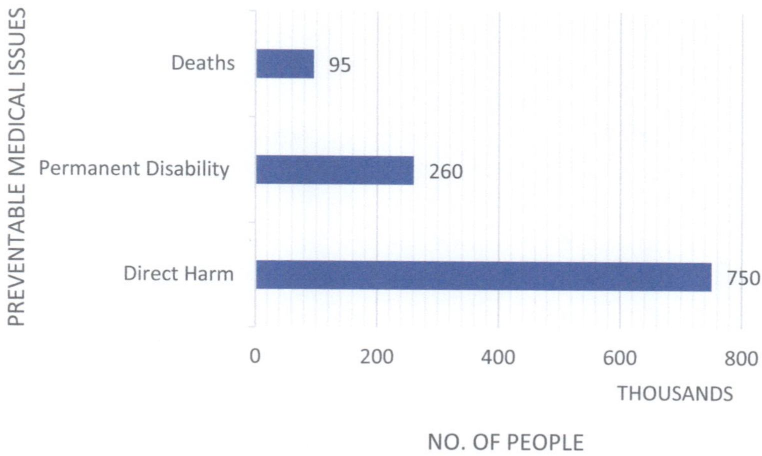

Fig. 5 Preventable medical issues by minimizing medical errors

for any disease specifically for a patient, depending on their health. Electronic health records (EHR) and the past records of the user can provide the best information in finding the drug. This would help diminish the adverse effects of the disease more effectively, in the given time frame. Furthermore, it can be used to assist in research-related activities, in finding the best research material and data on any given field to carry out research.

4 Artificial Intelligence in the Healthcare Administrative Perspective

The point of Artificial Intelligence being capable to provide the best data catering to the specific needs of the data point has been made vivid in the paper. This feature of Artificial Intelligence can be applied in designing systems that are efficient and simultaneously effective in their work. Hospital Administrators are using the techniques of machine learning to provide an optimal health plan for a patient, while at their stay in the hospital. The technique is often coupled with other modern technologies such as the Internet-of-Things (IoT). Harrow Council using the IBM Watson Care Manager is one of the many instances where AI is being employed to achieve the feat of efficiency in designing systems.

Administrators are helping patients out by creating the trend of telehealth, where a patient pays a virtual visit to a doctor. The technology is facilitated by the underlying learning algorithms. The industry is expected to grow rapidly, causing an impact on the healthcare sector and changing the way patients are treated. This application of Artificial Intelligence can also be replicated globally [15] (Fig. 6).

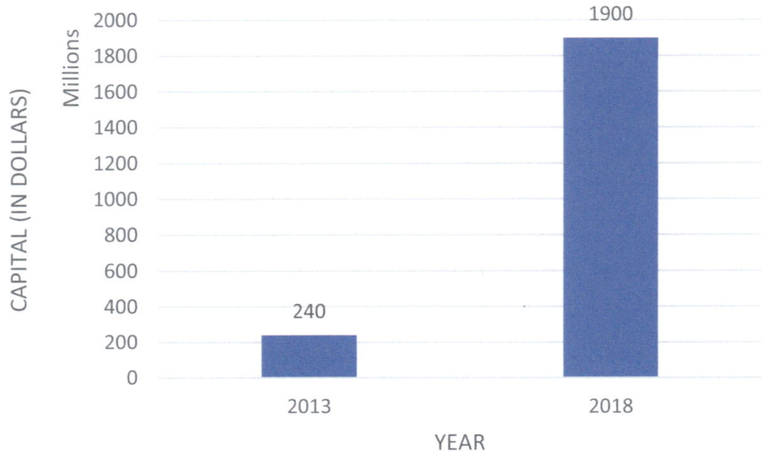

Fig. 6 Growth of the telehealth market

5 Conclusion

The technology of Artificial Intelligence though being experimented on vigorously and getting employed in several fields of the healthcare sectors, for every role of the healthcare sector, from patient to doctor to the administrator, still remains in its nascent stage. The future for learning from large datasets and predicting an ideal outcome with a 100 percent success rate will involve several disciplines like Internet-of-Things (to collect the EHR data from the patient), Cloud Computing (to efficiently store data and making it accessible 24 × 7), Big Data (to manage the data), and finally Artificial Intelligence (to learn from the data). The Healthcare Industry will see an exponential growth as new researches and techniques keep coming. Artificial Intelligence is facing legal issues of collecting the data, ethical issues like ensuring the privacy of the patient in the data to security issues of effectively and securely storing the data. Committees have been created to ensure that the data is used for purposes of research, and no one uses the modern techniques of machine learning for malicious purposes. Artificial Intelligence will struggle to establish trust among the masses, as every modern technique has to, but once the technique gets establishes and comes in use, it can prove to be a boon to the healthcare sector and help in saving umpteen lives.

References

1. Warren SM, Walter P (1943) A logical calculus of the ideas immanent in nervous activity. Bull Math Biophys
2. Alison C, Nigam HS (2017) Machine learning in healthcare. Key Adv Clin Inf

3. Yassaman K, Seyed AM (2017) A novel method for predicting kidney stone type using ensemble learning. Artif Intell Med
4. Derrick EW, James RW (2018) A machine learning approach for somatic mutation discovery. Sci Trans Med
5. Pranav R, Jeremy I (2017) Radiologist-level pneumonia detection on chest X-Rays with deep learning. Cornelll University Library
6. Stephen FW, Jenna R, Joe K, Jonathan MG (2017) Can machine-learning improve cardiovascular risk prediction using routine clinical data? PLOS One
7. IBM and harrow council to bring watson care manager to individuals in the UK (2016). http://www.harrow.gov.uk/news/article/397/
8. Dolores D (2017) Artificial intelligence for health and health care. JSR-17-Task-002
9. Bentley P, Ganesalingam J, Carlton JAL (2014) Prediction of stroke thrombolysis outcome using CT brain machine learning. Neuroimage Clin
10. Michael DA, Mona KG, Milan S (2010) Retinal imaging and image analysis. IEEE Trans Med Imaging
11. IBM is counting on its bet on Watson, and paying big money for it. The New York Times (2016). https://www.nytimes.com/2016/10/17/technology/ibm-is-counting-on-its-bet-on-watson-and-paying-big-money-for-it.html
12. Mortality Rate of India. IndexMundi (2017). https://www.indexmundi.com/g/g.aspx?c=in&v=26
13. Arthritis Research UK Introduces IBM Watson-powered virtual assistant to provide information and advice to people with arthritis (2017). https://ibm.com/press/pressrelease/51826.wss
14. World Health Organization. http://www.euro.who.int/en/health-topics/Health-systems/patient-safety/data-and-statistics
15. World Market for Telehealth. HIS Technologies (2014)

A Novel Approach for Stock Market Price Prediction Based on Polynomial Linear Regression

Jayesh Amrutphale, Pavan Rathore and Vijay Malviya

Abstract Every stock market investor wants to earn more profit from his/her investment. Investor tries different strategies to invest their money. Nowadays, many investors use computer algorithms based stock market prediction system to predict the future prices of stocks. Machine learning and artificial intelligence are one of the advanced and efficient techniques for stock price prediction. This paper will implement polynomial linear regression model and is compared with simple linear regression (SLR) machine learning model. The implementation and experimental results show that polynomial linear regression (PLR) model gives better prediction accuracy and results.

Keywords Stock market · Prediction system · Simple linear regression · Polynomial linear regression · Accuracy · Machine learning · Artificial intelligence · Mean abstract error

1 Introduction

Stock market is a public market, where company shares their company ownership with public. General public can buy of stock to become the part of that company. Usually, stock buyers invest their money to get profit from their investment. They usually expect that in future company profit will grow and they will get good returns on their investments. But companies' future growths are uncertain. Because it depends on many factors, like history of company, government policies, public opinion, market news, social media etc. [1, 2]. So it is difficult and challenging task to predict the future prices of any company stocks.

Now, stock market is one of the biggest parts of global economy. From the last few years, it is very trendy and popular research topic in the field of financial applications

J. Amrutphale (✉) · P. Rathore · V. Malviya
Malwa Institute of Technology, Indore, India
e-mail: jayesh.amrutphale@gmail.com

V. Malviya
e-mail: vijaymalviya@gmail.com

R. K. Shukla et al. (eds.), *Social Networking and Computational Intelligence*, Lecture Notes in Networks and Systems 100, https://doi.org/10.1007/978-981-15-2071-6_13

[3]. Many researchers are working to implement and proposed different types of algorithms and applications to predict stock prices.

There are many prediction systems working for stock market prediction. But none of them are best or worst, selection of prediction system based on accuracy, if accuracy of system is high than it is easy to select model for use [4].

Selection of accurate prediction model increases the probability of getting significant profits. Good prediction system reduces the risk of falling of stock prices. But stock market is always considered as a risky business [3].

Two types of analysis follow by investors one is fundamental and another next one is technical [5]. In fundamental analysis, investor determines the fundamental issues related to company like political environment, government policies, financial market situations etc. In technical analysis, they use some computational based prediction system to predict the future prices on the basis of some past data features.

The main motivation behind this work is that it is very critical and important to predict future prices of stocks for investor to reduce the risk factor and to get good returns on investment stock market is highly unpredicted market. So, it is a very challenging task to make a proper prediction model, which forecast the accurate predictions.

So in this work is going to implement and give a comparison study of experimental results of two different techniques of artificial intelligence (AI). One is simple linear regression, and another next one is our proposed (PLR) polynomial linear regression.

Our paper details are mentioned section wise. Section 2 has literature review for linear regression and stock prediction systems. Section 3 has details about simple linear regression (SLR) model. Section 4 contains the details about polynomial linear regression (PLR). Section 5 has implementation details of application developed. Different performance evaluation parameters are discussed in Sect. 6. Result analysis is discussed in Sect. 7, and finally, overall conclusion of research work is concluded in Sect. 8.

2 Literature Review

Iacomin [6] presented the study of various machine learning algorithms used in stock market prediction. Author represented artificial neural network algorithm with different feature selection. As a result of his study, he observed that support vector machine classifier with principal component analysis for feature selection can give good prediction values.

Waqar et al. [4] investigated the problem associated with high dimensionality in stock market. They applied principal component analysis abbreviation PCA algorithm along with linear regression. This PCA algorithm reduces the data redundancy and improved the performance of machine learning method. They compared accuracy with and without PCA. By the experimental results, they found that the PCA can enhance the performance of machine learning algorithm.

Dinesh et al. [3] proposed linear regression models to implement stock prediction model. They have experimented their model on forecasting behavior of TCS dataset. They predicted and compared their prediction based on linear regression, polynomial and radial basis function (RBF). They used 5 independent variables for analysis.

Shepon et al. [7] proposed and implemented a system that was used to calculate risk in buying the stocks. Proposed system calculates the higher risk rewards ratio and mange number of trades according to that.

Sasidhar et al. [5] proposed a linear regression model to predict the stock market prices using statistical modeling approach. They took open, close, high and low values of stocks as a independent variables.

There is lots of research work done by researchers in stock market prediction. A hybrid forecasting model is proposed by Ince et al. [8] in this authors combined independent component analysis (ICA) and kernel methods to develop prediction system. Impact of social media on stock market prices is also studied by many researchers [2, 9, 10].

3 Simple Linear Regression

SLR is generally based on the formula given in Eq. (1).

$$Y = b_0 + b_1 * X_1 \tag{1}$$

where Y is a dependent variable. Its value is always dependent on some independent variables.

X_1 is an independent variable. Value of this variable is directly taken from the real-world situations. In SLR, only single independent variable is used. This independent variable (X_1) is responsible to changes in the value of dependent variable Y. When multiple independent variables are used to evaluate dependent variable, it is called multiple linear regression (MLR).

b_1 is a coefficient for X_1 that mean it gives the impact on dependent variable, i.e., how much unit changes in X_1 change the unit of Y. b_1 is calculated by the formula given in Eq. (2).

$$b_1 = \frac{n\sum xy - \left(\sum x \sum y\right)}{n\sum x^2 - \left(\sum x\right)^2} \tag{2}$$

where n is number of observations and x and y are observation coordinates. b_0 is a constant term. It is like a base value of Y, i.e., value of dependent variable when independent variable value is zero. As shown in Fig. 1. It is calculated by formula given in Eq. (3).

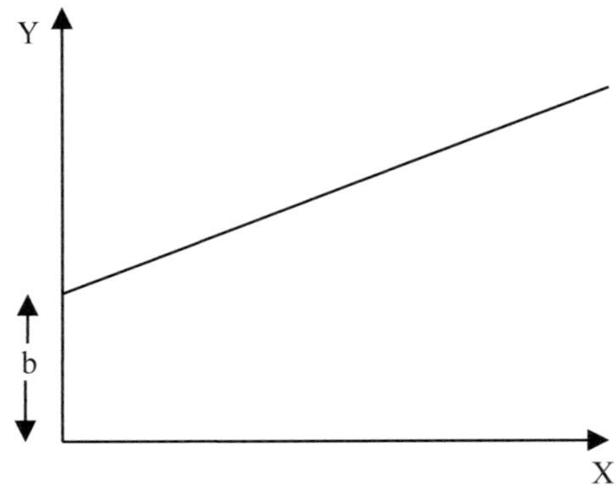

Fig. 1 Simple linear regression

$$b_0 = \frac{\sum y}{n} - b_1 \frac{\sum x}{n} \tag{3}$$

where n is number of observations and x and y are observation coordinates.

4 Polynomial Linear Regression

PLR prediction model is based on Eq. (4).

$$Y = b_0 + b_1 X_1 + b_2 X_1^2 + \cdots + b_d X_1^d \tag{4}$$

where Y is dependent variable similar to the SLR and X_1 is independent variable. But in PLR X_1 is presented in different powers. d represents the degree of polynomial.

Due to the coefficient (b_1, b_2, b_3 … b_d), it is called a linear regression. These coefficients are in linear in equation. Because ultimately coefficients are unknown and we have to find this unknown to prepare a model. These coefficients decide that which type of regression our model fits. It is also called a special case of multiple regression.

5 Implementation Details

For the implementation and experiment, we have used python programming language. Python provides great support for machine learning model implementation. Some important python models which are used in this implementation are as follows:

sklearn—This is a scikit-learn https://scikit-learn.org/ machine learning library, and this module has many machine learning inbuilt algorithms like linear regression, multiple regression, support vector machine etc.

matplotlib—This module is used to plot actual observation and prediction model in the form of graphs.
tkinter—This module is used to design GUI for application.
pandas—This module is used to read the input data from datasets.
numpy—It is used to perform numeric operations on datasets.

Figure 2 shows the GUI implemented for this application. In dataset selection section we have to give the name of dataset. There are two separate sections for SLR and PLR. After training the model, we get accuracy, mean absolute error (MAE), mean squared error (MSE) and root mean squared error (RMSE). PLR also required degree of polynomial as a input. Here, degree means number of X terms in polynomial equation. Model accuracy may vary according to variation take place in degree. So, we have to select best degree. Degree of polynomial may vary according to variation

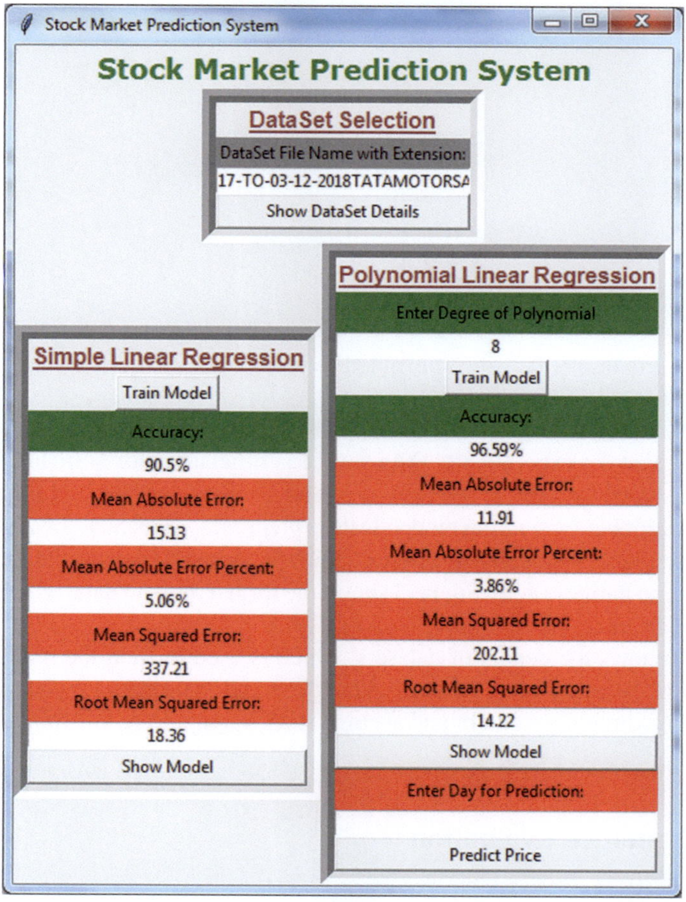

Fig. 2 GUI of stock market prediction system

Fig. 3 Input data format

Date	High Price
1	411.25
2	405.5
3	404
4	405.3
5	413.75
6	415.2
7	415.5
8	409.85
9	404.55
10	409
11	409.9
12	421.7
13	427.75
14	426.05
15	426.4
16	425.15
17	428.45
-	-
-	-
-	-

in pattern of datasets. It is recommended to test model with different degrees and select best one.

One year historical data (from December 04, 2017, to December 03, 2018) of stock market is taken for experiments. Data source is official site of NSE India (https://www.nseindia.com). Date is selected as an independent variable and high price as dependent variable for training model. Data is preprocessed manual for experiments. In preprocessing, unwanted features are removed from the records and date field is converted into numeric field. Figure 3 shows the format of input data, which is stored in csv file format.

Table 1 shows the details of datasets, which is used to train the models in this work. Total 256 days observations are taken. Observation summary is given in Table 1 including maximum, minimum and average stock value during this time period.

6 Performance Evaluation

For the implementation and experiment, we have used python programming language. Python provides great support for machine learning model implementation. Some important python models which are used in this implementation are as follows:

Table 1 Dataset details

S. No.	Company symbol	Max. price	Min. price	Avg. price
1	TATACOFFEE	173.4	94.7	125.28
2	TATACOMM	754	459	602.03
3	TATAGLOBAL	328.75	215.6	264.76
4	TATAINVEST	957.85	663.5	816.8
5	TATAMETALI	975.5	580.7	757.94
6	TATAMOTORS	443.5	169.75	308.76
7	TATAPOWER	101.8	62.4	81.16
8	TATASTEEL	793	505.75	620.9
9	TCS	3674.8	1734.9	2505.63

To evaluate the performance of implementation, accuracy and errors are calculated. Accuracy is calculated with reference to predicted value versus actual value. Three types of error rates are calculated for proposed prediction model.

a. Mean absolute error, i.e., MAE is the mean of the absolute value of the errors. It is calculated as shown in Eq. (5).

$$\frac{1}{n}\sum_{i=0}^{n}|\text{Actual} - \text{Predicted}| \tag{5}$$

b. Mean squared error, i.e., MSE is the mean of the squared errors and is calculated as shown in Eq. (6).

$$\frac{1}{n}\sum_{i=1}^{n}|\text{Actual} - \text{Predicted}|^2 \tag{6}$$

c. Root mean squared error, i.e., RMSE is the square root of the mean of the squared errors and formula is as shown in Eq. (7).

$$\sqrt{\frac{1}{n}\sum_{i=1}^{n}|\text{Actual} - \text{Predicted}|^2} \tag{7}$$

7 Result Analysis

Figure 4 shows the stock prices SLR model of TATA POWER, red dots represent the real observation and blue straight line represents the training model.

Figure 5 shows the stock prices PLR model of TATA POWER, red dots represent the real observation and green line represents the training model.

Similarly, other models are trained. Table 2 shows the experimental results of SLR model, in this accuracy of model and three different error rates (MAE, MSE and RMSE) are represented.

Table 3 shows the experimental results of PLR model. Table contains the model accuracy, selected optimum degree and error rates (MAE, MSE and RMSE).

As a result of both model (SLR and PLR) experimental results analysis, certain observations have been found.

Accuracy of PLR is far better than SLR. Figure 6 shows the comparison of accuracy of SLR and PLR.

MAR rate of PLR is low and in the acceptable range compared to SLR model. Figure 7 shows the comparison of MAR calculated by both SLR and PLR prediction model.

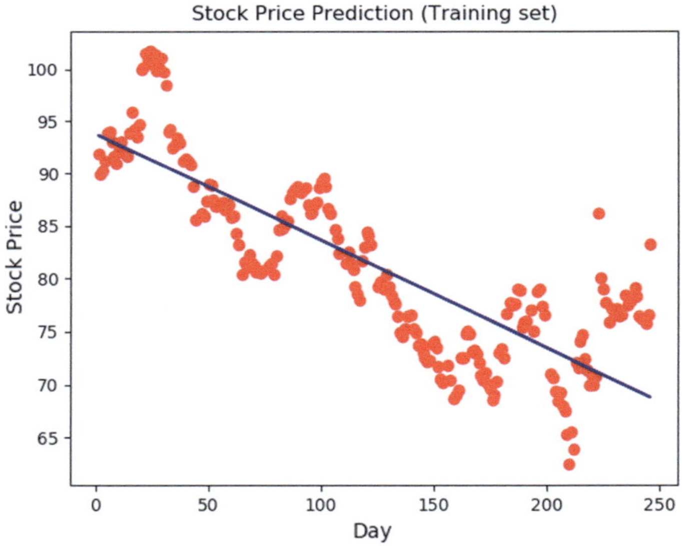

Fig. 4 SLR training model of TATA POWER

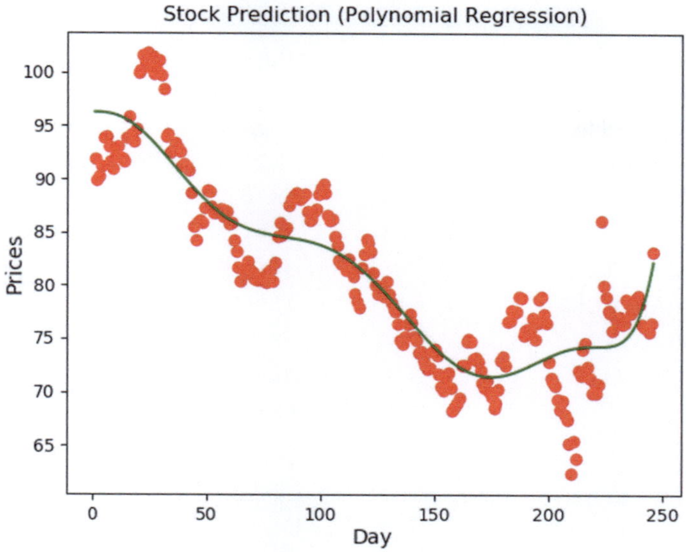

Fig. 5 PLR training model of TATA POWER

Table 2 Experimental results of SLR

S. No.	Company symbol	Accuracy (%)	MAE	MAE (%)	MSE	RMSE
1	TATACOFFEE	36.48	8.82	7.44	110.27	10.5
2	TATACOMM	76.30	19.18	3.26	531.06	23.04
3	TATAGLOBAL	75.51	8.59	3.32	129.22	11.37
4	TATAINVEST	34.19	37.86	4.71	2436.58	49.36
5	TATAMETALI	48.24	43.6	5.79	2500.3	50
6	TATAMOTORS	90.50	15.13	5.06	337.21	18.36
7	TATAPOWER	21.98	3.82	4.80	22.56	4.75
8	TATASTEEL	0.25	43.99	7.23	2639.87	51.38
9	TCS	46.13	326.19	13.13	204,660.22	452.39

Table 3 Experimental results of PLR

S. No.	Company symbol	Poly. degree	Accuracy (%)	MAE	MAE (%)	MSE	RMSE
1	TATACOFFEE	8	93.34	4.51	3.60	31.29	5.59
2	TATACOMM	6	93.50	12.29	2.04	247.52	15.73
3	TATAGLOBAL	9	85.93	8.71	3.29	120.77	10.99
4	TATAINVEST	8	68.75	26.83	3.28	1070.23	32.71
5	TATAMETALI	9	77.26	35.41	4.67	1902.28	43.62
6	TATAMOTORS	8	96.59	11.91	3.86	202.11	14.22
7	TATAPOWER	8	82.06	2.9	3.58	13.1	3.62
8	TATASTEEL	9	83.77	21.55	3.47	735.27	27.12
9	TCS	10	76.22	196.08	7.83	83,293.22	288.6

Fig. 6 Accuracy comparison of SLR and PLR model

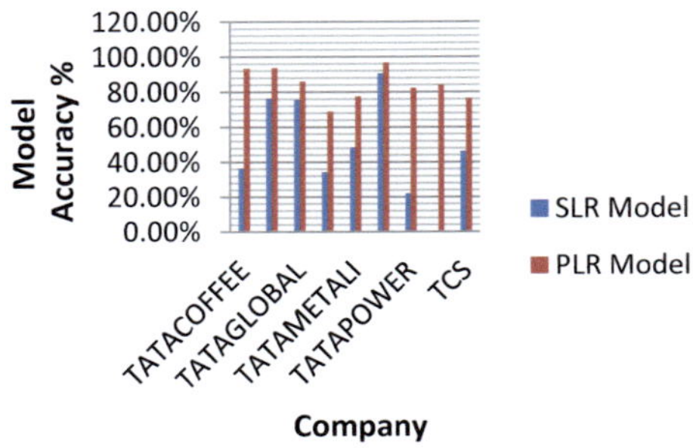

Fig. 7 MAE comparison of SLR and PLR model

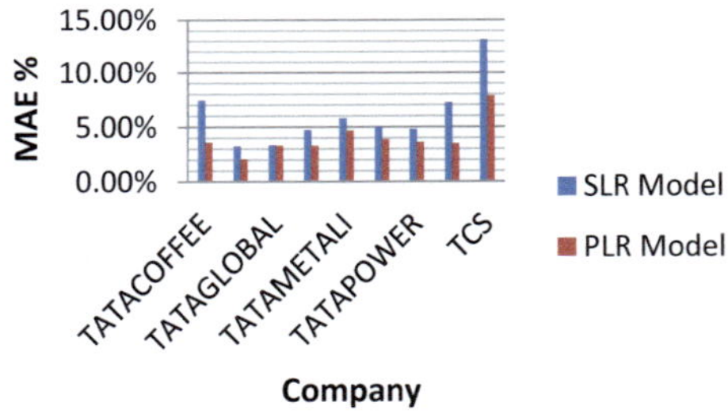

8 Conclusion

Stock market prediction system plays a significant role in investment planning to get better returns on investment. By this research work, it is found that our proposed (PLR) polynomial linear regression model is far better option than simple linear regression model to develop stock market prediction system. By the analysis of historical data of stock market, we found that stock market is fluctuating market. Many factors affect the stock market prices. So, it is very difficult to predict accurate value of stock prices using SLR model. But by the implementation and experimental results, we found that PLR model gives better prediction accuracy and results.

References

1. Yang B, Gong Z-J, Yang W (2017) Stock market index prediction using deep neural network ensemble. In: 36th Chinese control conference, 26–28 July 2017
2. Nguyen TH, Shirai K (2015) Topic modeling based sentiment analysis on social media for stock market prediction. In: Proceedings of 7th international joint conference on natural language processing, Beijing, China, 26–31 July 2015, pp 1354–1364
3. Bhuriya D, Sharma A, Singh U (2017) Stock market prediction using linear regression. In: International conference on electronics, communication and aerospace technology ICECA 2017
4. Waqar M, Dawood H, Shahnawaz MB, Ghazanfar MA, Guo P (2017) Prediction of stock market by principal component analysis. In: 13th international conference on computational intelligence and security
5. Bommareddy SR, Reddy KSS, Kaushik P, Vinay Kumar KV, Hulipalled VR (2018) Predicting the stock price using linear regression. Int J Adv Res Comput Sci 9(3):81–85
6. Iacomin R (2015) Stock market prediction. In: Proceedings of 19th international conference on system theory, control and computing (ICSTCC), 14–16 Oct, Cheile Gradistei, Romania, pp 200–205
7. Shepal Y, Yatish B, Rahul K, Anis S (2018) Stock market prediction. Int J Res Eng Appl Manag. Special issue—iCreate
8. Ince H, Trafalis TB (2017) A hybrid forecasting model for stock market prediction. Econ Comput Econ Cybern Stud Res 51(3):263–280
9. Oliveira N, Cortez P, Areal N (2017) The impact of microblogging data for stock market prediction: using twitter to predict returns, volatility, trading volume and survey sentiment indices. Expert Syst Appl 73:125–144
10. Sun A, Lachanski M, Fabozzi FJ (2016) Trade the tweet: social media text mining and sparse matrix factorization for stock market prediction. Int Rev Fin Anal. https://doi.org/10.1016/j.irfa.2016.10.009

Real-Time Classification of Twitter Data Using Decision Tree Technique

Shivam Nilosey, Abhishek Pipliya and Vijay Malviya

Abstract The data which comes from e-commerce site is unstructured text data. Text mining is becoming an important field in research for finding valuable information from unstructured texts. Data which contains an unstructured text which stores large volume of valuable information cannot exclusively be used for any process by computers. Therefore, we want a definite process strategies, techniques and algorithms in order to extract this meaningful information which is done by using text mining. Classification of these user opinions is an Information Extraction and Natural Language Processing task that classifies the user opinions into various categories like in the form of positive or negative. In these paper, we can build a classifiers based on SVM and decision tree classification algorithm to identify the opinions and classify them onto categories, and also, we can compute the performance measure of these classifiers and also compare the classification algorithms performance based on their accuracy, and we can say that decision tree perform better as compared with SVM. In this, we can also classify the real-time tweets review into various emotions and polarity through decision tree classification model.

Keywords Web data · Text mining · Data mining · R · Text mining techniques · SVM · Decision tree · Classification

1 Introduction

Sentiment analysis is principally involved with the identification and classification of opinions or emotions of every tweet. Sentiment analysis is broadly speaking classified within the 2 sorts; first, one could be a feature or side primarily sentiment analysis, and

S. Nilosey (✉) · A. Pipliya · V. Malviya
Malwa Institute of Technology, Indore, India
e-mail: shivam.nilosey@gmail.com

A. Pipliya
e-mail: aadeepipliya@gmail.com

V. Malviya
e-mail: vijaymalviya@gmail.com

R. K. Shukla et al. (eds.), *Social Networking and Computational Intelligence*, Lecture Notes in Networks and Systems 100, https://doi.org/10.1007/978-981-15-2071-6_14

also, the different is perspicacity-based sentiment analysis [1]. The tweets associated with picture show reviews come back beneath the class of the feature. Perspicacity primarily based SA [2] will the exploration of the tweets that are associated with the emotions like hate, miss, love, etc.

In general, varied symbolic techniques are accustomed to analyze the sentiment from the twitter knowledge. Thus in differently, we will say that a sentiment analysis could be a system or model that takes the documents that analyzed the input and generates an in depth document summarizing the opinions of the given input document. Within the pre-processing, we tend to are removing the stop words, white areas and continuance words. To properly classify the text [3], classification technique uses the coaching knowledge. So, this system does not need the info of words like employed in knowledge-based approach, and so, machine learning techniques are healthier and quicker.

The many strategies are accustomed extract the feature from the supply text. Feature extraction is finished in 2 sections: within the 1st phase, extraction of knowledge associated with twitter is finished, i.e., twitters-specific knowledge is extracted. Currently by doing this, the tweet is remodeled into traditional text. Within the next section, additional options are extracted and additional to feature vector [4]. Every tweet within the coaching knowledge is related to category label. This coaching knowledge is passed to completely different classifiers, and classifiers are trained. Then, check tweets are given to the model and classification is finished with the assistance of those trained classifiers. Thus, finally, we tend to get the texts that are classified into the positive, negative and neutral.

2 Literature Review

In [1], they planned a public sentiment analysis unit which collects the reviews and feedback reviews toward any products and merchandise and classify then into various polarities. In [4], they taken twitter website because on twitter websites millions of user can post their opinions toward any products and issues, so they can fetch the real-time tweets reviews from twitter website, and after fetching, they will pre-process the data by removing unwanted characters, numbers, urls, and redundant data is also been removed after that they can calculate or compute the sentiment score of the tweets.

In [2], they will collect the data from any review websites and then classify the review into positive or negative polarity. So, they will easily get the product quality and to find the product prons and cons so which will helps us to take a decision toward that products. There are many social sites and reviewer sites from where they can collect the data and started working on this data by applying various data mining techniques.

According to this paper [3], they proposed to find the sentiment based on the emoji because emoji can also express the sentiments toward any products'. Emojis are very short and can be easily used by user into day-to-day uses, so they can express their emotion with the help of various emoticons who they will collect the review which contains emojis, and then, they uses emoji-based lexicon dictionary; based on these, they will find the sentiment of user toward any topic or products' based on the emojis.

In [5], it is within the main used for text information analysis. The opinions on any product unit done supported the reviews provided by the users. Recently, social media users are abundant involved in usage of the images and videos rather than text messages. Since the sentiment analysis on such visual contents is improbably tedious, a fresh methodology is used therefore on try and image analysis. Throughout this paper, convolutional neutral networks (CNN) are used for this purpose. There unit a pair of approaches involved throughout this process: visual sentiment analysis with regular CNN and visual sentiment analysis with progressive CNN. The result is that sentiment analysis is quite useful for information analytics tasks which could be used for prediction and prognostication [6].

3 Problem Definition

Real-time sentiment analysis [7] provide better view to understand the real reviews, and based on these, a better decision taking system is developed. But there are many algorithms and techniques are present in machine learning, but the performance is different of all the algorithms. So a better algorithm provides better performance, so selection of the algorithm is very important.

4 Proposed Work

In this, we proposed SVM and decision tree classification algorithm to classify the reviews into various categories like positive or negative. We build classifiers based on these classification algorithms such as decision tree (CART), SVM (support vector machine) and trained these model on training data and then apply these model on test data and compute the performance measure of these classification algorithms (Fig. 1).

Step-1. First, we can collect the text dataset for classification.
Step-2. After collecting the dataset, we can pre-processed the data before apply classification algorithm. In this, we can label the dataset attributes and transform the data and also perform various text mining techniques like in this we find the term frequency.
Step-3. After that we can build classifier based on various classification algorithms and apply the train dataset to the models.
Step-4. Then, we can compute the performance of these classification algorithms on the test dataset.

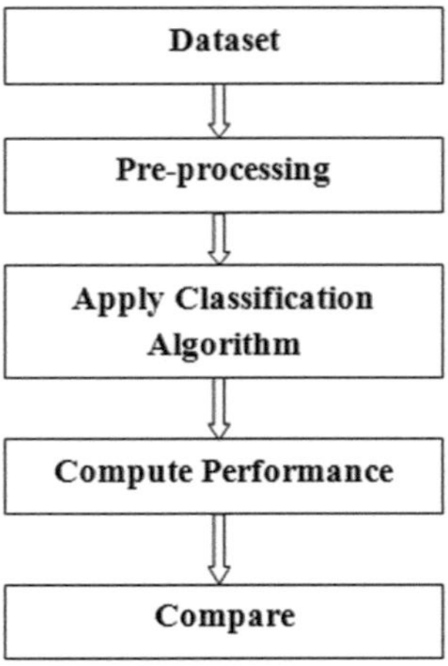

Fig. 1 Flow diagram

Step-5. After computing the performance measure, we can compare these algorithms based on their performance.

5 Experimental and Result Analysis

The experimental and result analysis is done by using Intel i5-2410M CPU with 2.30 GHz CPU along with 4 GB RAM, and the windows operating system is running. For result analysis, we use R and Rstudio for processing the data, and then, we load the tweets and performing sentiment analysis on that collected tweets. So, we can load the data, and Fig. 2 shows him data is loaded.

We can see that the dataset has 1150 observations and 2 variables. One of the variables is the tweet itself, and the other variable is the average sentiment about the tweet. After loading the dataset, we perform pre-processing the text, so we can first create a corpus of tweets, and so, we can see that the corpus has 1150 documents in it. Then, we create a function that will clean the corpus to convert all characters to lowercase, remove punctuation, remove any stop words and stem the document. After cleaning the tweets, we can look at the tweet, we can see that the tweets have been cleaned, and Fig. 3 shows the pre-processing the tweets and text.

After pre-processing the data, we build the model, so we can split the data into train data or test data, and then, we can take SVM and CART (decision tree) classification models to learn on train data, and then, we can compute the performance of these models. Let us evaluate these models and pick the best model for real-time classification. Figure 4 shows the training of these models, and Fig. 5 shows the evaluation measures of the models.

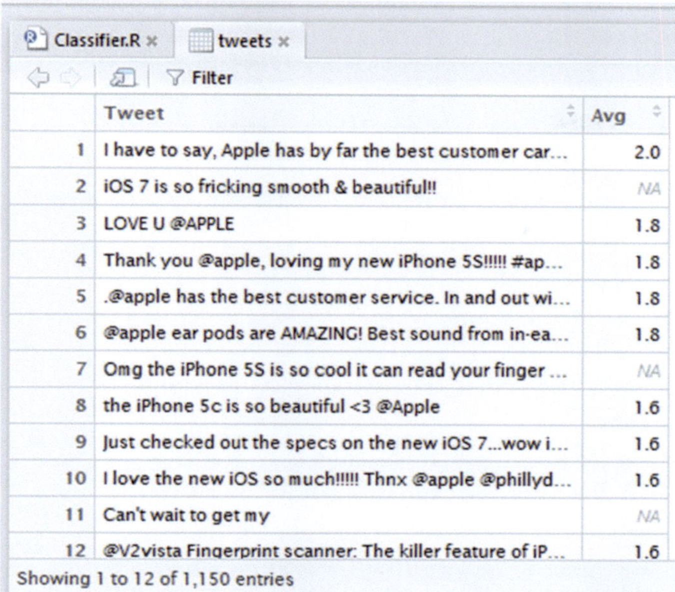

Classifier.R × | tweets ×

Filter

	Tweet	Avg
1	I have to say, Apple has by far the best customer car...	2.0
2	iOS 7 is so fricking smooth & beautiful!!	NA
3	LOVE U @APPLE	1.8
4	Thank you @apple, loving my new iPhone 5S!!!!! #ap...	1.8
5	.@apple has the best customer service. In and out wi...	1.8
6	@apple ear pods are AMAZING! Best sound from in-ea...	1.8
7	Omg the iPhone 5S is so cool it can read your finger ...	NA
8	the iPhone 5c is so beautiful <3 @Apple	1.6
9	Just checked out the specs on the new iOS 7...wow i...	1.6
10	I love the new iOS so much!!!!! Thnx @apple @phillyd...	1.6
11	Can't wait to get my	NA
12	@V2vista Fingerprint scanner: The killer feature of iP...	1.6

Showing 1 to 12 of 1,150 entries

Fig. 2 Load the data

```
> # View tweets
> tweet_corpus[[1]][1]
$content
[1] "I have to say, Apple has by far the best customer care service I have ever received
! @Apple @AppStore"

> clean_corpus <- function(corp){
+   corp <- tm_map(corp, content_transformer(tolower))
+   corp <- tm_map(corp, removePunctuation)
+   corp <- tm_map(corp, removeWords, c('apple', stopwords('en')))
+   corp <- tm_map(corp, stemDocument)
+ }
>
> # Apply function on corpus
> clean_corp <- clean_corpus(tweet_corpus)
>
> # Check a tweet from cleaned corpus
> clean_corp[[1]][1]
$content
[1] "say far best custom care servic ever receiv appstor"

>
```

Fig. 3 Pre-processing the data

```
# CART
set.seed(101)
tweet.cart <- train(Negative~., data=trainSparse, method='rpart',
                    trControl=control, metric=metric)


# SVM
set.seed(101)
tweet.svm <- train(Negative~., data=trainSparse, method='svmRadial',
                   trControl=control, metric=metric)
```

Fig. 4 Training of these models

```
Models: CART, SVM
Number of resamples: 10

Accuracy
          Min.   1st Qu.    Median      Mean   3rd Qu.      Max. NA's
CART 0.8292683 0.8704819 0.8957392 0.8897855 0.9126506 0.9397590    0
SVM  0.8433735 0.8662577 0.8727593 0.8705080 0.8795181 0.8915663    0

Kappa
          Min.   1st Qu.    Median      Mean   3rd Qu.      Max. NA's
CART 0.2347024 0.3424612 0.4403611 0.4650376 0.5890784 0.7296417    0
SVM  0.0000000 0.2244787 0.2779996 0.2559332 0.3360000 0.5133550    0
```

Fig. 5 Performance measure summary of the models

After computing the performance of the models, we can compare the accuracy of these models, and based on these comparisons, we pick the best models for classification. Figure 6 shows the comparison of the models.

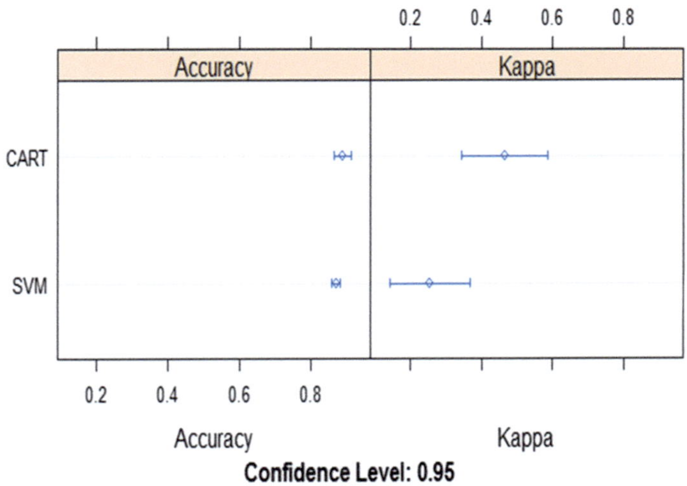

Fig. 6 Performance comparison

Observing the summary and graphs, we can see that CART (decision tree) model still has the highest accuracy of 0.8845 on test dataset and Fig. 7 shows the comparison of these models on the basis of accuracy. And after picking the best model, we can real-time classification on tweets data.

Based on comparison, we can say that decision tree perform better than SVM, so we can pick the decision tree for real-time classification of twitter data. So for collecting twitter tweets data, we need a r package called twitteR and ROAuth package to authenticate user consumer and token keys (Fig. 8).

For consumer key and access tokens, we need to create a twitter app through which we can generate our twitter keys and put into twitter_oauth and then we are

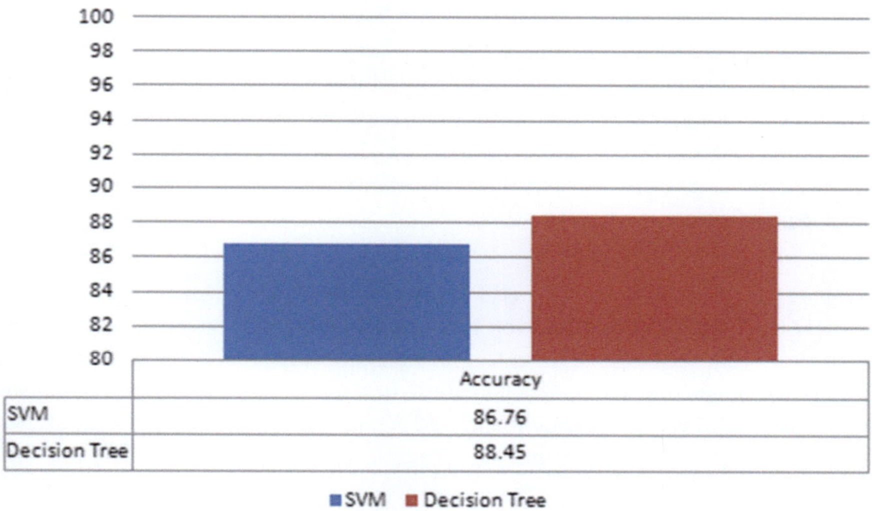

Fig. 7 Accuracy comparison

```
> library(twitteR)
> library(ROAuth)
>
>
> consumer_key<-'VE12EJOT2niBwUONTt7Q5sA4u'
> consumer_secret<- 'J5f0rVLmldAafwS4Pxm3eRyB537MMrrsXPdayANnpLyalfpCsY'
> access_token<- '772457331795726337-6IHz8xax4nKXgYfGy7qnZvYmayo0v41'
> access_secret<- 'xRc0OuVc7nRp2Wmq55Z2lzTCzYiDu2daRJNbX1ScOCcHn'
>
> consumer_key<-'VE12EJOT2niBwUONTt7Q5sA4u'
> consumer_secret<- 'J5f0rVLmldAafwS4Pxm3eRyB537MMrrsXPdayANnpLyalfpCsY'
> access_token<- '772457331795726337-6IHz8xax4nKXgYfGy7qnZvYmayo0v41'
> access_secret<- 'xRc0OuVc7nRp2Wmq55Z2lzTCzYiDu2daRJNbX1ScOCcHn'
>
> setup_twitter_oauth(consumer_key, consumer_secret, access_token,access_secret)
> setup_twitter_oauth(consumer_key, consumer_secret, access_token,access_secret)
[1] "Using direct authentication"
[1] "Using direct authentication"
>
>
> some_tweets = searchTwitter("oneplus6", n=1500, lang="en")
> some_tweets = searchTwitter("oneplus6", n=1500, lang="en")
```

Fig. 8 Data classification logic

storing to collects tweets on modi with frame size of 1000 and stored into some tweets variable.

Next pre-processing the data, we are applying tree classifier to classify the text into various sentiment emotions and sentiment polarities.

```
#Sentiment Analysis

# using sentiment package to classify emotions
emotions <- classify_emotion(clean_tweets, algorithm='tree')

# using sentiment package to classify polarities
polarities = classify_polarity(clean_tweets, algorithm='tree')
```

The decision tree algorithm classifies the text into various emotions shown in Fig. 9.

We can also classify the sentiment polarity of tweets text and which is shown in Fig. 10.

6 Conclusion

In this paper, we investigate various classification algorithms such as support vector machine (SVM) and CART (Decision Tree) to classify the user opinions into category positive or negative. In these, we can build a classifiers based on these classification algorithm compute the performance measure of these classifiers. And also compare performance of these classification and we found that decision tree performs better than SVM classification algorithm, so for real-time classification, we can pick the

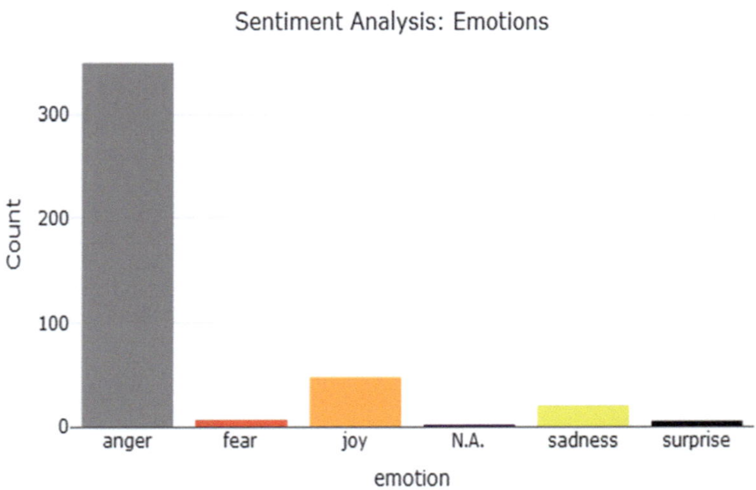

Fig. 9 Classification by emotions

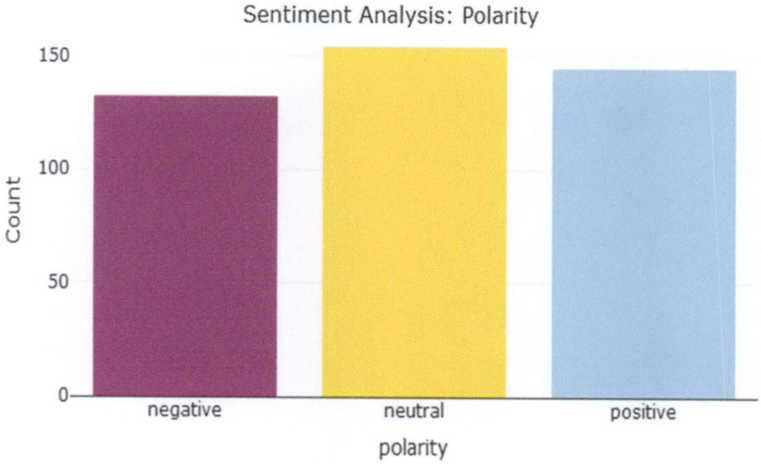

Fig. 10 Classification by polarity

decision tree classifier. In these, we can classify the real-time twitter data with the help decision tree classifier into various emotions and polarity.

References

1. Rangu C, Chatterjee S, Valluru SR (2017) Text mining approach for product quality enhancement. In IEEE 2017
2. Zhan J, Fang X (2015) Sentiment analysis using product review data. Springer, Berlin
3. Sluban B, Mozetič I, Smailović J, Novak PK (2015) Sentiment of emojis. Plos one
4. Islam MS, Ahmed F, Rahman RM, Anwar Hridoy SA, Tahmid Ekram M (2015) Localized twitter opinion mining using sentiment analysis. Springer
5. Jin H, Yang J, You Q, Luo J (2015) Robust image sentiment analysis using progressively trained and domain transferred deep networks. In: Association for the advancement of artificial intelligence
6. Mukherjee S (2012) Sentiment analysis-a literature survey. Indian Institute of Technology, Department of Computer Science and Engineering, Bombay
7. Riloff E, Wiebe J, Phillips W (2005) Exploiting subjectivity classification to improve information extraction. In: Proceedings of association for the advancement of artificial intelligence, pp 1106–1111

Dynamic Web Service Composition Using AI Planning Technique: Case Study on Blackbox Planner

Lalit Purohit, Satyendra Singh Chouhan and Aditi Jain

Abstract The dynamic composition of web services is an important research problem to offer value added services to the end user. As per the demands of the end user, the sequence in which services to be combined as well as participating services are to be decided at run-time. Planners based approach is useful to achieve the dynamic web service composition. Based on the functional parameters—input, output, precondition and effect, various AI planners achieve service composition differently. In this work, we present a AI planning-based dynamic web service composition approach using Blackbox planner. The experimental results show the effectiveness of the proposed approach.

Kewords Web service · Web service composition · AI planning · Planning domain definition language

1 Introduction

Web services are the web-accessible software components that allow machine-to-machine interaction. The web services can be published (in public registry), located and easily accessed over the web using XML-based standard protocols such as SOAP, WSDL and UDDI [1]. The application functionalities are encapsulated along with sources of information by the web service. They also enable programmatic access to applications available on the web. In the diverse software systems, the interoperability property offered by web services allows one system to exploit the functionality of the other. However, sometimes a single service when taken alone offers limited functionality and is insufficient to fulfil the requirements of the end user. By combining

L. Purohit (✉) · S. S. Chouhan · A. Jain
Shri Govindram Seksaria Institute of Technology and Science, Indore 452003, India
e-mail: purohitlalit@sgsits.ac.in

S. S. Chouhan
e-mail: schouhan@sgsits.ac.in

A. Jain
e-mail: jainaditi310@gmail.com

R. K. Shukla et al. (eds.), *Social Networking and Computational Intelligence*, Lecture Notes in Networks and Systems 100, https://doi.org/10.1007/978-981-15-2071-6_15

various web services together can perform the specified goal. Web service composition (WSC) is a mechanism of combining services with different functionalities together to represent any business functionality [2]. The web service composition can be categorized as static or dynamic, based on the time when web services are composed. A composition is said to be static composition when the composition of services is performed manually by user's intervention or by other methods, before the execution of the web service. In contrast, dynamic composition involves the ordering of tasks automatically and dynamically during the execution of web service so as to handle the highly flexible and dynamic environment of growing web repository, and to fulfil the user's request. However, the task of dynamic composition of web services is challenging and this paper focus on composition approach.

There are two aspect of web service behaviour—functional and non-functional. The input, output, precondition and effect (IOPE) parameters are useful in defining functional behaviour of a web service. The input parameter specifies functional behaviour of a web service can be defined using the input expected by the web service. Output parameter signifies output generated by the web service. The precondition which needs to be satisfied before service execution is specified using precondition parameter. The after effects upon execution of a web service are represented using effect parameter. Based on the IOPE parameters, the decision about dynamic composition of web services is affected.

The complexity of the process of dynamic WSC is mainly due to the following:

(i) The web repository contains large number of services which makes it difficult to search services in the huge repository.
(ii) The web services are dynamically created and updated, so the decisions should be made based on the recent information.
(iii) Entire failure of WSC can take place due to too many errors.
(iv) Upon increase of large number of web services, the efficiency of composition decreases.
(v) Fault issues such as incorrect order, poor response, unavailability and service incompatibility may occur [3].

Several techniques have been implemented in the past to achieve dynamic WSC, but have some limitations with them. HTN planning [2] does not assure to solve arbitrary planning problems. The model checking approach of AI planning [4] lacks on the basis of performance criteria. In OWLS-XPlan [5], the solution provided is not time efficient and it cannot handle non-determinism. Planning graph algorithm proposed for WSC comes with a redundancy problem. The Simplanner has scalability issues, and user's involvement is quite less in composition procedure.

The AI planning-based approach presented in this paper deals with the automatic and dynamic web service composition problem. In this work, we have used Blackbox AI planner [6] to achieve dynamic WSC.

The organization of the rest of the paper is as follows. Section 2 describes related work. Section 3 provides a detailed discussion on the use of AI planning for achieving dynamic WSC. Various experiments and results appear in Sect. 4 followed by conclusions in Sect. 5.

2 Related Work

This section provides a review of the state-of-the-art approaches and previous works in the field of dynamic composition of web services. Later in this section, the work carried out in AI planning for WSC is discussed.

In past years, a lot of researchers focused on dynamic web service composition problem. Several approaches have been proposed in the literature [7–12]. Most of the approaches for dynamic composition discussed in the literature are based on AI planning techniques such as HTN, Golog, model checking, Markov decision process [13], planning as graph [14] and linear logic [15]. It is observable that the problem of WSC is similar to the planning problem in the domain of AI planning.

Some of the techniques of AI planning for WSC and related to our proposed approach are discussed here. In [2], the composition of web services is done by using Hierarchical Task Network (HTN) planning. SHOP2, the domain-independent HTN planner performs the planning as to find the user's required goal. It uses some set of methods and proceeds the planning by continually decomposing the tasks into smaller subtasks until the composition plan contains only the initial task. It does have advantage in performance, but they have a limitation regarding the amount of domain knowledge. During the plan generation, only the information providing services are being executed, which itself is a limitation as world-altering effects if executed before the total plan generation may affect the results.

In [16], a new language called Golog is built on the top of situation calculus. The Golog is adapted for facilitating automatic composition of web services. In Golog, the request of the user and the constraints are presented by using the first-order language of situation calculus. The preconditions and conditional effects are computed with the formal methods. It also deals with the user's changing preferences during the composition.

In [4], planning with model checking, the technique based on verification is used. The planning is carried out by verifying the goal formula is true in particular model. To ensure the correctness of the composition plan, the true values of formulas are checked at each step. It provides advantages over other planning frameworks as it can deal with in partial observability situations and planning with uncertainty can manage non-determinism. The limitation is regarding the performance due to which its of no use in complex service composition.

In [5], the author proposed a WSC model based on OWL-S service composition planner OWLS-XPLAN. XPlan planner is built as a combination of Fast-Forward planner and HTN component. There are two steps followed to achieve the composition task. In the first step, the OWL2PDDL converter is applied on OWL and OWL-S to obtain PDDL. The OWL and OWL-S caries the details of domain ontology and web service descriptions, respectively. In the next step, XPlan uses these descriptions to generate a plan. The generated plan solves the given composition problem represented as planning problem. It lacks in non-determinism, and solution provided is not time efficient.

In [17], an approach based on a domain-independent AI planner, namely Simplanner is proposed for obtaining dynamic composition of web services. The work presented in this paper primarily focuses on determining the web services and perform the execution according to the order given in user's goal. The Simplanner takes the responsibility of keeping track of all state descriptions and also the effects of the actions. It handles situations when any change in state occurs and performs dynamic re-planning. It is responsive and fault-tolerant. The limitations of this approach are scalability and user's involvement is less in WSC.

3 AI Planning for Dynamic WSC

Dynamic composition involves the ordering of tasks automatically and dynamically during the execution of web service so as to handle the highly flexible and dynamic environment of growing web repository, and to fulfil the user's request. The problem of web service composition can be visualized as a planning problem from AI planning domain. The problem of web service composition typically involves three parameters of importance. (i) The data offered as input to the composite web service. (ii) The generated outputs, which reflect the desired results generated by the composite web service. (iii) A number representing the number of available web services to participate in composition to achieve a goal.

Definition: AI planning as Dynamic Web Service Composition
A classical AI planning problem is given as π = (P, I, A, G) where P represents the set of predicates (also called facts), I represents the initial state, the set of actions are represented as A and G is a goal state. In the context of web service, A indicates the set of functionalities/tasks provided by the web service. The overall objective of AI planning is to come up with a sequence/order of web services (i.e. a plan) along with ensuring that the end user request is also fulfilled by the plan. Consider for example the Travel Domain. The travel service uses three atomic services. Each one independently executes a task. Suppose, there are three tasks flight booking, hotel booking and booking car on rent executed by 3 services airplane service, hotel service and car rental service, respectively. These three services may dependent on each other and output of one service may act as input of other. Thus, obtaining a sequence of these web services dynamically requires automated planning.

Figure 1 shows the architecture of the proposed system to achieve dynamic WSC using AI planning. Firstly, the problem of WSC is described as a planning problem using Planning Domain Definition Language (PDDL), which works as an input to any state-of-the-art domain-independent planning system. The planner generates a plan that represents a composition of the functionalities provided by the web services. Many state-of-the-art AI planners are available which perform differently. For the case of WSC, we have considered Blackbox planner. The detailed discussion of Blackbox planner is as given below.

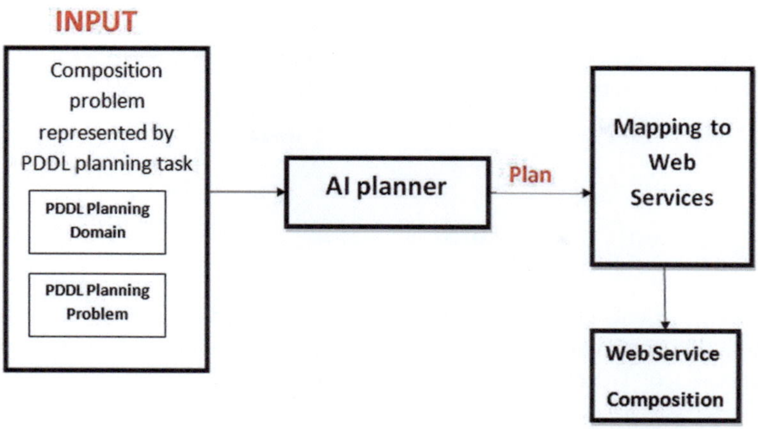

Fig. 1 Architecture for dynamic web service composition using AI planning

Planning Domain Definition Language (PDDL)

Planning Domain Definition Language (PDDL) is a standard encoding language which was originally proposed by Malik Ghallab and 1998 International Planning Competition (IPC) committee. This language was developed with a goal to standardize a language for planning domain. It is mainly inspired by STRIPS, as it also uses actions, preconditions, postconditions and effects. The applicability of the actions in state is described by precondition and postcondition. When the execution of these actions takes place in actual world, the impacts are represented by effects. Various planners exist which use PDDL as formal planning language. In PDDL language, definition of planning problem consists of two parts: (i) domain part and (ii) problem part.

Both the parts, typically are contained in separate files namely, domain file and problem file. This division of files allows the same domain file to be used for multiple problem files. The world described in the domain file does all the planning. The domain file consists of descriptions of the types of objects possible. The second part in the domain file describes a set of predicates. With the given set of objects, one can ground the given predicates and form a set of propositions which describe the state of the world. The third part consists of actions each has some parameters, a precondition and a effect. The precondition is the logical condition, which has the some predicates combined using standard first-order logic logical connectives. These preconditions should be satisfied before the service which is being requested. The effects are the logical conditions which become true after the execution of the services. They are generally the negations of the predicates defined in the preconditions.

Blackbox AI Planner

The Blackbox planning system unifies both the planning as graph and planning as satisfiability approach. The planner converts the problems described in STRIPS into boolean satisfiability problems and then solves it using various satisfiability engines. The front-end uses the graph plan system. It provides lot of flexibility. One can use walkSAT for 50 s, and if it fails, then can use satz for 900 s. This makes it capable of functioning efficiently over large range of problems. The Blackbox planner knows about the solvers and solvers know nothing about plans; hence, they are Blackbox to each other. For the detailed discussion, please refer to Ref. [6]

4 Experiments and Results

This section presents the experiments and results of the system discuss in the Sect. 3. To conduct experimentation, the problem from Travel Domain is considered and two different cases of Travel Domains are discussed.

4.1 Travel Domain

Travel Domain consists of various entities such as person and locations visited by that person. It has several number of locations like airport, hotel, restaurant, ATM, beach, national park and researchers home. The domain has following actions:

- goto hotel (p; x; y): person p from any location x goes to hotel y
- goto r home (p; x; y): person p from any location x goes to researchers home y
- goto restro (p; x; y): person p from any location x goes to restaurant y
- goto atm (p; x; y): person p from any location x goes to atm y
- goto beach (p; x; y): person p from any location x goes to hotel y
- goto park (p; x; y): person p from any location x goes to hotel y
- order food (p; x): person p orders food at location x
- withdraw money (p; x): person p withdraws money at location x.

With respect to the above-described domain description we have following cases:

4.2 Case I

In this case, person P is a student doing research work and wants to meet a researcher residing in another city. She plans to visit the researcher to explore her knowledge regarding her research. She reaches another city by flight and from airport she decides to go to any hotel. After reaching hotel, she gets ready immediately and went to researchers home by 11 AM. Her meeting with researcher finishes by 1 PM, she left her home and go to the restaurant to have lunch. At restaurant, she orders food and had it soon. Her return flight was at 10 PM in night so she plans to explore the famous places in the city. To visit the places, firstly she goes to the nearby ATM and withdraws some money. She is fond of water sports so visits a beach. At last in evening, she wishes for sightseeing at a national park in city.

Action 1	Action 2	Action 3	Action 4	Action 5	Action 6	Action 7	Action 8
Goto_hotel	Goto_r_home	Goto_restro	Order_food	Goto_atm	Withdraw_money	Goto_beach	Goto_park

With the help of the inputs, the initial and goal states, following plan is generated.

Similarly, we have generated various instances of Case I. The experimental results of Travel Domain performed on Blackbox planner are shown in Table 1. The problem instance is defined as (P, A, H, RH, R, AL, B, NP) where they represents person, airport, hotel, researchers home, restaurant, ATM location, beach and national park. We performed experiments on 10 different combinations of person and other locations. In the experiment, with the increasing number of combination we increased the number of objects. We have executed each of the combination 20 times on the planner and then calculated the average for them. In Table 1, the first column shows the problem instances combinations. Columns 2 and 3 represent the time to form the plan and elapsed time which is the time from start of an event to its finish. Column 4 shows the number of nodes created in the graph during the generation of the plan. The action variables and fluent variables are shown in columns 5 and 6. Column 7 represents number clauses.

Total number of actions in the plan also known as plan size in shown in column 9 and plan length is given in column 10.

From the above experiments and result shown in Table 1, we observed that as we increase the number of instances, the time to generate the plan also increases. The number of nodes created increases with the increasing number of objects of each instance.

4.3 Case II

In this case, person P is a student doing research work and wants to meet a researcher residing in another city. She plans to visit the researcher to explore her knowledge regarding her research. She reaches another city by flight and from airport decides to go to any hotel. Her meeting with researcher is fixed at 8 PM. Meanwhile, after reaching hotel at 10 AM, she gets ready and decides to visit the famous places in the city. To visit the places firstly she goes to the nearby ATM and withdraws some money. She is fond of water sports and visits one of the famous beach of the city. In evening, she wishes to see sunset from the famous national park and go to visit it. Afterwards, at 8 PM she went to researchers home and had her meeting. Later, the person went to the restaurant order her food and had dinner.

Table 1 Result of applying BlackBox planner in Travel Domain

Problem instance (P, A, H, RH, R, AL, B, NP)	Time in ms	Total elapsed time	Nodes created	Number of action variables	Number of fluent variables	Number clauses	Number of actions in the plan (plan Size)
(1, 1, 1, 1, 1, 1, 1, 1)	4.4	0.01	253	67	68	222	8
(1, 2, 2, 2, 2, 2, 2, 2)	9.1	0.01	515	133	110	728	8
(1, 3, 3, 3, 3, 3, 3, 3)	12.85	0.01	817	217	152	1954	8
(1, 4, 4, 4, 4, 4, 4, 4)	20.05	0.02	1159	319	194	4470	8
(1, 5, 5, 5, 5, 5, 5, 5)	30.6	0.03	1541	439	236	9026	8
(1, 6, 6, 6, 6, 6, 6, 6)	46.95	0.05	1963	577	278	16,552	8
(2, 2, 2, 2, 2, 2, 2, 2)	11.7	0.01	744	215	157	1422	14
(3, 3, 3, 3, 3, 3, 3, 3)	29.95	0.0315	1607	499	270	6496	20
(4, 4, 4, 4, 4, 4, 4, 4)	69.55	0.07	2962	973	407	22,584	26
(5, 5, 5, 5, 5, 5, 5, 5)	159.95	0.16	4929	1691	568	64,014	32
(6, 6, 6, 6, 6, 6, 6, 6)	347.9	0.35	7628	2707	753	155,962	41

Action 1	Action 2	Action 3	Action 4	Action 5	Action 6	Action 7	Action 8
Goto_hotel	Goto_atm	Withdraw_money	Goto_beach	Goto_park	Goto_r_home	Goto_restro	Order_food

Similarly, we have generated various instances of Case I. The experimental results of Travel Domain on Blackbox planner are shown in Table 1. The problem instance is defined as (P, A, H, RH, R, AL, B, NP) where they represents person, airport, hotel, researchers home, restaurant, ATM location, beach and national park. We performed experiments on 10 different combinations of person and other locations. In the experiment, with the increasing number of combination we increased the number of objects. We have executed each of the combination 20 times on the planner and then calculated the average for them. In Table 1, the first column shows the problem instances combinations. Columns 2 and 3 represent the time to form the plan and elapsed time which is the time from start of an event to its finish. Column 4 shows the number of nodes created in the graph during the generation of the plan. The action variables and fluent variables are shown in columns 5 and 6. Column 7 represents number clauses. Total number of actions in the plan also known as plan size is shown in column 9 and plan length is given in column 10.

Problem instance (P, A, H, RH, R, AL, B, NP)	Time in ms	Total elapsed time	Nodes created	Number of action variables	Number of fluent variables	Number clauses	Number of actions (plan size)
(1, 1, 1, 1, 1, 1, 1, 1)	4.05	0.0025	253	69	69	227	8
(1, 2, 2, 2, 2, 2, 2, 2)	8.2	0.01	502	145	120	753	8
(1, 3, 3, 3, 3, 3, 3, 3)	13.35	0.0115	793	241	171	2031	8
(1, 4, 4, 4, 4, 4, 4, 4)	20.45	0.0205	1126	357	222	4697	8
(1, 5, 5, 5, 5, 5, 5, 5)	31.9	0.0305	1501	493	273	9603	8
(1, 6, 6, 6, 6, 6, 6, 6)	49.35	0.0505	1918	649	324	17,817	8
(2, 2, 2, 2, 2, 2, 2, 2)	12.2	0.011	754	225	161	1433	14
(3, 3, 3, 3, 3, 3, 3, 3)	30.65	0.03	1641	529	279	65,527	22

(continued)

(continued)

Problem instance (P, A, H, RH, R, AL, B, NP)	Time in ms	Total elapsed time	Nodes created	Number of action variables	Number of fluent variables	Number clauses	Number of actions (plan size)
(4, 4, 4, 4, 4, 4, 4, 4)	75.1	0.0735	3040	1041	423	22,733	28
(5, 5, 5, 5, 5, 5, 5, 5)	171.05	0.1715	5077	1821	592	64,595	39
(6, 6, 6, 6, 6, 6, 6, 6)	368.68	0.37	7878	2929	789	157,697	46

5 Conclusion

The problem of dynamic Web Service Composition is important in the present scenario. In order to meet the dynamic demands of the end user, the dynamic web service composition plan needs to be generated. The AI planning technique is found suitable to solve this problem. In this work, we present dynamic web service composition using Blackbox planner. The problem from Travel Domain is considered for experimental purpose. Based on the results of experiments, it is observed that the Blackbox planner can be employed for obtaining composition.

As a future work, we will consider using cognitive parameters along with functional parameters during plan generation. We will also take up the comparative analysis with other state-of-the-art planners.

References

1. Zeng L, Ngu AH, Benatallah B, Podorozhny R, Lei H (208) Dynamic composition and optimization of web services. Distrib Parallel Databases 24(1–3):45–72
2. Rei-Marganiec S, Chen K, Xu J (2009) Markov-htn planning approach to enhance flexibility of automatic web services composition
3. Mustafa F, McCluskey T (2009) Dynamic web service composition. In: 2009 international conference on computer engineering and technology, vol 2. IEEE, pp 463–467
4. Bertoli P, Pistore M, Traverso P (2010) Automated composition of web services via planning in asynchronous domains. Artif Intell 174(3–4):316–361
5. Klusch M, Gerber A, Schmidt M (2005) Semantic web service composition planning with owls-xplan. In: Proceedings of the 1st international AAAI fall symposium on agents and the semantic Web, pp 55–62

6. Kautz H, Selman B (1998) Blackbox: a new approach to the application of theorem proving to problem solving. In: AIPS98 workshop on planning as combinatorial search, vol 58260, pp 58–60
7. Aggarwal R, Verma K, Miller J, Milnor W (2004) Constraint driven web service composition in meteor-s. In: Proceedings of 2004 IEEE international conference on Services computing, (SCC 2004), IEEE, pp 23–30
8. Casati F, Ilnicki S, Jin LJ, Krishnamoorthy V, Shan MC (2000) eFlow: a platform for developing and managing composite e-services. IEEE
9. Cui L, Li J, Zheng Y (2012) A dynamic web service composition method based on viterbi algorithm. In: 2012 IEEE 19th international conference on web services. IEEE, pp 267–271
10. Fujii K, Suda T (2006) Semantics-based dynamic web service composition. Int J Coop Inf Syst 15(03):293–324
11. Lecue F, Silva E, Pires LF (2008) A framework for dynamic web services composition. In: Emerging web services technology, vol II. Springer, Berlin, pp 59–75
12. Pires PF, Benevides MR, Mattoso M (2002) Building reliable web services compositions. In: Net. ObjectDays: International conference on object-oriented and internet-based technologies, concepts, and applications for a networked World. Springer, Berlin, pp 59–72
13. Doshi P, Goodwin R, Akkiraju R, Verma K (2005) Dynamic workflow composition: Using markov decision processes. Int J Web Serv Res (IJWSR) 2(1):1–17
14. Yan Y, Zheng X (2008) A planning graph based algorithm for semantic web service composition. In: 2008 10th IEEE conference on E-Commerce technology and the fifth IEEE conference on enterprise computing, E-Commerce and E-Services. IEEE, pp 339–342
15. Hao S, Zhang L (2010) Dynamic web services composition based on linear temporal logic. In: 2010 international conference of information science and management engineering (ISME), vol 1. IEEE, pp 362–365
16. McIlraith S, Son TC (2002) Adapting golog for composition of semantic web services. KR 2:482–493
17. Kuzu M, Cicekli NK (2012) Dynamic planning approach to automated web service composition. Appl Intell 36(1):1–28

A Study of Deep Learning in Text Analytics

Noopur Ballal and Sri Khetwat Saritha

Abstract Most of the data present today on Web is in the form of text. This text data is a rich source of information, but it is difficult to process because it is unstructured in nature. Various techniques have been developed in past to process this text data for information retrieval and text mining. Many machine learning techniques have been developed in past to drive valuable information from text data. However, these methods require a lot of preprocessing and feature extraction prior to processing which increases manual task. There are a lot of human interactions which not only consumes time but also makes the whole system where rigid and hard to generalize. Deep learning architectures mitigate these limitations. They require less human intervention and create better solutions than other machine learning techniques. Deep architectures are neural networks with multiple processing layers of neuron with each layer having a specific task. Various deep learning architectures have been developed till date with each having a specific task. These architectures have contributed significantly in the domain of text processing. Word2Vec and GloVe have helped to represent text in vector forms. Vector representation helps to process text easily. Architectures like CNN & autoencoders have helped to automate the task of feature extraction. On the other hand, architectures like LSTM and RNN are used for text processing. Text classification, document summarization, question answering, machine translation, caption generation, and speech recognition are some of the areas where improvements have been found due to deep learning architectures. This paper first introduces the text preprocessing techniques and feature extraction methods. Later on, a study of contributions of deep learning architectures in text processing is presented.

Keywords Feature extraction · Caption generation · Text classification · Machine translation · Text summarization

N. Ballal (✉) · S. K. Saritha
Department of Computer Science and Engineering, Maulana Azad National Institute of Technology, Bhopal, India
e-mail: ballalnoopur@gmail.com

S. K. Saritha
e-mail: sarithakishan@gmail.com

R. K. Shukla et al. (eds.), *Social Networking and Computational Intelligence*, Lecture Notes in Networks and Systems 100, https://doi.org/10.1007/978-981-15-2071-6_16

1 Introduction

Deep learning is one of the primary focuses of Data Sciences in recent times. With advances in social media platforms, the volume of data available on Web is continuously being increased. Many businesses are making use of this high volume data to derive insights and make decisions. Most of this data is in the form of text which is unlabeled and uncategorized. Various techniques have been developed in past for processing this text data. The primary advantage of using deep learning is that it can analyze unsupervised data [1]. Deep learning methods also have widespread applications in text processing like in caption generation, document summarization, text classification, question answering, and machine translation. However, before text is processed for these operations, it passes through a phase of preprocessing and feature extraction. Methods have been developed to utilize deep learning architectures for preprocessing and feature extraction. This paper first illustrates techniques in text preprocessing and feature extraction then summarizes the applications of deep learning architectures in various text processing domains.

2 Text Preprocessing

Before feeding text to neural network, it passes through several stages. First stage is sentence segmentation where sentence is broken into meaningful subsentences. This is followed by tokenization where words in text are identified. In third stage, part-of-speech tags are attached to words and lastly, lemmatization is done where the roots or concepts of words are identified. The words are represented in the form of vectors and then they are fed into neural network. The formation of words into vector is termed as word embedding. Word2Vec and GloVe are popular methods of word vector formulation. Skip-gram model used in these models helps in conversion of millions and trillions word corpus to vectors of reduced size [2].

3 Feature Extraction Using Deep Architectures

Data representation and feature extraction are the important aspects of machine learning. It helps in reducing the dimensionality of feature space. It removes uncorrelated extra features. Some traditional features extraction methods include mutual information, information gain, clustering methods, fusion methods and mapping methods. These methods involve some kind of human interaction. Deep learning methods help in automated feature extraction. The hierarchical nature of learning in deep learning networks motivates feature extraction. In deep architectures, the definition of higher level features is in terms of lower level features.

Convolutional neural network (CNN) is a popular architecture for image classification, but it can be used for extracting features automatically. Convolution neural network (CNN) is a form of artificial neural networks developed from the concept of visual cortex. It is used extensively in the field of image recognition, speech detection, and document classification. Text matrix is convolved using filters. Then, max pooling is applied to operate extractive vectors of every filter. Finally, digits are represented as filters and these filters are connected to represent sentence [3]. Multiple convolution and pooling layers can be applied. In some cases, convolution operations are followed by other architectures like long short-term memory (LSTM) which use the sentences created from CNN to make predictions or classification. In [3], it shows that this kind of architecture performs better than only LSTM.

Autoencoder is a feed forward network which is used to compress data. It has a hidden layer between input and output layers. The data is compressed (encoded) moving from input to hidden layer while it is decompressed (decoded) moving from hidden to output layer. Hidden layer outputs act as extracted features. A deep counterpart is a stacked autoencoder where each layer learns a compact representation of layer before it. Replacing input data with noised data creates a better learned autoencoder. Here, the output layer is the same. [4] This autoencoder has much generalization power. In [4], author introduces a contractive autoencoder which favors mapping that is more strongly contracting or decreasing. The objective is to keep output features unchanged when input features have disturbances. In [5], author introduces convolution autoencoder. Here, weights are shared among locations of all inputs so the spatial locality is preserved. In [3], the author lists some of feature extraction architectures which are a combination of deep learning and machine learning methods. In these methods, the conversion of high dimensional space to low dimensional space is done using deep learning architectures like autoencoders or convolutional neural networks and the feature extraction is done by traditional methods like clustering, Support Vector Machine, or principle component analysis.

Restricted Boltzmann machine (RBM) is another deep learning architecture which is used for feature extraction. It consists of two layers, visible and hidden. The restriction here is that there is no connection among visible layer units and hidden layer units. Visible layer is only connected to hidden layer. Each input x passed to each visible layer node is multiplied by a separate weight. This product is summed and added with a bias and passed to an activation function to produce a nodes output. In [3], experimental results prove that RBM gives better results as compared to SVM.

Deep belief network (DBN) consists of stack of multiple RBMs with full interconnections between two layers. In DBN, multiple layers of hidden units are used as feature detectors [6]. Text classification using DBN proves to be better than methods like SVM and k-nearest neighbors.

Recurrent neural networks (RNNs) process interrelated data. If the input depends on output of the previous layer, then recurrent neural networks are preferred. In these types of networks, there is a connection between output layer and input layer. RNNs have proved to be very effective for sequential inputs like speech and text, where there is a correlation between sentences.

4 Applications of Deep Learning in Text Processing

Deep learning architectures find applications in various domains like text processing, speech recognition, image processing, and automatic game playing. In Natural language processing, these architectures can be used in applications which are listed as [7].

4.1 Text Classification

The goal of text classification is to categorized documents or texts based on some topic or theme or label.

Sentiment analysis is a popular classification task where sentences are given emotional polarity of being "positive" or "negative." In [8], a comparison of sentiment analysis with SVM and CNN shows that CNN performs better than SVM. Authors in [9] used various deep learning methods CNN, RNN, regularization and optimization techniques on twitter data. A multilayer perceptron (MLP) model was also created. They found that accuracy increases in case of deep learning architectures as compared to MLP models. The experiments were performed on Tensorflow networks. Variants of CNN like CNN-static, CNN-non-static, CNN-rand, and CNN-multichannel were created in [10]. These were tested against other architectures like autoencoder, RNN, and SVM for various kinds of datasets like movie reviews, question–answer, etc. For movie review, CNN proved to be better than machine learning architectures and other deep learning methods. A new type of feed forward network called deep averaging network (DAN) for sentiment analysis was introduced in [10].

Spam filtering is another type of text classification task where emails are classified as spam or not. In [11], authors compare spam classifier created on deep learning architectures with traditional ones. A novel multimodal architecture is created which has CNN classifiers for image and text.

Age group classification is another implementation of deep learning architectures. Age is considered to be a parameter which influences the sentiment of user to a situation. In [12], deep convolutional neural networks were used to identify age group of around 7000 sentences. It reached a precision of around 0.95. A metric enhanced Sentiment Metric (eSM) was also introduced which indicated the influence of age group identification on sentiment analysis.

4.2 Document Summarization

Document summarization is creating a gist of document. It includes creating abstract or heading for a document. Extractive and abstractive are two types of summarization techniques. Extractive summaries are generated by extracting key features from text like key terms, numbers, and even texts. In this approach, the summaries generated may not be complete or meaningful.

On the other hand, abstractive summaries are created just like humans. They may or may not contain sentences from source text. The big challenge here is to identify the important concepts and compress them in lossless manner. Some of the important methods include sentence compression, fusion, and splitting [13]. In abstractive method, there are two techniques: semantic-based and structure-based. In structure-based approach, meaningful sentences are extracted from text and fed into predefined structures like templates, ontology trees, lead and body phrase structure, and rule-based structures [13].

In semantic-based approach, there are three stages: inputting the document, semantic representation of document, and feeding the document to natural language generation phase. The various models used are multimodal semantic model, semantic graph-based method, and information item-based method [13].

Advances in neural machine translation have helped to incorporate neural networks to contextual input encoder. A model based on attention-based encoder is proposed in [14]. In [15], the authors introduce improvements over attention-based encoder by capturing keywords, modeling rare and unseen words and capturing hierarchical structure of document. A new dataset is also introduced. Most of the summarization problems are dependent on human given features; however, in [16] a data-driven approach is proposed which does not depend on human given features.

4.3 *Question Answering*

In question answering type of implementation, a specific question is answered based on document of a text. Here, natural language questions are answered from a knowledge base. Tradition methods rely on human-trained features for this type of implementation however with deep learning architectures this dependence is mitigated. A novel method to select the answer sentence from a knowledge base is introduced in [17]. Here, distributed representation is used. Questions and answers are matched using their semantic encoding. In [18], the authors propose a multi-column convolutional neural network (MCCNN) which understands questions from different perspectives like answer path, answer context, and answer type. Then, candidate answers are ranked and the best answer is chosen. The experiments done in [18] show that MCCNN performs comparable and sometimes better than the baseline systems. A novel approach of converting summary sentences to context–query–answer triplets is introduced in [19]. A deep neural network model based on RNN is built to read and comprehend data.

4.4 Machine Translation

Machine translation is conversion of text from one language to another. Deep learning has been used recently in this kind of implementation. In [20], a model to translate English to French is developed. Here, a model for sequence to sequence learning is built. A sequence in English is converted to a vector of fixed dimensionality using multilayer LSTM. Decoding of this vector is done afterward using another deep LSTM which maps vector to a sequence in French. A joint model based on RNN is introduced in [21]. Authors in [22] identified that forming a fixed length vector is bottleneck and search for words which are imperative for predicting target words. A bidirectional RNN is used for this implementation.

4.5 Caption Generation

Caption generation is defined as identifying the topic or subject of the text or image. Given a document or image a heading or caption is generated. In image processing, contents of an image are identified and a text depicting these contents is generated automatically. It finds applications in various fields like guiding visually impaired, in field of education and faster and accurate image searches and indexing [23]. Traditional systems used rule-based methods where templates like <object, action, scene> were used and filled by sentence generators. However, sentences generated by this method were rigid and did not generalize well [24]. To overcome these limitations, deep neural architectures were used.

In [24], initially a CNN model is used for image classification. The output of last layer of CNN is fed as input to a RNN model which generates a caption. An attention-based extension to previous model is being proposed in [25]. When there is clutter in image, it becomes a difficult task to identify the salient features of image.

In [25], two models namely "Soft" and "Hard" are generated. "Soft" model is generated from back-propagation methods, and "Hard" is generated from reinforcement.

Authors in [26] take this one step forward and proposed a model to generate texts about what is going on in a video. LSTMs are used to form word sequences for a group of frames in video.

In [27], authors found that in LSTM-based systems, the sentence generators gradually loss information because the length of captions is large. To overcome this, they added a guide factor to LSTM cells which helps the sentence generators. The guides used were semantic information or the whole image itself.

In [28], authors take this further and add a topic-guided vector to the semantic and spatial representation of image and create sentences using LSTM models.

More detailed features from images are extracted by authors in [29] using Regional-CNN (R-CNN). Attributes of detected objects are predicted, and scene classification is also done using AlexNet CNN by the authors. Once these image details are extracted, they are fed in RNN models for caption generation.

Extensive amount of work is being done in the field of caption generation using deep neural networks which cater to their applications.

4.6 Speech Recognition

Speech Recognition is the process of converting speech into text. It is the problem of identifying what is being said in an audio file. RNNs are very effective for sequential data representation, and they are used for speech recognition as well. However, their performance is not at par with deep feed forward networks [30]. However, when they are combined with multilayer LSTM, they perform better [30].

5 Conclusion

Deep learning methods are being evolved continuously, and they are being applied extensively in various tasks. Text analysis is one of the fields which have found numerous implementations of deep learning architectures. These architectures have helped in overcoming the limitations of traditional machine learning methods. In comparison to most machine learning and feature extraction methods, deep learning architectures provide better solution. It automatically extracts insights from large and complex data. Architectures like CNN and autoencoders find implementation in feature extraction while architectures like RNN and LSTM found implement in predictive analysis and text generations. This paper illustrated how deep learning can be used in various areas of text processing. Before performing any kind of analysis on text, they need to be preprocessed so that the deep learning architectures can read and understand them clearly. A brief introduction of preprocessing methods is provided. After that, the implementations of deep learning architectures are listed. Firstly, the architectures which are used for feature extraction are listed. CNN and autoencoders are the most popular architectures for the same. However, other architectures like RBM and DBN can also find implementations for feature extraction. A brief introduction about these architectures is also provided. Later on, the application areas of text analysis are listed. It was found that RNN and LSTM are widely used for text generation purposes. The fields where they are used extensively are caption generation, summarization, machine translation, and question answering. For most of these applications, there can be an initial layer for feature extraction which is then followed by sentence creation. For the purpose of text classification, CNN is majorly used. Popular text classification applications like sentiment analysis and spam filtering have found drastic improvements when aided with deep learning architectures.

Text analytics have found advances because of deep learning. This is mainly because deep learning architectures are self learning and require minimum human interactions. A lot of work has been done, and new advances are introduced frequently. However, this field is still immature and extensive work is being carried out particularly in the field of data tagging, information retrieval, semantic indexing, high dimensionality, and streaming data analysis [1].

References

1. Najafabadi MM et al (2015) Deep learning applications and challenges in big data analytics. J Big Data. https://doi.org/10.1186/s40537-014-0007-7
2. Mikolov T et al (2013) Efficient estimation of word representations in vector space. arXiv: 1301.3781v3 [cs.CL] 7 Sep 2013
3. Shaheen F et al (2016) Impact of automatic feature extraction in deep learning architecture. In: 2016 International conference on digital image computing: techniques and applications (DICTA)
4. Rifai S et al (2011) Contractive auto-encoders: explicit invariance during feature extraction. In: Proceedings of the 28th international conference on machine learning, Bellevue, WA, USA
5. Masci J et al (2011) Stacked convolutional auto-encoders for hierarchical feature extraction. In: Conference paper, June 2011. https://doi.org/10.1007/978-3-642-21735-7_7. Source: DBLP
6. Deng L (2014) A tutorial survey of architectures, algorithms, and applications for deep learning. SIP 3(e2):1–29
7. www.machinelearningmastery.com
8. Reshma U et al (2018) Deep learning for digital text analytics: sentiment analysis. arXiv:1804. 03673v1 [cs.CL] 10 Apr 2018
9. Ramadhani AM (2017) Twitter sentiment analysis using deep learning methods. In: 2017 7th international annual engineering seminar (InAES), Yogyakarta, Indonesia
10. Iyyer M et al (2015) Deep unordered composition rivals syntactic methods for text classification. In: Conference paper, Jan 2015. https://doi.org/10.3115/v1/p15-1162
11. Seth S et al (2017) Multimodal spam classification using deep learning techniques. In: 2017 13th international conference on signal-image technology and internet-based systems (SITIS)
12. Guimaraes RG et al (2017) Age groups classification in social network using deep learning, digital object identifier. https://doi.org/10.1109/access.2017.2706674
13. Sunitha C et al (2016) Study of abstractive summarization techniques in Indian language. Procedia Comput Sci 87:25–31. https://doi.org/10.1016/j.procs.2016.05.121
14. Rush AM et al (2015) A neural attention model for abstractive sentence summarization. arXiv: 1509.00685v2 [cs.CL] 3 Sep 2015
15. Nallapati R et al (2016) Abstractive text summarization using sequence-to-sequence RNNs and beyond. arXiv:1602.06023v5 [cs.CL] 26 Aug 2016
16. Cheng J et al (2016) Neural summarization by extracting sentences and words. arXiv:1603. 07252v3 [cs.CL] 1 Jul 2016
17. Yu L et al (2014) Deep learning for answer sentence selection. arXiv:1412.1632v1 [cs.CL] 4 Dec 2014
18. Dong L et al (2015) Question answering over freebase with multi-column convolutional neural networks. In: Proceedings of the 53rd annual meeting of the association for computational linguistics and the 7th international joint conference on natural language processing, pp 260–269. Beijing, China, July 26–31
19. Mortiz K et al (2015) Teaching machines to read and comprehend. In: Advances in neural information processing systems 28 (NIPS 2015). arXiv:1506.03340
20. Suutskever I (2014) Sequence to sequence learning with neural networks. arXiv:1409.3215v3 [cs.CL] 14 Dec 2014

21. Auli M et al (2013) Joint language and translation modeling with recurrent neural networks. In: Proceedings of the 2013 conference on empirical methods in natural language processing, pp 1044–1054. Seattle, Washington, USA, 18–21 Oct 2013
22. Bahdanau D (2016) Neural machine translation by jointly learning to align and translate. arXiv: 1409.0473v7 [cs.CL] 19 May 2016
23. Mathur P et al (2017) Camera2Caption: a real-time image caption generator. In: International conference on computational intelligence in data science (ICCIDS)
24. Vinyals O et al (2015) Show and tell: a neural image caption generator. arXiv:1411.4555v2 [cs.CV] 20 Apr 2015
25. Xu K et al (2016) Show, attend and tell: neural image caption generation with visual attention. arXiv:1502.03044v3 [cs.LG] 19 Apr 2016
26. Venugopalan S et al (2015) Sequence to sequence—video to text, arXiv:1505.00487v3 [cs.CV] 19 Oct 2015
27. Jia X et al (2015) Guiding long-short term memory for image caption generation. In: IEEE international conference on computer vision
28. Zhu Z et al (2018) Topic-guided attention for image-captioning. In: International conference on image processing (IEEE)
29. Kinghorn P et al (2018) A region-based image caption generator with refined descriptions. Neurocomputing J
30. Donahue J et al (2017) Long-term recurrent convolutional networks for visual recognition and description. IEEE Trans Pattern Anal Mach Intell 39(4)

Image Segmentation of Breast Cancer Histopathology Images Using PSO-Based Clustering Technique

Vandana Kate and Pragya Shukla

Abstract Image segmentation has key influence in numerous medical imaging uses. An image segmentation model that is based on the particle swarm optimizer (PSO) is developed in this paper for breast cancer histopathology images of different magnification levels (40X, 100X, 200X and 400X), thus simplifying image representation and making it meaningful and easier for future analysis. As lower the magnification level, the bigger is the field of view and as it can provide greater detail, thus more time and care must be taken to use such images. Thus, finding a segmentation method that works equally well for all zoom levels of images is a big challenge. To explicate the better performance of the proposed method and its applicability on breast cancer images, results of applications and tests are augmented which shows PSO image clustering approach using intra-cluster distance as an optimization function, performs superior to cutting edge strategies, namely K-means and genetic algorithm (GA). The algorithms, when given specified number of clusters, find the centroids, thus grouping similar image primitives. The influence of different estimations of PSO control parameters on execution is additionally outlined.

Keywords Unsupervised clustering · Evolutionary algorithms · PSO · GA · K-means · Color segmentation

1 Introduction

Breast cancer is the most wide-spreading disease among lady in India and different nations, and thus, automated image analysis is playing a paramount role in potentially reducing the workload of pathologist and thus improving their quality of interpretation. Histopathology is the microscopic examination of the extracted biological tissues, conscientiously prepared into histological sections and stained utilizing histology stains (such as Immuno Histo Chemistry (IHC), hematoxylin and eosin (HE))

V. Kate (✉) · P. Shukla
Institute of Engineering and Technology, Indore, India
e-mail: vandana.kate@gmail.com

P. Shukla
e-mail: pragyashukla_iet@yahoo.co.in

R. K. Shukla et al. (eds.), *Social Networking and Computational Intelligence*, Lecture Notes in Networks and Systems 100, https://doi.org/10.1007/978-981-15-2071-6_17

to observe the appearance of diseased cells and tissues in very fine detail. Examination of stained tissues has been the hallmark of automatic computer diagnosis and cancer research for over a century. Heamatoxlin being a basic dye has an affinity for the nucleic acids of the cell nucleus and colors it blue or violet and in contrast eosin is an acidic dye with an affinity for the cytoplasmic components of the cell coloring it as pink [1]. This property can be used in breast cancer color segmentation of histological images which can be an important pre-processing step that will aid in identifying benign or malignant tumors [2] while maintaining high accuracy, and also reducing the costs for diagnosing. Cancer cells in general occurs in clusters and causes huge lumps of continually dividing cells that results in dense, overlapping cell regions. Thus, image clustering can play an important role as a preprocessing step in histopathology images prior to extracting histological structural properties, such as lymphocytes, cancer nuclei or glandular shape.

There is a wide variety of open access H&E stained histology image datasets for breast cancer (BC) that can be used as a benchmark dataset by researchers for algorithm assessment, e.g., UCSB Bio Segmentation [3], MITOS or mitosis detection etc. [4]. The proposed work has been experimented on BreakHis dataset [5] containing the benign and malignant subclasses.

The paper is composed in following areas. Section 2 describes review of previous work, Sect. 3 describes K-Means method, Sect. 4 widely used GA evolutionary technique, and Sect. 5 gives a detailed overview of PSO for image segmentation and its application to BC images. Section 6 demonstrates experimental results obtained. We conclude with summarizing methods used and suggesting future scope.

2 Literature Survey

A number of research studies have been applied to pre-process and segment the BC histopathology images to identify regions of interest for detection of the presence of cancer disease. Some of the work includes thresholding, clustering [6] evolutionary algorithm [7], etc., based methods for pre-processing cancer images. Xu et al. [8] proposed automatic image pixel clustering (e.g., hierarchical and partitioned) with the help of various differential algorithms that is used for finding naturally occurring clusters in images. Jung et al. [9] shades light on the watershed-based technique for segmentation of cervical and breast cell images and suggests a region merging and marker-controlled watershed method. Dundar et al. [10] explicated a system archetype that automatically discriminate among actionable subtypes (ADH & DCIS) and usual ductal hyperplasia (UDH) and accordingly pinpoints the breast tissues. This prototype estimates the tissues sample in computerized stained slides for a set cytological benchmark and then systematize them according to the image. Wienert et al. [11] worked on multi-objective clustering, thus integrating two or more-image segmentation algorithm for better results. Nucleus detection was seen as the cornerstone for a range of applications in automated assessment of histopathological images by Vink et al. [12]. The paper proposed an efficient machine learning-based nuclei detector for a large feature set with modifications in AdaBoost. Ghamisi et al. [13] proposed two novel methods for segmentation of images based on the Darwinian Particle Swarm Optimization (DPSO) and Fractional-Order Darwinian Particle Swarm Optimization (FODPSO) for determining the threshold on a given image. Exploratory outcomes demonstrate that the proposed strategies perform superior to different thresholding techniques when considering numerous criteria. Study made by Krishna et al. [14] develops Cuckoo Search (CS) algorithm-based image clustering model for the proper segmentation of breast histopathology images. Experimental results show that CS provides better quality segmented images compare to classical K-means algorithm by considering the computational time, fitness values and the values of quality parameters.

Evolutionary algorithms have been used in number of other domains such as Omran et al. [15] developed an image clustering method based on the particle swarm optimizer (PSO). The image classifier proposed here has been applied to synthetic, MRI and satellite images. Numerous validity test shows that the PSO image classifier performs better than state-of-the-art image classifiers (namely K-means, Fuzzy C-means, K-Harmonic means and genetic algorithms) in all measured criteria. The article by Samy et al. [16] gives an idea about image segmentation problem optimized by PSO (Particle Swarm Optimization) and some other methods to perform like (Hidden Markov Random Fields) and shows their evaluation. Van der Merwe et al. [17] explained about different techniques, which can be utilized to use Particle Swarm Optimization for data clustering. The usage of PSO for clustering was expounded in a great depth in this work.

3 K-means Clustering

Partitional clustering algorithms partition the dataset into a predefined number of clusters limiting certain criteria (e.g., a squared error function). Along these lines, they can be utilized to take care of optimization problem. The most broadly utilized partition algorithm is K-means. It is an unsupervised hard clustering method which allocates n data objects to predefined k number of clusters. The standard K-means clustering algorithm is summarized as

- The calculation inputs are the number of clusters K and the image to be processed. It begins with randomly initializing the K centroid vectors, where each centroid characterizes one of the clusters.
- As a subsequent stage, every datum point z_p is assigned to its nearest centroid, in light of the squared Euclidean distance calculated as—$d\left(z_{p,}c_j\right) = \sqrt{\sum_{k=1}^{N_d}\left(z_{pk} - c_{jk}\right)^2}$, i.e., data point zp is assigned to the centroid having $d(z_p, c_j)$ as minimum value, where c_j is the accumulation of centroids in set C and N_d signifies the input dimension (i.e., the number of parameters of each data point)

- In this progression, the centroids are recomputed. This is done by taking the mean of all data points assigned to that centroid's cluster as—$m_j = \left(\left(\sum_{\forall z_p inc_j} z_p\right)/n_j\right)$ where n_j is the number of data points in cluster j
- The algorithm iterates between steps two and three until a halting criteria is met (i.e., no data points change clusters, the sum of the separating distance is minimized, or some extreme number of iterations is reached).

4 Genetic Algorithm (GA)

Genetic Algorithm (GA) introduced by J. Holland in the 70's is an adaptive stochastic search procedure to find good solution for a given problem using principle of natural evolution by going through operations like inheritance, mutation, selection and crossover. In GAs, set of all candidate solutions is known as population represented as strings of 0's and 1's (called chromosomes or the genotype). Starting with some initial population, fitness function is computed for each individual working as a selection procedure for producing new set of offspring. This new population replaces the current population and is used as an input to the next iteration of the algorithm. GA when applied to images can greatly improve segmentation [18] and localization accuracy. GA mainly works in four stages

(i) *Population Initialization*—In this step, initial population of solutions composed of individuals representing different combinations of values of the segmentation parameters in $a * b*$ color space of $La * b*$ image representation is generated randomly.
(ii) *Evaluation of fitness*—For each candidate solutions, fitness function is evaluated to measure their cost value for selecting the best individuals for generating next population using proportional selection criteria called roulette wheel.
(iii) *Reproduction*—This procedure has four stages in particular—selection, crossover, mutation and accepting the solution.
    - *Selection*—The fittest members in the present population are selected in order to reproduce new solution. The fitness function for each individual measures the degree of color similarity in image obtained from the segmentation process.
    - *Crossover*—Two individuals are selected through selection operator to be used as parent, and random numbers of genes are swapped between them.
    - *Mutation*—It is done by flipping a bit randomly in the chromosome as crossover may stuck in local optima. After crossover and mutation, two new chromosomes are reproduced which can be included in new population
(iv) *Termination*—The algorithm proceeds till the maximum number of generations (50 generations) is reached, or a reasonable fitness value is attained.

5 Particle Swarm Optimization (PSO)

Particle Swarm Optimization is a nature inspired meta-heuristic, population-based optimization algorithm (i.e., it generates multiple solutions each iteration) for solving non-linear functions. PSO can be used as an unsupervised image segmentation approach to find best cluster centers [17, 19]. Same as genetic algorithm PSO is initialized with a population of random solutions called particles which are flown through search space [7]. The quality, or fitness, of a particle is measured using a fitness function which quantifies how close a particle is to the optimal solution. However, unlike GA each particle is also assigned a randomized velocity.

In PSO, each particle i has a position $x_i(t)$ at the time t in search space of all feasible solutions which change at time $(t+1)$ by a velocity $v_i(t)$ which is guided by—(i) Personal best fitness value achieved by individual particle so far also called $PBest$, (ii) Overall best coordinates obtained so far by any particle in the population also called $GBest$ and (iii) The present movement of particle, to decide its next move in the search space.

Digitized histopathology images are usually acquired and stored in RGB. As $La*b*$ color space is a non-linear transformation of RGB [20]. Clustering using PSO is done using $a*b*$ color space of $La*b*$ image representation model so that color variations can be compatible to visual perceptual differences, thus contributing for uniform color representation independent to illumination conditions. The global version of PSO image clustering algorithm is outlined beneath:

Initialize position $x_i(t)$ and velocity $v_i(t)$ of N number of particles at time t is represented as

$$x_i(t) = \left(x_{i1}, x_{i2}, \ldots x_{ij}, \ldots x_{ik}\right) \tag{1}$$

$$v_i(t) = \left(v_{i1}, v_{i2}, \ldots v_{ij}, \ldots v_{ik}\right) \tag{2}$$

where $1 \leq i \leq N$, and k is the number of clusters centroids. The personal best (*PBest*) coordinates of particle i at time t, i.e., $y_i(t)$ are given by-

$$y_i(t) = \left(y_{i1}, y_{i2}, \ldots y_{ij}, \ldots y_{ik}\right) \tag{3}$$

whereas the global best (*GBest*) coordinates of all particles at time t, i.e., $z(t)$ are given by-

$$z(t) = \left(z_1, z_2, \ldots z_j, \ldots z_k\right) \tag{4}$$

$PBest$ represented as $y_i(t)$ is updated overtime according to the following formula-

$$y_i(t+1) = \begin{pmatrix} y_i(t), & \text{if } f(x_i(t+1)) \geq f(y_i(t)) \\ x_i(t+1), & \text{if } f(x_i(t+1)) < f(y_i(t)) \end{pmatrix} \tag{5}$$

where f is the fitness function defined in Eq. (6) which is the sum of squared distance between the centroid and all points in the cluster also known as intra-cluster distance. This objective function is to be minimized as smaller value of f means that the clusters are more compact.

$$f(x_i) = \left(\max_{j=1tok} \sum_{\forall z_p \text{within clusters}} \big(d(z_p, x_{ij})\big)/n_j \right) \tag{6}$$

where n_j = number of data points in the cluster j $d(z_p, x_{ij})$ represents the Euclidean distance between pth pixel z_p and the centroid of ith cluster x_j. The $GBest - z(t)$ which belongs to—$(y_1(t), y_2(t) \ldots y_j(t) \ldots y_N(t))$ is calculated as

$$z(t) = \min\big(f(y_1(t)), f(y_2(t)), \ldots f\big(y_j(t)\big) \ldots f(y_N(t))\big) \tag{7}$$

The velocity $v_i(t) = (v_{i1}(t), v_{i2}(t), \ \ldots \ v_{ij}(t) \ldots v_{iN}(t))$ of particle i at the time t is updated utilizing the following formula

$$v_{ij}(t+1) = w * v_{ij}(t) + c_1 r_{1j}\big(y_{ij}(t) - x_{ij}(t)\big) + c_2 r_{2j}\big(z_j(t) - x_{ij}(t)\big) \tag{8}$$

where w is called the inertia factor ranging usually between [0.4, 0.9], c_1 and c_2 are learning or acceleration factors ranging between [2, 2.05], r_{1j} and r_{2j} are random variables between [0, 1], velocity v_{ij} ranging between [−10, 10] decides the maximum change one particle can take during one iteration to ensure convergence. The number of particles is typically range is [20, 40]. The position x_i of the particle i is updated using the following formula-

$$x_i(t+1) = x_i(t) + v_i(t+1) \tag{9}$$

6 Experimental Results

This section looks at the consequences of the PSO, GA and K-means clustering algorithms on breast cancer images of different magnification levels. The fundamental design is to look and compare the quality of the individual clustering results, where quality is estimated according to the intra-cluster distances (i.e., the distance between data points within a cluster). The goal is to minimize the intra-cluster distances.

Table 1 Performance after Clustering

Zoom	No. of Clusters	Best cost of		
		PSO	GA	K-means
40X	2	**1,746,608.398**	1.36E+07	1,746,710.791
100X		**2,008,151.243**	1.56E+07	2,008,646.352
200X		**1,908,583.23**	1.14E+07	1,908,946.072
400X		**2,131,501.279**	1.56E+07	2,141,910.537
40X	3	**1,386,903.287**	7.71E+06	1,408,158.705
100X		**1,495,000.911**	8.66E+06	1,503,576.076
200X		**1,463,996.716**	7.17E+06	1,482,194.405
400X		**1,432,407.867**	9.71E+06	1,451,232.095
40X	4	**1,176,889.575**	5.49E+06	1,204,008.175
100X		**1,225,362.333**	5.85E+06	1,264,246.678
200X		**1,240,853.225**	5.36E+06	1,242,788.597
400X		**1,060,704.462**	6.93E+06	1,170,704.462

Bold indicates better results

PSO and GA optimization method were run for over 100 iterations on breast cancer images of different magnification level (40X, 100X, 200X, 400X). Particles vary in the range of [20, 30, 40], and appropriate values of w, c_1 and c_2 were selected to ensure good convergence. Table 1 shows the comparison result of the various methods used, choosing random cluster number in the range [17, 20]. The results are also demonstrated using Figs. 1, 2 and 3 where all the result shows PSO as a better optimization algorithm in all the cases.

7 Conclusion

The paper compares evolutionary algorithms like PSO, GA with mostly used Image segmentation method called $KMeans$. The experiment shows that PSO performs well on $KMeans$ but gives comparable results with GA. As RGB model may not be considered as uniform color model, La$*b*$ color components of histological images were used as input features which can be further be tested on other model-based intermediate representations (such as SIFT, HMAX, Luv etc.) and color texture. Further features computed from segmented image can be classified using various techniques such as support vector machine (*SVM*), K nearest neighbor (*kNN*), or Bayesian classifier to know the state of disease.

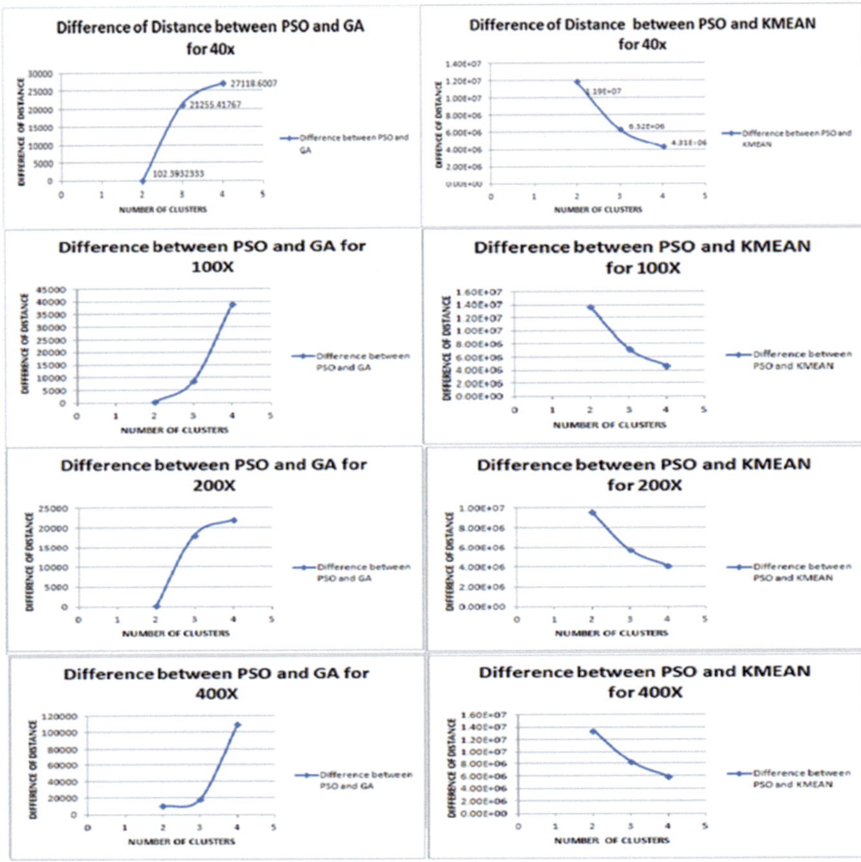

Fig. 1 Comparison graph for showing improvement of PSO over GA & K-Means for breast cancer images of high to low magnification level obtained through Selection of microscope objectives ranging from 40X to 400X

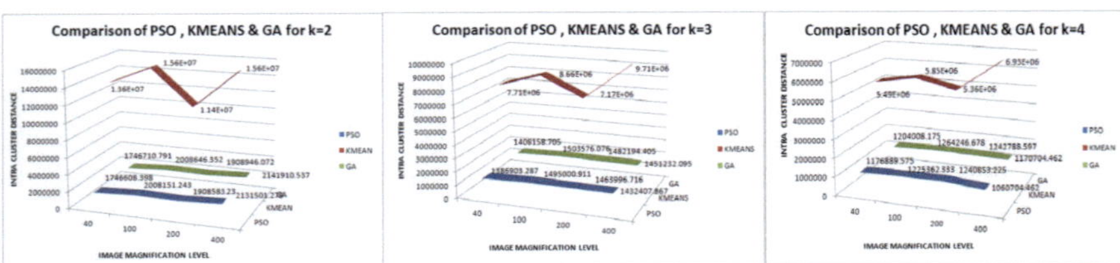

Fig. 2 Comparison graph showing improvement of PSO over GA & K-Means

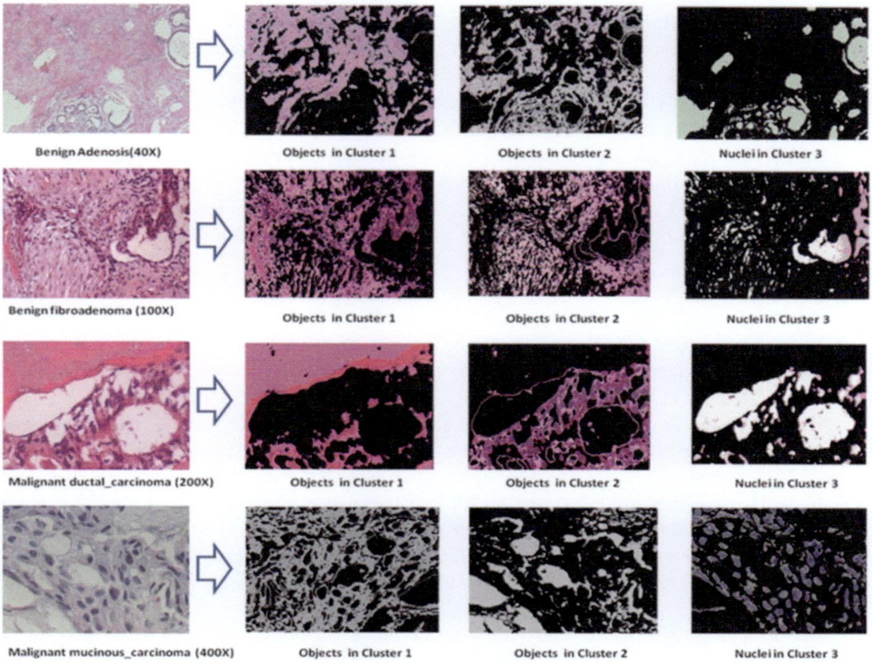

Fig. 3 Resulting clustered images using PSO for $k = 3$ when applied on BC 40X, 100X, 200X images separating blue/violet color stained nucleus and pink cytoplasm

Acknowledgements I would like to deeply express my thanks to my guide for giving valuable suggestions and her kind support.

References

1. Orlov NV, Chen WW, Eckley DM, Macura TJ, Shamir L, Jaffe ES, Goldberg IG (2010) Automatic classification of lymphoma images with transform-based global features. IEEE Trans. Inf. Technol. Biomed 14(4)
2. Tosta TAA, Neves LA, do Nascimento MZ (2017) Segmentation methods of H&E-stained histological images of lymphoma: a review. Inf Med 35–43
3. Gelasca, ED, Obara B, Fedorov D, Kvilekval K, Manjunath BS (2009) A biosegmentation benchmark for evaluation of bioimage analysis methods. BMC Bioinf 10(1)
4. MITOS, ICPR (2012) Contest, IPAL UMI CNRS Lab Std., [Online]. Available: http://ipal.cnrs.fr/ICPR2012/?q=node/5
5. Spanhol FA, Oliveira LS, Petitjean C, Heutte L (2016) A dataset for breast cancer histopathological image classification. IEEE Trans Biomed Eng 63(7):1455–1462
6. Xu R, Donald W (2005) Survey of clustering algorithms. IEEE Trans Neural Netw 16(3)
7. Ab Wahab MN, Nefti-Meziani S, Atyabi A (2015) A comprehensive review of swarm optimization algorithms. PLoS One 10.5

8. Xu Y et al (2012) Multiple clustered instance learning for histopathology cancer image classification, segmentation and clustering. In: 2012 IEEE conference on computer vision and pattern recognition (CVPR). IEEE
9. Jung C, Kim C (2010) Segmenting clustered nuclei using H-minima transform-based marker extraction and contour parameterization. IEEE Trans Biomed Eng 57(10):2600–2604
10. Dundar MM et al (2011) Computerized classification of intraductal breast lesions using histopathological images. IEEE Trans Biomed Eng 58(7):1977–1984
11. Wienert S, Heim D, Saeger K, Stenzinger A, Beil M, Hufnagl P, Klauschen F (2012) Detection and segmentation of cell nuclei in virtual microscopy images: a minimum-model approach. Sci Rep 2:503
12. Vink JP, Van Leeuwen MB, Van Deurzen CHM, De Haan G (2013) Efficient nucleus detector in histopathology images. J Microsc 249(2):124–135
13. Ghamisi P, Couceiro MS, Benediktsson JA, Ferreira NM (2012) An efficient method for segmentation of images based on fractional calculus and natural selection. Expert Syst Appl 39(16):12407–12417
14. Dhal KG, Fister I Jr, Das A, Ray S, Das S (2018) Breast histopathology image clustering using cuckoo search algorithm. Proceedings of the 5th student computer science research conference, pp 47–54
15. Omran M, Engelbrecht AP, Salman A (2005) Particle swarm optimization method for image clustering. Int J Pattern Recogn Artif Intell 19(03):297–321
16. Ait-Aoudia S, Guerrout EH, Mahiou R (2014) Medical image segmentation using particle swarm optimization. In: 2014 18th international conference on information visualisation (IV), pp 287–291
17. Van der Merwe DW, Engelbrecht AP (2003) Data clustering using particle swarm optimization. In: The 2003 Congress on evolutionary computation. CEC'03, vol 1, pp 215–220. IEEE
18. Bhandarkar SM, Hui Z (1999) Image segmentation using evolutionary computation. IEEE Trans Evol Comput 3(1)
19. Omran MGH, Salman A, Engelbrecht AP (2006) Dynamic clustering using particle swarm optimization with application in image segmentation. Pattern Anal Appl 8(4)
20. Sertel O, Kong J, Catalyurek UV, Lozanski G, Saltz JH, Gurcan MN (2009) Histopathological image analysis using model-based intermediate representations and color texture: follicular lymphoma grading. J Sig Process Syst 55(1–3)

Survey of Methods Applying Deep Learning to Distinguish Between Computer Generated and Natural Images

Aiman Meenai and Vasima Khan

Abstract With the advent of fake news and propaganda being spread throughout the Internet using forged or computer-generated images, it is important to evolve algorithms that are able to differentiate between computer-generated images and natural ones. In this paper, we provide a high-level summary of the methods proposed recently which classify images as computer generated or natural using deep learning concepts. We spelled out the pros and cons of each method and further suggested future research paths like building a standard computer generated (CG) versus natural images (NI) dataset targeting compressed images from heterogeneous sources which ensures that the dataset models real world well, testing proposed approaches in real-life conditions and trying out training classifiers using combination of features generated from various convolutional neural networks (CNNs) as opposed to a single neural network to assess its impact on the accuracy rate for classification of compressed images.

Keywords Convolutional neural networks · Image forensics · Natural image · Computer-generated image · Deep learning · Classification

1 Introduction

The advent of social media has changed the way we share and assimilate information. A sizable number of people rely on social media for checking news and other current affairs [1, 2]. Also, a lot of information online is exchanged in the form of images [3]. Images are known for infusing a sense of trust for the object it illustrates [4]. These facts put together make images a very powerful and efficient tool for spreading information as compared to text. This, in turn, entails that fake images can be used as a very efficient vehicle for spreading propaganda. Fake images can either be entirely

A. Meenai (✉)
UIT-RGPV, Bhopal, India
e-mail: jalaly.farah@gmail.com

V. Khan
SIRT, Bhopal, India
e-mail: aarish.azfar07@gmail.com

R. K. Shukla et al. (eds.), *Social Networking and Computational Intelligence*, Lecture Notes in Networks and Systems 100, https://doi.org/10.1007/978-981-15-2071-6_18

computer generated or produced by splicing, i.e., making a composite of a given number of images or created by tampering an original image in some other way, for example, been subjected to a copy move operation, i.e., coping a part of an image and pasting it on some other part of the original image [4]. Computer-generated images, especially, are getting more and more photorealistic [5]. As computer-generated (CG) images progressively become more photorealistic, they can potentially morph into a very viable weapon for spreading fake news. Studies show humans are not able to efficiently differentiate between computer-generated images and natural ones although they perform better after receiving targeted training [6]. It has also been demonstrated that humans have a bias and would rather assume that an image is a photograph when in doubt [6]. Hence, it is imperative to focus research on the task of distinguishing between photographs and photorealistic images. Methods to solve this problem are majorly based on two approaches [7]: acquisition process-based methods which use differences in generation process of CG and NI like [8–10] or statistical distribution-based methods which utilize usual image statistics like [11–14]. Some traditional methods employing machine learning have been in circulation since long. Recently, with the advent of deep learning, better accuracy is being achieved at the said task [15–17]. In this paper, we present a theoretical survey of the methods which use the concepts of deep learning to make a distinction between photorealistic and natural images (NI) and list the challenges that persist in this field.

The paper has been organized to give a basic understanding of the concepts currently being used to implement deep learning-based solutions for this problem. We start by explicitly defining the problem statement in Sect. 2, explore previous work in Sect. 3, then give an overview of the underlying concepts being used by the algorithms in section, discuss the high-level implementation of these algorithms theoretically in Sect. 4 and finally present our findings in Sect. 5. Currently, under development are arrhythmia monitors for ambulatory patients which analyze the ECG in real time [6, 15, 16]. Software QRS detectors typically include one or more of three different types of processing steps: linear digital filtering, nonlinear transformation and decision rule algorithms [17].

2 Problem Statement

For this paper, we define 'computer-generated (CG) images' as images created by the use of computer graphics. 'Natural images (NI)' are defined as photographs captured by a digital camera which have not been subjected to any processing. It should be noted that compressed photographs in jpeg format are also considered as NI since jpeg is the standard compression scheme for online uploads used by Twitter. Natural images which have been tampered with are not in the scope of our consideration. Distinguishing between CG and NI is considered as a classification problem in which given a test set $T = \{x^1; x^2 \dots x^n\}$ where each x^i in set T is an image, the algorithm has to accurately map the elements in T to one of the two target classes: CG or NI to give the output of the form $\{x^1, y^1; x^2, y^2 \dots x^n, y^n\}$ where x^i is the input image and y^i is the target class. Another format of accepted output can be a vector of the form $\{x^1, y_1^1, y_2^1; x^2, y_1^2, y_2^2; \ldots x^1, y_1^n, y_2^n\}$ where x^i is the input image, y_1^i is the probability of it belonging to the CG class, and y_2^i is the probability of it belonging to the NI class.

3 Previous Work

Previous work on detecting image forgeries or computer-generated images and separating them from natural images has relied on basic differences in the inherent characteristics of a photograph and photorealistic computer-generated image. Methods like the one proposed by Lyu and Farid [11] are based on the following two-phase implementation:

Phase 1. Identifying the property difference between CG and NI on the basis of which feature extraction is carried out from the given image data and the rest of the information is discarded. In the case of Lyu and Farid [11], the wavelet statistics are used as reliable features.
Phase 2 Training a classifier such as LDA or SVM on the features extracted in phase 1.

This approach presents problems like unnecessary discarding of additional information which might have been useful before training the classifier and the effort-intensive step of feature detection and extraction. Another apparent downside is that there is no mathematical way to determine whether the features we are using are optimal or not.

4 Methods Implementing Deep Learning to Distinguish Between CG and NI

4.1 Method I: Rahmouni et al. [15]

Rahmouni et al. [15] in the effort to eliminate the tedious feature extraction process and avoid other downsides listed in the previous section proposed using a convolution neural network to perform the classification task. Input images are divided into patches of resolution 100 × 100 each. The proposed architecture replaced the traditional approach as follows:

Filtering. The traditional manual filtering process is replaced by a set of N convolutional kernels of size containing $k \times k$ number of weights which are to be optimized. This is then subjected to rectified linear unit (ReLU) which introduces nonlinearity and composes the results.

Statistical feature extraction. This task is achieved by a custom pooling layer which takes the filtered images as input and can be implemented in the following two ways:

- ***Simple statistics***. estimates the mean, variance, maximum and minimum of the filtered images and gives a 4xNf feature as outputs to the classifier
- ***Histograms***. Makes a normalized histogram of pixel distribution and hence it is able to capture significantly more information than simple statistics. The authors implement an 11-bin histogram which passes an $11 \times N_f$ feature vector to the classifier.

Classification. A multilayer perceptron (MLP) is used for the classification task. It contains the following two layers:

The hidden layer has 1024 ReLU activated neurons and uses dropout [18] to prevent overfitting.

The output layer has two soft max activated neurons which map the input feature vectors to CG or NI classes. Then, a weighted voting scheme is applied on each patch of the image to calculate the classification strategy of the complete image.

$$y = \operatorname{sgn}\left(\sum_{i=1}^{Np} \log \frac{P(Y = 1|Xi = x_i)}{P(Y = -1|Xi = x_i)}\right)$$

where Y are the labels, Xi the patches, x_i the real observations and sgn(a) is a function that returns ± 1 according to the sign of a [15]. The network is optimized using Adam's algorithm. The authors tested the method on their own dataset collected from various sources and reached an accuracy of 93.2%, thus practically demonstrating that the performance of deep learning methods is considerably better than other tradition approaches. The tedious process of manual feature extraction was eliminated, and a higher accuracy was achieved (Fig. 1).

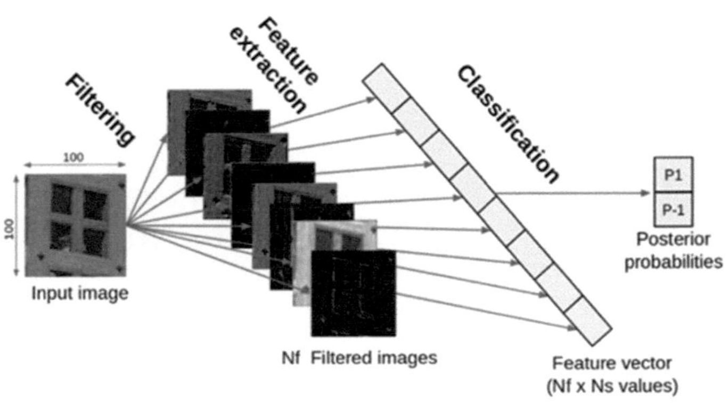

Fig. 1 The input image is filtered by a convolutional neural network, its statistical features are extracted, and then finally, the classifier uses the extracted features for classification [15]

4.2 Method II: Quan et al. [16]

Quan et al. [16] proposed a way to employ local to global strategy by training their CNN using a number of patches cropped randomly from each training image. The cropping of patched was done using maximal Poisson-disk sampling (MPS). The advantage of using MPS is that it completely covers the entire image which ensures that no important information is unnecessarily discarded. The proposed method divides images into patches of resolution 240 × 240. The architecture has the following three segments:

ConvFilter layer. Maps the input image patch to several feature images which is analogous to feature extraction phase. There are three convolutional groups:

- ***Convolution***. multidimensional linear operation produces multiple feature maps
- ***Batch Normalization***. makes the output to take on a unit Gaussian distribution
- ***ReLU Activation***. introduces nonlinearity
- ***Max-pooling layer***. performs a down sampling operation
- ***2 Fully Connected Layers***. Provides high-level reasoning. Implementing dropout during training prevents overfitting.
- ***Softmax layer***. Maps the output of the FC layers which is a high-level feature vector, to the output vector which contains the set of probabilities of the input image belonging to respective class labels.

$$J(\theta)^{(\text{data})} = -\frac{1}{N}\left[\sum_{i=1} N \sum_{j=1} K \emptyset\{y^i = j\}\log\frac{e^{a_j^i}}{\sum_{j=1}^{K} e^{a_j^i}}\right]$$

where N is the number of training samples, K is the number of categories, $\emptyset\{\ \}$ is the indicator function mapping truth value to a binary value, i.e., $\emptyset\{\text{True}\} = 1$ and $\emptyset\{\text{False}\} = 0$ [16]. The network was trained by minimizing the binomial logistic loss function. Reported average accuracy was 93.20% on the Colombia Photographic Dataset and the PRCG dataset [19] using the patch size 240 × 240. It is observed in general that deeper the neural network is the higher is the classification accuracy. Another observation is that even though the authors explicitly state that Colombia dataset, which is used to train their model, is more challenging to process since the origin of their images is heterogeneous, it is important to note that the said dataset was compiled in 2007, and the computer graphics technology has since made many advances. Hence, it is imperative to test the given algorithm on newly curated datasets too which majorly represent the levels of photorealism reached by present CGI techniques (Fig. 2).

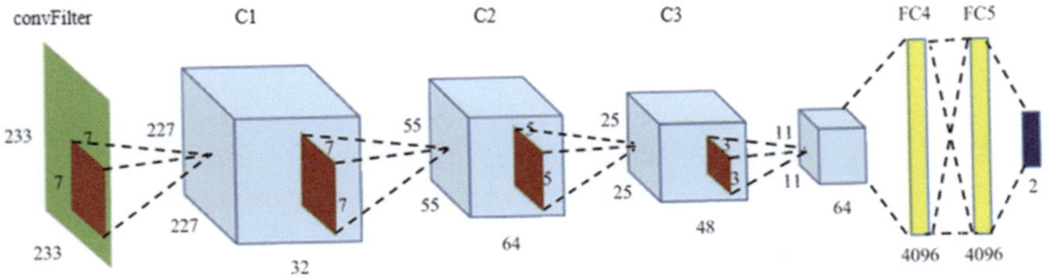

Fig. 2 The input image of size 240 × 240 × 3 is randomly cropped to the size of 233 × 233 × 3 represented by a green square. Each red square represents a convolutional kernel, and blue cuboids represent the feature maps. Kernel sizes are mentioned near the respective red squares. The feature images thus produced are subjected to fully connected layers (FC4 and FC5) providing high-level logic. The network terminated with a softmax layer which gives the output as a probability vector [16]

4.3 Method III: Rezende et al. [17]

Rezende et al. [17] put forward the idea that transfers learning can yield better results than building a network from scratch and initializing it with random weights. The transferred network outputs bottleneck features of input images. Using a classifier on top to train on the bottleneck features prevents overfitting. This is equivalent to replacing the last 1000 fc softmax layer of ResNet-50 with a SVM freezing the parameters of the convolutional layers during the training process. Transfer learning is based on the fact that the first few layers of any deep neural networks do not learn features specific to any dataset or task [20], and hence, we can copy the weights and architecture of a neural network trained on another dataset and completely different task and use it on our dataset. The proposed method inputs raw pixel values of 224 × 224 × 3 RGB images. Their architecture is explained in the following two phases:

Transfer learning. Consists of transferring the weights of the first 49 layers of ResNet-50. ResNet-50 was pre-trained on ImageNet dataset. The last fully connected softmax layer however is removed which was designed to map the input image to one of the 1000 output classes. The output activation map produced by the first 49 layers is referred to as bottleneck feature having a dimension of 2048 elements, which are then used to train the CG/NI classifier.
SVM classifier. The last softmax layer in ResNet-50 is replaced by a support vector machine which acts as a classifier and maps the input bottleneck features to either CG or NI class.

The network was trained and tested on the public dataset given by Tokuda et al. [21], giving an average accuracy of 94.1% on the said dataset. During pre-processing, the mean RGB value computed on the ImageNet dataset was subtracted from each pixel of the given images, as per the concept proposed by Krizhevsky et al. [22]. Also, since the first layers of the architecture were frozen and no fine tuning was performed, the training time was significantly less. There is still a scope for improvement as the accuracy can be further bettered by using a combination of bottleneck features which

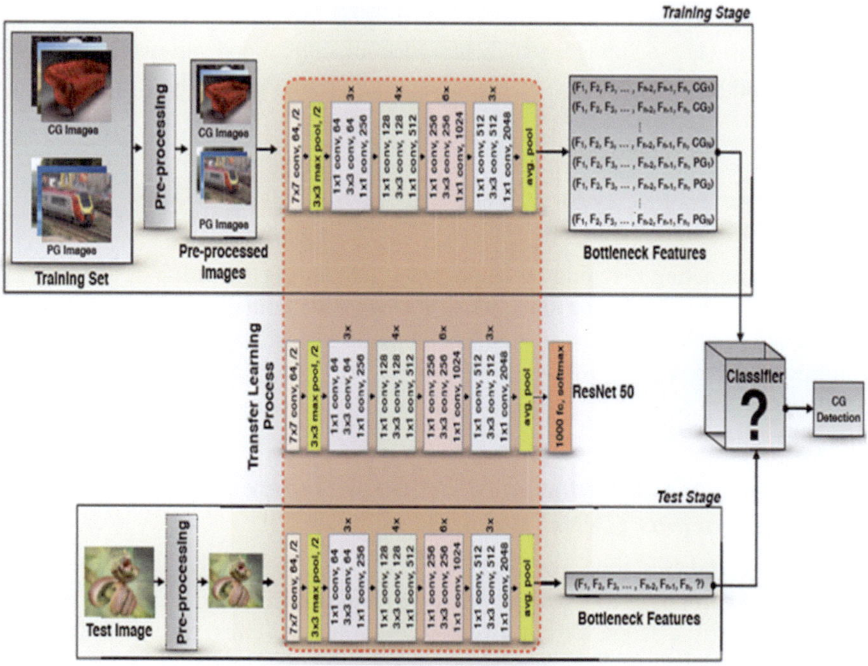

Fig. 3 The parameters of the first 49 layers of ResNet-50 used in the proposed architecture are analogous to feature extraction phase. This is an implementation of transfer learning. The extracted bottleneck features are used to train the SVM acting as a classifier [17]

have been extracted using different CNNs models. Also, the dataset used to train the proposed model only had 9700 images—a higher number of training examples might also result in better performance (Fig. 3).

5 Theoretical Findings

The first apparent conclusion of the findings seems to be that the neural network architectures perform significantly better than other state of the art approaches regarding the classification of CG and NI images. The following points need to be incorporated in the coming researches on the topic.

5.1 Building of Standard Dataset

As better techniques for creating more photorealistic content are introduced, it becomes more likely that CG images would be used to spread propaganda, and in turn, it becomes important for methods being proposed to classify them are tested on a standard dataset having images with heterogeneous sources and various compression techniques applied on them.

5.2 *Increased Focus on Classifying Compressed Images*

Approaches targeting reaching better accuracy rate for classification of compressed images need to be developed since most photographic content on the Web is compressed before uploading, and the present methods perform subpar when classifying compressed images.

5.3 *Testing of Combination of Features Generated from Various CNNs*

Every method proposed focuses on building a single CNN to extract feature vectors. Since there is still a scope for improvement in accuracy scores, especially for compressed images, using multiple CNNs to extract a combination of features and using them to train classifiers might prove useful.

6 Conclusion

In this paper, we provided a high-level summary of the methods recently proposed to classify images as computer generated or natural using deep learning concepts. We spelled out the pros and cons of each method and further suggested future research paths like building a standard CG versus NI dataset targeting compressed images from heterogeneous sources to model real world well, testing proposed approaches in real-life conditions and testing combination of features generated from various CNNs to test whether it increases accuracy rate for classification of compressed images.

References

1. Broersma M, Graham T (2012) Social media as beat: tweets as a news source during the 2010 British and Dutch elections. Journalism Pract 6(3):403–419
2. Hermida A (2010) Twittering the news: the emergence of ambient journalism. Journalism Pract 4(3):297–308
3. Duggan M (2013) Photo and video sharing grow online. Pew Res Internet Proj

4. Aldiri K, Hobbs D, Qahwaji R (2008) The human face of e-business: engendering consumer initial trust through the use of images of sales personnel on e-commerce web sites. Int J E-Bus Res (IJEBR) 4(4):58–78
5. Datta U, Sharma C (2013) Analysis of copy-move image forgery detection. Int J Adv Res Comput Sci Electron Eng (IJARCSEE) 2(8):607
6. Holmes O, Banks MS, Farid H (2016) Assessing and improving the identification of computer-generated portraits. ACM Trans Appl Percept (TAP) 13(2):7
7. Wu R, Li X, Yang B (2011) Identifying computer generated graphics via histogram features. In: 2011 18th IEEE international conference on image processing (ICIP). IEEE
8. Ng T-T, Chang S-F, Hsu J, Xie L, Tsui M-P (2005) Physics motivated features for distinguishing photographic images and computer graphics. In: Proceedings of ACM multimedia, pp 239–248
9. Dehnie S, Sencar T, Memon N (2006) Digital image forensics for identifying computer generated and digital camera images. In: Proceedings of IEEE ICIP, pp 2313–2316
10. Dirik AE, Bayram S, Sencar HT, Memon N (2007) New features to identify computer generated images. In: Proceedings of IEEE ICIP, vol IV, pp 433–436
11. Farid H, Lyu S (2003) Higher-order wavelet statistics and their application to digital forensics. In: CVPRW'03 conference on computer vision and pattern recognition workshop, 8. IEEE
12. Wang Y, Moulin P (2006) On discrimination between photorealistic and photographic images. Proc IEEE ICASSP II:161–164
13. Chen W, Shi YQ, Xuan G (2007) Identifying computer graphics using HSV color model and statistical moments of characteristic functions. In: Proceedings IEEEICME, pp 1123–1126
14. Sutthiwan P, Cai X, Shi YQ, Zhang H (2009) Computer graphics classification based on Markov process model and boosting feature selection technique. In: Proceedings IEEE ICIP, pp 2913–2916
15. Rahmouni N et al (2017) Distinguishing computer graphics from natural images using convolution neural networks. In: 2017 IEEE workshop on information forensics and security (WIFS). IEEE
16. Quan W et al (2018) Distinguishing between natural and computer-generated images using convolutional neural networks. IEEE Trans Inf Forensics Secur 13(11):2772–2787
17. De Rezende ERS, Ruppert GCS, Carvalho T (2017) Detecting computer generated images with deep convolutional neural networks. In: 2017 30th SIBGRAPI conference on graphics, patterns and images (SIBGRAPI). IEEE
18. Srivastava N, Hinton GE, Krizhevsky A, Sutskever I, Salakhutdinov R (2014) Dropout: a simple way to prevent neural networks from overfitting. J Mach Learn Res 15(1):1929–1958
19. Ng T-T, Chang S-F, Hsu J, Pepeljugoski M (2004) Columbia photographicimagesandphotorealisticcomputergraphicsdataset. ADVENT, Columbia University, Tech. Rep. 205-2004-5
20. Yosinski J, Clune J, Bengio Y, Lipson H (2014) How transferable are features in deep neural networks? Adv Neural Inf Process Syst 3320–3328
21. Tokuda E, Pedrini H, Rocha A (2013) Computer generated images versus digital photographs: a synergetic feature and classifier combination approach. JVCI 24(8):1276–1292
22. Krizhevsky A, Sutskever I, Hinton GE (2012) Imagenet classification with deep convolutional neural networks. Adv Neural Inf Process Syst 1097–1105

SVM Hyper-Parameters Optimization using Multi-PSO for Intrusion Detection

Dhruba Jyoti Kalita, Vibhav Prakash Singh and Vinay Kumar

Abstract Among all the classifiers available in recent time support vector machine (SVM) can be considered as one of the powerful classifiers. Support vector machine (SVM) modeled with properly tuned parameters can give significantly increasing accuracy rate for classification problem. Therefore, choosing the optimal values for the SVM hyper-parameters cannot be a mundane task. It is the most crucial and important task for SVM modeling. The basic emphasis of this paper is to optimize the values of C and γ which are two important kernel parameters of SVM which in turn can be used an intrusion detector in network. A set of properly optimized values of C and γ can increase the effectiveness and efficiency of SVM model. There are many state-of-the-art and metaheuristic techniques such as traditional grid search, gradient descent, genetic algorithm (GA), and particle swarm optimization (PSO). Among these, the most used optimization technique for SVM model selection is PSO. In our work, we have proposed a framework which uses a variant of PSO, multi-PSO for the selection of optimal values for the C and γ selection. The results show that it outperforms the other models for model selection in support vector machine.

Keywords Classifier · Support vector machine · Hyper-parameters · Particle swarm optimization

1 Introduction

Classification can be considered as a problem in machine learning which deals with identification of a category from a given set of categories for a new observation [1].

D. J. Kalita (✉)
Gaya College of Engineering, Gaya, India
e-mail: mestop12@gmail.com

V. Kumar
National Institute of Technology Jamshedpur, Jamshedpur, India
e-mail: vkumar.cse@nitjsr.ac.in

V. P. Singh
Motilal Nehru National Institute of Technology, Allahabad, Prayagraj, India
e-mail: vibhav@mnnit.ac.in

R. K. Shukla et al. (eds.), *Social Networking and Computational Intelligence*, Lecture Notes in Networks and Systems 100, https://doi.org/10.1007/978-981-15-2071-6_19

Table 1 List of Kernel functions

Kernels	Inner product
Linear kernel	$K(X_i, X_j) = X_i^T X_j$
Polynomial kernel	$K(X_i, X_j) = (\gamma X_i X_j + r)^u, u > 0$
Radial basis function kernel	$K(X_i, X_j) = \exp\left(-\gamma \lvert\lvert X_i - X_j \rvert\rvert^2\right), \gamma > 0$
Sigmoid function kernel	$K(X_i, Xj) = \tanh(\gamma X_i^T X_j + r)$

To solve this problem, one builds a mathematical model which can be termed as a classifier. A training set that contains observations then is fetched to the model, and the model learns from those observations. This kind of learning mechanism is called supervised learning. Support vector machine (SVM) is a classifier or classification model which can be designed based on supervised learning mechanism. It basically deals with binary classification problem.

In reality, support vector machine [2] can be considered as an optimization problem where we find an optimum hyper-plane which separates the positive and negative class. A kernel function is used in SVM model to map the original search space to a higher-dimensional search space.

Table 1 is a list of kernel functions that we can be used for mapping the original space to a higher-dimensional space. In this paper, we have used RBF kernel [3] because it has superiorities over the other kernel methods. Linear kernel can also be derived from RBF kernel. With a penalty parameter *C,* it can be adjusted to give same performance as RBF. In the same way, RBF kernel can also be tuned with certain parameters to make it behave like a sigmoid kernel. In polynomial kernel, the number of parameters is more than that of the RBF kernel. Also, RBF kernel has fewer numerical difficulties.

So our optimization problem basically includes two SVM hyper-parameters C and γ. Classical PSO (Particle Swarm Optimization) proposed by Kennedy and Eberhart [4] is a very popular nature inspired optimization technique that can be used to optimize C and γ. PSO is a population-based metaheuristic technique inspired by models of swarming and flocking [4]. The basic SVM model selection problem consists of two main parts—selection criterion and searching. Selection criterion is a function in terms of certain variables, which we need to maximize or minimize with respect to the values of those variables. These functions are called as objective functions. We can construct an objective function based on various factors of the problem domain. Vapnik [5] proposed a SVM formulation approach based on the radius margin bound. Likewise, some other factors span bound [6] and support vector count [1] can also be used very efficiently for the formulation of SVM. Other than all of the above, there are techniques like cross-validation and hold-out estimation to evaluate the effectiveness of the values of the parameters we have selected. In this paper, we have chosen the cross-validation error rate as the selection criterion.

Searching is the phase where we search for the values for the variables used in the objective functions which minimizes or maximizes those functions. There are a lot of searching methods that can be applied to find the optimum values of C and γ. Few of them are gradient descent [7–9], grid search [10–12], genetic algorithm (GA) [13–16], and CMA-ES [17].

In this proposed work, we have used a variant of PSO (multi-PSO) that first partitions the whole search space and deploy different swarm for each of the partition. The results have shown that the integration of multi-PSO with the SVM models improves the performance in terms of precision and recall.

2 Support Vector Machine (SVM)

Support vector machine (SVM) is a supervised machine learning models for classification, regression, and outlier detection proposed by Vapnik in [1] which is based on structural risk theory. In our work, we have considered only the binary classification problem through SVM.

Construction of a support vector machine can be considered as an optimization problem where an optimal hyper-plane is chosen that separates both the classes. Let us consider a training dataset

$$\{(X_i, y_i), i = 1 \ldots n\}, \quad X_i \in \mathbb{R}^m, \quad y_i \in \{+1, -1\}$$

If the training dataset can be separated linearly, then there exists $w \in \mathbb{R}^m$ and $b \in \mathbb{R}$ such that

$$w^T X_i + b > 0 \, \forall i, \quad \text{s.t. } y_i = +1$$

$$w^T X_i + b < 0 \, \forall i, \quad \text{s.t. } y_i = -1$$

So $w^T X_i + b = 0$ will be a separating hyper-plane. Finding an optimal hyper-plane in an infinite set of hyper-planes is the main motive for the formulation of support vector machine (SVM).

In the linear model of SVM, we normally scale w, b such that

$$w^T X_i + b \geq +1 \quad \text{if } y_i = +1$$

$$w^T X_i + b \leq -1 \quad \text{if } y_i = -1$$

For the above two equations, we can say that

$$y_i(w^T X_i + b) \geq 1, \quad \forall i \tag{1}$$

So when the training set is separable, any separating hyper-plane can be tuned to satisfy Eq. (1). In this formulation of SVM, a margin can be defined as the distance between $w^T X + b = +1$ and $w^T X + b = -1$ and the measurement of the distance between these two are $\frac{2}{||w||}$. So, intuitively we can say that more the margin better is the chance of correct classification of the new patterns. Following the above intuition, we can say that optimal hyper-plane is a solution of the following constrained optimization problem

$$\begin{aligned}&\text{minimize}\ \frac{1}{2}w^T w\\&\text{subject to}\ y_i\left(w^T X_i + b\right) \geq 1, \quad i = 1\ldots n\end{aligned}$$

To deal with a nonlinear separable problem in SVM modeling something called as "kernel trick" needs to be used. In this dual form of the SVM formulation, the only difference is that there is an upper bound on μ_i. Now to learn a nonlinear discriminate function, we need to use something called as kernel trick. Kernel trick transforms X_i (training dataset) into some high-dimensional space and learns a linear classifier there. This is equivalent to make a nonlinear classifier learning in lower-dimensional space.

Let,

$$\emptyset : \mathbb{R}^m \rightarrow \mathbb{R}^{m'}$$

is a mapping function to map a lower-dimensional space $\mathbb{R}^m$ into a higher-dimensional space $\mathbb{R}^{m'}$. Now the training set in this new space $\mathbb{R}^{m'}$ will be

$$\{(Z_i, y_i), i = 1\ldots l\}, \quad Z_i = \emptyset(X_i)$$

After this, we can find the optimal hyper-plane by solving the following dual problem

$$\text{maximizeq}(\mu) = \sum_{i=1}^{n} \mu_i - \frac{1}{2}\sum_{i,j=1}^{n} \mu_i \mu_j y_i y_j \emptyset(X_i)^T \emptyset(X_j)$$

$$\text{subject to}\ 0 \leq \mu_i \leq C, \quad i = 1\ldots n, \sum_{i=1}^{n} y_i \mu_i = 0$$

3 Particle Swarm Optimization (PSO)

The concept of particle swarm optimization is based on a social model. This model was simulation of movements of birds flock. Kennedy and Eberhart in the year 1995 came out with the conclusion that this model in fact can be used as an optimizer.

In a more simplified way, PSO can be defined as an evolutionary computation technique similar to genetic algorithm (GA) where initialization of a population of random solutions in the search space is the first step to get an optimal solution. Each potential solution is called as particle. PSO differs from the GA in the fact that particles are associated with movements and randomized velocity. In each time step, each particle's intention is to move toward two best positions *Pbest* and *Gbest*. *Pbest* is the local best position of individual particles so far, and *Gbest* is the global best position of any of the particles in the population so far.

Evaluating the objective function at a particular position, we can measure the fitness of that position. In this, *Pbest* is the memory associated with each of the particles and *Gbest* enables the information sharing between particles.

The movement of the particles toward *Pbest* and *Gbest* is controlled by their velocity. In each iteration, this velocity needs to be changed to reach the *Pbest* and *Gbest*. First the acceleration is calculated based on the above two best positions, and then this acceleration is added to the velocity. In PSO, constriction is one factor that needs to be taken care so that progressive slowdown of the particle's movement happens. The velocity which has been updated then is used to update the position of each particle from its current position to the new position.

Let $\boldsymbol{i}$ be a particle on the solution or search space. This $\boldsymbol{i}$ contains three components

(1) Particle's position vector $\vec{\boldsymbol{x}}_i$.
(2) Particle's own best solution found so far $\overrightarrow{\boldsymbol{pos}}_i$.
(3) Particle's velocity $\overrightarrow{\boldsymbol{vel}}_i$.

Apart from the above components, each particle also keeps track of the best solution found so far by any of its neighbors, i.e., the global best solution $\overrightarrow{pos}_{gb}$. Many different topologies can be used as explained in [18, 19]. Among them, the standard one is the global (*Gbest*). At each iteration, PSO algorithm updates the velocity of each particle and then their positions.

$$\overrightarrow{vel}_i = X\left[\overrightarrow{vel}_i + c_1\epsilon_1\left(\overrightarrow{pos}_{gb} - \vec{x}_i\right) + c_2\epsilon_2\left(\overrightarrow{pos}_i - \vec{x}_i\right)\right] \tag{2}$$

$$x_i = x_i + \overrightarrow{vel}_i \tag{3}$$

```
Algorithm 1: Standard PSO Algorithm

// Initialize all particles

FOR EACH particle i in S

Randomly initialize vel_i, x_i = pos_i

        g_best = arg max {f(pos_i)}

REPEAT

FOR EACH particle i in S do

update particle position x_i

according to equations (1) and (2)

// update the particle's best position

IF f(x_i) > f(pos_i) THEN

        pos_i = x_i

END IF

// update the global best position

IF    f(pos_i) > f(pos_gb) THEN

        pos_gb = pos_i

END IF

END FOR

UNTIL termination condition reached
```

4 Proposed Model

When we apply classical PSO for the selection of the optimal values for C and γ, we apply them on a particular range of values. The particles of the PSO can be treated as points on a two-dimensional space, i.e., a two-dimensional vector is the position of the particles on the search space. According to [18], we can take a standard range of the search space as $\left[2^{-5}, 2^{-15}\right]$ to $\left[2^{-15}, 2^{10}\right]$. In case of the classical PSO, we should initialize proper number of particles so that it covers the whole search space. A less or a very large number of particle's initialization may lead to improper selection of optimal values of the parameters C and γ. For this reason, we need to initialize an optimal number of particles within the search space.

When the number of particles becomes less, then the convergence rate of the particles becomes slow and it cannot cover the whole search space to find the optimal values for the C and γ. On the other hand, if the number of particles becomes more, the convergence process becomes fast and the collision rate of the particles becomes high.

Our proposed algorithm first partitions the whole search space to remove the above problem of particle initialization. The number of partitions depends upon the upper and the lower bound of the range of the search space. Figure 1 is depicting the partitions of the search space.

Inside each of the partitions, our algorithm then creates multiple swarms which contain a fixed number of particles. Doing this, we have decreased the probability of getting bad optimum values for the parameters C and γ. After doing this, our algorithm imposes searching of the optimum values within each of the partitions. Particles of a swarm will converge within the range of that partition itself which confirm no interference with the other swarms.

At end of the algorithm, we will get a list of optimum values for the SVM model. From this list of values, our algorithm chooses the best one for SVM modeling and model the classifier. Any optimization process is lead by an objective function. In our algorithm, we have taken the cross-validation error rate ϵ as the objective function which in turn is a function of (C, γ) and the dataset available in hand. Following is the objective function for our algorithm (Fig. 2).

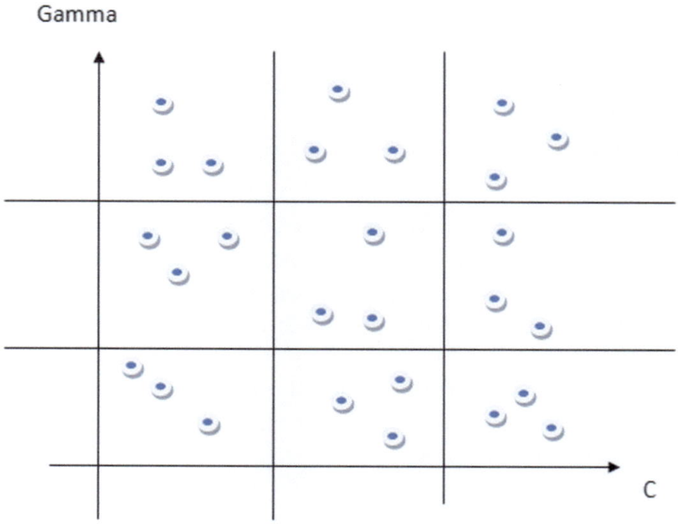

Fig. 1 Particle's initialization in each of the partitions

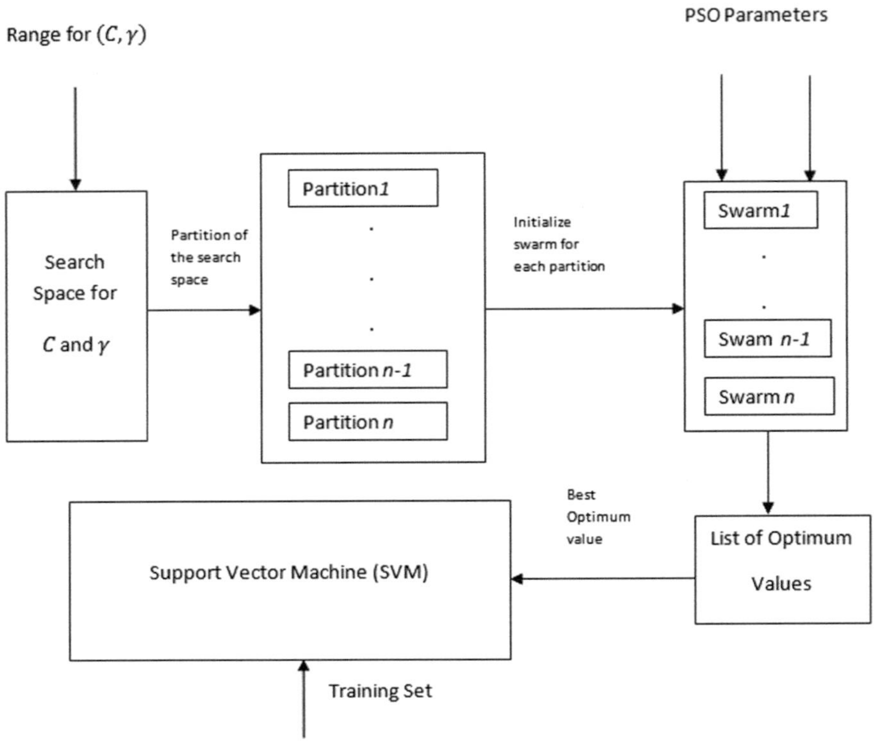

Fig. 2 General framework of our model

$$\text{minimize}_{C,\gamma}\epsilon((C,\gamma))$$

5 Result Analysis

The basic environment that we have used to carry out our experiment is python 2.7. Python is an interpreted language and gives a modular environment for programming. The modules or packages we have used for our experiments are as follows: scikit-learn-0.19.1, pandas-0.19.2, and numpy-1.13.2. For our experiment, we have used KDD cup 1999 dataset.

KDD cup 1999 dataset is normally used in intrusion detection system for learning. Intrusion detection is a classification problem which classifies network attacks into one of the four categories of attacks

- DOS: Denial-of-service attack,
- R2L: Unauthorized access from a remote machine,

- U2R: Unauthorized access to local (root) super user privileges,
- Probing: Surveillance and other probing.

For our purpose of working on a binary classification problem, we have modified the existing dataset to have two categories of attacks. We have taken DOS and probing as one class of attack and R2L and U2R as other categories of attack.

5.1 *Range of the Parameters* C *and* γ

We have defined the values of the parameters C and γ as the dimensions of the particles in the search space as explained in [18], i.e., we can say that a two-dimensional vector is the representation of the position of a particle. The lower bound of these dimensions is set to $\left[2^{-5}, 2^{15}\right]$ and the upper bound to $\left[2^{-15}, 2^{10}\right]$

5.2 *Results*

Figure 3 is showing an instance of our running algorithm which gives multiple optimum values for C and γ which we use to create a list of optimum values for the above-mentioned two parameters. Table 2 is an abstraction of the same result which is showing different optimum values for all the swarms or partition.

For each pair of (C, γ), Graph 1 are showing the precision-recall rate of SVM model. We have generated the entire precision-recall graph while executing our algorithm.

From the above observations, we have found that for the (C, γ) pair $(1, 0.000517980089733)$ and $(1, 0.000207762060518)$, the precision-recall rate is 99%. Among the list of values, $(2, 4.5350323811e - 05)$ is giving the lowest precision-recall rate. Now our algorithm keeps either $(1, 0.000517980089733)$ or $(1, 0.000207762060518)$ as the selected pair of C and γ to model the SVM.

To make our comparison, we have also run the classical PSO in the search space defined by the above range. In this case, in spite of doing any partition we initialize a single swarm in the search space and PSO returns a single pair of C and γ value. This pair is then used to build up our SVM model and the result have shown that it gives less precision-recall rate compared to the Partitioned model of PSO. Figure 4 is showing the result of the execution of PSO (Fig. 5).

Apart from the PSO, we have also compared our results with the three well-known optimization techniques—grid search, genetic algorithm, and gradient descent. The recorded precision-recall rate of each of the techniques has been shown in Graph 2a–c.

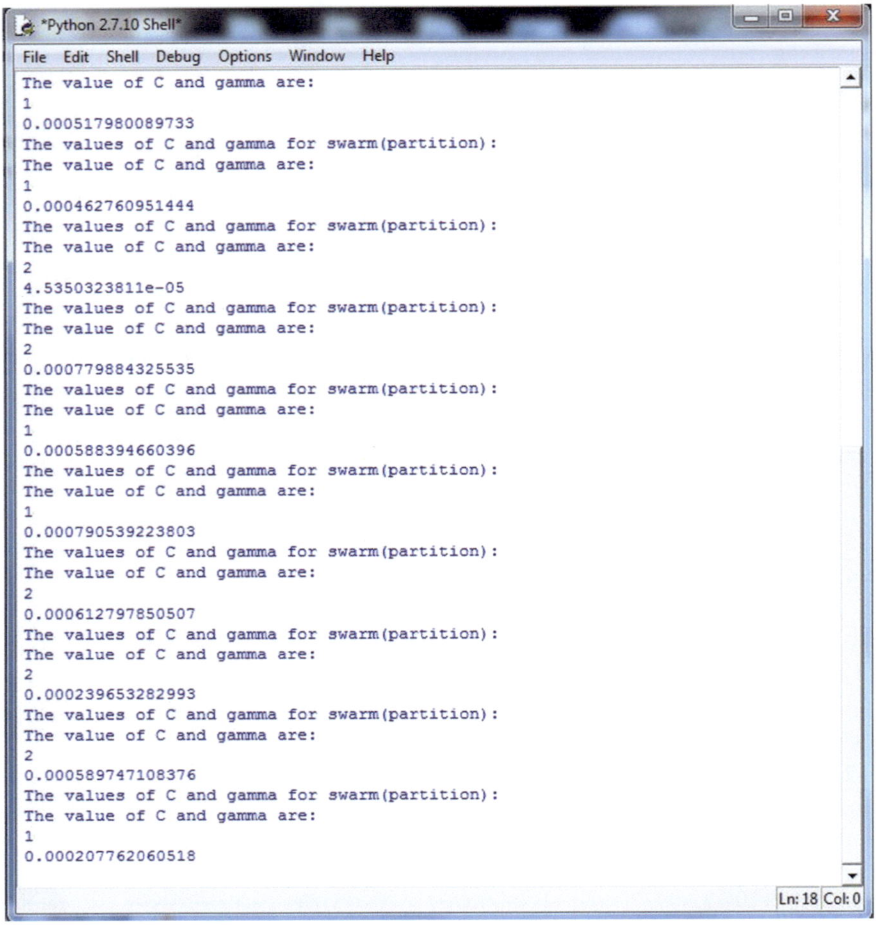

Fig. 3 Instance of the running algorithm

Table 2 Different values of C and γ for each swarm

Swarm number	c	γ
1	1	0.000517980089733
2	1	0.000462760951444
3	2	4.5350323811e−05
4	2	0.000779884325335
5	1	0.000588394660396
6	1	0.000790539223803
7	2	0.000612797850507
8	2	0.000239653282993
9	2	0.000589747108376
10	1	0.000207762060518

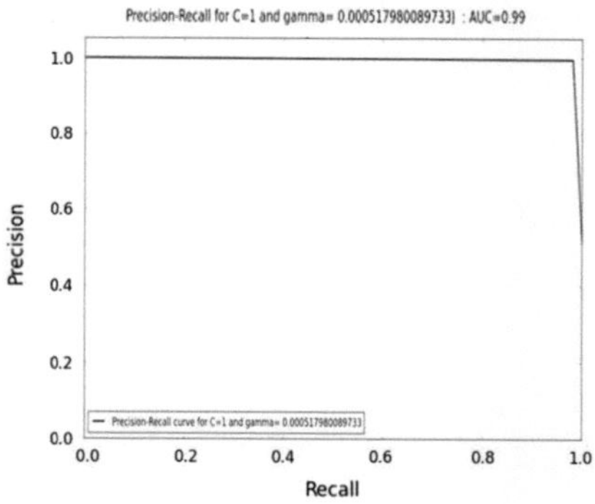

Graph 1(a)

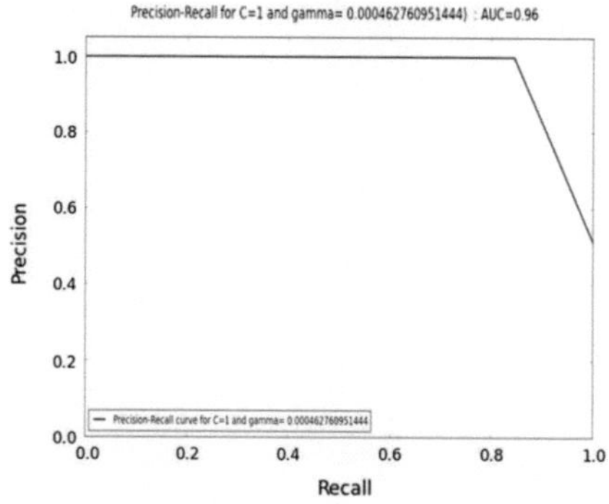

Graph 1(b)

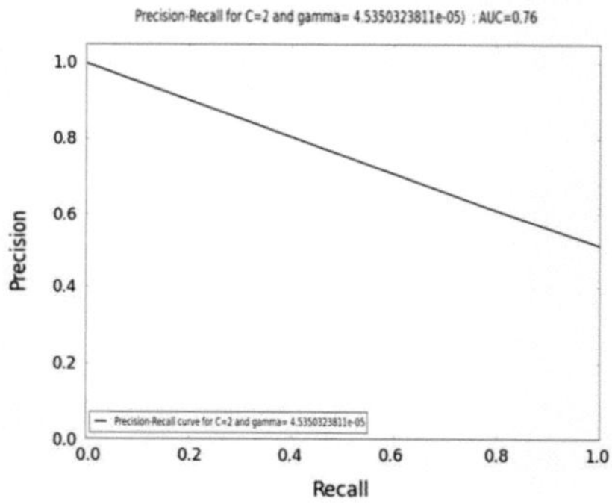

Graph 1(c)

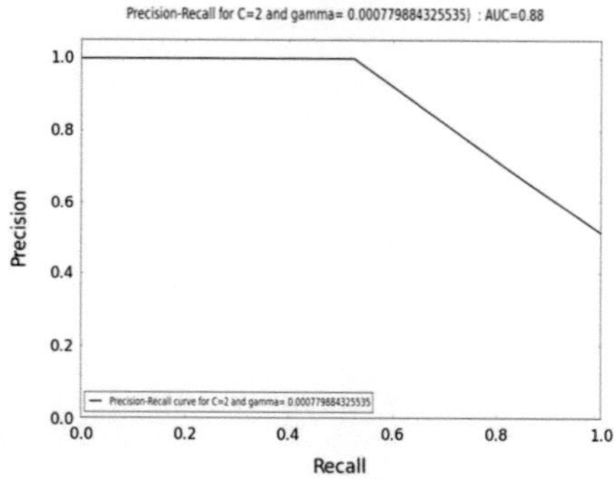

Graph1(d)

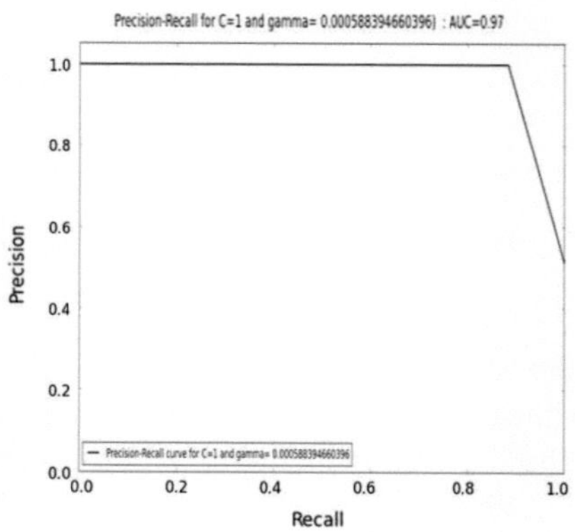

Graph 1(e)

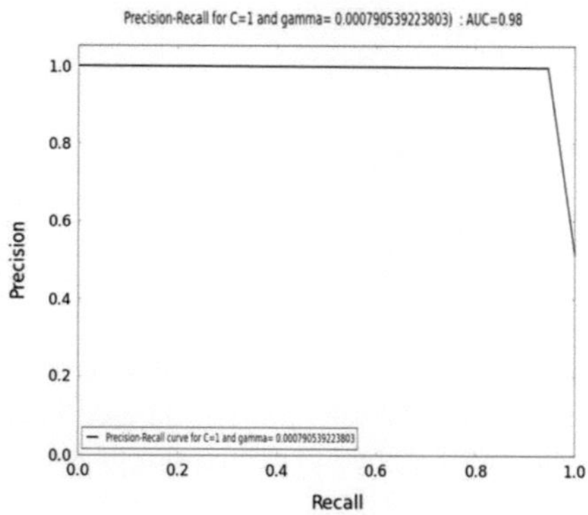

Graph 1(f)

Graph 1 Precision-Recall rate of SVM while using Multi-PSO to optimize C and γ

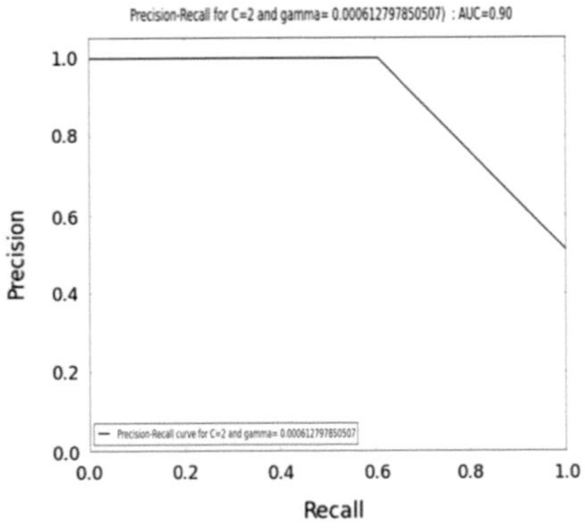

Graph 1(g)

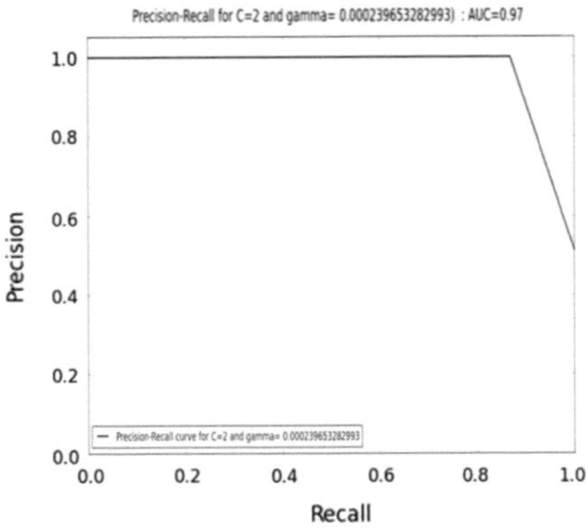

Graph1(h)

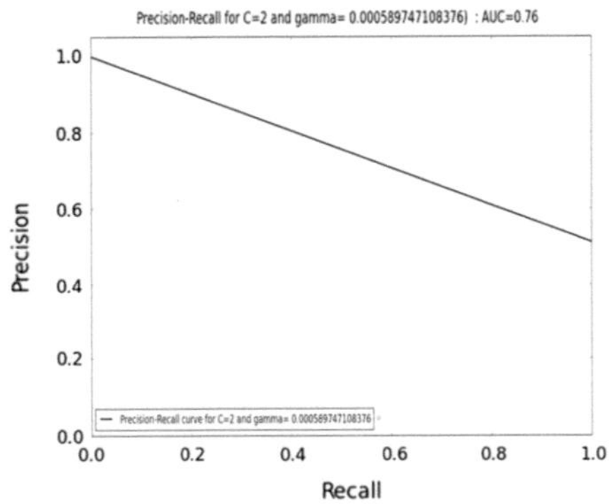

Graph 1(i)

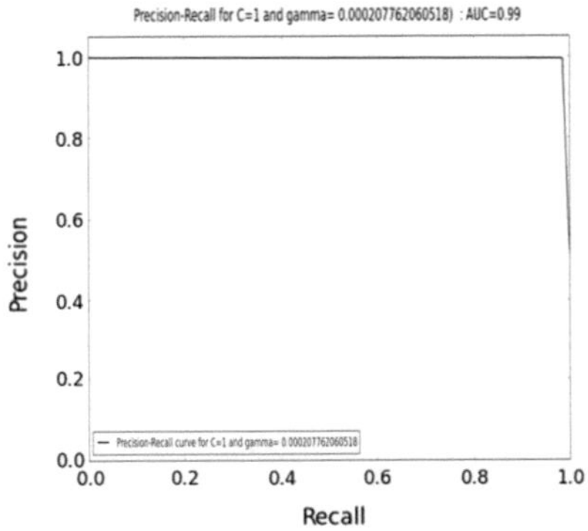

Graph 1(j)

Graph 1 (continued)

6 Conclusion

In this paper, we have proposed a model for the optimization of ***C*** and ***gamma*** RBF kernel parameters used in SVM modeling. An imbalance happens when the number of particles initialized in PSO is not proper. For the above problem, we have introduced a variant of PSO called multi-PSO which first partitions the search space for the ***C*** and ***gamma*** value and then deploys different swarms for each of the partition. Most of the applications of our model can be found in medical science, network security, text classification, image processing, software engineering, and many more. Further, there is a possibility of enhancement in our model where data is not static in nature, i.e., where data comes in batches and the time intervals between the data fetching is not uniform.

We have compared our results with well-known strategies such as classical PSO, grid search, genetic algorithm, and gradient descent. The results show an improvement in the precision-recall rate of SVM modeled using our strategy.

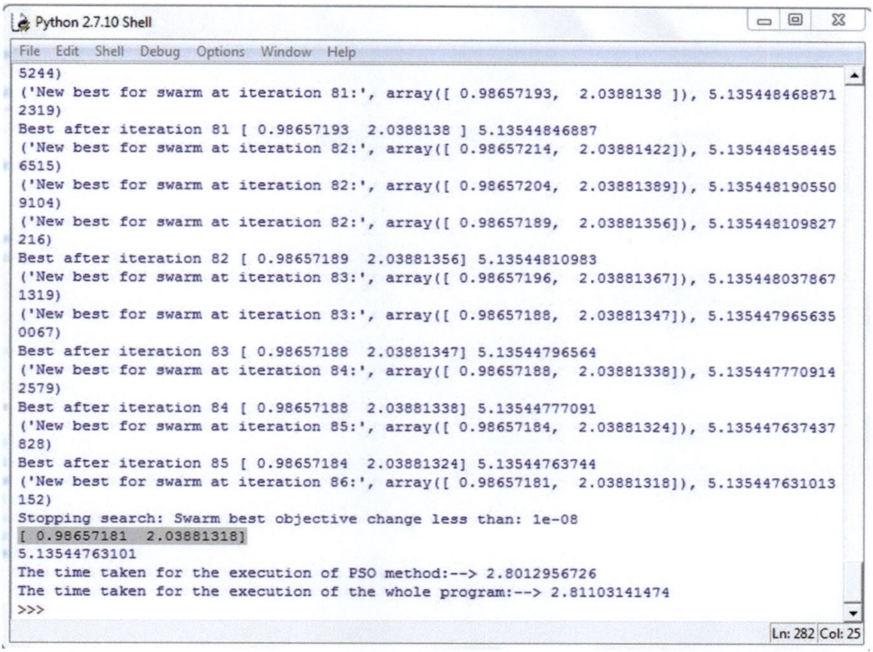

Fig. 4 Instance of execution of PSO

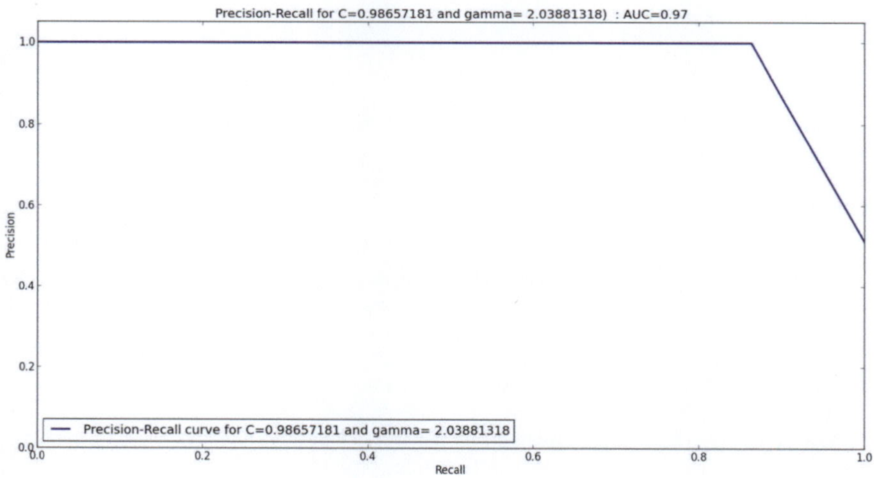

Fig. 5 Precision-recall rate of SVM using PSO

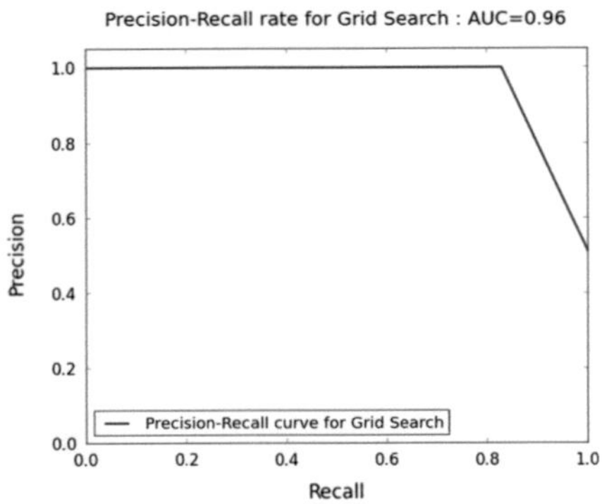

Graph2(a)

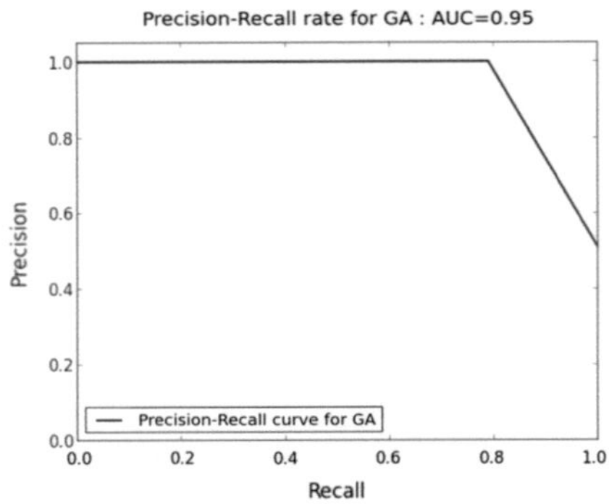

Graph2(b)

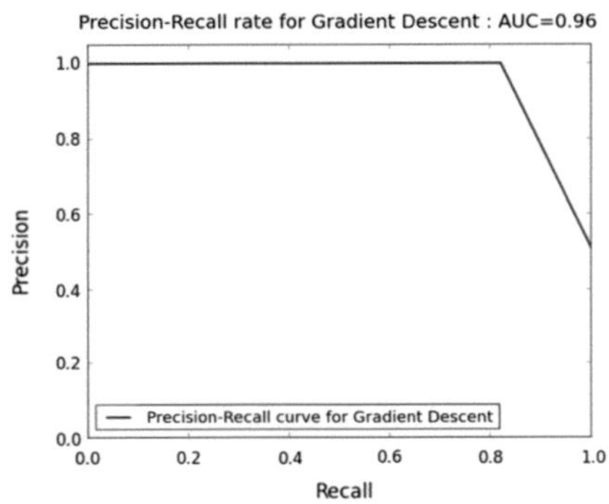

Graph 2(c)

Graph 2 Precision-Recall rate of SVM model while using Grid Search, GA and Gradient Descent

References

1. Vapnik V (1995) The nature of statistical learning theory. Springer, New York Google Scholar
2. Weston J (2014) Support vector machine. Tutorial http://www.cs.columbia.edu/~kathy/cs4701/documents/jason_svm_tutorial.pdf. Accessed, 10(0), 0-5
3. Bousquet O, Elisseeff A (2002) Stability and generalization. J Mach Learn Res 2:499–526
4. Kennedy J (2011) Particle swarm optimization. In: Encyclopedia of machine learning. Springer, Boston, MA, pp 760–766
5. Vapnik V (1998) Statistical learning theory new york. Wiley, NY
6. Chapelle O, Vapnik V (2000) Model selection for support vector machines. In: Advances in neural information processing systems, pp 230–236
7. Cristianini N, Shawe-Taylor J (2000) An introduction to support vector machines and other kernel-based learning methods. Cambridge University Press
8. Chapelle O, Vapnik V, Bousquet O, Mukherjee S (2002) Choosing multiple parameters for support vector machines. Mach Learn 46(1–3):131–159
9. Ayat NE, Cheriet M, Suen CY (2005) Automatic model selection for the optimization of SVM kernels. Pattern Recogn 38(10):1733–1745
10. Chang CC, Lin CJ (2011) LIBSVM: a library for support vector machines. ACM Trans Intell Syst Technol (TIST) 2(3):27

11. Hsu CW, Lin CJ (2002) A comparison of methods for multiclass support vector machines. IEEE Trans Neural Netw 13(2):415–425
12. Huang CM, Lee YJ, Lin DK, Huang SY (2007) Model selection for support vector machines via uniform design. Comput Stat Data Anal 52(1):335–346
13. Chunhong Z, Licheng J (2004) Automatic parameters selection for SVM based on GA. In: Fifth world congress on intelligent control and automation, 2004. WCICA 2004, vol 2. IEEE, pp 1869–1872
14. Cohen G, Hilario M, Geissbuhler A (2004) Model selection for support vector classifiers via genetic algorithms. An application to medical decision support. In: International symposium on biological and medical data analysis. Springer, Berlin, Heidelberg, pp 200–211
15. Suttorp T, Hansen N, Igel C (2009) Efficient covariance matrix update for variable metric evolution strategies. Mach Learn 75(2):167–197
16. Chatelain C, Adam S, Lecourtier Y, Heutte L, Paquet T (2007) Multi-objective optimization for SVM model selection. In: ICDAR, vol 1, pp 427–431
17. Igel C (2005) Multi-objective model selection for support vector machines. In: International conference on evolutionary multi-criterion optimization. Springer, Berlin, Heidelberg, pp 534–546
18. Kapp MN, Sabourin R, Maupin P (2012) A dynamic model selection strategy for support vector machine classifiers. Appl Soft Comput 12(8):2550–2565
19. Bai, Q. (2010) Analysis of particle swarm optimization algorithm. Comput Inform Sci 3(1):180

A Survey on SVM Hyper-Parameters Optimization Techniques

Dhruba Jyoti Kalita, Vibhav Prakash Singh and Vinay Kumar

Abstract Support vector machines can be considered as one of the most powerful classifiers. They are parameterized models build upon the support vectors extracted during the training phase. One of the crucial tasks in the modeling of SVM is to select optimal values for its hyper-parameters, because the effectiveness and efficiency of SVM depend upon these parameters. This task of tuning the values for the SVM hyper-parameters is called as the SVM model selection problem. Till now a lot of techniques have been proposed for optimizing values of hyper-parameters of SVM both in static and dynamic environment. Static environment is one where the knowledge about a problem does not change over time due to which static optimal values can be assigned to the hyper-parameters. On the other hand, due to the changing nature of the knowledge about a problem, in dynamic environment the optimization process has to be flexible enough to adapt the changes quickly. In dynamic environment, re-evaluation of the optimal values of the hyper-parameters is needed. This paper attempts to identify various optimization techniques used for SVM hyper-parameters tuning and recognize their pros and cons.

Keywords Classifier · Support vector machine · Model selection hyper-parameters · Optimization

D. J. Kalita (✉)
Gaya College of Engineering, Gaya, India
e-mail: mestop12@gmail.com

V. P. Singh
Motilal Nehru National Institute of Technology Allahabad, Prayagraj, India
e-mail: vibhav@mnnit.ac.in

V. Kumar
National Institute of Technology Jamshedpur, Jamshedpur, India
e-mail: vkumar.cse@nitjsr.ac.in

R. K. Shukla et al. (eds.), *Social Networking and Computational Intelligence*, Lecture Notes in Networks and Systems 100, https://doi.org/10.1007/978-981-15-2071-6_20

1 Introduction

Support vector machines (SVMs) are supervised machine learning models for classification, regression and outlier detection proposed by Vapnik in [1] which is based on structural risk minimization theory.

Construction of a support vector machine can be considered as an optimization problem where an optimal hyper-plane is chosen that separates both the classes. On the other hand, there is a dependency upon the kernel functions while constructing SVM for nonlinear dataset which maps from a lower dimensional space to a higher dimensional space. Kernel functions also use parameters that need to be tuned.

Let us consider a training dataset

$$\{(X_i, y_i), i = 1 \dots n\}, X_i \in \mathbb{R}^m, y_i \in \{+1, -1\}$$

If the training set is linearly separable, then there exists $w \in \mathbb{R}^m$ and $b \in \mathbb{R}$ such that

$$w^T X_i + b > 0 \, \forall i, \quad \text{s.t.} \ \ y_i = +1 \tag{1}$$

$$w^T X_i + b < 0 \, \forall i, \quad \text{s.t.} \ \ y_i = -1 \tag{2}$$

So $w^T X_i + b = 0$ will be a separating hyper-plane. Since there can exist infinitely many hyper-planes, in the formulation of the support vector machine our main motive is to find the optimal hyper-plane among these.

In the linear model of SVM, we normally scale w, b such that

$$w^T X_i + b \geq +1 \quad \text{if} \ \ y_i = +1 \tag{3}$$

$$w^T X_i + b \leq -1 \quad \text{if} \ \ y_i = -1 \tag{4}$$

For the above two equations, we can say that

$$y_i \left(w^T X_i + b \right) \geq 1, \quad \forall i \tag{5}$$

So, when the training set is separable, any separating hyper-plane can be tuned to satisfy the Eq. (5). In this formulation of SVM, a margin can be defined as the distance between $w^T X + b = +1$ and $w^T X + b = -1$ and the measurement of the distance between these two is $\frac{2}{||w||}$. So, intuitively we can say that more the margin better is the chance of correct classification of the new patterns. Following the above intuition, we can say that optimal hyper-plane is a solution of the following constrained optimization problem

$$\text{minimize} \frac{1}{2} w^T w$$

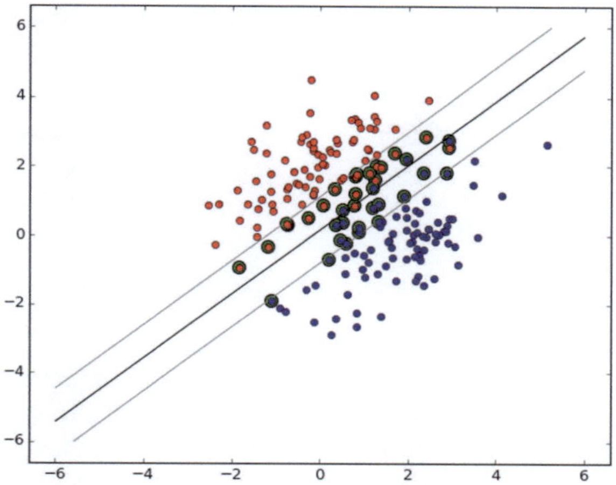

Fig. 1 Linearly separable data

$$\text{subject to } y_i\left(w^T X_i + b\right) \geq 1, \quad i = 1 \ldots n$$

The equivalent dual optimization for the same optimization problem is

$$\text{maximize} q(\mu) = \sum_{i=1}^{n} \mu_i - \frac{1}{2} \sum_{i,j=1}^{n} \mu_i \mu_j y_i y_j X_i^T X_j$$

$$\text{subject to } \mu_i \geq 0, \quad i = 1 \ldots n, \sum_{i=1}^{n} y_i \mu_i = 0$$

where $\mu_i, \ i = 1 \ldots n$ are Lagrangian multipliers that need to be computed for each of the training sample. This allows the SVM to select a part of the training samples with $\mu_i > 0$ which is used to define the decision boundary. These selected training samples are called support vectors (Fig. 1).

For the nonlinear dataset, the above formulation of the SVM model fails to detect an optimal hyper-plane. By introducing slack variables, we can deal with this problem. For the nonlinear dataset, we can model the optimization problem as follows

$$\text{minimize} \frac{1}{2} w^T w + C \sum_{i=1}^{n} \varepsilon_i$$

$$\text{subject to } y_i\left(w^T X_i + b\right) \geq 1 - \varepsilon_i, \quad i = 1 \ldots n$$

$$\varepsilon_i \geq 0, \quad i = 1 \ldots n$$

In this, we need to find the optimal values for the variables w, b, ε_i. In this ε_i are slack variables which measures the penalty that how much amount of constraint is not satisfied, i.e., it measures the extent of the violation of the optimal separation. If $\varepsilon > 0$, there is a margin error and when $\varepsilon_i > 1$, X_i is wrongly classified.

The dual optimization problem for the above can be now expressed as

$$\text{maximizeq}(\mu) = \sum_{i=1}^{n} \mu_i - \frac{1}{2} \sum_{i,j}^{n} \mu_i \mu_j y_i y_j X_i^T X_j$$

$$\text{subject to } 0 \leq \mu_i \leq C, \quad i = 1 \ldots n, \sum_{i=1}^{n} y_i \mu_i = 0$$

To deal with a nonlinear separable problem in SVM modeling, something called as "kernel trick" needs to be used. In this dual form of the SVM formulation, the only difference is that there is an upper bound on μ_i. Now to train a nonlinear discriminate function, we can use something called as kernel trick. This kernel trick is used to transform X_i (training dataset) into some high-dimensional space and learn a linear classifier there. This is equivalent to building a nonlinear classifier in the lower dimensional space (Fig. 2).

Let

$$\emptyset : \mathbb{R}^m \to \mathbb{R}^{m'}$$

is a function to map a lower dimensional space $\mathbb{R}^m$ to a higher dimensional space $\mathbb{R}^{m'}$. Now, the training set in this new space $\mathbb{R}^{m'}$ will be

$$\{(Z_i, y_i), \quad i = 1 \ldots l\}, \quad Z_i = \emptyset(X_i)$$

Then we can find the optimal hyper-plane by solving the following dual problem

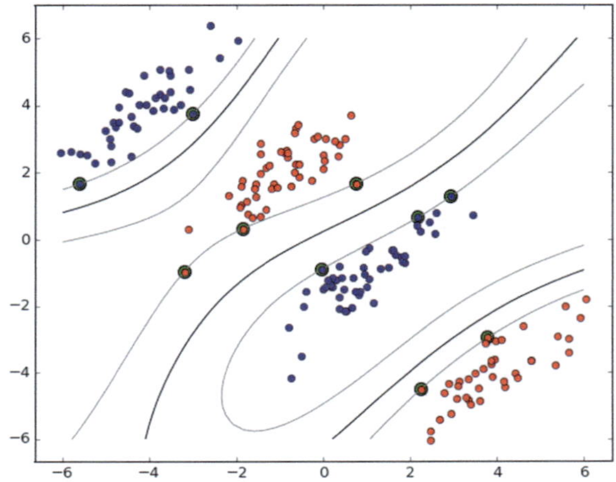

Fig. 2 Nonlinearly separable data

Table 1 Different kernel functions

Kernel functions	Inner producer
Linear kernel	$K(X_i, X_j) = X_i^T X_j$
Polynomial kernel	$K(X_i, X_j) = (\gamma X_i X_j + r)^u, u > 0$
Radial basis function kernel	$K(X_i, X_j) = \exp\left(-\gamma \lVert X_i - X_j \rVert^2\right), \gamma > 0$
Third-level heading	$K(X_i, Xj) = \tanh(\gamma X_i^T X_j + r)$

$$\text{maximizeq}(\mu) = \sum_{i=1}^{n} \mu_i - \frac{1}{2} \sum_{i,j=1}^{n} \mu_i \mu_j y_i y_j \emptyset(X_i)^T \emptyset(X_j)$$

$$\text{subject to } 0 \le \mu_i \le C, \quad i = 1 \ldots n, \sum_{i=1}^{n} y_i \mu_i = 0$$

The above problem is an optimization problem over $\mathbb{R}^{m'}$(with a quadratic cost function and linear constrains) irrespective of $\emptyset$ and m'.

Table 1 is a list of kernel functions that can be used for mapping the original space to a higher dimensional space. In this paper, we have used RBF kernel [1] because it has superiorities over the other kernel methods. Linear kernel can also be derived from RBF kernel. With a penalty parameter C it can be adjusted to give same performance as RBF. In the same way, RBF kernel can also be tuned with certain parameters to make it behave like a sigmoid kernel. In polynomial kernel, the number of parameters is more than that of the RBF kernel. Also, RBF kernel has fewer numerical difficulties.

Various optimization techniques for SVM hyper-parameters depend upon two basic factors.

- Selection criterion.
- Searching Methods used.

So, all the model selection techniques basically differ in these two aspects. Sections 2 and 3 identify various selection criterions and searching methods, respectively.

2 Various Selection Criterion

2.1 Leave-One-Out Bound

In leave-one-out procedure, a decision rule is constructed by eliminating one element from the training data at a time, and the removed element is then used for testing. So, one can test all the n elements or samples of the training dataset (with n different decision rules) [2]. Let

$$\mathcal{L}(x_1, y_1, \ldots x_n, y_n)$$

be the number of errors in the leave-one-out bound.

In [3] it has been explained that the leave-one-out procedure is capable of giving an almost unbiased estimation of the probability of the test error (Luntz and Brailovsky theorem).

$$Ep_{\text{error}}^{n-1} = E\left(\frac{1}{n}(\mathcal{L}(x_1, y_1, \ldots x_n, y_n))\right) \tag{6}$$

where p_{error}^{n-1} is the probability of the error generated during testing in the machine trained on $(n-1)$ data samples.

In SVM, one needs to perform leave-one-out procedure only for support vectors since removal of a point that is not support vector does not change the decision function.

This leave-one-out bound can be applied to SVM both in static and dynamic environments. In static environment, since the volume of data does not increase with time, the leave-one-out procedure can be used efficiently with less amount of computational time. On the other hand, in dynamic environment it takes more computational time as volume of data increases with time.

2.2 Support Vector Count

Removing a support vector which is not a support vector from the training set does not affect the solution computed by the support vector machine.

$$U_p = f^0(x_p) - f^p(x_p) = 0 \tag{7}$$

where x_p is a non support vector.

For the above reason, one can restrict the sum to support vectors and upper bound each term in the sum by 1 which gives the following bound on the number of errors made by the leave-one-out procedure [1].

$$T = \frac{N_{\mathrm{SV}}}{n} \tag{8}$$

where N_{SV} is the number of support vectors.

2.3 Radius-Margin Bound

The SVM formulation process is an optimization process that minimizes or maximizes an objective function. These objective functions can be designed using many factors. One of the factors is radius-margin bound. When we use both the radius and the margin in SVM error bound optimization then it provides a tighter error bound. Here, radius refers to the radius of the smallest sphere containing all data. In standard SVM [1], the optimization problem deals only with the margin, because in the feature space the smallest sphere which encloses the data is fixed. But in case of the linear transformations of the feature space the sphere and its radius changes. So, it is necessary to consider both the radius and the margin. In [4], Huyen Do and Alexandros Kalousis use both radius and margin to formulation of SVM [5].

For feature selection criterion used radius-margin bound ratio which can be defined as a minimization function as below

$$f(\sigma) = \frac{R^2}{\gamma}(\sigma) \tag{9}$$

Over σ where $\sigma \in \{0, 1\}^d$ which indicates the selection status of σ_i.

2.3.1 *Span* Bound

As explained by Vapnik and Chapelle [6], span of support vectors (SV) is a geometrical concept which influences the ability of generalization of support vector machines (SVM). This concept has a direct application in model selection.

In [5], Vapnik and Chapelle defined a set Λ_p which is constrained linear combinations of the points $\{x_i\}_{i \neq p}$ for any fixed support vector x_p

$$\Lambda_p = \left\{ \sum_{i=1, i \neq p}^{n} \lambda_i x_i : \sum_{i=1, i \neq p}^{n} \lambda_i = 1, \quad \text{and } \forall i \neq p, \alpha_i^0 + y_i y_p \alpha_p^0 \lambda_i \geq 0 \right\} \tag{10}$$

λ_i can be less than 0.

Span of the support vector X_p is a quantity which can be defined as the distance between X_p and the above set.

$$S_p^2 = d^2(X_p, \Lambda_p) = \min(X_p - X)^2 \tag{11}$$

3 Various Searching Techniques

3.1 Gradient Descent

Gradient descent [6–8] is a statistical searching technique which is frequently used for the optimization of parameters. Gradient descent is to optimize an objective function $f(\theta)$, where $\theta \in \mathbb{R}^n$ is the parameters of a model. Minimization in gradient descent is performed by changing the parameters in the opposite direction of the objective function $\nabla_\theta f(\theta)$ with respect to the parameters. The step's size to reach the local minima in gradient descent algorithm depends upon the learning rate η (Fig. 3).

Based on the amount of data available for computing the gradient of the objective, there can be three variants of gradient descent. They are

(1) *Batch Gradient Descent*:

To compute the gradient of the objective function with respect to the parameters θ batch gradient descent [8] considers the whole training dataset.

$$\theta = \theta - \eta.\nabla_\theta f(\theta) \tag{12}$$

Since for each update one has to consider the whole dataset to calculate the gradient, the batch gradient descent algorithm is very slow and when datasets are not fit into the memory then it is intractable. Batch gradient descent does not support the online updation of the model.

(2) *Stochastic gradient descent* (*SGD*):

Stochastic gradient descent [9, 10] is considered as a stochastic approximation of the gradient descent. SGD tries to find minimums or maximums by iteration. SGD performs the parameter updation by considering each training example x_i and label y_i

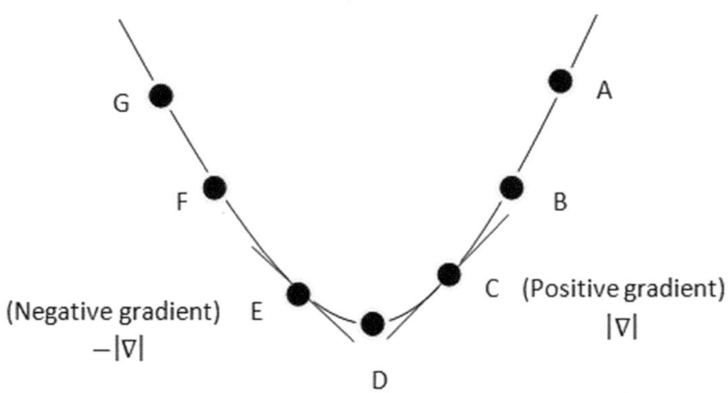

Fig. 3 Gradient descent searching

$$\theta = \theta - \eta.\nabla_\theta f(\theta; x_i, y_i) \tag{13}$$

SGD performs one update at a time, i.e., it does not recomputes gradients for similar examples before each parameter update. So, SGD is very fast which makes it feasible for the online updation of the model.

(3) Mini batch gradient descent:

A mini batch of n training samples is considered during the updation process in mini batch gradient descent.

$$\theta = \theta - \eta.\nabla_\theta f(\theta; x_{i:i+n}; y_{i:i+n}) \tag{14}$$

In reality, gradient descent is not a good option for model selection. Gradient descent uses a differentiable objective function with reference to the hyper-parameters and kernel functions. Multiple local minima in objective functions are also a hurdle for the gradient descent.

3.2 Grid Search

Grid search [11] is the most frequently used hyper-parameter optimization technique. It performs exhaustive search on a manually specified subset of the hyper-parameter space. Cross validation is used to evaluate the fitness of the parameter values in grid search. For optimizing the parameters like C, kernel and gamma in SVM classifier grid search is a good option.

A search in this manner consists of five basic components

- An estimator (regressor or classifier).
- Parameter space.
- Method for searching or sampling candidates.
- Cross-validation scheme.
- Score function.

The parameter space in an optimization process normally includes real valued or unbounded value spaces for the parameters to be tuned. For this reason, a manual setting of bounds and discretization is necessary in grid search.

3.3 Genetic Algorithms

Genetic algorithm [12–17] is an evolutionary algorithm (EA) that is based on natural evolution techniques such as inheritance, mutation, selection and crossover. GA mimics a natural selection process. Darwin's principle "survival of the fittest" is the main concept behind genetic algorithm. A population of candidate solutions to an optimization problem is first initialized in GA which in turn evolves toward better solutions. The basic steps in genetic algorithms are as follows

- Initialization: In this step, a random population of n chromosomes is generated or initialized which is a set of solutions for the problem.
- Fitness: A fitness function $f(x)$ evaluates the fitness for each chromosome in that population.
- Creation of the new population: A new population can be created using the following steps
- **Step 1**: Selection: Two parent chromosomes get selected from the populations based on the fitness value.
- **Step 2**: Crossover: Crossover happens on the parents to form new offsprings (children) with crossover probability. Without crossover, the offsprings will be the exact copy of the parents.
- **Step 3**: Mutation: A new offsprings get mutated at each locus (position in chromosome) with a mutation probability.
- **Step 4**: Accepting: New offspring is added to the new population.
- Replace: Above new generated population is then further used in next run of the algorithm.
- Test: After reaching the end condition, algorithm gets stopped and in the current population returns the best solution.
- Loop: Go To Step 2.

3.4 Covariance Matrix Adaptation Evolution Strategies (CMA-ES)

CMA-ES are methods for numerical optimization. They are stochastic, derivative-free methods. They are basically evolutionary algorithms and very much powerful for real-valued single-objective optimization [16, 18]. CMA-ES can be used for unconstrained or bounded constraint optimization problems, where search space dimensions are limited within hundred or three hundred. The CMA-ES is very advantageous because of its invariance properties. Two main invariance properties of CMA-ES are

- Invariance to order preserving (i.e., strictly monotonic) transformations of the objective function value (i.e., $||x||^2$ and $3||x||^{0.2} - 100$ are equivalent objective functions)
- Invariance to angle preserving (rigid) transformations of the search space (including rotation, reflection and translation), if initial search point is transformed accordingly.

Igel et al. in [16] have shown CMA-ES is capable to handling more kernel parameters.

3.5 *Particle Swarm Optimization*

The concept of particle swarm optimization came into existence due to the observations in a simulated and simplified social model. This simulation basically was that of the choreography of a bird flock. Kennedy and Eberhart in the year 1995 found that this model is an optimizer that can be used for optimization.

In a more simplified way, PSO can be defined as an evolutionary computation technique similar to genetic algorithm (GA) where initialization of a population of random solutions in the search space is the first step to get an optimal solution. Each potential solution is called as particle. PSO differs from the GA in the fact that particles are associated with movements and randomized velocity. In each time step each particle's intention is to move toward two best positions *Pbest* and *Gbest*. *Pbest* is the local best position of individual particles so far and *Gbest* is the global best position of any of the particles in the population so far.

Evaluating the objective function at a particular position we can measure the fitness of that position. In this, *Pbest* is the memory associated with each of the particle and *Gbest* enables the information sharing between particles.

The movement of the particles toward *Pbest* and *Gbest is* controlled by their velocity. In each iteration this velocity needs to be changed to reach the *Pbest* and *Gbest*. First, the acceleration is calculated based on the above two best positions and then this acceleration is added to the velocity. In PSO, constriction is one factor that needs to be taken care so that progressive slow down of the particles movement happens. The velocity which has been updated then is used to update the position of each particle from its current position to the new position (Fig. 4).

Let $\boldsymbol{i}$ be a particle on the solution or search space. This $\boldsymbol{i}$ contains three components

(1) Particle's position vector $\vec{\boldsymbol{x}}_i$.
(2) Particle's own best solution found so far $\overrightarrow{\boldsymbol{pos}}_i$.
(3) Particle's velocity $\overrightarrow{\boldsymbol{vel}}_i$.

Apart from the above components, each particle also keeps track of the best solution found so far by any of its neighbors, i.e., the global best solution $\overrightarrow{pos}_{gb}$. Many different topologies can be used as explained in [18–20]. Among them the standard one is the global (*gbest*). At each iteration, PSO algorithm updates the velocity of each particle and then their positions.

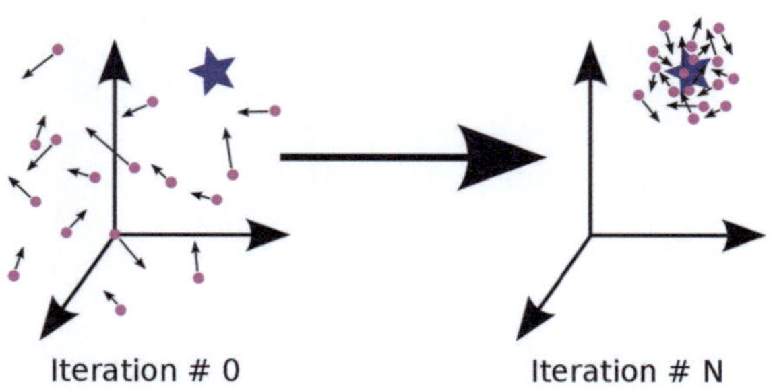

Fig. 4 Convergence in PSO

$$\overrightarrow{vel}_i = X\Big[\overrightarrow{vel}_i + c_1\epsilon_1\big(\overrightarrow{pos}_{gb} - \vec{x}_i\big) + c_2\epsilon_2\big(\overrightarrow{pos}_i - \vec{x}_i\big)\Big] \tag{15}$$

$$x_i = x_i + \overrightarrow{vel}_i \tag{16}$$

3.6 Hybrid-Optimization Techniques

Instead of using a single optimization technique for SVM hyper-parameters optimization we can also combine two or more optimization techniques[21–23] to increase the efficiency and effectiveness. Xiao et al. in [23] combined PSO and grid search to increase the efficiency of the parameter selection. They used this technique to tune C and γ parameters of SVM. They used a combination (C, γ) of a large step size with a range of $C = \left[2^{-16}, 2^{-15}, \ldots 2^{15}, 2^{16}\right]$ and range of $\gamma = \left[2^{-16}, 2^{-15}, \ldots 2^{15}, 2^{16}\right]$, where combined (C, γ) is a total of 1089. Then they used a combination (C, γ) of 2^2 steps to train different SVM, respectively, to get highest learning rate accuracy (C, γ) and applied PSO in the certain range of the (C, γ) neighborhood for more detailed search for highest learning accuracy.

Kapp et al. [21] in their paper have proposed a framework to perform optimization for SVM model selection problem in dynamic environments by combining a variant of grid search called adapted grid search and a dynamic particle swarm optimization technique. In this, they have implemented a change detection module which detects the changes in the environment and according to that perform searching at different levels. In real sense, the change detection module is responsible for monitoring the quality of the model selection process and avoiding "unnecessary" searching process. They have tried this framework for training SVM where data comes in batches and have shown its ability to cope up with the dynamic environments where optimum changes over time.

4 Conclusion

The performance of SVM classifier depends upon the parameters associated with it. These parameters are basically that of the kernel functions used by the SVM and of SVM itself. Finding an optimal value for the SVM parameters is called as SVM model selection where a model of SVM is built over the support vectors selected and optimal values of these parameters. So it is very much crucial to choose a proper optimization technique for SVM model selection.

All optimization techniques are associated with two basic aspects-selection criterion used and the searching method used. This paper identifies various selection criterion as well as searching methods used for SVM hyper-parameters optimization which are manifestations for various optimization techniques. This paper also identifies pros and cons of these techniques.

The basic limitation of most of the above-explained methods is that they can hardly cope up with the dynamic environment where data changes over time. So it is very much needed to create an algorithm which takes care of the data in the dynamic environment. Normally, to resolve this problem many of the authors have proposed hybrid system for model selection in SVM where they combine two or more techniques. So, the main challenge is to find a single algorithm which finds the optimum solution both in static and dynamic environment in less time.

References

1. Vapnik V (1995) The nature of statistical learning theory. Springer New York Google Scholar
2. Bousquet, O, Elisseeff A (2002) Stability and generalization. J Mach Learn Res 2:499–526
3. Vapnik V, Chapelle O (2000) Bounds on error expectation for support vector machines. Neural Comput 12(9):2013–2036
4. Do H, Kalousis A (2013) Convex formulations of radius-margin based support vector machines. In: International conference on machine learning, pp 169–177
5. Cristianini N, Shawe-Taylor J (2000) An introduction to support vector machines and other kernel-based learning methods. Cambridge University Press, Cambridge
6. Chapelle O, Vapnik V, Bousquet O, Mukherjee S (2002) Choosing multiple parameters for support vector machines. Mach Learn 46(1–3):131–159
7. Ayat N-E, Cheriet M, Suen CY (2005) Automatic model selection for the optimization of SVM kernels. Pattern Recogn 38(10):1733–1745
8. Wilson DRl, Martinez TR (2003) The general inefficiency of batch training for gradient descent learning. Neural Netw 16(10):1429–1451
9. Bottou L (2010) Large-scale machine learning with stochastic gradient descent. In: Proceedings of COMPSTAT' 2010. Physica-Verlag HD, pp 177–186
10. Hsu C-W, Lin C-J (2002) A comparison of methods for multiclass support vector machines. IEEE Trans Neural Netw 13(2):415–425
11. Huang C-M, Lee Y-J, Lin DKJ, Huang S-Y (2007) Model selection for support vector machines via uniform design. Comput Stat Data Anal 52(1):335–346
12. Chunhong Z, Licheng J (2004) Automatic parameters selection for SVM based on GA. In: Fifth world congress on intelligent control and automation, 2004. WCICA 2004, vol 2. IEEE, pp 1869–1872
13. Gilles C, Hilario M, Geissbuhler A (2004) Model selection for support vector classifiers via genetic algorithms. An application to medical decision support. In: International symposium on biological and medical data analysis. Springer, Berlin, pp 200–211
14. Chatelain C, Adam S, Lecourtier Y, Laurent L, Paquet T (2007) Multi-objective optimization for SVM model selection. In: ICDAR, vol 1, pp 427–431
15. Lessmann S, Stahlbock R, Crone SF (2006) Genetic algorithms for support vector machine model selection. IJCNN 6:3063–3069
16. Igel C, Hansen N, Roth S (2007) Covariance matrix adaptation for multi-objective optimization. Evol Comput 15(1):1–28
17. Friedrichs F, Igel C (2005) Evolutionary tuning of multiple SVM parameters. Neurocomputing 64:107–117
18. Janson S, Middendorf M (2004) A hierarchical particle swarm optimizer for dynamic optimization problems. In: Workshops on applications of evolutionary computation. Springer, Berlin, pp 513–524

19. Kennedy J (2011) Particle swarm optimization. In: Encyclopedia of machine learning. Springer, Boston, pp 760–766
20. Kennedy, J, Mendes R (2002) Population structure and particle swarm performance. In: Proceedings of the 2002 congress on evolutionary computation, CEC'02, vol. 2. IEEE, pp 1671–1676
21. Kapp MN, Sabourin R, Maupin P (2012) A dynamic model selection strategy for support vector machine classifiers. Appl Soft Comput 12(8):2550–2565
22. Weston J (1998) Support vector machine (and statistical learning theory) tutorial. NEC Labs Am 4
23. Xiao T, Ren D, Lei S, Zhang J, Liu X (2014) Based on grid-search and PSO parameter optimization for support vector machine. In: 2014 11th world congress on intelligent control and automation (WCICA). IEEE, pp. 1529–1533

Review of F0 Estimation in the Context of Indian Classical Music Expression Detection

Amit Rege and Ravi Sindal

Abstract The work addresses the need of fast and accurate F0 detection method for faithful transcription of Indian classical music. Three prominent F0 detection methods, viz. discrete Fourier transform (DFT), constant Q transform (CQT), and YIN algorithm are described and compared on the basis of accuracy and frame size against simulated signals of standard MIDI note frequencies. The same analysis is repeated on recorded data containing vocal recitals of eight notes from an octave in the equal tempered musical scale. That YIN method is most accurate and applicable for small frame size and is concluded.

Keywords F0 · Expression · Ornamentation

1 Motivation and Context

Music transcription has many applications in area of music information retrieval, query-by-humming (QBH), music-related equipment, and musicological analysis of different genres of music. One of the authors is a learner of music, and the other is keenly interested in it as a field of research. This is the motivation to research in the area of music signal processing.

The sound in a musical note is quasi-periodic signals, i.e., a superposition of piecewise periodic part and an aperiodic portion. It is clear from Fourier analysis that the periodic part can be expressed as sum of harmonically related sinusoids. The fundamental frequency of this representation characterizes the pitch of the sound. In going up in a musical scale, the fundamental frequency (F0) of the periodic part increases. The aperiodic portion leaves little effect on perception and therefore less important. Different notes in a musical system have a defined geometric relation between the fundamental frequencies of those (Fig. 1).

A. Rege (✉)
Medicaps University, Indore, India
e-mail: amit.rege@medicaps.ac.in

R. Sindal
IET Devi Ahilya University, Indore, India
e-mail: rsindal@ietdavv.edu.in

R. K. Shukla et al. (eds.), *Social Networking and Computational Intelligence*, Lecture Notes in Networks and Systems 100, https://doi.org/10.1007/978-981-15-2071-6_21

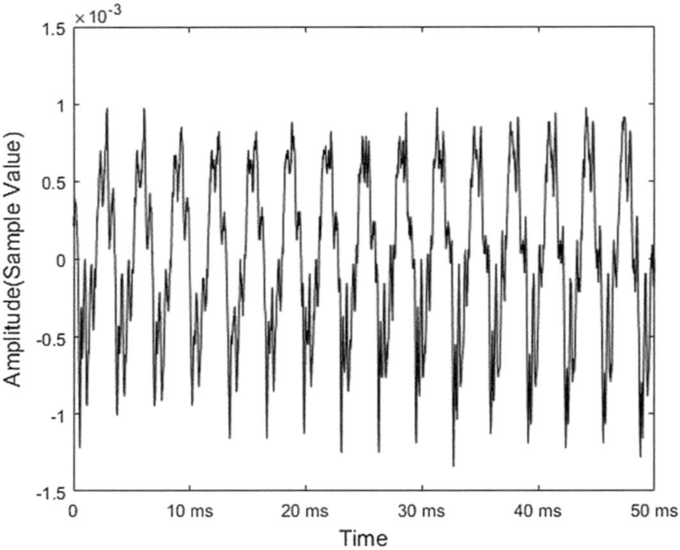

Fig. 1 Musical notes time domain description

Indian classical music (ICM) is sophisticated from the point of view of musicological[1] structure in the form of defined permutations and combinations of notes, largely only melodies. It has got two major schools, viz. Hindustani music, which is found in the northern region and Carnatic music, which is there in southern region of the subcontinent. Ancient Indian literature on music describes the presence of 22 *shrutis* (distinguishable tones) in one octave of a musical scale. Some of the tones prevailed over centuries leading to the presence of 12 notes in one octave. These had uneven musical intervals, i.e., ratios of frequencies. Later on, musical scale with equal musical intervals was adopted. This scale is called equal tempered scale and has equal musical intervals equal to one semitone and has 12 notes in one octave. Thus, the fundamental frequency increases by a factor of $2^{\frac{1}{12}}$ in going up by one semitone from a note. The scale with such relations is used worldwide nowadays.

It is worth stating here that music consists of both melody and harmony. Melody is characterized by the presence of only one note in a rendition at one time and a continuous change of that note over time in an articulated manner to give an esthetic sense to the listener. Harmony on the other hand is characterized by the presence of more than one note simultaneously in a well-defined and articulated manner to sound good to the listener.

The expressions, which are described as ornamentation in Western renditions, e.g., vibrato and glissando, have well-defined set of rules for improvising them in both vocal and instrumental renditions in ICM. The notes have well-defined fundamental frequencies which are standardized by MIDI. The expressions that are described as ornamentation in Western renditions, e.g., vibrato and glissando, have well-defined

[1]Musicology is the set of rules that are observed in particular genre of music.

set of rules for improvising them in both vocal and instrumental renditions in ICM. The notes have well-defined frequencies standardized by MIDI.[2] However, since the ornamentations are generally frequency-modulated expressions, accurate determination of the amount of deviation from the standard values is a must. Moreover, this process should be fast enough for faithful calculation of the F0 contours that characterize different musical expressions.

The literature on Indian musicology specifies names for some of the different types of musical expressions and defines rules of their improvisation in different *ragas*. For example, glissando, which is a continuum of the instantaneous frequency in the transition from one note to the next note, is called *meend* in Hindustani music and *gamaka* in Carnatic music. The usage of a particular type of glissando originating and ending at a particular set of notes characterizes a *raga*. Accurate determination of the type of expression contained would lead to betterment of the transcription system and would pave the way for the identification of sophisticated musical structures like *ragas* and *gharanas*.[3] Since the type of the expression is characterized by the pattern in which the underlying instantaneous frequency changes, accurate and fast determination of F0 is crucial for musical expression detection.

There have been numerous attempts at finding out the F0 for voiced part of speech signals. Vocal musical signals are always voiced, and some of the algorithms originally designed for speech worked well for music as well. Time domain methods for finding the pitch have their origin in autocorrelation-based methods [1], since the autocorrelation function is also almost periodic and is comparatively smoother to work than the original signal. Another time domain technique using average magnitude difference function (AMDF) is suggested in [2]. These methods work well for both instrumental and vocal inputs. In music note signals, the F0 may be feeble compared to the harmonics and even be altogether absent [3]. This makes frequency domain methods comparatively less accurate because it is difficult to track the F0.

Many efforts have been made to transcribe the complete note sequence in a monophonic recording as well as extracting the prominent melody from polyphonic sound and then the transcription task followed. Time–frequency domain analysis methods for finding pitch contour and, hence, the note sequence have been state of the art for quite a long time [4].

From the point of view of instrumental and vocal expression detection, work has been done to characterize and localize the type of expression in singing voice [5, 6] or the instrument being played [7]. The limitation of such approaches is that the methods and the corresponding works have not been able to characterize the expressions in a musicological perspective, and hence, sufficient emphasis on the exact nature of the expression has not been laid.

This article emphasizes the need for accurate and fast F0 estimation from the point of view of melody expression detection and evaluates three F0 estimation methods on the basis of accuracy and speed.

---

[2]Musical Instrument Digital Interface is a worldwide agreed upon set of protocols for operation of electronic music.

[3]This refers to the style of improvising a particular *raga*.

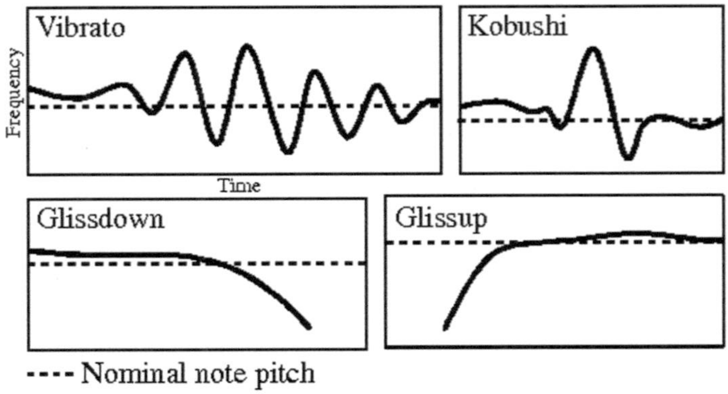

Fig. 2 Different types of musical expressions

Section 2 gives a brief description of different types of musical expressions that are improvised in vocal and instrumental renditions in both ICM and Western music. Section 3 describes the methods used in this work for the determination of F0. Section 4 gives details about the database that is used to evaluate the methods. Section 5 contains result and discussion. Section 6 discusses the conclusion that we draw from the work.

2 Different Types of Music Expressions

In the Indian art music as well as in Western classical music, different types of musical expressions are observed and therefore have been listed in the literature. Some of those are described next.

2.1 Glissando and Legato

The frequency slide of a continual nature in between two or more notes is called glissando. Exact difference between the terms glissando and legato could not be established [8]. A continual increment in F0 between two notes is termed as gliss-up, and decrement is termed as gliss-down by Ikemiya et al. [5] as shown in Fig. 2.

2.2 Vibrato

The modulation of the phonation frequency during a performed note around the defined frequency of the note in the manner described in the figure is known as

Vibrato [8]. The average rate of periodic variation of pitch is around 6 Hz and increases exponentially over the duration of a note.

2.3 Tremolo

The fluctuation of loudness in the singing voice is usually referred to as tremolo [8]. Sometimes the terms vibrato and tremolo are used to name the same ornamentation, but it is not correct to do so [9]. However, vibrato and tremolo could coincide in a singing voice.

2.4 Kobushi

Kobushi is a small deviation deliberated by the performer as shown [5]. In ICM, this expression is sometimes called a bracketed note if the upper and lower boundary is analogous to the preceding and following note frequencies of the notes used in the particular *raga*.

There are many other types of ornamentations that are specific to the ethnicity and culture. It is clear from the above discussion that for all kinds of frequency-modulated musical expressions, fast and accurate estimation of F0 is very important. The following section discusses the pitch extraction methods used for F0 estimation in this work in brief.

3 Pitch Extraction Methods Used

We use three methods for F0 detection and evaluate them on the basis of frame length and accuracy of determination of F0 against the frequencies standardized by MIDI. It is clear that the more accurate the algorithm is for the smaller frame length, the more suitable is the algorithm in the stated context.

3.1 Discrete Fourier Transform

We take the discrete Fourier transform of a frame of prescribed size and measure the F0 by picking the first peak after the DC component. This peak is, in general, the strongest. It can be picked by taking the first maxima after the DC component in the magnitude of Fourier transform or the power spectral density calculated by multiplying the transform sequence with its conjugate yielding the squared magnitude.

We use the latter approach. Discrete Fourier transform of a sequence is given by the following relation:

$$X(k) = \sum_{n=0}^{N-1} x(n)e^{\frac{j2\pi}{N}}$$

The basic limitation of Fourier transform, as discussed in the literature, is that the number of frequency bins is constant throughout the frequency range. The frequency bins are equally spaced from smallest to largest frequencies. If we take the frame length to be small that is we want high time resolution, the corresponding frequency resolution is very poor, and therefore, we cannot be sure about the exact value of the underlying frequency. On the other hand, if we take a large frame length, then the frequency resolution increases, but the time resolution is poor.

3.2 Constant Q Transforms

The above-stated problem can be solved by going for wavelet transform, which offer good frequency resolution for low frequencies and good time resolution for high frequencies. The ratio of center frequency to bandwidth is constant throughout. Therefore, we deliberately emphasize the name constant Q transform [4]. The constant Q transform is given by the following equation

$$X^{CQ}(k, n) = \sum_{j=n-N_k/2}^{n-N_k/2} x(j)a_k^*(j - n + N_k/2)$$

where $k = 1, 2, \ldots, K$ indexes the frequency bins of the CQT. The symbol $\lfloor . \rfloor$ denotes rounding toward negative infinity. The basis functions a_k are complex-valued waveforms, here also called time–frequency atoms. The definition of the same is as follows.

$$a_k(n) = \frac{1}{N_k}\omega\left(\frac{n}{N_k}\right)e^{-j2\pi n\frac{f_k}{f_s}}$$

where f_k is the center frequency of bin k, f_s denotes sampling rate, and $\omega(t)$ is a continuous window function for example Hann window or Blackman window, sampled at points determined by t. The window function is zero outside the range $t \in [0, 1]$

The window length $N_k \in \mathbb{R}$ in the above equations is real valued and is inversely proportional to f_k to ensure same Q factor for all frequency bins. Moreover in the CQT considered here, the center frequencies f_k obey the following equation.

$$f_k = f_1 2^{\frac{k-1}{B}}$$

where f_1 is the center frequency of the lowest frequency bin and B is the number of bins per octave.

In order to implement the transform, a computationally efficient method is obtained by using the following identity [4].

$$\sum_{n=0}^{N-1} x(n)a^*(n) = \sum_{j=0}^{N-1} X(j)A^*(j)$$

Further, in order to simplify the process and maintain the temporal kernel sparse, for computational efficiency throughout the frequency range, the authors in [4] process one octave at a time for CQT calculation. After computing the highest octave CQT bins over the entire signal, the input signal is low-pass filtered and down-sampled by factor two and same process is repeated to calculate the CQT bins for the next octave using the same DFT block size and spectral kernel. Figure 3 illustrates this process.

To calculate the fundamental frequency F0, from the so-obtained sparse time–frequency domain representation of the signal, the maximum value of the magnitude of the representation and the corresponding frequency index is collected and then converted to actual frequency with the help of the sampling rate.

This yields the value of instantaneous value of the F0 for the time frame being considered. For computation of the sparse time–frequency matrix representation, MATLAB toolbox published by the authors of [4] is used. We want to credit the authors for that.

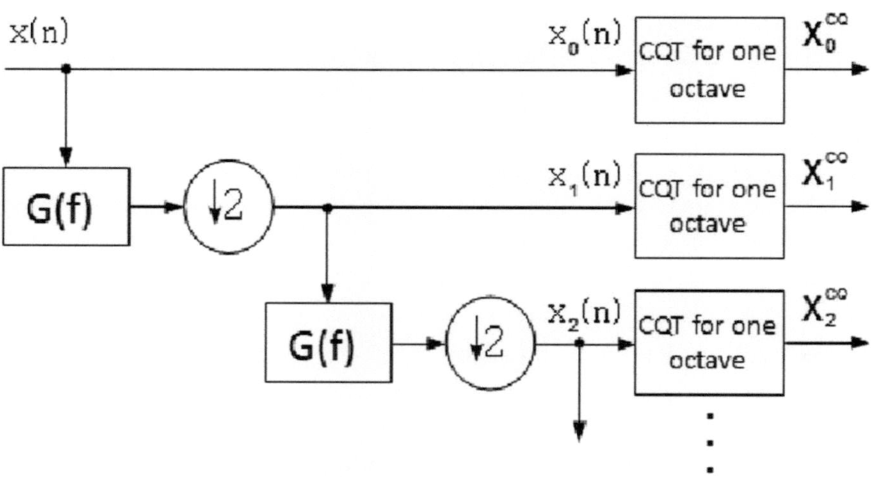

Fig. 3 Computation of CQT octave-by-octave

3.3 *YIN Algorithm*

This algorithm was proposed in [10] that is based on squared difference of the signal with delayed version of itself and then estimation of exact time period and hence frequency by going for parabolic interpolation of the curve hence obtained.

Steps of the algorithm as given by de Cheveigne and Kawahara [10] are as follows:

1. For a signal given by $x(t)$, the squared difference function $d_t(\tau)$ is calculated by the following equation where τ is the lag.

$$d_t(\tau) = \sum_{k=t}^{t+W-1} (x(k) - x(k+\tau))^2$$

2. Cumulative mean difference function $d'(\tau)$ is then derived using the following relation:

$$d'(\tau) = \begin{cases} 1, & \text{for } \tau = 0 \\ \frac{d_t(\tau)}{\left[\left(\frac{1}{\tau}\right)\sum_{j=1}^{\tau} d_t(j)\right]} & \text{otherwise} \end{cases}$$

3. Then, we search for the smallest τ for which a local minimum of $d_t'(\tau)$ is smaller than a given absolute threshold value κ. If no such value is found, then we search for the global minimum $d_t'(\tau)$ instead. We denote the found lag value with τ'.
4. We then interpolate the $d_t'(\tau)$ function values at abscissas $\{\tau' - 1, \tau', \tau' + 1\}$ with a second-order polynomial.
5. Then, we search for the minimum of the polynomial in the continuous range $(\tau' - 1, \tau' + 1)$ and denote the corresponding lag value with $\hat{\tau}$. The estimated F0 is then, $f_s/\hat{\tau}$.

This algorithm is also available with MATLAB package; however, for this work this is implemented by the authors in a simplistic way for the estimation of F0 of frames from the dataset.

4 Dataset for Experimentation

We experiment on the eight notes of C#3 major scale. This is the middle octave that is generally used in singing. The last note of the octave is double in frequency of the first note, i.e., the tonic. In Western terminology, the first note of a scale is called the tonic and the last note is called the octave. In ICM terminology, these are called *sa*, *re*, *ga*, *ma*, *pa*, *dha*, *ni*, *sȧ* provided we take the first note, i.e., C# as

Table 1 Standard notes used

Note	C#3	D#3	F3	F#3	G#3	A#3	C4	C#4
MIDI note number	61	63	65	66	68	70	72	73
Freq. (Hz)	277.1826	311.1270	349.2282	369.9944	415.3047	466.1638	523.2511	554.3653

the tonic or *sa*. The MIDI note numbers and standard frequencies for these notes are listed in Table 1. The pivot for these values comes from note number 69 that is conventionally assigned a frequency value of 440 Hz. A traversal of 12 semitones in forward direction leads to double of the F0, similarly while going back 12 semitone change will amount to halving the F0 of the note.

We create a database of sound files containing recorded monophonic notes. Three types of sound files, viz. vocal rendition, strings, and piano, are used to test the algorithms. We also simulate some sequences and use the algorithms with those. The simulated signals are considered the most accurate ones because otherwise there is no standard database available for reference MIDI notes. For all these variations, we experiment with the 8 notes of C# major scale. The sampling frequency used in all these calculations is 44,100 samples per second, to be at par with the standard CD recording sampling frequency.

4.1 Simulated Notes

Since the methods have to be tested on the basis of accuracy and speed (frame length), the first attempt made is to use simulated signals with the help of following formulae.

$$x(n) = \sin(2\pi f n / F_s)$$

So in the above formula we use the standard MIDI frequencies of the C# major scale as specified in Table 1 and use $F_s = 44{,}100$, which is the standard sampling frequency in audio recording.

4.2 Recorded Notes

Note renditions of a Casio synthesizer in strings and piano mode are recorded in the form of wav file at $F_s = 44{,}100$ samples per second, which is the default sampling frequency. Vocal renditions for the same notes in voice of a skilled vocalist are recorded, without any background. In the same manner, recording is done for piano mode synthesizer sounds with same notes. The part of piano recording taken is just following the onset of the notes. For all four kinds of dataset, a 2- or 3-s-long recording is extracted and analyzed with the three algorithms.

5 Results and Discussion

The mentioned data are analyzed using the three methods for finding out the F0. The error calculation for all of the three remaining is done against the standard values obtained from the standard frequencies up to 2 digits of precision. In case of both frequency domain methods, DFT and CQT, the octave error manifests in the form of large error in note frequency determination, which is corrected using octave correction [8].

Table 2 shows the percentage error in F0 estimation for recordings of strings notes against the standard frequencies. Similar observations are taken for vocal, piano notes, and simulated notes that are omitted for brevity. It can be seen that the percentage error follows least for YIN algorithm in spite of being a time domain approach.

6 Conclusion and Future Scope

The objective of this analysis has been to proceed in the direction of music expression detection by accurately estimating the instantaneous F0. It is concluded that YIN is the best algorithm from the point of view of both accuracy and speed, i.e., small size of the frame of observation. Octave errors also do not occur, and therefore, there is no need of probabilistic considerations.

It is hoped that if the shape of the temporal kernel of the CQT has a tailored shape according to the time domain profile of the musical signal, maintaining the properties desired from mathematical viewpoint, the frequency resolution can further be made better and faster for musical expression or ornamentation detection.

Acknowledgements The authors acknowledge the support provided by IET-DAVV for availing necessary infrastructure to carry out the research. Moreover, the authors of [4] are also acknowledged for providing beautiful toolbox.

Table 2 Percentage error in F0 estimation for strings recording

Notes	Percentage error in F0 estimation for different frame lengths														
	256			512			1024			2048			4096		
	DFT	CQT	YIN	DFT	CQT	YIN	DFT	CQT	YIN	DFT	CQT	YIN	DFT	CQT	YIN
C#3	24	7.9	0.0	6.8	0.8	0.0	6.8	0.0	0.0	16	0.0	0.1	16	0.0	0.1
D#3	11	3.4	0.2	11	3.9	0.2	14	3.9	0.2	14	3.9	0.1	3.5	0.0	0.2
F3	1.3	7.5	0.2	1.3	2.9	0.2	1.3	1.4	0.2	6.2	0.0	0.3	8.7	0.0	0.4
F#3	6.9	4.4	0.2	6.9	1.4	0.2	12	0.0	0.3	12	0.0	0.3	8.7	0.0	0.4
G#3	17	3.9	0.2	3.7	0.1	0.2	3.7	0.0	0.1	16	0.0	0.1	7.8	0.0	0.2
A#3	11	0.2	0.1	7.6	0.2	0.2	19	0.0	0.1	4.6	0.0	0.2	2.3	0.0	0.2
C4	1.2	2.3	0.1	17	2.3	0.1	17	0.4	0.1	12	0.0	0.2	10	0.0	0.2
C#4	6.8	11	0.2	6.8	0.4	0.2	23	0.4	0.3	16	1.4	0.2	3.9	0.1	0.2

References

1. Rabiner LR (1977) On the use of autocorrelation analysis for pitch detection. IEEE Trans Acoust Speech Sig Process 25(1)
2. Un CK, Yang SC (1977) A pitch extraction algorithm based on LPC inverse filtering and AMDF. IEEE Trans Acoust Speech Sig Process 25(6)
3. Gerhard D (2003) Pitch extraction and fundamental frequency: history and current techniques. Department of Computer Science, University of Regina, Regina, Saskatchewan, Canada
4. Schoerkhuber C, Klapuri A (2010) Constant-Q transform toolbox for music processing. In: 7th sound and music computing conference, Barcelona, Spain
5. Ikemiya Y, Itoyama K, Okuno HG (2014) Transcribing vocal expression from polyphosnic music. In: ICASSP, Florence, Italy
6. Sung D, Lee K (2014) Transcribing frequency modulated musical expressions from polyphonic music using HMM constrained shift invariant PLCA. In: Proceedings of tenth IEEE international conference on intelligent information hiding and multimedia signal processing (IIH-MSP)
7. Barbancho I, de la Bandera C, Barbancho AM, Tardon LJ (2009) Transcription and expressiveness detection system for violin music. In: IEEE International conference on acoustics, speech and signal processing (ICASSP), Taipei, Taiwan, pp 189–192
8. Klapuri A, Davy M (2006) Signal processing methods for music transcription. Springer, New York
9. Polrolniczak E, Kramarczyk M (2015) Computer assessment of tremolo feature in context of evaluation of singing quality. Signal processing: algorithm, architecture, arrangements and applications, Sept 2015 Poznan, Poland
10. de Cheveigne A, Kawahara H (2002) YIN, a fundamental frequency estimator for speech and music. J Acoust Soc Am 111(4):1917–1930

Classification and Detection of Breast Cancer Using Machine Learning

Rekh Ram Janghel, Lokesh Singh, Satya Prakash Sahu and Chandra Prakash Rathore

Abstract One of the major second largest issues found in the world is breast cancer. For increasing in long term, the continuance of women can be achieved by early and accurate diagnosis. Early diagnosis is the only remedy to prevent breast cancer. Since detection of this disease is a critical issue, this research work focuses on various machine learning methods which can assist doctors in giving promising results in a correct diagnosis of cancer. Thirteen machine learning models are employed and compared on the various measures. Wisconsin Breast Cancer Database (WBCD) dataset is employed in performing the experimentation, which is extract from UCI repository. AdaBoost, logistic regression and 1-NN machine learning models give promising accuracy of 98% in performing the experiment among all the models.

Keywords Machine learning · Breast cancer · WBCD · Diagnosis · Accuracy

1 Introduction

A huge number of women are at a possibility of breast cancer worldwide which is considered as leading cancer mortality rate as compared to another cancerous disease. As per the American Cancer Society (ACS), a uniformity was found in the occurrence of breast cancer in white women from 2007 to 2011 while in case of black women 3% increment is found in the related uniformity. Approximately, 40,730 deaths due to breast cancer have been evaluated by American Cancer Society (ACS) [1, 2].

R. R. Janghel (✉) · L. Singh · S. P. Sahu · C. P. Rathore
National Institute of Technology Raipur, Raipur 492001, India
e-mail: rrjanghel.it@nitrr.ac.in

L. Singh
e-mail: lokesingh@gmail.com

S. P. Sahu
e-mail: spsahu.it@nitrr.ac.in

C. P. Rathore
e-mail: Rathore_1st@yahoo.com

R. K. Shukla et al. (eds.), *Social Networking and Computational Intelligence*, Lecture Notes in Networks and Systems 100, https://doi.org/10.1007/978-981-15-2071-6_22

Among all living beings including humans, the cell is called as the basic biological unit. A cell throughout its life cycle develops in size, where it gathers the nutrients and then splits itself to generate two novice daughter cells according to the need of the body. New cells overcome the old cells as they get damaged. Cancer evolves when the life cycle of cell gets disturbed. When cells turn abnormal, old cells survive instead of being dead and new cells evolve while they are no longer required. These extra cells turn into tumors [3, 4].

When the tumor is found, it is required to be classified as a benign or malignant. This is examined as a binary classification problem in the field of machine learning. Diagnosis of breast cancer at previous stage for the effective therapy of cancer cells is the only remedy to reduce death rate due to breast cancer. Various machine learning, deep learning and computing methods exist which help the physician in diagnosing cancerous diseases [5, 6].

2 Related Work Heading

In [7, 8], an algorithm is designed by Cedeno et al. which is based on the biological metaplasticity property of neurons and Shannon's information theory and thus named artificial metaplasticity multilayer perceptron (AMMLP) algorithm. To model the metaplasticity artificially, this algorithm revises the weights at high priority. The algorithm is evaluated using the Wisconsin Breast Cancer Database (WBCD) at different measures like accuracy, sensitivity, specificity, etc., and achieves a classification accuracy of 99.26%.

In [3], Tiwari et al. have discussed GONN algorithm which categorizes breast cancer tumors into two classes as benign and malignant. To perform the experiment, WBCD is used which is taken from UCI machine learning repository. The performance is measured by various measures like accuracy, sensitivity, specificity, confusion matrix, ROC curves and AUC under ROC curves. The training and testing data is divided into the following ratios: 50–50, 60–40 and 70–30 and achieves an accuracy of 98.24, 99.63 and 100% toward corresponding ratios. Obtained results prove that algorithm outperforms with breast cancer database.

In research work proposed by Zaher et al. in [9], a computer-aided diagnosis (CAD) technique is used for diagnosing breast cancer which uses the unsupervised technique—deep belief network (DBN), based on supervised method, namely back-propagation. For performing the experiment, Wisconsin Breast Cancer Database (WBCD) is employed using the back-propagation neural network with Liebenberg–Marquardt learning function. The algorithm used gives an efficient accuracy of 99.68%.

In [7, 10], Janghel et al. have designed an ensemble model using MLP, RBF and LVQ methods for the detection of breast cancer. The designed model has achieved faster classification speed with incremental learning requiring minimum space. Various ensemble methods are employed for the evaluation of the model like the weighted averaging product, minimum and maximum integration methods. Minimum integration techniques outperform over the measures like accuracy, sensitivity and specificity.

In [10, 11], Janghel et al. have employed soft computing techniques for the detection of breast cancer disease. Several neural network architectures are compared to fulfill the need. Three different types of breast cancer databases are used for performing the experimentation which is gathered from UCI repository which is specially for machine learning. The following are the architectures used in building the model—Back-propagation algorithm, radial basis function networks, learning vector quantization, probabilistic neural networks, recurrent neural network and competitive neural network.

3 Methodology Used

3.1 Soft Computing Models

Adaptive Boosting (AdaBoost) Model

Boosting is one of the machine learning ensemble methodologies which constructs a strong classifier $H(x)$ using various weak classifiers $h_t(x)$ as stated in Eq. (1) [12].

$$H(x) = \text{sign} \sum_{t=1}^{T} (\alpha_t h_t(x)) \tag{1}$$

where T denotes various iterations, t denotes iteration counter and α_t is an evaluation of importance assigned to h_t which is stated in Eq. (2)

$$\alpha_t = \frac{1}{2} \ln\left(\frac{1 - \in_t}{\in_t}\right) \tag{2}$$

The data $(x_1, y_1), \ldots, (x_m, y_m)$ is considered as training dataset which is considered as input value where every value x_i € (belongs to) instance space X and y_i € (belongs to) classification label set Y. For poor or inadequate classifier, AdaBoost algorithm performs T times iterations where $t = 1, \ldots, T$; i.e., for each of the iteration t, set of weights $D_t(i)$ is adjusted for the training dataset. At the first step, overall weights are initialized to $1/m$, and as the repetition goes on, the weights of those instances which are incorrectly classified emphasize hardly on new instances which are inaccurately classified by the previous poor classifier. The integrity of the week classifier is evaluated by error $\in_t$ shown in Eq. (3),

$$\in_t = Pr_{i \sim D_t}[h_t(x_i) \neq y_i] = \sum_{i:h_t(x_i) \neq y_i}^{n} D_t(i) \tag{3}$$

C5.0 Model

It is **a decision tree algorithm of machine learning** which is based on entropy and information gain. The advantage of using C5.0 model is that it can easily deal with the data which has missing values and suffered from noise [13]. The algorithm works in four phases: (1) In the first phase, it checks for the base class; (2) in the second step, it generates decision tree with the help of training data; (3) it then searches the attribute which has the highest information gain; (4) for each $t_i \in$ D, decision tree is applied to describe its base class [14].

Cubist Committees (CC) model

The term committees are alike to boosting in machine learning which designs series of trees by adjustment in weights. Just like bagging take and average of predictions is calculated and get final results. As the trees are generated in sequence, the first tree is designed according to the cubist model technique. The remaining of the trees are then constructed by adjustment in weights [15].

Random Forest (RF) model

Random forest is a machine learning model designed by Leo et al. It is a group of treelike structured classifiers $\{h(x, \Theta k), k = 1, \ldots\}$ where $\{\Theta k\}$ denotes distributed random vectors. In this model, each tree emits a unit vote for the most prevalent class at input x00. The model gives better classification accuracy, and without discarding any variable it runs effectively on large datasets [16].

ANOVA Radial Basis Kernel SVM (KSVMARB) model

One of the machine learning classification model experiments for both the classification and regression tasks. To map original data from input space to feature space, nonlinear functions are used by creating hyperplanes. Using test dataset, predictions are made by evaluation to which category they belong. In [17], let $(x_i, y_i) 1 \leq i \leq l$ be the set of training instances where $x_i \in R^n$ corresponding to class labeled as $y_{i \in}\{-1, 1\}$. The decision function of SVM can be stated as

$$f(x) = \text{sign}\left(\sum_{i=1}^{l} \alpha_i^0 y_i K(X_i, X) + b\right), \tag{4}$$

Linear Regression (LR) model

It is a statistical technique of machine learning which comprises more than one independent variable which represents the relationship between dependent variable say y and independent variable say X. Binomial logistic regression gains its strength as it predicts outcome from a group of steady predictor variables [18]. The likelihood function can be stated as

$$l(\beta) = \sum_{i=1}^{n} [y_i \log(p_{i)} + (1 - y_i) \log(1 - p_i)] \tag{5}$$

where b represents a framework for regression, n denotes various responses and p_i is the probability of u.

Naive Bayes (NB) model

An effective superintended machine learning algorithm/model is used for classification with an assumption that existence of a particular feature in a class is not related to the existence of any other feature. Let Y be response attribute, and $X_1 \ldots X_n$ is the real invaluable attributes. The main objective is to train a classifier so as to obtain a probability distribution over response variable Y for each related case of X. The measured probability for the kth terms of Y in pursuance of Bayes rule is stated as [19]:

$$P(Y = y_k|X_1 \ldots X_n) = \frac{P(Y = y_k)P(X_1 \ldots X_n|Y = y_k)}{\sum_j P(Y = y_j)P(X_1 \ldots X_n|Y = y_j)} \tag{6}$$

Artificial Neural Network (ANN) model

ANN is a computational machine learning model. It has artificial neurons as the basic primary elements. Synapses refer to the small interconnection between neurons in the brain which are denoted by weights that regulate the input signals. Transfer function determines the nonlinear characteristic displayed by neurons. Transfer function transforms the neuron impulse which is evaluated as the weighted summation of input signals. Artificial neuron gains their learning ability by adjustment in weights according to the learning algorithm [13].

K-Means

K-Means is an unsupervised machine learning methodology used to perform clustering which separates a database into n groups. Initially, the process starts by choosing k number of clusters and iterations continue. Each sample is assigned to its neighbor cluster which is closer to it. Cluster centers are evaluated by finding the mean of corresponding samples. The process continues until the algorithm converges, i.e., when until the same result is achieved. Clusters are initialized by choosing the samples randomly. Euclidean distance or Manhattan distance is used in the evaluation of *K*-Means clustering [13].

3.2 Model Diagram

For performing the experiment, breast cancer dataset is used which is collected from UCI repository. The performance of the employed soft computing paradigms for the detection of disease identification is evaluated over various measures like sensitivity, specificity, accuracy, RMSE, etc. Normalization is performed over the dataset which makes the dataset efficient to work. The model diagram (Fig. 1) defines the complete process of the research flow.

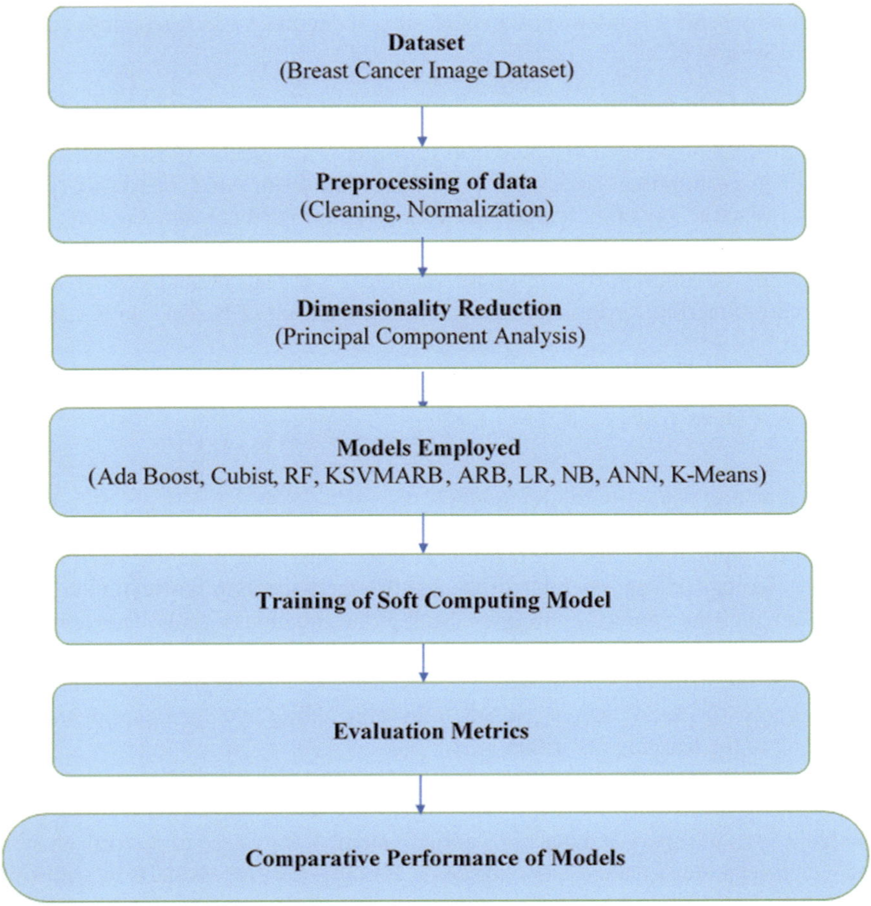

Fig. 1 Model diagram of research flow

3.3 Evaluation Measures

The performance of these soft computing models is evaluated over below-mentioned measures:

- **True Positive (TrP)**: the case when patient is actually suffering from breast cancer and model also classified as breast cancer.
- **False Positive (FaP)**: the case when patient is not suffering from breast cancer and model classified as breast cancer.
- **False Negative (FaN)**: the case when patient is suffering from breast cancer and model classified as no breast cancer.
- **True Negative (TrN)**: the case when patient is not suffering from breast cancer and model also classified as no breast cancer.

- **Accuracy**: It defines correctly diagnosed patients/people identified by the model out of all diagnosed people. It is defined in Eq. (7)

$$\text{Accuracy} = (\text{TrP} + \text{TrN})/(\text{TrP} + \text{TrN} + \text{FaP} + \text{FaN}) \tag{7}$$

- **Sensitivity/Recall/True-Positive Rate**: It can be defined as the ratio of amount of positive data items which are correctly predicted as positive corresponding to all positive data items. It is defined in Eq. (8)

$$\text{Sensitivity} = \text{TrP}/(\text{TrP} + \text{FaN}) \tag{8}$$

- **Specificity/False-Positive Rate/True-Negative Rate**: It can be defined as the ratio of amount of negative data items which are incorrectly predicted as positive corresponding to all negative data items. It is defined in Eq. (9)

$$\text{Specificity} = \text{TrN}/(\text{FaP} + \text{TrN}) \tag{9}$$

- **Negative Predictive Value (NPV)**: It defines the amount of accurately diagnosed healthy patient among all patients detected as healthy. It is defined in Eq. (10)

$$\text{NPV} = \text{TrN}/(\text{TrN} + \text{FaN}) \tag{10}$$

- **False-Negative Rate (FNR)**: It defines the amount of inaccurately diagnosed healthy patients among all breast cancer patients. It is defined in Eq. (11)

$$\text{FNR} = \text{FaN}/(\text{TrP} + \text{FaN}) \tag{11}$$

- **False Discovery Rate (FDR)**: It is calculated as the ratio of inaccurately diagnosed cancer patients among all patients diagnosed as suffering from cancer disease. It is defined in Eq. (12)

$$\text{FDR} = \text{FaP}/(\text{TrP} + \text{FaP}) \tag{12}$$

- **Root-Mean-Square Error (RMSE)**: It evaluates the mean significance of the error. It can be calculated as the square root of the mean of squared differences between the actual observed value and predicted value. It is defined in Eq. (13).

$$\text{RMSE} = \sqrt{\frac{1}{n}\sum_{i=1}^{n}\left(y_i - \hat{y}_i\right)^2} \tag{13}$$

3.4 Dataset Details

For experimentation, the Wisconsin Breast Cancer Database (WBCD) [20] is used in this research study, which is gathered from the UCI repository which classifies malignant (cancerous) from benign (non-cancerous) samples. Table 1 determines the description of the dataset.

During the experiment, 500 records are used for training the dataset and 69 records are chosen for the testing purpose. Table 2 gives a brief description of training and testing dataset.

Attribute Information

The WBCD dataset comprises total 32 attributes: ID, diagnosis, 30 real-valued input features. The following are the ten real-valued features evaluated for each cell nucleus (Table 3):

The above-mentioned features are evaluated from a digitized image of a fine needle aspiration (FNA) of a breast mass.

Table 1 Dataset description

Dataset used	Total attributes	Total instances	Total classes
Wisconsin Diagnosis Breast Cancer (WDBC)	32	569	2

Table 2 Training and testing dataset details

Dataset used	Data type	Total records
Breast cancer dataset	Training	500
	Testing	69

Table 3 Feature description

S. No.	Features	Description
1	ra	Average of distances from the center to points on the perimeter
2	te	Standard deviation of grayscale values
3	pe	Perimeter
4	ar	Area
5	sm	Local variation in radius lengths
6	cos	$\text{Perimeter}^2/\text{area} - 1$
7	con	The severity of concave portions of the contour
8	Concave	Number of concave portions of the contour
9	sy	Symmetry
10	Fractal	Coastline approximation − 1

4 Experimental Results

R language is used in this research work for performing the experiment. R packages are being utilized for the implementation of machine learning algorithms using the breast cancer dataset which is collected from UCI repository. Table 4 gives the description of packages of R, techniques and personalized parameter employed during designing of the soft computing models.

Table 5 describes the performance evaluation of 11 machine learning classifiers on training dataset which are being compared over various measures like sensitivity, specificity, TrP, FaP, FaN, TrN, NPV, FNR, FDR, RMSE and accuracy. Abbreviations are defined in Sect. 3.3.

Likewise, Tables 5 and 6 describe the performance evaluation of 11 machine learning classifiers on a testing dataset which are being compared over various measures like sensitivity, specificity, TrP, FaP, FaN, TrN, NPV, FNR, FDR, RMSE and accuracy.

Tables 5 and 6 represent the evaluation results of all classifiers, among which AdaBoost, logistic regression and 1-NN neighbor classifiers outperform. These evaluation results are measured by sensitivity, specificity and accuracy. It can be clearly seen from Table 5 that AdaBoost and logistic regression give 100% result over the sensitivity, specificity and accuracy measures while 1-NN gives 95% sensitivity, 100% specificity and 98% accuracy.

Figure 2 shows the comparative evaluation of all classifiers over average accuracy, sensitivity and specificity of the soft computing exemplars on training datasets, while Fig. 3 shows comparative evaluation of average RMSE of the soft computing exemplars on training datasets.

Figures 4 and 5 represent comparative evaluation of all classifiers over average accuracy, sensitivity and specificity of the employed soft computing exemplars on testing datasets along with the RMSE.

Table 4 R packages and methods

Models used	R package/technique	Personalized parameters
AdaBoost	Ada	Boosting iterations = 250 Complexity of trees = 0.01 Max. depth of trees = 30
C5.0	C50	Default parameter values taken
CT	Rpart	Default parameter values taken
RF	Random forest	No. of subtrees = 500 Randomly different sampled candidates at each split = 3
SVM	Ksvm	Kernel = vanilla dot
ANN	NET	No. of segments in hidden layer = 15 Max. number of likelihood iterations = 500
k-NN	k-NN	$k = 5$
1-NN	k-NN	$k = 1$

Table 5 Evaluation performance measures of the employed soft computing exemplars on training dataset

Method	TrP	FaP	FaN	TrN	Sensitivity	Specificity	NPV	FNR	FDR	RMSE	Accuracy
AdaBoost	218	0	0	366	1	1	1	0	0	0	1
C5.0	217	13	1	353	0.9954	0.9645	0.9972	0.0046	0.0565	0.1548	0.976027
Rpart	210	13	8	353	0.9633	0.9645	0.9778	0.0367	0.0583	0.1853	0.964041
Cubist	217	8	1	358	0.9954	0.9781	0.9972	0.0046	0.0356	0.1165	0.984589
Cubist_Committees	218	9	0	357	1	0.9754	1	0	0.0396	0.1031	0.984589
Random forest	218	0	0	366	1	1	1	0	0	0	1
Ksvm linear	212	11	6	355	0.9725	0.9699	0.9834	0.0275	0.0493	0.1706	0.97089
KsvmAnova radial basis	217	4	1	362	0.9954	0.9891	0.9972	0.0046	0.0181	0.0925	0.991438
Logistic regression	208	10	10	356	0.9541	0.9727	0.9727	0.0459	0.0459	0.1543	0.965753
Naïve Bayes	213	18	5	348	0.9771	0.9508	0.9858	0.0229	0.0779	0.1985	0.960616
Neural network	217	0	1	366	0.9954	1	0.9973	0.0046	0	0.0338	0.998288
K-nearest neighbor	214	10	4	356	0.9817	0.9727	0.9889	0.0183	0.0446	0.1548	0.976027
1-nearest neighbor	218	0	0	366	1	1	1	0	0	0	1

Table 6 Evaluation performance measures of the employed soft computing exemplars on testing dataset

Method	TrP	FaP	FaN	TrN	Sensitivity	Specificity	NPV	FNR	FDR	RMSE	Accuracy
AdaBoost	21	0	0	78	1	1	1	0	0	0	1
C5.0	20	4	1	74	0.9524	0.9487	0.9867	0.0476	0.1667	0.2247	0.949495
Rpart	20	4	1	74	0.9524	0.9487	0.9867	0.0476	0.1667	0.2152	0.949495
Cubist	19	2	2	76	0.9048	0.9744	0.9744	0.0952	0.0952	0.187	0.959596
Cubist committees	21	2	0	76	1	0.9744	1	0	0.087	0.1257	0.979798
Random forest	21	1	0	77	1	0.9872	1	0	0.0455	0.1005	0.989899
Ksvm linear	21	2	0	76	1	0.9744	1	0	0.087	0.1421	0.979798
KsvmAnova radial basis	21	3	0	75	1	0.9615	1	0	0.125	0.1741	0.969697
Logistic regression	21	0	0	78	1	1	1	0	0	5.7401	1
Naïve Bayes	21	2	0	76	1	0.9744	1	0	0.087	0.1421	0.979798
Neural network	20	2	1	76	0.9524	0.9744	0.987	0.0476	0.0909	0.1498	0.969697
K-nearest neighbor	20	1	1	77	0.9524	0.9872	0.9872	0.0476	0.0476	0.1421	0.979798
1-nearest neighbor	20	0	1	78	0.9524	1	0.9873	0.0476	0	0.1005	0.989899

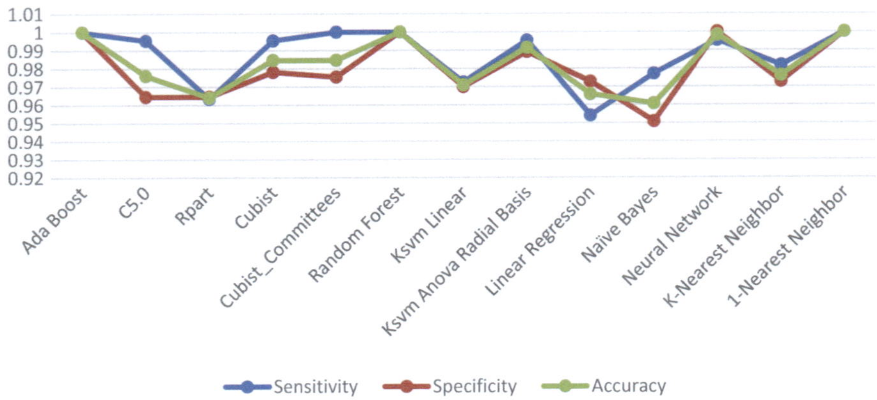

Fig. 2 Comparison of average accuracy, sensitivity, specificity of the soft computing models on training datasets

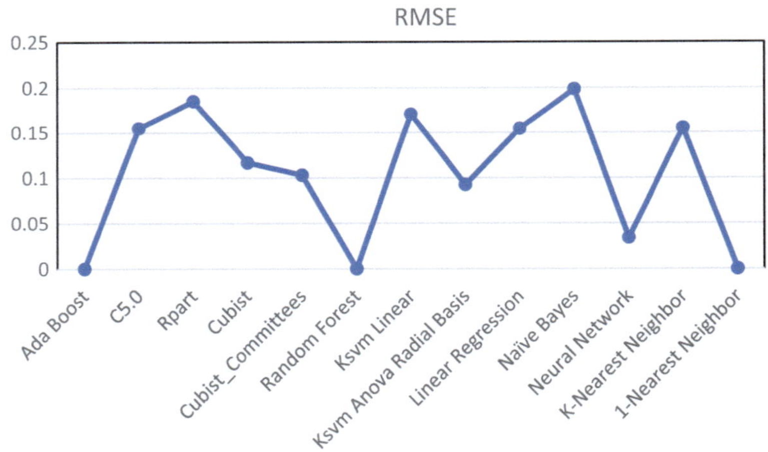

Fig. 3 Comparison of average RMSE of the soft computing exemplars on training datasets

5 Conclusion

Since breast cancer disease is hard to predict, this research work focuses on the key role of classification algorithms in predicting breast cancer diseases. Machine learning algorithms are being chosen as they gain their strength in classifying the instances correctly once the system is trained while requiring less human efforts and time. Thirteen models with their corresponding classifiers are used in conducting the research. For experimentation purpose, breast cancer database is chosen which is taken from UCI machine learning repository. Results showed that AdaBoost, logistic

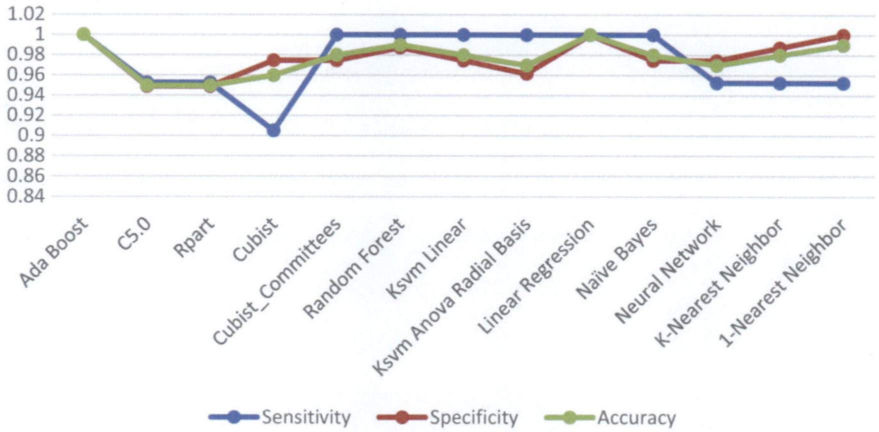

Fig. 4 Comparison of average accuracy, sensitivity, specificity of the soft computing models on testing datasets

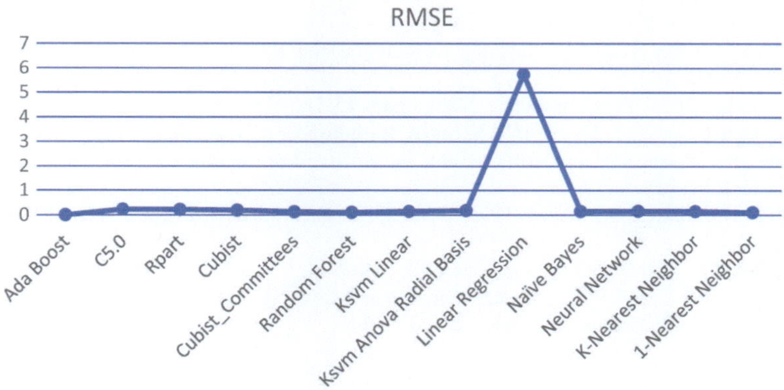

Fig. 5 Comparison of average RMSE of the soft computing models on testing datasets

regression and 1-NN neighbor classifiers perform well from training and testing also. Final results showed that 100% accuracy is achieved from classifiers.

References

1. Rouhi R, Jafari M (2015) Classification of benign and malignant breast tumors based on hybrid level set segmentation. Expert Syst Appl
2. Andina D (2011) Expert systems with applications WBCD breast cancer database classification applying artificial metaplasticity neural network. Expert Syst Appl 38:9573–9579
3. Bhardwaj A, Tiwari A (2015) Breast cancer diagnosis using genetically optimized neural network model. Expert Syst Appl

4. B. C. A. T. Review, Breast cancer, pp 79–81 (2010)
5. Azami H, Member S, Escudero J (2015) A comparative study of breast cancer diagnosis based on neural network ensemble via improved training algorithms, pp 2836–2839
6. Aswathy MA, Jagannath M (2016) Informatics in medicine unlocked detection of breast cancer on digital histopathology images: present status and future possibilities. Inf Med Unlocked pp 0–1
7. Janghel RR, Shukla A (2016) Expert system for breast cancer diagnosis using ensemble approach. Res Article 1(1):1–7
8. Jouni H et al (2016) Neural network architecture for breast cancer detection and classification, pp 4–8
9. Abdel-zaher AM, Eldeib AM (2016) Breast cancer classification using deep belief networks. Expert Syst Appl 46:139–144
10. Janghel RR, Shukla A, Sharma S, Gnaneswar AV (2014) Evolutionary ensemble model for breast, pp 8–16
11. Janghel RR, Shukla A, Tiwari R (2012) Hybrid computing based intelligent system for breast cancer diagnosis. Int J Biomed Eng Technol 10(1):1–18
12. Schapire RE (1999) A brief introduction to boosting. IJCAI Int Jt Conf Artif Intell 2(5):1401–1406
13. Hastie T, Tibshirani R, Friedman J (2009) Neural networks, pp 1–28
14. Galathiya A, Ganatra A, Bhensdadia C (2012) Improved decision tree induction algorithm with feature selection, cross validation, model complexity and reduced error pruning. Int J Comput Sci Inf Technol 3(2):3427–3431
15. Kuhn M, Coulter N (2012) Cubist models for regression. R Packag Vignette R Package Version 0.0 18(1992)
16. Breiman L (2001) Random forests. Mach Learn 45(1):5–32
17. Janghel RR, Shukla A, Rathore CP, Verma K, Rathore S (2017) A comparison of soft computing models for Parkinson's disease diagnosis using voice and gait features. Netw Model Anal Health Inf Bioinf 6(1):14
18. Nelder JA, Wedderburn RWM (1972) Generalized linear models. J R Stat Soc A 135(3):370–384
19. Mitchell TM (1997) No title. McGraw-Hill, New York
20. Areerachakul S et al (2012) Breast cancer diagnosis on three different datasets using multi-classifiers. Int J Comput Inf Technol 1(4):236–241

Data and Web Mining

Couplets Translation from English to Hindi Language

Anshuma Yadav, Rajesh Kumar Chakrawarti and Pratosh Bansal

Abstract We are a part of that society which has different-different cultures, languages, rituals, etc. There are some sentences which we used to say someone at particular situation or to motivate someone which we called as couplets. A couplet is a pair of consecutive lines of rhythm in poetry. Couplet usually consists of two consecutive lines that poem has the same rhythm. Couplets are balanced in length and word segmentation, in a manner that matching words in the two consecutive sentences match each other by observing certain limitation on semantic, syntactic, and lexical connection. In this research, authors have tried to translate few couplets from English to Hindi language and this kind of task comes under the umbrella of natural language processing. In continuation of the same, authors have used a statistical machine translation (SMT) approach for producing good results (of English to Hindi translation of couplets).

Keywords Machine translation (MT) · Language · Couplets · Statistical MT · Natural language processing (NLP)

1 Introduction

In today's scenario, machine translation is a difficult task because natural languages are very complex to translate; many words have different meanings and have their different probable solutions that are why we are proposing a type of system which

---

A. Yadav
Department of Computer Science and Engineering, SVIIT, SVVV, Indore, Madhya Pradesh 453111, India
e-mail: anshuma.yadav94@gmail.com

R. K. Chakrawarti (✉)
Department of Computer Engineering, IET, DAVV, Indore, Madhya Pradesh 452017, India
e-mail: rajesh_kr_chakra@yahoo.com

P. Bansal
Department of Information Technology, IET DAVV, Indore, Madhya Pradesh 452017, India
e-mail: pratosh@hotmail.com

R. K. Shukla et al. (eds.), *Social Networking and Computational Intelligence*, Lecture Notes in Networks and Systems 100, https://doi.org/10.1007/978-981-15-2071-6_23

converts the English couplets into Hindi language. The couplets may be an independent poem lines and might be a portion of other poems, like sonnets in Shakespearean poetry. If a couplet has a different meaning from the rest of the poem, and if it is independent, then it is called a closed couplet. If a couplet cannot have proper meaning without the rest of the poem, then it is called an open couplet. The consecutive lines (couplets) might have same but the meaning and the emotions may be different. In our Indian culture, we call couplets like Dohas, and in India, everyone is aware of couplets of Indian languages, but sometimes we do not know about the foreign couplets. These couplets sometimes may have deep and important meanings.

Foreign couplet example: "*Blessed are you whose worthiness gives scope, being had, to triumph, being lacked, to hope*." It means—you are very blessed that you have that much worthiness which gives you opportunities for win something had to be week. This is an example of couplet from foreign writer William Shakespeare. There are many more this kind of couplets given by foreign writers.

Translation of couplets is similar to translating one natural language into another natural language, but the target language should satisfy some linguistic constraints. For this kind of translation we want to use statistical machine translation for conversion of the language under development are arrhythmia monitors for ambulatory patients which analyze the ECG in real time [1–3]. Software QRS detectors typically include one or more of three different types of processing steps: linear digital filtering, nonlinear transformation, and decision rule algorithms [4].

2 Literature Survey

A. A research done by "*Long Jiang and Ming Zhou*" on Paper entitled "*Generating Chinese Couplets using a Statistical MT Approach.*"

   Authors [1] have presented a different approach to solve the problem of generating Chinese couplets. An SMT approach is proposed to make the "Second Sentence" for a "First Sentence" of a Chinese couplet. The system consists of a phrase-based SMT model for the generation of an N-best list of Second Sentence candidates, a set of linguistic filters to remove absolute candidates to meet the special constraints of Chinese couplets, and a discriminative re-ranking model incorporating multidimensional features to get better results.

B. A research done by "*H. S. Sreedeepa and Dr. Sumam Mary Idicula*" on paper entitled "Interlingual *based Sanskrit-English translation.*"

   Authors [3] have introduced a machine translation system for Sanskrit to English. In their proposed work, they use rule-based machine translation because it has interlingual approach. For their research work, they show the used algorithm, working of corpus, and also show interlingual representation of the sentences. They tokenize and f-structures to sentences created for the tokens. They also stated that the interlingual is a language independent approach and this approach can be used for any other natural language translation.

C. A research done by "*Franz Josef Och and Hermann Ney*" on paper entitled "*Alignment Tamplet approach to statistical Machine Translation.*"

Authors [4] have used statistical machine translation to perform different natural language translation. They perform German to English speech verbmobil task, and they also mentioned effects of various system components. For the other one French to English Canadian hansard task, they also conclude that alignment template gets better results.

3 Machine Translation

Machine translation is an automatic translation. It is a process, and this process takes place with the help of computer software which helps to translate one natural language to another natural language [5].

3.1 Machine Translation Process

Machine translations steps are shown in Fig. 1 follows

Step 1—Source Text

A source text is just a text from a particular language which we want to translate in targeted language.

Step 2—Morphological Analysis, Syntactic Analysis, and Semantic and contextual Analysis.

In this stage, three parts have been covered as follows: The first is morphological analysis, which dictates the word forms such as tense, number, and parts of speech; the second part is syntactic analysis, which dictates the word is subject or object; and the third and final part is contextual and semantic analysis, which dictates a correct version of a sentences from the result that is produced by the syntactic analysis. Semantic and syntactic are frequently executed at the same time and produce syntactic tree structure and semantic network.

Step 3—Internal Representation of source language translation to internal representation of target language.

In this stage, the overall representation of the source translation is performed in this phase.

Step 4—Contextual and Semantic generation and Syntactic generation.

In this stage, the semantic analysis composes meaningful representation and allots them into linguistic inputs. These analyzers use dictionaries and grammar to make

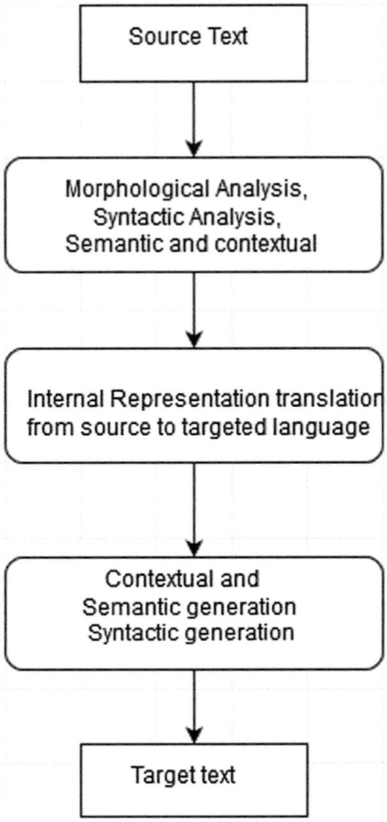

Fig. 1 A machine translation process

core meaning of words. The root of knowledge belongs to the meaning of words. Syntactic analysis determines the way words are grouped into category is called parts of speech, the way they grouped with their neighbors into phrases.

Step 5—Target text

In this stage, the final outcome of the previous stages of machine translation process came out.

3.2 Types of Machine Translation

Machine translation has different types of translation methods [2] which are as follows

1. Rule-based Machine Translation;

2. Example-based Machine Translation;
3. Statistical Machine Translation;
4. Neural Machine Translations;
5. Hybrid Machine Translation.

3.2.1 Rule-Based Machine Translation (RBMT)

RBMT is another method for language translation. Usually, this method uses a dictionary and grammar system. This method needs much awareness about the semantics of natural language.

Rule-based machine translation also has some types which are as follows

a. Transfer-based machine translation;
b. Interlingual;
c. Dictionary.

3.2.2 Example-Based Machine Translation (EBMT)

EBMT is a language translation method. Many times, it distinguishes by its use of a bilingual corpus with parallel texts.

3.2.3 Statistical Machine Translation (SMT)

SMT is a specific approach where the machine translation is happened on the basis of statistical model and facts are copied from the analysis of bilingual text corpora. Corpora are usually called treebanks and parsed corpora.

3.2.4 Neural Machine Translation

In this machine translation method, this method uses neural network prototype to learn and train statistical model for machine translation.

3.2.5 Hybrid Machine Translation

Hybrid machine translation is a method of machine translation which uses multiple machine translation methods in the single machine translation system.

4 Objectives

The objectives of this research are as follows:

a. To provide an accurate translation of couplets and also systematic result of couplets, we want to translate in Hindi in that way which helps to understand the meaning of a particular couplet.
b. To resolve the problem of translating ambiguous (which not having one clear meaning) sentences (couplets).
c. To provide knowledge to the world through our work. With help of this system, people get some knowledge about the foreign couplets in Hindi language.

5 Problem Domain

In a translation world, there were various systems and platforms are provided by much organization but there are not such system for translation of couplets or not that much accurate to translate couplet. The related example is shown in Fig. 2.

Existing System

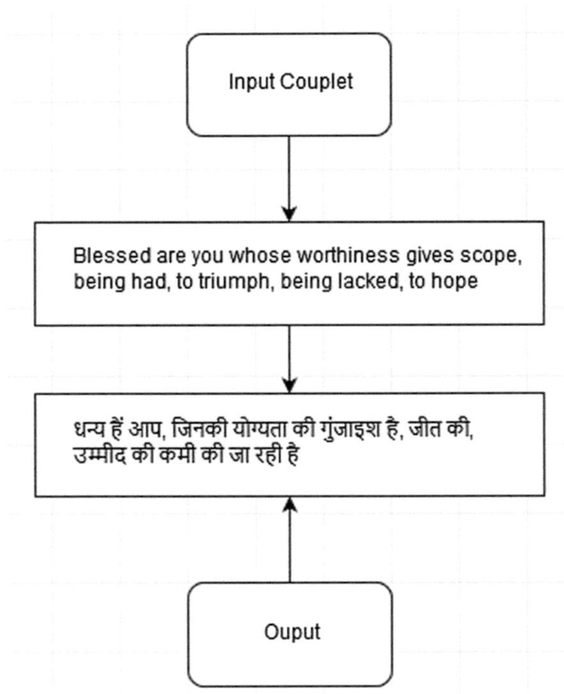

Fig. 2 An example of existing system

a. They do some mistake or the expected outcome comes with ambiguity.
b. Results are similar to the input but not acceptable.

Some translation services use their own immense databases, but seem to have limitations in dealing with some terminologies.

6 Solution Domain

To perform an accurate translation, we want to use SMT. Statistical machine translation is a specific approach where the machine translation takes place with the help of statistical model and facts are copied from the analysis of parallel text in corpora. Corpora are usually called treebanks and parsed corpora. Corpora are usually smaller, containing around 3 million words. With the help of this solution, I am providing much accurate solution than existing system.

In Fig. 3, the two models are working paralleled with the statistical analysis which is process the source text to the targeted text. This figure consists of three stages which are as follows

a. Insertion of source text.
b. Both the sides have their different models like they use translation model and language model. And both the model working in a statistical analysis.

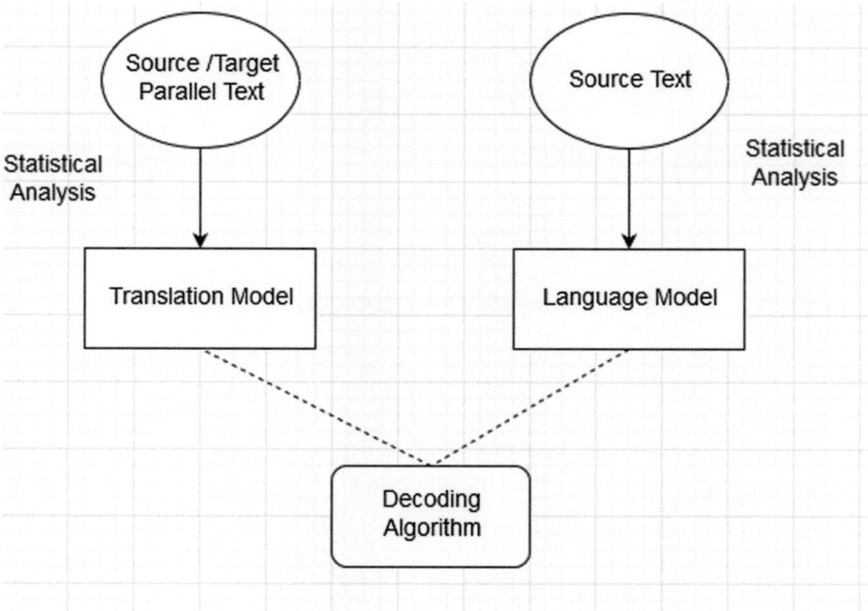

Fig. 3 Statistical machine translation (SMT) process diagram

c. In this phase of both the models, decoding algorithm is working for both the sides.

A System Architecture

This is a kind of system architecture. Right now, it is a very simple flow chart of system architecture. With the help of this flow chart, we can understand work flow easily. In Fig. 4 is giving us brief knowledge about the system working or a flow of a system which is starts from the input of a couplet till couplet translation from English to Hindi language.

a. In first step, we have to enter the couplet in the system.
b. In second step, couplet goes through a comparison phase where the words are going to be paired with their simple meaning with the help of English database.
c. In third step system, use the algorithm for couplet translation and processed this translation in our targeted language.

In final step, our expected outcome is process in targeted language.

These are some stages of the system which we are going to use in a translation.

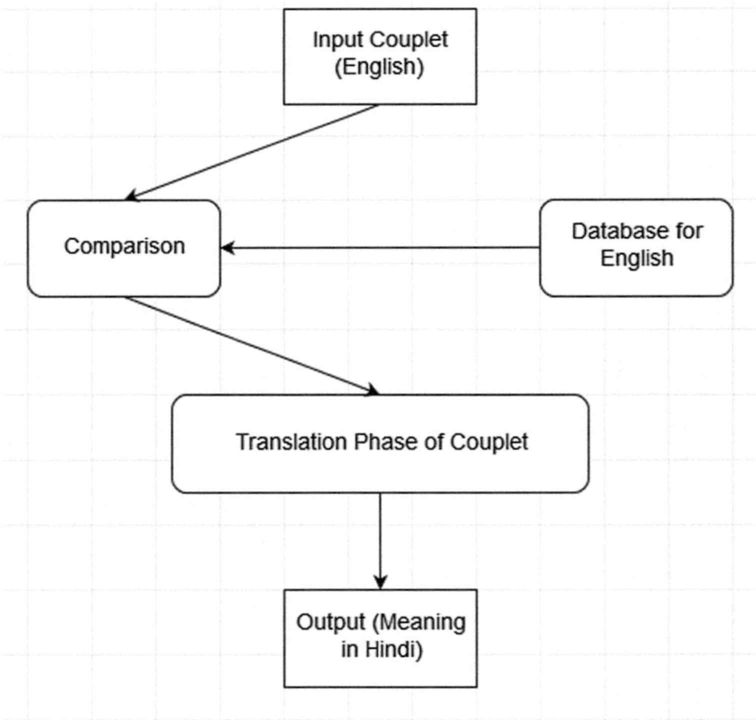

Fig. 4 System architecture

Table 1 Hardware/software requirement table

S. No.	Components	Specification
1	RAM	2 or 4 GB
2	Processor	Intel® Core™ i3–4005U@1.70 GHz
3	Operating system	Windows XP or higher version
4	Secondary storage	500 GB
5	Front end	Python
6	Back end	Microsoft SQL

7 System Domain

For the system domain, we analyze some tools and hardware requirements are shown in Table 1.

8 Application Domain

Our research work will be useful for students, authors, and those who want to connect with foreign legacy.

a. It will used in many fields like rural areas, people who only understand Hindi and wants to explore knowledge in case of wants to understand the meaning of the couplets.

It will also be used in academics for the students to connect with other culture, in literature for the authors, and in novels.

9 Benefits of SMT

The benefits of statistical machine translation (SMT) are as follows:

a. More effective use of human and data resource.
b. SMT gives a manner of automatically finding association between the properties of two languages from a parallel corpus.
c. SMT also uses less virtual space than any other models of machine translation, and it is easier to operate and train on smaller system.
d. It is more fluent translation system to use of language model.

With the help of SMT, we can pair different natural language very easily.

10 Expected Outcome

The expected outcome of the research will be as follows:

a. The proposed system helps to know about actual meaning of English couplets.
b. This system will help to overcome the problem of translating ambiguous sentences (couplets).

Acknowledgements I want to thank my guide, Rajesh Kumar Chakrawarti (Assistant Professor, Department of CSE, SVIIT, SVVV Indore and PhD Scholar, Department of Computer Engineering, IET DAVV Indore), and my special thanks to Dr. Pratosh Bansal (Professor, Department of IT, IET DAVV, Indore) for his guidance and support, which helped me throughout the work, and for giving me this opportunity to do this research and belief on me. I am also thankful to my department (CSE SVIIT, SVVV Indore).

References

1. Jiang L, Zhou M (2008) Generating Chinese couplets using a statistical MT approach. In: 22nd international conference on computational linguistics (Coling), Manchester
2. Chakrawarti RK, Mishra H, Bansal P (2017) Review of machine translation techniques for idea of Hindi to English idiom translation. Int J Comput Intell Res 13(5)
3. Sreedeepa HS, Idicula SM (2017) Interlingua based Sanskrit-English translation. In: International conference circuits powers and computing technology (ICCPCT), Kollam, India
4. Och FJ, Ney H (2004) The alignment template approach to statistical machine translation. Comput Linguist 30(4):417–449
5. Chakrawarti RK, Bansal P (2017) Approaches for Improving Hindi to English Machine Translation System. Indian J Sci Technol 10(16)
6. https://literarydevices.net/couplet/
7. http://language.worldofcomputing.net/machine-translation/machine-translation-process.html
8. www.draw.io (flowchart maker & online diagram software)

A Novel Approach for Predicting Customer Churn in Telecom Sector

Ankit Khede, Abhishek Pipliya and Vijay Malviya

Abstract Predicting customer churn in telecommunication industries becomes a most important topic for research in recent years. Because its helps in detection that client square measure possible to alter or cancel their subscription to a service. Analysis of knowledge that is extracted from medium corporations will helps to seek out the explanations of client churn and additionally uses the knowledge to retain the shoppers. So predicting churn is very important for telecom companies to retain their customers. So data mining techniques and algorithm play an important role for corporations in today's industrial conditions as a result of gaining a replacement customer's value is quite retentive the present ones. In this paper, we can focus on machine-learning techniques for predicting customer churn through which we can build the classification models such as SVM and random forest and also compare the performance of these models.

Keywords Churn prediction · Data mining · Telecom system · Customer retention · Classification system · Random forest · SVM

1 Introduction

In today's technical situations, data is an organism created by wholly completely diverse bases in many segments. However, it is uphill to abstract the useful information concealed in these data groups, they are treated suitably. Thus on dig up this unseen information, varied examines should to be completed victimization processing that comprises numerous ways [1].

A. Khede (✉) · A. Pipliya · V. Malviya
Malwa Institute of Technology, Indore, India
e-mail: Erankitkhede@gmail.com

A. Pipliya
e-mail: aadeepipliya@gmail.com

V. Malviya
e-mail: vijaymalviya@gmail.com

R. K. Shukla et al. (eds.), *Social Networking and Computational Intelligence*, Lecture Notes in Networks and Systems 100, https://doi.org/10.1007/978-981-15-2071-6_24

The churn analysis [2] goals to forecast clients unit around to stop employing an invention or facility among the shoppers. Today's inexpensive situations rectifier toward varied firm's commerce identical product at quite a similar service and merchandise quality.

With churn analysis [3], it stays attainable to exactly foretell clients stay planning to break mistreatment facilities or merchandise by assignment likelihood to every consumer. The study is often accomplished in line with client pieces and quantity of loss (monetary equivalent). Succeeding these studies, message with clients is often improved so as to steer the shoppers and growth consumer reliability. Operative selling operations for goal clients are often produced by hard churn rate or consumer abrasion. During this means, profit is often raised considerably or the attainable harm because of client loss is often reduced at constant rate. As a sample, if a facility supplier that contains overall of two million subscribers, improvements 750,000 new subscribers and lost 275,000 consumers; churn frequency is planned as 100%. The client churn rate contains vital result on monetary rate of the corporate. Thus, most of the businesses retain a watch on worth of the client at regular or three-monthly periods.

2 Related Work

According to Li et al. [4], from the beginning of the data mining [5] which is used to discover new knowledge's from the databases can help in various problems and in the business for their solutions. Telecom companies improve their revenue by retaining their customers. Customer churn in telecom sector is to leave a one subscription and join the other subscription. In these paper, they predict the customer churn by using various R packages and they created a classification model and they train by giving him a dataset and after training, they can classify the records into churn or non-churn and then they visualize the result with the help to visualization techniques [6]. In this, they are using logistic regression model and these models first train on training data after that they can test the model on test data to compute the performance measure of the classification model, so we can get the various parameters like true positive rate, false positive rate, and accuracy.

According to Wang et al. [7], telecom customer churn prediction is a cost-sensitive classification problem. Most of studies regard it as a general classification problem using traditional methods, that the two types of misclassification cost are equal. And, in aspect of cost-sensitive classification, there are some researches focused on static cost-sensitive situation. In fact, customer value of each customer is different, so misclassification cost of each sample is different. For this problem, we propose the partition cost-sensitive CART model in this paper [8]. According to the experiment based on the real data, it is showed that the method not only obtains a good classification performance, but also reduces the total misclassification costs effectively.

According to Dahiya et al. [9], customer churn plays an significant part in customer relationship management (CRM), and they are using various machine-learning algorithms to calculate customer churn and they found collective knowledge is best to predict client churn, however, around exist tons of issues like, however, they select the strategy of integration and the way to settle on the strategy, that makes the ultimate ensemble classifier. On the other hand, there is no good classifier, so it is also a main problem to chosen which classification algorithm is best for which situation. So we can consider various aspects like vertical and horizontal contrasts to find the best classifier to forecast the consumer churn in telecommunication sector [10].

3 Problem Definition

From the problems obligatory through market saturation and value implications, there has been associate identification of a desire for a computer based mostly churn prediction methodology that is capable of accurately distinctive a loss of client ahead, so proactive retention ways is deployed during a bid to retain the client. The churn prediction should be correct as a result of retention ways is pricey [11]. A limitation of current analysis is that alternative studies have focused virtually solely on churn capture, neglecting the problem of misclassification of non-churn as churn. Retention campaigns usually embrace creating service based mostly offers to customers during a bid to retain them [12].

4 Proposed Work

In the R [13] software development are wont to build prototypical for churn prediction. It's extensive used amongst theoreticians and information sappers for emerging applied mathematics package and information study. R is spontaneously obtainable and robust applied mathematics study tool that consumes not nonetheless been explored for structure models for churn prediction [14] (Fig. 1).

In this paper, we proposed different machine-learning algorithms to analyze customer churn analysis. Through which multiple different models are employed to accurately predict those churn customers in the dataset. These models are support vector machine and random forest.

Algorithm phases as follow:

1. Dataset: Telecom dataset is taken for predicting churn which is to identify trends in customer churn at a telecom company and the data which we have taken are in .csv format. The data given to us contains 7043 observations and 21 variables extracted from a datasets.
2. Data preparation: The dataset is no heritable and not pragmatic onto the churn estimate prototypes, thus we will be naming every attributes.

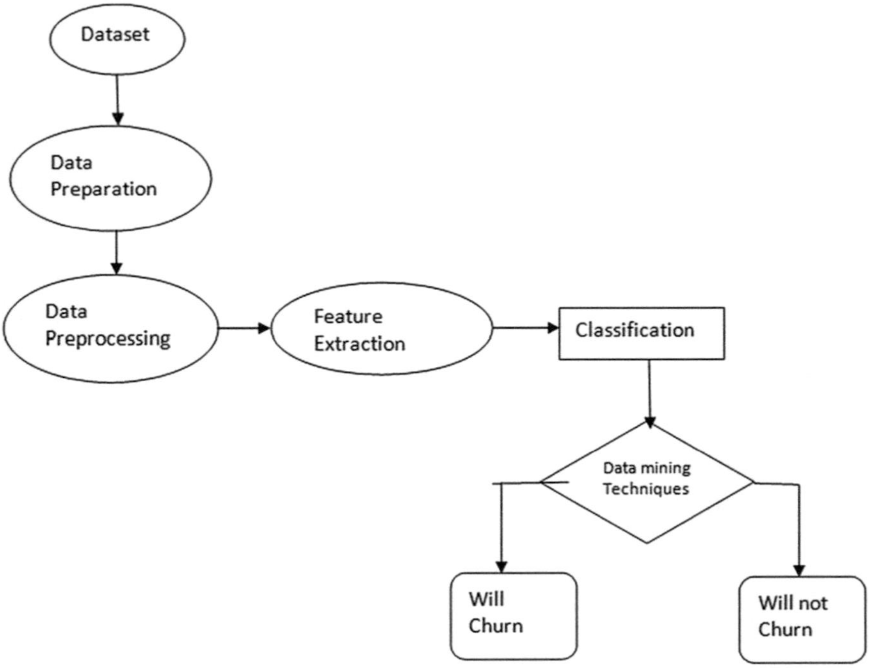

Fig. 1 Churn forecast structure

3. Data preprocessing: Information processing is that greatest significant introduce forecast models because the information contains of uncertainties, errors, redundancy, and transformation that must be cleansed beforehand.
4. Data Mining: The elements area unit is known for categorizing method.
5. Decision: Based on data extraction and classification models, we can take a decision whether the employee is churner or not.

5 Result Analysis

All the tests were done by an i5-2410M microprocessor @ 2.30 GHz CPU and 4 GB RAM running MS Windows. At the moment we will install R and Rstudio and then to identify trends in customer churn at a telecom company. The data given to us contains 7043 observations and 21 variables extracted from a data warehouse. These variables are shown in Fig. 2.

Now, we started to explore the data and cleaning the data for machine-learning models; we can explore the data by their multiple attributes such as average monthly charge of those who churned and average total charge of those who churned and also

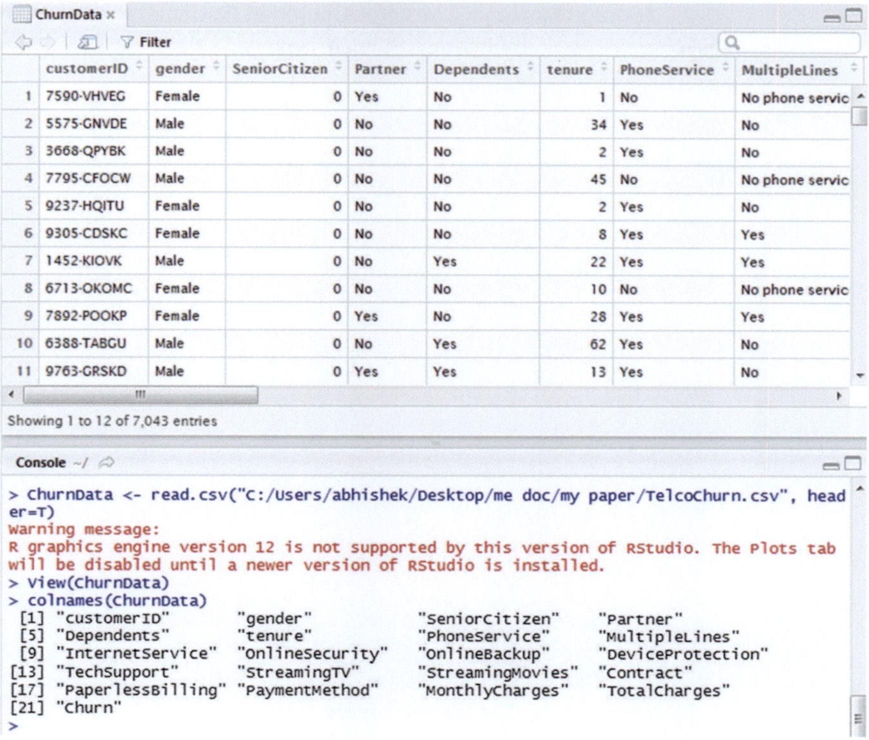

Fig. 2 Variables or sample values in datasets

visualize these exploring result such as average monthly charges by Internet service types for churned customer which is shown in Fig. 3.

After that, we can find the correlation between these categorical variables. Fig. 4 shows the relation between these variables.

These graphs explore relationships between categorical variables and which look like the only highly correlated variables are "streaming movies" and "streaming tv" which is expected.

Preparation for the Model Building

Now, we can build machine-learning models, such as SVM and random forest and then, we train these classifiers and after training, we can test these models and compare their performance. Before starting training, we can perform variable selection: We know that we should not include "streaming movies" and "streaming tv" in the same equation. Their correlation from the above section is fairly high.

Now, we can split the data into train (75%) and test (25%) datasets and then start learning of model on these train data and we can test the model on test dataset and compute model measures. Figure 5 shows the computing performance of these two machine-learning models.

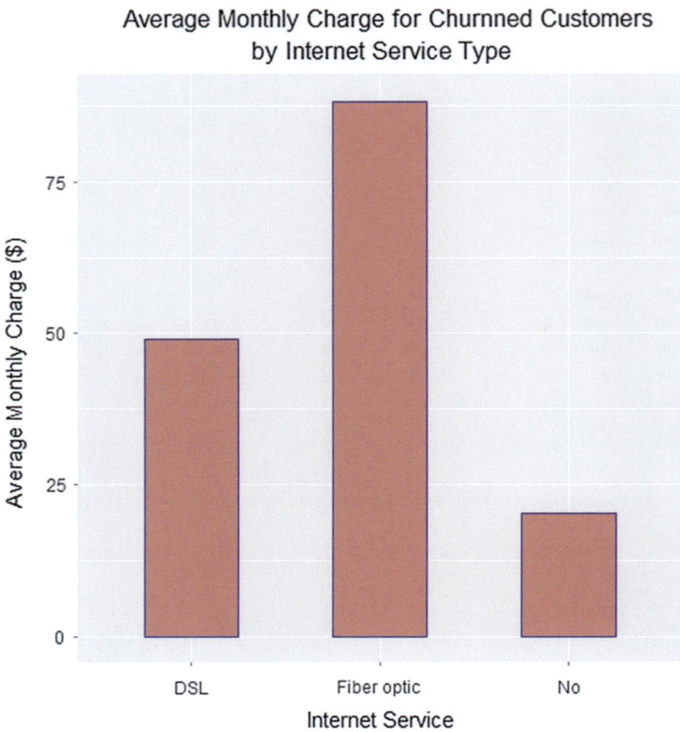

Fig. 3 Average monthly charges by Internet service types for churned customer

According to the AUC values which we have computed, the method that gives us the most accurate model is random forest with AUC value of 81.24%.

ROC curve shows the tradeoff between sensitivity and specificity: The measure of **sensitivity** is that the percentage of helpful samples that were properly classified. The live of **specificity** is the amount of negative samples that were properly classified.

AUC: The area underneath the curve. It is adequate to the chance that classifier can rank an indiscriminately selected helpful case over an indiscriminately chosen negative one. Figure 7 shows the AUC curve of these four models (Figs. 6 and 8).

According to the AUC curves, the method that gives us the most accurate model is random forest with AUC value of 81.24%.

Testing the Models

After the models get trained on the training datasets, we test these models on the test datasets. These models predict the labels as churn or not churn and compare these predicted outcomes to the actual outcomes and we get the performance measures of these models. The testing results of both the models are shown in Figs. 9 and 10.

Based on the AUC curve and the performance measures of both the models, we can compare the performance of the models and the conclusion of the performance as shown in Fig. 11.

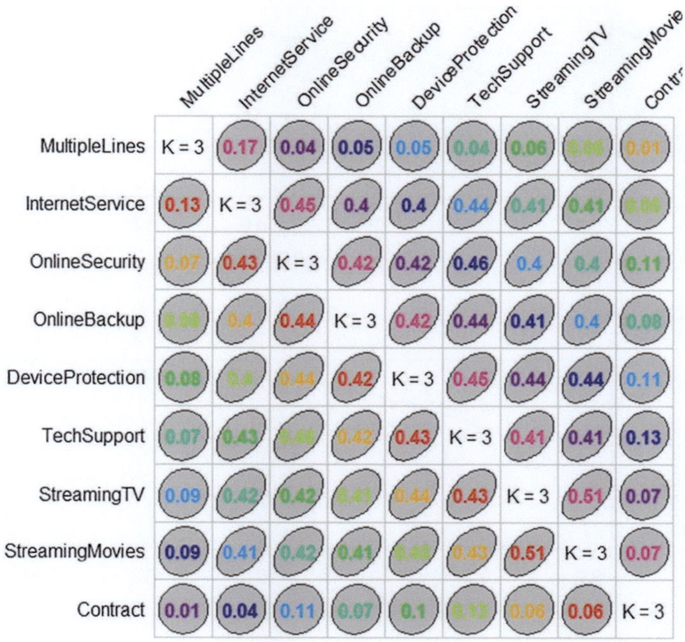

Fig. 4 Correlation between categorical variables

```
> ROCRperfwholeRandom <- performance(ROCRpredwholeRandom,'tpr','fpr')
> plot(ROCRperfwholeRandom)
> aucwholeRandom <- performance(ROCRpredwholeRandom,measure='auc')
> aucwholeRandom <- aucwholeRandom@y.values[[1]]
> aucwholeRandom
[1] 0.812689
> m <- matrix(c(aucSVM,aucwholeRandom),nrow=2,ncol=1)
> colnames(m) <- c("AUC Value")
> rownames(m) <- c("SVM","Random Forest")
> m
              AUC Value
SVM           0.7975093
Random Forest 0.8126890
> |
```

Fig. 5 AUC values of models

6 Conclusion

In order to retain existing customers, telecommunication suppliers have to be compelled to understand the explanations of churn, which may be realized through the data extracted from telecommunication data. In this paper, we train two machine-learning models which are SVM and random forest and we can say that random

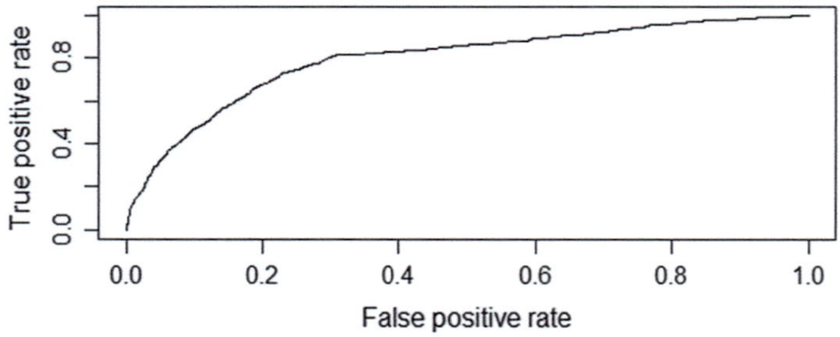

Fig. 6 AUC curves of SVM

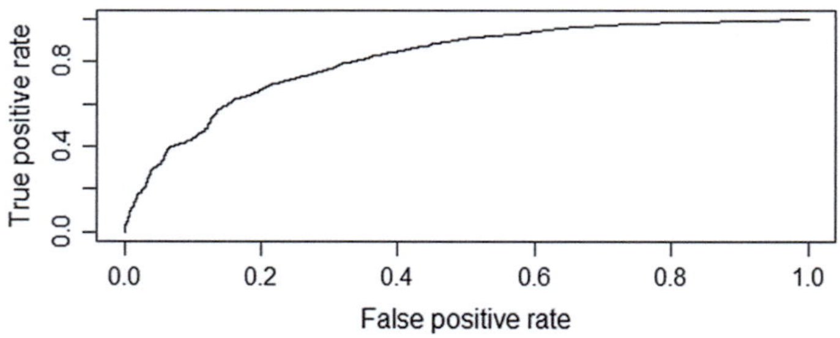

Fig. 7 AUC curve of random forest

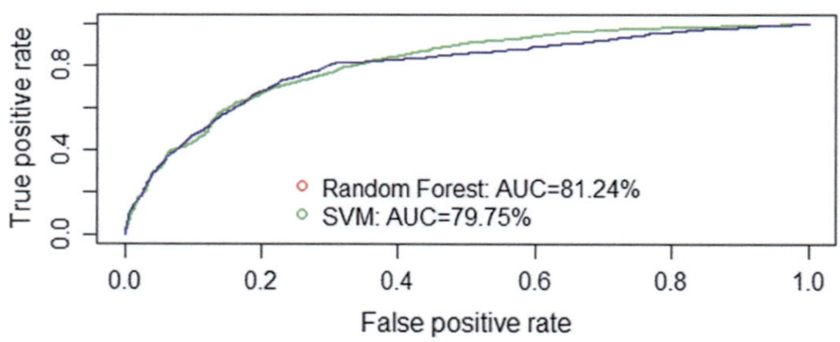

Fig. 8 AUC of both the models

```
> aucSVM
[1] 0.7975093
> predictions_step <- predict(trainSVMmodel1, Testset, type = "response")
>
>
> pred_step <- prediction(predictions_step, Testset$Churn)
> plot(performance(pred_step, "tpr", "fpr"), colorize = TRUE)
>
> roc_step <- roc(response = Testset$Churn, predictor = predictions_step)
> c <- coords(roc_step, "best", "threshold")
> c
  threshold specificity sensitivity
  0.1088537   0.7738187   0.7063830
```

Fig. 9 Performance measure of SVM model

```
> pred_step1 <- prediction(as.numeric(predictions_step1), as.numeric(Testset$Churn))
> plot(performance(pred_step1, "tpr", "fpr"), colorize = TRUE)
>
>
>
> roc_step1 <- roc(response = as.numeric(Testset$Churn), predictor = as.numeric(predict
ions_step1))
> c1 <- coords(roc_step1, "best", "threshold")
> c1
  threshold specificity sensitivity
  1.5000000   0.9047250   0.4382979
```

Fig. 10 Performance measure of random forest model

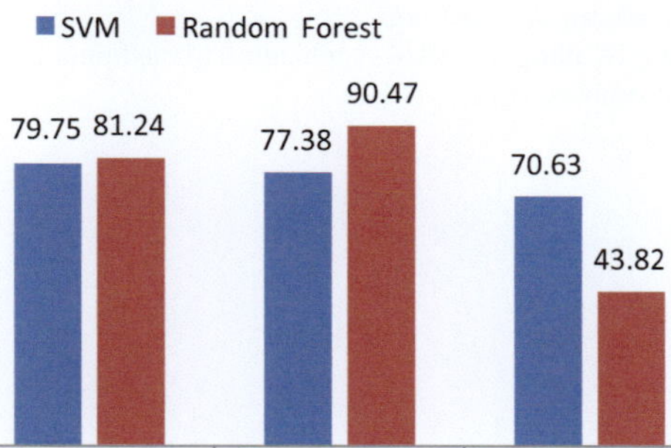

	Accuracy	Specificity	Sensitivity
SVM	79.75	77.38	70.63
Random Forest	81.24	90.47	43.82

Fig. 11 Performance on the models

forest performs better as compared to SVM because it provides better accuracy and specificity, but in terms of sensitivity, the SVM model performs better compared to random forest.

References

1. Kamalraj N, Malathi A (2013) A survey on churn prediction techniques in communication sector. IJCA 64(5)
2. Jadhav RJ, Pawar UT (2011) Churn prediction in telecommunication using data mining technology. IJACSA 2(2)
3. Dahiya K, Talwar K (2015) Customer churn prediction in telecommunication industries using data mining techniques- a review. IJARCSSE 5(4)
4. Li P, Li S, Bi T, Liu Y, Telecom customer churn prediction method based on cluster stratified sampling logistic regression. IEEE
5. Data Mining in the Telecommunications Industry||, Gary M. Weiss, Fordham University, USA
6. Kaur M et al (2013) Data mining as a tool to predict the churn behaviour among Indian bank customers. IJRITCC 1(9)
7. Wang C, Li R, Wang P, Chen Z (2017) Partition cost-sensitive CART based on customer value for telecom customer churn prediction. In: Proceedings of the 36th Chinese control conference 2017. IEEE

8. Praveen et al (2015) Churn prediction in telecom industry using R. IJETR 3(5). ISSN: 2321-0869
9. Dahiya K, Bhatia S (2015) Customer churn analysis in telecom industry. In: IEEE. 978-1-4673-7231-2/15
10. Khare R, Kaloya D, Choudhary CK, Gupta G, Employee attrition risk assessment using logistic regression analysis
11. Burez J, Van den Poel D (2009) Handling class imbalance in customer churn prediction. Expert Syst Appl 36(3)
12. Verbeke W, Dejaeger K, Martens D, Hur J, Baesens B (2012) New insights into churn prediction in the telecommunication sector: a profit driven data mining approach. Eur J Oper Res 218(1):211–229
13. R Data: http://cran.r-project.org/
14. Xia G, Wang H, Jiang Y (2016) Application of customer churn prediction based on weighted selective ensembles. IEEE

An Advance Approach for Spam Document Detection Using QAP Rabin-Karp Algorithm

Nidhi Ruthia and Abhigyan Tiwary

Abstract Document spam is the term which is related to the document copyright issue. It deals with the plagiarism of content from a genuine copy to another. Many researches are performed around the globe for the sake of improvement in different fields such as medical, technology, and agriculture The original content makes impact on the current scenario improvement. Many organization and individual make use of existing concepts to take credit of others' work in their profile. Document spamming is not a legal activity, and thus, there are many algorithms are derived by the research authors to avoid such spamming. Challenges behind such approach are processing the data and finding accuracy over the similarity detection. In this paper, a novel QAP-based Rabin-Karp algorithm is proposed. This approach is a combination of score computation using QAP functions and finally similarity measure computation using Rabin-Karp algorithm. The execution of experimental algorithm is performed using Java library along with sample documents. Algorithm is compared with traditional approach which shows the performance of proposed technique in terms of similarity measure, computation time, and throughput as parameter. The application found improvement, and hence, it shows the effectiveness of proposed approach.

Keywords Document spamming · Plagiarism · Rabin-Karp · Similarity measure · QAP optimization · Document processing

1 Introduction

Cyber law and plagiarism is defined as representing someone else words, thoughts, knowledge, methods, programs, etc., in our own name [1]. Plagiarism has a wider meaning by paraphrasing someone else text by replacing some data or method in our way is also a plagiarism. Copying someone's code is also stands with the plagiarism against the copyright programming content code [2]. It also violates the rule if you do

N. Ruthia (✉) · A. Tiwary
Department of Computer Science and Engineering, SIRTS, Sagar Group of Institute, Bhopal, India
e-mail: Nidhiruthia029@gmail.com

A. Tiwary
e-mail: Abhigyantiwary@gmail.com

R. K. Shukla et al. (eds.), *Social Networking and Computational Intelligence*, Lecture Notes in Networks and Systems 100, https://doi.org/10.1007/978-981-15-2071-6_25

not mention someone author name when you are copying their data on your own form of presentation. Plagiarism is difficult to deal when it comes to scientific research and engineering. It has also increased with the advent of easily availability of data over Internet.

As people start doing copy to someone else data and pasting with their own name, plagiarism detection techniques are applied by making difference between natural and programming languages. A dynamic structure of the content helps in detection of such scenario [3]. N-gram is an approach which works with the number of consequent pattern analysis while dealing with the sentence processing over documents [4].

There are various software for finding analysis such as viper, plagiarism detector, which makes use of permission-oriented plagiarism detection work solution [5], whereas in papers [6–8], they have discussed about the advantage and disadvantage of using the copied code. Author [9] shows about the text mining from the document and finding working of spam text detection, whereas in paper [10], text content spam detection based on the citation is discussed.

The remaining part of the paper is given in following 6 sections: Second section shows the literature survey study performed by other scholars, third section discuss with the problem identification, fourth section depicts the proposed methodology, fifth deals with the experimental setup with usage parameters, and section sixth deals with the result outcome observations, while the seventh provides conclusion of given paper.

2 Literature Survey

This section discusses about the previous work methods which shows their usability over the spamming content detection. Few surveys are depicted here which shows the sequence of latest work performed by previous authors.

In this paper [11], they have given an string-matching pattern match algorithm by using the byte processing of the data in the form of blocks, and thus, the block pattern identification and their comparison are getting performed by using the string pattern match technique. By the use of deep pattern and packet inspection, the identification over the document is getting performed by analyzing algorithm.

They use the winnowing idea [12] developed for fingerprint over the document, and putting the watermark sign of fingerprint on document for the analysis is given. Thus, the idea is to provide a unique identification to each document. Using which an ideal identification and delicacy detection can get perform with the unique fingerprint detection.

In this paper [13], SCAM (Standard Copy Analysis Mechanism) is given which is based on the standard protocol of copying data. The execution performance is done with accuracy and precision of analysis document.

In this paper [14], the new algorithm is named as multi-pattern parameterized shift-or (MPSO) string-matching algorithm and further extended MPSO which utilized the super alphabet concept for identification of the data algorithm. Multiple pattern analysis strategy is acquired by the given solution. A real-time HTTP usage and their traffic monitoring are getting performed in the paper which deals with the pattern match over Web links and their data [15].

Thus, the given work analysis shows the different pattern matching approaches over the time, which help in working with plagiarism analysis.

3 Problem Formulation

As per discussion of literature survey analysis, there are following major problems which are analyzed as follows:

1. Existing approach is limited to one way of feature analysis cost function calculation at single level.
2. JPlag is one dependency but it is missing some patch which gives false result while working with their tool.

   Integrity-based algorithm only works with exact content finding, which is the static content match-based approach. Thus, the accuracy of prediction is only on fixed text data match.
4. The existing algorithms are limited to traditional string search concept, whereas the proposed algorithm is well versed with multiple pattern search and analysis.
5. Existing approaches do not compute hash function values along with multiple score values, whereas the proposed solution finds its suitability over and finds hash function with QAP score generated.

Thus in order to work with given limitations and hybrid approach with the QAP, cost and weight measure along with Rabin-Karp are proposed in next section.

4 Proposed Methodology

In order to put an effective work toward the proposed system, an advance algorithm QAP-based Rabin-Karp approach is presented by work performed.

QAP Rabin-Karp Algorithm:

The proposed algorithm includes Jaccard cost function, weight measure of the input word and given source folder data, and frequency computation of the data words. Finally, evaluation of complete cost using summation of individual cost and then applying Rabin-Karp matching over the cost derived from the different entities. Finally, result evaluation using the major three parameters is proposed using the algorithm.

Following is the flow, and its execution description is provided which shows the algorithm working analysis.

Figure 1 shows the flow diagram of complete process which is taken for the execution.

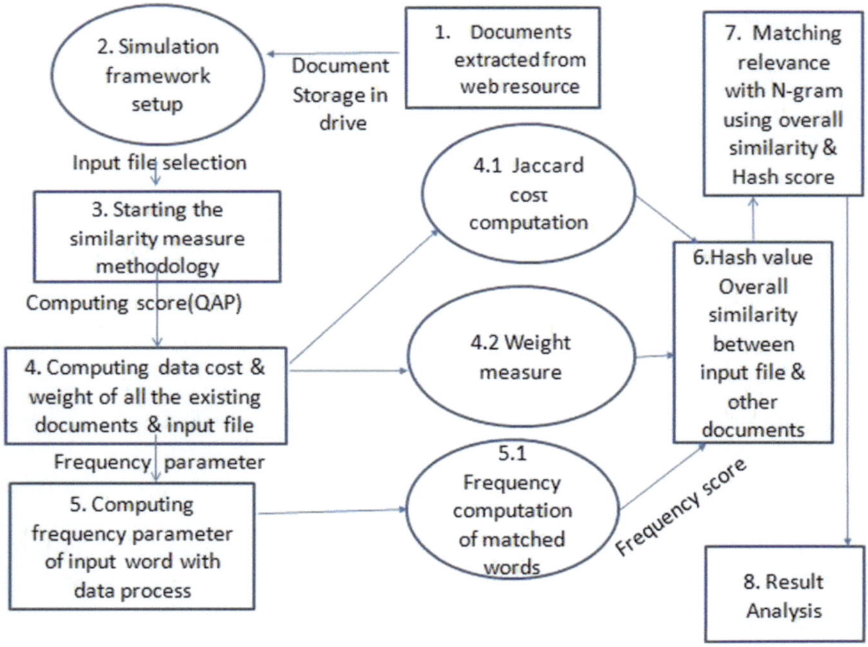

Fig. 1 Flow architecture of complete process algorithm

A detail execution and algorithm detail are given here in following execution algorithm steps which shows the sequence of given work:

1. Dataset document is extracted from the Web resource and stored over the local disk. Here, the document for evaluation is stored over the local storage.
2. Initializing Java library and components for the data processing.
3. Input file selection which is needed to match with existing documents.
4. Initializing the similarity matching method using QAP with its subcomponent function which returns Jaccard cost computation.
   Dij … Distance between *i* input and other document value

$$\textbf{Jaccard index} = (\text{complete set union})/(\text{one side set}) * 100 \; d(x, y) = 1 - j(x, y)$$

5. Weight measure of input string file and available document.
6. Computing word frequency between the input file and available document and returning frequency score.
7. Measuring overall cost of input file and available documents using summation of Jaccard cost, weight measure score, and frequency score.
8. Finding hash computation of summation value obtained from weight computation, frequency measure, and Jaccard cost.
   Summation of different weight obtained values:

$$(x)^n = \sum_{k=0}^{n} \binom{n}{k} F_c^k W_m^k J_c^k$$

Here, F_c is frequency count, W_m is weight measure, and J_c is Jaccard cost.
Hash computation of Summation values:

$$H = Hf(x)$$

9. Finally, computing the similarity measure using the overall computation hash score and compute the comparison parameters.
10. Exit.

Thus, the given steps are taken for simulation analysis and substeps for the proposed QAP Rabin-Karp algorithm which is efficient while compared with N-gram measure solution for plagiarism analysis.

5 Experimental Setup and Parameters

In order to work with the given approach and their simulation analysis, there are tools and programming language is considered where Java 8.0 is by using NetBeans 8.2 along with local file storage system of windows. The *PubMed* dataset is taken with different document numbers and size, on which experiment is performed. The experiment is performed with simulation comparison parameter as similarity measure, computation time, and throughput. Thus, the comparison performance shows the efficiency of the proposed algorithm over traditional analysis.

Jaccard Similarity: Document similarity measure is the average similar content unit found in the input unit and other source folder data. We have used Jaccard similarity to measure the most similar content using this formula:

$$\mathrm{JS}(\mathrm{doc1},\ \mathrm{doc2}) = \frac{|\mathrm{doc1} \cap \mathrm{doc2}|}{|\mathrm{doc1} \cup \mathrm{doc2}|}$$

Where doc1 is the suspicious document and doc2 is source folder.

Computation time: Computational time is the total execution time for the algorithm execution, which is calculated as difference between final finish times of algorithm subtracted from initialization time noted at the starting of algorithm.

Ct = Final time − Initialized time;

Throughput: Throughput is the measure of performance which computes as detection per unit of time.

Th = Number of detection/Unit time;

Table 1 Statically analysis of obtained result

Algorithms	Computation time (ms)	Similarity measure	Throughput
N-gram Measure	242.91	511.5	0.8
QAP Rabin-Karp	184.91	509.9	0.9

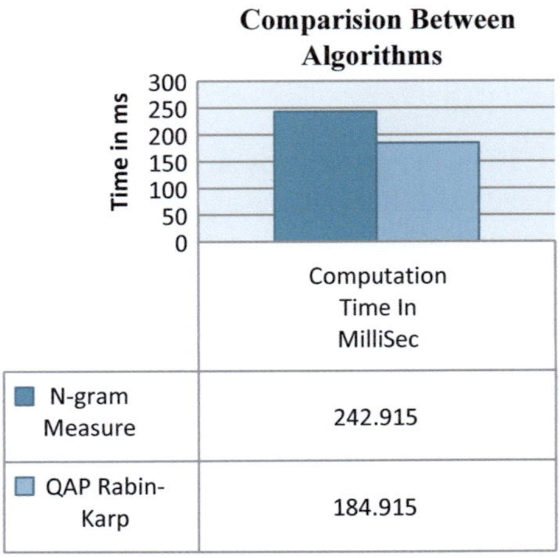

Fig. 2 Comparison analysis bar graph for computation time analysis

6 Result Analysis

As per the proposed QAP-based Rabin-Karp algorithm is implemented, also the existing solution of similarity measure is implemented with the given experimental setup scenario. There is result obtained in the term of similarity measure, computational time, and throughput. Following is the result computational analysis:

Statically Analysis: A tabular analysis of the value obtained from the experiment is shown and compared in Table 1 (Figs. 2, 3 and 4).

Thus, the result discussion showed the performance of proposed mechanism over the traditional approach. The given analysis of the result obtained shows the effectiveness of approach over existing solution.

7 Conclusion and Future Work

Plagiarism detection over the documents is a recent trend and challenging over the time. There are multiple techniques that use to reframing the data, and hence, plagiarism is avoided by the research group. Detection of spamming content is always needed a refinement which can improve the technique of existing scenario

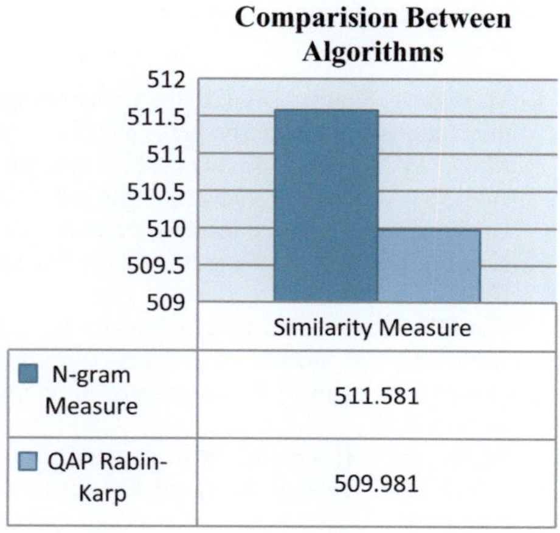

Fig. 3 Comparison analysis bar graph for similarity measure analysis

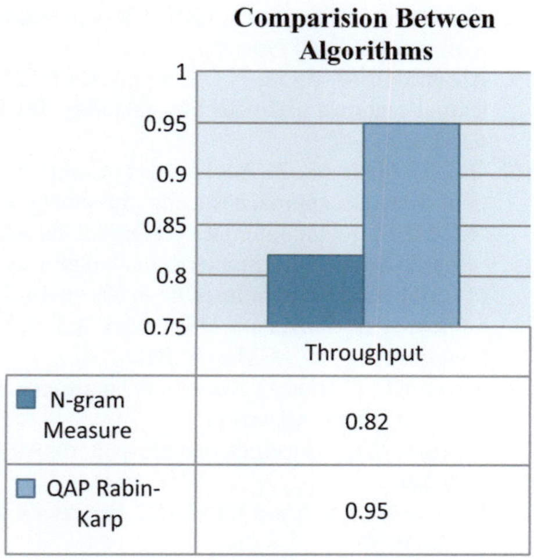

Fig. 4 Comparison analysis bar graph for throughput analysis

which deals with the document spamming. The traditional algorithms have limitation with high computation time and lower in efficiency. This paper presented a novel QAP-based Rabin-Karp algorithm for the data processing and finding the similarity between the documents. Score computation and matching similarity measure obtained from the proposed approach show the efficiency of the system. Algorithm computation parameter using the computation time, throughput, and similarity measure improvement shows the advancement of the proposed algorithm. A further improvement of using such algorithm with the real-time online tool with third-party permission access can be performed. Thus, a real-time cloud-based platform can be proposed for the further enhancement of the proposed system.

References

1. Mayank A, Sharma DK (2016) A state of art on source code plagiarism detection. In: 2nd international conference on next generation computing technologies (NGCT). IEEE
2. Mirza OM, Joy M, Cosma G (2017) Style analysis for source code plagiarism detection—an analysis of a dataset of student coursework. In: Proceedings of the 2017 IEEE 17th international conference on advanced learning technologies, pp 296–297
3. Kuo J-Y, Cheng H-K, Wang P-F (2018) Program plagiarism detection with dynamic structure. IEEE
4. Wang R, Utiyama M, Goto I, Sumita E, Zhao H, Lu B-L (2013) Converting continuous-space language models into N-gram language models for statistical machine translation. In: Proceedings of the 2013 conference on empirical methods in natural language processing, pp. 845–850
5. Vandana (2018) A comparative study of plagiarism detection software. IEEE, 11 Oct 2018
6. Pawelczak D (2018) Benefits and drawbacks of source code plagiarism detection in engineering education
7. Xylogiannopoulos K (2018) Text mining for plagiarism detection: multivariate pattern detection for recognition of text similarities. IEEE, 25 Oct 2018
8. Karnalim O, Sulistiani L (2018) Which source code plagiarism detection approach is more humane? IEEE, 01 Nov 2018
9. Alzahrani SM, Salim N, Abraham A (2012) Understanding plagiarism linguistic patterns, textual features, and detection methods. IEEE Trans Syst Man Cybern Part C (Appl Rev) 42(2):133–149
10. Gipp B, Meuschke N, Beel J (2011) Comparative evaluation of text and citation-based plagiarism detection approaches using guttenplag. In: Proceedings of the 11th annual international ACM/IEEE joint conference on Digital libraries. ACM, pp 255–258
11. Variable-Stride "Multi-pattern matching for scalable deep packet inspection". Nan Hua College of Computing Georgia Institute of Technology, nanhua@cc.gatech.edu
12. Schleimer S, Wilkerson DS, Aiken A (2003) Winnowing: local algorithms for document fingerprinting. In: ACM SIGMOD
13. Anzelmi D, Carlone D, Rizzello F, Thomsen R, Akbar Hussain DM (2016) Plagiarism detection based on SCAM algorithm
14. Prasad R (2017) An efficient multi-patterns parameterized string matching algorithm with super alphabet
15. Bremler-Barr A, Koral HY (2016) Accelerating multi-patterns matching on compressed HTTP traffic. IEEE

A Review on Enhancement to Standard K-Means Clustering

Mohit Kushwaha, Himanshu Yadav and Chetan Agrawal

Abstract In clustering, objects that have similar nature will lie collectively in the same group called cluster, i.e. same cluster and if they are of distinct nature, then they will be belonging to other cluster of similar nature. The standard k-means is a prime and basic procedure of the clustering but it suffers from some shortcomings, these are as follows (1) Its performance depends on initial clusters which are selected randomly in standard k-means. (2) The basic algorithm, standard k-means, has computational time of $0(NKL)$ where N represents the number of data points, K represents the number of distinct clusters and L is the number of different iterations that is time consuming or too much expensive. (3) The standard k-means algorithm also has the dead unit problems that result in clusters which contains no data points that is empty cluster. (4) In standard k-means if we do random initialization which causes to converse at local minima. Several enhancement techniques were introduced to enhance the efficiency of the basic k-means algorithm but most of the algorithms were focus only on one of the above drawbacks at a time. In this review paper, we consider initial centre as well as computational complexity problem along with dead unit problem in single algorithm.

Keywords Cluster analysis · Similarity proximity · Enhanced k-means · Standard k-mean

1 Introduction

Among various sub-fields of the artificial intelligence, one of the fields of artificial intelligence is machine learning. Machine learning is mainly concerned with

M. Kushwaha · H. Yadav (✉) · C. Agrawal (✉)
Computer Science and Engineering, RITS Bhopal, Bhopal, India
e-mail: himanshuyadav86@gmail.com

C. Agrawal
e-mail: chetan.agrawal12@gmail.com

M. Kushwaha
e-mail: mohitkushwaha786@gmail.com

R. K. Shukla et al. (eds.), *Social Networking and Computational Intelligence*, Lecture Notes in Networks and Systems 100, https://doi.org/10.1007/978-981-15-2071-6_26

the design, analysis, application and implementation of set of instructions which learn from training. Training the data is the procedure to feed data into the machine. Machine learning is further breaking down into two major fields that are supervised learning and unsupervised learning. In supervised learning, we have an input variable x_1 and an output variable y_1 and we have to map input and output variable. This mapping function is used to map new input variable to output variable. The above process is known as supervised learning because mapping function learns to make training dataset and it can be assumed to as a supervisor supervises the whole procedure from outside. Here, machine learning is a collaborative process which follows step-by-step learning, adjusts its result and it will stop when they accomplish a worthy dimension of exactness. The supervised learning can be further breaking down into classification and regression. On the other hands, the next major learning process is unsupervised learning, we have only input variable x_1 and no other variable like output variable as in supervised learning. It is known as unsupervised learning because here no supervisor and no exact outcomes. The algorithm remains to their own to find the data which have similar properties and to determine the structure in the data. Unsupervised learning problem breaks down into two other types, that is clustering and association. Human minds are skilled and learn the objects by dividing objects in to groups and allocate a particular object to appropriate class, for example, a young child can easily label objects in a photograph as the motor, building, plant and animal. Analysis of cluster is the way to determine these classes automatically from data without any previous knowledge about classes.

Analysis of clusters has several applications, such as in field of biology, clustering can be used to find collection of genes that have similar functions, in the area of medicine and psychology, cluster study can be used to detect temporal distribution and spatial distribution of several diseases. The algorithm of clustering is divided into major two categories hierarchical algorithm and partitioning algorithm [1]. In the first type of algorithm which is hierarchical clustering, either smaller sets are merged together to get the larger set or larger sets to divide to get smaller sets. From both the cases, we consider a dendrogram. Divisive clustering strategy uses a famous approach known as top-down approach in which larger ones are splitting into smaller ones and on the other hand, agglomerative clustering strategy follows the bottom-up approach in which smaller ones are combined together to become the larger one. In partitioning algorithm, each instance is to be found in only one cluster and these clusters are mutually exclusive as shown in Fig. 1 given below which show the type of clustering.

1.1 Hierarchical Clustering

Hierarchical clustering is another type of clustering in which we can select any one of the strategies from both bottom-up approach and top-down approach. In bottom-up strategy, each of the data point is firstly in a collective group of having similar property of its own and at each iteration, we determine the two closest clusters and

Fig. 1 Type of clustering

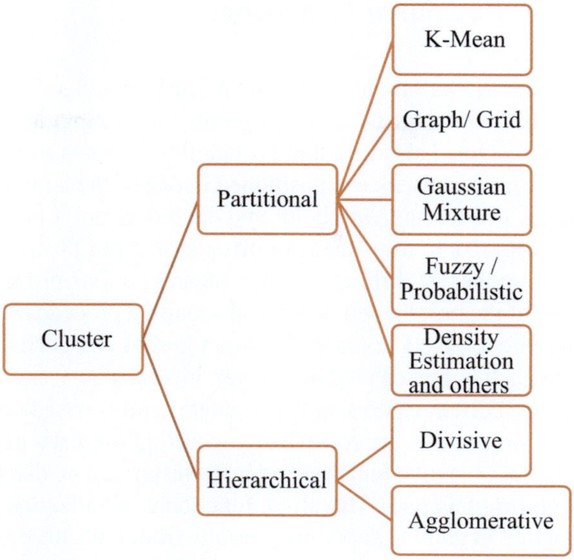

combine them into a single one. In divisive approach of hierarchical clustering, all the data points fall to a single cluster, to begin with, and we repeatedly split the cluster as we go along.

Hierarchical clustering algorithms repeat the cycle of either merging smaller clusters to form a larger one or dividing larger clusters into smaller ones. According to the both of above approach, it becomes a hierarchy of collection of similar groups that are of tree structure [2]. Another type of hierarchical clustering is agglomerative clustering in which it follows the bottom-up approach. In a bottom-up method, smaller groups are merged together to become a larger cluster. In divisive clustering approach, we follow the top-down approach. In top-down strategy, a larger group is divided into smaller similar group. Generally, a greedy approach is used to determine whether we should follow bottom-up approach or top-down approach. For numerical data, we generally use Euclidean distance, Manhattan distance and cosine similarity for metric of similarity, while for non-numerical datasets, we are using metric such as Hamming distance. In hierarchical clustering, actual observation is not necessarily the only matrix of distances is sufficient. The dendrogram is a visual picture of the group having similar properties, which displays the tree structure very clearly. The users can also determine various clustering depending on the category at which the dendrogram is cut.

1.2 Partitional Clustering

As we discuss about next type of clustering algorithm, partitional clustering algorithm first we create as many groups as possible and then calculate them on some parameters. This algorithm is sometimes known as the non-hierarchical algorithm. Here, each existence is positioned in one of the k mutually groups. Each group contains at least one data point and each data point belongs to exactly one group. In this approach, user needs to gives *k* as input to the procedure. Here, *k* represents the number of different groups having similar property. K-means strategy is also one of the best known partitional grouping procedures [2]. In k-means approach, we require to give a number of distinct groups and corresponding initial centre of clusters as the input. In external layer, k-means approach then allocate members based on the current centres and re-evaluate centres based on the existing members. These two procedures are recursively occurring until we get intracluster distance having similar objective function and intercluster having dissimilarity objective function in optimized manner. Therefore, functional initialization of centres is a very important feature in gaining superiority results from this clustering algorithm.

In the partitioning algorithm, each object is located in only one collective group and these clusters are mutually exclusive. Among many clustering algorithms, standard k-means is basic algorithm. It is introduced in 1967 by Mac Queen. K-means is easy to implement that become a popular among several fields. The basic and prime K-means process is a partitioning clustering in which data are divided into *k* groups and these *k* groups are mutually exclusive [3]. The basic *K*-means to have its own drawbacks such as its efficiency depends upon the selection of the initial centroid of the group, it has high computational time complexity of $O(NKL)$ where *k* is the no of groups or clusters, *K* is the no of data objects or data points and *L* denotes the number of different iterations, the standard k-means procedure also converses to local minima that sometimes gives unexpected results. In this paper, we are going to propose an enhanced k-means algorithm which will address three of its prime shortcomings, these are selection of random initial centre problem and its high computational time complexity to determine clusters and dead unit problem due to random selection of initial centroid. Most of the researchers address one of them at a time but in this paper, we are going to address all of them simultaneously.

2 Related Work

There have been done much related effort to reduce the shortcoming of the basic k-means algorithm. Another method to grouping the data points is the BIRCH method; in this approach, they introduce a CF-tree, which is a tree data structure proposed for a multiphase grouping algorithm. In this approach, they first scan the database to get an initial in storage as CF tree. And then, they use a random clustering procedure which is used for grouping the lower level nodes known as leaf nodes of the CF-tree cite [4].

Likewise, framework-based grouping system CLIQUE has been proposed for mining in high-dimensional information spaces. Info parameters are the span of the matrix and a worldwide thickness edge for the groups. The real distinction from all other bunching approaches is that this strategy likewise recognizes subspace of the most elevated dimensional with the end goal that high-thickness groups exist in those subspaces [5].

In this paper, author proposed that original k-means is not converse to the global minima but it converses to local minima. Some improvements are given but these improvements require many external inputs such as threshold values. Here, author proposes an algorithm in which they first normalize the data. Then partition them in k equal parts and take the median value as the initial centroid [6].

In this paper, author mainly addresses the higher time complexity problem. Like as in basic k-means algorithm, they compute the distance between the centre of each cluster and data points of cluster for each iteration that makes standard algorithm too expensive. In this paper, they use a special data structure to hold the previous steps data and uses that information for the successive steps. In this algorithm, they first calculate the distance between current data point and previous cluster centroid, if computed distance is less to the distance to the previous cluster then there is no requirement to compute its intracluster distance from other centroid.

In this paper, the author proposed that according to conditions, first compute the distance between the data points. And after that, find the closest points according to the similar behaviours. Finally, opt the actual centroid and get the improved outcomes [7].

Although basic k-means procedure is introduced 50 years ago but it is broadly used nowadays also because it is too difficult to organize generic grouping algorithm. In this paper, author also discussed the semi-supervised clustering that is in between supervised and unsupervised clustering. Author also identifies the major challenges to the clustering such as

(i) Feature selection
(ii) Data normalization
(iii) Define similarity
(iv) Selection of proximity function [8].

In this paper, author proposed a variant of k-means that is more efficient for multispectral image segmentation. The image having digital numbers of several spectral that sense particular wavelength. The author considers both spatial property as well as the spectral property of the image. The author encrypts the image for the above purpose [9] (Table 1).

Table 1 Comparative study

S. No.	Title of the research paper	Description
1	An efficient enhanced k-means clustering algorithm [3]	For every point of data, they calculate the distance to the closest group. After that they approach for next iteration, in this iteration they compute the distance to the previous closest cluster. If the distance which is computed from above iterations known as new distance which is less than or equal to the previous distance, the data points belongs to the group having similar data, and then there is no necessity to compute its distances of each data points to the other centre of clusters. This procedure reduces the lot of computational time which is essential to compute distances to $k - 1$ cluster centres
2	Efficient data clustering approach using very large databases [4]	Author proposed a BIRCH algorithm that is majorly essential for huge amounts of datasets. This algorithm becomes a large grouping problem controllable by concentrating on densely occupied portions, and using a compressed summary
3	Automatic subspace grouping of high-dimensional data for data mining applications [5]	Framework-based grouping system CLIQUE has been proposed for mining in high-dimensional information spaces. Info parameters are the span of the matrix and a worldwide thickness edge for groups. The real distinction from all other bunching approaches is that this strategy likewise identifies subspace of the most elevated dimensional with the end goal that high-thickness groups exist in those subspaces
4	Enhanced k-means clustering [6]	In this paper, author proposed that original k-means is not converse to the global minima but it converses to local minima. Some improvements are given but these improvements require many external inputs such as threshold values. Here, author proposes an algorithm in which we first normalize the data. Then partition them in k equal part and take the median value as the initial centroid
5	Improving the correctness and performance of the k-means clustering algorithm [7]	In this paper, author proposed that according to conditions, first find the distance among the various data points. Then after calculate the closest points (distance) according to the similar behaviour. At last, select the actual centroid and get the better result

(continued)

Table 1 (continued)

S. No.	Title of the research paper	Description
6	Contiguity-improved k-means clustering strategy for unsupervised multispectral image segmentation [8]	In this paper, author also discussed the semi-supervised clustering that is in between supervised and unsupervised clustering. Author also identifies the major challenges to the clustering such as feature selection, data normalization, define similarity and selection of proximity function which used
7	Enhanced k-mean clustering algorithm to decrease the number of steps and time complexity [9]	The image having digital numbers of several spectral that sense particular wavelength. The author considers both spatial property as well as the spectral property of the images. The author encrypts the image for the above purpose
8	Improving k-means clustering algorithm with improved initial centre [11]	This paper gives a technique to initiate the initial centre in k-means. Here, sort the data points and splitting them into k equal part and median value of each set is initial centroid

3 Standard K-Means Algorithm

The basic approach of grouping the data points having similar properties is k-means algorithm which is partitioning-based algorithm in which k represents the number of a different groups that is provided by user and then this algorithm randomly selects k initial centres from each group. In each iteration, data point is allocated to the closest centroid and then computes the new centroid of each group. This process will stop until no data point moves from one group to other it means that equilibrium condition set. Here, our aim is to decrease the intra group distance. It is calculated as:

$$\text{intra cluster distance} = \sum_{j=1}^{k} \sum\nolimits_{i\,\varepsilon\,c_j} \left((x_i)^2 - (c_j)^2 \right)$$

Here, c_j denotes the jth group. $X_i\ i\,\varepsilon\,c_j$ denotes the ith data points of jth group. For better clustering procedure, the intercluster distance should be more and intracluster distance should less. High intercluster distance shows that different clusters are more distinct and lower intracluster distance shows that object within the cluster is more similar.

Steps of basic k-means algorithm:

Let $A = a_1, a_2, \ldots, a_n$ are the data points,

$C = c_1, c_2, \ldots, c_k$ denotes the set of groups and k is the number of different groups.

(1) Initially, select k centre of each clusters randomly.
(2) Compute the distance between data point and their corresponding centroid of each cluster.
(3) Allocate the data points to appropriate group according to minimum distance.

Table 2 Different proximity

Proximity functions	Centroid	Objective
Manhattan	Median	To decrease the distance between data points and centroid of each clusters
Square Euclidean	Mean	To reduce the square of distinction between data items and its group's centroid
Cosine	Mean	To decrease the cosine similarity between data items and its group's centroid

(4) Recompute the new centroid of each clusters.
(5) If no data points were changing its group, i.e. stationary condition then stop otherwise go to step 3.

In basic k-means algorithm, we first select k initial centroid randomly and then required to allocate the given data points to these groups. To allocate data points, we required to define proximity function measure that specifies the notion of closeness. There are different types of proximity function such as Euclidean distance and Manhattan distance. In this paper, we are going to follow Euclidean proximity function for the notion of closeness. Different proximity is given in Table 2.

In this algorithm, many functions are used such as dist(a, b) denotes the distance between two points a and b. Here, we are using Euclidean distance function. This represents in the straight Table 2: Different Proximity function Centroid Objective Manhattan Median to decrease the distance between data points and cluster centroid Square Euclidean Mean to decrease the square of separation between data items and its group centroid Cosine Mean to minimize the cosine similarity between data items and its group centroid.

Line distance between two data points in Euclidean space. Let $m = (m_1, m_2, m_3, \ldots, m_n)$ and $l = (l_1, l_2, l_3, \ldots l_n)$ then Euclidean distance between these two point is given by

$$\text{dist}(l, m) = \sqrt{(l_1 - m_1)^2 + (l_2 - m_2)^2 + \cdots + (l_n - m_n)^2}.$$

In this algorithm, we use dist(C_j, O_i) which represent the distance between jth group's centroid and ith object. Mean (C_j) function is used to compute the mean of the jth group. For mean, we use the following function, let we have $x = (x_1, x_2, x_3, x_4, \ldots, x_n)$ and total data points are m.

$Mean = (x_1 + x_2 + x_3 + \cdots + x_n)/m$, min($r$, k) is a function that evaluates the minimum value in the rth array (Fig. 2).

Algorithm: Standard K-means Algorithm

Inputs:
K denotes the counts of groups or clusters
Manually feeds the centre for each group of Data points Outputs:
Groups or Cluster with data points
1: repeat

Fig. 2 K-means flow chart

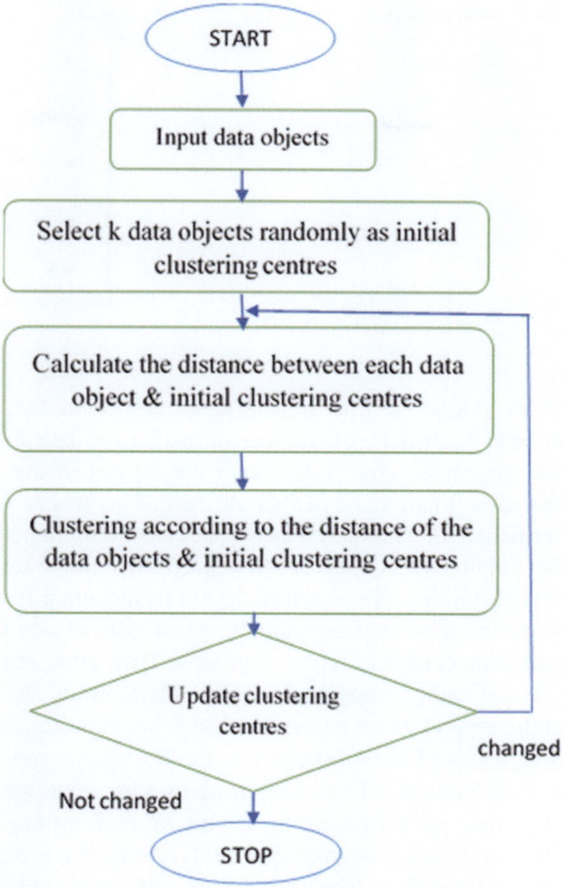


```
for l 1, N do
for m 1, K do
rm distance (Cm, Ol)
end for
end for
for j = 1 1, K do
OCj = Cj
Cj = compute mean (Cj)
end for
if then not equal (OC,C,K)
for i = 1 1, K do
clear Gi
end for
end if
until not equal (OC,C,K)
```

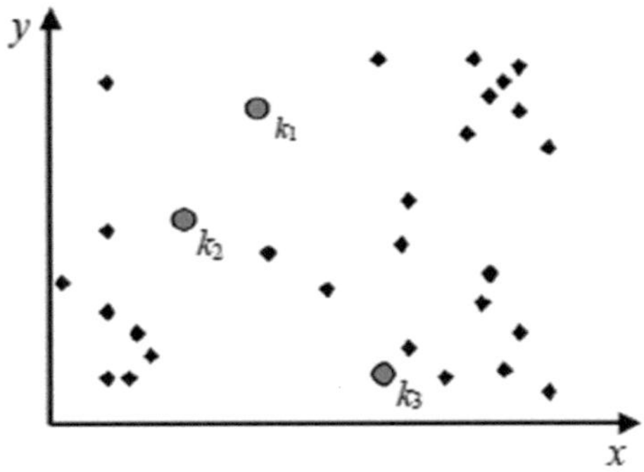

Fig. 3 Data points [3]

To address the first drawback, we first check that the data contain any negative value in its attribute; if yes, then normalize the data points. To achieve normalization, find largest negative value from the dataset in that respective attribute then add this value to all value of that attribute. This makes whole dataset to positive. This normalization is required as we are calculating the distance from the origin [10]. If there is no data contain negative magnitude in the respective attribute then no need to perform the normalization. In the forthcoming iteration, we sort the whole data point of dataset and separate the whole dataset into k equal individual parts. Here, k denotes counts of required groups. Then after, compute the result at the middle index of each groups, these intermediate results are treated as the initial centroid. In the second phase of the procedure, we want to decrease the computational time complexity of the procedure. In basic k-means approach, if any data point varies its initial position from one group to other, then we compute the closest distance of all data point from all the centres of each group. If N denotes the counts of all data points and K denotes the counts of different groups and L denotes the counts of different iterations, then basic k-means procedure has computational time complexity of $O(NKL)$. To decrease computational time complexity, we can take last iterations outcomes, for this we select to map the closest cluster distance for each and every data point. And after that if the new distance of centroid is less than or equal to the previous iteration distance then the data point remains in the group having similar property, so there is no requirement to compute its distance from other group's centroid. This will work well because in the k-means grouping, group are spherical in shape [2] (Figs. 3, 4 and 5).

3.1 Application of Standard K-means

1. E-retailers or online business organizations have overwhelmed the retail e-commerce. They are knocking out drawing offers and rebates. They plan to move from debate prompted comfort or separation drove offering over a period of time. Some of them have been bound to begin the excursion. One of the large

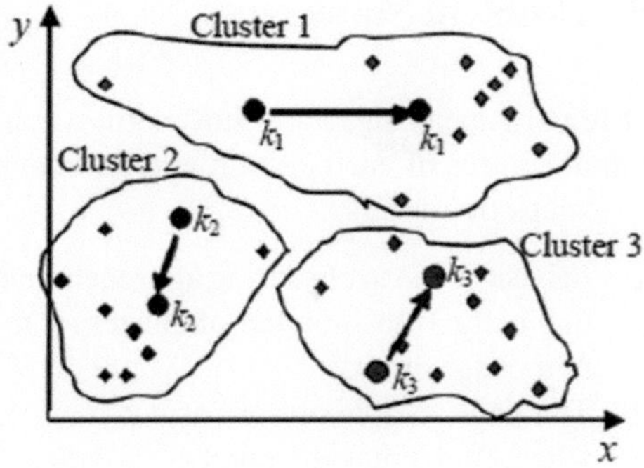

Fig. 4 Distribution over these centroid [3]

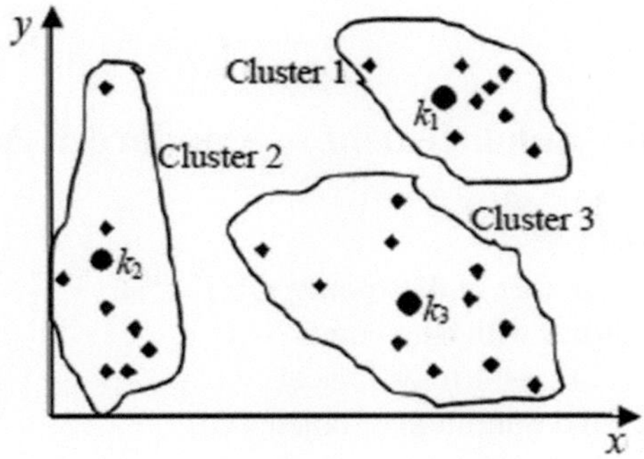

Fig. 5 Final cluster with centroid [3]

areas for retailers needed to section the clients in light of client spend designs and comprehend value which affects the capability of the clients.

2. Student performance measurement by using basic k-means approach, we groups the students according to their marks and compute the performance index for a group. If student performance index is low in a semester, then it means that we should give some remedial classes to this student so that student can improve their performance.

4 Objectives

Many enhancements techniques were introduced to improve the performance of the basic k-means algorithm, but most of these algorithms address only one of them at a time. In this paper, we consider the following problems in one algorithm.

1. Initial centroid of clusters
2. Computational time complexity
3. Intracluster distance
4. Dead unit problems.

5 Problem Statements

Original *k*-mean algorithm suffers from many shortcomings such as it depends upon initial centres of each cluster, and its time complexity also depends on the number of groups or clusters.

1. In basic k-mean clustering approach, select initial centres for each cluster as input manually. If the number of clusters is too many, then it is more overhead to feed in the algorithm.
2. The time complexity of basic k-mean is more or too expensive that is $O(NKL)$ where N denotes the number of data points, K denotes the number of groups or clusters and L denotes the number of different iterations.

6 Enhancement to Standard K-Mean Algorithm

```
INPUT:
K-number of data points OUTPUT:
Cluster with data points
for i ← 1 to N do next
find minimum of objects Oi
for loop end
if Oi < 0 then
for i ← 1 to N do
Oi = Oi – minimum
for loop end
end if
sort algorithm (heap sort) (Oi)
step size = N/K
for i ← 1 to K, C = step size/2 do
Ci = Oc
C = C + step size
for loop end
recurse
for i ← 1 to N do
for j ← 1 to K do
rj ← compute new distance (Cj, Oi)
for loop end
Parent node [i] = minimum (row; K i)
Object[i] = index of (Parent node [i]; row; K)
add in the Group
for loop end
for j = L ← 1 to K do
```

```
25: OCj = Cj
26: Cj = compute mean (Cj)
27: end for loop
28: if not equal (OC, C, K) then
29: for loop i = L ← 1 to K do
30: empty Gi
31: end for loop
32: end if condition
33: until not equal (OC,C,K)
```

In this enhancement procedure, we will decrease the computational complexity of basic k-mean approach by applying some modifications.

(1) First change is to compute the initial centroid for cluster automatically which is done by step 1–8 in above algorithm. It will take $O(N)$ time complexity.
(2) Then sort the data points and for sorting, we are using heap sort. Heap sort has $O(N\log N)$ time complexity in best case, average case as well worst case and then separate them in k equal parts. This approach will take time of $O(k)$ in step 9.
(3) If points remain in the same group then it will take complexity of $O(1)$ and if point do not remain in the same group, then it will take $O(k)$ in step 10–15.
(4) From step 16 to 33, we compute new distance which take $O(NK\log L)$
(5) So total computational complexity of the whole algorithm is $O(NK\log L)$ + $O(N\log N)$ which is approximately equal to $O(\mathbf{NKlogL})$.
(6) Standard k-means methodology gives $O(NKL)$ and this proposed enhanced approach gives $O(NK\log L)$ which is comparatively less time complexity than basic k-mean.

7 Conclusion

In this research, we will get the improved time complexity which is $O(NK\log L)$. To achieve this time complexity, first we analyse the standard k-mean algorithm and their drawbacks with their performance. It may give same time complexity when data points are less but in case of large number of data points, enhanced k-mean algorithm performs better than standard k-mean algorithm. The huge time consumed by standard k-mean to achieve initial centroid and distance calculation between centroid and their data points which is overcome by using heap sort and new distance calculation algorithm.

8 Research Scope

In order to overcome the limitation of the previous applied algorithm mentioned in the above comparative analysis of various research papers. Implementation of this algorithm on different huge datasets to analyse different populations, poverty, illiteracy, e-commerce application like amazon, etc.

References

1. Datta S, Datta S (2003) Comparisons and validation of statistical clustering techniques for micro array gene expression data. Bioinformatics 19(4):459–466
2. Jain AK, Dubes RC (1988) Algorithms for clustering data. Prentice-Hall, Inc.
3. Fahim AM, Salem AM, Torkey FA, Ramadan MA (2006) An efficient enhanced k-means clustering algorithm. J Zhejiang Univ-Sci A 7(10):1626–1633
4. Birch ZT (1996) An efficient data clustering method for very large databases. In: Zhang T, Ramakrishnan R, Livny M (eds) Proceedings of the 1996 ACM SIGMOD international conference on management of data (SIGMOD'96). ACM, New York, pp 103–114
5. Rakesh A, Gehrke J, Dimitrios G, Raghavan P (1998) Automatic subspace clustering of high dimensional data for data mining applications, vol 27. ACM
6. Campos MM, Milenova BL, McCracken MA, Enhanced k-means clustering, September 152009. US Patent 7590642
7. Nazeer KAA, Sebastian MP (2009) Improving the accuracy and efficiency of the k-means clustering algorithm. In: Proceedings of the world congress on engineering, vol 1, pp 1–3
8. Theiler JP, Gisler G (1997) Contiguity-enhanced k-means clustering algorithm for unsupervised multispectral image segmentation. In: Optical science, engineering and instrumentation' 97. International Society for Optics and Photonics, pp 108–118
9. Rauf A, Sheeba SM, Khusro S, Javed H (2012) Enhanced k-mean clustering algorithm to reduce number of iterations and time complexity. Middle-East J Sci Res 12(7):959–963
10. Tian Z, Ramakrishnan R, Birch ML (1996) An efficient data clustering method for very large databases. In: ACM sigmod record, vol 25. ACM, pp 103–114
11. Yedla M, Pathakota SR, Srinivasa TM (2010) Enhancing k-means clustering algorithm with improved initial center. Int J Comput Sci Inf Technol 1(2):121–125

A Review on Benchmarking: Comparing the Static Analysis Tools (SATs) in Web Security

Rekha Deshlahre and Namita Tiwari

Abstract In this present IOT (Internet of things) era, strong security in a Web application is critical to the success of your online presence. Security importance has grown on a vast scale among Web application. Static analysis tools (SATs) are currently useful tools for developers to explore the vulnerabilities present in the initial source code in a Web application. The aim of the SAT is to improve the effectiveness and usefulness of the source code. There are many SATs are present in this era. However, different tools provide different results according to the complexity of the source code underneath analysis and the application scenario. To compare tool abilities, benchmarking is used on SATs. Benchmarks are used for comparing and accessing different system codes and components. Thus, while reporting the alarm information to the tools, vulnerability missing causes a problem and gives the result as a poor infrastructure of the source code. Benchmark is used to address the limitation of the SATs. However, present benchmarks have strict representative restrictions, disregarding the specificity of the domain, where the tools under the benchmarking will be used. In this paper, benchmark is introduced to compare and access static analysis tools (SATs) in terms of their vulnerability detection capabilities for security. Benchmark uses four real-life development scenarios, including workload with different goals and constraints.

Keywords Static analysis tools (SATs) · Benchmarking · OWASP · SAMATE · Security metrics · Vulnerability detection

1 Introduction

Rapid development of Web services requires new technologies with an additional feature. Many of the business and organizations are a useful Web application to communicate with their consumers. The Web application has an exceptional ability

R. Deshlahre (✉) · N. Tiwari (✉)
Maulana Azad National Institute of Technology, Bhopal, Madhya Pradesh, India
e-mail: rekhadeshlahre999@gmail.com

N. Tiwari
e-mail: namita_tiwari21@rediffmail.com

R. K. Shukla et al. (eds.), *Social Networking and Computational Intelligence*, Lecture Notes in Networks and Systems 100, https://doi.org/10.1007/978-981-15-2071-6_27

of rapid deployment and direct accessibility to the millions of users at the same time instance. So, the demand for the new application with advanced and complex features is increasing constantly [1]. Hence, application development is often carried out under strict time and schedule. Thus, most of the vulnerabilities and bugs are found under Web applications that reflect the result of the organization with their market share's losses, product demands, financial loss, brand damage, organization liability problems, etc. In the last few years, several initiatives have been introduced to provide security in software development life cycles. Every initiative introduces a new feature in different software development phases to improve the security of the software in development life cycle before executing it. Recurrently, many tools are available to support security issues rising in software development. Static analysis tools (SATs) are one of the most important activities that discover vulnerabilities in an early stage.

In fact, research estimates show half of the security vulnerabilities are detected by static analysis tools (SATs). Static analysis has some limitations to cover all vulnerabilities, such as missing application flaw because of the complex structure, raising a false alarm, and missing vulnerabilities. Therefore, each tool gives different results according to their time complexity and analysis of the source code. Thus, selecting SATs that best fit for your perspective application is a difficult task.

Benchmarking is an approach that helps to select the standby SATs by comparing their testing result and behavior during testing relevant application. However, the presently available SATs benchmark is restricted, well-known software assurance metric and tool evaluation (SAMATE) project from NIST [2] and the OWASP benchmark for security BSA [3]. These benchmarks also not give the best result for real-life applications and also lacking the ability to identify the context are noncritical or critical applications that shape the produced result. This paper suggests to focus on building benchmarks to evaluate vulnerabilities of SATs at every critical level of applications. Differing from SAMATE and BSA, we propose to apply workload on the real application that has most of the vulnerabilities. This convinces that SATs are testing the way how source code builds and gives the complexity of the code and also addresses the vulnerabilities presented in the code. Additionally, by inspecting the use of scenarios of Web application (a scenario used for vulnerability detection that based on the criticality level of the Web application) in benchmarking, two methods are used: First is to apply metrics in application and second is to compose workload. Application scenarios have criticality levels; according to that criticality, several metrics are used in between different scenarios, one main metrics and tiebreaker metrics for each level of scenarios, where the main metrics is used for ranking the tools and the tiebreaker mentions the ties between two and more scenarios. The workload is considered as a characteristic set of vulnerable source code for each of the scenarios that must follow the standard agenda to assign source code to scenarios according to code criticality levels. The application (source code) quality measurement is based on a quality model standard ISO/IEC 9126 [2], that is based on a set of SCMs, which are associated with nonfunctional demands and it can be obtained before running the application. In this approach, build a benchmark for ranking the SATs on WordPress plugins [1]. WordPress plugins are widespread content management system which is used on Web services.

The summary of the paper [1] is as follows:

(1) Design a benchmark for SAT evaluation that tends to detect the vulnerabilities present in the applications, considering different rank metrics and workload that include vulnerable application characteristic scenarios with different criticality levels.
(2) Collecting vulnerable application to build workload and characterizing them by a vulnerable or non-vulnerable line of code and assign them to the scenarios.
(3) According to the software quality of application, process of assigning them to scenarios is done.
(4) The rank evaluation obtained with the help of OWASP's BSA benchmarking and SAMATE methodologies.

2 Background Work of Benchmark

This partition presents a review of SATs that tend to detect the vulnerabilities of applications, and how they are compared with existing benchmarks. Software quality model is also included to show how the workload is applied to scenarios.

2.1 Benchmarks

Standard, or a set of standard, that use as a set of reference for evaluating performance or level of quality of the tools that access and compare the performance of the tools and show the difference in result that such process is called by benchmark. It typically includes three main constituents:

(1) Mechanism and order for benchmark execution.
(2) Workload is a set of characteristic test cases of benchmarking SATs.
(3) Metrics compare tool fitness under benchmarking for their motives.

The workload is the constituent of benchmarking that is mainly used on it and helps to find the result. Thus, workload verifies the following properties:

Representativeness: Workload that applied on the benchmark should be typical of the domain and also be influenced by the variety and the test case size. Benchmark should give a related result to the placed users that are planned for use.

Comprehensiveness: Workload should exercise the major features used by the desired domain and that attributes should be maintained according to the test case usage.

Focus: In case of characterizing the target in benchmarking, workload should be centralized.

Configurability: Workload should be user-friendly, in case of user want to customize workload according to their requirement.

Scalability: Workload should be scalable in the feature. Test case number and complexity can be increased and decreased.

2.2 Vulnerability Detection of SATs

OWASP top 10 of Web application security has top 10 vulnerabilities occurred in applications. XSS (cross-site scripting) and SQL (SQL injection) are most of the exploited ones. An XSS attack is the JavaScript injection in vulnerable applications, and SQL is the source code injection that affects or changes SQL query that is sent to the ending database [4]. These attacks are very threatening; since they can manipulate the database, they are allowed to perform unwarranted action like accessing privileged database account, updating the unauthorized data, injecting malware and virus, and performing operation behalf of the legal user when they are allowed to access their account and can impersonate the legal user. Most of the SATs add a feature to detect XSS and SQLi. SATs are scanning source code before executing it and detect most of the vulnerabilities in the development phase. However, SATs are also limited in nature and can include some vulnerability itself because some of the programming structure is difficult to analyze. Therefore, SAT developer assuming by producing a nearby solution leads to a false alarm and missing vulnerabilities.

Different tools give different results according to their algorithm complexity because their development technologies differ. Resulting false alarm tacks longer time to examine the problem that may lead to exploit more vulnerabilities.

2.3 SAT Benchmark

To evaluate SATs, metrics (recall, coverage, discrimination, precision, F-measure) are used that help to improve the effectiveness of SATs. To calculate metrics, three test case characteristics are used such as relevance, statistical significance, and ground truth. These are publically not available, and creating them needs lots of effort. Thus, what we can do is to find test case combination of those characteristics: production software (statistical significance and relevance), common vulnerability enumeration (CVE) software, and synthetic test cases (ground truth and statistical significance). OWASP BSA [3] and SAMATE project from NIST [2] are two SAT benchmarks. SAMATE project creates a methodology to show the weakness of SATs and against it find the capacity through tool metrics, development of tool function specification, and test suites. C/C++, PHP, and Java are the test suites of the workload, and variety of test case is included on applications. For evaluating the tools metrics are used that show the tools are a recall, false-positive rate, and precision. To examine the SAT services, BSA from OWASP is used to check the accuracy, speed, and coverage.

And to examine the SAT effectiveness, BSA and SAMATE benchmarks need manual work: in the synthetic workload running the SATs can detect the vulnerabilities, there a common format for the result so SATs result should be in that format, also there a target result for test cases so the resultant one must be compared with it and evaluation metrics will compute. A synthetic workload is the main restriction of the SAMATE and BSA, which is mostly collected by small-scale test cases with few numbers of programming constructs. In benchmarks, propose a workload which is controlled software in between production. The workload is characterizing vulnerability and non-vulnerability in an application.

2.4 *Quality of Software*

In traditional software quality model lack of ability to fully access Web application quality, since Web application becomes essential for the user. Thus, in Web application, many developers build tools to improve the quality characteristics of the software and also proposed the extended ISO 1926 software quality model with additional quality for the Web application such as security, reusability, popularity, and scalability. The static analysis community identifies that source code evaluation is harder. The NIST workshop on software measurement and metrics for vulnerability detection or reducing it advised that code should be easily persuaded or controlled. It is used to make tools less vulnerable and less complex code which gives the better result [5].

3 Benchmarking Approach

Our benchmarking proposition is a specification form approach which defines the target tools have to achieve the functions of the benchmarks. The main idea is with the use of real-world vulnerable software, run the target tools and collect all the vulnerabilities and check their correctness [6]. With the use of metrics, obtain each scenario rank and detect their capabilities. Using heterogeneous component, high quality of application can be built, and to make all SAT feasible, diversity of vulnerability is used. This paper shows that workload affected by the benchmark domain includes the Web application class selection. Also, the SATs can detect the vulnerability classes. Since workload is not fulfilled all criteria at each case, thus balancing those criteria with the help of determining workload strength and weakness to make workload result better, their multiple workloads needed. Therefore, the specific scenarios of workload are defined to build workload representation of vulnerabilities for real-world software [7]. Our approach consisted of four components, and these are described as follows:

3.1 Scenarios

According to the security issues, scenarios should be based on the requirement of the application of the organization with the use of available resources in the development phase. Therefore, all of the presented resources need to make sure to check all the vulnerability presented in the SAT. In this approach, according to the criticality, level of workload scenarios is evaluated. Scenario names are mentioned according to Antunes et al. [8]. Such as,

(1) **Highest quality**: Missing all of the vulnerability can give the huge problem to the application criticality. The aim of this task is to report much false alarm with SAT highest number of vulnerability.
(2) **High quality**: Criticality level of application is not highest but have few of vulnerabilities and some false alarm, so the main task of this scenario is to still give the highest number of vulnerability with not much false alarm there false alarm will be very less.
(3) **Medium quality**: During the process of reducing the false alarm, vulnerability will be missing. Thus, the task of these scenarios is to return very less amount of false alarm and vulnerability can be negligible or skipped.
(4) **Low quality**: Vulnerability detection is very important for the workload. During the strict restriction of vulnerability detection, everyone will be detected. The goal of this scenario is to report the lowest number of SAT false alarm with reporting vulnerabilities.

With the help of these scenarios, developers can easily make decisions because it can control the static analysis expected outcomes that suited in scenarios.

3.2 Metrics

Metrics are the best appropriate proposal for the vulnerability detection it helps to give a comparison between the benchmarks and can rank them. Metrics are used between each of the scenarios to detect the vulnerabilities in each stage there is one main metrics to provide ranking order of tools and there is a tiebreaker metrics that perform an operation between two tools which gave the same result or tie between them. However, here, operation of the metrics is to perform the vulnerability detection and metrics is very effective in between scenarios of SAT, considering that in workload a number of P (positive, vulnerability) instances are less frequent than the N (negative, non-vulnerability) instances. Then, the metrics F-measure and recall are taken focus on all P instances, markedness and informedness is focus on all N instances and all P instances, and precision has focused on all N instances, it means TP (true positive, vulnerability classified correctly), FN (false negative, vulnerability incorrectly classified) give vulnerability missing alarm, FP (false positive, non-vulnerability incorrectly classified) give false alarm, TN (true negative, non-vulnerability correctly classified).

(1) **Precision**: Since N instance is more frequent than P instance, it gives less FP report because precision is focused on all N instances. This metrics is used as a tiebreaker metrics for the list of the tools that report the exact number of the vulnerability, and for SAT, best are with the highest precision.

$$\text{precision} = \frac{\text{TP}}{\text{TP} + \text{FP}}$$

(2) **Recall**: Since N instance is more frequent than P instance recall use of all the P instances, it means true-positive vulnerabilities are classified correctly. SAT ranking reports the highest quality scenarios require the highest number of TPs, and there is a tie in high quality and medium quality.

$$\text{Recall} = \frac{\text{TP}}{\text{TP} + \text{FN}}$$

(3) **F-measure**: F-measure is measure, which weighs recall higher than precision and the measure, which weighs recall lower than precision, and it means F-measure is a harmonic means of recall and precision. This shows that TPs have twice-weighed of the FPs, and this kind of case is suitable for medium-quality scenarios where instead of fixing more vulnerability it prefers to fix less vulnerability.

$$\text{F-Measure} = \frac{2 * \text{TP}}{2 * (\text{TP} + \text{FP} + \text{FN})}$$

(4) **Informedness**: Since N instance is more frequent than P instance, every FP decreases proportion of metrics by $1/N$ and every TP increases proportion of metrics by $1/P$, however, that shows that SAT can predict the outcome of the TN and FP. Since SAT predicts the priority of the outcome of vulnerabilities or at the time they are not much FP report so that is the result of high-level scenarios.

$$\text{Informedness} = \frac{\text{TP}}{\text{TP} + \text{FN}} + \frac{\text{TN}}{\text{FP} + \text{TN}} - 1$$

$$\text{Informedness} = \text{Recall} + \text{inverse Recall} - 1$$

(5) **Markedness**: Metrics identified how many numbers of negative and how many numbers of positive proportion of vulnerabilities are presented and find the sum of those proportions. That is, the reason metrics are called a marker. Based on

the precision focus only *N* instances on FPs reporting of SATs, the vulnerability reporting of SAT is better in terms of reporting FPs. This case satisfies the low-quality scenarios where there are no FP addressing resources. And in the inverse precision, *P* instances of FNs reporting are done. Thus, SAT's metrics with the same inverse precision, SAT vulnerabilities reporting are to be ranked. That is the low-quality scenario.

$$\text{Markedness} = \frac{\text{TN}}{\text{FN} + \text{TN}} + \frac{\text{TP}}{\text{FP} + \text{TP}}$$

$$\text{Markedness} = \text{Precision} + \text{inverse} \;\; \text{Precision}$$

3.3 *Workload*

The workload is a vast set of real application of various sizes. A workload is created according to the industry requirement to find the vulnerability presentation. But workload has a drawback to not to exist and create, and its impractical task ingests vast resources. For reducing this problem, propose a method to combine different SAT results, and for finding vulnerabilities and non-vulnerabilities of software, show manual review [9]. The process proposed has three stages which are discussed are as follows:

3.3.1 Phase 1 Source Code Selection for Vulnerable Application

This method follows several steps for the selection of a characteristic set of vulnerable application for workload define. Steps include as follows:

(1) From benchmarking domain selecting application that contains source code because for vulnerability detection, SATs require source code.
(2) According to the benchmark domain, vulnerability classes will be chosen.
(3) From the selected application, all their vulnerabilities will be selected from their data repository.
(4) Selecting vulnerability can show that application is affected by the attack. Mean vulnerability with proof of concept (PoC).
(5) From source code repository, applications with vulnerabilities have to download.

NIST benchmarking and OWASP benchmarking have vulnerability representativeness since it occurs in real application that makes the great methodology.

3.3.2 Phase 2 Allocating Application in the Scenario

To create a workload, in each scenario, needs to assign vulnerable application representatives. This takes two steps:

(1) Rating of application: ISO/IEC 1926 and SCMs is a standardized model of application rating. SCM measures the properties of a software product.
(2) Applications for scenarios: This schema shows mapping the rating of each scenario. The scenarios vary from the level of criticality (low to high) in descending order, and the rating is in quality order in ascending order. Thus, there is some mapping rating for each scenario. If the mapping rating of the scenario is high, then it is the highest quality scenario, and if the mapping rating is low, then it is a low-quality scenario. Like that mapping, the rating was done for high-level and medium-level scenarios also by using the one-value interval to map the rating of each scenario quality.

3.3.3 Phase 3 Non-vulnerability and Vulnerability Identification

For SAT evaluation, need to identify the LOC (line of code) is vulnerable or not means which LOCs contain the *P* instances and which contain *N* instances. *P* instance LOCs are called as VLOCs (vulnerable line of code), and *N* instance LOCs are called as NVLOCs (non-vulnerable line of code). This is a tough task to identify the VLOCs and NVLOCs in such a large code for the security testing persons. And the result accuracy will also not high. To identify this problem, some procedure is to be followed.

(1) **VLOC characterization**: Workload is the collection of the vulnerabilities, because it is an initial source code that comes from the user database. The best approach to find the VLOCs is to run the workload in several SATs, and for the result confirmation, manual review can be used. To find the VLOCs, vulnerabilities of the workload will be detected by running the SATs in the selected applications and the result of SATs was combined and also review the result manually to identify it is FP (non-vulnerability) or TP (vulnerability).
(2) **NVLOC characterization**: The NVLOCs are depended on the FP report of the tools which are used in the process. Therefore, if the FPs are less in the tools, report the set size will be small. Thus, the values of the metrics are depending on how many NVLOCs are obtained. The effective way of NVLOC identification is to find the difference between LOCs and VLOCs. However, the result shows NVLOCs in LOCs. In such case, metrics have very small effect by FPs.
(3) **Achieve NVLOCs and VLOCs**: To achieve NVLOCs and VLOCs following steps are included:

(1) Select a set of SATs with the configuration setting definition that can identify the list of NVLOCs and VLOCs.
(2) Running SATs for vulnerability detection on the specified workload application. Result shows list of VLOCs.
(3) Perform a manual review to verify the vulnerability of tools. Also, differentiate them such as NVLOCs or VLOCs.
(4) Build the list of VLOCs by combining the vulnerability and the initial list.
(5) Build a set of NVLOCs by combining the LOCs and the FPs of the tools. But the VLOCs are excluded from the LOCs.
(6) Characterize the list of the VLOCs with some information like vulnerability classes, list of LOCs, initial code, and vulnerability files.

It is a fact that the process of NVLOC and VLOC extraction also leaves some undetected vulnerability during benchmark execution. This needs to identify the manual review for the FP and TP classification. This process constantly updates the NVLOCs and VLOCs that can change the rank of the SATs and the value of the metrics which also required to update. The best approach to overcome this problem is to use the various SATs in the LOC characterization.

4 Conclusion

In this paper, for the vulnerability detection, selection of SATs was a big problem. This paper shows those problems in Web applications and proposes the benchmarking evaluation design according to the criticality levels of SATs. This approach organizes the workload in scenarios according to the level of criticality increase. For each scenario, different metrics are to be used for the tool ranking. With the use of manual review of the workload, characterize the NVLOCs and the VLOCs of the SAT LOCs.

References

1. Nunes P, Medeiros I, Fonsica JC, Neves N, Correia M, Vieira M (2018) Benchmarking static analysis tools for web security. IEEE Trans Reliab 67(3):1159–1175
2. (Online) Available: http://samate.nist.gov/, Accessed on: Jun 6 2015
3. Online Available: http://www.owasp.org/index.php/benchmark, Accessed on 10 Apr 2016
4. Avizienis A, Laprie J-C, Randell B, Landwehr C (2004) Basic concepts and taxonomy of dependable and secure computing. IEEE Trans Depend Secur Comput 1(1):11–33
5. Okum V, Guthrie WF, Gaucher R, Black PE (2007) Effect of static analysis tools on software security: preliminary investigation. In: Proceedings of ACM workshop quality, protection, pp 1–5
6. Albreiki HH, Mahmoud QH (2014) Evaluation of static analysis tools for software security. IEEE, pp 93–98
7. Delaitre A, Stivalet B, Fong E, Okun V (2015) Evaluation bug finders: test and measurement of static code analysis. In: Proceedings of 1st international workshop complex faults failure large software system, pp 14–20

8. Antunes N, Vieira M (2015) On the metrics for benchmarking detection tools. In: Proceedings of IEEE/IFIP 45th annual international depend system network, pp 505–516
9. Masood A, Java J (2015) Static analysis for web service security-tools and techniques for a secure development life cycles. IEEE, pp 1–6

Farmer the Entrepreneur—An Android-Based Solution for Agriculture End Services

Jayashree Agarkhed, Lubna Tahreem, Summaiya Siddiqua and Tayyaba Nousheen

Abstract Rapid growth of mobile telephony and the introduction of mobile-enabled information services to provide ways to improve information dissemination to the knowledge-intensive agriculture sector and also help to overcome information asymmetry existing among the group of farmers. The agriculture role in the society and economy of various countries is above the per capita income. The enormous amount of population survives on their little source of farming for their living and the state of being comfortable. All the farmers overlook unusual challenges frequently because of the absence of approach to knowledge about prices and trades.

Keywords Agriculture · Farmer friend · Crop · Fertilizer

1 Introduction

Many of applications are Android-based with exploration in technology and trend carried out in the today's world. Android applications lack to satisfy requirements of agricultural field. The proposed work provides an optimal solution to this by providing information regarding the crops, their market rates and other farming information to the farmers. Mobile apps in the arena of agriculture can be the best option to increase countries agriculture production. The inventions in technology in agriculture domain are not getting to the farmers because of either most of them are illiterates or due to unawareness of the location from where they can have information.

J. Agarkhed · L. Tahreem (✉) · S. Siddiqua · T. Nousheen
P.D.A College of Engineering, Kalaburagi 585102, Karnataka, India
e-mail: lubnatahreem789@gmail.com

J. Agarkhed
e-mail: jayashreeptl@yahoo.com

S. Siddiqua
e-mail: Summaiyatech@gmail.com

T. Nousheen
e-mail: Tayyabanousheen081@gmail.com

R. K. Shukla et al. (eds.), *Social Networking and Computational Intelligence*, Lecture Notes in Networks and Systems 100, https://doi.org/10.1007/978-981-15-2071-6_28

Farmers are capable with a low-cost smartphone and the particular software to gain facilities which could not be available in their hands before.

This paper is organized in five sections; Sect. 1 outlines the introduction of the service work, and Sect. 2 represents the related work. Section 3 represents system architecture and proposed algorithm. In Sect. 4, results are discussed with the simulation environment and snapshots. Section 5 concludes.

2 Related Work

The review study provides an overview of different methods and techniques involved in the growth of an Android application for agriculture-based proposed solutions and services.

Patil and Nikam [1] proposed—Automation in Farming using the Android Application—a new architecture for remote control of agriculture devices. The author presented automated controller using Bluetooth devices and microcontroller. The project works automatically and hence reduces the man power. The main idea behind the proposed architecture is to design system, which would be used as a platform which provides the services needed to perform remote control of agricultural devices. The farmer will be able to on/off the irrigation device, cultivation device, seeding device, decide the pesticides proportion and monitor the farming activities remotely. This system provides reminder to the user so that their farming activities will take place on time and also provide all online information about any particular crop.

Sharma et al. [2] proposed—E-agro Android Application—the software application is used to maintain progress of farmers. Huge amount of farmer are unable to take decisions relating to in choosing the right fertilizer, pesticide and time to do particular farming actions. This application is extremely beneficial to avoid these problems. For each type of crop, a fertilizer program is recorded. As per schedule, the farmers will get reminders about fertilizer, herbicide, pesticide and weather alerts if a crop surpasses its temperature range depending on the sowing date of a crop. For efficient and stable farming, the system merges the Internet and mobile communication systems with GPS. The author has made the brief study of some common problems faced by the farmers across the nation.

Santosh and Sudarshan [3] proposed—A Modern Farming Techniques using Android Application—with the various ways in which a farmer can yield better crop with better way of cultivation of crops and merchandise. The mobile computing (MC)-based application assists farmers for better cultivation. This system not only provides weather updates but also helps assist farmers to get market updates. The framework uses MC, which in effect, puts power consumption issue into a farmer's hand.

Geo et al. have proposed [4]—Study and Implementation of Agricultural SMS Management System— an short message service (SMS) for agricultural management utilizing GSM modem for message transmission and reception. The proposed system maintains SMS in database tables. Users can send request of query to the platform in short messages through a mobile phone.

Singhal et al. [5] have introduced—Krishi Ville Android founded rectification for Indian agriculture industry— mobile-based application known as—Krishi Ville—availing farming recreation. An application considers agriculture properties and its improvement with proper updates of weather conditions.

Yukikazu et al. [6] have introduced—ifarm evolution of Web-based mode cultivation and expense orientation for agriculture—to facilitate farm management. The working mode of application was structured as of smartphone application with Web browsers support integrated with cloud server. The Web browser assists farmer to fetch data from cloud allowing sharing of data with main office. Farmers on farm can cheaply consign to work plans, either field source into the cloud system, or share them with main office. Farmer at main office examines the data in the cloud system, calculates farming costs and forms work plans based on their investigation.

Jagyasi et al. [7] proposed event-based experiential computing in agro advisory system for rural farmers. The assortment and impersonation of agriculture events, peroration of the agricultural experiences and a system to browse through the history of agriculture intellect are the central endowment in the present work.

Umadikar et al. [8] present a functioning—A functioning personalized ICT-based agriculture advisory system: Implementation—that has been built with the aim of bridging information gaps between farmers and agriculture knowledge workers. The works discussed only the benefits of using an ICT approach to provide personalized agricultural advisories, and this paper covers details of the technology implementation, presents of brief summary of the impact analysis carried out with the farmers registered into our system and discusses new features that could make the system more effective.

De Oliveira et al. [9] proposed development of an agricultural management information system based on open-source solutions. The author has proposed an detail data procedure planned toward the management of agriculture resources, with the name—Agrifootprint system. The system supported geographical information designed using Web technology and agile to provide user-friendly process for merging and control of data from various agriculture enterprises.

Narkhedel [10] proposed Kisan Monitoring System Focused on Android-based Application. An Android-based agriculture system namely Kisan monitoring system with the proper framework and also the database used for the system. The Android application framework is designed to simplify the reuse of components.

Mittal and Mehar [11], proposed—How Mobile Phones Contribute to Growth of small Farmers? Evidence from India—the discussed the use of mobile phone to provide information to farmer as reply to their query such as farm income and cost of manufacture in India. In order to improve the incomes of number of small farmers and their productivity, the author has focused on complete aim and consequences.

Patel and Patel [12] proposed—survey of Android apps for agriculture sector—the purpose of this paper is to explore how Android apps of agriculture services have impacted the farmers in their agricultural interest. The author surveys the certain

Android app in the different area of agriculture and found that all apps are having dissimilar features.

The author of—Socio-Economic Impact of Mobile Phones on Indian Agriculture—[13] has investigated a series of questions to design a system that assist information service providers, mobile operators and for policy makers. The primary issue in the current system is interpretation of farm-gate price dispersion for a staple grain and money in cash crop and to correlate the observable or not to be observed attribute with mobile phone coverage. The start of cell phone may decrease farmers search cost across markets as compared to personal travel. Further, it may provide price data from medium and large agricultural markets, but it does not cover data from small (e.g., less than 20 traders) and remote agricultural markets. A second issue is that the current dataset focuses on producer prices.

Indian farmer never uses proper management and marketing for crop cultivation. Absence of co-ordination beside the agricultural worth series from farm inputs to farm operation, the cost of production increases and lowers revenue for farmers is a further problem. And for getting advice, farmers need to visit local agriculture office or visit other expert farmer. If there is proper management and organization, then that will show very good result in crop yields and hence gives more income to farmers [14, 15].

The outcome of survey of existing system from the review study shows the following open issues left unattended.

- Existing apps do not provide necessary information at a stretch.
- Apps that are already in use are static.
- Present apps do not give the information about transportation and storage efficiency.
- Farmers are unable to make decisions about whether to sell their crops or not.
- Does not analyze the current situation.
- English is generally supported which majority of farmers does not understand.
- Poor performance.
- Some apps support newer Android versions only.

3 System Architecture

An extensive literature survey for agricultural services using latest technology is done. This includes the overall block diagram of the system, various modules of the proposed system, their working flows, algorithms and flow charts.

Example Domain

The proposed system Farmer Friend—an Android app consists of two entities or modules namely admin and user. Admin is the person who will be having an modules namely admin and user. Admin is the person who will be having an authorized access to the application. The duties of the admin are to update the crop price and to post

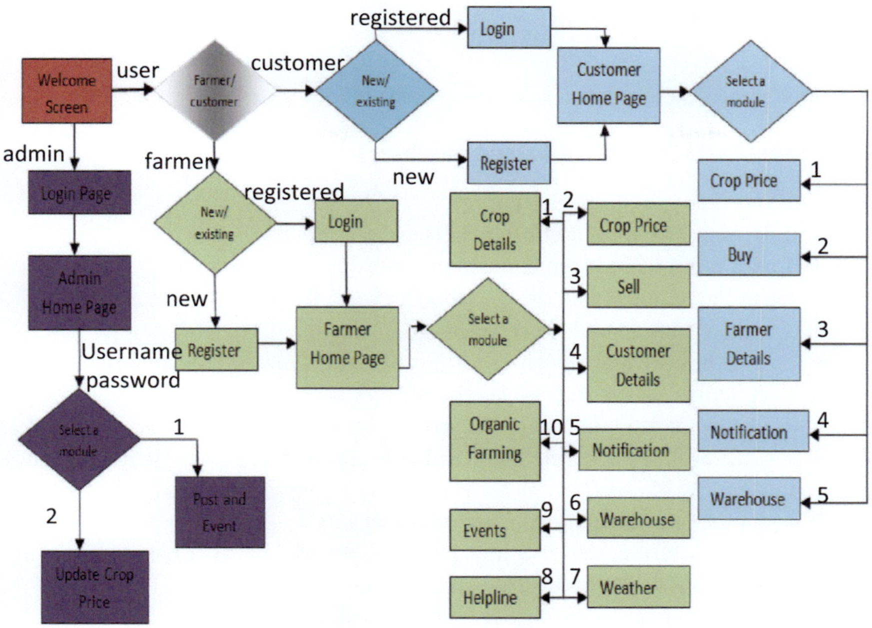

Fig. 1 Block diagram representation of the proposed system

the events for the farmers if any. A unique username and password are given to admin. Users are mainly the farmers, but sometimes customers may also need to use the application to buy the crops or to get the information about crops prices, warehouses, etc. So, users are either farmers or customers. The proposed system has been represented with the block diagram as shown in Fig. 1.

They are two key modules in the system

(1) Admin, (2) User.

Admin and user can login to get their interested information by entering mobile number and password if he/she is existing in the system through login page. Admin can update the information whenever needed. Users are either farmer or customer that they will have the secured access to information. If there is a new user, it should first register. The farmer information like farmer name, phone number, location of farmer, etc., provided through registration form to allow them to have the access to app and make use of it. There is a list of modules such as crop details, crop price, sell, customer details, notification, warehouse, weather, helpline, events, organic farming that can be accessed by farmer user, and the modules of customer user are crop details, crop price, buy, farmer details, notification, warehouse, weather, helpline, events and organic farming. Crop price module will provide the list of crops with their current market rates according to quality.

3.1 Proposed Algorithm

The algorithm of the proposed system is shown below.

Algorithm: FARMER_FRIEND (farmer/user)

Step 1: Select the type of user (admin/user)
Step 2: if the user is admin, go to step 3 else go to step 9
Step 3: if admin has valid username and password, go to step 4
Step 4: select the required module (price/post event) from admin home page
Step 5: if update price module is selected, go to step 6
Step 6: Update the crop price by entering the crop name, max price, min price, variety and click on update button then go to step 32
Step 7: if post-event module is selected, go to step 8
Step 8: post the event by entering in an available edit box and click on update button then go to step 32
Step 9: select the user category (farmer/customer)
Step 10: if user is farmer and is a new member, register but if it has login details, then go to step 11 else go to step 26
Step 11: select the required module from the farmer home page
Step 12: if the crop details module selected, go to step 13
Step 13: A list of crops is displayed, choose the required crop, crop information will be displayed. Click home button to move to the crop list then go to step 32
Step 14: if the crop price module selected, go to step 15
Step 15: A screen with an edit box and a button is displayed. Enter the crop name in edit box and click the search button. The price of the entered crop will be displayed then go to step 32
Step 16: if the organic farming module is selected, a list of organic farming techniques is displayed, choose the required technique, information about the selected technique is displayed. Click home button to move to the technique list then go to step 32
Step 17: if the weather module is selected, weather information is displayed then go to step 32
Step 18: if the events module is selected, events posted by admin are displayed then go to step 32
Step 19: if the warehouse module is selected, select the warehouse type, type of good to be stored in warehouse and the rent range per square feet per day. Warehouses satisfying the mentioned needs will be displayed then go to step 32
Step 20: if the helpline module is selected, helpline contact numbers and details of agriculture officers are displayed. Hold the number for some time to make a call then go to step 32
Step 21: if the sell module is selected, go to step 22
Step 22: A form for selling the crop opens. Fill the form by entering appropriate particulars. Click update button then go to step 32
Step 23: if the notification module is selected, go to step 24

Step 24: A notification about the customers interested in buying a particular crop the farmer has registered is displayed. If the farmer is interested to sell the crop, click SMS button, enter the details of crop in the form that opens, click the send button and then, go to step 32
Step 25: if the customer details module is selected, customer information is displayed then go to step 32
Step 26: if the user is customer and is new member, first register but if it has login details, then select the required modules from the customer home page
Step 27: follow the steps from step 11 to step 20 to access some of the customer modules which are similar to farmer modules and to access remaining customer modules go to step 28
Step 28: if the buy module is selected, a form for buying the crop opens. Fill the form by entering appropriate particulars. Click update button then go to step 32
Step 29: if the notification module is selected, go to step 30
Step 30: A notification about the farmers interested in selling a particular crop the customer has registered is displayed. If the customer is interested to buy the crop, click SMS button, enter the details of crop in the form that opens, click the send button and then, go to step 31
Step 31: if the farmer details module is selected, farmer information is displayed then go to step 32
Step 32: Click back button to exit the module.

4 Results and Discussions

The proposed Android application is compared with the existing applications by checking number of end services supported by each app, and a conclusion is drawn that the proposed application tends to provide majority of end services and helps

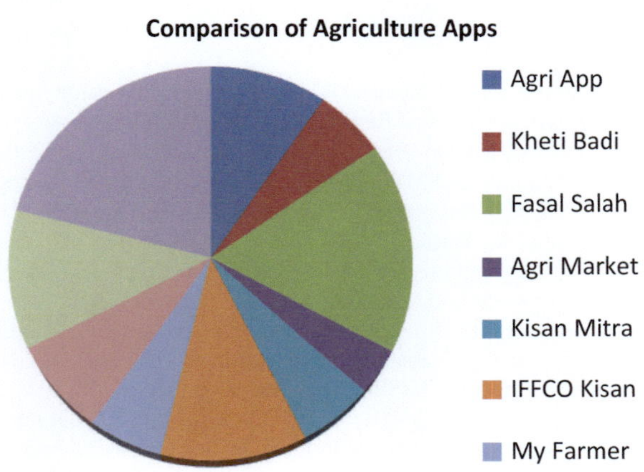

Fig. 2 Pie chart showing performance of farmer friend in providing end services

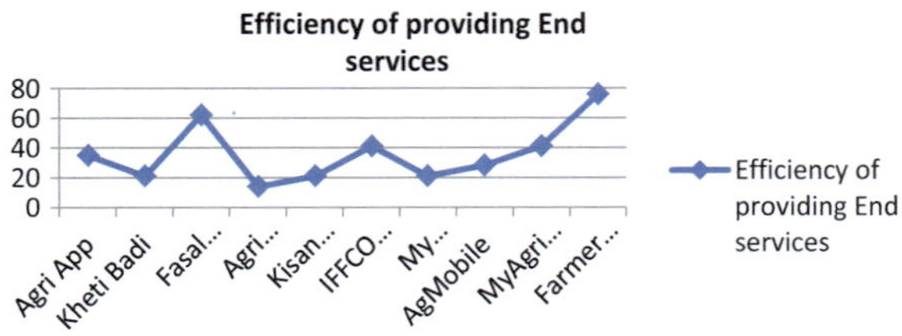

Fig. 3 Graph showing the farmer friend with approximately 91% efficiency

farmers to improve the crop growth and crop selling by providing single source of information (Figs. 2 and 3).

5 Conclusion

This kind of system can help in providing helpline to farmers in an innovative way, and definitely, it would be beneficial in the growing and cultivation of the crops. Another thing that can bring more usability of the proposed app is the server. An admin module can be replaced by a server. Where a person operating the server can make the price and event updates, additionally the contents such as crop information, details of organic farming techniques can be changed whenever needed that is the application can be made dynamic and hence could be served as the real-time dynamic project and will definitely help farmers in almost all stages of crop growing, cultivation and selling.

References

1. Patil SS, Nikam VD (2016) Automation in farming using android application. In: Proc. International Conference on Recent Innovations in Engineering and Management, pp. 572–576
2. Sharma S, Patodkar V, Simant S, Shah C, Godse S (2015) E-agro android application (integrated farming management systems for sustainable development of farmers). Int J Eng Res Gen Sci 3(1)
3. Santosh KG, Sudarshan GG (2015) A modern farming techniques using android application. Int J Innov Res Sci, Eng Technol (IJIRSET) 4(10):1–8
4. Gao W, Zhang G, Jiang X, Wang Q, Yu L, Lu L, Li J (2009, July) Study and implementation of agricultural SMS management system. In: 2009 International conference on information technology and computer science, vol 1. IEEE, pp 468–471
5. Singhal M, Verma K, Shukla A (2011, December) Krishi Ville—android based solution for Indian agriculture. In: 2011 Fifth IEEE international conference on advanced telecommunication systems and networks (ANTS). IEEE, pp 1–5
6. Yukikazu Murakmi Dept of Electrical and computer Engineering, Kagawa National College of Technology, Japan "ifarm evolution of web based mode cultivation and expense orientation for agriculture". INSPEC Accession Number:14650001
7. Jagyasi BG, Pande AK, Jain R (2011, October) Event based experiential computing in agro-advisory system for rural farmers. In: 2011 IEEE 7th international conference on wireless and mobile computing, networking and communications (WiMob). IEEE, pp 439–444

8. Umadikar J, Sangeetha U, Kalpana M, Soundarapandian M, Prashant S, Jhunjhunwala A (2014, August) Mask: a functioning personalized ICT-based agriculture advisory system: implementation, impact and new potential. In: 2014 IEEE region 10 humanitarian technology conference (R10 HTC). IEEE, pp 121–126
9. De Oliveira THM, Painho M, Santos V, Sian O, Barriguinha A (2014) Development of an agricultural management information system based on open source solutions. Procedia Technol 16:342–354
10. Narkhede S (2016) Kisan monitoring system focused on android based application. Int Res J Eng Technol 3(2):964–968
11. Mittal S, Mehar M (2012) How mobile phones contribute to growth of small farmers? Evidence from India. Q J Int Agric 51(8922016–65169):227
12. Patel H, Patel D (2016) Survey of android apps for agriculture sector. Int J Inf Sci Tech 6:61–67
13. Mittal S, Gandhi S, Tripathi G (2010) Socio-economic impact of mobile phones on Indian agriculture (No. 246). Working paper
14. Agarkhed J (2017) Agricultural applications using IoT based WSN. Int J Curr Adv Res (IJCAR) 6(9):6325–6329
15. Agarkhed J, Kashika (2016) iFriendly: a WSN based system technique for precision agriculture. In: 3rd International conference on microelectronics, circuits and systems, Micro2016. July 9th–10th 2016 in Science City, Kolkata, India, pp 272–282

Face Recognition Algorithm for Low-Resolution Images

Monika Rani Golla, Poonam Sharma and Jitendra Madarkar

Abstract Recently, in the field of face recognition, deep learning has achieved great success. FaceNet, one of the Google's deep learning frameworks, obtained nearly 100% accuracy in recognizing high-resolution faces. But, its performance on low-resolution images is unsatisfactory. Here, to justify the above statement, the performance of FaceNet has been evaluated on four well-known facial databases. For improving the performance, the face recognition algorithm for low-resolution images has been designed. Specifically, the algorithm performs image enhancement using a customized error back propagation network, and then the nearest patterns (NP) have been extracted and fed as embeddings to FaceNet. NP is a modified version of dual-cross patterns face descriptor that encodes the texture information of each image pixel in eight directions as unique features. The experimentation done on various databases such as LFW, ORL, EYB, and AR. The results of proposed approaches with NP have shown better performance than the other face descriptors like DCP, LBP, and LTP. Moreover, better results have been obtained on shrinking the Inception-Resnet-v1 network, which resulted in the fast convergence of the FaceNet.

Keywords Face recognition · Low resolution · FaceNet · Local binary pattern · Face descriptor

1 Introduction

In today's digital world, every single activity of the public is being monitored with the help of the video surveillance system. If any undesirable activity happens, then there comes the need of performing face recognition on the captured image in order to find

M. R. Golla · P. Sharma · J. Madarkar (✉)
Department of Computer Science and Engineering, Visvesvaraya National Institute of Technology, Nagpur, India
e-mail: jitendramadarkar475@gmail.com

M. R. Golla
e-mail: gmonikarani@gmail.com

P. Sharma
e-mail: dr.poonamasharma@gmail.com

R. K. Shukla et al. (eds.), *Social Networking and Computational Intelligence*, Lecture Notes in Networks and Systems 100, https://doi.org/10.1007/978-981-15-2071-6_29

the source (person) of such activity. Face recognition has attracted much attention of researches due to the security concern and becomes one of the most important applications. Several existing methods [1–9] have been proposed to resolve the issue of face recognition, such as linear variation, undersampled, occlusion, illumination, pose, expression, and low resolution. Some issues like low resolution need to explore more to gain high accuracy.

It has been observed that the distance between the camera and capturing object plays a vital role in face recognition. As distance between camera and object increases, it leads to reduced the object size in an image, and due to a few pixels of the object in the image, it loses the discriminative feature of the object. The low-resolution issue affects the performance of face recognition. Recently, the deep learning-based method [10] and face image descriptor methods are being used in recent research of face recognition.

The face recognition process is carried out in four processes: image acquisition, pre-processing, feature extraction, and classification. A classification and feature extraction play a major role in the recognition process. Feature extraction extracts a discriminative feature from the given images, and the classification process compares a test image features with the gallery training samples features.

Still, the face recognition have some challanges [11] and it needed to explore such as low resolution, undersampled, and pose variation. In this paper, the deep module has been used for face recognition and used descriptor features.

The rest of this paper is organized as follows: Sect. 2 reviews the significant issue of the existing work on face recognition. Section 3 provides a brief introduction to some of the state-of-the-art face descriptors that are used in face recognition. Additionally, their main drawbacks are also discussed. Section 4 provides a proposed methodology that explains about the four modules such as image resizing, image enhancement, feature extraction, and face recognition which are used in the design of the algorithm. Section 5 has shown the experimental results of the proposed approaches. Section 6 concludes the conclusion with the scope of future work.

2 Literature Survey

The classification process of the face recognition application must need the same resolution of training samples and probe sample but in the most practical scenario, it might be difficult to collect the same resolution of samples. In that case, it needs to bring in common resolution or convert high resolution into low resolution (LR) or vice verse. Several methods are used to bring different images in common resolution, such as scaling, inter-resolution, and downscaling. Aforementioned methods lose discriminative feature after changing the resolution of images. The first method converts low-resolution images into high-resolution (HR) images. The second method projects LR and HR space into common subspace with mapping methodologies. The third method converts high image resolution into low resolution, but it is not an efficient mechanism with a performance perspective.

A reconstruction method of upscaling approach has shown better performance over the interpolation technique. In the reconstruction method [12], divide the LR and HR images into a number of patches, where each patch contributes to a single model. In the case of low-resolution images, 20–30% accuracy is achieved by an interpolation method; whereas, 55–62% accuracy is achieved by the reconstruction-based method. Still, the results of the aforementioned algorithm are not up to the benchmark.

Shi et al. [13] introduced coupled mappings technique which projects LR and HR images into common feature space. Embedded space is beneficial for different resolution images, and this space is being used to measure the similarity of the different resolution images. The mapping preserves the most discriminative features of the image while making certain the consistency between HR and LR samples. This method helps to avoid the matrix singularity issue, and hence, provide consistent result. The principal component analysis (PCA) [14] method has been used to remove outlier in the inputted LR and HR face images, and it helps to enhance the performance of face recognition on LR images but it leads to high computational cost. The downscaling approach is used in [15], and it helps to reduce the size or dimension of the images, but reduced image loses discriminative feature, and this feature is not a good choice for face recognition.

The most straight forward super-resolution (SR) [16] method is used to solve low-resolution issue. This method first transfers LR images into HR space and then performs the classification task. Recently, several promising methods have been proposed for the issue of low resolution [17–20]. However, these methods have some limitation and did not explore fully. To improve the accuracy of low resolution images, a face recognition needed dismatched resolution based method.

3 Face Descriptors

Since decades, the face description-based face recognition has drawn attention for extracting patterns and projecting them as features. This is because of its capability of capturing each and every small appearance details. The well-known face descriptors such as dual-cross patterns (DCP), local binary patterns (LBP), local ternary patterns (LTP) and Gabor wavelets have their own advantages and disadvantages.

3.1 Local Binary Patterns

The LBP [21] extracts the texture information of the image, and it also invariant to the illumination effect. LBP is simple to implement and requires less computational cost, which makes it affordable to analyze the images. In LBP, the eight neighbors of each pixel are considered for sampling. LBP has high discriminative power for texture classification. Thus, a 3×3 matrix for each pixel is considered, and the pixel value gets compared with all of the neighboring pixel values in the matrix. If the value of a center pixel is greater than the value of compared neighbor pixel, then it is encoded as "0," otherwise encoded as "1" using the equation below.

$$\mathrm{LBP}(x_c, y_c) = \sum_{n=0}^{7} s(i_n - i_c)2^n$$

where

$$s(x) = \begin{cases} 1, & x \geq 0 \\ 0, & x < 0 \end{cases}$$

Several research papers have used LBP features in face recognition application. Most of the components of face are relative uniform; it has potential to improve the robustness of the underlying descriptors of face component.

3.2 *Local Ternary Patterns*

LBP is not robust to an outlier of the uniform region. The aformentioned drawback is resolved by an extended version of LBP called local ternary patterns [21], which is less sensitive and more discriminant to noise in uniform regions. Also, the central pixel can be easily influenced by noise. Since LBP could not give better patterns if noise images are to be considered, its extension known as LTP makes sure that comparison takes place with the help of threshold t as shown in the equation below.

$$\mathrm{LTP}(x_c, y_c) = \sum_{n=0}^{7} s'(i_n - i_c)3^n$$

where

$$s'(x, t) = \begin{cases} 1, & x \geq t \\ 0, & |x| < t \\ -1, & x \leq -t \end{cases}$$

LTP is encoded into three values $\{-1, 0, 1\}$, where 0 value is assigned to a neighboring pixel if its value is within a threshold range of central pixel, 1 is assigned to far greater value, and -1 is assigned to far smaller number. LTP is more powerful than LBP and also more discriminant and less sensitive to noise. But here, we need to set the threshold value manually. The threshold in LTP is not data-adaptive and does not robust to noise pixel.

3.3 Dual-Cross Patterns

DCP [22] samples two pixels in each of the eight directions for each pixel in an image. Thus, it makes sure that four patterns are designed in each direction and encodes using the below equation.

$$\mathrm{DCP}_i = S\left(I_{A_i} - I_O\right) \times 2 + S\left(I_{B_i} - I_{A_i}\right), \quad 0 \leq i \leq 7$$

where

$$S(x) = \begin{cases} 1, & x \geq 0 \\ 0, & x < 0 \end{cases}$$

In order to reduce the sampling size, the eight codes are divided into two subsets. The odd and even subsets are formed; if the distance between the pixels is maximum, then maximum Shannon entropy is achieved. The entropy feature represents the randomness of gray level distribution. If the entropy is high, the gray levels are distributed randomly throughout the image.

$$\mathrm{DCP} = \left\{ \sum_{i=0}^{3} \mathrm{DCP}_{2i} \times 4^i, \sum_{i=0}^{3} \mathrm{DCP}_{2i+1} \times 4^i \right\}$$

Upon dividing an image into non-overlapping grids of regions, the LBP, LTP, and DCP capture the texture information of each cell of an image and concatenate the histograms of those regions. These are taken as features, and thus, face recognition can be done with the help of any of the existing similarity matching techniques.

4 Proposed Methodology

The proposed methodology is a four-stage process as depicted in Fig. 1. The four stages are image resizing, image enhancement, feature extraction, and face recognition. Each stage has its own importance to make a better feature of low-resolution images and extract discriminative features that are needed for face recognition process.

4.1 Image Resizing

Since our focus is on low-resolution images, the images of high resolution have been resized down to low resolution. The MATLAB function, imresize() which uses

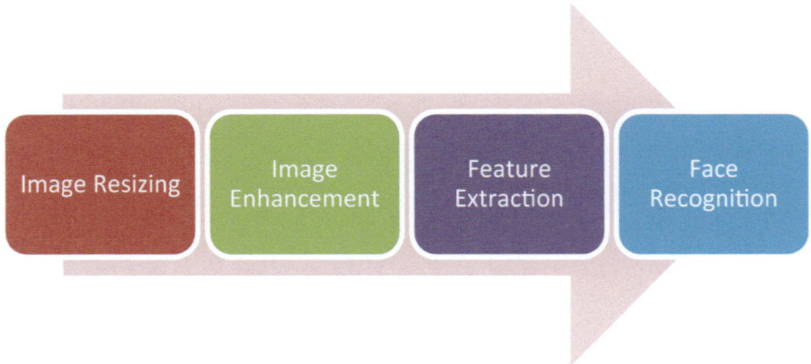

Fig. 1 Face recognition stages of the proposed methodology

nearest neighbor interpolation, has been used for resizing. The algorithm of image resizing is presented below Algorithm 1.

Algorithm 1: Image Resizing algorithm

```
for each folder, N in the LFW dataset
  | for each file, j in the N
  |   | im= imread(path[j]);       //read the image in the specified path of the file
  |   | imr= imresize(im, [15 15]);       //image resizeto 15×15 image size
  | end for
end for
```

4.2 Image Enhancement

The image enhancement helps to improve the quality of the given images for further analysis, and there is a need for performing image enhancement on the images. The goal of this stage is to avoid unwanted noise in the input images. For this purpose, the multilayer neural network made up of three layers has been built. The error back propagation algorithm, which is presented in Algorithm 2 was used for learning and updating the weights by applying error-correcting learning rule and stochastic gradient descent optimizer.

In the case of 15 × 15 resolution images, the network consists of 225 nodes in both the input and output layers each. And the hidden layer constitutes 25 nodes after performing trial and error. Only the subjects whose number of images is more than eight have been considered. Each pixel value of an image corresponding to each node in the input layer has been fed to the network. The parameters like weights and biases of the layers were initialized with random values (between −1 and 1). The initialization of the learning rate with 0.95 and the momentum with 0.95 has been done. And these two parameters vary during training by using step decay function after each epoch. Here, the weights are learned for each subject separately, and while testing if noise image is given, then it gets inclined to any of the trained images of the corresponding subject.

Algorithm 2: Error back propagation algorithm:

1. Let the learning rate be η, momentum be μ
2. Initialize the weights to some random numbers.
3. **repeat**
4. **for** each training sample, **do**
5. Input the training sample to the network and compute the outputs.
6. **for** each output unit k
7. $\delta_k = (t_k - O_k)O_k(1 - O_k)$
8. **end for**
9. **for** each hidden unit h
10. $\delta_h = O_h(1-O_h)\sum_{k_{outputs}} w_{kh}\sigma_{kh}$
11. **end for**
12. Update each network weight w_{ji}:
13. $w_{ji} = w_{ji} + \mu\, \Delta w_{ji}$ where $\Delta w_{ji} = \eta\, \delta_j x_{ji}$
14. **end for**
15. **until** a termination condition(no. of epochs completes) is met

The step decay of the parameters helped in the convergence of the algorithm without encountering the saturation problem. It drops the learning rate by a factor every few epochs. We have used the step decay function that drops the learning rate by half every 10 epochs. And the same is applicable for momentum.

The number of epochs used for training was 2500 since the error values obtained around this epoch number were less than 0.01 which is approximately zero (convergence). During the network training phase, the weights' updating takes place when all the images of a subject were fed once (epoch). This goes on for 2500 times, and then the resulted network with updated weights has been used for testing purpose. In the testing phase, each image of the subject was fed, and the corresponding predicted (enhanced) image was obtained from the output layer. These two phases were performed on each subject of the dataset, and then these enhanced images of the entire dataset were given as input to the following module.

4.3 Feature Extraction

After performing the image enhancement, local sampling has been performed on the resulted images. Since the components of the face extend horizontally or vertically and converge diagonally to the ends, the sampling is done in 0°, 45°, 90°, 135°, 180°, 225°, 270°, and 315° directions. In local sampling, for each pixel O, i.e., $M_{(x,\,y)}$ where

x and y are the coordinates of the image say I, the sampling has been done on the local 16 neighborhood pixels: $\{I_{(x-2,\, y-2)}, I_{(x-1,\, y-1)}\}$; $\{I_{(x-2,\, y)}, I_{(x-1,\, y)}\}$; $\{I_{(x-2,\, y+2)}, I_{(x-1,\, y+1)}\}$; $\{I_{(x,\, y+2)}, I_{(x,\, y+1)}\}$; $\{I_{(x+2,\, y+2)}, I_{(x+1,\, y+1)}\}$; $\{I_{(x+2,\, y)}, I_{(x+1,\, y)}\}$; $\{I_{(x+2,\, y-2)}, I_{(x+1,\, y-1)}\}$; and $\{I_{(x,\, y-2)}, I_{(x,\, y-1)}\}$ which covers eight directions that corresponds to the extension directions of major facial textures. The resulted sample points are denoted as $\{A_0, B_0; A_1, B_1; \ldots; A_7, B_7\}$.

The nearest patterns in each of the eight directions have been encoded, and these were combined to form the NP codes as shown in the below equation.

$$\mathrm{NP}_i = T\big(L\big(M_{B_i}, M_{A_i}\big), L\big(M_{A_i}, M_{O_i}\big)\big), \quad 0 \le i \le 7$$

where

$$L(x, y) = \begin{cases} 1, x \ge y \\ 0, x < y \end{cases} \text{ and } T(x, y) = \begin{cases} 3, x = 1 \wedge y = 1 \\ 2, x = 1 \wedge y = 0 \\ 1, x = 0 \wedge y = 1 \\ 0, x = 0 \wedge y = 0 \end{cases}$$

and M_O, M_{Ai}, and M_{Bi} are the grayscale values of pixels O, A_i, and B_i, respectively.

From Fig. 2, it is clear that the first four directions NP codes are reflecting the same result as remaining four directions NP codes. Hence, the nearest four directions NP codes are considered for further process. Number of NP codes reduces, hence reduces the time complexity of NP to the same level as LBP. This results in the subset $\{\mathrm{NP}_0, \mathrm{NP}_1, \mathrm{NP}_2, \text{and } \mathrm{NP}_3\}$. The NP descriptor of image pixel O is calculated by

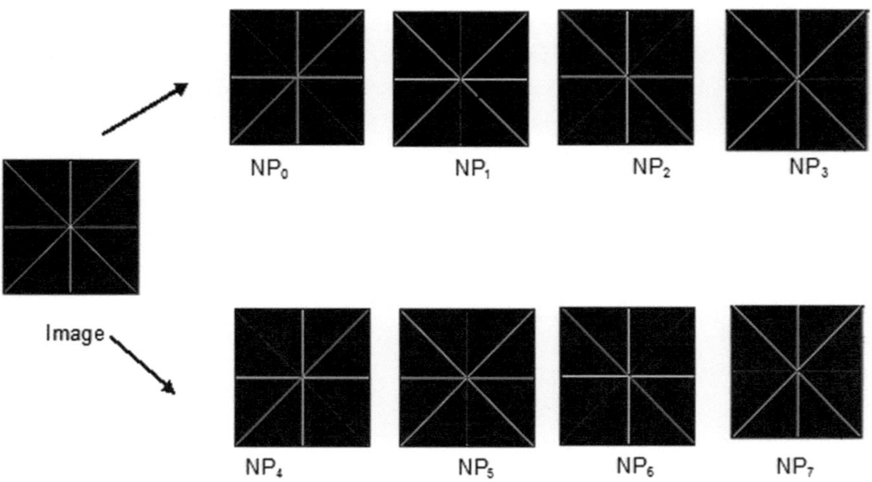

Fig. 2 NP codes of an image for each of the eight directions

$$\text{NP} = \sum_{i=0}^{3} \text{NP}_i \times 4^i$$

After encoding the image pixel, the image is divided into the number of grid of non-overlapping region. Histograms of NP codes are computed in each region (block size of 9) by using histogram() method of numpy package in Python. The following formula is for building histogram

$$H(k) = \sum_{x=1}^{n} \sum_{y=1}^{m} f(E_L(x, y), k), \quad k \in [0, P]$$

$$f(u, v) = \begin{cases} 1, & u = v \\ 0, & \text{otherwise} \end{cases}$$

where P is the largest NP pattern value, $n \times m$ is image size, and $E_L(x, y)$ is an encoded image.

Thus, all the histograms of each region are concatenated and given as features to the deep learning framework, FaceNet for face recognition purpose. And the NP encoded images are fed as inputs to FaceNet.

4.4 Face Recognition

The above resulted concatenated histograms of DCP codes from the images were given as embeddings (features) to FaceNet architecture, i.e., Inception-Resnet-v1 model. From the Inception-Resnet-v1 architecture, the Reduction-A, Inception-Resnet-B, and Reduction-B layers are removed in order to make the training and testing image sizes compatible. This resulted in fast convergence of the network.

After embeddings are learned, we have evaluated our model on LFW dataset. Nine training splits are used to learn the optimal L2 distance threshold. Classification (same or different) is then performed on the tenth test split. The person is recognized by the minimum (threshold value) Euclidean distance between two embeddings. Otherwise, the embeddings are recognized as of different persons.

Here, face recognition accuracy is calculated for the given number of samples that are correctly classified true positives (TP) and true negatives (TN) and is evaluated by the formula:

$$\text{Accuracy} = \frac{TP + TN}{\text{total no. of samples}}$$

Table 1 FaceNet performance on LFW database

Resolution	Descriptor				
	FaceNet	LBP	LTP	DCP	NP
160 × 160	99.2	99.2	99.2	99.2	99.2
100 × 100	97.4	98.7	98.7	99.1	99.1
80 × 80	84.6	95.3	96.4	98.4	99.2
60 × 60	80.9	83.2	83.9	84.6	98.5
40 × 40	78.5	78.8	79.3	80.5	98.5
20 × 20	64.6	72.1	72.3	77.7	81.3
10 × 10	58.2	66.3	66.9	68.1	71.6

5 Experimental Results and Discussions

In this paper, experimentation is carried out on LFW, ORL, AR, and EYB benchmarking face databases to evaluate the performance of proposed approach, LBP, LTP, and DCP. For experimentation, the radius of neighbors for LBP and LTP (with a threshold, 1.5) is taken as 1 and for both DCP and NP inner radius is 1 and outer radius is 2. To evaluate that our proposed method is not sensitive to noise, we have tested by adding Gaussian noise into the test samples.

5.1 LFW Database

From Table 1, it is clear that the NP descriptor produces overall best results when compared to other descriptors upon evaluating on LFW database. However, the effect of NP is not much on high-resolution images as they have already been recognized well enough by FaceNet. More than 15% accuracy has been achieved by NP with FaceNet approach than FaceNet without any descriptor approach.

5.2 ORL Database

The ORL database contains 40 classes or individual and each individual having ten images that differ in illumination, pose, and expression from each other. Since the images' size is 112 × 92, they are resized to 100 × 100 image size in order to make them appeal as a square matrix. The training is performed on eight images of each individual, and then the performance is tested on the remaining two images of the same individuals. For histogram representation, each encoded image has been

Table 2 FaceNet performance on ORL database

Resolution	Descriptor			
	LBP	LTP	DCP	NP
100 × 100	98.3	98.3	98.3	98.3
80 × 80	94.3	95.4	98.1	98.2
60 × 60	80.6	82.8	83.6	83.9
40 × 40	76.9	77.3	81.5	82.4
20 × 20	71.9	73.3	75.5	79.8
10 × 10	64.8	65.9	67.1	70.2

Table 3 FaceNet performance on AR database

Resolution	Descriptor			
	LBP	LTP	DCP	NP
80 × 80	98.7	98.7	98.7	98.7
60 × 60	87.1	88.9	90.6	93.5
40 × 40	80.3	82.6	84.6	87.9
20 × 20	76.5	78.3	79.1	82.7
10 × 10	70.3	72.2	77.0	80.6

divided into regions of 10 × 10 size. Due to various effects, LTP showed better performance than LBP. DCP performed better than LTP. Furthermore, the NP improved the accuracy rate by nearly 8% compared to LBP as observed from Table 2.

5.3 *AR Database*

The AR database consists of 76 men images and 60 women images with the difference in illumination, expressions, and occlusions. These images are also resized from 92 × 92 to 80 × 80 image sizes. From Table 3, it has been observed that the accuracy obtained for low-resolution images of the AR database is higher when compared with that of LFW database. This is due to its training set which is much less than LFW and slightly more than ORL. The obtained accurate rate of the proposed approach is 10% more than that of LBP.

5.4 *EYB Database*

The EYB database has 38 individuals with pose and illumination differences. Firstly, these images are resized to 80 × 80 images, and then training has been performed. In

Table 4 FaceNet performance on EYB database

Resolution	Descriptor			
	LBP	LTP	DCP	NP
80 × 80	92.4	92.4	95.3	95.8
60 × 60	87.1	88.9	90.6	93.5
40 × 40	84.4	85.0	87.5	89.3
20 × 20	81.3	82.2	84.1	84.7
10 × 10	75.3	76.2	78.0	82.3

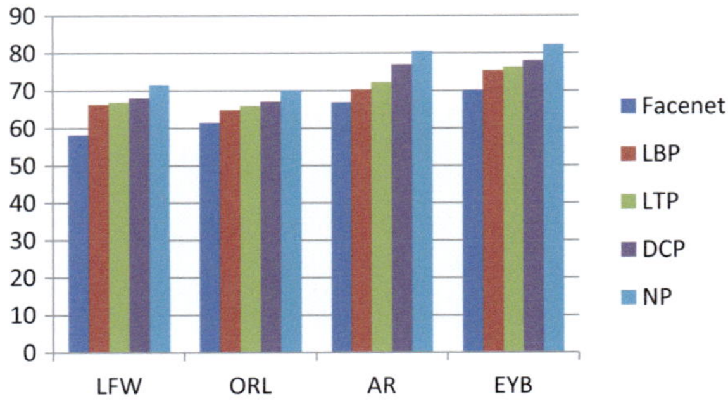

Fig. 3 Comparison of NP-based FaceNet performance with that of LBP, LTP, DCP, and FaceNet on LFW, ORL, AR, and EYB databases in terms of accuracy (percentage)

this experiment, out of 73 samples of an individual 53 samples are taken as training set and remaining 20 samples as a testing set. Due to less variation of the images, the accuracy obtained is the highest compared to the other databases. The proposed NP result increased by more than 7% compared to that of LBP as observed from Table 4.

The results of these FaceNet-based experiments with LBP, LTP, DCP, and NP on the four face database are summarized in Fig. 3. Here, only the 10 × 10 image size has been considered since the results obtained are worth noting for least resolution images. The accuracy list of the conventional FaceNet on 10 × 10 resolution images on each of LFW, ORL, AR, and EYB databases is 58.2, 61.5, 66.9, and 70.2%; whereas, the corresponding accuracy list of the proposed system is 71.6, 70.2, 80.6, and 82.3%. It has been observed that EYB database obtained the highest accuracy while ORL databases obtained the lowest accuracy compared to all other databases. This is due to the number of training samples and variations range of illumination, pose, expression, and occlusion.

6 Conclusion

The classification task becomes very challenging due to the loss of important features in low-resolution images. Here, after evaluating, the performance of FaceNet on low-resolution face images provoked the necessity of designing the proposed methodology. The image enhancement performed with EBPA helped in removing noise from the images. NP, the proposed descriptor, encoded the texture information of an image uniquely. Our proposed approach achieved more than 80% accuracy by using NP features as embeddings in FaceNet and by feeding the NP encoded image as FaceNet's input image too. Moreover, the shrinking of Inception-Resnet-v1 architecture resulted in the fast convergence of the network along with less time complexity.

The future work in this NP-based FaceNet can be attained in the following aspects:

- The techniques like similarity feature-based selection [21] can be used that helps in dimension reduction of embedding and also improves the performance.
- Some new features of NP-like multi-scale block [17] can be experimented for achieving better accuracy.

References

1. Sharma P, Yadav RN, Arya KV (2014) Pose invariant face recognition using curvelet neural network. IET Biom 3(3):128–138
2. Turk Matthew, Pentland Alex (1991) Eigenfaces for recognition. J Cogn Neurosci 3(1):71–86
3. Ahonen T, Hadid A, Pietikainen M (2006) Face description with local binary patterns: application to face recognition. IEEE Trans Pattern Anal Mach Intell 28(12):2037–2041
4. Bartlett MS, Movellan JR, Sejnowski TJ (2002) Face recognition by independent component analysis. IEEE Trans Neural Netw 13(6):1450–1464
5. Sharma P, Arya KV, Yadav RN (2013) Efficient face recognition using generalized mean wavelet neural network. Signal Process 93(6):1557–1565. SCI Journal (impact factor 2.238)
6. Phillips PJ (1999) Support vector machines applied to face recognition. In: David AC (ed) Proceedings of the 1998 conference on advances in neural information processing systems II, 1999. MIT Press, Cambridge, MA, USA, pp 803–809
7. Sharma P, Arya KV, Yadav RN (2011) Extraction of facial features using higher order moments in curvelet transform and recognition using generalized mean neural networks. In: International conference on soft computing for problem solving at IIT Roorkee, vol 131, pp 717–728, 20–22 Dec 2011
8. Sharma P, Yadav RN, Arya KV (2016) Face recognition from video using generalized mean deep learning neural network. In: 2016 4th international symposium on computational and business intelligence (ISCBI). Olten, pp 195–199
9. Arya KV, Upadhyay G, Upadhyay S, Tiwari S, Sharma P (2016) Facial recognition using histogram of Gabor phase patterns and self organizing maps. In: 2016 11th international conference on industrial and information systems (ICIIS). Roorkee, pp 883–889
10. Ghazi MM, Ekenel HK (2016) A comprehensive analysis of deep learning based representation for face recognition. In: 2016 IEEE conference on computer vision and pattern recognition workshops (CVPRW). Las Vegas, NV, pp 102–109
11. Kurmi US, Agrawal D, Baghel RK (2014) Study of different face recognition algorithms and challenges. Int J Eng Res 3:112–115. https://doi.org/10.17950/ijer/v3s2/216
12. Li Z, Hou Y, Liu H, Li X (2014) Very low resolution face reconstruction based on multi-output regression. In: IEEE workshop on electronics, computer and applications. Ottawa, ON, pp 74–77
13. Shi J, Qi C (2015) From local geometry to global structure: learning latent subspace for low-resolution face image recognition. IEEE Signal Process Lett 22(5):554–558

14. Kaur R, Himanshi E (2015) Face recognition using principal component analysis. In: IEEE international advance computing conference (IACC). Bangalore, pp 585–589
15. Xu Y, Jin Z (2008) Down-sampling face images and low-resolution face recognition. In: 3rd international conference on innovative computing information and control. Dalian, Liaoning, pp 392–392
16. Raghavendra R, Raja KB, Yang B, Busch C (2013) Comparative evaluation of super-resolution techniques for multi-face recognition using light-field camera. In: 2013 18th international conference on digital signal processing (DSP). Fira, pp 1–6
17. Jia K, Gong SG (2005) Multi-modal tensor face for simultaneous super-resolution and recognition. In: Proceedings of IEEE 10th international conference on computer vision (ICCV), vol 2. Beijing, China, pp 1683–1690
18. Choi JY, Ro YM, Plataniotis KN (2009) Color face recognition for degraded face images. IEEE Trans Syst Man Cybern, Part B, Cybern 39(5):1217–1230
19. Biswas S, Bowyer KW, Flynn PJ (2012) Multidimensional scaling for matching low-resolution face images. IEEE Trans Pattern Anal Mach Intell 34(10):2019–2030
20. Ren CX, Dai DQ, Yan H (2012) Coupled kernel embedding for low-resolution face recognition. IEEE Trans Image Process 21(8):3770–3783
21. Tran CK, Lee TF, Tuan CC, Lu CH, Chao PJ (2013) Improving face recognition performance using similarity feature-based selection and classification algorithm. In: 2013 Second international conference on robot, vision and signal processing. Kitakyushu, pp 56–60
22. Ding C, Choi J, Tao D, Davis LS (2016) Multi-directional multi-level dual-cross patterns for robust face recognition. IEEE Trans Pattern Anal Mach Intell 38(3):518–531

A Cognition Scanning on Popularity Prediction of Videos

Neeti Sangwan and Vishal Bhatnagar

Abstract There is an explosive growth of the information on the web which leads to the online competition for the attention of the viewers. Among the large amount of the videos, only some of the videos become popular while the rest of the videos remain unknown. To anticipate the prevalence of the published video is a challenging task. It is noticed that some portion of the published visual content gains the significant popularity and the rest of the content is viewed by small number of the viewers. This largely depends upon intrinsic and extrinsic factors such as content included, relevancy to the users that influence the popularity is difficult. So, the popularity prediction becomes an active area of research. Various prediction methods are given by different researchers. In this paper, we have studied the various factors, tools and challenges involved in prediction process of the video popularity.

Keywords Popularity · YouTube · Correlation · Prediction · Video

1 Introduction

In the recent years, the popularity of the online platforms such as YouTube, Facebook, Twitter or Instagram increases to ease the users to share their content with other people. All the groups of people, business and organizations are benefitted from the advantages of publishing in the social media. Different individuals get popular using social media. Companies may find the product that attracts more attention from the customers, decide intelligently and manage their resources accordingly to increase the revenue. Among the different formats used by users, video content becomes more popular. It is noticed that some portion of the published visual content gains

N. Sangwan (✉)
GGS Indraprastha University, Dwarka, India
e-mail: neetisangwan@gmail.com

MSIT, New Delhi, India

V. Bhatnagar
Ambedkar Institute of Advanced Communication Technologies and Research, New Delhi, India
e-mail: vishalbhatnagar@yahoo.com

R. K. Shukla et al. (eds.), *Social Networking and Computational Intelligence*, Lecture Notes in Networks and Systems 100, https://doi.org/10.1007/978-981-15-2071-6_30

the significant popularity and the rest of the content is viewed by small number of the viewers.

To forecast the prevalence of the published video is a challenging task. Determining the intrinsic and extrinsic factors such as content included, relevancy to the users that influence the popularity is difficult. So, the popularity prediction becomes an active area of research. Various prediction methods are given by different researchers. In general, popularity is defined in the terms of the interactions made by audience on a platform. In this paper, some light has been thrown on the current scenario of the research in the field. Inputs of the paper to the current field are as follows:

i. List the recent advancements in the area;
ii. List the various challenges;
iii. List various tools for popularity prediction released between 2009 and 2018.

Rest of the paper includes research work carried in the field discussed in the Sect. 2. Section 3 provides the issues present in the field. Section 4 provides the major developments and automated tools related to the field. Finally, Sect. 5 concludes the paper.

2 Literature Review

Li et al. [1] proposed a cloud transcoder that performed video transcoding in the cloud to fill the format-resolution gap. Borghol et al. [2] developed a method to assess the effect of various hidden factors such as previous popularity, age and publishers on the popularity of the video. Wang et al. [3] explored the relation between video links that are exchanged in the microblogging system and number of sight count of the video.

Avramova et al. [4] monitored 50 videos from YouTube for five months and came up with a model for progression of the video fame based on the accumulative distribution of the video popularity traces, i.e., demand of a particular trace until any time instant. Luo et al. [5] provided fast-decaying popularity evolved for thirty days on the dataset of news and music. Luo et al. [6] demonstrated the qualitative difference in popularity decay for 10 news and 10 video clips. Tang et al. [7] provided the knowledge about synthetic media server workload with the help of real media server log trace. Cha et al. [8] suggested the power-law and exponential cut-off-based popularity unevenness. Crane and Sornette [9] found out that 90% of the videos did not get so much attention and only 10% of the videos have significant power-law behavior. Authors grouped the videos according to the type of disturbance that affects the interests of the viewers and the degree of the criticality. Cha et al. [10] analyzed the internal properties of the user-generated videos that affect the static popularity distribution and dynamic popularity evolution of UGC videos. Szabo and Huberman [11] presented a method that predicts the popularity instantly as the content published and found the high correlation between popularities at initial time and later. Niu et al. [12] exploited the time series analysis techniques for online population evolution and gave a novel approach to foresight the individual contribution with respect to popularity progression. Niu et al. [12] also addressed the various issues in the performance and demand prediction of the videos. Figueiredo et al. [13] analyzed the popularity patterns of three different video datasets and found that the top videos got sudden bursts of the popularity. Famaey et al. [14] determined two schemes to forecast the popularity on the basis of the multimedia data caching to improve the hit ratio. Borghol et al. [15] proposed a model for finding the fame of recently published YouTube videos and concluded that the popularity is highly non-stationary.

3 Challenges in the Video Popularity Prediction

There are many challenges in exploiting popularity of the videos quickly, accurately and efficiently.

- Prediction should be fast to minimize the miss rate of the number of views of a video.
- Accuracy in prediction cannot be ignored.
- Prediction needs to be scalable so it can handle video workloads at a global scale like Facebook.
- Trends in other media (radio, television, newspapers) also have a significant role in predicting the online popularity of the content.
- Number of followers and friends that reflect the social network of the publisher also have a significant impact on the future popularity of the online content.
- Capturing the pertinence of the video to the viewers and finding the relationship between the content and real world are difficult and complex.

4 Major Recent Developments and Automated Tools

The recent development and opportunities in the video popularity prediction have been listed in this section. Advancement in the field of video popularity prediction from 2009 to 2018 is stated in Table 1. As per the findings, major advancements in the field are:

(i) Temporal evolution in the prediction;
(ii) Introduction of cloud transcoder;
(iii) Use of transfer learning framework;
(iv) Stochastic fluid model for prediction is proposed;
(v) Different visualization systems are launched.

Different tools for the video popularity prediction are explored that are developed in the recent history. Various tools that are available for the video analysis are listed in Table 2. Recent tools for the prediction are built on different platforms. These tools

Table 1 Advancement in the field of video popularity prediction

Year	Description of advancement to popularity prediction	Aim
2009 [4]	Power-law or exponential distribution	To represent popularity dynamics
2009 [10]	Power law with exponential cut-off	A way to determine the pattern of video popularity
2011 [12]	Introduce a time series-based scheme to find instantaneous online population and peer upload contribution	To forecast server bandwidth demand for capacity planning and quality control
2011 [15]	Model for three-phase characterization of popularity dynamics	To generate synthetic data matching characteristics for user-generated video services
2011 [14]	Forecast in the environment based on the realistic multimedia data caching	To improve the cache hit rate in multimedia content delivery networks
2011 [15]	Log-normal distribution	A way to determine the pattern of video popularity
2012 [1]	Cloud transcoder is introduced	To minimize the resolution differences between the mobile devices and online videos
2012 [3]	Relationship between video propagation in the microblogging system and its popularity	To enhance the prediction accuracy of the videos by using the various microblogging intrinsic parameters
2012 [16]	Usage of geographic properties of videos for popularity analysis	To present the videos to the users according to their geographic parameters
2013 [2]	Popularity prediction method based on content-agnostic parameters for clone videos is introduced	To assess the effect of different parameters on the video popularity
2013 [17]	Diffusion of the videos in the network by using the inter-arrival time between the requests made for a video and virus propagation model	To improve the popularity prediction ability of the videos
2013 [18]	SoVP framework	To overcome the limitations like predicting early peaks and later burst of video accesses in conventional models
2013 [19]	Transfer learning framework	To learn topics from social streams for improved popularity prediction
2013 [20]	Models based on multivariate linear regression and MRBF	To lower the prediction errors
2013 [21]	Temporal evolution	To determine popularity dynamics
2014 [22]	Degree of skewness	To improve predictability by caching the videos
2014 [22]	Gamma distributions	A way to determine the pattern of video popularity

(continued)

Table 1 (continued)

Year	Description of advancement to popularity prediction	Aim
2015 [23]	Dual sentimental Hawkes process (DSHP)	A topology independent framework to correlate view counts in initial and later time
2015 [24]	Use visual sentiment and content features	To improve the prediction over the baseline models
2015 [25]	Predictions in terms of number of solicitations	To determine the content that should be cached
2016 [26]	Stochastic fluid model	To identify the different factors to determine the video popularity evolution
2016 [27]	Hawkes process	To provide feature-based and generative-based methods for better prediction
2016 [28]	Transductive multimodal learning approach for micro-videos	To represent the micro-videos more efficiently.
2017 [29]	Model based on support vector regression	Improvement in prediction
2018 [30]	Hawkes intensity process insights explorer (HIPie)—an interactive visualization system	Provides open-source visualization platform for video popularity prediction
2018 [31]	Graph model based on log-normal distribution and power-law distributions	Determine user relationships and various social structures
2018 [32]	P2P-VoD streaming system using mesh-based network	To minimize the delay, bandwidth consumptions and transferring costs
2018 [33]	Spatial characterization using pareto-principle video request similarity calculation	To determine the contents to be cached in a particular location
2018 [34]	Multifactor differential influence (MFDI) prediction model	To predict top-N popular videos

provide the ease of use and conserve the time and efforts for the video popularity prediction.

Figure 1 shows the pictorial representation of the advancements in the field of video popularity prediction. This graph depicts the publication trends and tools developed in the field during recent period of time.

5 Conclusions

It is an uphill struggle task to predict the popularity of the published video. It is noticed that some portion of the published visual content gains the significant popularity and

Table 2 Recent tools used in video popularity prediction

Year	Tool name	Description
2017	Pliers (python)	For extraction of the features from multimodal stimuli, identify the objects or faces and transcribe speech in audio or video file
2016	Extract video-frames	To extract frames from a video using openCV
2017	Fast video feat	To extract motion features or motion vector from video compression information
2017	Yt8m-feature-extractor	Extract features from video files as the format in YouTube 8 M
2018	EDCAV2018 (audio project)	Audio classification for audio files
2018	Ibeaucourt/object detection	Apply tensorflow object detection on the input video stream
2018	Ricoms	Feature extraction from images and videos
2006	VideoIQ	It basically observes all videos on YouTube and finds the most popular keywords among them. And it is certified by YouTube
2014	TubeBuddy	It is an optimization tool and follows four steps: automate, optimize, promote and connect. It also helps in managing the YouTube channel
2008	SocialBlade	To make the statistical charts and graph corresponding to the data from YouTube to map the growth
2006	TubeTracker	Provide tools and dashboards for YouTube channels and optimize everything
2011	JVZoo	It tracks the traffic of audience, analyzes and compares views. It also helps channel to increase viewers
2014	Vidooly	Online video analytics software for advertising and media business
2011	YouTube analytics	It is a tool provided by YouTube to every channel and used to count the views and uses graph for comparison
2011	Channel meter	It is a tool that use to count view and compare the video with most viewed video and uses different graphs for comparison as well as for progress
2012	Tubular labs	For global ranking

the rest of the content is viewed by small number of the viewers. This largely depends upon intrinsic and extrinsic factors such as content included and relevancy to the users that influence the popularity. So, the popularity prediction becomes an active area of research. Various prediction methods are given by different researchers. This paper focused on the prominent issues, recent development and tools available in the field of video popularity prediction. We found that in spite of the different models and correlation techniques in prediction, there are many points to be explored. Some of the future directions that need to be work upon include exploring more intrinsic and extrinsic factors of the videos that affect the prediction performance of the videos.

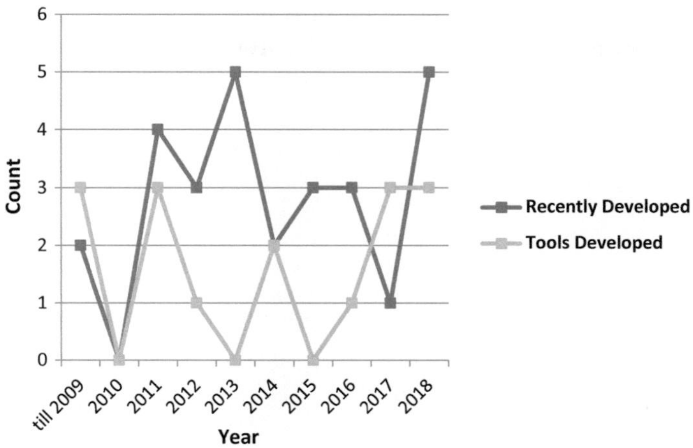

Fig. 1 Recent publications and tools developed for predicting video popularity

References

1. Li Z, Huang Y, Liu G, Wang F, Zhang ZL, Dai Y (2012) Cloud transcoder: bridging the format and resolution gap between internet videos and mobile devices. In: Proceedings of the 22nd international workshop on network and operating system support for digital audio and video, pp 33–38
2. Borghol Y, Ardon S, Carlsson N, Eager D, Mahanti A (2012) The untold story of the clones: content-agnostic factors that impact YouTube video popularity. In: Proceedings of the 18th ACM SIGKDD international conference on knowledge discovery and data mining, pp 1186–1194
3. Wang Z, Sun L, Wu C, Yang S (2012) Guiding internet-scale video service deployment using microblog-based prediction. In: INFOCOM, 2012 proceedings IEEE, pp 2901–2905
4. Avramova Z, Wittevrongel S, Bruneel H, De Vleeschauwer D (2009) Analysis and modeling of video popularity evolution in various online video content systems: power-law versus exponential decay. In: First international conference on evolving internet. INTERNET'09, pp 95–100
5. Luo JG, Tang Y, Zhang M, Yang SQ (2007) Characterizing user behavior model to evaluate hard cache in peer-to-peer based video-on-demand service. In: International conference on multimedia modeling, pp 125–134
6. Luo J-G, Zhang Q, Tang Y, Yang S-Q (2009) A trace-driven approach to evaluate the scalability of P2P-based video-on-demand service. IEEE Trans Parallel Distrib Syst 20(1):59–70
7. Tang W, Fu Y, Cherkasova L, Vahdat A (2007) Modeling and generating realistic streaming media server workloads. Comput Netw 51(1):336–356
8. Cha M, Kwak H, Rodriguez P, Ahn YY, Moon S (2007) I tube, you tube, everybody tubes: analyzing the world's largest user generated content video system. In: Proceedings of the 7th ACM SIGCOMM conference on internet measurement, pp 1–14
9. Crane R, Sornette D (2008) Robust dynamic classes revealed by measuring the response function of a social system. Proc Natl Acad Sci 105(41):15649–15653
10. Cha M, Kwak H, Rodriguez P, Ahn Y-Y, Moon S (2009) Analyzing the video popularity characteristics of large-scale user generated content systems. IEEE/ACM Trans Netw 17(5):1357–1370
11. Szabo G, Huberman BA (2008) Predicting the popularity of online content. Available SSRN 1295610

12. Niu D, Liu Z, Li B, Zhao S (2011) Demand forecast and performance prediction in peer-assisted on-demand streaming systems. In: INFOCOM, 2011 proceedings IEEE, pp 421–425
13. Figueiredo F, Benevenuto F, Almeida JM (2011) The tube over time: characterizing popularity growth of youtube videos. In: Proceedings of the fourth ACM international conference on web search and data mining, pp 745–754
14. Famaey J, Wauters T, De Turck F (2011) On the merits of popularity prediction in multimedia content caching. In: International symposium on integrated network management (IM), 2011 IFIP/IEEE, pp 17–24
15. Borghol Y, Mitra S, Ardon S, Carlsson N, Eager D, Mahanti A (2011) Characterizing and modelling popularity of user-generated videos. Perform Eval 68(11):1037–1055
16. Brodersen A, Scellato S, Wattenhofer M (2012) Youtube around the world: geographic popularity of videos. In: Proceedings of the 21st international conference on World Wide Web, pp 241–250
17. Nwana AO, Avestimehr S, Chen T (2013) A latent social approach to youtube popularity prediction. In: Global communications conference (GLOBECOM), 2013 IEEE, pp 3138–3144
18. Li H, Ma X, Wang F, Liu J, Xu K (2013) On popularity prediction of videos shared in online social networks. In: Proceedings of the 22nd ACM international conference on information and knowledge management, pp 169–178
19. Roy SD, Mei T, Zeng W, Li S (2013) Towards cross-domain learning for social video popularity prediction. IEEE Trans Multimed 15(6):1255–1267
20. Pinto H, Almeida JM, Gonçalves MA (2013) Using early view patterns to predict the popularity of youtube videos. In: Proceedings of the sixth ACM international conference on web search and data mining, pp 365–374
21. Ahmed M, Spagna S, Huici F, Niccolini S (2013) A peek into the future: predicting the evolution of popularity in user generated content. In: Proceedings of the sixth ACM international conference on web search and data mining, pp 607–616
22. Tatar A, De Amorim MD, Fdida S, Antoniadis P (2014) A survey on predicting the popularity of web content. J Internet Serv Appl 5(1):8
23. Ding W, Shang Y, Guo L, Hu X, Yan R, He T (2015) Video popularity prediction by sentiment propagation via implicit network. In: Proceedings of the 24th ACM international on conference on information and knowledge management, pp 1621–1630
24. Fontanini G, Bertini M, Del Bimbo A (2016) Web video popularity prediction using sentiment and content visual features. In: Proceedings of the 2016 ACM on international conference on multimedia retrieval, pp 289–292
25. Ben Hassine N, Marinca D, Pascale M, Barth D (2015) Machine learning and popularity prediction of a video content. In: The 4th international conference on performance evaluation and modeling in wired and wireless networks (PEMWN)
26. Wu J, Zhou Y, Chiu DM, Zhu Z (2016) Modeling dynamics of online video popularity. IEEE Trans Multimed 18(9):1882–1895
27. Mishra S, Rizoiu MA, Xie L (2016) Feature driven and point process approaches for popularity prediction. In: Proceedings of the 25th ACM international on conference on information and knowledge management, pp 1069–1078
28. Chen J, Song X, Nie L, Wang X, Zhang H, Chua TS (2016) Micro tells macro: predicting the popularity of micro-videos via a transductive model. In: Proceedings of the 2016 ACM on multimedia conference, pp 898–907
29. Trzciński T, Rokita P (2017) Predicting popularity of online videos using support vector regression. IEEE Trans Multimed 19(11):2561–2570
30. Kong Q, Rizoiu MA, Wu S, Xie L (2018) Will this video go viral? Explaining and predicting the popularity of Youtube videos. arXiv Prepr arXiv1801.04117
31. Jia AL, Shen S, Li D, Chen S (2018) Predicting the implicit and the explicit video popularity in a user generated content site with enhanced social features. Comput Netw 140:112–125
32. Marza V, JadidiNejad A, et al (2018) A novel caching strategy in video-on-demand (VoD) peer-to-peer (P2P) networks based on complex network theory. J Adv Comput Res 9(1):17–27

33. Yan H, Liu J, Li Y, Jin D, Chen S (2018) Spatial popularity and similarity of watching videos in large-scale urban environment. IEEE Trans Netw Serv Manag
34. Tan Z, Zhang Y (2018) Predicting the top-N popular videos via a cross-domain hybrid model. IEEE Trans Multimed

Review on High Utility Rare Itemset Mining

Shalini Zanzote Ninoria and S. S. Thakur

Abstract Today's era is an era of data. Hence, the most attentive field of research and study is to collect and select the data rapidly as per the need nowadays. Data mining is the field which helps industries to overcome the problem of data extraction. Association rule mining (ARM) is one of the major techniques of data mining which identifies the itemsets appear frequently in the dataset known as frequent itemset and generates the association rules. This helps in decision making. The extension of traditional association rule mining has come up with the concept of utility which should be considered while mining; hence, utility mining aims to identify the itemsets which not only have the frequent occurrences but also have considered the utility of the itemset. High utility itemsets mining can be used as an efficient method to discover interesting patterns. Rare items are items that appear fewer frequently in a database. High utility itemsets can be frequent or rare. Even rare itemsets can also be of high or low utility. In this paper, a literature survey of various research works has been discussed on High Utility Rare Itemset Mining. We have thoroughly surveyed High Utility Rare Itemset Mining methods and applications here. Also, we have focused on some open research issues to represent future challenges in this domain.

Keywords Data mining · Association rule mining · Frequent itemsets · High utility itemsets · Rare itemsets

1 Introduction

During the last few years, data mining, which also known as KDD, i.e., knowledge discovery in databases, has come up as a prominent research area aiming to extract

S. Z. Ninoria (✉) · S. S. Thakur
Department of Mathematics and Computer Science, RDVV, Jabalpur, Madhya Pradesh, India
e-mail: shalini.ninoria@gmail.com

S. S. Thakur
e-mail: samajh_singh@rediffmail.com

Department of Applied Mathematics, Jabalpur Engineering College, Jabalpur, Madhya Pradesh, India

R. K. Shukla et al. (eds.), *Social Networking and Computational Intelligence*, Lecture Notes in Networks and Systems 100, https://doi.org/10.1007/978-981-15-2071-6_31

significant hidden information from the data [19]. Data mining has been used in various domains and considered as an algorithmic process that takes data as input and produces patterns, such as clusters, classification rules, association rules, etc. as output [59]. Data mining tasks can be split into two groups, descriptive mining and predictive mining. Clustering, Association Rule Discovery, Sequential Pattern Discovery are descriptive mining techniques which are used to locate human-interpretable patterns that describe the data. The predictive mining techniques cover the techniques like classification, regression, deviation detection which uses some variables to predict unknown or future values of other variables [45]. The most important technique in which a lot of work has been done is frequent itemset mining (FIM) [5, 21, 26] and high utility itemset mining [36, 37, 50, 57] separately. FIM mines those itemsets from the database which are frequent. In FIM, the technique only considers the frequency of the occurrences of the itemset and does not consider the profit value or quantity value associated with the items. High utility itemset mining behaves as an extension to FIM as overcome from this limitation and considers the utility of the itemset along with its frequency for HUIM. Most of the studies in data mining have primarily concentrated on frequent itemsets and production of association rules from them. Their common property is that they all extract frequent itemsets. The infrequent or rare itemsets have also to be investigated because such itemsets also hold significant information just as frequent itemsets do. The organization of the paper is as follows. Section 2 comprehensively discusses preliminaries. Section 3 discusses Rare Itemset Mining schemes. Section 4 discusses state-of-the-art literature regarding High Utility Rare Itemset Mining. Moreover, Sect. 5 gives details about the related work regarding High Utility Rare Itemset Mining, later Sect. 6 discusses open research issues and significant applications in terms of future perspective. Section 7 concludes the whole paper and depicts the future work.

2 Preliminaries

2.1 Data Mining

Data mining is a method of determining hidden patterns and information from the existing data. The structure of the data may be scattered or not uniformly structured for processing. Hence, in data mining, we also need to primarily concentrate on cleansing the data. The process of cleansing data can be either made by using tools such as ETL [41]. Data mining emerged in the 1990s and has a big impact on business, industry, and science. Data mining has been used for many years by many fields such as businesses, scientists, governments, etc. It is used to filter through volumes of data such as population data and marketing data to generate market research reports. Data mining generally involves four classes of tasks [13]: Classification which arranges the data into predefined groups. Other is clustering, it is like classification but the groups are not predefined; hence, the algorithm groups similar items together; next can be regression, it attempts to find a function which models the data with the least error. The most popular is association rule mining, searching for relationships between variables. Han et al. [20] presented data mining functionalities include data characterization, data discrimination, association analysis, classification, clustering, outlier analysis, and data evolution analysis. Data mining is the procedure of applying these methods to data with the intention of discovering hidden patterns [23]. There are various application areas where data mining techniques are tremendously used. Financial data collected in the banking and financial industry is often relatively complete and reliable, which facilitates systematic data analysis and data mining. Usual cases comprise classification and clustering of customers for targeted marketing, detection of money laundering, and other financial crimes as well as design and construction of data warehouses for multidimensional data analysis. The retail industry is also a major application area for data mining as it collects huge amounts of data on customer shopping history, consumption, and sales to discover customer buying behavior, to find out customer purchasing pattern and to forecast customer consuming trends [27].

2.2 *Association Rule Mining*

Association rule mining is a technique for extracting the pattern which has the association between them in the huge database. For example, in a computer shop, if a customer purchases a computer, he is likely to purchase a pen drive or software with varying possibilities. Association rules are formed by examining information for nonstop if/then examples and utilizing the criteria support. Support is a term which shows how frequently the things show up in the database [51]. The concept of association rules for discovering regularities between products in large databases has introduced by Agrawal and Srikant [5, 9, 19]. Mining association rules can be divided into two steps: the first is generating frequent itemsets and the second is generating association rules. Association rule mining works on two major parts. The first part is to find all itemsets with adequate supports and the second part is to generate association rules [38]. In the traditional association rules mining [6, 53], minimum support threshold and minimum confidence threshold values are assumed to be available for mining frequent itemsets, which is hard to be set without specific knowledge; users have difficulties in setting the support threshold to obtain their required results. To use association rule mining without support threshold [4, 7, 10], another constraint such as similarity or confidence pruning is usually introduced. Association rule mining is all about finding all rules whose support and confidence exceed the threshold, minimum support, and minimum confidence values. Setting the support threshold too large would produce only a small number of rules or even no rules to conclude. In that case, a smaller threshold value should be guessed (imposed) to do the mining again, which may or may not give a better result, as by setting the threshold too small, too many results would be produced for the users, too many results would require not only very long time for computation but also for screening these rules.

2.3 *Frequent Itemset Mining*

The task of discovering frequent itemsets in databases was introduced by Agrawal and Srikant [5]. Discovering associations between items are helpful to understand customer behavior. For example, a retail store manager can use these facts to take strategic marketing decisions such as copromoting products or putting them closer on the shelves [14]. The first algorithm has been proposed named Apriori for finding the frequent itemsets occurred in the database [34]. Apriori algorithm is useful for searching the association rules among items in market-basket data [1, 54]. Association rules use two main constraints, i.e., minimum support and minimum confidence. The mining process is divided into two subprocesses. First to find the itemsets those are having occurrences in the database more than or equal to the minimum support value and later in the second phase generating association rules which satisfies the minimum confidence [40]. Hence Apriori is a bottom-up, breadth-first search algorithm. As Apriori holds the downward closure property, only candidate frequent itemsets, whose all subsets are also frequent, are generated in each database scan. There are various variants of Apriori developed such as, [34, 48, 55] have presented [18]. Table 1 gives some remarkable variations of frequent itemset mining.

3 Rare Itemset Mining

The task of frequent itemset mining has various applications, but it can also be viewed with limitations in terms of the assumptions that it makes. These limitations have generated the need of extension to FIM. One of the remarkable limitations of traditional FIM is that it assumes that all items are equal. But in real-life applications, items are often different from each other [34]. In actual situations, some items naturally have more chance of being frequent than others. This shows the way to generate the rare

Table 1 Some remarkable variations of FIM [14]

Algorithm	Type of search	Database representation
Apriori [5]	Breadth-first (candidate generation)	Horizontal
Apriori—TID [20]	Breadth-first (candidate generation)	Vertical (TID-lists)
Eclat [64]	Depth-first (candidate generation)	Vertical (TID-lists, diffsets)
FP-growth [21]	Depth-first (pattern-growth)	Horizontal (prefix-tree)

item problem [34], which means that some items are much less likely to appear in frequent itemsets than others.

Only frequency consideration in traditional techniques is not sufficient. Mining rare patterns from databases have always been overlooked and giving more emphasis on frequent one. The unknown and unusual patterns are proficient in discovering hidden useful information from databases in various domains of applications.

For this issue, many researchers have developed different algorithms to find frequent itemsets using multiple minimum support thresholds, such as MSApriori [34], CFPGrowth [22], and CFPGrowth++ [24]. Liu et al. [34] noted that some individual items can have low support and even they cannot be part of the associations generated by Apriori, even though they have very high confidence. This problem is solved by specifying the constraint that the frequent items can have higher minimum support and the rare items can have lower minimum support [34]. Later on, significant effort has been made by Lin et al. [32] and Poovammal and Ponnavaikko [47], to alleviate the problems in the previous work, for which they added the measure of lift or conviction, which obtained the minimum support dynamically from the item support [32, 47]. Relative Support Apriori Algorithm (RSAA) is proposed by Yun et al. [59] for generating rare itemset rules without the need to specify the support threshold by the user. This algorithm allocates high support threshold for items with low frequency and low support threshold for items with high frequency [59]. Apriori-inverse algorithm by Koh and Rountree [25] has been proposed. In this work, the algorithm is used to generate rules that may contain items over the maximum support threshold called perfectly sporadic rules. This algorithm is much faster than Apriori in finding perfectly rare itemsets that are a subclass of rare itemsets containing itemsets whose all subsets are rare [5]. Szathmary et al. [52] have proposed the algorithm Apriori-rare which finds all minimal rare itemsets. This algorithm finds out two sets of items. One is maximal frequent itemset (MFI) and another is minimal rare itemset (mRI). An itemset is referred to as a MFI if it is frequent but not all its supersets; similarly, an itemset is referred to as a mRI if it is rare but all its proper subsets are not. It also finds the generator which generates the frequent itemsets (FIs) [52]. Adda et al. described ARANIM algorithm for Apriori-rare and Non-Present Itemset Mining to mine rare and non-present itemsets in [31]. In the proposed approach, the technique is same like Apriori and the mining idea is that if the itemset lattice representing the itemset space presented in classical Apriori approaches is traversed in a bottom-up manner, then the equivalent properties for the Apriori exploration is provided to discover rare itemsets [3]. Apriori-rare is an alteration of the Apriori algorithm which is used to mine frequent itemsets. To extracts all rare itemsets from minimal rare itemset (mRIs), a prototype algorithm called—A Rare Itemset Miner Algorithm—(Arima) was proposed in [55]. Arima generates the set of all rare itemsets and splits those into two sets: the set of rare itemsets with zero support and others with nonzero support. If an itemset is rare, then any extension to that itemset will also result in a rare itemset [52]. Adda et al. proposed a framework which is used to represent different categories of interesting patterns. A common framework was presented to mine patterns based on the Apriori approach. The generalized Apriori framework was represented to mine rare itemsets. The Apriori algorithm, called

AfRIM for Apriori-rare itemset to mine rare itemsets, was proposed which performs a level-wise search. The backward traversal method is used with a property that leads to prune out potentially non-rare itemsets in the mining process. This includes an antimonotone property and a level-wise exploration of the itemset space [2]. A Rare Itemset Mining Algorithm (Arima) presented by Adda et al. [2].

4 High Utility Rare Itemset Mining

A utility mining is an apparent topic in data mining. The main focus in the field of utility mining is not only FIM but also the consideration of utility. Practically, it has been found that the utility is of great interest in the industry if considers with high utility itemsets [39]. Due to the important limitations of traditional FIM algorithm, several extensions of FIM have been proposed; some of the most important ones are the following.

Weighted itemset mining is an extension of frequent itemset mining where weights are coupled to each item to indicate their relative importance [60, 62, 63] with the objective to find itemsets that have a minimum weight. The infrequent weighted itemsets are a popular variation to this problem [8]. The major extension of weighted itemset mining has arisen in terms of high utility itemset mining (HUIM) where not only weights are considered but also purchase quantities or utility in transactions [15, 28, 33, 35, 37, 61, 64]. In traditional FIM, either an itemset appears in a transaction or not. In HUIM, the utility or quantity is also indicated in transactions. For example, a transaction could specify that a customer has bought two desktops and one pen drive. In HUIM, weights are used to indicate how much profit is generated by each unit sold of a product. The objective of HUIM is to find all itemsets that have a utility higher than a given threshold in a database (i.e., itemsets generating a high profit). Plenty of work has been done on HUIM so far like two-phase algorithm [37] where an upper-bound, called the TWU is used to reduce the search space, tighter upper-bounds on the utility is introduced so that the algorithm be able to prune a larger part of the search space and improve the performance of HUIM algorithms [15, 29, 61, 64] current fastest HUIM algorithm is EFIM [64], shelf-time periods of items [16], discount strategies applied in retail stores [30] discover the top-k most profitable itemsets [11, 56], etc. Table 2 presents the example of transaction database with a total utility which have been calculated with the help of Table 3 profit table.

Let us have a closer look at the basic definitions related to utility of items in a dataset.

$I = \{i1, i2, i3, \ldots, im\}$ each item has external utility as pr(i_p, Itemset X of length k is $X = \{i1, i2, i3, \ldots, ik\}$) where for $j \in 1, \ldots, k$, $ij \in I$, a transaction database $D = \{T1, T2, \ldots, Tn\}$, every item I_p in a transaction T_d has a quantity $q(i_p, T_d)$ associated with it.

Definition 1: The utility of an item I_p in a transaction T_d is the product of the profit of the item and its quantity in the transaction

i.e., $u(i_p, T_d) = q(i_p, T_d) * \text{pr}(i_p)$.

Table 2 Transaction table with utility

TID	Transaction	TU
T1	(C:5) (D:20)	70
T2	(C:1) (F:40)	42
T3	(A:1) (B:1) (C:2) (G:10)	20
T4	(A:1) (B:1) (C:2)	10
T5	(A:5) (C:10)	45
T6	(B:1) (C:1) (E:1)	5
T7	(B:1) (C:1) (E:1) (G:10)	15
T8	(B:1) (C:1) (E:1) (H:1)	6
T9	(C:10) (E:10)	40
T10	(A:1) (B:1) (C:1)	8

Table 3 Profit table

Item	A	B	C	D	E	F	G	H
Profit	5	1	2	3	2	1	1	1

Definition 2: The utility of an itemset X in a transaction T_d is denoted as $u(X, T_d)$ and defined as

$$\sum_{x \subseteq T_d \wedge i_p \in X} u(i_p, T_d).$$

Definition 3: The utility of a transaction T_d is denoted as $\sum_{i_p \in T_d} u(i_p, T_d)$.

Definition 4: The utility of an itemset X in database D is denoted as $u(X)$ and defined as

$$\sum_{x \subseteq T_d \wedge T_d \in D} u(X, T_d).$$

Definition 5: An itemset is called a high utility itemset if its utility is no less than a user-specified minimum threshold denoted by min_util.

Definition 6: The support of an itemset X denoted by sup(X) is the number of transactions in database D

which contain itemset X.

Definition 7: An itemset X is called a rare itemset, if sup(X) < max_sup_threshold.

If max_sup_threshold = 2, Table 4 shows the rare itemsets of above example database.

High utility rare itemsets fall below a maximum support value but above a user provided high utility threshold min_util [44]. Jyothi et al. [43] proposed influential work in an extension of high utility work in terms of a two-phase algorithm which

Table 4 Rare itemset table

Itemsets	List of rare itemsets
1-itemset	{D},{F},{H}
2-itemset	{AG},{BH},{CD},{CF},{CH},{EG},{EH}
3-itemset	{ABG},{ACG},{BCH},{BEH},{BEG},{CEG} {CEH}
4-itemset	{ABCG},{BCEG},{BCEH}

Table 5 High utility rare items

Itemsets	List of high utility rare itemsets
1-itemset	{D},{F}
2-itemset	{CD},{CF}
3-itemset	{Ø}
4-itemset	{Ø}

is used to find high utility rare itemsets from transaction databases. The two-phase work accomplished as the rare itemsets is mined in the first phase and utility of rare itemsets are calculated in the second phase and high utility rare itemsets. Jyothi et al. [46] proposed an approach for handing profitable transactions along with high utility rare itemsets from a transaction database. A UP-rare growth, which uses a UP-tree data structure to find high utility rare itemsets, has been further proposed by Vikram et al. in [17].

If min_util = 30 and max_sup_threshold = 2, Table 5 shows the high utility rare itemsets in the example database.

The objective of HUIM is to find all itemsets that have a utility higher than a given threshold in databases. Utility mining is an apparent topic in data mining. The main focus is not to find the only FIM but also consider utility. Including unusual and rare patterns makes the approach more efficient and useful. Hence, High Utility Rare Itemset Mining has the objective to satisfy the below condition,

$$\textbf{Max\_sup} > \textbf{HURI} < \textbf{Min\_util}$$

5 Related Work

In conventional pattern mining, the main target is to find frequent patterns and associations between the items. But in many applications, some items appear more frequently in the data, while others rarely appear. Hence, the concept of rare itemset mining is introduced. Nowadays, utility mining is a very significant association rule-mining paradigm. Adda et al. [2] introduce a good initial model of utility itemset mining, where a utility table *UT* < *I*, *U* > is defined by items *I* and their utilities *U* computed for each transaction and termed local utility of a transaction. Utility mining approach was enhanced by Yao and Hamilton [58]. Some utility approaches have considered performance enhancements to enable handling of large candidate sets too. Rare itemsets provide very valuable information in real-life applications such as security, business strategies, biology, medicine, supermarket, etc. Adda et al. [2] show that normal behavior is very frequent, whereas abnormal or suspicious behavior is less frequent. For example, one can consider a database where the behavior of people in sensitive places such as airports or shopping complexes is recorded, and if those behaviors are modeled, it is normal that the common normal behaviors can be represented by frequent patterns and but the uncommon behavior can be considered suspicious and represented by rare patterns. Rare itemsets contain items of high utility and may appear rarely in transactions or datasets. High utility frequent itemsets contribute the most to a predefined utility, objective function, or performance metric [12]. There are some different approaches to discover rare association rules. The most prominent way is Apriori which can directly be applied by setting the minimum threshold (minsup) to a low value. This leads to a combinatorial explosion in the number of patterns, most of them frequent with only a small number of them actually rare. Shankar et al. [49] presented an algorithm Fast Utility Mining (FUM), which finds all high utility itemsets within the given utility constraint threshold. The authors also suggested methods to generate, i.e., High Utility and High Frequency itemsets (HUHF), High Utility and Low Frequency itemsets means (HULF), Low Utility and High Frequency itemsets, i.e., (LUHF) and Low Utility and Low Frequency itemsets as (LULF) using a combination of FUM and Fast Utility Frequent Mining (FUFM) algorithms [49]. A significantly different approach is Apriori-inverse; it was proposed by Koh and Rountree [25]. It involves the modification of the Apriori algorithm to use the infrequent itemsets during rule generation. A change makes here use of the maximum support measure, instead of the usual minimum support, to generate candidate itemsets, i.e., only items with a lower support than a given threshold are considered. Szathmary et al. [52] represented a notified algorithm for computing all rare itemsets by splitting the rare itemset mining task into two steps. The first step is to the identification of the minimal rare itemsets, and in the second step, the minimal rare itemsets are processed in order to restore all rare itemsets. Apriori-rare is an alteration of the Apriori algorithm used to mine frequent itemsets. Apriori-rare produces a MRM, i.e., a set of all minimal rare generators that correspond to the itemsets usually pruned by the Apriori algorithm when seeking for frequent itemsets. "A Rare Itemset Miner Algorithm (ARIMA)" was proposed to regain all rare itemsets from minimal rare itemset (mRIs). It splits the generated itemsets into two sets: the set of rare itemsets having zero support and the set of rare itemsets with nonzero support. A totally different approach to all these algorithms presented demands developing new algorithms to tackle these new challenges. Firstly consider Apriori-inverse [25], which can be seen as a more intricate variation of the traditional Apriori algorithm. The main idea is that given a user-specified maximum support threshold, MaxSup and a derived MinAbsSup value, a rule X is rare if $\text{Sup}(X) < \text{MaxSup}$ and $\text{Sup}(X) >$ MinAbsSup. Adda et al. [2] proposed a framework to represent different categories of interesting patterns and then instantiate it to the specific case of rare patterns. A generic framework, called AfRIM for Apriori-rare itemset, was presented to mine patterns based on the Apriori approach. The comprehensive Apriori framework was instantiated to mine rare itemsets. The resulting approach is Apriori-like where the itemset lattice representing the itemset space in classical Apriori approaches is traversed on a bottom-up manner, equivalent properties to the Apriori exploration of frequent itemsets are afforded to mine rare itemsets. Pillai and Vyas [44] presented a new foundational approach to temporally weighted itemset utility mining. Further, a conceptual model was presented by Pillai and Vyas [42, 43], which allows the development of an efficient and applicable algorithm to real-world data and captures real-life situations in fuzzy temporal weighted utility association rule mining. HURI algorithm considers the utility of itemsets other than the frequency of items in the transaction set. The utility of items is determined by considering factors such as profit, sale, temporal aspects, etc. of items. By using HURI, high utility rare itemsets can be generated based on minimum threshold values and user preferences.

6 Open Research Issues and Applications

6.1 Open Research Issues

As we can see that the itemset mining or pattern mining problems have been an active and prominent research topic for more than 20 years. Utility mining can be seen as the most significant area of research in this era, and there are still vast opportunities for research in this area. We here provide some key types of research opportunities in this field:

(a) Novel algorithms: The very opening research opportunity can be applying existing pattern mining algorithms like HUIM, rare itemset mining, HURI, etc. in new ways in terms of application domains. In particular, the use of pattern mining methods in emerging research areas such as social network analysis, the Internet of Things, sensor networks provides several novel possibilities in terms of applications.
(b) Enhancing the performance: As we can see the pattern mining process is quite time-consuming, this can be an important problem especially for new extensions of the pattern mining problem like high utility itemset mining amalgamation with soft computing, uncertain HURI, etc. which have been less explored.
(c) More complex and meaningful patterns mining: Another research opportunity is to develop utility mining algorithms that can be applied on complex types of data, i.e., extending high utility mining to consider more complex data.
(d) Applications of rare techniques: The major research opportunity can be seen in the application areas like medical diagnosis, intrusion detection system, credit card fraud detection system, Web mining, hardware fault detection, drug study, etc. Researchers can have the opportunities to work in these various application areas and produce good and useful results for decision making in future.

6.2 Applications in Future Prescriptive

There are most of the important areas in real life where the rare patterns have been found to be more desirable and vital compared to the frequent patterns. The applications here discuss in terms of the future prescriptive with future directions for handling significant real-life application in the area of rare pattern mining research.

(a) Abnormalities identification in biological dimensions: The biological datasets can be taking up to identify the rare gens presence which can result in major disease identification or major abnormalities in the high dimensional biological databases in the field of bioinformatics.
(b) Study of drug reactions in pharmaceuticals: In pharmaceuticals, the drug doses and contents are well studied for the effectiveness; in that case, the adverse reaction which can be rarely found may identify and used for the future study and findings.
(c) Web usage: With the concern of security and fraud detection, the rare and suspicious behaviors can be studied and addressed while browsing Web applications. The rare behavior of website visitors can be mined and studied to find the suspects and improve the securities.
(d) Cancer detection in medical diagnosis: If the case of mammogram images seen, only the small fragment in the entire image indicates the cancerous pixels which can be found in terms of rare patterns.

7 Conclusion and Future Work

7.1 Conclusion

This comprehensive study gives significant approaches in the field of utility mining in terms of High Utility Itemset Mining and High Utility Rare Itemset Mining. High Utility Rare Itemset (HURI) Mining discovers itemsets from a database which have their support less than a given frequency threshold and utility no less than a given minimum utility threshold. Identifying high utility rare itemsets from a database can assist in better business decision making. Table 6 gives the summary of remarkable contributions reviewed in this paper.

This survey will be useful for developing new efficient and optimize techniques in future in the field of utility mining. The open research opportunities in this field can be explored which presented in this paper. The researchers can get the motivation in the field on rare itemset mining through the application areas which have been also discussed in this paper.

Table 6 Conclusions of remarkable contributions reviewed

Sr.No	Algorithm	Proposed By	Year	Noteworthy contribution	Shortcoming
1	Apriori algorithm	Agrawal et al.	1994	Produce frequent pattern with single support and confidence	Generation of enormous candidate itemset. Multiple databases scanning. Time-consuming
2	Sampling-based algorithm	Toivionen	1996	Data reduction techniques have applied using sampling	Works in phases
3	MS Apriori algorithm	Liu et al.	1999	Multiple support framework used	Assignment of support values individual to each item
4	FP tree algorithm	Han et al.	2000	Only two phases of scanning used. It refers to FP tree structure only rather scanning of whole database	Expensive tree data structure require which can amplify memory consumption to some extent
5	OOA algorithm	Chan et al.	2003	Mining the top-K utility frequent closed patterns	Antimonotone property used in Apriori algorithm which does not hold good for utility mining
6	Apriori-inverse algorithm	Koh et al.	2005	Generate perfectly sporadic items. Faster than Apriori to generate rare items	Generate rules that contain items over the maximum support threshold
7	ARIMA algorithm	Szathmary et al.	2007	Took Apriori with single support base and generate zero support rare and nonzero support rare both	Spend a lot of time for looking frequent and rare both itemsets

(continued)

Table 6 (continued)

Sr.No	Algorithm	Proposed By	Year	Noteworthy contribution	Shortcoming
8	AFRIM algorithm	Adda et al.	2007	Perform level-wise search top to bottom and prune non-rare items	Maintenance of candidate list
9	HURI algorithm	Pillai et al.	2011	Consider both frequency and utility and generate high utility rare items	Multiphase working can be time-consuming
10	UP-rare growth algorithm	Goyal et al.	2017	Used a tree structure and generate rare items	Maintenance of tree structure

7.2 *Future Work*

In the future studies, we planned to explore the existing approaches with complex database such as noisy database and develop the novel approaches with enhanced performance. The proposed work will deal with noisy data with uncertainties. Also, the predefined threshold can be replaced by dynamic allocation. Individual ranks can be assigned to the items based on their support value only minimal rare will be considered so that the time consumption can be reduced.

References

1. Abaya SA (2012) Association rule mining based on Apriori algorithm in minimizing candidate generation. Int J Sci Eng Res 3(7):1–4
2. Adda M, Wu L, Feng Y (2007) Rare itemset mining. In: IEEE sixth international conference on machine learning and applications, ICMLA 2007, pp 73–80
3. Adda M, Wu L, White S, Feng Y (2007) Pattern detection with rare item-set mining. arXiv preprint. arXiv:1209.3089
4. Agrawal R, Mannila H, Srikant R, Toivonen H, Verkamo (2007) AI Fast discovery of association rules. Adv Knowl Discov Data Min 12(1):307–328
5. Agrawal R, Srikant R (1994) Fast algorithms for mining association rules. In: Proceedings of 20th international conference very large data bases VLDB, vol 1215, pp 487–499
6. Ashrafi MZ, Taniar D, Smith K (1994) A new approach of eliminating redundant association rules. In: International conference on database and expert systems applications. Springer, Berlin, Heidelberg, pp 465–474S
7. Ashrafi MZ, Taniar D, Smith K (2007) Redundant association rules reduction techniques. Int J Bus Intell Data Min 2(1):29–63
8. Cagliero L, Garza P (2014) Infrequent weighted itemset mining using frequent pattern growth. IEEE Trans Knowl Data Eng 26(4):903–915

9. Chui CK, Kao B, Hung E (2007) Mining frequent itemsets from uncertain data. In: Pacific-Asia conference on knowledge discovery and data mining. Springer, Berlin, Heidelberg, pp 47–58
10. Dimitrijevic M, Bosnjak Z, Subotica S (2010) Discovering interesting association rules in the web log usage data. Interdiscip J Inf Knowl Manag 5:191–207
11. Duong QH, Liao B, Fournier-Viger P, Dam TL (2016) An efficient algorithm for mining the top-k high utility itemsets using novel threshold raising and pruning strategies. Knowl Based Syst 104:106–122
12. Erwin A, Gopalan RP, Achuthan NR (2007) A bottom-up projection based algorithm for mining high utility itemsets. In: Proceedings of the 2nd international workshop on integrating artificial intelligence and data mining Australian computer society Inc., vol 84, pp 3–11
13. Fayyad U, Piatetsky-Shapiro G, Smyth P (1996) From data mining to knowledge discovery in databases. AI Mag 173:37
14. Fournier-Viger P, Lin JC, Vo B, Chi TT, Zhang J, Le HB (2017) A survey of itemset mining. Wiley Interdiscip Rev: Data Min Knowl Discov 7(4):e1207
15. Fournier-Viger P, Wu CW, Zida S, Tseng VS (2014) FHM: faster high-utility itemset mining using estimated utility co-occurrence pruning. In: International symposium on methodologies for intelligent systems. Springer, Cham, pp 83–92
16. Fournier-Viger P and Zida S (2015) FOSHU: faster on-shelf high utility itemset mining with or without negative unit profit. In: Proceedings of the 30th annual ACM symposium on applied computing, pp 857–864
17. Goyal V, Dawar S, Sureka A (2015) High utility rare itemset mining over transaction databases. In: International workshop on databases in networked information systems. Springer, Cham, pp 27–40
18. Grahne G, Zhu J (2005) Fast algorithms for frequent itemset mining using fp-trees. IEEE Trans Knowl Data Eng 17(10):1347–1362
19. Gyenesei A (2000) Mining weighted association rules for fuzzy quantitative items. In: European conference on principles of data mining and knowledge discovery. Springer, Berlin, Heidelberg, pp 416–423
20. Han J, Pei J, Kamber M (2011) Data mining: concepts and techniques. Elsevier
21. Han J, Pei J, Yin Y (2000) Mining frequent patterns without candidate generation. ACM Sigmod Rec ACM 29(2):1–12
22. Hu YH, Chen YL (2006) Mining association rules with multiple minimum supports: a new mining algorithm and a support tuning mechanism. Decis Support Syst 42(1):1–24
23. Kantardzic M (2011) Data mining: concepts, models, methods, and algorithms. Wiley
24. Kiran RU, Reddy PK (2011) Novel techniques to reduce search space in multiple minimum supports-based frequent pattern mining algorithms. In: Proceedings of the 14th international conference on extending database technology ACM, pp 11–20
25. Koh YS, Rountree N (2005) Finding sporadic rules using Apriori-inverse. In: Pacific-Asia conference on knowledge discovery and data mining. Springer, Berlin, Heidelberg, pp 97–106
26. Leung CK, Khan QI, Li Z, Hoque T (2007) CanTree: a canonical-order tree for incremental frequent-pattern mining. Knowl Inf Syst 11(3):287–311
27. Li Y (2011) Data mining: concepts, background and methods of integrating uncertainty. In data mining
28. Lin CW, Hong TP, Lu WH (2011) An effective tree structure for mining high utility itemsets. Expert Syst Appl 38(6):7419–7424
29. Lin CW, Hong TP, Lu WH (2009) The Pre-FUFP algorithm for incremental mining. Expert Syst Appl 36(5):9498–9505
30. Lin JC, Gan W, Fournier-Viger P, Hong TP, Tseng VS (2016) Fast algorithms for mining high-utility itemsets with various discount strategies. Adv Eng Inform 30(2):109–126
31. Lin JC, Gan W, Fournier-Viger P, Hong TP, Tseng VS (2015) Mining potential high-utility itemsets over uncertain databases. In: Proceedings of the ASE bigdata and social informatics 2015 ACM
32. Lin WY, Tseng MC (2006) Automated support specification for efficient mining of interesting association rules. J Inf Sci 32(3):238–250

33. Lin YC, Wu CW, Tseng VS (2015) Mining high utility itemsets in big data. In: Pacific-Asia conference on knowledge discovery and data mining. Springer, Cham, pp 649–661
34. Liu B, Hsu W, Ma Y (1999) Mining association rules with multiple minimum supports. In: Proceedings of the fifth ACM SIGKDD international conference on Knowledge discovery and data mining ACM, pp 337–341
35. Liu M, Qu J (2012) Mining high utility itemsets without candidate generation. In: Proceedings of the 21st ACM international conference on information and knowledge management, ACM, pp 55–64, 29 Oct 2012
36. Liu Y, Liao WK, Choudhary A (2005) A fast high utility itemsets mining algorithm. In: Proceedings of the 1st international workshop on Utility-based data mining ACM, pp 90–99
37. Liu Y, Liao WK, Choudhary A (2005) A two-phase algorithm for fast discovery of high utility itemsets. In: Pacific-Asia conference on knowledge discovery and data mining. Springer, Berlin, Heidelberg, pp 689–695
38. Mahgoub H, Rosner D (2006) Mining association rules from unstructured documents. In: Proceedings of 3rd international conference on knowledge mining, pp 167–172
39. Ninoria SZ, Thakur SS (2017) Study of high utility itemset mining. Int J Comput Appl 175(4):43–50
40. Patel AM, Bhalodiya D (2014) A survey on frequent itemset mining techniques using GPU. Int J Innov Res Technol 1(5)
41. PhridviRaj MS, GuruRao CV (2014) Data mining–past, present and future–a typical survey on data streams. Procedia Technol 12:255–263
42. Pillai J, Vyas OP, Muyeba M (2013) Huri–a novel algorithm for mining high utility rare itemsets. In: Advances in computing and information technology. Springer, Berlin, Heidelberg, pp 531–540
43. Pillai J, Vyas OP (2011) High utility rare item set mining (HURI): an approach for extracting high utility rare item sets. J Futur Eng Technol 7(1)
44. Pillai J, Vyas OP (2010) Overview of itemset utility mining and its applications. Int J Comput Appl 5(11):9–13
45. Pillai J, Vyas OP (2013) Transaction profitability using HURI algorithm [tphuri]. Int J Bus Inf Syst 2(1)
46. Pillai J, Vyas OP (2011) User centric approach to itemset utility mining in market basket analysis. Int J Comput Sci Eng 3(1):393–400
47. Poovammal E, Ponnavaikko M (2009) Utility independent privacy preserving data mining on vertically partitioned data 1. Int J Comput Sci 5:666–673
48. Savasere A, Omiecinski ER, Navathe SB (1995) An efficient algorithm for mining association rules in large databases. Georgia Institute of Technology 502
49. Shankar S, Babu N, Purusothaman T, Jayanthi S (2009) A fast algorithm for mining high utility itemsets. In: Advance computing conference, IEEE international, pp 1459–1464
50. Shie BE, Tseng VS, Yu PS (2010) Online mining of temporal maximal utility itemsets from data streams. In: Proceedings of the 2010 ACM symposium on applied computing ACM, pp 1622 –1626
51. Shridhar M, Parmar M (2017) Survey on association rule mining and its approaches, pp 129–135
52. Szathmary L, Napoli A, Valtchev P (2007) Towards rare itemset mining. In: Tools with artificial intelligence 19th IEEE international conference, vol 1, pp 305–312
53. Tang P, Turkia MP (2006) Parallelizing frequent itemset mining with FP-trees. In: Computers and their applications, pp 30–35
54. Thakur SS, Ninoria SZ (2017) An improved progressive sampling based approach for association rule mining. Int J Comput Appl 165:7
55. Toivonen H (1996) Sampling large databases for association rules. VLDB, vol 96, pp 134–145
56. Tseng V, Wu CW, Fournier-Viger P, Philip SY (2016) Efficient algorithms for mining top-k high utility itemsets. IEEE Trans Knowl Data Eng (1)
57. Tseng VS, Shie BE, Wu CW, Philip SY (2013) Efficient algorithms for mining high utility itemsets from transactional databases. IEEE Trans Knowl Data Eng 25(8):1772–1786

58. Yao H, Hamilton HJ, Geng L (2006) A unified framework for utility-based measures for mining itemsets. In: Proceedings of ACM SIGKDD 2nd Workshop on Utility-Based Data Mining, pp 28–37
59. Yun H, Ha D, Hwang B, Ryu KH (2003) Mining association rules on significant rare data using relative support. J Syst Softw 67(3):181–191
60. Yun U, Leggett JJ (2005) WFIM: weighted frequent itemset mining with a weight range and a minimum weight. In: Proceedings of the 2005 SIAM international conference on data mining, pp 636–640
61. Yun U, Ryang H, Ryu KH (2014) High utility itemset mining with techniques for reducing overestimated utilities and pruning candidates. Expert Syst Appl 41(8):3861–3878
62. Yun U (2007) Efficient mining of weighted interesting patterns with a strong weight and/or support affinity. Inf Sci 177(17):477–99
63. Yun U (2009) On pushing weight constraints deeply into frequent itemset mining. Intell Data Anal 13(2):359–383
64. Zida S, Fournier-Viger P, Lin JC, Wu CW, Tseng VS (2015) EFIM: a highly efficient algorithm for high—utility itemset mining. In: Mexican international conference on artificial intelligence. Springer, pp 530–546

A Study on Impact of Team Composition and Optimal Parameters Required to Predict Result of Cricket Match

Manoj S. Ishi and J. B. Patil

Abstract Cricket is getting a huge amount of popularity all around the world. More than a hundred countries now becomes part of cricket playing nations. Currently, the selection of the players and prediction of winner of a match becomes a challenging task. Performance of player is measure with factors such as current form, opposite team, venue, strike rate, etc. These parameters are considered for the selection of eleven player squad. The aim of forming squad is to get best playing eleven out of number of available players to form a balanced team. With the selection of players, there is also a need for finding the right set of parameters for the winning a match. Batsman, bowler, weather condition, venue are some of the factors that have effects on the outcome of the match. It becomes very difficult to show dominance over team in their home conditions but some team shows complete dominance all around the world. For deciding teams and predicting the result of match, there is a need to find new parameters. In this paper, we will study the methods previously used to get a balanced squad and outcome of the match.

Keywords Team formation · Winning prediction · Classifiers · Venue · Machine learning

1 Introduction

After soccer cricket becomes a very popular game all around the world. In the sixteen century, the game of cricket was started. After the end of the eighteenth century, cricket becomes an international sport. First international game was played in 1877 in the longer format called as test cricket. Cricket is played between two teams, eleven players, and with bat and ball. Every team gets an opportunity to bat once in the game to score runs, while other teams' tries to oppose that team from free

M. S. Ishi (✉) · J. B. Patil
Department of Computer Engineering, R. C. Patel Institute of Technology, Shirpur, Maharashtra, India
e-mail: ishimanoj41@gmail.com

J. B. Patil
e-mail: jbpatil@hotmail.com

R. K. Shukla et al. (eds.), *Social Networking and Computational Intelligence*, Lecture Notes in Networks and Systems 100, https://doi.org/10.1007/978-981-15-2071-6_32

scoring of runs. Players are assigned with a roll like batsmen, bowler, all-rounder, wicketkeeper and captain. For the training and monitoring, an extra person is assigned with team known as coach of team and considers as experts of cricket. For each nation, cricketing boards are formed. After the formation of international cricket council, the shorter format of a game was evolved. One day international and Twenty 20 cricket produced result in single day to decide the winner of match to make the game more interesting. Twenty 20 game is design to match timing with other game. One of the major events in the game of cricket is world cup. This event takes place once in four years [1]. ICC was formed in 1909 by South Africa, Australia and England. ICC observes the performance of team in all formats to defined ranking metrics for team. Price is given to team which maintains rank one in particular format. Points are awarded to team on the basis of their win, and if team wins against team having higher rank then more points are awarded. If team loses match against low-rank team then points are deducted from team. Table 1 shows the current rank metrics of the team in ODI format. Currently, team England is at the first position in one day international format with 126 points, while India at second position with 124 points. In the multination tournament like world cup, champion's trophy team with higher ranks gets direct participation. The associate team or teams not having higher ranks in ICC ranking need to play qualifier rounds to get shortlisted for multination tournament [2].

Players are assigned with special roles such as batsmen, bowlers, all-rounder's and wicketkeeper. Batsmen divided as top order batsmen, middle order batsmen, and lower order batsmen. Fast bowlers and spinners are the category of bowlers. All-rounder's considered as batting all-rounder if he is dominant in batting as compared to bowling. Bowling all-rounder is dominant in bowling performance. The wicketkeeper is having the job of batting along with fielding. Batsmen performance is measured with batting average, strike rate and high score. Economy, strike rate and bowling average is used to find the talent of bowler. Each team tries to get the right composition of players to improve their winning chances.

Table 1 ICC ODI ranking

Rank	Team	Matches	Ratings
1	England	55	126
2	India	64	124
3	New Zealand	49	110
4	South Africa	49	110
5	Pakistan	47	102
6	Australia	43	100
7	Bangladesh	39	93
8	Sri Lanka	57	78
9	West Indies	40	72
10	Afghanistan	36	67

In this paper, we will study various method used for the team formation and parameters required to decide the winner match. In Sect. 2, research direction or problems related to crickets are discussed. Section 3 consists of literature survey on the existing approaches used for team and winner prediction. In Sect. 4, we conclude with problem identification from the previous method.

2 Research Direction in Cricket

Currently, in the game of cricket, the problems occur when the games get interrupted due to weather condition and crowd problem. Team formation and winner prediction are also one of the biggest issues in the game of cricket. Lastly, the fantasy league also becomes a research problem in cricket [3].

Resetting of Targets: Matches are interrupted due to many reasons, such as bad light, weather condition and lastly crowd problem. In that case, to decide the winner of match, Duckworth and Lewis designs D/L method which helps to get winner of match without favoring particular team. Before D/L method, average run rate, most productive overs without or with discount, parabola curve, Clarke curve and V. Jaydevan are used for target resetting on the basis of run rate as a main parameter. Run rate is considered as an important parameter in all methods. ICC adopts Duckworth–Lewis method which considers net run rate and wickets fall as main parameter. Resource table is designed for resetting target. Resource table values are calculated on the basis number of overs remaining and number of the wickets fallen. Duckworth–Lewis method fails to provide good result for high scoring match. So Stern provides modification to existing method. It provides fair target for avoiding biasing toward one team. This method is designed such that with pocket calculator results can be provided without complex calculations. The Duckworth–Lewis method removes the disadvantage of all methods, still it favors team batting second, especially in Twenty 20 format of game where only 20 overs are assign per side [1, 4–6]. Jaydevan considers only runs remaining to get an end result of match in case of interrupted game [7].

Team Formation: Cricket is a game of statistics where a number of records are form and broken in every match. One of the main aspects of cricket is to form the balanced squad for team. Player's performance is accounted with batting average, strike rate, bowling average, economy rate, consistency, recent form players, etc. Team management, coach, captain need to consider this parameter for the formation of optimum side to get positive outcome of the match [3].

Winning Prediction: Every team tries to enhance their chance of winning in the cricket. Number of cricket league plays all around the world, where number of franchises are involved. They are investing a huge amount of money to select quality player to increase their chances of winning the title. Numerous parameters are affecting the outcome of cricket match such as run rate, venue, home/away game and team composition. For the tour to any nation or participating in any league, teams need to find out strong point to get result of match in their favor. This result in finding optimal parameters required for winning the match in all formats of the game [3].

In this paper, we will study the number of existing work done on team composition and predicting outcome of match.

3 Related Work

Selecting the balanced team in the game of cricket with right mixing of players is very difficult task for the team management. Following works are done by the authors to provide solution to from the balanced team.

Siripurapu et al. have proposed the technique for selection of balanced site in the game of cricket [8]. To make this process easy and simple, an adaptive neuro-fuzzy inference model is developed that considers the number of parameter for a player. Player data and ratings are provided to android application that does the task of formation of team using fuzzy rules. The principles of fuzzy logic and neural network together provide more benefits. Principal of both fuzzy logic and neural network are combined together into adaptive neuro-fuzzy inference system (ANFIS) to get benefits under single umbrella. If-then rules using fuzzy logic are act as backbone of this system. Player's performance is calculated on an average of previous five performances against the respective team. If-then rules consider the parameters like matches played, innings played, batting or bowling average, strike rate, wickets taken, recent form of player, team strength and opposite team strength. On the basis of these parameters, players are rated as very high, high, moderate, low and very low. If players are good in all parameter then it is rated as very high. After the implementation model, author still thinks that to add more parameter and fuzzy rules to get more clarity to system.

Passi and Pandey have described that player selection in cricket is more important task [9]. The performance of player is predicted as how many runs score by batsman and how many wickets taken by bowler with minimum run given to opposite team. They use naïve Bayes, random forest, SVM and decision tree classifiers to generate prediction model. The performance of player is weighted according to analytic hierarchy process. It is used to set priorities and decision making to form the team. The ranking is decided by optimal score and weight using AHP method for the formation of team. They conclude that random forest classifier was accurate for the problem. They want to elaborate his work for other formats of game like test cricket, T20 cricket. This module is designed for ODI matches. They also are focusing to improve the accuracy of classifier for ODI matches.

Saikia et al. have presented the approach for balanced squad with experts like bowler, batsman, wicketkeeper and all-rounder [10]. They measure the performance of player using single numeric value, which calculates on the basis of player's cricketing efficiency. Normalization method is used to measure the performance of batting, bowling and wicket keeping in the similar range. To measure the performance of player on the basis of their area range of 0–1 is used. Some parameters related to player act as positive parameter and some of the parameter acts as negative. Hence, the author decides to maintain balance of positive and negative parameter. After normalization weights are assigned on the basis of different factors. Weight acts as multiplication factor to normalized value and the result is combined in linear fashion to get composite index. Correlation analysis is among factors are performed under different skills of the players. In the final step, player identification is done under expertise for team formation. At last, author suggests that designing a tool for selection of player performance measure of a player is needed before using data.

Agrawal et al. have introduced the cricket team prediction statistical modeling approach using Hadoop [11]. As the cricket data is available in huge amount, Hadoop is selected for the formation of team. Batting and bowling scores are calculated using factor such as overall status, year-wise status, opposition-wise status, location-wise status. Hadoop and Hive framework provide an easy way for computing the analysis. Calculation for running algorithm is done with MapReduce jobs, where data is analyzed by splitting into number of independent chunks. It is processed in parallel proposed by map task to produce desired result. For increasing the accuracy of prediction, medical fitness data of player is required. Players who are not fit discarded from training data so that only fit player's data can be considered for prediction.

Sharma et al. have provided a new way to select team for T20 matches [12]. Cricket batsmen are analyzed using the ordered weighted averaging operator. They apply OWA method with respect to various attributes of cricket players. Author considered ten important attributes of batsmen. They consider the attribute like number of matches played, innings, number of not out innings, total run scored, highest score, strike rate, number of fours and sixes, number of fifties and hundreds. OWA aggregation operator checks aggregate performance score to from the team. OWA operator is the aggregation of different decision-making problems. To find the score of particular batsman, OWA calculates the weight of the player using linear programming model called as minimax disparity model. This model works for only batsmen selection. This model can be extended for selection of bowlers by measuring their attributes.

Ahmad et al. have used machine learning techniques [13]. According to them finding rising stars is main need of today's era. In this method, bowling as well as batsman is considered for team formation. To classify the players generative and discriminative algorithms of machine learning are used. Generative variables generate observable data values from latent variable provided. In discriminative unknown variable depends upon known variable. Bayesian network, naïve Bayesian, support vector machine and classification and regression tree are used to predict the rising stars in the game of cricket. They conclude that naïve Bayesian is better than other method. The weighted average and performance evolution of players are used to define rankings on the basis of co-players, teams and opposite teams. In this work, it is found that generative classifiers performance is better as compared to other classifiers. Chi-squared statistics, information gain and gain ratio are also considered for checking relative importance of features for the prediction of rising stars. In this technique, opposite team strength, number of hundreds, number of fifties, number of three or five wicket haul, home or away game parameters are not considered. This method can also be extended for other sports.

Harsha Perera et al. have presented a simulated annealing method for team formation in T20 cricket [14]. Simulated annealing algorithm is based upon probabilistic search algorithm. They used three components namely batting order, bowling order and team selection. In the first phase of algorithm, data is searched to consider distant search but latter on search is restricted to nearest neighbor. This algorithm evaluates the combinatorial place. It spends more time on regions to form the promising team. This approach is strictly based on analysis of data, and it avoids the opinion, tradition and folklore. They also focused on batting and bowling order of team. Systematic approach is provided for experiment with team lineup to shuffle the player's throughout the inning. This method considers only recent form of player. If the player is not playing cricket due to injury then medical problem may ignore the player for selection.

Ahmed et al. have proposed Web-based model for identification of cricket talent and selection method [15]. Web-based system works on OWA aggregation operator, relative fuzzy linguistic quantifier and normalized adequacy coefficient used to identify weakness in player and select most talented players from the available number of players. They use 28 parameters like speed, catching ability, self-control, decision making, etc. These 28 parameters are classified into three types as physical/motor abilities, cognitive/physiological abilities and anthropometric abilities. The OWA aggregation operator provides mean for aggregation. Relative quantifiers are found by relative fuzzy linguistic quantifier. Normalized adequacy coefficient calculates the difference between two elements called as hamming distance which helps to make decision-making process. In this method, the questionnaire is sent to experts with 28 parameters through multiple choice questions. If certain parameters are not included then response from experts are collected to improve questionnaire for next iteration. The problem with this approach is batting, bowling and fielding performance is not considered. Web-based system is designed accessible to cricket selectors at respective places. The players are labeled as very much talented, much talented, moderately talented and not talented. This system also helps in the identification of weaker area for the improvement. If someone excess value of threshold, then cricket selectors get alert form system. For making this module more successful, the involvement of more cricket experts is needed. Feature of alerting needs to be added into this system to make it more efficient.

Score and winning prediction are also one of the biggest issues in the cricket. Strategies are defined on the basis of inning number.

Mustafa et al. have predicted the winner of match using crowd opinion on social network [16]. They collect the tweets from twitter to predict the winner of match. Three different methods are used which depends upon the tweets before the match, sentiments of fans toward each team and fans prediction of score on twitter. These three methods combine together to decide the winner of match before the game starts. Opinion mining technique is applied for classification on the basis of sentiment. Sentiment analysis is identified with the guessing of opinion. The opinion may be in the format of sentence, document or feature and labeled as positive, negative and neutral. Support vector machine, naïve Bayes and logistic regression are used to train and evaluation of parameters. They finally conclude that SVM is better than other approaches with respect accuracy. In the training phase, tweets collection, representing features and training the classifiers these steps are performed. Testing phase consists of collecting tweets, representing feature, prediction of hypothesis and evaluation. Tweets are collected with the help of number of hashtags used for that team on official cricket Web page. This method can be applied to other methods after selection of right attributes.

Pathak and Wadhwa have proposed a technique to predict the outcome of ODI cricket using classification techniques [17]. They developed a tool called as cricket outcome predictor (COP) which decides the probability of win/loss. The modern classification techniques consists of naïve Bayes, support vector machine and random forest algorithm are used. Comparative study on their performances and outcomes is performed. For the evaluation of performance, kappa statistics and balanced accuracy are used. Higher kappa denotes better classification. Balanced accuracy finds mean, specificity, sensitivity and average accuracy obtained for other class. They conclude that naïve Bayesian is better than other approaches with respect to balanced accuracy. For extension to this method, more classifiers can be added because computation part is kept separate. This method can be extended to other formats of cricket. By adding more parameters, prediction accuracy can be improved. This approach can be used for classification of other sports like football and baseball with a different implementation.

Peter O Donoghue has included wicket loss and risk taken to predict the winner of the match [18]. The purpose of this method to study what optimal strategy is used in one day international cricket for the balance of number of run score to wickets fallen. In the first module, they study on effect of net run rate. In the second module, the study of wicket loss is done on number of runs scored. The second module is check on first innings as well as second inning and compare with each other. This method observed that during first half of inning less number of runs scored with less wickets are fallen. In the second half, the team tries to raise run rate due to that more wickets are fallen. To justify the study simulation is performed on the data. This method explains that keeping more wickets at the end of inning with less number of runs scored not awarding any extra points for the team. In the future, more variables need to add along with observed run rates and dismissal rates to occur. The area needs to improve are: ball line, ball length, fielding position, field position and types of short selected. This method can also be extended for women cricket, domestic and junior cricket performance.

Asif et al. have designed dynamic logistic regression model for the win probability of winning the game in one day international cricket [19]. This model is dynamic because parameters are allowed to change while game is in progress. Logistic regression model is used to reduce number of parameters for producing stable and forecast probabilities. For the identification of variable, cross-validation technique is used. They divide variable into two subparts: pre-match variable and in-play variable. The factors considered for the prediction are home advantage, toss, day–night effect, team quality and form. These variables are treated as pre-match variable. For the calculation of past performance of team, ICC rankings are considered. Team performance is measured with difference of rank between team. For getting more accurate form of a team, instead of considering ICC ranking previous five results of teams need to consider. In-play variables are number of runs scored or runs remaining depend upon inning number, number of wicket lost and number of balls remain. This model summarized the condition of match on the basis of number of runs scored, number of wickets lost and number of balls remain. These estimated coefficients are allowed to move smoothly in match. Leave one out cross-validation process used for testing of the model. For reducing the computing time for LOOCV considers over-by-over data instead of ball-by-ball data. In the future, pitch effect can be consider according to high scoring or low scoring games. This module can also use to develop ranking system for player or team in ODI cricket.

Jhawar and Pudi have proposed predict the outcome of cricket match using supervised learning approach with respect to team composition [20]. Team strength relative to opposite team is the most important factor to predict the winner of match. Individual batting and bowling performance are considered along with career statistics and recent performance of the player. Selection of players are the most difficult task, selections are performed on the basis of match conditions, venue, opponent team, etc. Team composition changes with match condition, venue, opponent team, injury and retirement from cricket. They used *k*-nearest neighbor algorithm to provide a better result as compared to other classifiers. Batsmen and bowler strength are measured with overall score. Highest score indicates that batsmen or bowler gets chance in every match. If the score is low then it indicates that batsmen or bowler not getting more chance to bat or bowl. For the prediction of winner of match venue, outcome of toss and relative strength are also considered in this method. In the future, more parameter may be added to improve the efficiency of method. Timing of match (day/night) and state of match at current instant may be considered as further work. This method provides promising result with simple features.

Shankaranarayanan et al. have discussed prediction system using combination of nearest neighbor clustering algorithm and linear regression model [21]. Supervised and unsupervised learning algorithms are applied on cricket data set. They extract the feature like home or away game, power play, target, performance, wickets and overs. These matches require an instantaneous record to predict the remaining game up to certain point. The problem of prediction is breakdown into two problems home runs and non-home runs for prediction. Weighted combination of instantaneous feature and past data is considered to predict and simulate the result. Two separate modules are defined for home runs and non-home runs using historical data. Attribute bagging algorithm and ridge regression algorithms are used to predict runs scored in the innings. Data variation was handled by regression algorithm, similar pattern of run scoring is identified using attribute bagging algorithm. This module uses historical feature and instantaneous feature. In historical feature, average run scored, number of wicket lost, frequency of all out, average run conceded in inning, average of wickets taken of opposite team and frequency of taking all wickets of opposite team was consider. In instantaneous feature home/away game, power play, target, batsmen performance feature, snapshot of game are considered. Historical and non-historical features are applied to nearest neighbor clustering method and attribute bagging method. Attribute bagging performs better than nearest neighbor. In the future, wicket prediction, avoiding data sparsity, bowler features can be used to improve prediction accuracy.

Akhatar and Scarf have proposed forecast match outcome in test cricket, session by session when game is in play [22]. Multinomial logistic regression model is used to forecast the outcome of the match. These models can help captain to decide certain aggressive or defensive strategies for the next session of matches. Probabilities of win, loss and draw provided session by session in this approach. Pre-match effect and in-play effects are considered in the analysis. Pre-match effect consists of parameter such as ground condition, advantage of home game and toss outcome. Lead taken or score, number of overs used and remaining, wickets remaining and run ret are considered for in-play effect. The outcome of study indicates that lead is not having at much effect on match result, but it can effective in later stage of game. Strength of team before match, effect of ground and home condition also have an impact on prediction. Wickets are important through the complete session of the match. Strength of team is calculated using ICC rating difference and wins percentage difference. This model is designed for test cricket for providing information at the start of each session to decide the plan for upcoming session. Session by session planning on the basis of team strength decides to predict the winner of match in case of test cricket.

Kaluarachchi and Varde have designed classification-based tool to predict the outcome in ODI cricket [23]. Bayesian classifier is used to predict the outcome of the match using factors affecting the end result of match. Crick AI tool is designed by considering the factor like home game, day/night effect, toss and inning. Naïve Bayesian, C4.5, bagging and boosting algorithms are considered for the classification using machine learning technique. Comparative study of classifiers is performed and then result is produced. They used association rule mining, clustering and classification for this tool. For the implementation of association rule mining, apriori algorithm is used for setting the rule. They finally conclude that naïve Bayes works best over other algorithms. Naïve Bayes algorithm handles a numerous number of independent variables. Classification techniques are compared using receiver operator curve, root mean squared error. Highest ROC and lowest MSE are provided with better learning techniques. This method says that toss is not having too much impact on the end result of match. Mixed outputs are produced by toss attribute. Crick AI module provides the prediction of match before game starts. ICC rankings also added to determine the strength of team. This study shows that classification is best approach for prediction. In the future, the previous match result of opponent team can be added, number of known batsmen from both teams can also considered for improving prediction. Finally, this module can also extend to another outdoor game such as baseball, depends on the application, dataset and parameters of that particular sport format

4 Conclusion

In this paper, we have performed an extensive literature survey of existing techniques and approaches used for team and winning prediction. We have study number of parameters required for prediction of team and outcome of match. We also study the impact of home or away game, power play, target, performance, wickets, overs, venue in the game of cricket. Naïve Bayes, support vector machine, k-nearest neighbor algorithms providing better result for classification of data. Cricket is game of statistics, so there is need to improve the accuracy of classifiers. Classifiers providing better result as compare to other technique. At the end, we summarized that there is a need of number of new parameters to be added in the existing approach to make prediction more accurate.

References

1. Jewson J, French S (2018) A comment on the Duckworth–Lewis–Stern method. J Oper Res Soc 69:1160–1163
2. https://www.icc-cricket.com/rankings/mens/team-rankings/odi
3. Swartz TB (2016) Research directions in cricket. Department of Statistics and Actuarial Science
4. Stern SE (2016) The Duckworth-Lewis-Stern method: extending the Duckworth-Lewis methodology to deal with modern scoring rates. J Oper Res Soc 67:1469–1480
5. Duckworth FC, Lewis AJ (2004) A successful operational research intervention in one-day cricket. J Oper Res Soc 59:749–759
6. Duckworth FC, Lewis AJ (2004) A fair method for resetting the target in interrupted one-day cricket matches. J Oper Res Soc 49:220–227
7. Jaydevan V (2014) VJD method software. http://jayadevan.yolasite.com/cricket.php. Accessed 1 Jan 2015
8. Siripurapu N, Mittal A, Mukku RP, Tiwari R (2018) Intelligent system for team selection and decision making in the game of cricket. In: Smart computing and informatics, pp 467–474. Springer, Berlin
9. Passi Kalpdrum, Pandey Niravkumar (2018) Predicting players' performance in one day international cricket matches using machine learning. Int J Data Mining Knowl Manag Process (IJDKP) 8:19–36
10. Saikia H, Bhattacharjee D, Radhakrishnan UK (2016) A new model for player selection in cricket. Int J Perform Anal Sport 16:373–388
11. Agarwal S, Yadav L, Mehta S (2017) Cricket team prediction with hadoop: statistical modeling approach. In: 5th International conference on information technology and quantitative management (ITQM2017), pp 525–532. Elsevier, Amsterdam
12. Sharma SK, Gholam Amin R, Gattoufi S (2012) Choosing the best Twenty20 cricket batsmen using ordered weighted averaging. Int J Perform Anal Sport 12:614–628
13. Ahmad H, Daud A, Wang L, Hong H, Dawood H, Yang Y (2017) Prediction of rising stars in the game of cricket. IEEE Transact 5:4104–4124
14. Perera H, Davis J, Swartz TB (2016) Optimal lineups in Twenty20 cricket. J Stat Comput Simul 86:1–13
15. Ahamad G, Kazim Naqvi S, Sufyan Beg MM, Ahmed T (2015) A web based system for cricket talent identification enhancement and selection. In: The 2015 international conference on soft computing and software engineering SCSE 2015, pp 134–142. Elsevier, Amsterdam

16. Mustafa RU, Saqib Nawaz M, Ikram Ullah Lali M, Zia T, Mehmood W (2017) Predicting the cricket match outcome using crowd opinions on social networks: a comparative study of machine learning methods. Malays J Comput Sci 30:63–76
17. Pathak N, Wadhwa H (2016) Applications of modern classification techniques to predict the outcome of ODI cricket. In: 2016 International conference on computational science, pp 55–60. Elsevier, Amsterdam
18. O'Donoghue P (2017) Wicket loss and risk taking during the 2011 and 2015 cricket world cups. Int J Perform Anal Sport 16:80–95
19. Asif M, McHalec IG (2016) In-play forecasting of win probability in one day international cricket: a dynamic logistic regression model. Int J Forecast 32:34–43
20. Jhawar MG, Pudi V (2016) Predicting the outcome of ODI cricket matches: a team composition based approach. In: European conference on machine learning and principles and practice of knowledge discovery in databases (ECML-PKDD 2016), Aug 2016
21. Sankaranarayanan VV, Sattar J, Lakshmanan LVS (2014) Auto-play: a data mining approach to ODI cricket simulation and prediction. In: 14th SIAM international conference on data mining, SDM 2014, vol 2, pp 1064–1072
22. Akhtar S, Scarf P (2012) Forecasting test cricket match outcomes in play. Int J Forecast 28:632–643 Elsevier
23. Kaluarachchi A, Varde AS (2010) CricAI: a classification based tool to predict the outcome in ODI cricket. In: 5th international conference on information and automation for sustainability. IEEE

Using Analytic Hierarchal Processing in 26/11 Mumbai Terrorist Attack for Key Player Selection and Ranking

Amit Kumar Mishra, Nisheeth Joshi and Iti Mathur

Abstract This article having a significant analysis of the Mumbai terrorist attack held on November 26, 2011, using a decision based on multi-criteria which are a potential approach for the social network analysis. The data source for this analysis is a dossier report on 26/11 Mumbai attack submitted to the Ministry of External Affairs that was published in the year 2009 and many more articles. This report gives the complete details about this tragic event consisting number of terrorist involved in India as well as from Pakistan, points where they had done operation, their communications in between, number of casualties, etc. When law enforcement agencies want to analyze any terrorist attack concerning key players involved in the attack, investigators should consider more than one criterion or factors that may also inconsistent and contradictory. Therefore, key player selection and ranking of terrorist nodes are a multi-criteria decision-making issue. AHP resolves this multi-criteria decision-making issue. This study reveals the key players and ranked them accordingly, involved in 26/11 Mumbai terrorist attack using the analytic hierarchy process (AHP).

Keywords Social network analysis (SNA) · Investigative data mining (IDM) · Terrorist network mining (TNM) · AHP · Centrality measures

1 Introduction

From the last few decades, India is facing a major issue of terrorism. Terrorism is not only threats to human life as well as it is potential break in the development of any country. Terrorism has become a worldwide issue nowadays [1]. This problem is not only strengthening its hold in India but also taking a vivid form at the international level. The whole India shocked on November 26, 2011, after the Mumbai Terrorist attack has taken place at different locations of Mumbai. It was carried out by 10 militants linked from Lashkar-e-Taiba, a terrorist group based on Pakistan. In this

A. Kumar Mishra (✉) · N. Joshi · I. Mathur
Department of Computer Science and Engineering, Banasthali Vidyapith, Vanasthali, Rajasthan, India
e-mail: amitmishra.mtech@gmail.com

R. K. Shukla et al. (eds.), *Social Networking and Computational Intelligence*, Lecture Notes in Networks and Systems 100, https://doi.org/10.1007/978-981-15-2071-6_33

terrorist attack, we lost 260 peoples and wounded at least 308 persons. National Investigation Agency and various researchers had submitted their report on 26/11 Mumbai attack [2–5].

Relationships and the information flow within or outside the world of the organization can be easily identified and quantified by using the social network analysis (SNA). Researchers used this area of research for the Investigative Data Mining (IDM) to analyze the terrorist social network. This concept of terrorist network analysis is popularly known as terrorist network mining (TNM). It is important for the smooth operation to frame the deductive argument and to show the relationship among organized crime and terrorist networks. It can be represented as a network with nodes and links where nodes represent terrorist's links that represent associations or relationship among them [6].

After the study, TNM will give us an answer to the following questions:

1. Who was the most centralized terrorist of 26/11 Mumbai attack?
2. How many numbers of subgroups existed?
3. What was the pattern of interactions within subgroups?
4. What was the architecture of the relationship?
5. What was the flow of information sharing?

The 26/11 Mumbai attack was planned and executed by the militants of Lashkar-e-Taiba—a terrorist group based on Pakistan. Ten terrorists (Table 1) started from Karachi on November 22, 2008, toward Mumbai, India. They had ammunition, pistols, hand grenades, bag with IED, etc. They reached a point four nautical miles off Mumbai at 1600 h on November 26, 2008. Ismil Khan was decided as a leader of this group. This group divided into five subgroups to attack five prime locations of Mumbai named the Taj Mahal Hotel, CST railway station, the Oberoi Trident Hotel, Café Leopold, and the Nariman (Chabad) house Jewish community center [5]. During the attack, terrorists were connected with other terrorists in Pakistan (Table 2) using mobiles and satellite phones, which have been received by National Investigation Agency after a complete operation against terrorists and intercepted the communication between them during attack, which were in progress. Terrorists were killed more than 165 innocent peoples and many others injured in this whole tragic attack. Nine terrorists were killed out of ten. One terrorist named Mohammed Ajmal Amir Kasab has been arrested. Kasab was interrogated by the Mumbai crime branch and reveals all the facts of 26/11 Mumbai attack and its Pakistan connections.

Thomas Saaty in the year 1980 introduced a concept of making an efficient decision using multiple criteria known as analytic hierarchy process (AHP). Using AHP, decision-makers can derive the priorities for each option available and make the best

Table 1 26/11 Mumbai attack handlers operated from Pakistan

S. No.	Terrorist name	Place
1	Abu kafa	Pakistan
2	Wassi	Pakistan
3	Zarar	Pakistan

Table 2 26/11 Mumbai attack terrorist operated at Mumbai, India

S. No.	Terrorist name	Place (in India)
1	Mohammad Aimal Amir Kasab	CST railway station
2	Ismail Khan	CST railway station
3	Babar Imran	Nariman House
4	Abu Umar	Leopold Cafe and Bar, Taj Mahal Hotel
5	Shoaib	Taj Mahal Hotel
6	Nazir	Nariman House
7	Hafiz Arshad	Leopold Cafe and Bar, Taj Mahal Hotel
8	Javed	Taj Mahal Hotel
9	Abdur Rehman	The Oberoi Trident Hotel
10	Fahadulla	The Oberoi Trident Hotel

decision out of them. It follows the concept of reducing the complexity of complex decision by series of pairwise comparisons and finally builds a synthesized result. AHP works on both subjective and objective aspects of a decision.

Motive behind this analysis is a significant and quantitative assessment of 26/11 Mumbai attack, and article organized as follows:

Section 2: Values and description of data set under consideration.
Section 3: Measurement of measures and visualization of terrorist network.
Section 4: Ranking algorithms to rank nodes [7].
Section 4: Results and analysis.

2 Data Set

Data set of 26/11 Mumbai attack is based on the Mumbai terrorist attacks 2008 India Ministry of External Affairs Dossier [6] and news reports [8, 9]. Ten terrorists operated in India distributed in five subgroups, simultaneously three other persons come in light as per report those were having in continue touch with these terrorists from Pakistan and giving them instructions (Tables 1 and 2) [9].

At first, an adjacency matrix is created on the basis of communication between terrorists in India as well as from Pakistan. From the Ministry of External Affairs report, they have found terrorists were in continuing in touch with each other as well as handlers in Pakistan. This communication has been intercepted by the intelligence team. Table 3 represents the adjacency matrix A_{ij} of relationship (communication/interaction) between handlers in Pakistan and attackers in India during the attack. If any node from caller had called to any node at receiver or any node (either from caller or receiver) operated together during attack, then respective matrix value is 1; otherwise, the value is 0 as per Eq. 1.

Table 3 Data set of terrorist network of 26/11 Mumbai attack based on their interaction during attack [10]

	Abu kafa	Wassi	Zarar	Haifz Arsad	Javed	Abu Shoaib	Abu Umar	Abdur Rehman	Fahadullah	Baba Imran	Nazir	Ismil Khna	Md. Ajmal kasab
Abu Kafa	0	1	1	0	0	0	0	0	0	0	0	0	0
Wassi	1	0	1	1	0	0	1	0	0	1	1	0	0
Zarar	1	1	0	0	0	0	0	0	0	0	0	0	0
Hafiz Arshad	0	1	0	0	1	1	1	0	0	0	0	0	0
Javed	0	0	0	1	0	0	1	0	0	0	0	0	0
Abu Shoaib	0	0	0	1	1	0	1	0	0	0	0	0	0
Abu Umar	0	1	0	1	1	1	0	0	0	0	0	0	0
Abdur Rehman	1	0	0	0	0	0	0	0	1	0	0	0	0
Fahadullah	0	0	1	0	0	0	0	1	0	0	0	0	0
Baba Imran	0	1	0	0	0	0	0	0	0	0	1	0	0
Nazir	0	1	0	0	0	0	0	0	0	1	0	0	0
Ismil Khan	0	0	0	0	0	0	0	0	0	0	0	0	1
Md. Ajmal Kasab	0	0	0	0	0	0	0	0	0	0	0	1	0

$$A_{ij} = \begin{cases} 0, & \text{if No interaction} \\ 1, & \text{if Interaction done} \end{cases} \tag{1}$$

3 Measures of Social Network Analysis for TNM

This study measures the relationship and information flow between terrorist, group and other information processing units. The analysis can give us a brief idea about the path of information flow, who is the most influential person, which subgroup of people are more effective as compared to other subgroups, degree of interaction, etc. (Fig. 1).

3.1 Measure Based on Location (Influence, Prestige, or Control)

Directional Relations

Centrality

In the expansion of directional relation, centrality is an important measure that is based on degree centrality (in-degree, out-degree), closeness centrality, betweenness centrality, and information centrality at the node level.

Degree centrality:

Degree centrality measures the score of the importance of any node within the network by determining the number of direct contacts, which indicates the quality of

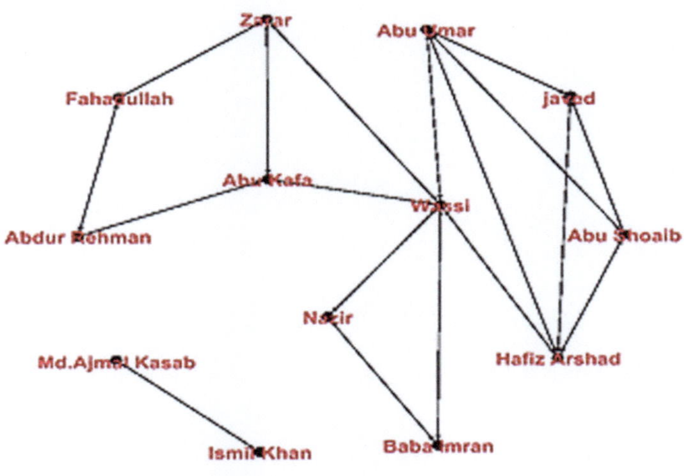

Fig. 1 26/11 Mumbai attack terrorist network

a member's interconnectedness inside a network. A node with a higher degree of centrality considered as a highly important node. It will have a higher impact on other nodes within network. Consider a network that has N nodes. The degree of centrality of node I C_i^D is defined as:

$$k_i = \sum_{j=1}^{N} A_{ij} \tag{2}$$

where
k_i is the degree of node i.

$$A_{ij} = \begin{cases} 1, & \text{if } i \text{ and } j \text{ is connected} \\ 0, & \text{otherwise} \end{cases}$$

$$C_i^D = \frac{k_i}{N-1}, \quad 0 \le C_i^D \le 1 \tag{3}$$

where $N - 1$ is the normalization factor which will limit the degree measurement between 0 and 1.

Figure 2 shows the degree distribution of 26/11 terrorist network, and different color schemes were used to represent the degree distribution. Wassi has a higher degree since he had 12 direct communication links with others. Table 4 shows the ranking of each node within the network according to the number of direct communication link with other actors within the network.

In-degree $\left(C_i^{D-}\right)$: Number of edges incoming at node *I,* i.e., the node whose distance to *I* is one (Fig. 3 and Table 5).

Out-degree $\left(C_i^{D+}\right)$: Number of edges outgoing from node *I,* i.e., node which has a distance one from node *i* (Fig. 4 and Table 6).

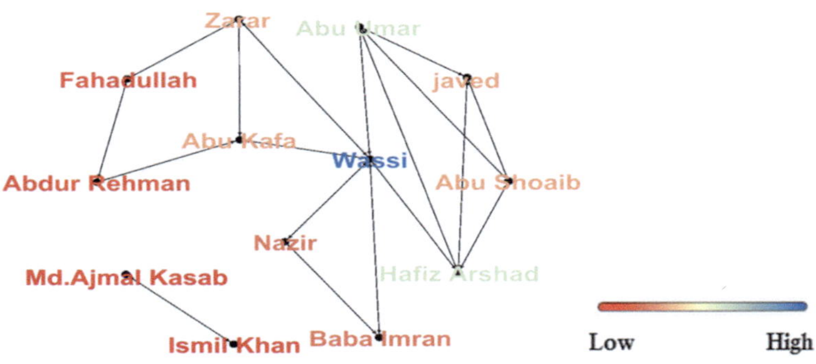

Fig. 2 Degree distribution graph

Table 4 Degree distribution table

Nodes	Abu kafa	Wassi	Zarar	Haifz Arsad	Javed	Abu Shoaib	Abu Umar	Abdur Rehman	Fahadullah	Baba Imran	Nazir	Ismil Khna	Md. Ajmal kasab
ID's	1	2	3	4	5	6	7	8	9	10	11	12	13
C_i^D	5	12	5	8	5	5	8	3	3	4	4	2	2
[0,1]	0.385	0.924	0.385	0.616	0.385	0.385	0.616	0.231	0.231	0.308	0.308	0.154	0.154
Rank	3	1	3	2	3	3	2	5	5	4	4	6	6

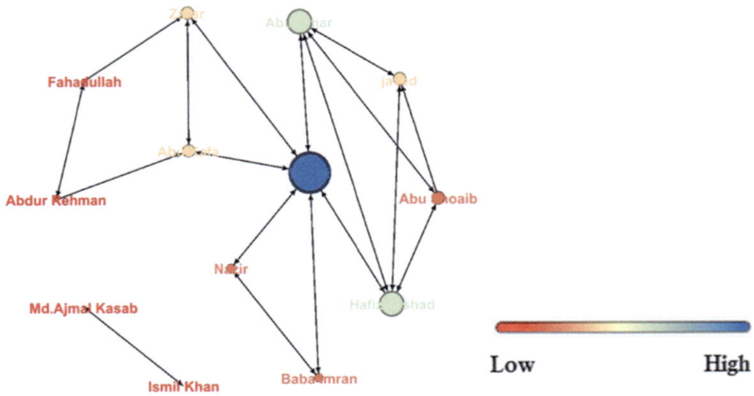

Fig. 3 In-degree distribution graph

Closeness Centrality:

Closeness centrality measures the score of a node which gives the idea of the speed of communication. Nodes with a short distance to the other nodes can spread information within network with a high speed and in an efficient way. The node which is closest to all other nodes will have highest closeness centrality value. This measure gives the idea to find a person in the terrorist network that can spread information in very less time to all other terrorists within network. If d_{ij} is the minimum number of edges between node *I* and node *j*, the closeness centrality $\left(C_i^C\right)$ of node *I* can be measured as: (Fig. 5 and Table 7)

$$C_i^C = \frac{N-1}{\sum_{j=1}^{|N|} d_{ij}}, \quad 0 \le C_i^C \le 1 \tag{4}$$

Betweenness Centrality:

Betweenness centrality measures a score, and how many times a node comes in the path of maximum possible pair's shortest path. It is considered as any node that is well connected and plays an important role within network if it comes in between the shortest path of as possible pairs of other nodes. According to Freeman [11], the betweenness centrality for a node *I* is calculated as:

$$C_i^B = \frac{1}{(N-1)(N-2)} \sum_{j,k \in G, j \ne k \ne i}^{|N|} \frac{n_{jk}(i)}{n_{jk}}, \quad 0 \le C_i^B \le 1 \tag{5}$$

where n_{jk} is number of the shortest path between node *j* and node *k* and $n_{jk}(i)$ is the number of shortest path in which node *I* comes in-between node *j* and node *k* (Fig. 6 and Table 8).

Table 5 In-degree distribution table

Nodes	Abu kafa	Wassi	Zarar	Haifz Arsad	Javed	Abu Shoaib	Abu Umar	Abdur Rehman	Fahadullah	Baba Imran	Nazir	Ismil Khna	Md. Ajmal kasab
ID's	1	2	3	4	5	6	7	8	9	10	11	12	13
C_i^{D-}	3	6	3	4	3	2	4	1	1	2	2	1	1
[0,1]	0.231	0.462	0.231	0.308	0.231	0.154	0.308	0.077	0.077	0.154	0.154	0.077	0.077
Rank	3	1	3	2	3	4	2	5	5	4	4	5	5

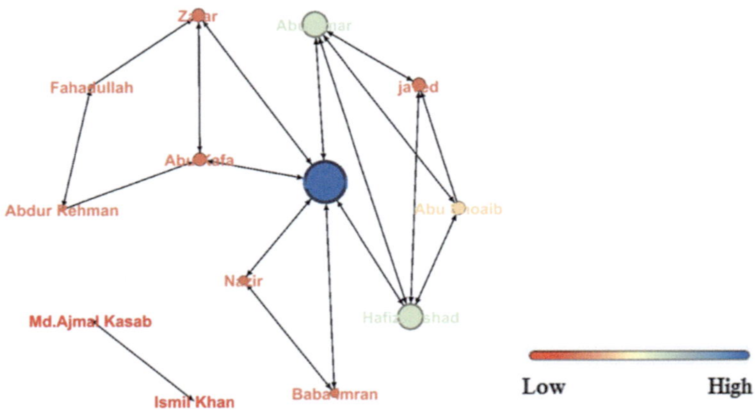

Fig. 4 Out-degree distribution graph

Eigenvector Centrality:

The idea behind the eigenvector centrality is to represent a node that has greater centrality if it is connected to more interconnected node and having less centrality value if it is connected to less interconnected node. For a node n_i, eigenvector centrality $\left(C_i^E\right)$ is calculated as [12]:

$$C_i^E = \frac{1}{\lambda}\left(\sum_{n_t \in N(n_i)} n_t\right) \tag{6}$$

where $N(n_i)$ set of nodes connected to n_i, and λ is a constant (Fig. 7 and Table 9).

Katz centrality is one of the generalized forms of eigenvector centrality; it measures the centrality value on the basis of its neighbor nodes. Katz centrality for node I will be measured as [13]:

$$\sigma_K(n_i) = \alpha \sum_{j=1}^{|N|} N_{i,j}(n_i + 1) \tag{7}$$

where α is a attenuation factor in (0,1) (Table 10).

Node ranking algorithms

PageRank:

Suppose a directed network has N nodes. PageRank of each node can be measured recursively as [14]:

Table 6 Out-degree distribution table

Nodes	Abu kafa	Wassi	Zarar	Haifz Arsad	Javed	Abu Shoaib	Abu Umar	Abdur Rehman	Fahadullah	Baba Imran	Nazir	Ismil Khna	Md. Ajmal kasab
ID's	1	2	3	4	5	6	7	8	9	10	11	12	13
C_i^{D+}	2	6	2	4	2	3	4	2	2	2	2	1	1
[0,1]	0.154	0.462	0.154	0.308	0.154	0.231	0.308	0.154	0.154	0.154	0.154	0.077	0.077
Rank	4	1	4	2	4	3	2	4	4	4	4	5	5

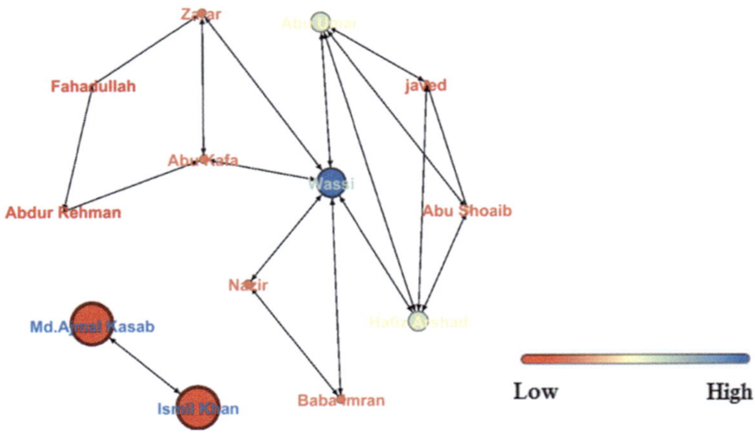

Fig. 5 Closeness centrality distribution graph

$$\mathrm{PR}_i^{(t+1)} = \alpha \sum_{j:k_j^{\mathrm{out}}>0} A_{ji} \frac{\mathrm{PR}_j^{(t)}}{k_j^{out}} + \frac{\alpha}{N} \sum_{j:k_j^{\mathrm{out}}=0} \mathrm{PR}_j^{(t)} + \frac{1-\alpha}{N} \tag{8}$$

where

A_{ji} is adjacency matrix between node j and i.
k_j^{out} is out-degree of node j.
α is the teleportation parameter ($\alpha = 0.85$).
t is iteration number.

The concept of PageRank is to traverse the graph by following directed edges with α probability or to teleport to a new node with $1-\alpha$ probability (Fig. 8 and Table 11).

Analytic Hierarchy Process (AHP):

The analytic hierarchy process (AHP) is a process to make the best decision among set of alternatives using multi-criteria. AHP makes its decision in three consecutive steps:

(1) Computing the criteria weight vector for each criterion

Importance weight for each criterion computes by creating a pairwise comparison matrix A for each criterion. In this problem, eight criteria have been chosen to make decision. Matrix A is 8×8 real matrix. Each a_{ij}^{th} element of matrix A shows the importance of ith criteria relative to jth criteria as shown in Table 1 [15, 16].

Table 7 Closeness centrality distribution table

Nodes	Abu kafa	Wassi	Zarar	Haifz Arsad	Javed	Abu Shoaib	Abu Umar	Abdur Rehman	Fahadullah	Baba Imran	Nazir	Ismil Khna	Md. Ajmal kasab
ID's	1	2	3	4	5	6	7	8	9	10	11	12	13
C_i^C	0.5	0.8	0.5	0.667	0.445	0.471	0.667	0.385	0.385	0.5	0.5	1	1
Rank	4	2	4	3	6	5	3	7	7	4	4	1	1

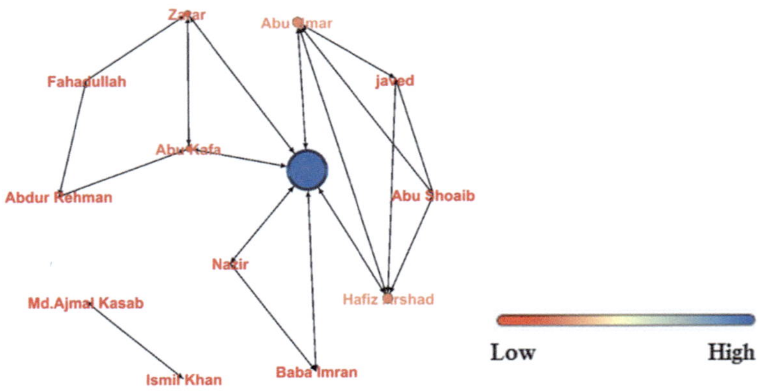

Fig. 6 Betweenness centrality distribution graph

Once the pairwise comparison matrix **A** built a normalized pairwise comparison matrix, $\mathbf{A}_{\text{norm}}$ is computed as (Table 12):

$$\bar{a}_{ij} = \frac{a_{ij}}{\sum_{i=1}^{m} a_{ij}}$$

Now by the averaging entry of each row of $\mathbf{A}_{\text{norm}}$, a criteria weight vector $\boldsymbol{w}$ measured as:

$$w_i = \frac{\sum_{i=1}^{m} \bar{a}_{ij}}{m}$$

where m is the total number or criteria (Tables 13 and 14).

(2) Computing the matrix of alternative scores

The pairwise comparison matrix of alternatives is a $n * n$ matrix B, where n is the total number of alternatives available. Each entry B_{ij} of matrix $B^{(k)}$ is an importance weight of ith alternative with respect to jth alternative for each criterion m, k = 1 … m. These pairwise comparison matrices of alternatives were used to derive alternative score matrix of $n * m$ real matrix S.

Now for each matrix $B^{(k)}$, it divides each entry by the sum of the same column. This step derives the average value on each row. It calculates the score $s^{(k)}$, $k = 1 \ldots m$. Finally, the score matrix S generated as (Table 15).

Table 8 Betweenness centrality distribution table

Nodes	Abu kafa	Wassi	Zarar	Haifz Arsad	Javed	Abu Shoaib	Abu Umar	Abdur Rehman	Fahadullah	Baba Imran	Nazir	Ismil Khna	Md. Ajmal kasab
ID's	1	2	3	4	5	6	7	8	9	10	11	12	13
C_i^B	0.057	0.394	0.057	0.095	0	0	0.095	0.004	0.004	0	0	0	0
Rank	3	1	3	2	5	5	2	4	4	5	5	5	5

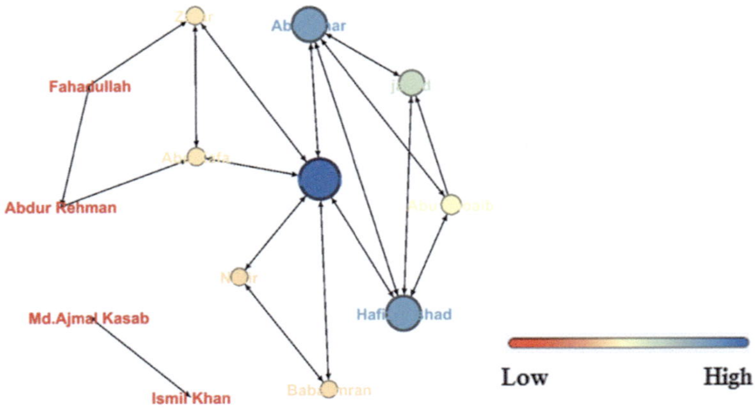

Fig. 7 Eigenvector centrality distribution graph

(3) Ranking the option

If v is global score of each alternative, then it will measure as:

$$v = S * w \tag{9}$$

where S is an alternative score matrix, w is weight vector for each criterion (Table 16).

4 Results and Analysis

The experimental result of AHP in 26/11 Mumbai attack terrorist nodes is given in Fig. 9.

Result shows Wassi was the main leader with the highest score; he was operating from Pakistan, and all terrorist subgroups in India were directly reporting to him. He was having the highest number of communication link among all 13 terrorists. Then the rest of the ranking graphically displayed in Fig. 9.

Figure 10 shows the graph of score for each criterion chosen for the evaluation of ranking of each terrorist involved in 26/11 Mumbai attack. These values represent the score in normalized [0,1] interval later on converted in idealized way 0–1.

Figure 11 represents the graph showing the criteria score for each terrorist involved in the attack. Criteria scores were ranged between 0 and 1, and each terrorist has a measured value for each criterion. This is useful to decide the position of terrorists within the network.

Table 9 Eigenvector centrality distribution table

Nodes	Abu kafa	Wassi	Zarar	Haifz Arsad	Javed	Abu Shoaib	Abu Umar	Abdur Rehman	Fahadullah	Baba Imran	Nazir	Ismil Khna	Md. Ajmal kasab
ID's	1	2	3	4	5	6	7	8	9	10	11	12	13
C_i^E	0.439	1	0.439	0.851	0.624	0.486	0.851	0.011	0.011	0.415	0.415	0.011	0.011
Rank	5	1	5	2	3	4	2	7	7	6	6	7	7

Table 10 Katz centrality distribution table

Nodes	Abu kafa	Wassi	Zarar	Haifz Arsad	Javed	Abu Shoaib	Abu Umar	Abdur Rehman	Fahadullah	Baba Imran	Nazir	Ismil Khna	Md. Ajmal kasab
ID's	1	2	3	4	5	6	7	8	9	10	11	12	13
$\sigma_K(n_i)$	0.287	0.374	0.287	0.324	0.290	0.264	0.324	0.221	0.221	0.263	0.263	0.221	0.221
Rank	4	1	4	2	3	5	2	7	7	6	6	7	7

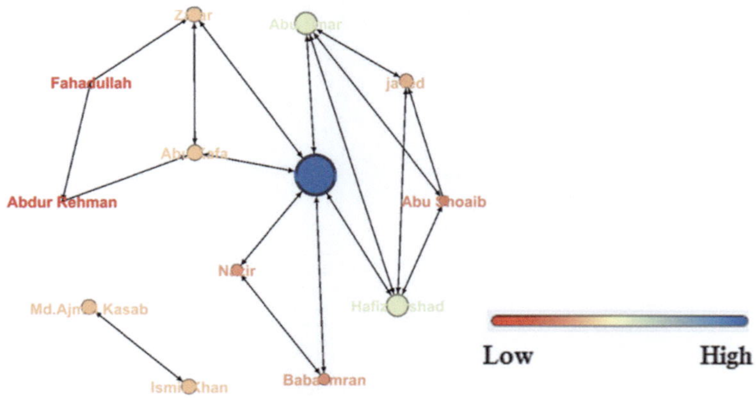

Fig. 8 PageRank distribution graph

Figure 12 shows the ranking of each terrorist according to each measure under consideration for rank evaluation. It shows AHP is the best technique out of them to evaluate the ranking of each terrorist more precisely and accurately.

5 Conclusion

The analytic hierarchy process (AHP) is an efficient process to rank the nodes in social network data. As a social network of terrorist has multiple terrorist nodes within the network, it is a needy task to identify the key players who influence the other nodes within the network most. So it would easy for the investigation agencies to find the leaders and the persons who played the other important roles within network to conduct any Terrorist activity. AHP uses multi-criteria to make a super decision on any given options, so it is a more accurate strategy compared to the traditional concept of deciding ranking on the basis of centrality measure which is not enough to decide. This paper gives the brief procedural introduction of an analytic hierarchical process to find the ranking and position of each terrorist who are all involved in 26/11 Mumbai attack. In the future, it needs more efficient methods that can make its decision on the basis of network processing, multi-attributes or on the basis of the ideal solution, and so better execution efficiency can also be achieved.

Table 11 PageRank distribution table

Nodes	Abu kafa	Wassi	Zarar	Haifz Arsad	Javed	Abu Shoaib	Abu Umar	Abdur Rehman	Fahadullah	Baba Imran	Nazir	Ismil Khna	Md. Ajmal kasab
ID's	1	2	3	4	5	6	7	8	9	10	11	12	13
$PR_i^{(t+1)}$	0.079	0.179	0.079	0.107	0.073	0.057	0.107	0.021	0.021	0.065	0.065	0.077	0.077
Rank	3	1	3	2	5	7	2	8	8	6	6	4	4

Table 12 Saaty's 1–9 scale for AHP performance

Intensity of importance	Definition	Explanation
1	Equal importance	Two activities contribute equally to the objective
3	Moderate importance	Experience and judgment slightly favor one over another
5	Strong importance	Experience and judgment strongly favor one over another
7	Very strong importance	Activity is strongly favored and its dominance is demonstrated in practice
9	Absolute importance	Importance of one over another affirmed on the highest possible order
2, 4, 6, 8	Intermediate values	Used to represent compromise between the priorities list above
Reciprocal of above nonzero number	If activity i has one of the above nonzero numbers assigned to it when compared with activity j, then j has the reciprocal value when compared with i	

Table 13 Criteria pairwise comparison matrix

CI: 0.0380, CR: 0.0271, λ: 8.2660

Criteria preferences	Degree	Eigenvector	In-degree	Out-degree	Closeness	Betweenness	Katz	PageRank
Degree	1	1/4	2	2	1/2	1/4	1/4	1/5
Eigenvector	4	1	4	4	2	1/2	1/2	1/3
In-degree	1/2	1/4	1	1	1/3	1/4	1/4	1/5
Out-degree	1/2	1/4	1	1	1/3	1/4	1/4	1/5
Closeness	2	1/2	1/3	1/3	1	1/2	1/2	1/3
Betweenness	4	2	4	4	2	1	2	1/2
Katz	4	2	4	4	2	1/2	1	1/2
PageRank	5	1/3	5	5	1/3	2	2	1

Table 14 Criteria preference matrix

Result	
Degree	0.0521
Eigenvector	0.1339
In-degree	0.0379
Out-degree	0.0379
Closeness	0.0928
Betweenness	0.1978
Katz	0.1658
PageRank	0.2817

Table 15 Alternative pairwise comparison matrix

Criteria preferences	Degree	Eigenvector	In-degree	Out-degree	Closeness	Betweenness	Katz	PageRank
Abu Kaahfa	0.0816	0.0522	0.0845	0.0507	0.0536	0.0931	0.0744	0.0995
Wassi	0.2024	0.2284	0.2054	0.2110	0.1330	0.2128	0.2255	0.2158
Zarar	0.0816	0.0522	0.0845	0.0507	0.0536	0.0931	0.0744	0.0995
Hafiz Arshad	0.1344	0.1620	0.1371	0.1427	0.0872	0.1447	0.1589	0.1494
Javed	0.0816	0.1134	0.0845	0.0507	0.0235	0.0329	0.1099	0.0443
Abu Shoaib	0.0816	0.0786	0.0498	0.0907	0.0343	0.0329	0.0498	0.0207
Abu Umer	0.1344	0.1620	0.1371	0.1427	0.0872	0.1447	0.1589	0.1494
Abdul Rehman	0.0311	0.0211	0.0294	0.0507	0.0167	0.0572	0.0207	0.0149
Fahadullah	0.0311	0.0211	0.0294	0.0507	0.0167	0.0572	0.0207	0.0149
Baba Imran	0.0493	0.0334	0.0498	0.0507	0.0536	0.0329	0.0326	0.0299
Nasir	0.0493	0.0334	0.0498	0.0507	0.0536	0.0329	0.0326	0.0299
Ismail Khan	0.0208	0.0211	0.0294	0.0290	0.1935	0.0329	0.0207	0.0658
Ajmal Amir Kasab	0.0208	0.0211	0.0294	0.0290	0.1935	0.0329	0.0207	0.0658

Table 16 Alternative score matrix

Consistency ratio (CR): 0.0213

Alternatives rankings with structure	Degree	Eigenvector	In-degree	Out-degree	Closeness	Betweenness	Katz	PageRank	Result
Abu Kaahfa	0.0043	0.0070	0.0032	0.0019	0.0050	0.0184	0.0123	0.0280	0.0801
Wassi	0.0105	0.0306	0.0078	0.0080	0.0123	0.0421	0.0374	0.0608	0.2095
Zarar	0.0043	0.0070	0.0032	0.0019	0.0050	0.0184	0.0123	0.0280	0.0801
Hafiz Arshad	0.0070	0.0217	0.0052	0.0054	0.0081	0.0286	0.0263	0.0421	0.1444
Javed	0.0043	0.0152	0.0032	0.0019	0.0022	0.0065	0.0182	0.0125	0.0639
Abu Shoaib	0.0043	0.0105	0.0019	0.0034	0.0032	0.0065	0.0083	0.0058	0.0439
Abu Umer	0.0070	0.0217	0.0052	0.0054	0.0081	0.0286	0.0263	0.0421	0.1444
Abdul Rehman	0.0016	0.0028	0.0011	0.0019	0.0015	0.0113	0.0034	0.0042	0.0280
Fahadullah	0.0016	0.0028	0.0011	0.0019	0.0015	0.0113	0.0034	0.0042	0.0280
Baba Imran	0.0026	0.0045	0.0019	0.0019	0.0050	0.0065	0.0054	0.0084	0.0362
Nasir	0.0026	0.0045	0.0019	0.0019	0.0050	0.0065	0.0054	0.0084	0.0362
Ismail Khan	0.0011	0.0028	0.0011	0.0011	0.0180	0.0065	0.0034	0.0185	0.0526
Ajmal Amir Kasab	0.0011	0.0028	0.0011	0.0011	0.0180	0.0065	0.0034	0.0185	0.0526

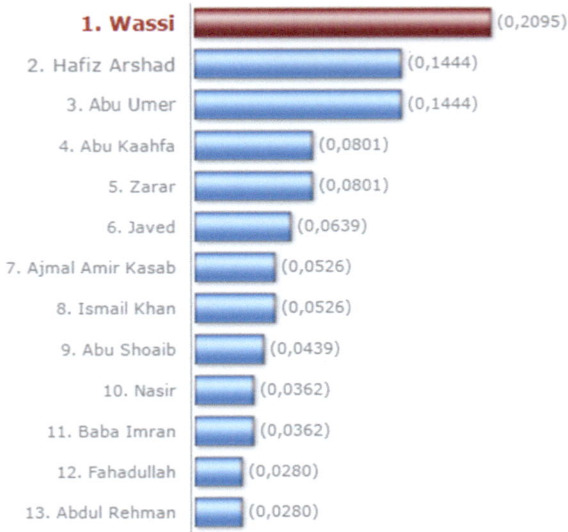

Fig. 9 Ranking of 26/11 Mumbai terrorist attack

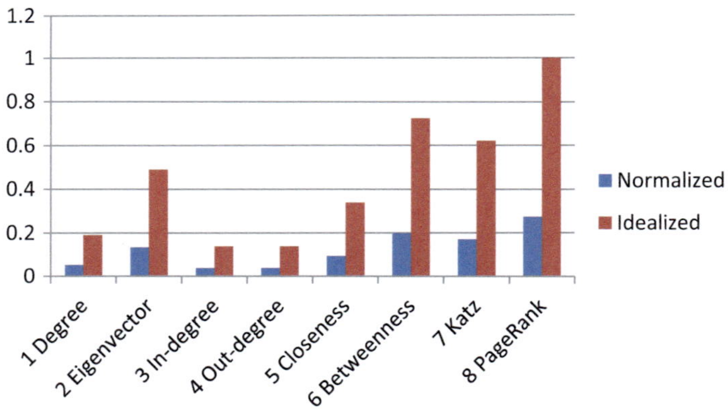

Fig. 10 Criteria score for evaluation

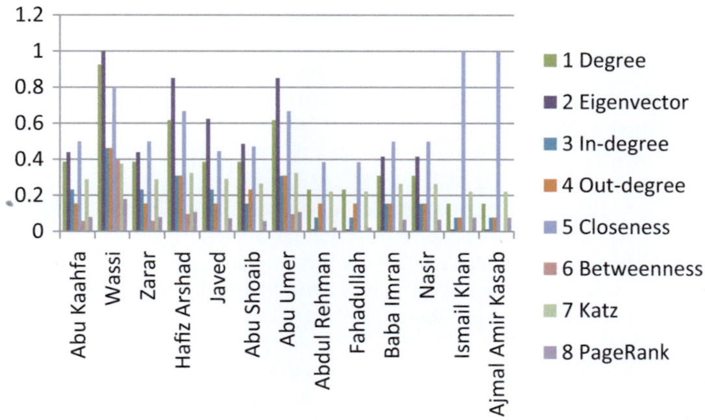

Fig. 11 Criteria alternative pairwise comparison

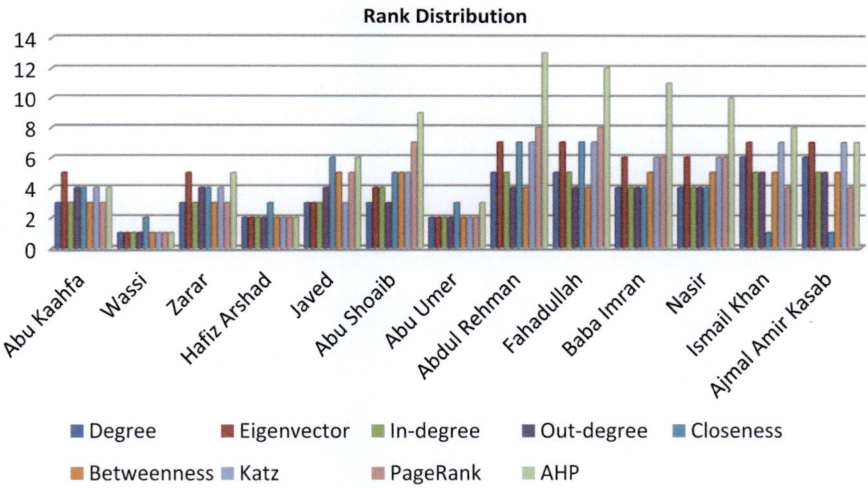

Fig. 12 Rank distribution of all measures under consideration

References

1. Fenstermacher L, Rieger KT, Speckhard A (2010) Protecting the homeland from international and domestic terrorism threats. White Paper: Counter Terrorism, p 178
2. Acharya A, Mandal S, Mehta A (2009) Terrorist attacks in Mumbai: picking up the pieces. International Centre for Political Violence and Terrorism Research S. Rajaratnam School for International Studies Nanyang Technological University, Singapore
3. Gunaratna R (2009) Mumbai investigation: the operatives, masterminds and enduring threat. UNISCI Discussion Paper, no 19, p 142
4. Onook O, Agrawal M, Rao R (2010) Information control and terrorism: tracking the Mumbai terrorist attack through twitter

5. Azad S, Gupta A (2011) A quantitative assessment on 26/11 Mumbai attack using SocialNetwork Analysis. J Terrorism Res V2(I2):4–14
6. A report on Mumbai attack (2009) Mumbai terrorist attack (26–29 Nov 2008). Govt. Of India
7. Marsden P (2015) Network centrality, measures of International Encyclopedia of the Social & Behavioral Sciences, pp. 532–539
8. Magnier M, Sharma S (2008) India terrorist attacks leave at least 101 dead in Mumbai. Los Angeles Times. p A1. Retrieved 28 Nov 2008
9. Masood S (2009) Pakistan announces arrests for Mumbai Attacks. Inf Syst Front 13(1):33–43. (New York Times. Retrieved 12 Feb 2009)
10. Chaurasia N, Tiwari A (2014) On the use of brokerage approach to discover influencing nodes in terrorist networks. In: Social networking: mining, visualization, and security, 1st edn, vol 65, Intelligent Systems Reference Library 65. Springer, Cham, pp 271–295. https://doi.org/10.1007/978-3-319-05164-2
11. De S, Dehuri S (2014) Machine learning for auspicious social network mining. In: Social networking: mining, visualization, and security, 1st edn, vol 65. Intelligent systems reference library 65. Springer, Cham, pp 45–83. https://doi.org/10.1007/978-3-319-05164-2
12. Freeman L (1977) A set of measures of centrality based on betweenness. Sociometry 40:35–41. https://doi.org/10.2307/3033543
13. Bonacich P (2007) Some unique properties of eigenvector centrality. Soc Netw 29:555–564
14. Katz L (1953) A new status index derived from sociometric analysis. Psychometrika 18(1):39–43
15. Mariani MS, Medo1 M, Zhang Y-C (2015) Ranking nodes in growing networks: when PageRank fails. Sci Rep 5(Article number: 16181):1–10. https://doi.org/10.1038/srep16181
16. Saaty TL, Rogers PC, Pell R (1980) Portfolio selection through hierarchies. J Portfolio Manag 6(3):16–21. https://doi.org/10.3905/jpm.1980.408749

A Comprehensive Study of Clustering Algorithms for Big Data Mining with MapReduce Capability

Kamlesh Kumar Pandey, Diwakar Shukla and Ram Milan

Abstract Big data mining is modern scientific research, which is used by all data related fields such as communication, computer, biology, geographical science, and so on. Basically, big data is related to volume, variety, velocity, variability, value, veracity, and visualization. Data mining technique is related to extract needed information, knowledge and hidden pattern, relations from large datasets with the heterogeneous format of data, which is collected by multiple sources. Data mining have classification, clustering, and association techniques for big data mining. Clustering is one of the approaches for mining, which is used for mine similar types of data, hidden patterns, and related data. All traditional clustering data mining approaches, such as partition, hierarchical, density, grid, and model-based algorithm, works on only high volume or high variety or high velocity. If we Apply the traditional clustering algorithms for big data mining then these algorithms will not work in the proper manner, and they need such clustering algorithms that work under high volume, high variety and high velocity. This paper presents the introduction to big data, big data mining, and traditional clustering algorithms concepts. From a theoretical, practical, and existing research perspective, this paper categorized clustering framework based on volume (dataset size, dimensional data), variety (dataset type, cluster shape), and velocity (scalability, time complexity), and presented a common framework for scalable and speed-up any type of clustering algorithm with MapReduce capability and shown this MapReduce clustering framework with the help of *K*-means algorithm.

Keywords Big data · Big data mining · Clustering · Clustering algorithm · MapReduce framework

K. K. Pandey (✉) · D. Shukla · R. Milan
Department of Computer Science and Applications, Dr. Harisingh Gour Vishwavidyalaya, Sagar 470003, Madhya Pradesh, India
e-mail: kamleshamk@gmail.com

D. Shukla
e-mail: diwakarshukla@rediffmail.com

R. Milan
e-mail: rammilan.in@gmail.com

R. K. Shukla et al. (eds.), *Social Networking and Computational Intelligence*, Lecture Notes in Networks and Systems 100, https://doi.org/10.1007/978-981-15-2071-6_34

1 Introduction

Currently, big data and its related services like cloud computing, Internet of things, social network, sensor network, medical applications, enterprise management, collective intelligence, smart grid, and data center, etc., are given and processed data very fast. For example, Google search engine processes hundreds of petabyte data query and Facebook generated 10 petabyte data such as text, video, images, and audio per day [1]. A lot of surveys discussed by researchers about data growing, one survey says according to Dobre et al. (2014), 2.5 quintillion bytes (2,500,000 terabytes) data generated every day all over the world [2]. According to the International Data Corporation report our world will be set to reach 44 zettabyte by 2020, which is ten times double of 2013. In 2017, Sivarajah and Kamal given two reports, according to his first report every day the world produces around 2.5 quintillion bytes of data, where 90% of data is unstructured and only 10% of data is structured, and according to the second report till 2020, 40 zettabyte or 40 trillion gigabytes of data will get generated, imitate, and consumed [3]. One report says, Facebook has 600 million active users are sharing more than 30 million data items like photos, notes, blogs, posts, Web links, and news, and Youtube has 490 million visitors worldwide spend approx 2.9 billion hours each month and it uploads 24 h of video every minute [4].

The main aim and objective of this paper is identifying a clustering algorithm for big data based on existing research outcomes and gave the common scalable and speed-up framework for clustering algorithm with the MapReduce approach under the big data mining. This paper has some important sections for big data mining and its clustering algorithms. The first section defines the introduction about what is big data, big data characteristics, big data mining, and how to differ data mining techniques from big data mining techniques on the context of database management. The second section defines the classification of the clustering algorithm with respect to big data mining. The third section describes the comparative analysis of clustering algorithms on the basis of three dimensions of big data and tries to find out which clustering algorithm is suitable for big data mining. The fourth section presented the scalable and speed-up the clustering framework. This presented framework have capability of executing the existing clustering algorithm with the MapReduce approach.

1.1 Big Data

Big data is not related to the only huge amount of data but it is also related to variety, velocity, variability, value, veracity, and visualization. Laney et al. (2001) is given volume, variety, and velocity characteristics of big data. These three dimensions are known as main characteristics or three V's and other V's is known as supportable characteristics of big data. In the present time, total of seven V's are available for big data [5]. In general, any dataset has met any two or three characteristics proposed by Laney et al. are known as big data because big data commonly buildup high volume

Fig. 1 Seven V's characteristics for big data

of different types and formats of data which is gathered from multiple sources at high speed [6]. Figure 1 shows all characteristics of big data [7].

First V's of big data is volume, which refers to the size of the dataset with respect to multiple terabytes or petabytes data scale. Second V's of big data is variety, which refers to the different types of data with respect to the homogenous format like structured, unstructured, and semi-structured data with multiple sources. The third of big data is velocity, which refers to the speed of data generation, mining, and analysis of real-time or non-real-time data set. Fourth V's of big data is variability, which refers the continuing changes of data or schema per second, minutes, and hours. Fifth V's of big data is value, which defines the attributes of big data which respect to data mining or data analysis. Six V's of big data is veracity, which represents the accuracy and quality of mined and analyzed data. Seven V's of big data is visualization, which represents the data under the stable and readable according to the end users [3–5].

1.2 Big Data Mining Technique

Data mining, statistical analysis, and machine learning are the best approach for extracting information about big data. Big data mining is the process of extracting for the needed information or knowledge in the huge volume with high velocity and high variety of data. Big data mining techniques are helpful for finding hidden patterns and relationships from big data. Every big data mining technique must be meet any two V's of three dimensions of big data with suitable data management framework, effective preprocessing steps, advanced parallel, distributed environment, highly scalable strategies, and intelligent user interaction [8, 9].

Traditional databases such as relational database systems are not suitable for big data because it holds only structured data, small volume, homogenous data, but big data have structured, unstructured, and semi-structured data. Big data is always controlled, processed, storage, and operated by NOSQL databases, which is handled by the Hadoop or similar framework. Hadoop has used HDFS and MapReduce function. HDFS used for the big data storage, and MapReduce function used for big data mining with parallel and distributed computing [1, 4, 10]. Big data are stored in the form of document-oriented, column-oriented, graph-based, and key-value [11]. MapReduce techniques are inspired by the map and reduce function. The idea of map function is breakdown to a task into possible phases and executes these phases in parallel order without disturbing any phases. Map function also assigns appropriate key/value pairs in every data. Reduce function collects all map results and combining all values based on the same key and given the final result of the MapReduce computational task. This concept reduces the computational time for big data mining [10, 11].

Big data mining techniques include basic three types of algorithm, the first approach is the supervised algorithm, which refers to models from labeled high volume dataset and used to predict the label of unlabeled heterogeneous data, for example, classification algorithm. The second approach is the unsupervised algorithm, which discovers hidden structure, patterns, and relation from unlabeled heterogeneous high volume dataset, for example, clustering and last approach are association mining or frequent itemsets mining, which discover dependencies, relations, and correlations among a high variety of data on large volume dataset [12].

2 Clustering Algorithms Techniques for Big Data Mining

In this section, the paper describes what is clustering and their classification for big data mining. The Clustering algorithm classifies heterogeneous and homogenous data into the different groups based on similarity or dissimilarity. These data groups are known as a cluster. The main purpose of clustering is to find out all hidden relations inside of cluster members or between every cluster for unlabeled data. Big data mining and machine learning viewpoint clustering are known as the unsupervised classification of hidden patterns such as data items, observations, and feature vectors result as so on into suitable groups or clusters [13]. In general, the clustering algorithm is categorized into five groups as a partition, hierarchical, density, grid, and model-based algorithms for clustering framework. Table 1 shows the summarization of all clustering algorithms [14–16].

Table 1 Summarization of clustering algorithm framework based on three-dimensional properties of big data

Clustering categories	Clustering algorithm	Volume		Variety		Velocity	
		Dataset size	Uses high-dimensional data	Dataset type	Cluster shape	Scalability	Complexity/time
Partition-based algorithms	*K*-mean	Large	No	Numerical	Convex	Medium	$0\ (knt)$/low
	K-medoid	Small	Yes	Categorical	Convex	Low	$0(k(n - k)^2)$/high
	k-modes	Large	Yes	Categorical	Convex	Medium	$0(n)$/Medium
	PAM	Small	No	Numerical	Convex	Low	$0\ (k^3 * n^2)$/high
	CLARA.	Large	No	Numerical	Convex	High	$0(ks^2 + k(n - k))$/medium
	FCM	Large	No	Numerical	Convex	Medium	$0(n)$/low
	CLARANS	Large	No	Numerical	Convex	Medium	$0(n^{2)}$/low
Hierarchical-based algorithms	BIRCH	Large	No	Numerical	Convex	High	$0(n)$/low
	CURE	Large	Yes	Numerical	Arbitrary	High	$0(n^2 \log n)$/low
	ROCK	Large	No	Numerical/Categorical	Arbitrary	Medium	$0(n^2 \log n)$/high
	Chameleon	Large	Yes	All data	Convex	High	$0(n^2)$/high
	ECHIDNA	Large	No	Multivariate data	Convex	High	$0(nb(1 + \log_b m)$/high
	WARDS	Small	No	Numerical	Arbitrary	Medium	–/low
	SNN	Small	No	Categorical	Arbitrary	Medium	$0(n^2)$/low
	CACTUS	Small	No	Categorical	Arbitrary	Medium	$0(cn)$/low
	GRIDCLUST	Small	No	Numerical	Arbitrary	Medium	$0(n)$/low

(continued)

Table 1 (continued)

Clustering categories	Clustering algorithm	Volume		Variety		Velocity	
		Dataset size	Uses high-dimensional data	Dataset type	Cluster shape	Scalability	Complexity/time
Density-based algorithms	DBSCAN	Large	No	Numerical/spatial	Arbitrary	Medium	$0(n\log n)$/medium
	OPTICS	Large	No	Numerical	Arbitrary	Medium	$0(n\log n)$/medium
	DBCLASD	Large	No	Numerical	Arbitrary	Medium	$0(n\log n)$/medium
	DENCLUE	Large	Yes	Numerical	Arbitrary	Medium	$0(\log\|d\|)$/medium
	STING	Large	No	Spatial	Arbitrary	High	$0(n)$/low
	SUBCLU	Large	Yes	Numerical	Arbitrary	Medium	–
	GDBSCAN	Large	No	Numerical	Arbitrary	Medium	–
Grid-based algorithms	Wave Cluster	Large	No	Numerical	Arbitrary	Medium	$0(n)$/low
	CLIQUE	Large	No	Numerical	Convex	High	$0(n + k^2)$/low
	FC	Large	Yes	Numerical	Arbitrary	Low	$0(n)$/high
	OptiGrid	Large	Yes	Spatial data	Arbitrary	Low	$0(nd)$ to $0(nd - \log n)$/low
	MAFIA	Large	No	Numerical	Arbitrary	High	$0(c^p + p^n)$/medium
	ENCLUS	Large	No	Numerical	Arbitrary	High	$0(nd + m^d)$/medium
	PROCLUS	Large	Yes	Spatial	Arbitrary	Medium	$0(n)$/low
	ORCLUS	Large	Yes	Spatial	Arbitrary	Medium	$0(d^3)$/low
	STIRR	Large	No	Categorical	Arbitrary	Medium	$0(n)$/low
	BANG	Large	Yes	Numerical	Arbitrary	Medium	$0(n)$/low

(continued)

Table 1 (continued)

Clustering categories	Clustering algorithm	Volume		Variety		Velocity	
		Dataset size	Uses high-dimensional data	Dataset type	Cluster shape	Scalability	Complexity/time
Model-based algorithms	EM	Large	Yes	Spatial data	Arbitrary	Medium	$0(knp)$/low
	COBWEB	Small	No	Numerical	Arbitrary	Medium	$0(n^2)$/low
	SOM	Small	Yes	Multivariate data	Arbitrary	Low	$0(n^2m)$/high
	CLASSIT	Small	No	Numerical	Arbitrary	Medium	$0(n^2)$/low
	SLINK	Large	No	Numerical	Arbitrary	Medium	$0(n^2)$/low

2.1 Partition-Based Algorithms

Partitioning clustering algorithm divided the similar heterogeneous and homogenous data object into k partitions based on the objective function. Parameter k is a user define constant, which defines the number of clusters. Each partition group has a similar type of data and it has always met two properties, the first property describes each cluster have must be at least one data object, and the second property says each data object must belong to exactly one cluster. Generally, heuristic-based iterative optimization algorithm is used to partition-based algorithms. Some partition-based algorithms are K-mean, K-medoids, K-modes, PAM, CLARA, FCM, and CLARANS [13–21].

2.2 Hierarchical-Based Algorithms

Hierarchical clustering algorithm divides the similar data object by creating a hierarchy of clusters based on the medium of proximity, and proximities can be obtained by the intermediate data nodes for using distance matrix. Generally, the connectivity-based approach used to hierarchical clustering. The hierarchical clustering algorithm is classified in the agglomerative (bottom-up) or divisive (top-down) groups. In the agglomerative hierarchical method, all data objects are organized into self-cluster and first be selected any data object for each cluster and merge to the two or more successively objects until the k cluster are not established based on the minimum, average or maximum distance. In divisive hierarchical clustering, all data objects are organized into one cluster and recursively splitting the most fitting cluster until the k cluster is not established. Some hierarchical-based algorithm is BIRCH, CURE, ROCK, Chameleon, ECHIDNA, WARDS, SNN, CACTUS, and GRIDCLUST [13–21].

2.3 Density-Based Algorithms

Density clustering algorithm divided the similar heterogeneous and homogenous data object based on their regions of density, connectivity, and boundary which is growing too high density and clusters can be found of arbitrary shapes and data are categorized into the core, border, and noise points. Generally, natural protection against outliers-based algorithm used to the density clustering algorithm. Some density-based algorithms are DBSCAN, OPTICS, DBCLASD, DENCLUE, STING, SUBCLU, and GDBSCAN [13–21].

2.4 Grid-Based Algorithms

Grid clustering algorithm divided the similar heterogeneous and homogenous data object in the form of cells or grid structure for using subspace and hierarchical clustering techniques. This algorithm maps the infinite amount of data object to finite numbers of grids cluster in less computation time. Grid size must be less than used database size and grid size also define grid clustering algorithms performance. Some grid-based algorithms are Wave Cluster, CLIQUE, FC, OptiGrid, MAFIA, ENCLUS, PROCLUS, ORCLUS, STING, and BANG [13–21].

2.5 Model-Based Algorithms

Model clustering algorithm divided the similar heterogeneous and homogenous data objects based on statistical methods, conceptual methods, neural network approach, and robust clustering methods. Generally, neural network approach and statistical methods are the most popular approach to this clustering algorithm. Most probably algorithm is mixture density algorithm is EM, the conceptual clustering algorithm is such as COBWEB, neural network-based algorithm is SOM, CLASSIT, and SLINK [13–21].

3 Comparative Analysis of Clustering Algorithms Techniques for Big Data Mining

In the present time, various clustering algorithms are available for big data mining. Every algorithm has own merits and demerits based on their working process with similarity and dissimilarity function. Clustering algorithms are only suitable for big data mining, if they have met any two V's of big data such as volume, variety, and velocity. The aim of this section is to identify the clustering algorithm for big data mining based on some clustering criterion. This clustering criterion is related to volume, variety, and velocity.

Volume of big data defines the ability of a clustering algorithm works under a huge amount of the data. A considerable criterion of identifying volume with respect to clustering algorithms are two ways, first is the size of the dataset, and the second is high dimensionality. Variety of big data defines the clustering algorithm handles different types and format of heterogeneous and homogenous data. A considerable criterion of identifying variety with respect to clustering algorithms are two ways first is a type of data set and second is cluster shape. Velocity defines the ability of a clustering algorithm with respect to high-speed data generation and mining with streaming data. The run-time complexity and scalability are considerable criteria of identifying velocity with respect to clustering algorithms. Table 1 shows summarization and identifying clustering algorithms for big data mining using volume, variety, and velocity clustering criterion [13–21].

4 Scalable and Speed-up Clustering Framework for Big Data Mining

In many cases, theoretical-based analysis on existing work is not sufficient for deciding the clustering algorithm for big data mining but it identified which clustering algorithm is scalable and works under big data mining. In general, according to big data concepts Table 1 shows, if any clustering algorithm handles huge datasets with high dimensional, scalability, and heterogeneous data in the form of arbitrary shape so they are suitable for big data mining [20, 21]. Designing of big data clustering algorithm has must be the capability for parallel and distributed computing because Hadoop or other frameworks used this type of capability with the MapReduce approach [10, 11, 22].

A lot of researchers are discovered better algorithms for each clustering algorithm with a specific criterion such as scalability, large data set, high-dimensional data, parallel execution, and so on. In this section, paper presents the common scalable and speed-up clustering framework for all existing clustering algorithms, which is executed by the MapReduce model, and the execution of this clustering framework shown the help of existing *K*-means algorithm. The flow chart of scalable and speed-up clustering framework shows in Flow chart 1.

According to the above clustering framework flow chart, in the first step, we choose any clustering algorithm based on big data criteria, which is shown in Table 1. The second step is to identify (key, value) pair because MapReduce processed data as (key, value) pair where every key is store unique index of data value for high-speed computation. In the third step, map function defines the working of a chosen clustering algorithm for using data as (key, value) pair. In the fourth step, reduce function combines the output of all map functions and maps to each clustering result of each map function result with help of chosen clustering algorithm concept. In the final step, reduce function given the needed result of the clustering algorithm.

The execution of the presented clustering framework is shown the *K*-means clustering algorithm because *K*-means have the capability for executing the large data set with scalability. *K*-means algorithm is the top second algorithm for the data mining technique. *K*-means used distance function such as Euclidian distance for creating cluster because *K*-means create a cluster based on means between data points [22–24]. Implementation of *K*-means clustering algorithm based on presented clustering framework under the big data mining shown in Table 2.

In this paragraph, we take the power dataset for executing on *K*-mean clustering algorithm based on this clustering MapReduce approach. The power dataset is consists of 512,320 real data points with seven dimensions [25]. In this experiment

Flow chart 1 Flow chart of scalable and speed-up clustering framework

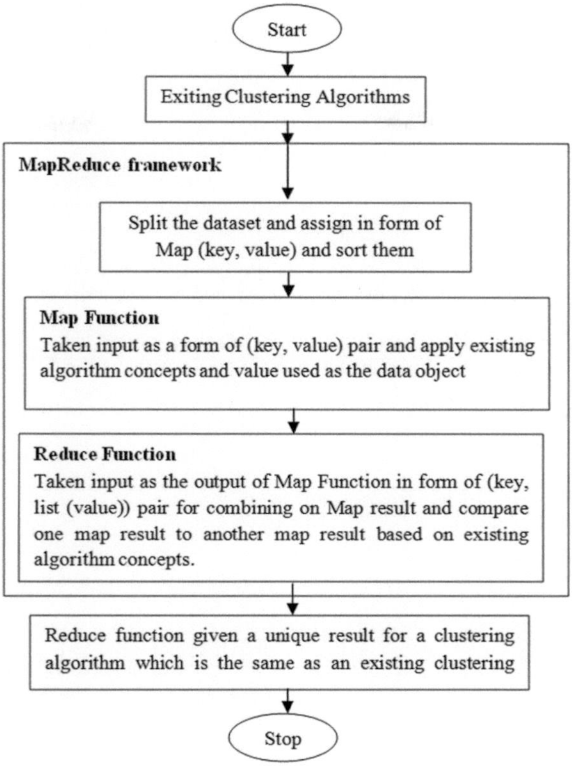

we used Hadoop with ten nodes cluster, and our system is configured with an Intel I3 processor, 4 GB DDR3 RAM, 320 GB hard disk, and operating system used is Windows 7. Here we show only execution time of existing *K*-mean (using Euclidian distance function) and MapReduce-based *K*-Mean clustering algorithm in Table 3.

5 Conclusion

This paper reviewed the basic and core idea of big data, big data mining, clustering algorithms with respect to big data characteristics, and defined which traditional clustering algorithms are suitable for big data mining. The first section provides information about big data and big data mining, big data databases and mining process of big data. The second section defines the concept of all existing clustering classification such as partition, hierarchical, density, grid, and model-based clustering. This section is very helpful for identifying clustering algorithm groups based on their working processes. The third section is given categorization or summarization of the clustering algorithm based on three-dimensional properties of big data such

Table 2 MapReduce clustering framework-based on *K*-mean algorithm

Steps	Clustering Algorithm for MapReduce based Clustering Framework
Step 1	Input : D Dataset, K number of Cluster Start 1. Choose k initial cluster center/mean from the set D 2. for i=1 to D.length 3. calculate distance of data from each cluster center 4. assign D_i to the nearest K cluster center 5. End for Stop
Step 2	Input: Map (key, value) = D key – index/ offset of data D value- content of Dataset D
Step 3	**Map function** Start 6. For i=1 to key.length 7. calculate distance of data value from each cluster center 8. assign D_i to the nearest K cluster center 9. End for Stop
Step 4	**Reduce function** 10. Map to each Map result to another Map result and store (key, list(value)) 11. For i=1 to key.length 12. calculate distance of data value from each Map result cluster center 13. assign D_i to the nearest K cluster center 14. End for Stop
Step 5	Taken final result from Reduce function

Table 3 Running time of *K*-mean algorithm

Algorithm	Execution time in second
K-mean (Existing)	649
K-mean (MapReduce-based)	140

as volume, variety, and velocity. This section is very helpful for identifying the clustering algorithm for big data mining. The fourth section presents common clustering framework for the existing clustering algorithm for using MapReduce approach. The MapReduce approach gives the scalable and speeds-up capability for any clustering algorithms. The working process of the presented framework is described with the help of existing *K*-mean algorithm. The presented clustering framework shows MapReduce based *K*-means algorithm takes less time as compared to the traditional *K*-means algorithm.

References

1. Chen M, Mao S, Liu Y (2014) Big data a survey. Mob Netw Appl 19(2):171–209. https://doi.org/10.1007/s11036-013-0489-0
2. Rouhani S, Robbie S, Hamidi H (2017) What do we know about the big data researches? A systematic review from 2011 to 2017. J Decis Syst 26(4):368–393. https://doi.org/10.1080/12460125.2018.1437654
3. Sivarajah U, Kamal MM (2017) Critical analysis of Big Data challenges and analytical methods. J Bus Res 70:263–286. https://doi.org/10.1016/j.jbusres.2016.08.001
4. Gole S, Tidke B (2015) A survey of Big Data in social media using data mining techniques. Proc IEEE ICACCS. https://doi.org/10.1109/ICACCS.2015.7324059
5. Gandomi A, Haider M (2015) Beyond the hype: Big Data concepts methods and analytics. Int J Inf Manag 35(2):137–144. https://doi.org/10.1016/j.ijinfomgt.2014.10.007
6. Wasastjerna MC (2018) The role of big data and digital privacy in merger review. Eur Compet J 14(2–3):417–444. https://doi.org/10.1080/17441056.2018.1533364
7. Pandey KK (2018) Mining on relationship in big data era using Apriori algorithm. In: Proceedings of NCDAMLS, pp 55–60. ISBN: 978-93-5291-457-9
8. Che D, Safran M, Peng Z (2013) From big data to big data mining challenges issues and opportunities. LNCS, vol 7827, pp 1–12. https://doi.org/10.1007/978-3-642-40270-8_1
9. Li N, Zeng L, Qing H, Zhongzhi S (2017) Parallel implementation of apriori algorithm based on MapReduce. In: Proceedings of 13th IEEE ACIS international conference on SEAIPDC. https://doi.org/10.1109/snpd.2012.31
10. Elgendy N, Elragal A (2014) Big data analytics a literature review paper. LNAI, vol 8557, pp 214–227. https://doi.org/10.1007/978-3-319-08976-8_16
11. Ozkose H, Ari ES, Gencer C (2015) Yesterday, today and tomorrow of big data. Proc Soc Behav Sci 195:1042–1050. https://doi.org/10.1016/j.sbspro.2015.06.147
12. Apiletti D, Baralis E, Pulvirenti F, Cerquitelli T, Garza P, Venturini L (2017) Frequent itemsets mining for big data: a comparative analysis. Big Data Res 9:67–83. https://doi.org/10.1016/j.bdr.2017.06.006
13. Jain AK, Murty MN, Flynn PJ (1999) Data clustering a review. ACM Comput Surv 31(3):264–323. https://doi.org/10.1145/331499.331504
14. Nagpal A, Jatain A, Gaur D (2013) Review based on data clustering algorithms. In: Proceedings of IEEE ICT, pp 298–303. https://doi.org/10.1109/cict.2013.6558109
15. Berkhin P (2006) A survey of clustering data mining techniques. In: Teboulle M (eds) Group Multidimens Data 25–71. https://doi.org/10.1007/3-540-28349-8_2
16. Mann AK, Kaur NB (2013) Review paper on clustering techniques. Global J Comp Sci Tech Soft Data Eng 13(5)
17. Shirkhorshidi AS, Aghabozorgi S, Wah TY, Herawan T (2014) Big data clustering: a review. LNCS, vol 8583, pp 707–720. https://doi.org/10.1007/978-3-319-09156-3_49
18. Xu D, Tian Y (2015) A comprehensive survey of clustering algorithms. Ann Data Sci 2(2):165–193. https://doi.org/10.1007/s40745-015-0040-1
19. Oyelade J, Aromolaran O, Itaewon I, Uwoghiren E, Oladipupo F, Ameh F, Adebiyi E, Achas M (2016) Clustering algorithms their application to gene expression data. Bioinf Biol Insights 10:237–253. https://doi.org/10.4137/BBI.S38316
20. Fahad A, Alshatri N, Tari Z, Alamri A, Khalil I, Zomaya AY, Foufou S, Bouras A (2014) A survey of clustering algorithms for big data taxonomy and empirical analysis. IEEE Trans Emerg Top Comput 2(3):267–279. https://doi.org/10.1109/tetc.2014.2330519
21. Pandove D, G.S.: A comprehensive study on clustering approaches for big data mining. In: IEEE 2nd ICECS, pp 1333–1338. https://doi.org/10.1109/ecs.2015.7124801
22. Sardar TH, Ansari Z (2018) Partition based clustering of large datasets using MapReduce framework: an analysis of recent themes and directions. Fut Comput Inf J 3(2):247–261. https://doi.org/10.1016/j.fcij.2018.06.002
23. Macqueen J (1967) Some methods for classification and analysis of multivariate observations. In: Proceedings of 5th BSMSP, vol 1, pp 281–297

24. Sinha A, Jana PK (2018) A hybrid MapReduce-based k-means clustering using genetic algorithm for distributed datasets. J Supercomput 74(4):1562–1579. https://doi.org/10.1007/s11227-017-2182-8
25. Berard A, Hebrail G (2013) Searching time series with hadoop in an electric power company. In: Proceedings of BDSHSMASPMA, pp 15–22. https://doi.org/10.1145/2501221.2501224

Parametric and Nonparametric Classification for Minimizing Misclassification Errors

Sushma Nagdeote and Sujata Chiwande

Abstract Parametric classification fits the parametric model to the training data and interpolates to classify the test data, whereas nonparametric methods like regression tree and classification trees use different techniques to determine classification. The classification process can be of two types: supervised and unsupervised. In supervised classification, training data are used to design the classifier. Bayes's rule, nearest neighboring rule, and perceptron rules are few widely used supervised classification rules. For unlabeled data, the process of classification is called clustering or unsupervised classification. This paper proposes a wrapper-based approach for pattern classification to minimize the error factor. Techniques, such as Bayes's classification, *K*-NN classifier, and NN classifier, are used to classify the patterns using linearly separable, linearly nonseparable, and Gaussian sample dataset. These methods classify the data in two stages: training stage and prediction stage. In this paper, we will be using parametric and nonparametric decision-making algorithm as we know the statistical and geometric properties of the patterns under study.

Keywords Parametric · Nonparametric · Maximum margin hyperplane · Soft margin · Wrapper

1 Introduction

The purpose of this paper is to use parametric and nonparametric classification techniques to minimize misclassification errors. The innovation in this design is to use wrapper approach. In this approach, set of input patterns are used as a template and a feature set is obtained from feature extractor for an input template. We consider set of data points that are to be classified into one of the four classes using hyperplane classifier. Classifiers such as Bayes's classifier, *K*-NN classifier, and NN classifier are used as hyperplane classifier. We consider datasets that are linearly separable,

S. Nagdeote (✉)
Department of Electronics Engineering, Fr. CRCE, Mumbai, India

S. Chiwande
Department of Electronics and Telecommunication Engineering, YCCE, Nagpur, Nagpur, India
e-mail: sujata_chiwande28@yahoo.co.in

R. K. Shukla et al. (eds.), *Social Networking and Computational Intelligence*, Lecture Notes in Networks and Systems 100, https://doi.org/10.1007/978-981-15-2071-6_35

linearly nonseparable, and Gaussian distributed dataset. For obtaining 100% of correct classification, we have to maximize the margins from the separating hyperplane to the nearest data points. Hence, we have to find maximum margin hyperplane. The data points which are closest to the hyperplane are known as support vectors. The number of classes is linearly separable which means the classes do not intersect, and we would be able to separate all four classes. Separable data can be learned without forcing errors in learning process. In linearly nonseparable case, the classes overlap or classes are intersecting. They cannot be learned without forcing errors in learning process. The soft margin is created when two classes overlap each other and is controlled by penalty parameter. Within soft margin, all misclassified data lie. The Gaussian distributed dataset is governed by probability density function. The classes are assumed to be distributed about means and variances. Usually, all statistical learning techniques are learning from the data. The attributes can be quantitative or categorical that we desire to speculate depending on the feature set. A training dataset is used to detect the result and attributes for a given set of data. By making use of this data, we can build statistical model and prediction model which is beneficial for predicting the output of test objects. The formulation of learning problem is generally based on two types: First is regression estimation and second is classification. The first problem is for minimizing the risk function with squared error loss function, and the second problem is to find a loss function which minimizes misclassification error.

2 Related Work

There are various works proposed on parametric and nonparametric classification. The author in paper [1] proposed fully automated chromosome classification algorithm. He adopted K-NN and maximum likelihood estimation for classification of chromosomes for M-FISH images. For $K = 7$, they achieved highest classification accuracy with K-NN method. In paper [2], the author proposed tree representation for different classifiers. They have adopted training set method and 10-cross-fold method for different classifiers and also compared their results. To classify the data when they used training set method, multilayer perceptron performs better than logistic regression classifier. The author in paper [3] discussed the formulation of labeled and unlabeled data that incorporates constrained data for model fitting. They modeled constraint information by principle of maximum entropy. Their strategy allows to handle constraint violations and soft constraints and to speed up the optimization process. This proposed algorithm is computationally efficient and generates groupings when compared to other methods. In paper [4], comparison of different NN methods was done. They divided different NN methods into structured and structure-less based category. Based on these categories, different NN algorithms are adopted and their results were compared. In the thesis [5], geometrical models for classification were proposed. A one-class classification using geometrical model is proposed. A one-class problem is modeled by random projections in 2D. This proposal was based on kernelization of approximate polytope ensemble (KAPE) and non-convex APE (NAPE). A kernel function is used in SVM to enable linear classifier to model nonlinear separable data to improve the original implementation of APE and NAPE. In paper [6], convolutional neural network was discussed. Different methods are classified on the basis of how segmentation is performed that is semantic segmentation, instance segmentation, and hybrid approaches. One approach is based on low-rank representation which captures only the global structure of data, and a graph construction method [7] was proposed. They also proposed low-rank and local linear graph which takes local and global structure into consideration with better outcome compared to traditional method. In paper [8], the author presented probabilistic principal geodesic analysis (PGA) model on Riemannian manifolds which is automatic dimensionality selection with Bayesian formulation of PGA. They also developed different parameters that use different algorithms to integrate the posterior distribution of latent variables. In paper [9], algorithms for analyzing planar shapes and closed curves were proposed. They used Surrey fish database. They demonstrated interpolation and extrapolations of shape changes, clustering of objects in low-quality images. A model on the planar shape space [10] was proposed. Nonparametric Bayesian approach is developed for compact metric spaces and manifolds. Gibbs sampling methods were proposed for posterior computation. The researcher of this paper applied Gibbs Sampling to the problems in density estimation and classification with shape based predictors.

3 Existing Methodology

The flow chart for the existing methodology in Fig. 1 is shown below. The process consists of three basic steps:

3.1 Pattern Extraction and Analysis

The first step of the methodology is pattern analysis and extraction as shown in Fig. 1. In this conventional approach, the patterns are extracted based on low-level homogeneity metric for the pixels. The color is used as the homogeneity metric.

Fig. 1 Existing methodology

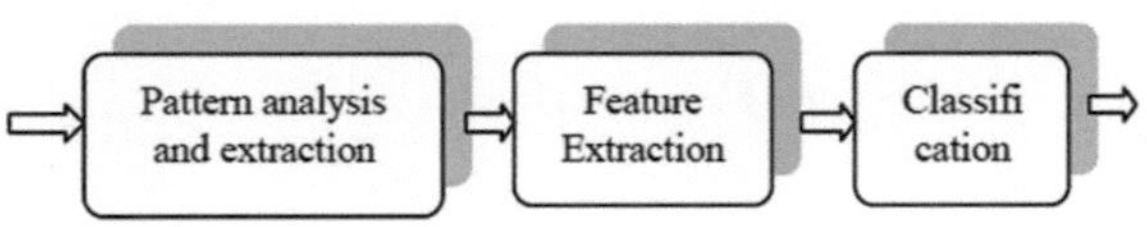

3.2 Feature Extraction

The features of the extracted patterns are obtained. The feature extraction algorithm is run on the extracted patterns as shown in Fig. 1. A special form of feature extraction is feature selection which selects a subset of given measurement as features. It reduces the dimensionality of input vectors to the classifier by mapping from n-dimensional space to lower-dimensional feature space. The feature selected should differentiate an object belonging to different classes.

3.3 Classification

A classifier is built from training data for classification as shown in Fig. 1. For training data, we use different measurements for patterns. The classifier classifies data in two stages: training stage and prediction stage.

4 Proposed Method

See Fig. 2.

4.1 Proposed Approach Requirements

(1) Extract patterns.
(2) Create pattern database.
(3) Extract the features of the patterns using principal component analysis (PCA).
(4) Create feature database.
(5) For a given query extract pattern. Extract the features of the pattern.
(6) Find out the best classification algorithm to obtain the best match with existing similarity metric using wrapper method.
(7) The ground truth of the test data and the result of classifier is as shown in Fig. 2.

4.2 Wrapper Framework

The wrapper framework processes each pattern independently. Small patterns are first tested with training database to check if any of them matches to an object of interest. Then remaining more complex patterns are compared where a variety of combinations of patterns are compared to see if any of the combinations matches

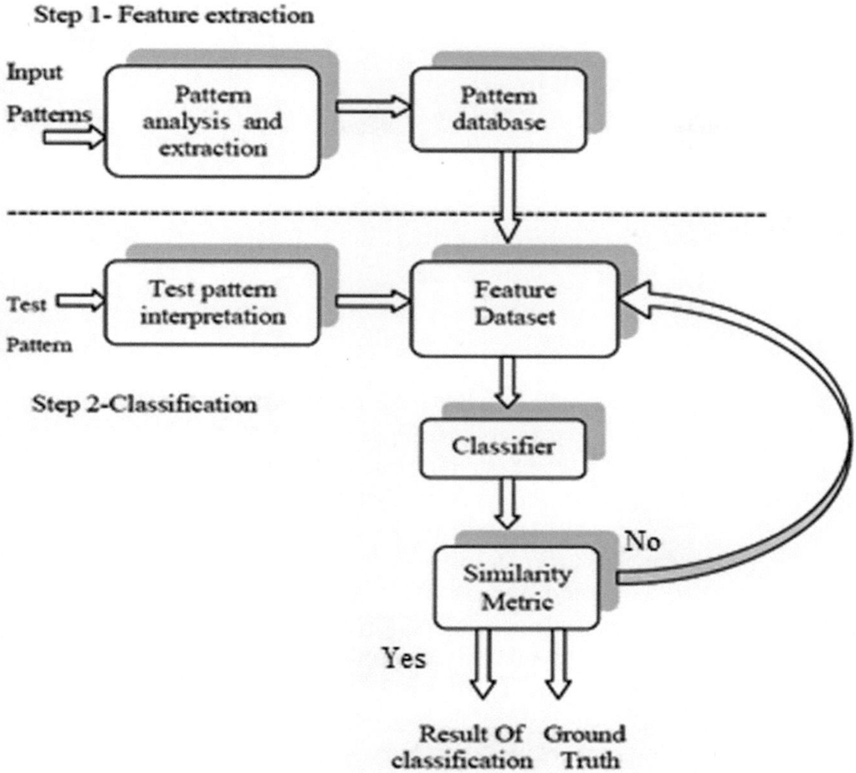

Fig. 2 Wrapper-based approach for pattern classification

with the pattern in training database. The first task in wrapper processing is feature extraction for each pattern. The proposed feature extraction algorithm used is principal component analysis which gives best result as compared to boundary descriptor. The purpose of PCA is to minimize the large dimensionality of data. The features extracted are mean, eigenvalues from covariance matrices, and eigenvectors. We calculate mean, eigenvalues and eigenvectors from covariance matrices: Mean can be calculated as

$$m = \frac{1}{p} \times \text{sum} \tag{1}$$

where $T = 2\text{D}$ matrix. Assume all P patterns in training database which has same size of $M \times N$, so the length of column vector in training database is $M \times N$.

Now calculate the deviation of each pattern from mean pattern and merging all centered pattern. Hence, $B = T - m$, where B is matrix of centered pattern.

$$C = B^* B' \tag{2}$$

where C = covariance matrix.

$$L = B''^* B \tag{3}$$

where L = surrogate of covariance matrix C, and diagonal elements of the surrogate and covariance matrix are eigenvalues. To calculate eigenvectors, take the product of centered pattern vector B and eigenvalues.

$$\text{Eigen}_{\text{vectors}} = B \times \text{Eigen}_{\text{vectors}} \tag{4}$$

In training stage, features are extracted for each pattern in training set. Let α_1 training set for pattern 1. To extract the features of $\alpha_{1,}$ first convert the input vector φ_1 by integrating each M row in a one-dimensional vector. For each training pattern $\alpha_{i,}$ feature vectors ¥$_i$ are obtained and stored. In testing stage, the feature vector ¥$_j$ of the test pattern α_j is computed. To correctly identify the test pattern α_j, the similarity metric between feature vector ¥j and all feature vectors in training set is computed. If $i = j$, then the pattern is correctly classified; otherwise, j is misclassified.

The framework of wrapper approach works by assuming a pattern class C to be the true class and computes the similarity metric of the extracted pattern to that class. The task of region combination is performed for every class C, and at the end of processing of all pattern classes, the class C gives the highest probability $P(\alpha_i|C)$. The class with highest probability is selected where set$\{\alpha_i\}$ is the subset of region that comprises the best classification for kth iteration. If the probability for class C is low, then the region of interest or pattern does not belong to the class.

5 Classification Techniques for Minimizing Errors

5.1 Bayes's Classifier

Based on the general theory, three classifiers such as Bayes's classifier, K-NN classifier, and NN classifier are used. Classification is generally a supervised learning for which the labels of true class for the data points are given in the training data. The setup for supervised learning is first to generate training data. The feature vector is obtained from the training data. Next is to obtain the response variable and finally form a predictor to predict the response variable based on feature vector. The predictor divides the feature vector space into collection of small regions, and each of these regions is labeled by one class. Bayes's classification rules estimate the probabilities of occurrence of various attributes for various classes in a training set.

In the process of classification, we want to minimize the probability of misclassification. To do so, we have to calculate the loss function. This function is denoted by

L which is the square of error function. The error function is the deviation between the actual and predicted outcome.

$$L(x, x_d) = (x - x_d)^2 \tag{5}$$

$$R(x_d) = E[L(x, x_d)] \tag{6}$$

where R is the risk factor associated with the prediction, and $\boldsymbol{x} - \boldsymbol{x_d}$ can be positive, negative, or zero.

$$R(x_d) = \sum_{x_i > x_d}(x_i - x_d)P(x_i) + \sum_{x_i < x_d}(x_i - x_d)P(x_i) + \sum_{x_i = x_d}(x_i - x_d)P(x_i) \tag{7}$$

$$\frac{\mathrm{d}R}{\mathrm{d}x_d} = -\sum_{x_i > x_d} P(x_i) + \sum_{x_i < x_d} P(x_i) = 0 \tag{8}$$

To minimize the risk factor, we should select x_d such that $P(x_i > x_d) = P(x_i < x_d)$. Now let us estimate the loss for classifying patterns into classes. Let L_{kj} be the loss associated with misclassifying a pattern x to class C_j when the actual class should be C_k. The minimum risk can be given by,

$$\sum_{k=1}^{k} L_{kj}\, p(x/C_k) < \sum_{k=1}^{k} L_{ki}\, p(C_k/x) \tag{9}$$

If we assume that loss of misclassification is 1 and that of correct classification is 0, then according to Bayes's theorem the above equation reduces to

$$p(x/C_k)P(C_K) > p(x/C_j)P(C_j) \quad \text{for all} \quad i \neq j \tag{10}$$

Equation 10 is the rule for minimizing the probability of misclassification.

5.2 K-*NN Classifier*

This algorithm is very simple and fast and is also used as nonparametric classification. This algorithm is used to find the nearest neighbor of all the neighbors. In K-NN, k is the number of training samples compared. When $k = 1$, then the algorithm is called 1-nearest neighbor algorithm. For better classification, k should have higher values and is input from the user depending on the design choice. To obtain optimal value of k, cross-validation technique is used. Cross-validation involves dividing the training data into groups or folds of equal size. 90% of the data is used to train the model and remaining to validate it. The error rate is computed on validation data. This step is repeated for each fold. A validation set for each fold is obtained and then averaged out as result. The error rate for Bayes's classification technique of a pattern x to its more likely class C_m.

$$P(E|x) = 1 - P(C_m|x) \tag{11}$$

The net probability of error is given by,

$$P(E|x)\text{NN} = 1 - \sum_{i=1}^{k} P(C_i|x)^2 = 1 - P(C_m|x)^2 - \sum_{i \notin m} P(C_i|x)^2 \tag{12}$$

Using these two equations, we get nearest neighbor classifier error rate which is less than the difference between twice the Bayesian error rate and its square.

$$P(E|x)\text{NN} \le 2P(E|x)_{\text{Bayes}} - P(E|x)^2_{\text{Bayes}} \tag{13}$$

5.3 Neural Network Classifier

This classifier has three layers: the input layer, hidden layer, and output layer. It processes information by learning. A set of training patterns are given at the input nodes, and their response is observed at the output nodes. Based on this response, a decision rule is generated which determines the characteristics of input pattern. The network is said to be learned when all the input pattern characteristics are determined. After training, they are used to classify unknown patterns of similar types. In the absence of exact match in training database, a different pattern with closest match is associated with unknown input. Let d and o be the desired output and actual output respectively for each training pattern. The error of classification is measured as the difference between desired and actual output. Let w_i be the weight with arbitrary random value and x_i be the input pattern. This error is used to modify the weights of the network. For all possible inputs, the aim is to minimize the error e. Hence, the weights are updated during training phase. These weights should satisfy minimum squared error criterion.

$$e = \sum_{j=1}^{m} \left(o_j - d_j\right)^2 \tag{14}$$

For pth input pattern x, the error can be computed as

$$e = \sum_{j=1}^{m} \left(o_{pj} - d_{pj}\right)^2 = \left\| o_p - d_p \right\| \tag{15}$$

In vector notation,

$$o = Wx \tag{16}$$

$$o_j = w_0 x_0 + \cdots + w_{nj} x_n = \sum_{i=0}^{n} w_{ij} x_i \tag{17}$$

Partially differentiating e with respect to w, we get

$$\Delta w_{ij} = (o_j - d_j) x_i \tag{18}$$

To minimize error e, the weights are updated according to

$$\Delta w_{ij} = \beta\big(o_j - d_j\big) x_i \tag{19}$$

5.4 *Figures and Tables*

See Figs. 3, 4, 5, 6, and 7 and Tables 1, 2, 3, 4, and 5.

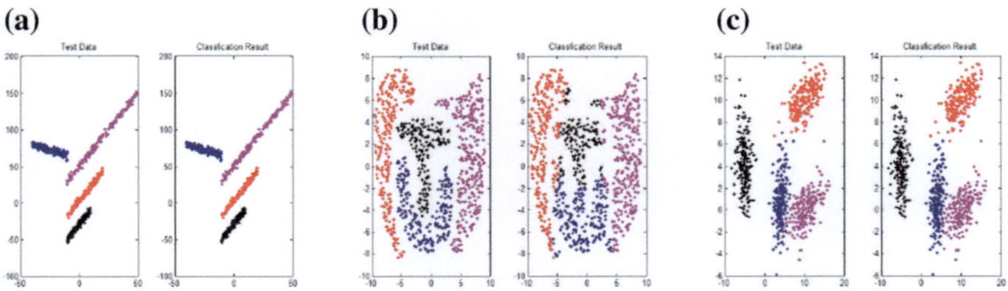

Fig. 3 Bayesian Ground Truth of test data and classification result for **a** linearly separable dataset, **b** linearly nonseparable dataset, **c** Gaussian distributed dataset

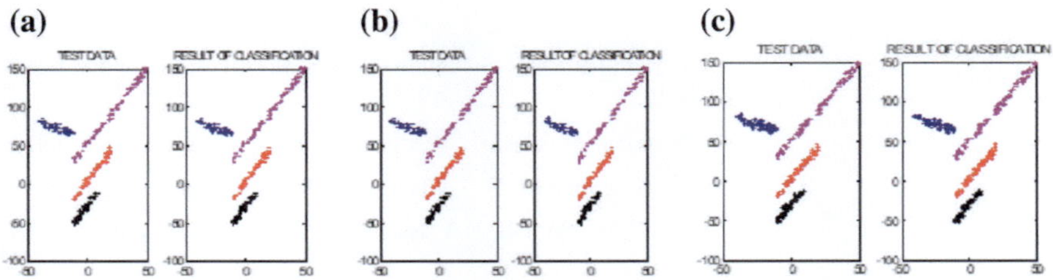

Fig. 4 *K*-NN Ground Truth of test data and classifier result for linearly separable dataset. **a** 1-NN, **b** 3-NN, **c** 5-NN

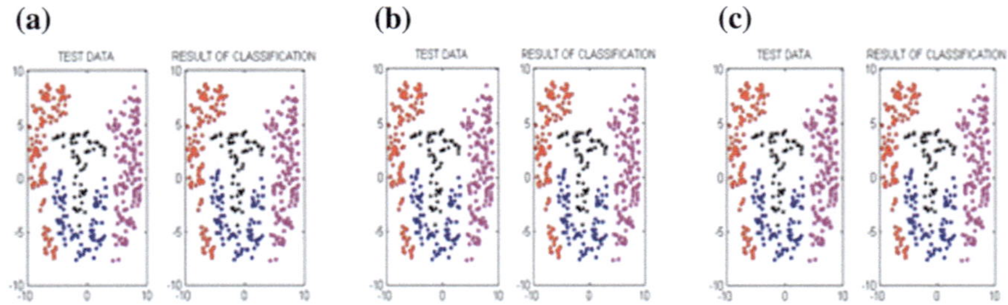

Fig. 5 *K*-NN Ground Truth of test data and classifier result for linearly nonseparable dataset. **a** 1-NN, **b** 3-NN, **c** 5-NN

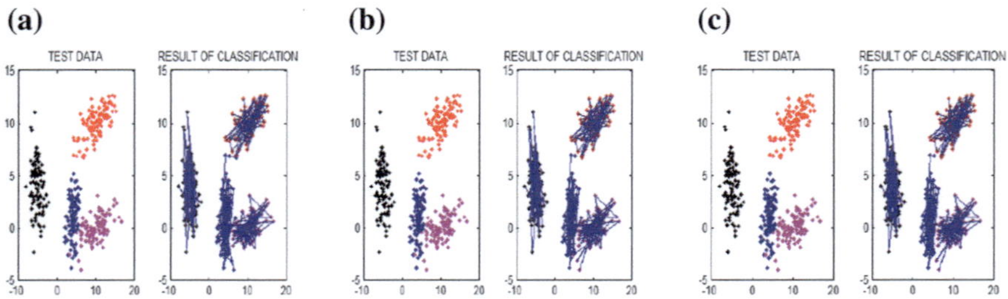

Fig. 6 *K*-NN Ground Truth of test data and classifier result for Gaussian distributed dataset. **a** 1-NN, **b** 3-NN, **c** 5-NN

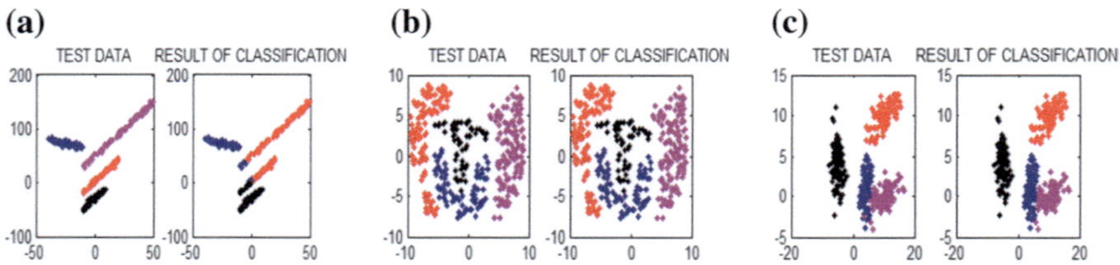

Fig. 7 NN Ground Truth of test data and classifier result for **a** linearly separable dataset, **b** linearly nonseparable dataset, **c** Gaussian distributed dataset

Table 1 Error rate for Bayes's classifier

Bayes's classifier		
Linearly separable	Linearly nonseparable	Gaussian
Training dataset: 800	Training dataset: 800	Training dataset: 800
Test dataset: 300	Test dataset: 300	Test dataset: 300
Misclassified: 0	Misclassified: 93	Misclassified: 9
Error rate: 0.00%	Error rate: 31.00%	Error rate: 3.00%

Table 2 Error rate for *K*-NN classifier for linearly separable dataset

K-NN classifier result for linearly separable dataset		
1-NN classifier	3-NN classifier	5-NN classifier
Training dataset: 800	Training dataset: 800	Training dataset: 800
Test dataset: 300	Test dataset: 300	Test dataset: 300
Misclassified: 0	Misclassified: 0	Misclassified: 0
Error rate: 0.00%	Error rate: 0.00%	Error rate: 0.00%

Table 3 Error rate for *K*-NN classifier for linearly nonseparable dataset

K-NN classifier result for linearly nonseparable dataset		
1-NN classifier	3-NN classifier	5-NN classifier
Training dataset: 800	Training dataset: 800	Training dataset: 800
Test dataset: 300	Test dataset: 300	Test dataset: 300
Misclassified: 0	Misclassified: 0	Misclassified: 0
Error rate: 0.00%	Error rate: 0.00%	Error rate: 0.00%

Table 4 Error rate for *K*-NN classifier for Gaussian distributed dataset

K-NN classifier result for Gaussian distributed dataset		
1-NN classifier	3-NN classifier	5-NN classifier
Training dataset: 800	Training dataset: 800	Training dataset: 800
Test dataset: 300	Test dataset: 300	Test dataset: 300
Misclassified: 6	Misclassified: 5	Misclassified: 3
Error rate: 2.00%	Error rate: 1.50%	Error rate: 1.00%

Table 5 Error rate for three-layer NN classifier

Three-layer NN classifier result		
Linearly separable	Linearly nonseparable	Gaussian
Training dataset: 800	Training dataset: 800	Training dataset: 800
Test dataset: 300	Test dataset: 300	Test dataset: 300
Learning completed in: 30th iterations	Learning completed in: 28th iterations	Learning completed in: 23rd iterations
Misclassified: 89	Misclassified: 3	Misclassified: 1
Error rate: 29.67%	Error rate: 1.00%	Error rate: 0.33%

6 Conclusion

The classification decisions are taken such that error rate is minimum. In this approach, feature extraction process is wrapped inside the classifier for better result of classification. This approach was applied for linearly separable, nonseparable, and Gaussian distributed dataset. Our result shows that linearly separable dataset for Bayesian classifier as shown in Table 1, and linearly separable as shown in Table 2, and linearly nonseparable as shown in Table 3 for K-NN classifier shows 100% correct classification which indicates the classes do not intersect, and we are able to separate all classes without error. The error rates are very high for Bayesian classifier when nonseparable dataset is used which indicates classes intersect each other and a soft margin is created where all misclassified data lie. The K-NN classifier error rate reduces when $K = 5$, for Gaussian dataset as shown in Table 4. Error rate for three-layer NN classifier is 29.67% for linearly separable dataset and 0.33% for Gaussian dataset as shown in Table 5. Hence, results show that the three-layer NN classifier with Gaussian sample dataset gives 99.67% accuracy for correct classification. The Fig. 3 shows the Ground truth of test data and the best classification result for Bayesian classifier for all the three datasets, the error rate is calculated depending on the misclassification and is shown in Table 1. The Figs. 4, 5 and 6 shows the ground truth of test data and best classification for K-NN classifier for linearly separable, nonlinearly separable and gaussian distributed dataset for $K = 1, 3, 5$ and the error rate is calculated depending on the misclassification and is shown in Tables 2, 3 and 4. Figure 7 shows the ground truth of test data and the classification results for all three datasets for NN classifier and the error rate is calculated depending on the misclassification.

References

1. Sampat MP, Bovik AC, Aggarwal JK, Castleman KR (2005) Supervised parametric and non parametric classification of chromosome image. Pattern Recogn 38(8):1209–1223. https://doi.org/10.1016/j.patcog.2004.09.010
2. Kumar Y, Sahoo G (2012) Analysis of parametric and non parametric classifiers for classification techniques using WEKA. Int J Inf Technol Comput Sci 7:43–49. https://doi.org/10.5815/ijitcs.2012.07.06
3. Lange T, Law MH, Jain AK, Buhmann J (2005) Learning with constrained and unlabelled data. IEEE Comput Conf Soc Comput Vis Pattern Recogn 1:730–737. https://doi.org/10.1109/CVPR.2005.210
4. Bhatia N, Vandana (2010) Survey of nearest neighbour techniques. Int J Comput Sci Inf Secur 8(2)
5. Fernandez AP (2015) Geometrical models for time series analysis. 18 ECTS thesis in artificial intelligence, Oct 2015
6. Aimal H, Rehman S, Farooq U, Ain QU, Riaz F, Hassan A (2018) Convolutional neural network based image segmentation: a review. In: Proceeding, vol 10649. Pattern recognition and tracking. SPIE 2018. https://doi.org/10.1117/12.2304711

7. Ahmadi SA, Mehrshad N, Razavi SM (2018) Semisupervised graph-based hyperspectral images classification using low-rank representation graph with considering the local structure of data. J Electron Imag 27(6):063002, 13 Nov 2018. https://doi.org/10.1117/1.jei.27.6.063002
8. Zhang M, Thomas Fletcher P (2013) Probablistic principal geodesic analysis. In: Advances in neural information processing systems, Jan 2013
9. Klassen E, Srivastava A, Mio M, Joshi SH (2004) Analysis of planar shapes using geodespic paths on shape spaces. IEEE, June 2004. https://doi.org/10.1109/tpami.2004.1262333
10. Bhattacharya A, Dunson DB (2010) Non parametric bayesian density estimation on manifolds with applications to planar shapes. Biometrika 97(4):851–865

IoT

A Review on IoT Security Architecture: Attacks, Protocols, Trust Management Issues, and Elliptic Curve Cryptography

Lalita Agrawal and Namita Tiwari

Abstract Internet of Things has become popular for the industrial and commercial environment and also for research purpose by providing a large number of applications that facilitate human daily and social life more comprehensible and convenient. Internet of Things amalgamates various kinds of sensors, actuators, and physical object to communicate with each other directly even in the absence of a human. For the formation of the IoT-equipped devices and network, security and trust among various parties are major concerns. Here, we discuss the security and privacy available at each layer of the IoT model and also the limitation and future direction that can be implemented to enhance IoT security architecture. However, trustworthy channel must be necessary for secure communication, and here, we also discuss elliptic curve light-weighted encryption technology used for trust management in IoT.

Keywords Internet of Things (IoT) · Security · Attacks · Protocol · Trust management · Elliptic curve cryptography (ECC)

1 Introduction

The word Internet of Things has gained attention across the world through its sustainable contribution to industrialization, commercial environment, and technical domain. Developing an intelligent, well-organized, and cost-effective IoT-based infrastructure is the central attraction of the researchers. The word 'Internet' in IoT refers to a universal interconnection of computer networks providing heterogeneous information and various transmission facilities using standardized communication protocols [1]. The word 'Things' in IoT is an integration of various kinds of virtual and physical sensors, actuators, smart devices, automobiles, physical objects, and embedded system [1]. Amalgamation of these things in structured and systematic

L. Agrawal (✉) · N. Tiwari
Maulana Azad National Institute of Technology, Bhopal, India
e-mail: lalitaagrawalmits@gmail.com

N. Tiwari
e-mail: namita_tiwari21@rediffmail.com

R. K. Shukla et al. (eds.), *Social Networking and Computational Intelligence*, Lecture Notes in Networks and Systems 100, https://doi.org/10.1007/978-981-15-2071-6_36

way has a sensory capability to autonomously assemble heterogeneous data, examine and monitor those data to extract functional and serviceable information used in miscellaneous IoT-based applications like smart city, smart healthcare system, public and defense services, smart surveillance system, asset tracking system, smart transportation, smart farming, etc. [2, 3].

There may have many issues while deploying IoT-based application in a distributed environment at a larger scale that could be a major concern. Some of them are enormously discussed in this paper. Security is one of the major concerns in case of IoT-based environment that can harm system anytime, anywhere. Everyone should aware of that for prevention purpose. For a system to be secure, it should satisfy some data security criteria like data confidentiality, data integrity, and data availability [4]. IoT embedded system that satisfies all security goals is not presented till now. The existing security issues like authorized access, trust management, tracking of intruders, public user data privacy, prevention from malicious insiders, and security against various attacks present at different layers of IoT architecture are obligatory [5]. Researches are still going on to satisfy all security goals. Hence, IoT security and trust awareness have a promising future scope.

2 Security at IoT Architecture

Trust is immensely considerable insecurity demonstration of IoT network system. The level of trust is assessable attribute to distinguish the security parameter in IoT infrastructure available at different layers. Trust parameters define objects that belief each other in any communication network. IoT architectural model has three layers, so trust responsibility at each layer is presented in this section and what are the future proposals to strengthen the existing techniques is also discussed in this section.

2.1 Physical Layer

For the realization of the IoT system model, sensor nodes are placed at the bottom layer. All the sensors, transducer, actuators, and physical objects are the basis of this layer [1]. Connected sensing unit has the capability to gather and derive information continuously in real time and remotely from larger data [4]. The gathered information is examined under various benchmarks because vulnerable data is caused by attackers. If collected information is harmed by any adversary due to lack of reliable encryption mechanism, it cannot be further processed by upper-layer services [6]. Hence, reliability of collected data at sensor node is a must for next layer trustworthy IoT environment as well as for user acceptance.

2.2 Network Layer

This layer has two sub-layers: information processing layer and service-oriented layer. These two layers have different responsibilities.

2.2.1 Information Processing System Layer

This layer processed the data taken from sensor nodes in a valuable and functional form by an available information processing system, and then processed data is kept into storage awareness units. From there, it is transmitted to the service layer [4, 5].

2.2.2 Service Layer

This layer is responsible to perform some activity based on computerized outcome from processed data. The major concern of this layer is that all the activities have the same service type. Finally, it transmits the services to the application layer [4].

2.3 Application Layer

This layer is also known as user acceptance layer since it provides a real-life application to end users like a business oriented and commercial application, etc. [5]. Final trust characteristics are influenced by many security primitives. Table 1 conjunctly depicts attacks and challenges presented at each layer of the IoT layered model.

In this paper, our major concern is trustworthy communication between devices and human, so the whole system should be tied together for reliable data transmission and reliability must also be presented at individual IoT layer.

Since IoT network is a network of capability constraint equipped sensors that provide the facility of a device to device communication [1]. Therefore, IoT architecture has some unique functional requisite and constraint at its each layer that makes IoT

Table 1 Attacks at IoT security layered architecture [1, 4, 5]

IoT layers	Attacks affecting IoT layered security architecture
Physical layer	Unauthorized access, spoofing, man-in-the-middle (MitM) attack, eavesdropping
Network layer	Denial of service (DoS) attack, man-in-the-middle (MitM) attack, sleep deprivation attack, malicious code injection, unauthorized access, Sybil attack
Application layer	Spear-phishing attack, malicious code injection, denial of service (DoS) attack, sniffing attack

Table 2 Conventional network protocol versus IoT network protocol [1, 5, 7]

Layers	Conventional network protocol	IoT network protocol
Physical layer	IEEE802.2, IEEE802.3, IEEE802.11, point to point protocol (PPP), link control protocol (LCP)	IEEE802.15.4, IEEE802.11AH, wireless HART, DASH 7, Z-WAVE, LTE-A
Network layer	IP (IPv4.IPv6), ARP, RARP, ICMP, IGMP, ICMPv6, TCP, SCTP, DCCP, IPsec, WTLS	Datagram transport layer security (DTLS), routing protocol for low power and lossy network (RPL), cognitive (CoRPL), channel aware routing protocol (CARP)
Application Layer	HTTP, FTP, DNS, DHCP, SMTP, TELNET, OFTP, XML, BGP, SSH, SNMP	Constrained application protocol (CoAP), data distribution service (DDS), extensible messaging and presence protocol (XMPP)

network differs from the traditional network. The security requirement, protocol, and standard of IoT network and traditional network are incompatible at each layer and shown in Table 2.

3 Trust Requirement and Techniques

3.1 Trust

Well-known term trust, security, and privacy are substantially interdependent to each other in the IoT communication model [8]. Trust estimation is a perception that how much devices and human begins are reliable on one another [9]. Over the last 8 years, many researchers have presented contrasting trust augmented model. Thus, we have concluded many trust-related open challenges presented at each layer of IoT security architecture shown in Fig. 1. One of them we have described extremely further.

3.2 Trust Issues and Requirement

- To assemble application-specific data, trustworthy communication must take place between sensor nodes for accurate, reliable, confidential security perspective [5].
- Integration of IoT application service must assure individual requirement also such as ‘only’ determined and personalized [6].
- Existing data perception trust solutions are too heavy or complicated for capability-constrained wireless sensors to adopt. So the development of lightweight encryption technology for IoT devices, elliptic curve cryptography, has been proposed that is adaptable by IoT network discussed in later section [10].

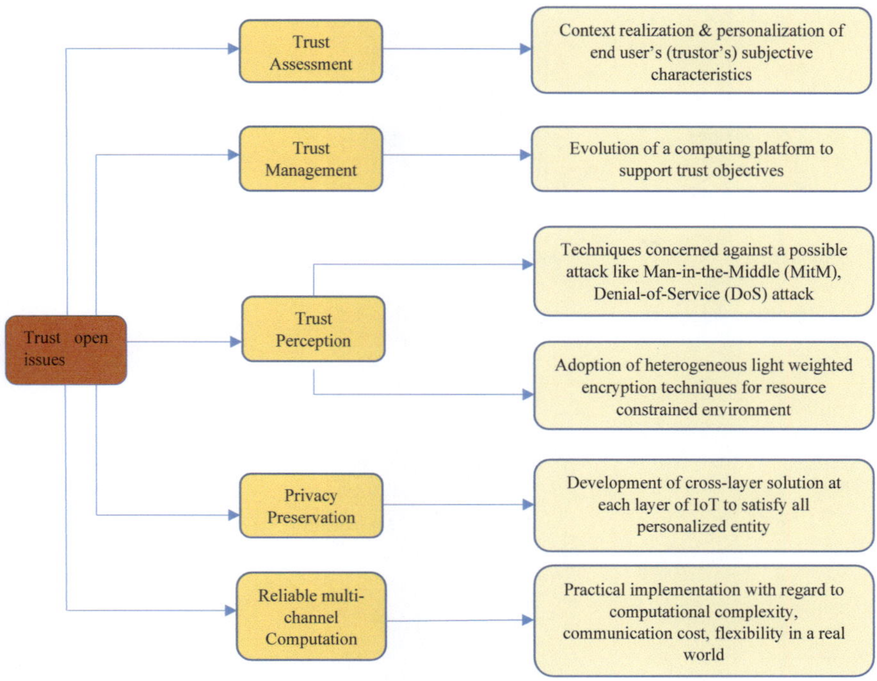

Fig. 1 Trust-related open issues for the Internet of Things [6, 8, 9]

- Developing phenomenon for sensor nodes to execute automatically intelligent activity based on gathered and observed data by realistic hardware implementation [9].

4 Illustration of IoT Architecture Based on Realistic Application

4.1 Smart Surveillance System

Safety and security are imperative for the development of any institutes like educational, government, etc. IoT-based sustainable, economic, eco-friendly security system collaborated with automated surveillance can enhance the traditional system. An IoT-based solution that effectively tracks and monitors access and intrusion, facilitates prevention of property loss, and upgrades the institute to national and international norms. The security architecture of a system is made up of three modules or phases: (1) intrusion detection, (2) access control, and (3) asset tracking. A smart surveillance system is clearly delineated in Figs. 2 and 3.

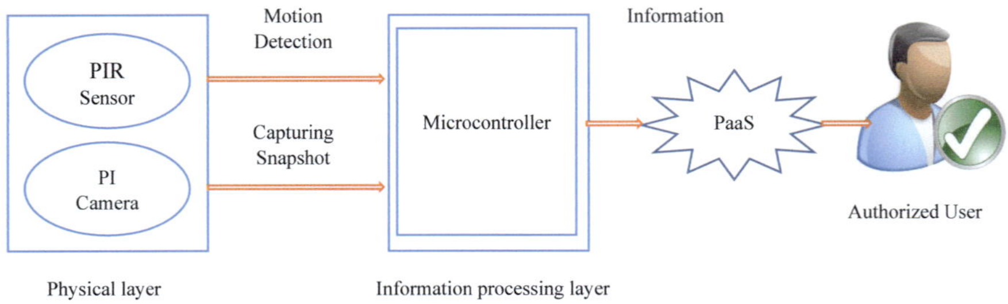

Fig. 2 Intrusion detection

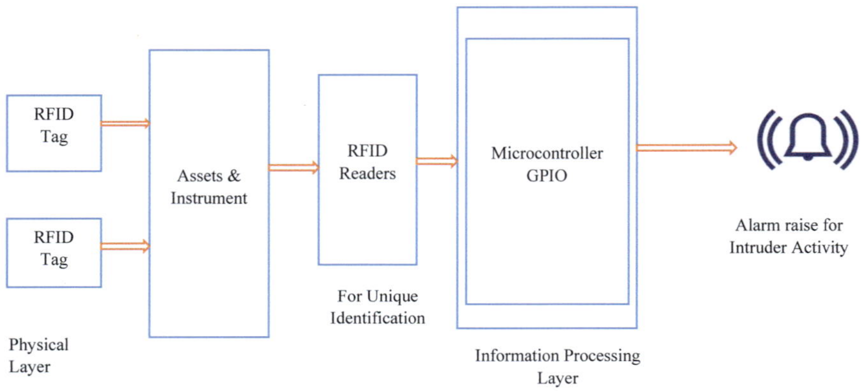

Fig. 3 Access control and asset tracking

4.2 Intrusion Detection

This module equipped with HC-SR501 Pyro-electric Infrared PIR Motion Proximity Sensor Module, Infrared Camera Module V Supports Night Vision 8 MP, and preferable microcontroller [11]. PIR sensor labeled pin (VCC, GND, OUT) and camera module are connected to hardware appropriately for capturing snapshot during any intruder's mobility and then send to authenticate authority using any platform as a services (PaaS) communication platform (like Twilio, etc.).

4.3 Access Control and Asset Tracking

RFID-based tracking system (RFID tags and readers) can be deployed to provide the facility for access control and asset tracking from unauthorized prevention [11, 12]. Expensive property can also be equipped with RFID tags (used for unique identification) so that any movement can be tracked through RFID readers [12, 13].

RFID labeled pins (SDA, SCK, MOSI, MISO, RST, GND, 3.3v) are integrated with suitable microcontroller GPIO's pins.

5 Elliptic Curve Cryptography

For secure data transmission, the security mechanism is used for data/message prevention from intruders and malicious insiders. Cryptographic algorithms are one security technique where encryption algorithms are used to encrypt heterogeneous data to achieve data confidentiality, integrity, authentication, and access control. The traditional public-key cryptographic system (here, two keys are involved: One is a public-key and another one is private key) has a larger key size because of that it becomes expensive and complicated for capability constraint tiny sensor nodes that have limited processing power, memory capacity, resource availability, etc. [1, 6]. IoT bottom most layer (physical layer) is made up of heterogeneous capability constraint sensors, which require a lightweight encipherment technique. For which cryptographic algorithm based on elliptic curves is a most prominent alternative approach called elliptic curve cryptosystem (ECC) [10, 14]. These are like mathematical framework or simulator used to simulate existing encryption algorithm with smaller key size.

Elliptic curves are represented by Weierstrass cubic equation [15]: $\mathbf{y^2 + b_1xy + b_2y = x^3 + a_1x^2 + a_2x + a_3}$.

Elliptic curves are defined over three sets for cryptographic purposes [15].

1. Elliptic curves over real numbers *R*,
2. Elliptic curves over GF(*P*),
3. Elliptic curves over GF(2^n).

Elliptic curve equation over *R* and GF(*p*) is as follows: $\boldsymbol{y^2 = x^3 + ax + b}$

Elliptic curve equation over GF(2^n) is as follows: $\boldsymbol{y^2 + xy = x^3 + ax + b}$

Elliptic curve equation used for cryptosystem must have three distinct roots (either real or complex) that is why elliptic curve must satisfy its non-singular characteristic, i.e., $\mathbf{4a^3 + 27b^2 \neq 0}$. An elliptic curve is a set of points, and it has basically two operations: (1) point addition and (2) point doubling that is performed to find another point on a curve. Let us assume $P(x_1, y_1)$ and $Q(x_2, y_2)$ are two points on curve and $P + Q = R$, where R is another point on the curve, and its coordinates are found using point addition operation in this way. Firstly, we have to find slope $\lambda = y_2 - y_1/x_2 - x_1$, and then its coordinates are $x_3 = \lambda^2 - x_1 - x_2$ and $y_3 = \lambda(x_1 - x_2) - y_1$. Point doubling operation is like $P + P = R$, where slope is calculated as $\lambda = 3x_1^2 + a/2y_1$ and coordinates are $x_3 = \lambda^2 - x_1 - x_2$ and $y_3 = \lambda(x_1 - x_3) - y_1$. These characteristics of elliptic curves form a basis for lightweight encryption [14, 15].

Elgamal encryption technique and digital signature scheme can be simulated using elliptic curves. ECC-based Elgamal has easier calculation and less computational cost (exponentiation and multiplication operations are replaced by multiplication and addition operation, respectively) than original technique providing same level or kind security services that are data authentication, data integrity, and confidentiality [15, 16]. Hence, elliptic curve-based cryptosystem has a smaller key size of 160 bits than original asymmetric key cryptosystem like RSA, Elgamal, and DSS that have 1024 bits key size [15]. These characteristics of ECC make it acceptable for light-weighted IoT environment [16]. Lightweight encryption using attribute-based encryption without bilinear pairing and identity-based encryption have some ongoing research that has a prominent subsequent scope for the Internet of Things [10].

6 Conclusion

This literature survey presents IoT layered security paradigm in detail. The attack and threat presented at each layer are demonstrated in clear-cut tabular form. Here, we have also differentiated IoT network protocols and standard with conventional security network protocols. The importance of trust, privacy, and secrecy is demonstrated with regard to the IoT environment. Trust-related open challenges at each layer are presented in figure clearly. IoT security layered architecture is illustrated with the help of realistic smart surveillance system.

References

1. Alaba FA, Othman M, Hashem IAT, Alotaibi F (2017) Internet of things security: a survey. J Netw Comput Appl 88:10–28
2. Atzori L, Iera A, Morabito G (2010) The internet of things: a survey. Comput Netw 54:2787–2805
3. Fraga-Lamas P, Fernández-Caramés TM, Suárez-Albela M, Castedo L, González-López M (2016) A review on internet of things for defense and public safety. Sensors MDPI
4. Farooq MU, Waseem M, Khairi A, Mazhar S (2015) A critical analysis on the security concerns of internet of things (IoT). Int J Comput Appl (IJCA) 111(7)
5. Kouicem DE, Bouabdallah A, Lakhlef H (2018) Internet of things security: a top-down survey. Comput Netw 141:199–221
6. Yan Z, Zhang P, Vasilakos AV (2014) A survey on trust management for internet of things. J Netw Comput Appl 42:120–134
7. Granjal J, Monteiro E, Silva JS (2015) Security for the internet of things: a survey of existing protocols and open research issues. IEEE Commun Surv Tutor 17(3):1294–1312
8. Daubert J, Wiesmaier A, Kikirasa P (2015) A view on privacy & trust in IoT. In: IEEE International Conference on Communication (ICC), London, UK
9. Suryani V, Selo, Widyawan (2016) A survey on trust in internet of things. In: 8th International Conference on Informational Technology and Electrical Engineering (ICITEE), Yogyakarta, Indonesia
10. Yao X, Chen Z, Tian Y (2015) A lightweight attribute-based encryption scheme for the internet of things. Futur Gener Comput Syst 49:104–112
11. Anwar S, Kishore D (2016) IOT based smart home security system with alert and door access control using smart phone. Int J Eng Res Technol (IJERT) 5(S12). ISSN: 2278-0181
12. He D, Zeadally S (2015) An analysis of RFID authentication schemes for internet of things in healthcare environment using elliptic curve cryptography. IEEE Internet Things J 2:72–83

13. Chatzigiannakis L, Vitaletti A, Pyrgelis A (2016) A privacy-preserving smart parking system using an IoT elliptic curve based security platform. Comput Commun 89–90:165–177
14. Martínez VG, Encinas LH, Ávila CS (2010) A survey of the elliptic curve integrated encryption scheme. J. Comput Sci Eng 2
15. Forouzan BA (2007) Cryptography and network security. Tata McGraw-Hill
16. Liu Z, Grobschadl J, Hu Z, Jarvinen K, Wang H, Verbauwhede I (2017) Elliptic curve cryptography with efficiently computable endomorphisms and its hardware implementations for the internet of things. IEEE Trans Comput 66(5)
17. Chernyshev M, Baig Z, Bello O, Zeadally S (2018) Internet of things (IoT): research, simulators, and testbeds. IEEE Internet Things J 5:1637–1647

A Comprehensive Review and Performance Evaluation of Recent Trends for Data Aggregation and Routing Techniques in IoT Networks

Neeraj Chandnani and Chandrakant N. Khairnar

Abstract Internet of things (IoT) is a ubiquitous network which supports and offers a system that observes and manages the physical world through the aggregation, filtering, and investigation of generated data using IoT devices. Aggregation of data and routing of nodes in IoT devices are always challenging tasks. A well-organized data aggregation and routing of nodes is necessary factor for successful placement and use of IoT devices. IoT devices usually share large amount of data that can be converted into information. The information is aggregated to enhance the overall efficiency of the IoT network. Data aggregation is the process in which information is collected and expressed for the purpose of statistical analysis. Routing in the IoT network plays a vital role. IoT devices act as routers for sending information to the gateways. The routing of data affects the power consumption of progressing IoT devices. For these reasons, aggregation of data and routing of nodes are important for IoT devices. This paper conveys and evaluates comparison on current data aggregation and routing techniques of IoT devices. Ad hoc On-demand Distance Vector (AODV) routing protocol is simulated for ten different mobility conditions, and its performance is observed in respect of throughput, delay, and packet delivery ratio.

Keywords IoT devices · Data aggregation · Centralized data aggregation · Cluster-based data aggregation · Tree-based data aggregation · Cluster and non-cluster-based routing · Secure routing · AODV

N. Chandnani (✉)
Devi Ahilya University, Indore 452001, Madhya Pradesh, India
e-mail: chandnani.neeraj@gmail.com

Military College of Telecommunication Engineering, Mhow 453441, Madhya Pradesh, India

C. N. Khairnar
Faculty of Communication Engineering, Military College of Telecommunication Engineering, Mhow 453441, Madhya Pradesh, India
e-mail: cnkhairnar@gmail.com

R. K. Shukla et al. (eds.), *Social Networking and Computational Intelligence*, Lecture Notes in Networks and Systems 100, https://doi.org/10.1007/978-981-15-2071-6_37

1 Introduction

Internet of things (IoT) is a unique example which constantly propagates and attracts the next-generation statistics and communication architectures. Evaluation of protocol stack in an IoT environment, criteria that influence application layer execution, and a possible trade-off between packet delivery ratio, delay, and throughput is discussed in reference [1]. IoT supports collection of information and forwards it to each individual node through the communication link. In IoT device-based network, lifetime increase is often critical challenge. For this concern, aggregation of data is an effective method to increase the transmission rate in IoT devices which in turn reduce the data redundancy, increase network lifetime, and also reduce energy consumption [2]. For efficient aggregation of data, mixed integer programming formulations and algorithms are introduced. Data are aggregated using nodes during transmission, such that given data are combined based on the functions, namely sum, count, and average. After aggregation, node sends single packet which represents the data aggregation. In reference [3], two cases are performed; the first case is named 1 K; in this destination node represents the fog gateways, and it is not significant which gateway gathers information from given dimension. In the second case, nK, destination acts as actuator which performs some action based on dimensions. A novel Lightweight Compressed Data Aggregation (LWCDA) algorithm is presented to increase the lifetime of IoT network. The LWCDA algorithm arbitrarily splits the complete network system into the non-overlapping groups for aggregation of data. In reference [4], the non-overlapping cluster offers two advantages such as energy efficiency and low complexity. Energy efficiency is achieved as each node sends a measurement to the head of the cluster. Highly sparse matrix is introduced to reduce the complexity in the network. A Recursive Principal Component Analysis (R-PCA)-based data aggregation algorithm is introduced in the IoT system [5]. The R-PCA configuration is based on the cluster-based data analysis which aggregates the unnecessary data and detects the irregularities in meantime. The constraints in the RCA are repeatedly reorganized to adapt the changes in the IoT network. Spatially correlated IoT device data gathered from the cluster members are aggregated by taking the key components. The functional and non-functional constraints are discussed for aggregation of data in IoT devices [6] which is based on the active designing-oriented Quality of Experience (QoE) constraint. At first, knowledge model of similarities between service categories is created. The aggregation complications in the service types are planned to the dynamic programming problem based on the correlation among service configuration. Service selection is achieved using the semantic similarity computing method. Energy-efficient link stable routing introduces conservation of energy in the IoT devices [7] to provide network stability for enhancing its lifetime. In order to check the exactness of the introduced routing, two processes are held. At first, analytical models are introduced for network stability and residual energy of the route. Secondly, optimal route selection algorithm is used for residual energy of route, link stability, and route distance. A heterogeneous IoT Routing Decision-making mechanism which is three dimensional is based on the Cellular Address (RDCA) is presented in [8] to establish communication model. At first, RDCA establishes nodes of network data forwarding based on the cellular automata and average of node received signal strength. Secondly, optimization of route and control using IPv6 addressing is done in RDCA. Finally, route decision-making algorithm is determined by the cellular address. The congestion and interference-aware energy-efficient routing method are introduced for optimum routing in IoT which is called survivable path routing [9]. The introduced protocol works in the IoT networks with more traffic because several sources attempt to send their data to the destination at same instant of time. To select the next hop for routing, algorithm uses the three factors. They are signal-to-interference ratio, signal-to-noise ratio of the network link, survivability factor path from next hop node to destination, and congestion level at next hop node. A new method based on the tree routing protocol is introduced in reference [10]. The introduced protocol begins from the local best which start the process of saturating to obtain a spanning tree. While performing spanning tree formation, local leaders' values will be routed. The best value is selected, if two spanning trees meet each other. The selected value tree is continued to process, while other tree stops working. This algorithm provides low energy consumption in IoT network. A multi-objective Fractional Gravitational Search Algorithm (FGSA) is discussed for effective routing in IoT [11]. To increase the lifetime of IoT network, the FGSA is introduced which finds the optimum cluster head for IoT network model. FGSA selects cluster head by computing fitness function which considers the following metrics, such as distance, delay, network link lifetime, and energy which is known as multi-objective FGSA. An energy-efficient Fractional Gravitational Gray Wolf Optimization (FGGWO) algorithm is introduced for routing in reference [12]. In this method, two processes are held. At first, cluster heads are nominated by FGSA. After choosing cluster head, FGGWO algorithm checks for the best optimal multi-path from source to destination. The proposed optimal path selection algorithm is a combination of FGSA and GWO algorithms. The proposed FGGWO method offers optimum routes for transmission with capability to perform in restraint problems.

The data aggregation and routing in IoT networks should be done in such a manner so that there is minimal loss of data through transmission, and the routing algorithm should have less delay and more packet delivery ratio and throughput.

The rest of this paper is organized as follows: Sect. 2 describes the recent data aggregation techniques. Section 3 describes the recent routing techniques. Section 4 describes the performance metrics parameters. Section 5 evaluates performance of AODV routing protocol under different mobility conditions which are not invoked in the algorithms of Sect. 4, Table 1. Finally, paper ends with results, findings, and conclusion.

Table 1 Comparison of techniques

Author	$S_{m,n}$	$R_{m,n}$	D	T_m	$P_{m,n}$	$\mathrm{MUI}_{m,n}$	$C(n)$	$A(n)$
Bhandari et al. [13]	✓	✗	✓	✗	✓	✗	✗	✓
Ghate [14]	✓	✓	✗	✓	✗	✗	✓	✗
Lu et al. [15]	✓	✗	✓	✗	✓	✗	✓	✗
Ko et al. [16]	✓	✗	✓	✗	✓	✓	✓	✗
Li et al. [17]	✓	✗	✗	✓	✗	✓	✗	✗
Abdul-Qawy [18]	✓	✗	✓	✗	✗	✓	✗	✗
Preetha et al. [19]	✓	✗	✓	✗	✓	✓	✓	✗
Chhabra et al. [20]	✓	✗	✓	✗	✓	✗	✓	✗
Sathish et al. [21]	✗	✓	✗	✗	✓	✗	✓	✗
Khan et al. [22]	✓	✗	✓	✓	✗	✓	✗	✓
Hasan et al. [23]	✗	✓	✗	✗	✓	✗	✓	✗
Hamrioui et al. [24]	✓	✗	✗	✓	✓	✗	✓	✗
Nguyen et al. [25]	✗	✓	✓	✗	✓	✓	✗	✓
Ma et al. [26]	✓	✗	✓	✗	✓	✗	✓	✗
El Hajjar et al. [27]	✓	✓	✓	✗	✓	✗	✓	✓
Mick et al [28]	✓	✗	✓	✓	✗	✓	✓	✗

2 Recent Trends in Data Aggregation Techniques

In this section, data aggregation techniques in the IoT devices are briefly explained. Data aggregation techniques are divided into three types, cluster head-based data aggregation, tree-based data aggregation, and centralized data aggregation. These three techniques are described in the following subsections.

2.1 Cluster Head-Based Data Aggregation Schemes

In cluster-based data aggregation scheme shown in Fig. 1, IoT devices form clusters and a cluster head (CH) is chosen by suitable algorithm and then cluster head aggregates data from the cluster members and sends it to the gateway device in the IoT network. Clustered IoT devices send the required information to the cluster head which in turn sends to the gateway. There is no direct connection between the IoT devices and gateway. They communicate through the cluster head.

To avoid latency in the IoT networks, priority-based channel access and aggregation of data scheme are discussed in reference [13]. A priority-based channel access scheme is deployed at cluster head to reduce the channel access latency. In this, preemptive M/G/I queuing model is proposed which separates the high- and low-priority queue before transferring aggregated data packets to the gateway device. In

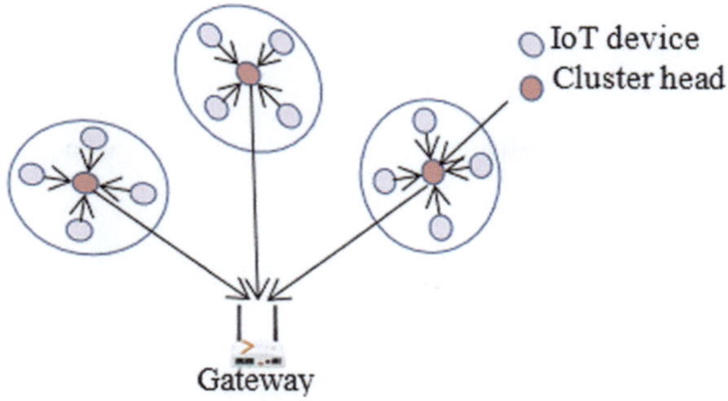

Fig. 1 Cluster-based data aggregation

this work, two levels of data aggregation scheme are introduced. The aggregation of data without prioritization, data packets from the IoT sensor nodes arrive at CH are in queue. The aggregation of data with prioritization model, M/G/I queuing model, holds priority class of data packet. The data packets with ith priority have arrival rate $i = \{1, 2, \ldots P\}$ and follow the Poisson distribution. The aggregation of data scheme based on priority is discussed in reference [14]; it contains two aggregations of data schemes. In first scheme, if the class label for input data is already known, values are extracted using the machine learning techniques or filtering techniques. Threshold values are monitored, and priority levels are decided based on the ranking and weights. The cluster head forwards data based on the priority on immediate factor or aggregates the data more if priority is less. In second scheme, output class labels are not known; in this case, unsupervised machine learning algorithms such as clustering algorithms are implemented to construct clusters of data values based on the similarities. A Lightweight Privacy-Preserving Data Aggregation (LPDA) scheme is introduced for IoT network in reference [15]. The proposed algorithm consists of four parts. They are system initialization, IoT device report generation, fog device data aggregation, and control center report reading and analytics. In system initialization phase, trusted authority assigns the key materials to all entities. In report generation phase, all the IoT devices report their sensed data using two steps. Reported data are aggregated using the fog devices. In data aggregation process above, four steps are performed. At last, control center runs the received data after verification process.

In cluster head-based data aggregation scheme, each cluster head sends the data to gateway. So, the probability of data loss is minimum in case of link failure between the cluster head and gateway.

2.2 *Centralized Data Aggregation Schemes*

In centralized data aggregation scheme, shown in Fig. 2, IoT devices send the required information to intermediate nodes which in turn send it to header node. Header node finally sends the data to gateway. All the intermediate nodes communicate to gateway through a single header node.

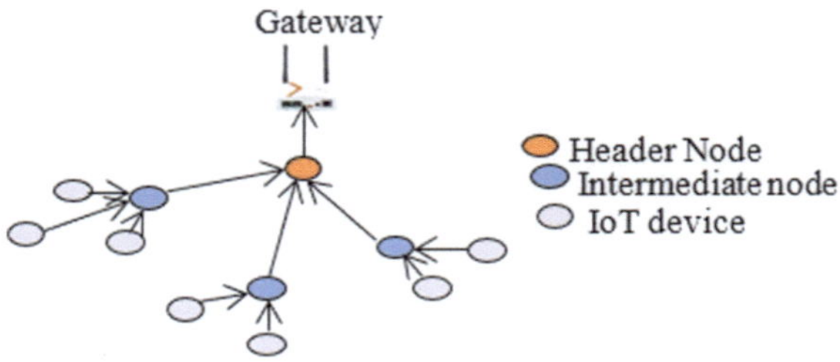

Fig. 2 Architecture for centralized data aggregation

Data acquisition-based IoT devices are considered in reference [16] which gathers data and sends it to the IoT gateway. In this method, consistency is definite and energy-efficient sleep scheduling method is proposed along with aggregation of data. Markov decision process (MDP) is used to find the best sleep duration of IoT device and data aggregation duration in IoT gateway device. The duration for aggregation of data is selected by the appropriate value. The sleep duration for IoT devices and data aggregation duration for IoT gateway device jointly determine the reflection on data consistency. The proposed MDP has five submodels. They are decision epoch, action, state, transition probability, and reward and cost functions. The modern cryptography and permutation circuits are combined for aggregation of data in IoT devices [17]. The protocol offers privacy in aggregation of data scheme for many applications such as wireless sensor networks, smart grids, and mobile health. The proposed protocol consists of three phases such as initialization, client to server, and server to client. The scalable energy efficient scheme (SEES) is proposed for efficient data aggregation in IoT network [18]. A multi-stage weighted linear combination method is introduced which is based on the election heuristic (MSWE) that considers the number of static and active parameters. A Minimum Cost Cross-layer Transmission (MCCT) method is proposed for data distribution from lowest part of layer to base station at the top most layer.

In the architecture for centralized data aggregation scheme, all sensor nodes direct their data to the intermediate node. Intermediate node sends data packets to the header node which aggregates the data. Header node sends single aggregated packet to gateway device.

2.3 *Tree-Based Data Aggregation Schemes*

In this scheme, shown in Fig. 3, the IoT devices send the required information to their aggregator node which aggregates the data of all its IoT devices. All the aggregator nodes finally direct the data to gateway. All IoT devices communicate to gateway through Aggregator node.

Tree-based data aggregation scheme for aggregation of data in IoT network is described in reference [19]. A reinforcement learning-based fuzzy interference system is proposed for efficient data collection in each cluster in the IoT network. The

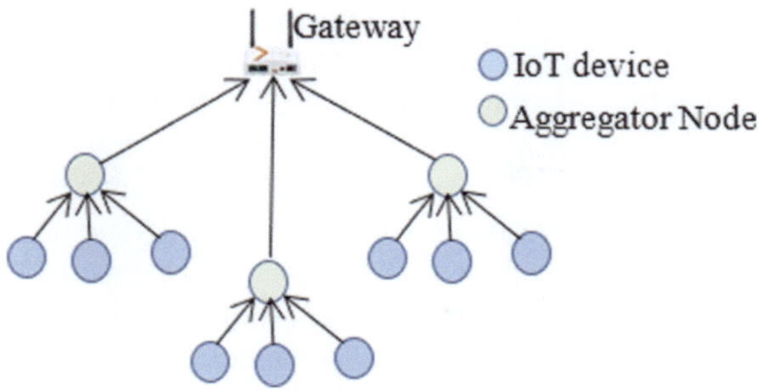

Fig. 3 Architecture for tree-based data aggregation

tree-based cluster is formed based on the density of the IoT network. Neighborhood overlap is one of the metrics to gauge degree of shared network between end hubs of a connection. The action pair value symbolizes the long period expected reward for every state and action pair. The achievable state and action state indicate the best policy that to learn fuzzy combination rules.

In the architecture of tree-based data aggregation technique in the IoT network, source node directs their data to the root node which aggregates the child node data. Aggregated data packet in the root node is directed to the gateway device.

3 Recent Trends in Routing Techniques

Routing in IoT is important for energy consumption, delay, and network lifetime. This section describes recent techniques that are used for efficient routing in IoT network.

3.1 Cluster-Based Routing Schemes

In this scheme, routing is done based on the cluster-based methods in IoT network. An energy-efficient routing protocol is introduced for optimum routing. To save the energy of the IoT devices, the routing procedure in reference [20] first aggregates the devices into bunches. Formation of cluster is based on the features such as data length, distance between the source and destination in the network, and data sensed in the surrounding area in the current period. After formation of cluster, cluster head is nominated and focused acyclic graph is plotted with all cluster heads, as a node. Edges in the graph are characterized as the communication intent from transmitter to the receiver. In this scheme, sleep scheduling is also proposed to save the energy in IoT devices. Pragmatic two-layer IoT architecture in reference [21] for efficient routing uses heuristic-based clustering algorithm for formation of cluster in two-layer network which works from bottom to top. An IoT node has the data of other nodes that are within its transmission range based on these nodes, formation of cluster takes place. To carry on task, graph-based clustering algorithm method is used. This algorithm chooses the cluster head from the clusters in the sensing layer. In this cluster formation, two parameters, i.e., residual energy and number of neighbors of a node, are considered based on these clusters formed. Routing is performed on the formed clusters. Cluster head selects the optimum route for transmission from source to destination. Data are transmitted through the neighbor nodes in the cluster. An energy optimization-based Modified Percentage LEACH protocol is presented for effectual routing in reference [22] for IoT network. The protocol reduces the energy consumption by minimizing communication between cluster head node and sink node. To communicate the nodes with sink node, first-order radio model is used. Nodes are arranged in the initialization phase. Each node itself forms cluster, in which each cluster is headed by a cluster head. Cluster head is nominated as per the probability. If the probability of a node is 1, then it is designated as cluster head or else cluster member. In communication phase, intra-cluster communication takes place in which cluster members are communicating between each other. During intra-cluster communication, single hop routing is used. Multi-hop routing is performed during the inter cluster communication.

3.2 Non-clustering-Based Routing Schemes

In this section, swarm intelligence algorithm-based routing methods are discussed. A bio-inspired Particle Multi-swarm Optimization algorithm (PMSO) is used for routing in [23]. To satisfy the QoS parameters in routing PMSO is used which build, improve, and choose *k*-disjoint network paths. The multi-swarm method finds the optimum guidelines in multi-path routing by interchanging information from the all position in the network link. The algorithm provides fast retrieval from path failure with the help of the objective functions. The operative values of these objective functions are optimized which are selected at each node to construct the *k*-disjoint multi-path. Ant colony optimization (ACO) algorithm is used for optimum routing in the IoT network [24]. The algorithm considers the following parameters for efficient route selection such as mobility, energy, and path length. The technique is called Efficient IoT Communication-based Ant System (EICAntS). EICAntS subordinates to each link between the source and destination. The group of attributes determines the quality that is length, energy factor and stability factor. The packet from source node reached at destination node; after receiving packets from source node, destination node computes the global factor, and then directs it to the source node. Source node updates the previous value of global factor in its routing table. Energy Harvesting Aware Routing Algorithm (EHARA) is described in [25] for improvement in lifetime of the IoT network. The algorithm uses new constraint, namely energy back off which is combined with the IEEE802.15.4 CSMA/CA mechanism. The energy estimation model is proposed for arrival of collected energy at nodes. The algorithm combines the energy back off and energy estimation process for efficient routing in the IoT network. By combining different energy harvesting algorithms, the method increases the nodes lifetime and Quality of Service of the network.

3.3 Secure Routing Schemes

In this section, recent trends in security-based routing schemes are discussed. A hierarchical clustering network topology is introduced in reference [26] for security-efficient routing. The protocol in the topology provides security against the black hole attack. The multi-hop routing protocol for low power and lossy networks (RPL) is proposed which establish different paths in clusters during route selection process. The top layer of the network is the gateway node, middle layer is the cluster head node, and bottom layer is the common node. Cluster head is nominated based on the arbitrary value. The nodes in the upper layer broadcast the Destination advertisement object Information Object (DIO) message which contains the identity value of Destination Advertisement Object (DAG), rank value of the cluster head, and IP of the neighbor nodes. Based on DIO information, common node selects the parent node. After selection of parent node, common node marks the IP address in default routing table as the parent IP address. Based on the selected neighbor and parent nodes, common node selects the route path. The modification of RPL protocol is introduced for security-efficient routing in IoT network in reference [27]. The protocol introduces the SISLOF which ensures only nodes share the information of suitable key in routing table. This process ensures the network nodes to be connected in secure way. The protocol modifies the RPL protocol messages. The approach finds the secured link between any node and its candidate parent node to form the secure RPL routing table. It reduces the number of nodes that are excluded because of insecure link. A new Lightweight Authentication and Secured Routing is discussed in reference [28] to achieve efficient and secure routing in IoT network. The protocol has three steps, discovery of network and its validation, standard node (SN) validation, and key delivery and path advertisement. In first step, SN discovers the neighbor node who asks Island Manager (IM) to validate the network to new SN. In second step, SN authenticates itself to the IM and gets the keys essential to broadcast the route. At last, SN advertises its route in the network. The route is then forwarded hop by hop to the anchor node using SetNext message. The resulting route is similar to route obtained in the RPL protocol.

4 Performance Metrics

In this section, the parameters that are needed to compare various techniques in references [13–28] pertaining to data aggregation and routing are explained.

4.1 Synchronization

Synchronization is defined as the active connection between source and destination nodes. It is expressed as

$$S_{m,n} = \delta(m, n) \tag{1}$$

where $\delta(m, n) = 0$ means nodes '*m*' and '*n*' share the active connection and 1 represents vice versa.

4.2 Link Reliability

Link reliability is stated as the number of data packets exchanged between the source and destination nodes within the particular interval. It can be expressed as

$$R_{m,n} = \frac{1}{2}\left(\frac{1}{d_p(m, n)} + \frac{1}{d_{\text{Nei}}(n) - 1}.\sum_{i=1, i\neq m}^{d_{\text{Nei}}(n)}\left(1 - \frac{k_{\text{min}/m}}{k(n, i)}\right)^2\right) \tag{2}$$

where d_p represents the number of packets received in '*n*' node from '*m*' node in the last interval. $d_{\text{Nei}}(n)$ denotes the number of neighbors known to *n* node. $k_{\text{min}/m}$ indicates distance between the '*n*' and its closest neighbor excluding '*m*'.

4.3 End-to-End Delay

End-to-end delay is stated as the time taken by the nodes to transfer data packets from source to destination. End-to-end delay is directly related to the number of hops at nodes. More number of hops led to an increased delay.

4.4 Traffic

Traffic in the network is defined as the total number of data packets that are transmitted in the given period. It can be expressed as

$$T_m = \frac{1}{b_M(m)} \sum_{i=0}^{N_{\text{active}}(m)-1} b_i \tag{3}$$

where $b_M(m)$ represents the maximum data rate guaranteed by node m. b_i indicates the rate of active connection involved in m. N_{active} indicates total number of active connections involved in the m.

4.5 Power

It is the power required to transfer data packet from source to destination nodes. It is expressed as

$$P_{m,n} = \left(\frac{t(m,n)}{t_m}\right)^{\propto} \tag{4}$$

where $t(m,n)$ represents distance between the node nodes 'm' and 'n'. t_m indicate maximum distance between the 'm' and 'n'. $\propto$ is the path loss extent.

4.6 Autonomy

Autonomy is stated as ratio of energy needed to transmit data packets to the total energy. It can be expressed as below

$$A(n) = 1 - (\text{Residual energy}/\text{Total energy}) \tag{5}$$

4.7 Multi-user Interference (MUI)

MUI is defined as the potential impact transmission from source node to the sink node through neighbor of source node.

$$\text{MUI}_{m,n} = \frac{1}{d_{\text{Nei}}(m) - 1}.\sum_{i=1,i\neq n}^{d_{\text{Nei}}(m)} \left(1 - \frac{k_{\min/n}}{k(m,i)}\right)^2 \tag{6}$$

where d_{Nei} is the number of neighbors known to node m. i is generic neighbor, excluding node n. $k_{\min/n}$ is the distance between m and its closest neighbor, excluding n.

4.8 Coexistence

Coexistence is defined as the ratio of measured external interference to the maximum interference. It can be expressed as below

$$C(n) = \frac{\text{Measured External Interference } (n)}{\text{Maximum Interference}(n)} \tag{7}$$

Table 1 shows the comparison of different techniques in references [13–28] based on the above-discussed parameters. From the table, it is observed that different authors have considered different performance metric parameters for the performance evaluation of the aggregation of data and routing techniques.

5 AODV Routing Protocol Performance Evaluation

This section describes the AODV routing protocol performance assessment on throughput, delay, and packet delivery ratio for ten different mobility conditions given in Table 2. Ad hoc On-demand Distance Vector (AODV) is a reactive protocol which obtains route from source to destination only when data transmission is recommended. Thus, it reduces burden and has less traffic as only required routes are maintained. It uses node sequence numbers to check efficiency of route. It supports unicast and multi-cast communications and has bidirectional links. It uses routing tables to store unicast and multi-cast routes. There are two phases for finding valid routes from source to destination and taking care of link breakage; these are Route Discovery and Route Maintenance. The packets in AODV are Route Request (RREQ), Route Reply (RREP), Route Error (RERROR), and Hello Packets (HELLO), as shown in Fig. 4.

Table 2 Node mobility conditions, route changes, and link changes

Node mobility conditions	Node	Route changes	Link changes
1	0	6	1
2	1	6	2
3	2	3	0
4	3	9	1
5	4	6	0
6	5	1	1
7	6	3	2
8	7	5	1
9	8	8	1
10	9	1	1

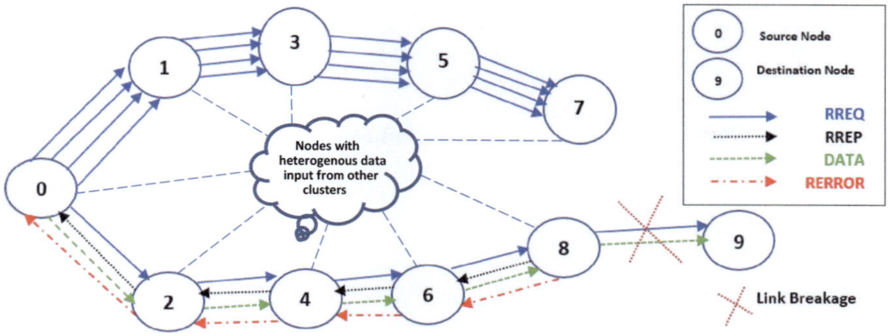

Fig. 4 Packets' transmission in AODV protocol

The comparison of different data aggregation and routing techniques is listed in Table 1 by various authors considering the performance metric parameters only. These *authors do not invoke mobility conditions* to nodes in a network. To understand the AODV protocol behavior for various mobility conditions given in Table 2 and its impact on important performance metric parameters, such as throughput, delay, and packet delivery ratio has been simulated in this paper using Network Simulator-2 (NS-2).

In NS-2, simulation of the network with ten wireless mobile nodes is considered, where each node may set input in heterogenous fashion. Node 0 is the source node, and node 9 is the destination node. Node 0 initially checks its routing table for effective route from itself to the node 9. If it did not find any valid route, then it starts the route discovery phase. In route discovery phase, it generates RREQ packets and broadcast it to the network. Figure 4 shows all the typical phases of data transmission of AODV protocol.

After discovery of a valid route from source to destination, node 8 sends RREP packet to the node 0 (source). This is the acknowledgment phase and is known as route detection phase of the AODV protocol. After the confirmation of valid route, node 0 (source) sends data to node 9 (destination). This phase is called data transfer phase which is executed after the acknowledgment from the destination node. If there is no valid route found during the discovery phase or there is link breakage or the node is shifted from its position, then RERROR packet is transferred from node 8 to source node to stop the data transmission. This phase is known as route maintenance phase of the protocol.

AODV routing protocol is simulated using NS-2 *invoking ten different mobility conditions,* and its performance is evaluated for throughput, delay, and packet delivery ratio. Table 3 shows the simulation parameters considered for the performance evaluation of AODV protocol for ten different mobility conditions listed in Table 2.

The following graphs are obtained as the results of the simulation on NS-2 which shows the interpretation of AODV protocol in respect of delay (in seconds), throughput (in kbps), and packet delivery ratio (in %) represented on *Y*-axis and simulation number on *X*-axis for ten different node mobility conditions (Graphs 1, 2, and 3).

Table 3 Simulation parameters

Parameters	Value
Simulation area	800 × 800 m
Number of nodes	10
Speed type	Constant
Min speed	10 m/s
Max speed	15 m/s
Average speed	12.33 m/s
Simulation time	10 s

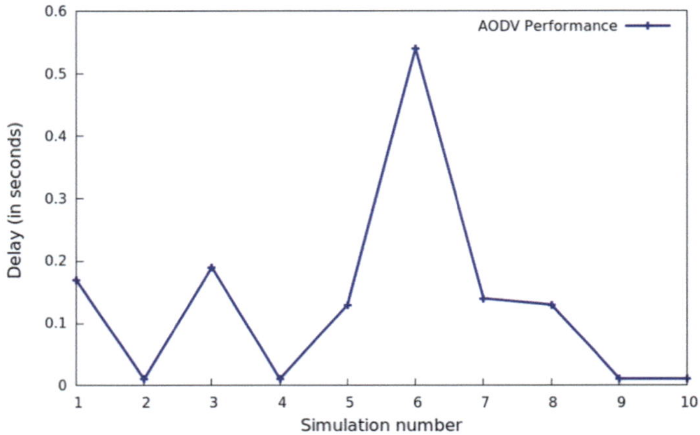

Graph 1 AODV performance on delay (*Y*-axis) for ten different mobility conditions (*X*-axis)

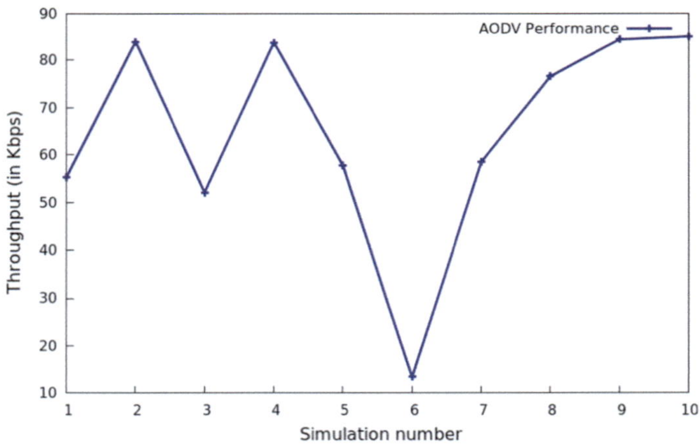

Graph 2 AODV performance on throughput (*Y*-axis) for ten different mobility conditions (*X*-axis)

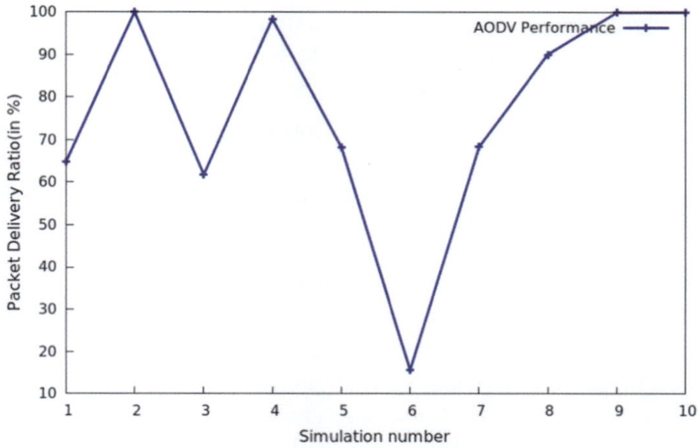

Graph 3 AODV performance on packet delivery ratio (*Y*-axis) for ten different mobility conditions (*X*-axis)

From the graphs, it is observed that mobility condition number 6 (Node 5: route changes: 1 link change: 1) should be avoided from network to prevent data loss. For the remaining node mobility conditions, the network performance is found to be comparatively non-degraded. Table 4 shows the summary of the simulation results.

From Table 4, it is observed that the maximum packet delivery ratio (100%) and throughput (85.05 kbps) are achieved for mobility condition number 10, and minimum delay (0.01 s) is for the mobility condition numbers 2, 4, 9, and 10, respectively, and minimum packet delivery ratio (15.69%) and throughput (13.42 kbps)

Table 4 Summary of the simulation results obtained using NS-2

Simulation number/mobility conditions	No. of packets sent	No. of packets received	Packet delivery ratio (%)	Throughput (kbps)	Delay (s)
1	1600	1037	64.81	55.26	0.17
2	1600	1599	99.94	83.98	0.01
3	1600	989	61.81	52.07	0.19
4	1600	1573	98.31	83.77	0.01
5	1600	1092	68.25	57.76	0.13
6	1600	251	15.69	13.42	0.54
7	1600	1094	68.38	58.5	0.14
8	1600	1440	90	76.58	0.13
9	1600	1599	99.94	84.39	0.01
10	1600	1600	100	85.05	0.01
Average	1600	1227.4	76.713	65.078	0.134

are for mobility condition number 6 with maximum delay of 0.54 s. The average packet delivery ratio, throughput, and delay are 76.713%, 65.078 kbps, and 0.134 s, respectively.

6 Conclusion

In this paper, the recent trends in data aggregation and routing techniques for conservation of energy and improved lifetime of network in the IoT network have been discussed. The data aggregation process is described in three parts, i.e., tree-based data aggregation, centralized data aggregation, and cluster-based data aggregation schemes. Centralized and cluster-based aggregation of data schemes plays major part in the aggregation of data. Tree-based aggregation of data scheme contributes weak in data aggregation process.

Routing process is described under three categories, i.e., cluster-based routing, non-cluster-based routing, and security-based routing schemes. Security-based routing schemes provide secure data transmission in IoT device networks. The paper discusses the parameters that are related to energy consumption, network lifetime increase, and reduced delay in the IoT network. These parameters are synchronization, link reliability, power, traffic, coexistence, autonomy, MUI, and delay. Finally, the proposed work compares all the data aggregation and routing techniques discussed with respect to the considered parameters given in Table 1, and the performance of AODV routing protocol is evaluated *invoking ten different mobility conditions*. Thus, in conclusion the suitable mobility conditions are to be followed so that data transmission and routing are not affected in the IoT network.

References

1. Karamitsios K, Orphanoudakis T (2017) Efficient IoT data aggregation for connected health Applications. In: IEEE symposium on computers and communications (ISCC), https://doi.org/10.1109/ISCC.2017.8024685
2. Pourghebleh B, Navimipour NJ (2017) Data aggregation mechanisms in the internet of things: a systematic review of the literature and recommendations for future research. Elsevier J Netw Comput Appl 97: 23–24. https://doi.org/10.1016/j.jnca.2017.08.006
3. Fitzgerald E, Pióro M, Tomaszewski A (2018) Energy-optimal data aggregation and dissemination for the internet of things. IEEE Internet Things J 5(2):955–969. https://doi.org/10.1109/JIOT.2018.2803792
4. Amarlingam M, Mishra PK, Rajalakshmi P, Channappayya SS, Sastry CS (2018) Novel light weight compressed data aggregation using sparse measurements for IoT networks. Elsevier J Netw Comput Appl 121: 119–134. https://doi.org/10.1016/j.jnca.2018.08.004
5. Yu T, Wang X, Shami A (2017) Recursive principal component analysis based data outlier detection and sensor data aggregation in IoT systems. IEEE Internet Things J 4(6): 2207–2216. https://ieeexplore.ieee.org/xpl/tocresult.jsp?isnumber=8172502, https://doi.org/10.1109/jiot.2017.2756025

6. Jia B, Hao L, Zhang C, Zhao H, Khan M (2018) An IoT service aggregation method based on dynamic planning for QoE restraints. Springer J Mob Netw Appl 24(1):25–33. https://doi.org/10.1007/s11036-018-1135-7
7. Kumar K, Kumar S (2018) Energy efficient link stable routing in internet of things. Springer Int J Inf Technol 10(4):465–479. https://doi.org/10.1007/s41870-018-0141-0
8. Wang Y, Tian Y, Miao R, Chen W (2018) Heterogeneous IoTs routing strategy based on cellular address. In: IEEE international conference on smart internet of things (SmartIoT), pp 64–69. https://doi.org/10.1109/smartiot.2018.00021
9. Elapp M, Chinara S, Parhi DR (2018) Survivable path routing in WSN for IoT applications. Elsevier J Pervasive Mob Comput 43: 49–63. https://doi.org/10.1016/j.pmcj.2017.11.004
10. Bounceur A, Bezoui M, Lounis M, Euler R, Teodorov C (2018) A new dominating tree routing algorithm for efficient leader election in IoT networks. In: IEEE annual consumer communications & networking conference (CCNC). https://doi.org/10.1109/CCNC.2018.8319292
11. Dhumane AV, Prasad RS (2017) Multi-objective fractional gravitational search algorithm for energy efficient routing in IoT. Springer J Wirel Netw 25(1):399–413. https://doi.org/10.1007/s11276-017-1566-2
12. Dhumane AV, Prasad RS (2018) Fractional gravitational grey wolf optimization to multi-path data transmission in IoT. Springer J Wirel Pers Commun 102(1):411–436. https://doi.org/10.1007/s11277-018-5850-y
13. Bhandari S, Sharma SK, Wang X (2017) Latency minimization in wireless IoT using prioritized channel access and data aggregation. In: IEEE global communications conference. https://doi.org/10.1109/glocom.2017.8255038
14. Ghate VV, Vijayakumar V (2018) Machine learning for data aggregation in WSN: a survey. Int J Pure Appl Math 118(24). https://doi.org/10.1016/j.inffus.2018.09.013
15. Lu R, Heung K, Lashkari AH, Ghorbani AA (2017) A lightweight privacy-preserving data aggregation scheme for fog computing-enhanced IoT. IEEE Access 5: 3302–3312. https://doi.org/10.1109/access.2017.2677520
16. Ko H, Lee J, Pack S (2017) CG-E2S2: consistency-guaranteed and energy-efficient sleep scheduling algorithm with data aggregation for IoT. Elsevier J Futur Gener Comput Syst 92:1093–1102. https://doi.org/10.1016/j.future.2017.08.040
17. Li R, Sturtivant C, Yu J, Cheng X (2018) A novel secure and efficient data aggregation scheme for IoT. IEEE Internet Things J. https://doi.org/10.1109/JIOT.2018.2848962
18. Abdul-Qawy ASH, Srinivasulu T (2018) SEES: a scalable and energy-efficient scheme for green IoT-based heterogeneous wireless nodes. Springer J Ambient Intell HumIzed Comput, 1–26. https://doi.org/10.1007/s12652-018-0758-7
19. Preetha SSL, Dhanalakshmi R, Kumar R (2018) An energy efficient framework for densely distributed WSNs IoT devices based on tree based robust cluster head. Springer J Wirel Pers Commun 103(4): 3163–3180. https://doi.org/10.1007/s11277-018-6000-2
20. Chhabra A, Vashishth V, Khanna A, Sharma DK, Singh J (2018) An energy efficient routing protocol for wireless internet-of-things sensor networks. Networking and Internet Architecture. https://arxiv.org/abs/1808.01039. Downloaded on 29 Nov
21. Kumar JS, Zaveri MA (2018) Clustering approaches for pragmatic two-layer IoT architecture. Wirel Commun Mob Comput, pp 1–16. https://doi.org/10.1155/2018/8739203
22. Khan FA, Ahmad A, Imran M (2018) Energy optimization of PR-LEACH routing scheme using distance awareness in internet of things networks. Springer Int J Parallel Program, pp 1–20. https://doi.org/10.1007/s10766-018-0586-6
23. Hasan MZ, Al-Turjman F (2017) Optimizing multipath routing with guaranteed fault tolerance in internet of things. IEEE Sens J 17(19): 6463–6473. https://doi.org/10.1109/jsen.2017.2739188
24. Hamrioui S, Lorenz P (2017) Bio inspired routing algorithm and efficient communications within IoT. IEEE Netw J 31(5):74–79. https://doi.org/10.1109/MNET.2017.1600282
25. Nguyen TD, Khan JY, Ngo DT (2018) A distributed energy-harvesting-aware routing algorithm for heterogeneous IoT networks. IEEE Trans Green Commun Netw 2(4): 1115–1127. https://doi.org/10.1109/tgcn.2018.2839593

26. Ma G, Li X, Pei Q, Li Z (2017) A security routing protocol for internet of things based on RPL. In: IEEE international conference on networking and network applications. https://doi.org/10.1109/NaNA.2017.28
27. El Hajjar A, Roussos G, Paterson M (2017) Secure routing in IoT networks with SISLOF. In: IEEE global internet of things summit (GIoTS). https://doi.org/10.1109/GIOTS.2017.8016278
28. Mick T, Tourani R, Misra S (2017) Laser: lightweight authentication and secured routing for NDN IoT in smart cities. IEEE Internet Things J 5(2):755–764. https://doi.org/10.1109/JIOT.2017.2725238

An Efficient Image Data Encryption Technique Based on RC4 and Blowfish Algorithm with Random Data Shuffling

Dharna Singhai and Chetan Gupta

Abstract In this work, an able safety structure has been proposed which is efficient in image data security as well as the image loss is minimum. In our approach, we have used the combination of RC4 and Blowfish algorithms. Blowfish is suitable for image data for which the key does not vary for a single cycle. It is not suitable where more compact ciphers are needed. Due to the advantages of RC4 substitution in different rounds and Blowfish algorithm for the image data, we have used the combination for image data security. Bit shuffling has been performed for the bit randomization and XOR operations, so that the proper bit shuffling is possible. Then, for comparative analysis, peak signal-to-noise ratio (PSNR) and mean square error (MSE) cover have been designed. It provides the comparative and analytical comparison based on the different images. Our results show that it is efficient in terms of information loss, MSE, and PSNR values.

Keywords Encryption · Decryption · PSNR · MSE · XOR · RC4

1 Introduction

Picture facts' safety is an important part of security nowadays [1]. It is the need of current age data gathering and security mechanism for the aspects of all other related data aggregations and communication on the Internet and offline system [2–5]. In this regard, encryption and decryption standards provide an insight in the way of achieving better security [6–9]. It can be helpful in several aspects of data hiding, shuffling, and preventing it [10, 11]. Thus into the resultant section it look at information Encryption scheme used pro print encryption [12]. The moreover utter on critical edges are used like a touch of image encryption by their inspirations intrigue with preventions [13]. There are different essential techniques which are unavoidable cryptography, for example, private key cryptography and hash [14]. In private key

D. Singhai (✉) · C. Gupta
Department of Computer Science and Engineering, SIRTS, Bhopal, India
e-mail: dharnasinghai1992@gmail.com

C. Gupta
e-mail: chetangupta.gupta1@gmail.com

R. K. Shukla et al. (eds.), *Social Networking and Computational Intelligence*, Lecture Notes in Networks and Systems 100, https://doi.org/10.1007/978-981-15-2071-6_38

cryptography, a single key is remaining for both encryption and unwinding. This obliges wander as a last resort part pass on offering a copy the key be strike in be passed swear off an ensured channel to the going with individual [15]. Private key estimations are indestructible level and appropriately recognized in contraption. Along these lines, they are well truly for mass estimation encryption. In tremendous please of the in and not in objective encryption depend upon plaintext, encryption estimation, key and unscrambling figuring. The plaintext is the size ahead requiring the encryption check. The encryption check is the estimation uses to keep scheduled manage the information new arrival plaintext to form composed. The puzzle key be an in every way that really matters indistinct to rebuke of the encryption figuring and of the plaintext and it is assistant of the encryption's wellsprings of data add up [16, 17].

In this document, a well-organized safety approach along with the performance comparison of the proposed approach has been discussed and analyzed in a comparative way.

2 Literature Survey

In 2018, Guo et al. [18] suggested that the watermarking and data hiding techniques may vary the design of data embedded systems. They have suggested a new rounding/truncation error (RTE) domain for the data embedding. It shows the rounding error. They have proposed this for the data hiding and authentication purpose.

In 2018, Singh et al. [19] discussed regarding the impact of currency counterfeiting. They have suggested an authentication system. Their system has the compact and adheres the properties of hardware components easily. Security thread and latent images are the main features of this system. It also extracts, encodes, and enhances the security parameters and features. For classification, they have used SVM model.

In 2018, Shrestha et al. [20] discussed the drawback of manual attendance system. They have suggested that it is very time-consuming. They have also highlighted the impact of authentication in case of RFID-based attendance system. They have used image processing techniques with the automatic face recognition system. They have suggested that their system has the capability of identifying the intruders. Unknown faces detected are called intruders. So in their system, it is efficient in assuring the security of the students.

In 2018, Dhanva et al. [21] discussed the cheque truncation system (CTS). It eases the clearing system by the use of MICR codes and image transferring of the images. There is a security concern in case of data passing because of the medium of transfer. They have suggested the chances of attacks like message modification and denial of service which can be possible. They have suggested that the SVD watermarking and AES encryption may be a solution for the data security. They have also applied skew correction techniques.

In 2018, Chen et al. [22] discussed the role of edge detection. They have also suggested these things in terms of cloud computing also. They have proposed an approach for edge detection based on the Sobel edge detector. They have used Gaussian filtering and Sobel operator. They have used homomorphism property. They have obtained the image edge and then performed the encryption procedure on this.

In 2018, Yang et al. [23] discussed the electronic health records (EHRs). They have used visual cryptography (VC) and secret sharing with password of practitioners. It has been used for the distributed system. They have also used "openEMR" for the performance evaluation.

In 2018, Chumuang et al. [24] proposed the data matching with histogram shaping for the image enhancement. Their images used for experimentation are of 1280 × 720 pixel size. It is collected from the CCTV surveillance video system. They have used JPG images and surveillance video of AVI.

In 2018, Li et al. [25] discussed image forensics. They have suggested that the previous approach only considers binary classification and it can lead to the irrelevant features. So they have suggested the need of multiple image operations identifications. They have analyzed the local pixel for the better detection and classification. They have proposed a compact universal feature set. Their results show the improvement in terms of effectiveness and universality.

In 2018, Usman et al. [26] proposed a steganography approach for the medical data security. They have used swapped Huffman tree coding. It has been used for the lossless compression. For cover images, they have used only edge regions. They have suggested that the results are efficient in confidentiality and secrecy.

In [27], this writer planned an able figure cryptography algorithm through by encryption among steganography. In this used for RC4 stream cipher and RGB pixel shuffling through steganography in by hash-least significant Bit (HLSB) is apply. For secret image, PSNR is infinity and MSE is 0. In support of cover image, PSNR is about 63 db and MSE is about 0.03.

3 Proposed Work

In proposed work, a well-organized safety framework has been proposed which is efficient in image data security as well as the image loss is minimum. Figure 1 shows the working procedural flowchart which can be helpful in understanding the approach.

In our approach, we have used the combination of Rivest's cipher (RC4) and Blowfish algorithms. The major advantage in RC4 algorithm is that it is faster for the streaming function and Blowfish is suitable for image data for which the key does not variable for a single cycle. It is not suitable where more compact ciphers are needed. Due to the advantages of RC4 and Blowfish algorithms for the image data, we have used the combination for image data security. Bit variations and shuffling operations are also applied for the bit randomization so that the proper bit shuffling is possible. Then, histogram comparison has been provided for the image data for the different phases. Then, the information loss is calculated based on the entropy values. Then, for comparative analysis, peak signal-to-noise ratio (PSNR) and mean

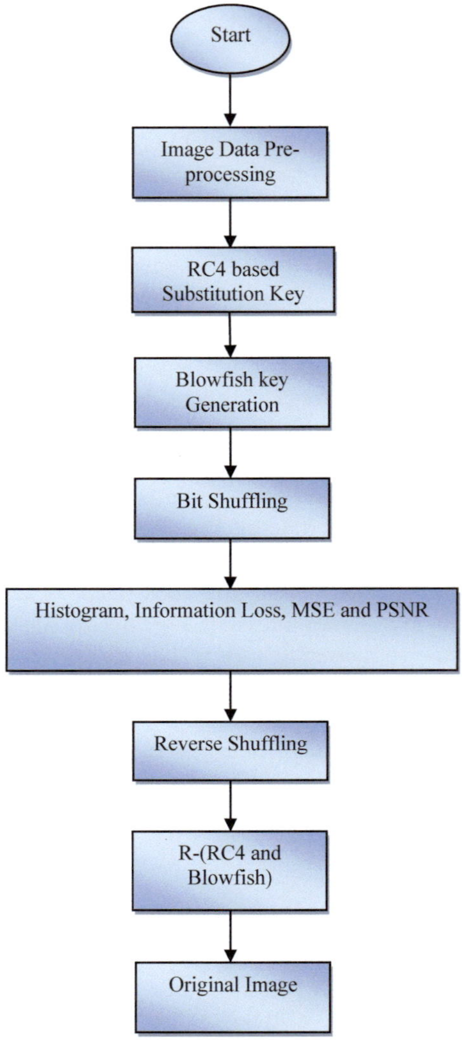

Fig. 1 Flowchart of proposed work

square error (MSE) have been calculated. It provides comparative and analytical comparison based on the different images.

Algorithm of Proposed work

Step 1: Accept the images for the input.
Step 2: Convert it into the algorithm compatible language, which is in our case in temporary array.
Step 3: Perform substitution through RC4 algorithm. Key generation after the substitution round is completed.

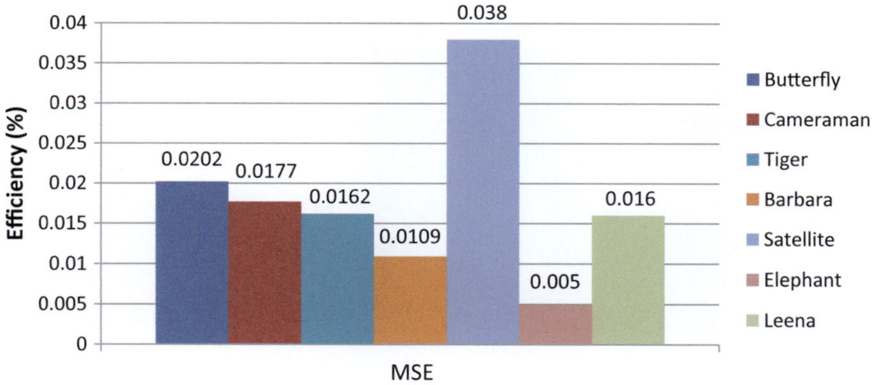

Fig. 2 MSE comparison for images

Step 4: The arranged substituted array has been inputted to Blowfish algorithm for the further encryption and key generation.
Step 5: Bit-wise random shuffling has been provided.
Step 6: Performance analysis based on the histogram, entropy, MSE, and PSNR values.
Step 7: Decryption process has been started.
Step 8: Time computation performances have been evaluated.

4 Results

Here we have presented the outcome which is used to obtain and discuss the comparison on the basis of various parameters. Figure 2 shows the MSE comparison of different images. Figure 3 shows the PSNR evaluation. Figure 4 shows the encryption time comparison used for images (RC4). Figure 5 shows the encryption time comparison for images (Blowfish). Figure 6 shows the decryption time for images in case of RC4 and Blowfish algorithms. Figure 7 shows the MSE comparison. It clearly indicates that the proposed approach is superior in terms of MSE. Figure 8 shows the PSNR comparison for images. It shows the performances in case of proposed work is better (Fig. 9).

5 Conclusion

In this proposed work, a well-organized image security framework had been developed. In this paper, RC4 and Blowfish algorithms are used for encryption to do the proper shuffling between pixels to create higher variation. In proposed work, two

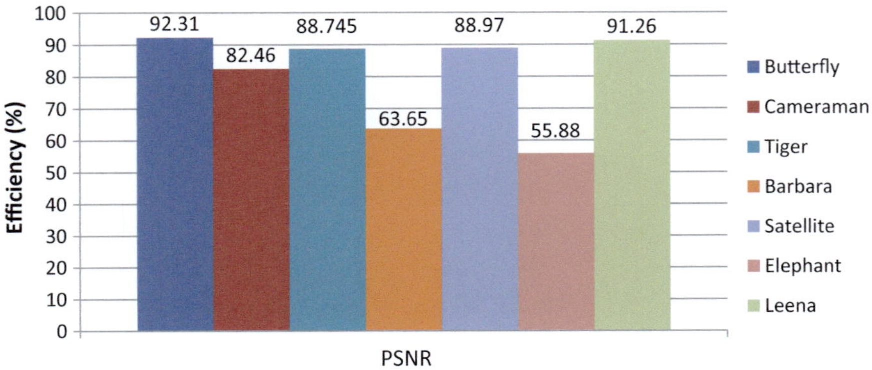

Fig. 3 PSNR comparison for images

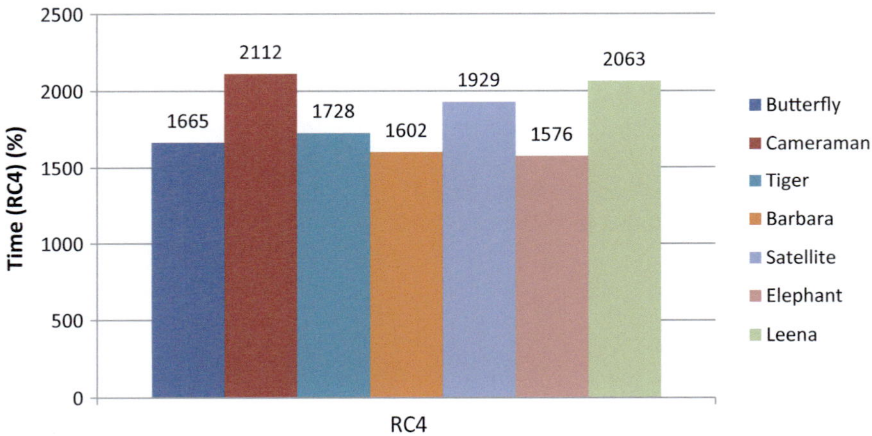

Fig. 4 Encryption time comparison for images (RC4) (ms)

random keys are generated to provide the more security to the data which we used during the decryption of image. We had also calculated the encryption and decryption time that show the effectiveness of our approach. The results obtained clearly indicate that the proposed approach has the better PSNR, MSE, and entropy in comparison to the previous method.

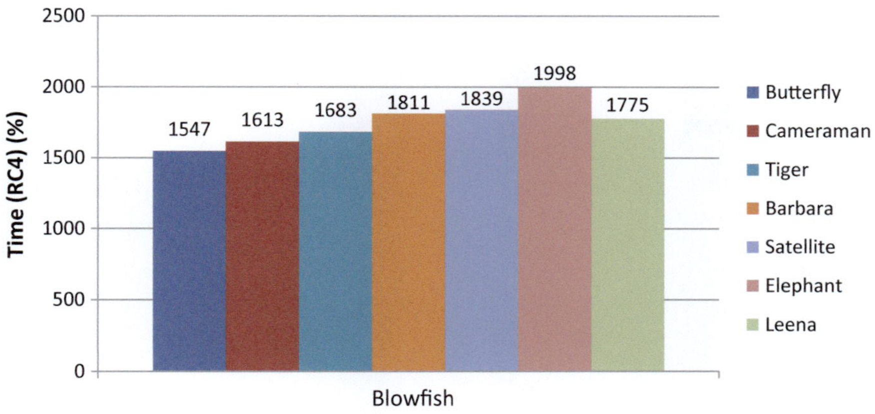

Fig. 5 Encryption time comparison for images (Blowfish) (ms)

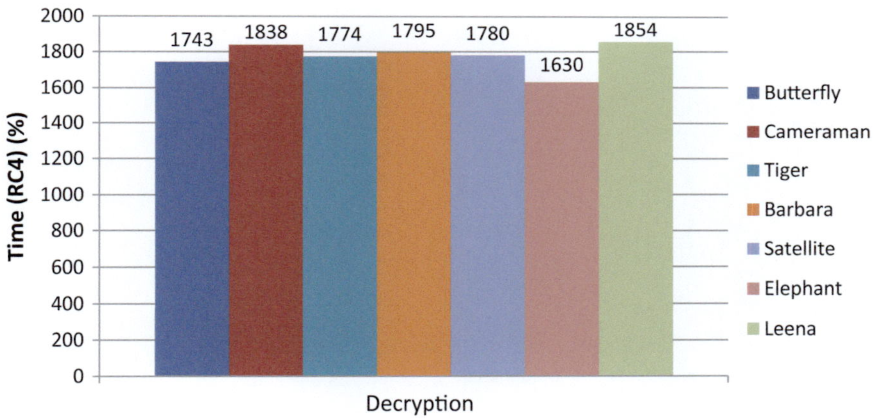

Fig. 6 Decryption time for images (ms)

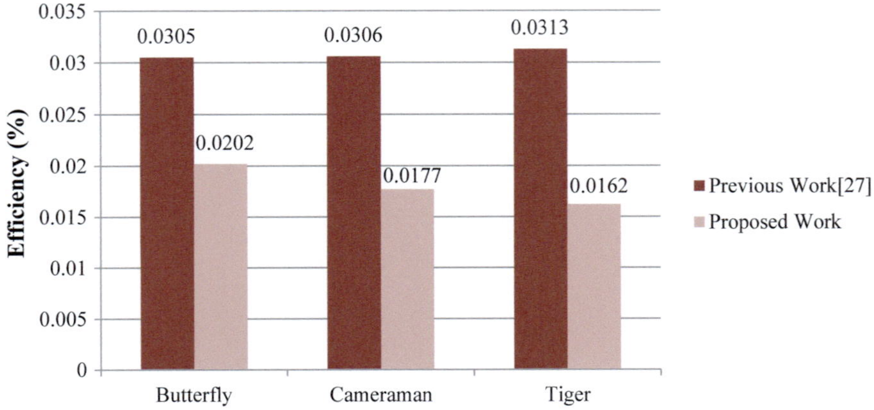

Fig. 7 MSE comparison from previous work [27]

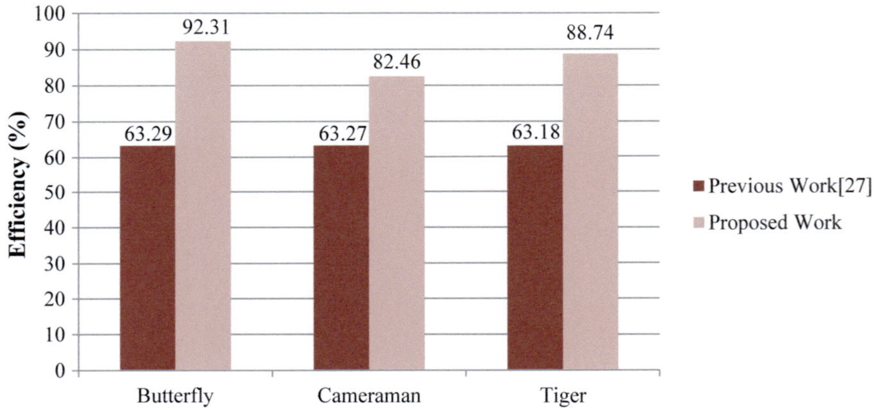

Fig. 8 PSNR comparison for images

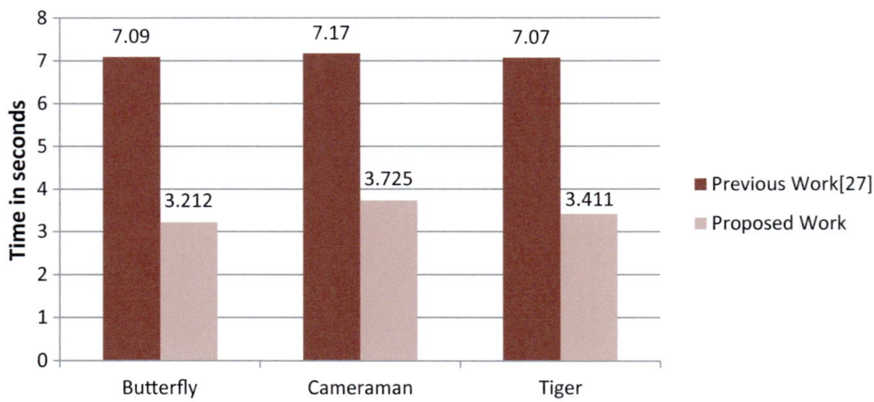

Fig. 9 Time comparison for images (time in second)

References

1. Wen W, Zhang Y, Fang Y, Fang Z (2016) A novel selective image encryption method based on saliency detection. In: Visual communications and image processing (VCIP), IEEE, 27–30 Nov 2016
2. Chatterjee A, Dhanotia J, Bhatia V, Rana S, Prakash S (2017) Optical image encryption using fringe projection profilometry fourier fringe analysis and RSA algorithm. In: 14th india council international conference (INDICON), IEEE, 2017
3. Dragoi IC, Coltuc D (2018) Reversible data hiding in encrypted color images based on vacating room after encryption and pixel prediction ioan. In: 25th IEEE (ICIP), 2018
4. Zhang Y, Li X, Hou W (2017) A fast image encryption scheme based on AES. In: 2nd international conference on image vision and computing (ICIVC), IEEE, 2017
5. Bhowmick A, Sinha N, Arjunan RV, Kishore B (2017) Permutation-substitution architecture based image encryption algorithm using middle square and RC4 PRNG. In: International conference on inventive systems and control (ICISC), IEEE, 2017
6. Gayathri SP, Sajeer M (2017) Chaotic system based image encryption of biometric characteristics for enhanced security. In: International conference on circuit, power and computing technologies (ICCPCT), IEEE, 2017
7. Nayak P, Nayak SK, Das S (2018) A secure and efficient color image encryption scheme based on two chaotic systems and advanced encryption standard. In: International conference on advances in computing, communications and informatics (ICACCI), IEEE, 2018
8. Guo L, Li J, Xue Q (2017) Joint image compression and encryption algorithm based on SPIHT and crossover operator. In: 14th international computer conference on wavelet active media technology and information processing (ICCWAMTIP), IEEE, 2017
9. Karthick S, Sankar SP, Prathab TR (2018) An approach for image encryption decryption based on quaternion fourier transform. In: International conference on emerging trends and innovations in engineering and technological research (ICETIETR), IEEE, 2018
10. Zhang Y, Zhang Q, Liao H, Wu W, Li X, Niu H (2017) A fast image encryption scheme based on public image and chaos. In: International conference on computing intelligence and information system (CIIS), IEEE, 2017
11. Joshy A, Baby KXA, Padma S, Fasila KA (2017) Text to image encryption technique using RGB substitution and AES. In: International conference on inventive computing and informatics (ICICI), IEEE, 2017
12. Takkar P, Girdhar A, Singh VP (2017) Image encryption algorithm using chaotic sequences and flipping. In: International conference on computing, communication and automation (ICCCA), IEEE, 2017
13. Chen D, Chen W, Chen J, Zheng P, Huang J (2018) Edge detection and image segmentation on encrypted image with homomorphic encryption and garbled circuit. In: International conference on multimedia and expo (ICME), IEEE, 2018
14. Lin J, Luo Y, Liu J, Bi J, Qiu S, Cen M, Liao Z (2018) An image compression-encryption algorithm based on cellular neural network and compressive sensing. In: 3rd international conference on image, vision and computing (ICIVC), IEEE, 2018
15. Zou Z (2018) A novel image encryption method based on modular matrix transformation and coordinate sampling. In: International conference on applied system invention (ICASI), IEEE, 2018
16. Chuman T, Sirichotedumrong W, Kiya H (2018) Encryption-then-compression systems using grayscale based image encryption for JPEG images. Trans Inf Forensics Secur, IEEE 2018
17. Kovalchuk A, Lotoshynska N (2018) Elements of RSA algorithm and extra noising in a binary linear-quadratic transformations during encryption and decryption of images. In: Second international conference on data stream mining & processing (DSMP), IEEE, 2018
18. Guo Y, Cao X, Wang R, Jin C (2018) A new data embedding method with a new data embedding domain for JPEG images. In: 2018 IEEE fourth international conference on multimedia big data (BigMM), IEEE, pp 1–5, 13 Sep 2018

19. Singh M, Ozarde P, Abhiram K (2018) Image processing based detection of counterfeit indian bank notes. In: 9th international conference on computing, communication and networking technologies (ICCCNT), IEEE, pp 1–5, 10 Jul 2018
20. Shrestha R, Pradhan SM, Karn R, Shrestha S (2018) Attendance and security assurance using image processing. In: Second international conference on computing methodologies and communication (ICCMC), IEEE, pp 544–548, 15 Feb 2018
21. Dhanva K, Harikrishnan M, Babu PU (2018) Cheque image security enhancement in online banking. In: Second international conference on inventive communication and computational technologies (ICICCT), IEEE, pp 1256–1260, 20 Apr 2018
22. Chen D, Chen W, Chen J, Zheng P, Huang J (2018) Edge detection and image segmentation on encrypted image with homomorphic encryption and garbled circuit. In: IEEE international conference on multimedia and expo (ICME), IEEE, pp 1–6, 23 Jul 2018
23. Yang D, Doh I, Chae K (2018) Secure medical image-sharing mechanism based on visual cryptography in EHR system. In: 20th international conference on advanced communication technology (ICACT), IEEE, pp 463–467, 11 Feb 2018
24. Chumuang N, Ketcham M, Yingthawornsuk T CCTV based surveillance system for railway station security. In: International conference on digital arts, media and technology (ICDAMT), IEEE, pp 7–12, 25 Feb 2018
25. Li H, Luo W, Qiu X, Huang J (2018) Identification of various image operations using residual-based features. IEEE Trans Circuits Syst Video Technol 28(1):31–45
26. Usman MA, Usman MR (2018) Using image steganography for providing enhanced medical data security. In: 15th IEEE annual 2018 consumer communications & networking conference (CCNC), IEEE, pp 1–4, 12 Jan 2018
27. Abood MH (2017) An efficient image cryptography using hash-LSB steganography with RC4 and pixel shuffling encryption algorithms. In: annual conference on new trends in information & communications technology applications (NTICT), IEEE, pp 86–90, 7 Mar 2017

IoT Devices for Monitoring Natural Environment—A Survey

Subhra Shriti Mishra and Akhtar Rasool

Abstract Environment plays the most important role for any human or living being. Monitoring various perspectives of our surrounding has turned out to be one of the most vital parts of recent science and technology. One of the advancing and pacing technologies is Internet of Things. The rapid development in the field of wireless sensors and networks has made it possible for IoT to grow as one of the most revering fields of study. These IoT devices can not only automate things but also help with the improvement of health conditions by monitoring the deformities in the environment. Visualising the pollution caused in air, water, soil and other elements of nature is the vital need of today's world. This paper offers a survey on the devices that are being used to detect and monitor the pollution caused by human activity and other natural phenomena.

Keywords Internet of Things-IoT · Heating · Ventilation and air conditioning-HVAC · Heated metal oxide sensor Cell-HMOS · Gas sensitive Semiconductor-GSS

1 Introduction

An in and around analysis of the environment gives us an overview of how it has changed from being a beautiful entity to a human intervened zone. Air, water, soil and every attribute present in the domain is being haunted with different level and types of pollutants. We as humans have made the surrounding a less habitable place to live in. The sufferings of the thoughtless behaviour started by us will not only last us but will continue on till each and every person works towards combat of the pollution around us. The aware humans have started different drives to make the locality a better place to live in. Technological advancements can be optimally made use of and a visible change can be brought about. Revolution in technology

S. S. Mishra · A. Rasool (✉)
Maulana Azad National Institute of Technology, Bhopal, Madhya Pradesh, India
e-mail: akki262@gmail.com

S. S. Mishra
e-mail: subhrashriti@rediffmail.com

R. K. Shukla et al. (eds.), *Social Networking and Computational Intelligence*, Lecture Notes in Networks and Systems 100, https://doi.org/10.1007/978-981-15-2071-6_39

is increasing day by day with the introduction of Internet of Things, Web-based approaches like cloud and other amenities, machine learning methods, etc. This can bring about a huge change in the way the environment can be helped to rehabilitate and rapidly as well. These techniques are being widely used for monitoring health, industrial monitoring, home monitoring, etc. The pollutant levels in the resources of the environment can also be monitored using IoT devices and other emerging technologies. Environmental parameters are defined as biotic and abiotic elements that structure the ecosystem and affect its functionalities. Biotic elements include all living organisms divided into plant and animal community. Abiotic factors include a portion of non-living chemical or physical parameters that consist of: temperature, light, humidity, minerals, water, soil, sound, etc.

This paper provides a brief survey of the devices that have been introduced to monitor the levels of pollutions in various entities of the environment. It also talks about the factors on which the pollution depends and increases or decreases it. The devices used to detect such factors and their effects can also be envisaged in this paper. These devices can be manufactured and made use of for lots of applications like air pollution monitoring, ecology, climate change observation, water quality monitoring, soil quality monitoring. These monitoring ways give a large amount of information that has to be stored and processed in order to come up with accurate analysis and reports which helps a quick access to the environmental menace definitively [1].

2 IoT Devices Used in Environment

It has been seen in the recent past that IoT devices have acquired a wide market with the introduction of wireless sensor devices. A mesh of independent nodes wireless sensor network contains various sensors that help with collection and transmission of data. Application of these WSNs has been wide-ranged starting from healthcare, accident detection, RFID, farming to pollutant detections, etc. For a long time, now various organisations and institutes have been making use of various kinds of sensors but with the advancement of the Internet of Things it has made the job of sensors along with growth of these devices to a totally new height.

Internet of Things platforms functions and delivers valiant kinds of intelligence and information using a variety of sensor detectors. These make it accessible for collection of data, shoving it and distributing it over an interconnected network of devices. The huge amount of data being collected makes the autonomous working of these devices possible that makes the whole ecosystem smarter everyday [2]. In this, there are some vital sensors discussed.

Temperature sensors: These are defined as devices being used to calculate the quantity of heat energy which permits to sense an alteration in the physical temperature from a specific source which then changes the data for either a user or a device, is called temperature sensor.

A variety of devices have made use of such sensors. But, with the arrival of Internet of Things, there has been more use of these sensors.

Proximity sensor: These kinds of sensors used in devices detect whether an object is present or absent nearby, or its properties, and changes that to signals that is easily readable by the operator or an electronic device without getting in connection with them. Retail industries make use of proximity sensors largely used, in order to detect motion and the relationship of the customers and the products that interest them. The user gets immediate notification about special offers, discounts of proximal products.
Pressure sensor: In a pressure sensor, the devices convert pressure to signal showing electric energy properties. Here, the amount is decided by the level of pressure applied.
Water quality sensor: In water issuing systems primarily water quality measurement detectors are being used to sense the standard of water as well as ion monitoring.
Chemical sensor: The main goal of chemical sensors is to detect changes in chemical properties in liquid or air.
Gas sensor: Though similar to the chemical sensors, gas sensors are specially being used to track the presence and retention of various gasses in the atmosphere. Industries like manufacture, agriculture and healthcare, sensing of poisonous or inflammable gasses in chemical labs, coal, oil and gas mines, plastic, rubber, petrochemical industries, etc.

Gas detector, catalytic bead sensor, ozone monitor, air pollution sensor, breathalyser, electrochemical gas sensor, hydrogen sensor, hygrometer, nitrogen, oxide sensor, etc. are gas sensors that are generally available.
Smoke sensor: The device that identifies smoke and their level which includes airborne particulates and gasses is a smoke sensor.

Common types of smoke sensors are smoke, flame and gas are detected by smoke sensors in their surrounding field. There are two methods this can be done, one optically and the other by physical process.

- Photoelectric or Optical smoke Sensor: The principle behind optical smoke sensors is the usage of trigger involving the light scatter.
- Ionisation smoke sensor: Ionisation is the main principle according to which ionisation smoke sensor works, to detect particles causing an alarm.

Infrared sensors: Infrared sensors are made use of in order to sense certain attributes of its environment by effusing or sensing infrared radiation. Heat being discharged by a device can also be detected. Various IoT projects like in Healthcare for measurement of blood pressure and blood flow, smartwatches, smartphones, Infrared vision, automotive blind-angle detection, breath analysis, optical measuring devices, non-contact-based temperature measurements.
Level sensors: The level and amount of liquids, fluids, etc. can be measured by level sensor in open as well as closed systems. Used in industries working with liquid materials like fuel, juice, alcohol as well as recycling industries use these sensors to know about the liquid assets present.
Image sensors: Image having optical properties is converted to electronic signals using these sensors for visualising and storage of files on electrical medium.

Charge-coupled device (CCD) and complementary metal-oxide semiconductor (CMOS) imagers are two major types of image sensors with usage of different types of images.

Motion detection sensors: Movement of objects or human beings or any type of physical motion is detected by an electronic device motion sensor which transforms motion into a signal with electric energy. Used in automatic doors, smart camera, automatic parking systems, security industries, etc. these sensors are being used extensively.

Accelerometer sensors: An electrical output is obtained by conversion from mechanical motion created due to an inertial force measured with the aid of this sensor which acts as a transducer and measures the acceleration of an object.

Gyroscope sensors: The angular velocity or angular rate can be measured by gyro sensors. Using this application, object orientation can be monitored.

Humidity sensors: Measures the quantity of water vapour present in the atmosphere.

Optical sensors: An optical sensor calculates the physical quantities like rays of light and transforms it into electrical signals that are made conveniently accessible by an operator or an electronic device. They can be used practically for computing different things at the same time like light, electricity, etc.

Some other sensors used more often are Passive IR (to detect human presence in the room), one temperature sensor (LM35 to collect room temperature) and one LDR (to detect light intensity near room window). A cloud-based metering service named Smart-me [3] has been introduced. With the help of electronic gadgets like tablets, smartphones, computers controlling and monitoring of meters like gas meters, electricity meters, water meters, heat meters, etc. from anywhere.

FilesThruTheAirTM [4] gives a thorough scope of environmental monitoring things including PC having a remote data logging sensor or remote cloud management, handheld multi-functional smart thermometers, data loggers with USB and indicative data loggers for integration into other available instruments. The speciality of FilesThruTheAirTM products is they have Wi-Fi and touch screen alongside a range of Wi-Fi data logging detectors and smart thermometers.

3 IoT Devices for Monitoring Air Pollution

Today the main pollutants present in the environment are ozone, carbon monoxide, sulphur dioxide, nitrogen dioxide, particulate matter, dispersion of organic matter, carbon dioxide, etc.

Major ozone detection sensors are:

3.1 Electrochemical Ozone Sensors

Ozone gas gets diffused into a cell with electrodes and electrolytes through a porous membrane in an electrochemical ozone sensor. With the increase in the existence of ozone, there is a proportional increase in the electrical signals. Interpretation of these signals and display of ozone concentration in PPM (parts per million) are done by the monitor.

3.2 Semiconductor-Based Ozone Sensors

Heated Metal Oxide Sensor Cell (HMOS)
Gas Sensitive Semiconductor (GSS)

The basic principle on which the HMOS sensor works is heating a mini substrate on high temperature of around 300 °F/149 °C. The substrate becomes very responsive to ozone at this temperature and shows a transformation in resistance which gets to be proportional to the quantity of ozone that touches its surface. An ozone level in PPM or PPB gets displayed on a monitor that interprets the change in resistance [5].

With the introduction of a global network of hardware devices like the uRADMonitor detection of different kinds of physical or chemical pollutants are possible. Measurement of barometric pressure, air temperature, dust concentration, humidity, VOC as well as alpha, beta and gamma radiation is made possible by current detectors. The latest uRADMonitor model D uses the BME680 sensor from Bosch to gauge air's condition. The BME680 amalgamates best-in-class air pressure, humidity and ambient air temperature with a gas sensor, sensing works within a single package [6].

Shinyei PPD42NS was decided to be used as a module for sensing the pollution as it costs less and is dependable and durable. By using a lens, photodiode and an LED this sensor helps with the measurement of opaqueness in ambient air. The low pulse occupancy (LPO) shown by the sensor is obtained by reading the opaqueness and by calculating the no. of particles per 0.01 cubic ft.

The next major gas that causes air pollution is carbon monoxide. Because of its properties like no colour, smell or taste carbon monoxide (CO) is named as "silent and invisible killer". CO is released into the atmosphere when there is burning of gasoline, oil, charcoal, wood, propane or natural gas.

The aim of the paper et al. [7] is to use an IoT device with MQ7 Arduino sensitive to CO to get the pollution generated by public transport. The amount of carbon monoxide released is detected once in (say 15 km) and also the location of automobile is used for determining the locality which is polluted the most. When the vehicle causes higher level of pollutants, a simple notification service (SNS) is sent to the smartphone. CO concentration can be detected by the MQ-7 somewhere from 20 to 2000 ppm as it has quick reaction time and high sensitivity.

Sulphur dioxide (SO_2) is a colourless, extremely toxic gas with a strong odour. Significant proportion of sulphur dioxide is emitted by burning of coal and petroleum, from thermal power plants, oil refineries, acid plants, volcanoes, smelters, organic waste, incinerators, etc. [8]. Industrial Scientific introduced portable gas detection devices like Tango™ TX1 & GasBadge® Pro single-gas sensors, the Ventis™ Pro Series, Ventis™ MX4, MX6 iBrid™ multi-gas detectors, as well as the Radius™ BZ1 Area Monitor to save workers from sulphur dioxide subjection, along with other hazardous gases that might be present. In this paper, a sensor principled on a porous Au–solid polymer electrolyte sensing electrode that is in immediate touch with the gas-containing air is used for sensing of concentrations of sulphur dioxide in the low-ppb range in the atmosphere. This can be manipulated to work even for approx 1 ppb (Fig. 1).

Nitrogen Oxides (NO_2): Killer Gas:
A reddish-brown corrosive gas generally categorised by nitric oxide (NO) plus nitrogen dioxide (NO_2) or combustion of atmospheric nitrogen and organic nitrogen in ambient air because of automobile exhaust. Nitrogen dioxide sensor has been detailed in Fig. 2 [9].
Particulate matter:
Generally, generated by solid bulk material breakdown or condensation these are the liquid or solid contents in the shape of mist and dust fumes.

SULFUR DIOXIDE SENSOR HEAD 0-100PPM

Sensor Code	ESP / ESP2
Range	0-100ppm
Sensor Type	GSE
Minimum Detection Limit	0.4ppm
Accuracy of Factory Calibration	<±0.5 ppm 0-5ppm <±10% 5-100ppm
Resolution	0.1ppm
Response Time	30 Seconds
Temp	0 to 40°C
Relative Humidity	15 to 90%

SULFUR DIOXIDE SENSOR HEAD 0-10PPM

Sensor Code	ESO / ESO2
Range	0-10ppm
Sensor Type	GSE
Minimum Detection Limit	0.04ppm
Accuracy of Factory Calibration	<±0.05 ppm 0-0.5ppm <±10% 0.5-10ppm
Resolution	0.01ppm
Response Time	60 Seconds
Temp	0 to 40°C
Relative Humidity	15 to 90%

NITROGEN DIOXIDE SENSOR HEAD 0-1PPM

Sensor Code	ENW / ENW2
Range	0-1ppm
Sensor Type	GSE
Minimum Detection Limit	0.005ppm
Accuracy of Factory Calibration	<±0.02 ppm 0-0.2 ppm <±10% 0.2-1 ppm
Resolution	0.001ppm
Response Time	30 Seconds
Temp	0 to 40°C
Relative Humidity	15 to 90%

CARBON MONOXIDE SENSOR HEAD 0-100PPM

Sensor Code	ECN / ECN2
Range	0-100ppm
Sensor Type	GSE
Minimum Detection Limit	0.2ppm
Accuracy of Factory Calibration	<±1 ppm 0-10 ppm <±10% 10-100ppm
Resolution	0.1ppm
Response Time	30 Seconds
Temp	0 to 40°C
Relative Humidity	15 to 90%

CARBON MONOXIDE SENSOR HEAD 0-25PPM

Sensor Code	ECM / ECM2
Range	0-25ppm
Sensor Type	GSE
Minimum Detection Limit	0.05ppm
Accuracy of Factory Calibration	<±0.5 ppm 0-5ppm <±10% 5-25ppm
Resolution	0.01ppm
Response Time	60 Seconds
Temp	0 to 40°C
Relative Humidity	15 to 90%

CARBON MONOXIDE ANALYZER MODULE 0-25PPM

Sensor Code	AQM ECM
Range	0-25 ppm
Sensor Type	GSE
Noise /ppm	0.020
Lower detectable limit /ppm	0.040
Precision	<0.050 ppm 0 to 2 ppm<3% of reading above 2 ppm

CARBON DIOXIDE SENSOR HEAD 0-5000PPM

Sensor Code	CE
Range	0-5000ppm
Sensor Type	NDIR
Minimum Detection Limit	20ppm
Accuracy of Factory Calibration	<±20ppm + 5%
Resolution	1ppm
Response Time	120 Seconds
Temp	0 to 40°C
Relative Humidity	10 to 95%

METHANE SENSOR HEAD 0-10000PPM

Sensor Code	MT / MT2
Range	0-10000ppm
Sensor Type	GSS
Minimum Detection Limit	10ppm
Accuracy of Factory Calibration	<±20ppm + 15%
Resolution	1ppm
Response Time	60 Seconds
Temp	0 to 40°C
Relative Humidity	10 to 90%

PARTICULATE MATTER SENSOR HEAD PM10 / PM2.5

Sensor Code	PM
Range	0.000 to 1.000 mg /m^3
Measurement Parameters	$PM_{2.5}$ and PM_{10}
Sensor Type	Laser particle counter
Minimum Detection Limit	0.001 mg/m3
Accuracy of Factory Calibration	± (0.002 mg/m3 + 15 % of reading)
Resolution	0.001 mg/3
Response Time	5 Seconds
Temp	0 to 40°C
Relative Humidity	0 to 90% non-condensating

Fig. 1 Various sensing devices for air pollution

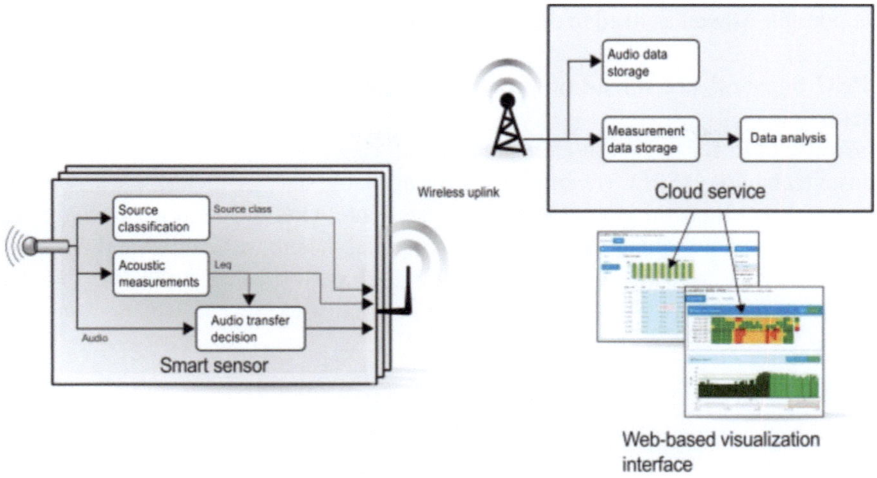

Fig. 2 Noise detection device

Particulate Matter Sensor SPS30 for HVAC and air standard measurement applications

A recent technological development in optical PM sensors is marked by SPS30 particulate matter sensor. It uses Sensirion's novel contamination-resistance technology and laser scattering.

PM2.5 and PM10 can be detected by the SPS30 by enabling the execution of fresh air quality monitoring devices that intercept air pollution damage.

For suspended particulate matter:

PM2.5/PM10 Particle Sensor Analogue Front-End for Air Quality Monitoring Design designed by Texas Instruments—An analogue front-end (AFE) solution for measuring PM2.5 and PM10 particulate matter (PM) is provided by this T1 Design. In order to convert the analogue output into a particle size and for measuring the concentration, a sample software algorithm is provided. Along with software and hardware design files, mulberry pollen, Arizona dust, and cigarette smoke test results were also provided [10].

Methane

Methanogenesis is the process which produces methane around landfill sites. In order to eradicate suffocation and uncontrolled explosion, monitoring at landfills and the surrounding locations is important. Aeroqual manufactures monitors to detect portable methane as shown in Fig. 2.

Atmospheric Carbon Dioxide

Since industrial revolution because of human activity, there is a significant rise in atmospheric CO_2. Industrial processes like combustion of fossil fuels, transportation and electrical generation are main sources of carbon dioxide causing ocean acidification, global warming, etc. Aeroqual manufactures carbon dioxide sensors which are portable as well as fixed to detect the amount of CO_2 in the atmosphere as shown in Fig. 2.

SPEC Sensors' are ultra-low power, high performance, and small-low form factor

Basically used for industrial and safety monitoring screen-printed electrochemical sensor technology (SPEC) revolutionises the available state-of-the-art. These sensors are basically ultra-thin, can be easily integrated into wireless, are portable and have good networking making them ideal for industrial, health, residential and environmental monitoring which gives the benefit of high performance, small size and low cost [11] (Fig. 1).

4 IoT Devices for Monitoring Noise Pollution

Noise is primarily an unwanted sound (>90 dB), a type of pollution that damages the environment and living beings in a vast way. Research shows that humans can endure up to 85 dB. In [12] a real-time, noise pollution detector is proposed where noise can be detected at specific locations and a time-dB Graph [12].

The World Health Organization (WHO) has declared noise pollution as the second cause of ailment due to environmental reasons, after air pollution.

Grove Sound Sensor detects sound strength in the surrounding based on LM386 amplifier and an electret microphone. The output obtained in this module is analogue and Seeeduino can easily sample and test it. The average sound pressure level can be calculated using the value given is the equivalent continuous sound level, with A weighting (LeqA) in a period of time and is used for controlling noise at workplace and the streets [13].

In [14], the noise monitoring system proposed consists of smart sensors which are connected through wireless link to the cloud service. This smart sensor comprises of a single-board computer with a wireless transmission unit and a measuring microphone. Within the measuring time segment, predominant sources of noise are sensed and a weighted 10 min corresponding sound pressure level (Lp,A,600 s) values are calculated continuously. Classification of the sound sources can be utilised to discover the noise source probabilities within the analysis segment.

High-end acoustic hardware manufacturers use noise measuring detectors with active self-testing abilities. Charge Injection Calibration (CIC) is used in the field by Brüel and Kjaer to assess the calibre of an environmental microphone [4]. G.R.A.S makes use of SysCheck, that transmits an extra signal with a recognised voltage level and frequency into the microphone pre-amplifier. Capacitance and resistance of the microphone are measured here [7]. Electrostatic actuator which is placed on a microphone tablet is used in Norsonic's environmental microphones [15].

In [16], they have used a microphone-speaker closed system that helps with measuring the frequency acknowledgement in it and uses this calculation to guess its condition [17].

In [18], a normal Electret Condenser microphone with Arduino is proposed to try and measure the noise pollution level in dB approximately. A "Sound Meter" android application is used to check for correct results. With the advancement of computers and information technology, there has been significant rise in the use and development of electronic devices in medical sciences, and with the unfolding of IoTs, the medical IoT has slowly but steadily penetrated itself into lives of individuals. The medical Internet of Things has been seen as a way in which technology has helped in embedding wireless sensors in medical equipments which then gets combined with the World Wide Web and interacts with patients, hospitals and medical equipments to make use of the new development in the model of modern medical.

5 IoT Devices for Monitoring Water Pollution

In [19], the network of storage and pipelines of the standard water supply system of an urban municipality is taken into account. The suggested system is designed in such a way that it is fixed to intermediate tanks placed underground or ground level reservoirs, covered or opened so that any amount of adulteration can be tracked prior to the time it reaches the consumer side. The system contains an amalgamation of sensors that work in conjunction to come up with a sensing system that can nearly sense almost any kind of contamination that is existing in the water that penetrates into the tank. An alert system and a "Flush On Request" mechanism is proposed to take out the impurities out of the storage tank.

I. pH Sensor

The pH sensor is used to sense the nature (i.e. acidic or basic) of the water present in the tank which includes pH level detection along with sending a corresponding electrical signal to the micro-controller. This is then compared with default values pre-programmed into the micro-controller. The micro-controller uses the 16-bit LCD to display the pH level of the water.

II. Laser Sensor

The laser circuit is designed to yield a stable voltage fed to the micro-controller. If there are impurities present in the water, a change input is given to the micro-controller. The supply to the micro-controller is cut if there are any solid impurities.

III. IR Circuit

Any colour change in water is detected using the IR circuit. The reflection of normal water through the IR circuit is set as the default. If there is a change in the colour of the water from the reflection of normal water, the output generated by the IR circuit alters. This change is detected by the micro-controller and a signal assigned if any colour change is detected is sent to the LCD.

IV. Solenoid Valve

The proposed system uses two solenoid valves in our device. One is used to control the water input into the tank and the other to drain the impure water if necessary. The solenoid valves with GSM module are commanded by the micro-controller.

In [20], they have approached to come up with a sensor for real-time and in-pipe monitoring, evaluation of water standard on the go and to measure the quantity of water transported. A sensor array is developed based on some specific features, with several microsystems for analogue signal logging, conditioning, processing, and presentation of data as well as testing of congregation of heavy metals and Escherichia coli in the water.

Smart Water Sensors to estimate water quality in the sea, lakes and rivers
For simplification of remote water monitoring in 2014, Libelium launched a wireless sensor platform Smart Water. First of its kind Waspmote Smart Water is an autonomous water quality-sensing platform which communicate with the cloud for real-time monitoring. Dissolved oxygen, pH, salinity, temperature, turbidity and dissolved ions (fluoride, calcium, nitrate, chloride, iodide, fluoroborate, ammonia, magnesium, perchlorate, potassium, sodium, etc. are some of the water quality parameters measured.

This platform being ultra-low-power sensor node is designed to be used in rugged environments and deployment in hard-to-access locations in Smart Cities. Smart Water is better with efficiency, accuracy, and low functional costs. It can be used in municipalities for its reliability, autonomous nature and flexibility [21].
Application of Smart Water:

- Potable water monitoring: The existence of micro-organisms in sewage, runoff or factory discharge is detected by the change in dissolved oxygen level. The right healthiness of drinking water is indicated by Turbidity being below 1 NTU.
- Chemical leakage sensing: Chemical spills result in small DO or drastic pH values.
- Swimming pool remote measurement: There should be control over the limit of pH, oxidation-reduction potential (ORP) and Chloride levels swimming pools.
- Level of pollution in sea: Temperature, oxygen, salinity, pH, and nitrates are measured to verify the quality of seawater.
- Corrosion and limescale accumulation prevention: They get deposited in water treatment devices and dishwashers and can be stopped by controlling the hardness of water.
- Fish Farming: By measuring pH, DO, NH_4, NO_3-, NO_2- and temperature the water conditions of aquatic animals such as crayfish, snails, shrimps, fish or prawns in containers can be calculated [40].

New sensor detects contaminants in water in real time:
In [22], researchers like Harold Hemond from MIT and Centre for Environmental Sensing and Modelling (CENSAM) have instituted a sensor named LED-induced fluorescence (LEDIF). Its features include low-cost, compact, compatible with multiple platforms, is capable of 3-D mapping and needs to withdraw only 10–20 ml of water for measurement and detection. This device is able to triumphantly detect six

substances simultaneously while in trial phase, some are fluorescence, absorbance and scattering. These applications help in safeguarding water resources from oil spills, industrial pollution, and dangerous algae growth.

6 IoT Devices for Monitoring Soil Pollution

For scrutinising big areas and measuring soil contamination at high temporal and spatial intervals proximal and remote sensing techniques are essential. An inimitable data stream detecting soil contaminants can be provided by newly developed satellites. Spectroscopy methods have been used to evaluate chosen soil contaminants together with potentially toxic elements and petroleum hydrocarbons from reflectance information [39].

7 IoT Devices for Measuring Other Environmental Factors

- **Temperature**

1. Negative Temperature Coefficient thermistor (NTC): Change in resistance precisely due to variation in temperature working on the principle of thermal sensitivity.
2. Resistance Temperature Detector (RTD): Also called as resistance thermometer an RTD works on correlating the resistance of the RTD element with temperature.
3. Thermocouple: Proportionate transformation in temperature of varying voltage is reflected between two points where metals are attached.
4. Semiconductor-based sensors: Change in temperature can be measured using the temperature-sensitive voltage vs current characteristics. Placed on an IC, it is a semiconductor-based temperature sensor.

Humidity

The relative humidity (moisture and temperature) in the atmosphere is sensed, measured and reported by the humidity sensor. Humidity or dew detectors use capacitive calculation, which is principled on electrical capacitance. A non-conductive polymer film is sandwiched between two metal plates in this sensor. The voltage in the middle of the two plates changes due to the moisture collected on this film and is converted to digital values to show the status of moisture in the air.

Three basic types of humidity sensors are capacitive, resistive and thermal.

Capacitive: A thin piece of metal oxide is laid between two electrodes to measure the relative humidity in a capacitive humidity sensor. With the change of the air's relative humidity, the electrical capacity of the metal oxide changes.

Resistive: The principle of measurement of electrical impedance of atoms by utilising ions in salts is the base of resistive humidity sensors.

Thermal: Based on humidity of atmosphere, two thermal detectors conduct electricity. Of these two, one is encased in dry nitrogen whereas another is used to measure ambient air and their difference gives humidity.

HDC2080 (http://www.ti.com/product/HDC2080)

Integrated humidity and temperature sensor, this provides with high accuracy as well as very low power consumption in small DFN package.

HDC2010 (http://www.ti.com/product/hdc2010)

Low Power Humidity and Temperature Digital Sensor

HDC1080 (http://www.ti.com/product/hdc1080)

Low Power, High Accuracy Digital Humidity Sensor along with Integrated Temperature Sensor in dust-resistant package.

Pressure

Piezoelectric pressure sensors can further be classified according to whether the crystal's electrostatic charge, its resistivity, or its resonant frequency electrostatic charge is measured. Depending on which phenomenon is used, the crystal sensor can be called electrostatic, piezoresistive, or resonant.

Types of Pressure Sensors

Based on the type of applications they are used in, pressure sensors can be categorised into many types. However, following are most common types of pressure sensors that have been widely used:

1. **Strain Gauge Type:** A voltage would be generated proportional to every deviation from the normal balanced condition, so every single compression or expansion movement of the diaphragm will produce an output indicating a change in pressure conditions. Since resistance change is the main cause for potential difference, these sensors are also termed as piezo-resistive type of pressure sensors.
2. **Capacitive Pressure Sensor:** These sensors, though much ineffective at high temperatures, are widely used at ambient temperature range due to their linear output.
3. **Piezoelectric Pressure Sensor:** When subjected to mechanical pressure, Piezoelectric crystals formulate a potential difference (i.e. voltage is induced across the surfaces). Voltage is produced when a pressure is generated by the sensor as it pushes the attached shaft down that pressurises the crystal.

8 Conclusion

The Internet of Things is nothing but an integration of sensors attached to various objects with the Internet in order to provide data to the Internet along with using the already available data from the Internet. This to and fro relationship can extensively be experimented for the betterment of human health. The various devices explained above (which are available) are helping with the measurement of various pollutants in the environment. The main aim of this work was to give a comprehensive view of IoT usage in pollution detection and to report the extensive range of gadgets and tools available as well as proposed. In this survey paper, we have given priority to both research works and commercial devices to study and investigate the currently available and future technologies.

References

1. Environmental Monitoring through Embedded System and Sensors
2. https://www.finoit.com/blog/top-15-sensor-types-used-iot/
3. "FilesThruTheAir," FilesThruTheAir, [Online]. Available: https://www.filesthrutheair.com. [Accessed 15 02 2017]
4. Smart pollution detection and tracking system embedded with AWS IOT cloud. Int J Adv Res Comput Sci Softw Eng
5. https://pubs.acs.org/doi/abs/10.1021/ac9812429
6. https://www.bosch-sensortec.com/bst/applicationssolutions/iotsmarthome/overview_iot-smarthome
7. Solomon GM, Campbell TR, Feuer GR, Masters J, Samkian A, Paul KA (2001) No breathing in the aisles, diesel exhaust inside school buses. Report prepared by the natural resources defense council and coalition for clean air. http://www.nrdc.org/air/transportation/schoolbus/schoolbus.pdf
8. https://www.aeroqual.com/product/nitrogen-dioxide-sensor-0-1ppm
9. https://www.aeroqual.com/product/sulfur-dioxide-sensor-0-10ppm
10. https://www.digikey.in/en/product-highlight/s/spec-sensors/carbon-monoxide-sensorsutm_adgroup=General%26mkwid=s6OMlQvGj%26pcrid=275741081675%26pkw=%26pmt=b%26pdv=c%26productid=%26%26gclid=CjwKCAjwwo7cBRBwEiwAMEoXPIVDZx2Kri83tOQws2pMjfjtm1ZV4cUZt1hbSTTJeBPVjKtAKZVW4hoCYigQAvD_BwE
11. Pollution control using internet of things (IoT). In: 2017 8th annual industrial automation and electromechanical engineering conference (IEMECON)
12. http://www.libelium.com/new-sound-level-sensor-to-control-noise-pollution/
13. Environmental noise monitoring using source classification in sensors
14. https://www.ncbi.nlm.nih.gov/pmc/articles/PMC3892865/ Active self-testing noise measurement sensors for large-scale environmental sensor networks
15. Chung WY, Oh SJ (2006) Remote monitoring system with wireless sensors module for room environment. Sens Actuators B Chem 113(1):64–70. ISSN 0925-4005. https://doi.org/10.1016/j.snb.2005.02.023
16. Tsow F, Forzani E, Rai A, Wang R, Tsui R, Mastroianni S, Knobbe C, Gandolfi AJ, Tao NJ (2009) A wearable and wireless sensor system for real-time monitoring of toxic environmental volatile organic compounds. Sens J IEEE 9(12):1734–1740, Dec 2009. https://doi.org/10.1109/jsen.2009.2030747
17. https://circuitdigest.com/microcontroller-projects/arduino-sound-level-measurement

18. Advanced water impurity detection system. Int J Innov Res Sci Eng Technol. https://www.ijirset.com/upload/2017/ncacces/45_NCACCES_TK003-007-PC03_m.pdf
19. Automated sensor network for monitoring and detection of impurity in drinking water. Int J Res Appl Sci Eng Technol (IJRASET) Syst. https://www.ijraset.com/fileserve.php?FID=1615
20. http://news.mit.edu/2014/new-sensor-detects-contaminants-in-water-in-real-time
21. https://www.researchgate.net/publication/323689485_Monitoring_of_Selected_Soil_Contaminants_using_Proximal_and_Remote_Sensing_Techniques_Background_State-of-the-Art_and_Future_Perspectives
22. http://www.libelium.com/smart-water-sensors-to-monitor-water-quality-in-rivers-lakes-and-the-sea/#!prettyPhoto
23. "smart-me," smart-me, [Online]. Available: https://smart-me.com/Description/HowItWorks.aspx. [Accessed 10 02 2017]
24. Doraiswamy P, Davis WT, Miller TL, Fu JS, Lam YF (2005) Measuring air pollution inside and outside of diesel truck cabs. Report prepared for US EPA by University of Tennessee, Knoxville TN. http://www.epa.gov/smartway/documents/publications/incabairquality-110405.pdf
25. https://www.ozonesolutions.com/journal/2013/ozone-sensors-technology-comparison/
26. https://www.safewise.com/resources/carbon-monoxide-detectors-guide
27. http://www.ti.com/lit/ug/tidub65c/tidub65c.pdf
28. Real-time air quality monitoring system for Bangladesh's perspective based on internet of things. In: 3rd international conference on electrical information and communication technology (EICT), Khulna, Bangladesh, 7–9 Dec 2017
29. Monitoring pollution: applying IoT to create a smart environment. In: International conference on electrical and computing technologies and applications (ICECTA)
30. IoT device used for air pollution campaign to encourage cycling habit in inverleith neighborhood. In: 2017 international conference on information management and technology (ICIMTech)
31. Distributed system as internet of things for a new low-cost, air pollution wireless monitoring on real time. In: IEEE/ACM 19th international symposium on distributed simulation and real time applications
32. Development of an IoT-based atmospheric environment monitoring system. In: 2017 international conference on information and communication technology convergence (ICTC)
33. Design of air quality meter and pollution detector. In: 2017 8th annual industrial automation and electromechanical engineering conference (IEMECON)
34. Smart IoT based system for vehicle noise and pollution monitoring. In: International conference on trends in electronics and informatics ICEI, 2017
35. Smart industry pollution monitoring and controlling using LabVIEW based IoT. In: 2017 3rd international conference on sensing, signal processing and security (ICSSS), IEEE
36. Identifying high pollution level regions through a terrestrial mobile monitoring system. In: 2017 region 10 symposium (TENSYMP), IEEE
37. https://www.skyfilabs.com/project-ideas/noise-pollution-detector
38. Monitoring of selected soil contaminants using proximal and remote sensing techniques: background, state-of-the-art and future perspectives computing and communications and IEEE internet of things and IEEE cyber, physical and social computing, IEEE, 2053–2058 Aug 2013

Suspicious Event Detection in Real-Time Video Surveillance System

Madhuri Agrawal and Shikha Agrawal

Abstract In today's generation, video security is becoming more important in the real-world applications because of the happening of suspicious events in our surroundings, and the safety and security in public places have become a priority. Video surveillance system might be used for enhancing the security in various areas such as offices, mall, theater, organizations, analysis of athletic events, content-based image storage and retrieval and many more. This paper focused research on the automatic analysis of suspicious event detection in the real-time video surveillance system and provided a recommendation on how it can be monitored automatically.

Keywords Video surveillance · Suspicious event detection · Video monitoring · Video security · Anomaly detection

1 Introduction

Video security is becoming more important nowadays not only in the security monarchy but also in our daily life in restaurants, malls, banks, universities, colleges, schools, etc., due to the suspicious events in our surroundings [1]. The video surveillance system is a process of analyzing and monitoring of video sequences. The main purpose behind this is to monitor and identify the information from a video sequence like checking the behavior, activities and other valuable information through the system [2]. The applications of video surveillance are used for monitoring the actual time for an event and to detect the suspicious actions from that event [3]. A 'suspicious event' is an activity that occurs sometimes or which has not been viewed before or an event which is not predictable. The major challenge in suspicious event detection is the proper classification between a suspicious event and a normal event because most of the scene related to the normal event while resting of the scene related to the suspicious event which constitutes a small portion of the whole scenes.

M. Agrawal (✉) · S. Agrawal (✉)
UIT, RGPV, Bhopal, Madhya Pradesh, India
e-mail: madhuriagrawal2000@gmail.com

S. Agrawal
e-mail: shikha@rgtu.net

R. K. Shukla et al. (eds.), *Social Networking and Computational Intelligence*, Lecture Notes in Networks and Systems 100, https://doi.org/10.1007/978-981-15-2071-6_40

The video surveillance system has low intelligence, cannot make decisions. Data mining applications can help us to make decision automatically to recognize video solutions, which improves the intelligence of video surveillance systems applications. Anomaly detection is the analysis of video data and identifies the suspicious event by using the techniques of data mining. It can be used to detect a deviation in behavior or actions and need a model to automate the decision-making process. Mostly, suspicious event detection methods depend on the anomaly detection techniques. There are three types of anomaly detection methods: supervised, semi-supervised and unsupervised methods. In the anomaly detection method, a model is developed by normal events using unsupervised learning. By using this model, if an event deflects from its normal, result is called a suspicious event.

Not only the decision-making process but also sentiment analysis as well should be automated. The automatic identification of human mind states especially private state of mind like emotions, sentiments, behaviors, beliefs and opinions is the sentiment analysis. It classifies data broadly into three categories that are a positive, negative and neutral category. It determines the sentiment polarity of the data, and it will help to find a deviation in behavior. Today, sentiment analysis is restricted to text-based only. So, it is extremely recommended from multiple modalities to mine emotions, opinions and identify sentiments. Visual data within a video deliver the facial expressions, which are used to analyze the private state of the user mind. For sentiment analysis video, data may be a fair source but a major challenge has to be solved of articulating the opinions which vary from one person to another.

To tell the truth, video surveillance allows to monitor and provide confidence of security but it is also true to say "open street video surveillance is adaptable but not a universal cure." When a video surveillance system along with various preventative measures is bundled in a package works effectively. Media may play a role in advertising a system which helps in expanding public knowledge and minimizing terror of crime. It may help to maintain prolong criminal vigilance and provide a feeling of public security at public places.

A good example for video surveillance published in July 2014 Mumbai city, to monitor cleanliness the ministry of Railways directs compulsory checking of contractual labor presence as well cleaning railway stations at regular intervals, will be monitored through closed circuit television cameras. The footage will act as proof of the work done and is monitored by the station master as well by senior officers by using Internet Protocol (IP)-based cameras. Another one example is from Panaji, Goa, in July 2014 "On Goa beaches, tourists are under closed circuit television surveillance." Entire security arrangements on the beaches area are reviewed and patrolled by India Reserve Battalion (IRB). Even Metro may adopt a security camera system with artificial intelligence in near future. To strengthen safety measures at educational institutes, the University Grants Commission (UGC) has directed universities and colleges across the country in July 2015, to install closed circuit cameras on their campuses. The UGC directed them to identify the busiest spots on the university campus and bring them under surveillance.

In India, closed circuit television cameras are going to be installed daily in large numbers, but the problem is either there is no monitoring or they are going to be monitored manually and as if monitoring is done by some manpower, having a disadvantage that any person can not able to do twenty-four cross seven hours in a day monitoring. Even if we make it possible, due to security issues, there is a requirement of research in this field to implement such a method that performs video surveillance automatically not manually. There are so many places where closed circuit television cameras are installed, but due to the lake of manpower, the monitoring is not going to be implemented. For this, it is a need for an automotive decision-making system by which closed circuit television has to be monitored.

2 Challenges in Real-Time Video Surveillance System

There are so many challenges in the real-time video surveillance system [4–7]. Few have mentioned below but not limited to the given list.

1. Preprocessing.
2. Feature extraction.
3. Less available training data set.
4. Detection of one or more moving objects from a video sequence.
5. To detect the topological changes of a moving region.
6. Segmentation of moving objects.
7. Less-efficient video tracking algorithms.
8. Foreground and background detection.
9. Human re-identification problem.
10. Identification of anomaly patterns.
11. Differentiation between normal behavior and suspicious behavior.

3 Related Work

A video process is represented by scene, shot and frame which is hierarchical structure units. A shot can be continuous action images or video frames. It has two types abrupt or gradual. Video shot is the smallest indexing unit in which no changes are perceived in the scene content. It is indexed by the temporal and spatial features which are represented by keyframes [8]. For identifying moving object detection and tracking in videos, several image features are used as color, shape, texture, contours and motion to track moving objects. There are several important methods, such as first, background subtraction, which is advanced of traditional method frame differencing. Second, mean shift uses the clustering concepts for the segmentation of images or identifying features. Third, mean shift filtering, it is a smoothing method having a non-parametric model. Fourth, continuously adaptive mean shift, which is used for identifying tracking. It is the successor of the mean shift. Fifth, optical flow which highlights brightness based on the pixel level and sixth trackers: sparse and dense based on optical flow for detection [8].

For abandoned object detection, tracking-based approach is unreliable due to occlusions, change in light and other factors which make it complex. To handle such complexity, lots of improvement is implemented on shadow removal, fragment reduction, light changing adaptation and stable update rate with different frame rates in the video stream. To detect abandoned and removed objects, a matching method is modeled by using three Gaussian mixtures [9].

For multiple objects detection, tracking is handled by using an adaptive Gaussian mixture model, tracking algorithm and particle filtering which uses fuzzy techniques for feature estimation. It can handle color video image sequences [10]. Nighttime bright objects can be identified by image segmentation using an automatic multilevel histogram threshold to extract object which is efficient in nighttime illumination conditions [11]. People counting can be done based on information provided by an overhead stereo system. An extended particle filter is used to detect and track human motion. For robustness, 3-D measurement is taken, the *k*-means algorithm is applied for the number of iterations, and then, trajectory generation can be used for counting people in different directions even in occlusion [12].

In video surveillance systems, behavior recognition may play a vital role, but it is ambiguous and uncertain. Analysis of human behavior is very difficult because the same behavior may have different meanings depending upon different situations. For effective results, behavior recognition should be clubbed with the uncertainty reasoning model with low computational complexity and robustness in nature [3].

In behavior analysis, the identification, tracking and monitoring activities of humans, objects, motion, for instance, help in understanding and enhancing new applications of surveillance through indoor and outdoor monitoring for security [13]. Moreover, understanding of non-verbal behaviors is a challenging task to forecast the everlasting values of emotion at each moment in time. For this reason, it requires human emotion recognition, gesture recognition and facial recognition all has to be analyzed.

Chao et al. [14] proposed a model to recognize emotion states from audio and visual modalities by utilizing deep belief network. It encodes the information in features and predicts the decision from various modalities and emotion sensual context information which improves the efficiency of key points in the method. In the past few years back, it is performed only on limited controlled lab data. For automatic emotion recognition in a wild sample, SVM-Hidden Markov model is used for training and classification which involves a blend of optical flow, facial features, audio features and Gabor filtering which produce a better result in overall accuracy parameter [15]. Emotion recognition can be done through Ekman emotion classification. It works in sentence level. It employs word lexicon, lexicons of emotions, common abbreviations and some set of rules. It is having high accuracy [2]. It recognizes the type of feeling such as happiness or sadness.

In human emotion recognition, emotions are recognized through facial expressions as well in sentiment analysis, and sentiments are also recognized through facial expressions. Moreover, if sentiment analysis has been added to it, may provide more accurate analysis [16]. Fully automated facial expression detection and classification can be done by face detector and classifier with support vector machine [17].

In gesture recognition, a template-based approach provides accurate performance. It measures the similarity between a template set and a user gesture by using the nearest neighbor classifier in Gestimator, which combines shape and stoke-based similarity in sequential classification [18]. It is important because people communication is possible only with the gesture.

Looking toward different aspects of the problems associated with video surveillance, it is very much required and important to preserve the privacy of a human being. Privacy is a human right and protection of privacy, while monitoring video surveillance is the current need for real-time video surveillance. Privacy has to be maintained in the video surveillance which requires privacy-preserving mechanisms. There are various mechanisms given by different researchers which are concentrated in providing specific part modification, using sizeable pixels to look like a black box for hiding specific part, using encryption and decryption with video modification techniques [19]. This privacy-preserving mechanisms must have to ensure that the privacy of innocent and authorized human is always to be protected, and suspicious human has to be revealed in any event detection.

A real-time video surveillance system is a good resource for monitoring a wide range of peculiar crimes in open places such as parking areas, public parks, streets in city centers, stations, etc. It is a solution to the crime but with another point of view for own personal privacy, it is invading. This requires some constitutional and privacy concerns for the surveillance system. The main aim of video surveillance is to reduce suspicious events by monitoring. This operation of monitoring can be handled by either police operators or civilian operators of the local authority, but for police, it is hard to involve with the local authority for security reasons, and for civilians, it is not trustworthy. So, integrated of both two may provide a good solution. Command and control system should be in hands of police. Actually, direct communication links from the video surveillance control location to local police will work effectively.

4 Suspicious Event Detection Through Anomaly Detection

Existing techniques are grouped into different categories depending upon the approach used in techniques to differentiate between normal and suspicious event. It aims to detect previously unobserved events in the video. Nature of input data is an event which is a vector. It can be defined using a set of dimensions. Each vector may consist of a single attribute or more than one attribute [20]. Attribute nature determines the technique which is used in the application. There are three types of anomalies. First, point anomaly based on only one feature. Second, a conditional anomaly. It is used majorly in the spatial data set and time-series data set, and third is collective anomaly using an entire data set. The data label which linked with vector decides whether the event is normal or suspicious. Based on labels, anomaly detection can be divided into three different types: supervised having training data set, semi-supervised used for normal class and unsupervised required no training data set. The aspects in which manner they are reported are the output of anomaly. That is scores and labels.

Table 1 Strengths and weakness of various techniques

Author	Techniques	Strengths	Weakness
Candamo et al. [3]	► For motion detection, background subtraction, optical flow and temporal differencing ► For object classification: shape and motion-based classification ► Color-based tracking	► Human behavior recognition for the prevention of incidents ► Behaviors for event and activity are divided into four groups	► Poor resolution ► Inefficient hardware availability
Chao et al. [14]	► Multi-scale temporal modeling ► Regression model using deep belief network	► Shape and appearance features are used ► Using temporal pooling function to include dynamic information in the feature level	► With an increase in window size, initially results increased and then keep stable at 13
Krishna et al. [1]	► Classification is performed using the SVM-Hidden Markov model ► For emotion detection blend of optical flow, Gabor filtering is used	► Data set consists of movie clips making a real-life scenario	► Less accuracy of only 20% because of the unavailability of fiducial points in facial regions
Pradeep et al. [21]	► Anomaly detection using unsupervised learning ► For moving object detection, background modeling (static or dynamic) using recursive and non-recursive techniques	► Proposes video surveillance systems architecture which is a combination of image processing within the digital video through the networked surveillance system ► For moving object detection, concluded that detection of moving pixels is easy and fast in the temporal differencing method	► For moving object detection, the temporal differencing method fails to detect the changes in the two or more consecutive frames and leaves some hole in the foreground objects when moving object stops
Little et al. [22]	► For classification modern learning, histograms are used as input to SVM with radial basis function	► Developed integrating generic and specific analysis tool ► An interaction between monitors of VSS and ML experts	► Low temporal segmentation of video on performance

For video surveillance, different authors focused on various techniques with their strengths and weakness which are tabulated as follows (Table 1).

The main aspect of real-time video surveillance systems includes the identification of human from the other objects by using size, shape or movement. Second, detection of the background scene and foreground objects in case of motion also. Third, track multiple objects in a group of people. Fourth, prepare a model for each person to identify the person after occlusion. Fifth, detect and track main body parts like the head, hands and legs of a person. Sixth, classify an object from the other object. It may be a single object or multiple objects along with tracking. Seventh, object recognition from one frame to another. Eighth, the count number of people in the event. Ninth, perform analysis of the recognized event by applying sentiment analysis, gesture recognition, emotion recognition, facial expression detection and human behavior recognition methods. Tenth, identify the suspicious event from the normal events and finally buzz an alarm if suspicious event identified for a security alert.

5 Conclusion and Future Enhancement

Now, video surveillance systems have little intelligence. As the human mind is having high quality to take a decision, thus to monitor the system we required people. In India, video surveillance systems are going to be installed daily in large numbers, but the problem is either there is no monitoring, or they are going to be monitored manually, and as if monitoring is done by some man power, has a drawback that any person cannot able to do monitoring twenty-four cross seven hours in a day. Even if we make it possible, due to security issues, there is a requirement of research in this field to implement such a method that performs video surveillance automatically not manually. There are so many places where video surveillance systems are installed, but due to the lake of manpower, the monitoring is not going to be implemented. Due to the high demand for intelligence, there is a need for an automotive decision-making systems in video surveillance.

References

1. Popoola OP, Wang K (2012) Video-based abnormal human behavior recognition-A review. IEEE Trans Syst Man Cybernetics Part C Appl Rev 42(6):865–878
2. Krcadinac U, Pasquier P, Jovanovic J, Devedzic V (2013) Synesketch: an open source library for sentence-based emotion recognition. IEEE Trans Affect Comput 4(3):312–325
3. Candamo J, Shreve M, Goldgof DB, Sapper DB, Kasturi R (2010) Understanding transit scenes: a survey on human behavior-recognition algorithms. IEEE Trans Intell Transp Syst 11(1):206–224
4. Kalaiselvan C, SivananthaRaja A (2012) Investigation on tracking system for real-time video surveillance applications. In: CUBE 2012, ACM, Pune, Maharashtra, India, pp 108–112
5. Kao LJ, Huang YP (2011) An efficient strategy to detect outlier transactions for knowledge mining. IEEE, pp 2670–2675

6. Li Y, Wu Z, Karanam S, Radke RJ (2014) Real-world re-identification in an airport camera network. ACM, ICDSC'14, Venezia Mestre, Italy
7. Liu H, Schneider M (2011) Tracking continuous topological changes of complex moving regions. ACM, SAC'11, Taichung, Taiwan, pp 833–838
8. Karasulu B, Korukoglu S (2013) Moving object detection and tracking in videos. Springer, Performance Evaluation Software, pp 7–30
9. Tian Y, Feris RS, Liu H, Hampapur A, Sun MT (2011) Robust detection of abandoned and removed objects in complex surveillance videos. IEEE Trans Syst Man Cybernetics Part C Appl Rev 41(5):565–576
10. Thomas V, Ray AK (2011) Fuzzy particle filter for video surveillance. IEEE Trans Fuzzy Syst 19(5):937–945
11. Chen YL, Wu BF, Huang HY, Fan CJ (2011) A real-time vision system for nighttime vehicle detection and traffic surveillance. IEEE Trans Ind Electron 58(5):2030–2044
12. García J, Gardel A, Bravo I, Lázaro JL, Martínez M (2013) Tracking people motion based on extended condensation algorithm. IEEE Trans Syst Man Cybern Syst 43(3):606–618
13. Ahmad I, He Z, Sinica A, Sun MT (2008) Special issue on video surveillance. IEEE Trans Circuits Syst Video Technol 18(8):1001–1005
14. Chao L, Tao J, Yang M, Li Y, Wen Z (2014) Multi-scale temporal modeling for dimensional emotion recognition in video. ACM, AVEC'14, Orlando, Florida, USA, pp 11–18
15. Krishna T, Rai A, Bansal S, Khandelwal S, Gupta S, Goyal D (2013) Emotion recognition using facial and audio features. ACM, ICMI'13, Sydney, Australia, December 9–13
16. Poria S, Cambria E, Howard N, Huang GB, Hussain A (2015) Fusing audio, visual and textual clues for sentiment analysis from multimodal content. Elsevier, Neurocomputing 50–59
17. El Meguid MKA, Levine MD (2014) Fully automated recognition of spontaneous facial expressions in videos using random forest classifiers. IEEE Trans Affect Comput 5(2):141–154
18. Ye Y, Nurmi P (2015) Gestimator—shape and stroke similarity based gesture recognition. ACM, ICMI 2015, Seattle, WA, USA, pp 219–226
19. Zhang P, Thomas T, Emmanuel S, Kankanhalli MS (2010) Privacy-preserving video surveillance using pedestrian tracking mechanism. ACM, MiFOR'10, Firenze, Italy, pp 31–36
20. Chandola V, Banerjee A, Kumar V (2009) Anomaly detection: a survey. ACM Comput Surv 41(3):1–58
21. Verma KK, Kumar P, Tomar A (2015) Analysis of moving object detection and tracking in video surveillance system. In: 2nd international conference on computing for sustainable global development, IEEE, pp 1758–1762
22. Little S, Clawson K, Nieto M (2013) An information retrieval approach to identifying infrequent events in surveillance video. ACM, ICMR'13, Dallas, Texas, USA., 16–20 April 2013, pp 223–230

Time Moments and Its Extension for Reduction of MIMO Discrete Interval Systems

A. P. Padhy and V. P. Singh

Abstract This paper extends the reduction procedure available for reducing the order of single-input-single-output (SISO) discrete interval system (DIS) to the reduction of multi-input-multi-output (MIMO) DISs. The methodology utilizes the time moments (TMs) for deriving the numerator of model. The paper also provides the methods available for calculating the different expressions for TMs. For analyzing this technique, a two-input-two-output (TITO) DIS is considered for reducing the order of DIS. The whole study is performed in two phases. The denominator of discrete interval TITO model is obtained by clustering the poles of the given higher-order DIS at first. Then, the numerators of discrete interval TITO model are derived by matching the (TMs). The results are presented in the form of time-domain responses. Also, a comparative analysis is carried out using these responses.

Keywords Model order reduction · Discrete interval system · Padé approximation · Multi-input-multi-output system · Dominant poles

1 Introduction

The mathematical analysis and design of controllers for a plant having higher-order transfer function are complex tasks. In such cases, it is necessary to reduce the order of the system using appropriate order reduction techniques. Several order reduction techniques have been proposed earlier for both non-interval continuous and discrete-time systems [1–3].

The performance of physical systems is affected by various types of uncertainties. Uncertainties in the system may arise due to lack of knowledge about the parameter variations, noise, disturbances, etc. For uncertain parameter variations, it is better to obtain transfer function with coefficients in certain ranges rather than fixed

A. P. Padhy (✉) · V. P. Singh
National Institute Technology, Raipur, India
e-mail: adityappadhy@gmail.com

V. P. Singh
e-mail: vinaymnnit@gmail.com

R. K. Shukla et al. (eds.), *Social Networking and Computational Intelligence*, Lecture Notes in Networks and Systems 100, https://doi.org/10.1007/978-981-15-2071-6_41

coefficients. However, systems with such kind of coefficients are termed as interval systems.

Numerous order reduction techniques are available for reducing the order of continuous interval systems (CISs). Out of which, the most popular method is Routh-Padé approximation [4] where the Routh table and expansion coefficients (like time moments) are utilized to obtain the model of CIS. This technique is a direct extension of Routh-Padé approximation earlier proposed for fixed-coefficient systems [5]. Another important work appeared for reduction of CISs is gamma-delta method [6] which is the extended form of alpha-beta method [7] proposed for fixed-coefficient systems. In continuation of works proposed in [4] and [6], Sastry et al. [8] suggested that only gamma table is sufficient to obtain the model of CISs. Furthermore, many improved and mixed techniques are reported in the paper for order reduction of CISs. The important methods of them are the factor division method [9] and Kharitonov's polynomial method [10].

Many methods [11–16] are proposed for reducing the order of DISs. But, few of them are extensions of methods proposed for CISs. The work presented in [11] is a direct extension to DISs. Similarly, extension from CISs to DISs using gamma-delta technique via bilinear transformation is noticed in [6]. Recently, Choudhary and Nagar [12] proposed a improved method for order reduction for uncertain z-domain systems using conventional Routh array. Another methodology is used in [13] where Routh array is employed to form two tables for order reduction. Furthermore, the works [14, 15, 17] present a mixed methods for diminishing the order of DISs.

Among the numerous techniques proposed in the literature, time moment matching (TMM)-based methods grabbed greater attention due to its computational simplicity. In [18], new expressions for derivation of (TMs) and Markov parameters (MPs) are presented for reduction of (CISs). However, in [19], authors tried to simplify the computation of TMs by using direct series expressions. Recently, in [20], Singh et al. proposed mathematical expressions for calculating the (TMs) and (MPs) in which there is no need of denominator inversion of the system transfer function. Similarly, the TMs for DISs are proposed in [21].

In this present work, the reduction technique [16, 21] available for SISO discrete interval system is extended for the reduction of MIMO DISs. For this study, a TITO system is considered for order reduction and system is analyzed in two steps: At first, the denominator of discrete interval model (DIM) is computed by clustering of poles of the given higher-order DIS and secondly the numerators of discrete interval model are calculated by matching the TMs of system and model. The results obtained are presented in the form of time response of various systems and models. A comparative analysis is also performed using time-domain responses.

In this article, Sect. 2 describes the problem formulation. Time moments of interval systems are illustrated in Sect. 3, clustering of poles method is discussed in Sect. 4. A MIMO test system is provided in Sect. 5 to validate the above technique. Finally, the conclusion of this paper is given in Sect. 6.

2 Mathematical Formulation

2.1 For Continuous Interval Systems

The higher-order MIMO interval systems (ISs) can be represented as

$$G(s) = \begin{bmatrix} G_{11}(s) & G_{12}(s) \\ G_{21}(s) & G_{22}(s) \end{bmatrix} \tag{1}$$

The transfer functions are given by

$$G_{11}(s) = \frac{[p_0^-, p_0^+]_{11} + [p_1^-, p_1^+]_{11}s + \cdots + [p_{n-1}^-, p_{n-1}^+]_{11}s^{n-1}}{[q_0^-, q_0^+]_{11} + [q_1^-, q_1^+]_{11}s + \cdots + [q_n^-, q_n^+]_{11}s^n} \tag{2}$$

$$G_{12}(s) = \frac{[p_0^-, p_0^+]_{12} + [p_1^-, p_1^+]_{12}s + \cdots + [p_{n-1}^-, p_{n-1}^+]_{12}s^{n-1}}{[q_0^-, q_0^+]_{12} + [q_1^-, q_1^+]_{12}s + \cdots + [q_n^-, q_n^+]_{12}s^n} \tag{3}$$

$$G_{21}(s) = \frac{[p_0^-, p_0^+]_{21} + [p_1^-, p_1^+]_{21}s + \cdots + [p_{n-1}^-, p_{n-1}^+]_{21}s^{n-1}}{[q_0^-, q_0^+]_{21} + [q_1^-, q_1^+]_{21}s + \cdots + [q_n^-, q_n^+]_{21}s^n} \tag{4}$$

$$G_{22}(s) = \frac{[p_0^-, p_0^+]_{22} + [p_1^-, p_1^+]_{22}s + \cdots + [p_{n-1}^-, p_{n-1}^+]_{22}s^{n-1}}{[q_0^-, q_0^+]_{22} + [q_1^-, q_1^+]_{22}s + \cdots + [q_n^-, q_n^+]_{22}s^n} \tag{5}$$

where $[p_i^-, p_i^+]$ for $i = 0, 1, \ldots, (n-1)$ and $[q_i^-, q_i^+]$ for $i = 0, 1, \ldots, n$ are the interval coefficients. The power series expansions of the DISs, Eqs. (2)–(5), around $s = 0$ can be written as

$$G_{kl}(s) = (T_0)_{kl} + (T_1)_{kl}s + \cdots + (T_i)_{kl}s^i + \cdots \text{ (expansion around } s = 0) \tag{6}$$

where k and l represent subscript of individual transfer functions, i.e. for G_{11}, the values of k and l becomes 1 and 1, respectively. $(T_i)_{kl} = \left(\left[T_i^-, T_i^+\right]\right)_{kl}$ for $i = 0, 1 \ldots$ represent the TMs of the ISs.

Further, the reduced-order model (ROM) becomes

$$\widehat{G}(s) = \begin{bmatrix} \widehat{G}_{11}(s) & \widehat{G}_{12}(s) \\ \widehat{G}_{21}(s) & \widehat{G}_{22}(s) \end{bmatrix} \tag{7}$$

The transfer functions are provided by

$$\widehat{G}_{11}(s) = \frac{\left[\widehat{p}_0^-, \widehat{p}_0^+\right]_{11} + \left[\widehat{p}_1^-, \widehat{p}_1^+\right]_{11}s + \cdots + \left[\widehat{p}_{m-1}^-, \widehat{p}_{m-1}^+\right]_{11}s^{m-1}}{\left[\widehat{q}_0^-, \widehat{q}_0^+\right]_{11} + \left[\widehat{q}_1^-, \widehat{q}_1^+\right]_{11}s + \cdots + \left[\widehat{q}_m^-, \widehat{q}_m^+\right]_{11}s^m} \tag{8}$$

$$\widehat{G}_{12}(s) = \frac{\left[\widehat{p}_0^-, \widehat{p}_0^+\right]_{12} + \left[\widehat{p}_1^-, \widehat{p}_1^+\right]_{12} s + \cdots + \left[\widehat{p}_{m-1}^-, \widehat{p}_{m-1}^+\right]_{12} s^{m-1}}{\left[\widehat{q}_0^-, \widehat{q}_0^+\right]_{12} + \left[\widehat{q}_1^-, \widehat{q}_1^+\right]_{12} s + \cdots + \left[\widehat{q}_m^-, \widehat{q}_m^+\right]_{12} s^m} \tag{9}$$

$$\widehat{G}_{21}(s) = \frac{\left[\widehat{p}_0^-, \widehat{p}_0^+\right]_{21} + \left[\widehat{p}_1^-, \widehat{p}_1^+\right]_{21} s + \cdots + \left[\widehat{p}_{m-1}^-, \widehat{p}_{m-1}^+\right]_{21} s^{m-1}}{\left[\widehat{q}_0^-, \widehat{q}_0^+\right]_{21} + \left[\widehat{q}_1^-, \widehat{q}_1^+\right]_{21} s + \cdots + \left[\widehat{q}_m^-, \widehat{q}_m^+\right]_{21} s^m} \tag{10}$$

$$\widehat{G}_{22}(s) = \frac{\left[\widehat{p}_0^-, \widehat{p}_0^+\right]_{22} + \left[\widehat{p}_1^-, \widehat{p}_1^+\right]_{22} s + \cdots + \left[\widehat{p}_{m-1}^-, \widehat{p}_{m-1}^+\right]_{22} s^{m-1}}{\left[\widehat{q}_0^-, \widehat{q}_0^+\right]_{22} + \left[\widehat{q}_1^-, \widehat{q}_1^+\right]_{22} s + \cdots + \left[\widehat{q}_m^-, \widehat{q}_m^+\right]_{22} s^m} \tag{11}$$

where m is represented as the order of model such that $m < n$ and $\left[\widehat{p}_i^-, \widehat{p}_i^+\right]$ for $i = 0, 1 \ldots, (m-1)$ and $\left[\widehat{q}_i^-, \widehat{q}_i^+\right]$ for $i = 0, 1, \ldots, m$ are the interval coefficients.

The power series expansion of the models, Eqs. (8)–(11), around $s = 0$ can be written as

$$\widehat{G}_{kl}(s) = \left(\widehat{T}_0\right)_{kl} + \left(\widehat{T}_1\right)_{kl} s + \cdots + \left(\widehat{T}_i\right)_{kl} s^i + \cdots \text{(expansion around } s = 0) \tag{12}$$

where k and l represent subscript of individual transfer functions. $\left(\widehat{T}_i\right)_{kl} = \left(\left[\widehat{T}_i^-, \widehat{T}_i^+\right]\right)_{kl}$ for $i = 0, 1 \ldots$ denote the TMs of the ISs.

2.2 *For Discrete Interval Systems*

Consider a stable MIMO discrete interval transfer function

$$H(z) = \begin{bmatrix} H_{11}(z) & H_{12}(z) \\ H_{21}(z) & H_{22}(z) \end{bmatrix} \tag{13}$$

The transfer functions are written as

$$H_{11}(z) = \frac{\left[e_0^-, e_0^+\right]_{11} + \left[e_1^-, e_1^+\right]_{11} z + \cdots + \left[e_{n-1}^-, e_{n-1}^+\right]_{11} z^{n-1}}{\left[f_0^-, f_0^+\right]_{11} + \left[f_1^-, f_1^+\right]_{11} z + \cdots + \left[f_n^-, f_n^+\right]_{11} z^n} \tag{14}$$

$$H_{12}(z) = \frac{\left[e_0^-, e_0^+\right]_{12} + \left[e_1^-, e_1^+\right]_{12} z + \cdots + \left[e_{n-1}^-, e_{n-1}^+\right]_{12} z^{n-1}}{\left[f_0^-, f_0^+\right]_{12} + \left[f_1^-, f_1^+\right]_{12} z + \cdots + \left[f_n^-, f_n^+\right]_{12} z^n} \tag{15}$$

$$H_{21}(z) = \frac{\left[e_0^-, e_0^+\right]_{21} + \left[e_1^-, e_1^+\right]_{21} z + \cdots + \left[e_{n-1}^-, e_{n-1}^+\right]_{21} z^{n-1}}{\left[f_0^-, f_0^+\right]_{21} + \left[f_1^-, f_1^+\right]_{21} z + \cdots + \left[f_n^-, f_n^+\right]_{21} z^n} \tag{16}$$

$$H_{22}(z) = \frac{[e_0^-, e_0^+]_{22} + [e_1^-, e_1^+]_{22} z + \cdots + [e_{n-1}^-, e_{n-1}^+]_{22} z^{n-1}}{[f_0^-, f_0^+]_{22} + [f_1^-, f_1^+]_{22} z + \cdots + [f_n^-, f_n^+]_{22} z^n} \tag{17}$$

where $[e_i^-, e_i^+]$ and $[f_i^-, f_i^+]$ are the interval coefficients.

Further, the transfer function of ROMs can be considered as

$$\widehat{H}(z) = \begin{bmatrix} \widehat{H}_{11}(z) & \widehat{H}_{12}(z) \\ \widehat{H}_{21}(z) & \widehat{H}_{22}(z) \end{bmatrix} \tag{18}$$

The transfer functions are represented as

$$\widehat{H}_{11}(z) = \frac{\left[\hat{e}_0^-, \hat{e}_0^+\right]_{11} + \left[\hat{e}_1^-, \hat{e}_1^+\right]_{11} z + \cdots + \left[\hat{e}_{m-1}^-, \hat{e}_{m-1}^+\right]_{11} z^{m-1}}{\left[\widehat{f}_0^-, \widehat{f}_0^+\right]_{11} + \left[\widehat{f}_1^-, \widehat{f}_1^+\right]_{11} z + \cdots + \left[\widehat{f}_m^-, \widehat{f}_m^+\right]_{11} z^m} \tag{19}$$

$$\widehat{H}_{12}(z) = \frac{\left[\hat{e}_0^-, \hat{e}_0^+\right]_{12} + \left[\hat{e}_1^-, \hat{e}_1^+\right]_{12} z + \cdots + \left[\hat{e}_{m-1}^-, \hat{e}_{m-1}^+\right]_{12} z^{m-1}}{\left[\widehat{f}_0^-, \widehat{f}_0^+\right]_{12} + \left[\widehat{f}_1^-, \widehat{f}_1^+\right]_{12} z + \cdots + \left[\widehat{f}_m^-, \widehat{f}_m^+\right]_{12} z^m} \tag{20}$$

$$\widehat{H}_{21}(z) = \frac{\left[\hat{e}_0^-, e_0^+\right]_{21} + \left[\hat{e}_1^-, \hat{e}_1^+\right]_{21} z + \cdots + \left[\hat{e}_{m-1}^-, \hat{e}_{m-1}^+\right]_{21} z^{m-1}}{\left[\widehat{f}_0^-, \widehat{f}_0^+\right]_{21} + \left[\widehat{f}_1^-, \widehat{f}_1^+\right]_{21} z + \cdots + \left[\widehat{f}_m^-, \widehat{f}_m^+\right]_{21} z^m} \tag{21}$$

$$\widehat{H}_{22}(z) = \frac{\left[\hat{e}_0^-, \hat{e}_0^+\right]_{22} + \left[\hat{e}_1^-, \hat{e}_1^+\right]_{22} z + \cdots + \left[\hat{e}_{m-1}^-, \hat{e}_{m-1}^+\right]_{22} z^{m-1}}{\left[\widehat{f}_0^-, \widehat{f}_0^+\right]_{22} + \left[\widehat{f}_1^-, \widehat{f}_1^+\right]_{22} z + \cdots + \left[\widehat{f}_m^-, \widehat{f}_m^+\right]_{22} z^m} \tag{22}$$

where m denotes the order of model such that $n > m$ and $\left[\hat{e}_i^-, \hat{e}_i^+\right]$ and $\left[\widehat{f}_i^-, \widehat{f}_i^+\right]$ are the interval coefficients.

3 Time Moments for Interval System

For interval system, TMs are broadly classified into two categories, i.e. continuous and discrete TMs.

3.1 Continuous Interval Systems

3.1.1 Time Moments Derived Using Gamma-Delta Coefficients

For stable linear invariant continuous interval systems Eqs. (2)–(5), the first three time moments $(T_1)_{kl}$, $(T_2)_{kl}$ and $(T_3)_{kl}$ are derived using gamma-delta method [6] are represented as

$$\left.\begin{aligned} (T_1)_{kl} &= \tfrac{\delta_1}{\gamma_1} \\ (T_2)_{kl} &= \tfrac{\delta_1}{\gamma_1^2} + \tfrac{1}{\gamma_1}\tfrac{\delta_2}{\gamma_2} \\ (T_3)_{kl} &= \delta_1\left(\tfrac{1}{\gamma_1^3} - \tfrac{1}{\gamma_1^2\gamma_2}\right) - \tfrac{\delta_2}{\gamma_1^2\gamma_2} + \tfrac{1}{\gamma_1\gamma_2}\tfrac{\delta_3}{\gamma_3} \end{aligned}\right\} \tag{23}$$

The interval parameters $\delta_1, \delta_2, \ldots, \delta_n$ and $\gamma_1, \gamma_2, \gamma_3, \ldots, \gamma_n$ are computed from the Routh table.

3.1.2 Time Moments Derived by Sastry and Rao [18]

Time moments of [18] of system transfer functions are given as

In general, for $i = 1, 2, \ldots$

$$[T_i]_{kl} = c\left\{\frac{p_i}{q_0} \times \hat{q}_0 + \frac{p_0}{q_0}\left\{\hat{q}_i - \frac{q_i}{q_0} \times \hat{q}_0\right\} + \sum_{j=1}^{i-1}(-1)^j\left\{\frac{q_j}{q_0} \times \hat{p}_j - \frac{p_j}{p_0} \times \hat{q}_j\right\}\right\} \tag{24}$$

where, $p_i = \left[p_i^-, p_i^+\right]$, $q_i = \left[q_i^-, q_i^+\right]$ are the interval coefficients.

$c = \left\{\frac{C_0^- + C_0^+}{2}\right\}$; with $\left[C_0^-, C_0^+\right] = \frac{\left[p_0^-, p_0^+\right]}{\left[q_0^-, q_0^+\right]}$ and $\hat{q}_j = \left[\hat{q}_j^-, \hat{q}_j^+\right]$ denote the coefficients of reduced order denominator.

3.1.3 Time Moments Derived Using Direct Series Expansion [19]

The TMs of CIS are obtained using direct series expression method.

$$\left.\begin{aligned} T_0 &= \left[\varepsilon_0^-, \varepsilon_0^+\right] \\ T_1 &= \left[\varepsilon_0^-, \varepsilon_0^+\right]\left[\varepsilon_1^-, \varepsilon_1^+\right] \\ &\vdots \\ T_x &= \prod_{i=0}^{x-1}\left[\varepsilon_i^-, \varepsilon_i^+\right] \end{aligned}\right\} \tag{25}$$

where $T_0, T_1, \ldots, T_x$ denotes the TMs of CIS and the parameters $\left[\varepsilon_0^-, \varepsilon_0^+\right]$, $\left[\varepsilon_1^-, \varepsilon_1^+\right], \ldots, \left[\varepsilon_i^-, \varepsilon_i^+\right]$ are computed using direct series expansion method [19].

3.1.4 Time Moments Derived Using Singh et al. [20]

Recently, in [20], authors proposed a simple expressions of TMs of CSIs Eqs. (2)–(5) which are

$$(T_x)_{kl} = \frac{p_x}{q_0} - \sum_{i=0}^{x-1} \frac{p_{x-i} T_i}{q_0}, \quad x = 0, 1, 2, \ldots \tag{26}$$

where, $T_i = \left[T_i^-, T_i^+\right]$ for $i = 0, 1, \ldots$ are the TMs of the IS.

3.2 *Discrete Interval Systems*

3.2.1 Time Moments Derived Using Direct Series Expansion

The TMs of higher-order DISs, Eqs. (14)–(17) computed using direct series expansion method [21] are

$$\left.\begin{aligned} \left[t_0^-, t_0^+\right] &= \left[\varepsilon_0^-, \varepsilon_0^+\right] = \left[\frac{E_0^-}{F_0^+}, \frac{E_0^+}{F_0^-}\right] \\ \left[t_1^-, t_1^+\right] &= \left[\varepsilon_0^-, \varepsilon_0^+\right].\left[\varepsilon_1^-, \varepsilon_1^+\right] \\ &= \left[\frac{E_0^-}{F_0^+}, \frac{E_0^+}{F_0^-}\right].\left[\frac{E_1^-}{F_0^+} - \frac{F_1^+}{F_0^-}, \frac{E_1^+}{E_0^-} - \frac{F_1^-}{F_0^+}\right] \\ &\vdots \\ \left[t_x^-, t_x^+\right] &= \prod_{i=0}^{x} \left[\varepsilon_i^-, \varepsilon_i^+\right] \end{aligned}\right\} \tag{27}$$

$$\left[T_x^-, T_x^+\right] = \begin{cases} \left[t_x^-, t_x^+\right] & \text{for } x = 0 \\ (-1) \sum_{j=1}^{x} \frac{1}{j!} (T_s)^j \beta_{xj} \left(t_j^-, t_j^+\right) & \text{for } x = 1, 2 \ldots \end{cases} \tag{28}$$

where $\left[E_0^-, E_0^+\right]$ and $\left[F_0^-, F_0^+\right]$ are interval coefficients in p-domain,

$$\beta_{xj} = \begin{cases} \beta_{x-1,j-1} + j\beta_{x-1.j} & \text{for } x > j \\ 0 & \text{for } x < j \end{cases}$$

$$\beta_{xx} = \beta_{x1} = 1$$

Similarly, the TMs of the DIS model are given by

$$\left[\hat{T}_x^-, \hat{T}_x^+\right] = \begin{cases} \left[\hat{t}_x^-, \hat{t}_x^+\right] & \text{for } x = 0 \\ (-1)\sum_{j=1}^{x} \frac{1}{j!}(T_s)^j \beta_{xj}\left(\hat{t}_j^-, \hat{t}_j^+\right) & \text{for } x = 1, 2 \ldots \end{cases} \tag{29}$$

where T_s represents the sampling time and β_{xj} are constants and can be computed using the algorithm [22].

4 Clustering of Poles

Cluster center is computed by pole clustering method in which poles are accumulated on the basics of the poles relative distance and a desirable order of model through a modeling process. Each pole is substituted by an individual or a pair of complex conjugate pole (CCP). Nevertheless, separate cluster center (CC) can be created with real and complex conjugate values of poles. In pole clustering method, model is obtained when poles on the imaginary axis are held back.

The CC is given as

$$\phi^c = \left\{\left(\sum_{i=1}^{y}\left(\frac{1}{\alpha_i}\right)\right) \div y\right\}^{-1} \tag{30}$$

where ϕ^c is CC of y real poles $(\alpha_1, \alpha_2, \ldots, \alpha_y)$ of the higher-order discrete interval systems.

CCs for CCPs is in the form of $\phi^R \pm j\phi^I$ expanded from y pairs of CCPs are represented as:

$$\left.\begin{aligned} \phi^R &= \left\{\left(\sum_{i=1}^{y}\left(\frac{1}{\alpha_i^R}\right)\right) \div y\right\}^{-1} \\ \phi^I &= \left\{\left(\sum_{i=1}^{y}\left(\frac{1}{\alpha_i^I}\right)\right) \div y\right\}^{-1} \end{aligned}\right\} \tag{31}$$

To generate reduced order polynomial for denominator of lower order model calculated as:

Case 1: If all pole CCs are real, then the denominator of mth-order model can be computed as

$$D_m(z) = \left(z - \phi_1^c\right)\left(z - \phi_2^c\right)\ldots(z - \phi_m) \tag{32}$$

where $\phi_1, \phi_2 \ldots \phi_m$ are 1st, 2nd, … mth pole CCs respectively.

Case 2: If all pole CC are complex conjugates, then the denominator of the mth-order model can be calculated as

$$D_m(z) = \big(z - (\phi_1^R + j\phi_1^I)\big)\big(z - (\phi_1^R - j\phi_1^I)\big)\ldots \\ \left(z - \left(\phi_{m/2}^R + j\phi_{m/2}^I\right)\right)\left(z - \left(\phi_{m/2}^R - j\phi_{m/2}^I\right)\right) \tag{33}$$

Case 3: If $(m - 2)$ is the real CC with a pair of CCP CC then mth-order model can be obtained as

$$D_m(z) = (z - \phi_1)(z - \phi_2)\ldots \\ \big(z - \phi_{(m-2)}\big)\big(z - (\phi_1^R + j\phi_1^I)\big)\big(z - (\phi_1^R - j\phi_1^I)\big) \tag{34}$$

By retaining the dominant and cluster poles of the system, the denominator of the reduced-order model is calculated and the first m time moments of the system is compared with the model to obtain the numerator.

5 Test System

Consider a MIMO transfer function

$$H(z) = \begin{bmatrix} H_{11}(z) & H_{12}(z) \\ H_{21}(z) & H_{22}(z) \end{bmatrix} \tag{35}$$

where the transfer functions are given as

$$H_{11}(z) = \frac{[8, 10] + [3, 4]z + [1, 2]z^2}{[0.8, 0.85] + [4.9, 5]z + [9, 9.5]z^2 + [6, 6]z^3} \tag{36}$$

$$H_{12}(z) = \frac{[8, 10] + [3.5, 4]z + [1.5, 2]z^2}{[0.8, 0.85] + [4.9, 5]z + [9, 9.5]z^2 + [6, 6]z^3} \tag{37}$$

$$H_{21}(z) = \frac{[8.2, 10] + [3.1, 4]z + [1.2, 2]z^2}{[0.8, 0.85] + [4.9, 5]z + [9, 9.5]z^2 + [6, 6]z^3} \tag{38}$$

$$H_{22}(z) = \frac{[8.5, 10] + [3.5, 4]z + [1.5, 2]z^2}{[0.8, 0.85] + [4.9, 5]z + [9, 9.5]z^2 + [6, 6]z^3} \tag{39}$$

The denominator polynomial of Eqs. (36)–(39) is

$$D(z) = [0.8, 0.85] + [4.9, 5]z + [9, 9.5]z^2 + [6, 6]z^3 \tag{40}$$

The poles of the characteristic equation, computed using [23], are

$$\alpha_1 = [-0.534, -0.268]$$
$$\alpha_2 = [-0.712, -0.536]$$

$$\alpha_3 = [-0.853, -0.720] \quad (41)$$

As poles Eq. (41) are real, thus, using Eq. (30) the CC is computed, by grouping α_2 and α_3 is

$$\phi^c = [-0.7766, -0.6147] \quad (42)$$

and keeping the dominant pole α_1, denominator polynomial, using Eq. (32), is computed as

$$\left.\begin{aligned} D_m(z) &= (z-\alpha_1)(z-\phi^c) \\ &= (z-[-0.5340, -0.2680])(z-[-0.7766, -0.6147]) \\ &= [1, 1]z^2 + [0.8827, 1.3106]z + [0.1647, 0.4147] \end{aligned}\right\} \quad (43)$$

The numerators of the transfer function are calculated by matching the two TMs of the DIS and model

For H_{11}

The first and second TMs of the DIS, as given by Eq. (28) are

$$\left.\begin{aligned} [T_0^-, T_0^+]_{11} &= [0.5621, 0.7729] \\ [T_1^-, T_1^+]_{11} &= [0.7022, 1.3266] \end{aligned}\right\} \quad (44)$$

and first and second TMs of the discrete interval model, as given by Eq. (29) are

$$\left.\begin{aligned} [\widehat{T}_0^-, \widehat{T}_0^+]_{11} &= \left[\frac{\hat{e}_0^-}{\hat{f}_0^+}, \frac{\hat{e}_0^+}{\hat{f}_0^-}\right]_{11} \\ [\widehat{T}_1^-, \widehat{T}_1^+]_{11} &= \left[\frac{\hat{e}_0^-}{\hat{f}_0^+}, \frac{\hat{e}_0^+}{\hat{f}_0^-}\right]_{11} \cdot \left[\frac{\hat{e}_1^-}{\hat{f}_0^+} - \frac{\hat{e}_1^+}{\hat{f}_0^-}, \frac{\hat{e}_1^+}{\hat{e}_0^-} - \frac{\hat{f}_1^-}{\hat{f}_0^+}\right]_{11} \end{aligned}\right. \quad (45)$$

By matching the TMs of the DIS and model, $\left[T_0^-, T_0^+\right]_{11} = \left[\widehat{T}_0^-, \widehat{T}_0^+\right]_{11}$ and $\left[T_1^-, T_1^+\right]_{11} = \left[\widehat{T}_1^-, \widehat{T}_1^+\right]_{11}$ the second-order transfer function becomes

$$\widehat{H}_{11}(z) = \frac{[1.3030, 2.0583] + [-0.4757, 0.2286]z}{[0.1647, 0.4147] + [0.8827, 1.3106]z + [1, 1]z^2} \quad (46)$$

The step response of system and model is constructed using Kharitonov polynomials of transfer function Eq. (36) and model Eq. (46) is shown in Fig. 1.

For H_{12}

The first and second TMs of the DIS, as given by Eq. (28) are

$$\left.\begin{aligned} [T_0^-, T_0^+]_{12} &= [0.7877, 1.2541] \\ [T_1^-, T_1^+]_{12} &= [0.6088, 0.7729] \end{aligned}\right\} \quad (47)$$

and first and second TMs of the discrete interval model, represented in Eq. (29) are

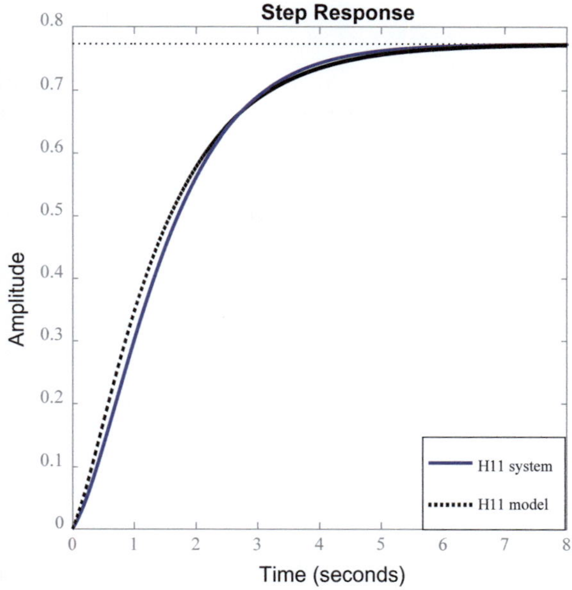

Fig. 1 Step responses of H_{11} DIS and model

$$\begin{aligned} [\widehat{T}_0^-, \widehat{T}_0^+]_{12} &= \left[\frac{\hat{e}_0^-}{\hat{f}_0^+}, \frac{\hat{e}_0^+}{\hat{f}_0^-}\right]_{12} \\ [\widehat{T}_1^-, \widehat{T}_1^+]_{12} &= \left[\frac{\hat{e}_0^-}{\hat{f}_0^+}, \frac{\hat{e}_0^+}{\hat{f}_0^-}\right]_{12} \cdot \left[\frac{\hat{e}_1^-}{\hat{f}_0^+} - \frac{\hat{e}_1^+}{\hat{f}_0^-}, \frac{\hat{e}_1^+}{\hat{e}_0^-} - \frac{\hat{f}_1^-}{\hat{f}_0^+}\right]_{12} \end{aligned} \tag{48}$$

By matching the TMs of the system and model, $\left[T_0^-, T_0^+\right]_{12} = \left[\widehat{T}_0^-, \widehat{T}_0^+\right]_{12}$ and $\left[T_1^-, T_1^+\right]_{12} = \left[\widehat{T}_1^-, \widehat{T}_1^+\right]_{12}$ the second-order transfer function represented as

$$\widehat{H}_{12}(z) = \frac{[1.5224, 2.3931] + [-0.7340, 0.0600]z}{[0.1647, 0.4147] + [0.8827, 1.3106]z + [1, 1]z^2} \tag{49}$$

The step response of system and model is constructed using Kharitonov polynomials of transfer function Eq. (37) and model Eq. (49) is shown in Fig. 2.

For H_{21}

The first and second TMs of the DIS, as given by Eq. (28) are

$$\left.\begin{aligned} [T_0^-, T_0^+]_{21} &= [0.5854, 0.7729] \\ [T_1^-, T_1^+]_{21} &= [0.7467, 1.3024] \end{aligned}\right\} \tag{50}$$

and first and second TMs of the discrete interval model, represented in Eq. (29) are

$$\begin{aligned} [\widehat{T}_0^-, \widehat{T}_0^+]_{21} &= \left[\frac{\hat{e}_0^-}{\hat{f}_0^+}, \frac{\hat{e}_0^+}{\hat{f}_0^-}\right]_{21} \\ [\widehat{T}_1^-, \widehat{T}_1^+]_{21} &= \left[\frac{\hat{e}_0^-}{\hat{f}_0^+}, \frac{\hat{e}_0^+}{\hat{f}_0^-}\right]_{21} \cdot \left[\frac{\hat{e}_1^-}{\hat{f}_0^+} - \frac{\hat{e}_1^+}{\hat{f}_0^-}, \frac{\hat{e}_1^+}{\hat{e}_0^-} - \frac{\hat{f}_1^-}{\hat{f}_0^+}\right]_{21} \end{aligned} \tag{51}$$

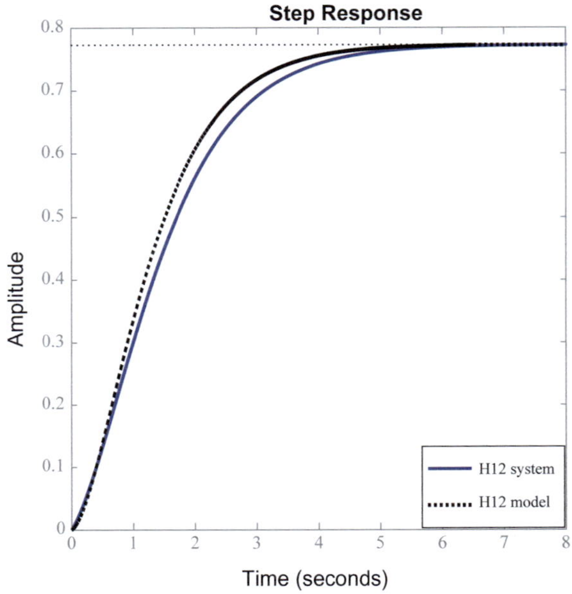

Fig. 2 Step responses of H_{12} DIS and model

By matching the TMs of the system and model, $[T_0^-, T_0^+]_{21} = \left[\widehat{T}_0^-, \widehat{T}_0^+\right]_{21}$ and $[T_1^-, T_1^+]_{21} = \left[\widehat{T}_1^-, \widehat{T}_1^+\right]_{21}$ the second-order transfer function becomes

$$\widehat{H}_{21}(z) = \frac{[1.4364, 2.5572] + [-0.9619, 0.146]z}{[0.1647, 0.4147] + [0.8827, 1.3106]z + [1, 1]z^2} \tag{52}$$

The step response of system and model is constructed using Kharitonov polynomials of transfer function Eq. (38) and model Eq. (52) is shown in Fig. 3.

For H_{22}

The first and second TMs of the DIS, as given by Eq. (28) are

$$\left.\begin{array}{l} [T_0^-, T_0^+]_{22} = [0.5854, 0.7729] \\ [T_1^-, T_1^+]_{22} = [0.7467, 1.3024] \end{array}\right\} \tag{53}$$

and first and second TMs of the discrete interval model, represented in Eq. (29) are

$$\begin{array}{l} [\widehat{T}_0^-, \widehat{T}_0^+]_{22} = \left[\frac{\hat{e}_0^-}{\hat{f}_0^+}, \frac{\hat{e}_0^+}{\hat{f}_0^-}\right]_{22} \\ [\widehat{T}_1^-, \widehat{T}_1^+]_{22} = \left[\frac{\hat{e}_0^-}{\hat{f}_0^+}, \frac{\hat{e}_0^+}{\hat{f}_0^-}\right]_{22} . \left[\frac{\hat{e}_1^-}{\hat{f}_0^+} - \frac{\hat{e}_1^+}{\hat{f}_0^-}, \frac{\hat{e}_1^+}{\hat{e}_0^-} - \frac{\hat{f}_1^-}{\hat{f}_0^+}\right]_{22} \end{array} \tag{54}$$

By matching the TMs of the system and model, $[T_0^-, T_0^+]_{22} = \left[\hat{T}_0^-, \hat{T}_0^+\right]_{22}$ and $[T_1^-, T_1^+]_{22} = \left[\hat{T}_1^-, \hat{T}_1^+\right]_{22}$ the second-order transfer function becomes

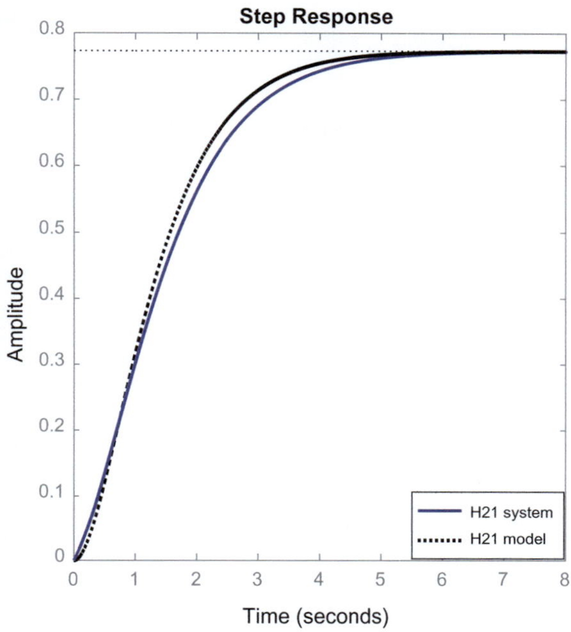

Fig. 3 Step responses of H_{21} DIS and model

$$\widehat{H}_{22}(z) = \frac{[1.6069, 2.3030] + [-0.5797, -0.0245]z}{[0.1647, 0.4147] + [0.8827, 1.3106]z + [1, 1]z^2} \tag{55}$$

The step response of system and model is constructed using Kharitonov polynomials of transfer function Eq. (39) and model Eq. (55) is shown in Fig. 4.

From the curves shown in Figs 1, 2, 3 and 4, it is clear that the obtained second-order models Eqs. (46), (49), (52), and (55) are the better approximation of transfer functions given in Eqs. (36)–(39).

6 Conclusion

In this work, a method of MOR for multi-input-multi-output (MIMO) DISs is presented. Clustering of poles technique is used to determine the denominator of different transfer functions of lower order MIMO DIS. However, numerators of different transfer functions of lower order model are obtained by matching of TMs. Various expressions of TMs are also represented in this paper. To demonstrate the proposed technique, a third-order two-input-two-output (TITO) discrete interval test system is reduced into a first-order MIMO model.

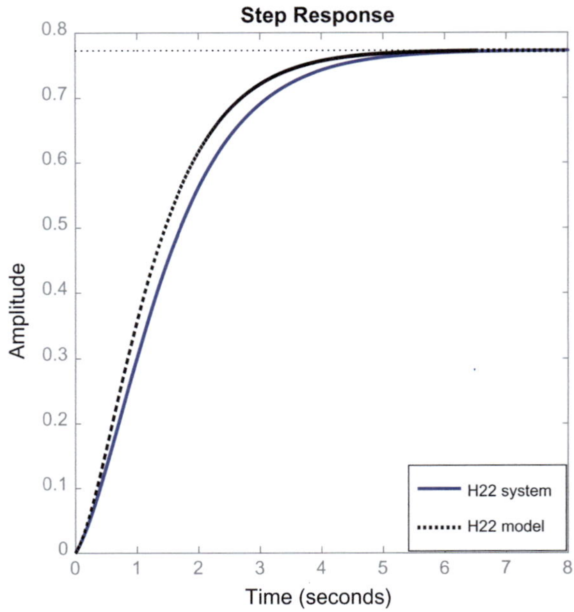

Fig. 4 Step responses of H_{22} DIS and model

Acknowledgements The work is supported by SERB, DST, GOI (ECR/2017/000212).

References

1. Fortuna L, Nunnari G, Gallo A (2012) Model order reduction techniques with applications in electrical engineering. Springer Science & Business Media
2. Pan S, Pal J (1995) Reduced order modelling of discrete-time systems. Appl Math Model 19:133–138
3. Deepa S, Sugumaran G (2011) Model order formulation of a multivariable discrete system using a modified particle swarm optimization approach. Swarm Evol Comput 1:204–212
4. Bandyopadhyay B, Ismail O, Gorez R (1994) Routh-pade approximation for interval systems. IEEE Trans Autom Control 39:2454–2456
5. Shamash Y (1975) Model reduction using the routh stability criterion and the padé approximation technique. Int J Control 21:475–484
6. Bandyopadhyay B, Upadhye A, Ismail O (1997) /spl gamma/-/spl delta/Routh approximation for interval systems. IEEE Trans Autom Control 42:1127–1130
7. Hutton M, Friedland B (1975) Routh approximations for reducing order of linear, time-invariant systems. IEEE Trans Autom Control 20:329–337
8. Sastry G, Rao RR, Rao PM (2000) Large scale interval system modelling using routh approximants. Electron Lett 36:768–769
9. Kumar Dk, Nagar S, Tiwari J (2011) Model order reduction of interval systems using modified Routh approximation and factor division method. In: 35th national system conference (NSC)
10. Potturu SR, Prasad R, (2017) Reduction of interval systems using Kharitonov's polynomials and their derivatives. In: 2017 6th international conference on computer applications in electrical engineering-recent advances (CERA), pp 445–449

11. Sastry G, Rao PM (2003) A new method for modelling of large scale interval systems. IETE J Res 49:423–430
12. Choudhary AK, Nagar SK (2017) Novel arrangement of routh array for order reduction of z-domain uncertain system. Syst Sci Control Eng 5:232–242
13. Choudhary AK, Nagar SK (2018) Order reduction in z-domain for interval system using an arithmetic operator. Circuits Syst Signal Process, pp 1–16
14. Choudhary AK, Nagar SK (2018) Model order reduction of discrete-time interval system based on Mikhailov stability criterion. Int J Dyn Control, pp 1–9
15. Choudhary AK, Nagar SK (2018) Model order reduction of discrete-time interval systems by differentiation calculus. Autom Control Comput Sci 52:402–411
16. Singh V, Chandra D (2012) Reduction of discrete interval system using clustering of poles with padé approximation: a computer-aided approach. Int J Eng, Sci Technol 4:97–105
17. Padhy AP, Singh VP, Pattnaik S (2018) Model reduction of multi-input-multi-output discrete interval systems using gain adjustment. Int J Pure Appl Math 119(12):12721–12739
18. Sastry G, Rao GR (2003) Simplified polynomial derivative technique for the reduction of large-scale interval systems. IETE J Res 49:405–409
19. Singh VP, Chandra D (2010) Routh-approximation based model reduction using series expansion of interval systems. In: 2010 international conference on power, control and embedded systems (ICPCES), pp 1–4
20. Singh V, Chauhan DPS, Singh SP, Prakash T (2017) On time moments and markov parameters of continuous interval systems. J Circuits Syst Comput 26:1750038
21. Singh VP, Chandra D (2011) Model reduction of discrete interval system using dominant poles retention and direct series expansion method. In: 2011 5th international power engineering and optimization conference (PEOCO), pp 27–30
22. Berge C (1971) Principles of combinatorics
23. Ismail O, Bandyopadhyay B, Gorez R (1997) Discrete interval system reduction using pade approximation to allow retention of dominant poles. IEEE Trans Circuits Syst I Fundam Theory Appl 44:1075–1078

Human Activity Recognition Using Smartphone Sensor Data

Sweta Jain, Sadare Alam and K. Shreesha Prabhu

Abstract The hot topic in recent times is recognition of human activities through a smartphone, smart home, remote monitoring and assisted healthcare. These fall under ambient intelligent services. This also includes recognition of simple activities like sitting, running and walking, and more research is being held for semi-complex activities such as moving upstairs and downstairs, running and jogging. Activity recognition is the problem of predicting the current action of a person by using the motion sensors worn on the body. This problem is approached by using supervised classification model where a model is trained from a known set of data, and a query is then resolved to a known activity label by using the learned model. The exigent issue here is whether how to feed this classification model with a set of features, where the input provided is a raw sensor data. In this study, three classification techniques are considered and their accuracy in predicting the correct activity. In addition to the systematic comparison of the results, a comprehensive evaluation of data collection and some preprocessing steps are provided such as filtering and feature generation. The results determine that feeding a support vector machine with an ensemble selection of most relevant features by using principal component analysis yields best results.

Keywords Human activity · Smartphone sensors · Walking · Running

1 Introduction

Nowadays, smartphones have occupied most of the human life. There are various application domains which contain activity recognition technology such as health

---

S. Jain (✉) · S. Alam · K. S. Prabhu
Maulana Azad National Institute of Technology, Bhopal, India
e-mail: shwetaj82@yahoo.com

S. Alam
e-mail: itisalam@gmail.com

K. S. Prabhu
e-mail: shreeshaprabhu@gmail.com

R. K. Shukla et al. (eds.), *Social Networking and Computational Intelligence*, Lecture Notes in Networks and Systems 100, https://doi.org/10.1007/978-981-15-2071-6_42

and elder care or sportive motion tracker devices. There are various studies which have used accelerometer sensors earlier [1, 2]. The studies can be outlined as the data was collected by developing an Android application with a simple interface. Through this application, raw sensor data was collected. The data is preprocessed by applying various filters, and appropriate features were extracted. Further, various classification models were used to classify data into different classes.

In this paper, the generation of the dataset from the smartphone sensors has been described, and then, using the data generated from the smartphones to predict various human activities such as running, walking, jumping and sitting is also described. In previous works, the activities such as running, walking are predicted [1]. As suggested by [1], in this work, jumping has been included in the new activity.

In this work, we are dealing with sensor data obtained from Android smartphone. From the data, relevant features were extracted for the classification. Three different training algorithms have been used to predict the activities, and the precision and recall of each activity are analyzed. The training algorithms used are—SVM, MLR and J48.

The following section deals with the technology and methodology used in this project.

1.1 Android Smartphones

Android smartphones are becoming the bread butter of our daily life. They are used in every phase of our daily routine. They can be brought in use to determine our daily activities. Sensors have been used in smartphones since they were invented. From the advent of microphones and touch key to the invention of accelerometers, gyroscope, GPS, etc., sensors have played a vital role in the formation of smartphones.

The Android sensor framework allows accessing many types of sensors such as hardware-based and software-based sensors. Hardware-based sensors are physical components which derive their raw data by directly measuring specific environmental parameters, such as acceleration, geomagnetic field strength or angular change, whereas, software-based sensors (linear acceleration sensor and the gravity sensor) are not physical devices, although they imitate hardware-based sensors.

The Android coordinate system uses the standard three-axis coordinate system (x, y and z). Here, accelerometer and gyroscope are used.

The accelerometer and gyroscope readings are used for dataset generation. The data generated is used for training/testing purposes.

1.1.1 Accelerometer

It adapts the orientation change when the position of the device is changed. In smartphones, one can feel accelerometer's use while tilting our screen (changes orientation:

landscape, portrait), playing games, detecting drops, etc. Through the Android sensor framework, raw data can be collected which would be represented in the form of vectors.

1.1.2 Gyroscope

It tracks rotation or twist, and it is primarily used for navigation and measurement of the angular rotational velocity. It measures the phone's rotation rate by detecting the roll, pitch and raw motions of the smartphones along the x, y and z-axis, respectively. The raw data obtained from the gyroscope is the rate of rotation in rad/sec (unit) around each physical axis (x, y and z).

2 Related Works

Human activity recognition is the most researched topic in today's research arena. And the topic of sensor-based activity recognition is not new. Much of the work is being done in analyzing the performance of various classification methods such as Naïve Bayes, decision tree, support vector machine, multinomial logistic regression and neural networks. In [3], a two-layer model is developed in which multi-component Gaussian mixture model and Markov models were combined to classify a range of user activity states, including sitting and walking. More researchers are diverted toward wearable sensor data. In [1], a system is developed that uses phone-based accelerometers to collect a raw data and draw useful patterns from this data. However, much of the researches are done without keeping postural transitions and the processes between the two activities, into consideration. In [4], a human activity recognition system based on feature selection techniques is developed. It also deals with finding most important features for activity recognition. In [5], the multi-sensor system is used to study the effectiveness of activity classifiers when the position of sensors is varied. It helps to decide the best position to place sensors and also compares the tradeoff between the sensor position and classification performance. Other works focus on the various applications that can be built on sensor-based activity recognition. In [6], a standard dataset is collected and made available to the public in UCI machine learning website. The dataset collected in [6] consists of data from 30 volunteers performing six different daily life activities.

3 Feature Generation

The data obtained from the smartphones is the raw data, and it needs to be converted into some desirable format before it can be used for model training and testing. Feature generation is the process of finding out the useful information from the raw data that may be beneficial for our model generation. The raw data is converted into some structured format in this process. The raw data is converted into features in such a way that it does not lose any significant information present in the data that may be beneficial to us.

3.1 Data Collection

In order to collect data for various activities such as walking, jumping, sitting and running, an Android application was needed which can collect the sensor data and user's information.

An Android application with the simple interface was developed. The application collects the tri-axial acceleration from the accelerometer and angular velocity from the gyroscope. Also, for labeling the data with activity labels, the information about the activity being performed is also taken. Some information about the user is also collected. User's information included their name, age and gender. The sensor data recorded the x, y and z-axis readings collected at a frequency of 50 Hz. The application stores the accelerometer and gyroscope readings along with their timestamps in a file on the Android system itself. This data can be then taken out of the phone for use in model training and testing. Using this application, four subjects were tracked for different activities for two days.

3.2 Smoothing and Filtering

The raw data contains some unnecessary information such as the noise if the pocket is loose, the phone may also move in the pocket. Both accelerometer and gyroscope available in a smartphone do not provide many accurate readings. Acceleration obtained from accelerometer has gravity component in it. To avoid the errors in the model due to the presence of the noise in it, digital filters are applied on the raw data. First of all, a median filter with a width size of 128 is applied on the data for smoothing. After filtering, the gravity component in the acceleration is separated using low-pass filter with cutoff frequency 0.3 Hz, and body acceleration component is obtained. Jerk of body acceleration and angular acceleration is also calculated. The magnitude of each of these signals is also calculated using Euclidian norm.

3.3 Generating Useful Features

The data obtained from the smartphones is a time series data. Time series data so obtained is sampled into 2.56 s fixed-width sliding windows. Each window will have

128 readings of each signal. Two consecutive windows overlap 50% with each other. Readings of a window cannot be directly used as features for classification purpose. In this paper, eight features from time domain of each signal are taken.

1. Arithmetic mean of values.
2. Standard deviation of values.
3. Median absolute deviation.
4. Largest value (Maximum).
5. Smallest value (Minimum).
6. Interquartile range.
7. Average sum of the squares (Energy).
8. Entropy of the signal.

In each window, there are five basic signals—body acceleration, body acceleration jerk, gravity, angular velocity and angular acceleration, each reading of signal has four values—one corresponding to each of *XYZ* directions and one value corresponding to magnitude making a total of 20 signals. From each signal, eight features are extracted making a total of 160 features for each window.

4 Experiments

The dataset so created is used for training the classifiers. Features in the dataset are reduced using PCA. Three different classifiers are trained. Trained classifiers are then tested.

4.1 Training Set and Test Set

The dataset is randomly shuffled. Two independent sets are created from the shuffled dataset. First 70% of the shuffled dataset is taken as the training set. Last 30% of the shuffled dataset is taken as the test set. Models are trained with the training set and tested with the test set. Shuffling, training and testing are repeated for five times for each classifier.

4.2 Data Normalization

The different features present on the data may have different order of magnitudes, and it may be the case that a feature having larger magnitude may have very less variance in it. For this purpose, z-score normalization is performed on the data. Mean and standard deviation of training dataset are used for normalizing both training and test examples.

4.3 Dimensionality Reduction

The number of features generated from the dataset is 200, so if all features are used for the model training and testing, it may require a large time on an ordinary machine, also, some of the features may carry a very little/less information, or they may have very little variance in it. For this purpose, the number of features in the dataset is reduced. For reducing the number of features, principal component analysis (PCA) is used.

The principal components of X are the eigenvectors of XXt [7] PCA can be used to examine the variances SY associated with the principal components. If X has m principal components, then the large variances are associated with first k principal components for some k, $k < m$ [7].

First, the PCA is applied on the training set, number of features corresponding to 95% variance in data is calculated, and the number of features comes out to be 45.

Using first k principal components, training examples and test examples are projected into k dimension space.

4.4 Training the Classifiers

Training set is used to train three classifiers namely J48, SVM and MLR.

1. **J48**
   J48 classifier is an implementation of C4.5 decision tree. J48 classifier is an open-source implementation in Java.
2. **Support Vector Machine (SVM)**
   In this experiment, one-vs-one implementation of SVM from *R* programming language is used. Support vectors are found for each pair of classes and for a test example, class predicted based on voting among all the classes.
3. **Multinomial Logistic Regression (MLR)**
   In this experiment, one-vs-all logistic regression is used for multiclass classification. For a test example, class with highest confidence is predicted. All the classifiers are tested with test set.

5 Results

The results from the testing are an average of five runs of the models on the training data.

Table 1 gives the percentage recall of each activity for each of the training algorithms. Simple activities like walking and sitting have better recall than complex activities like jumping and running. It can be clearly observed from the analysis of

Table 1 Percentage recall of activities

Activity	% of records correctly predicted		
	J48	SVM	MLR
Walking	100	99.88	99.54
Jumping	95.34	99.04	89.45
Running	98.43	100	92.03
Sitting	99.63	100	98.09

Table 2 Overall accuracy of the models

Training algorithm	Accuracy
J48	99.62
MLR	98.21
SVM	99.90

the results that while JUMPING is the very new activity, its recall is comparable to other activities, and it shows that our dataset generation is quite homogeneous and error free.

The reasons for good recall in model can be attributed to:

- The number of classes in the dataset is very less (only four), so the dataset is not very complex. Due to reduced complexity in the dataset, the decision boundary can easily separate classes.
- In the case of sitting, the data has almost no variance, so separation of sitting class from other classes is very easy.
- In case of jumping, the recall is higher because for the training examples corresponding to jumping, variance is mainly along any one axis, whereas other two axes almost remain constant.
- In case of running and walking, these are also very easily separable, walking has almost no acceleration, while in case of running, some acceleration is present. So, this fact is also very helpful in finding decision boundaries in the dataset.

Table 2 gives the accuracy of the different models on the testing data, from the table, it can be easily observed that the models are highly accurate, and thus, it also reflects the quality of the dataset.

6 Conclusion

In this project, an Android application was developed to collect data for predicting human activities. To test the quality of dataset generated, the same dataset was used to do the prediction of activity labels using SVM, MLR and J48 algorithms. From the results, it is found that all the three models give a very good prediction accuracy and recall for the given dataset.

From Table 1, it can be seen that the support vector machine(SVM) gives the best recall in case of jumping as compared to other models such as MLR and J48. The recall obtained for jumping by using SVM is 99.04% which is comparable to 99.88% in walking, 100% in running and 100% in sitting. The accuracy of the models is 99.90% for SVM, 98.21% for MLR and 99.62% for J48. It can be concluded that the dataset generated is quite standard and can be used for prediction of human activities.

7 Future Work

The human activity recognition is widely researched topic; however, there are still quite a few improvements and additions that can be done. The following are some areas that can be considered for the future research works.

1. Dataset generation and recognition of complex activities such as swimming, cycling and dancing.
2. Due to high computing power of the Android devices, the activity recognition system can be implemented solely on the Android devices to perform online learning.
3. A large dataset can be generated and made available so that no problem is faced while training and testing.

References

1. Wu Z, Zhang S, Zhang C Human activity recognition using wearable devices sensor data
2. Anguita D, Ghio A, Oneto L, Parra X, Reyes-Ortiz JL Human activity recognition on smartphones using a multiclass hardware-friendly support vector machine. In: Ambient assisted living and home care
3. DeVaul RW, Dunn S (2001) Real-time motion classification for wearable computing applications. Project Paper
4. Zhang M, Sawchuk AA (2011) A feature selection-based framework for human activity recognition using wearable multimodal sensors. In: Proceedings of the 6th international conference on body area networks. ICST (Institute for Computer Sciences, Social-Informatics and Telecommunications Engineering)
5. Maurer U, Smailagic A, Siewiorek D, Deisher M (2006) Activity recognition and monitoring using multiple sensors on different body positions. In: International workshop on wearable and implantable body sensor networks (BSN'06)
6. Anguita D, Ghio A, Oneto L, Parra X, Reyes-Ortiz J. L (2013) A public domain dataset for human activity recognition using smartphones. In: 21th European symposium on artificial neural networks, computational intelligence and machine learning, ESANN 2013, Bruges, Belgium, 24–26 April 2013
7. Shlens J (2014) A tutorial on principal component analysis. arXivn preprint arXiv:1404.1100
8. Brown M, Deitch T, O'Conor L Activity classification with smartphone data

9. Pai A, Nachum O, Kanter M (2013) Activity classification using smartphone accelerometer data
10. Sunny JT et al Applications and challenges of human activity recognition using sensors in a smart environment. Int J 2: 50–57
11. Lara OD, Labrador MA (2013) A survey on human activity recognition using wearable sensors. IEEE Commun Surv Tutor 15(3):1192–1209
12. Olguın DO, Pentland AS (2006) Human activity recognition: accuracy across common locations for wearable sensors. In: Proceedings of 2006 10th IEEE international symposium on wearable computers, Citeseer, Montreux, Switzerland

Novel Software Modeling Technique for Surveillance System

Rakesh Kumar, Priti Maheshwary and Timothy Malche

Abstract A technological revolution Internet of things (IoT) describes the forthcoming of communications and computing. Development of IoT from wireless sensors to nanotechnology in many important fields depends on dynamic technical innovation. In IoT, each object will be tagged for identifying, monitoring, automating and controlling. This research paper describes novel software development technique-based surveillance system that leads to the energy efficiency concept based on IoT technology. PIR sensor will capture the movement in the given time and sends the signal to the microcontroller which leads to activate the camera to capture and also sends the alert to mail and mobile of the user. User is facilitated to control the working from remote location. This system will be more secure and flexible with change request incorporated with fast development.

Keywords IoT · Energy efficiency · PIR · Camera · Secure

1 Introduction

Internet of things (IoT) is a phenomenon which connects a variety of things where everything has ability to communicate. There are many application areas of the IoT, but security is the main aspect like IoT-based surveillance system. In many environments, surveillance systems are widespread and common in current era. Smart surveillance has been a key component in ensuring security at airports, banks, casinos, institutions, governments' agencies, businesses and even in schools for ensuring security surveillance systems are the key component. In crime prevention, resolution and protection, surveillance system is an important emerging tool.

R. Kumar (✉) · P. Maheshwary · T. Malche
Computer Science and Engineering, Rabindranath Tagore University, Bhopal, India
e-mail: rakeshmittan@gmail.com

P. Maheshwary
e-mail: pritimaheshwary@gmail.com

T. Malche
e-mail: timothy.malche@gmail.com

R. K. Shukla et al. (eds.), *Social Networking and Computational Intelligence*, Lecture Notes in Networks and Systems 100, https://doi.org/10.1007/978-981-15-2071-6_43

In this current era, need of smart cameras is to cope with serious problem like high processing demand due to increasing resources and gigantic video data management. When processing is pushed to the system edge resulting in maximum processing by the sensor, then scalable and efficient solutions are achieved. Research papers reviewed by the researcher results that the time stamping based systems for surveillance is not in common practice and use. As there are basically software development methodologies like lightweight software development methodologies and heavyweight software development methodologies are in fame. In heavyweight also known as traditional methodologies like waterfall model, the development goes in a sequence like a waterfall; i.e., the previous step ends, and next step starts. If there a change is incorporated with these developments which are heavyweight methodology–based, then the whole process will be again implemented from the beginning stage. There was some relax if the development is done with the lightweight software development methodologies; here, the small chunks are released in a short time frame so that customer relies that the work is going on and the involvement of the customer is aggressively involved. In these methodologies, the technical team retains a lot of pressure instead of the development team of heavyweight methodologies. But when there is product development related to Internet of things, then there are no such methodologies which can cover the issues which are generated during the IoT-based product development. These issues are virtualization, safety, data collection, data management and ownership, scalability and regulation, integrability, interoperability and composability, trust, privacy, identity management and security. So there is a need of methodology having software method and development techniques in such a way that the IoT-based development should be secure, fast and change-oriented which is a demand. Here in this research paper, researcher proposed a novel software development technique-based surveillance system which will cover the issues related to IoT.

2 Literature Review

Mukesh Kumar Saini in his article "From Smart Camera to SmartHub: Embracing Cloud for Video Surveillance" investigated the reason behind that why smart cameras are used at pinpoint level commercially and found that to achieve scalability, there additional constraints are put in smart cameras to improve quality. Researcher following a cloud based hub architecture, a proposed cloud entity and SmartHub the constraints can be relaxed results a solution which is scalable with minimum quality constraints. Researcher proposed a framework for designing SmartHub system for placement of given camera [1]. Saber Talari in his research paper "A Review of Smart Cities Based on the Internet of Things Concept" provides a review on smart city concept with their different applications, benefits and advantages along with the possible IoT technology introduction with capabilities [2]. Faisal Qureshi in his research paper "Smart Camera Networks in Virtual Reality" demonstrated a smart camera network which results in an extensive coverage of a large virtual public space having static and active simulated video surveillance cameras as a model. The presented model is governed by local decision making at each node along with intra node communication [3]. Yu Shi in his paper "Smart Cameras: A Review" described the smart camera technologies and applications. Researcher analysis the smart camera rapid growth along with categories and their system architectures. Researcher after examine intelligent algorithms, features and applications of smart camera [4]. Sushma. N. Nichal in her research paper "Raspberry Pi Based Smart Supervisor using Internet of Things (IoT)" implemented in an ARM Linux embedded platform-based digital video monitoring system which is real time in nature along with data control and acquisition [5]. N. Sugumaran in his research paper "Smart Surveillance Monitoring System using Raspberry Pi and PIR Sensor" described the efficient use of PIR sensor and Raspberry Pi controller in design of surveillance system which is smart in nature for mobile to increase mobile technology usage for home security. According to researcher proposed system captures information and transmit them on mail [6]. Chinmaya Kaundanya and Omkar Pathak in their research paper "Smart Surveillance System using Raspberry Pi and Face Recognition" proposed a system which will monitor and capture image when there is detection of motion. The captured image is then checked for the faces, and if the face detected is not stored in database then an alert is sent with the help of the face recognition alerts [7]. Lubos Ovseník in his research paper "Video Surveillance Systems" focused on the surveillance applications in the detection, tracking and classifying targets. In this paper, researcher described object modeling and activity analysis along with change detection. Design of proposed video surveillance system has been described by the researcher [8].

3 Methodology

The proposed novel software development technique is used to develop the surveillance system. The novel software development technique consists of three phases.

First Phase: Requirements are collected and categorized according to the defined categories like database-oriented, feature-oriented, error-oriented or others related. Here in this product, the requirements are:

- Camera should capture data in a time period; this requirement is feature- and database-oriented.
- Capture data should be saved; this requirement is database-oriented.
- Image capture should be any activity recognized by the sensor in given time frame; this requirement is feature-oriented.
- In required time period, the alert should be sent to user; this requirement is database and feature-oriented.
- The product should be Wi-Fi enabled; this requirement is feature-oriented.
- Results can be viewed anywhere in the world.

Second Phase: In this phase, the first process is the proper agile modeling implemented while it is not a thorough process but effective for documentation and modeling. In IoT-based development, the documentation is a major issue that is solved by this modeling. In this process, the empirical approach has been implemented because the proposed system should be developed in shorter time frame as a working module. Here, the Scrum an agile method is being used in a revised framework for better development. In proposed framework, Scrum consists of the following artifacts.

- Product backlog
- Sprint backlog
- Burn down chart
- Dashboard
- Quality assurance engineer (QAE).

Product backlog contains the product requirement list from customer in prioritized manner. Sprint backlog contains the list of requirement for the next sprint in prioritized manner. Burn down chart contains the sprint progress report. A burn down chart is a graphical representation of pending work versus time. Dashboard shows the status of sprint backlog items like not done, in progress and done. Quality assurance engineer role is activated from the beginning of the sprint instead of at the end of project like waterfall model. Quality assurance engineer incorporated quality thinking by uncovering business logics in project to ensure quality product development. Product backlog contains the requirement derived from the customer stories which are as follows:

- Hardware Requirement: In hardware, the microcontroller, sensors, actuators, connecting technology and power supply are required. Here, researcher uses Raspberry Pi 3 instead of other microcontrollers, because according to requirement this microcontroller is best suitable because of its feature like Wi-Fi enabled and Ethernet as a connecting technology, primary memory, storage capacity and easy implementation, and instead of NodeMCU and Arduino Uno microcontrollers. In sensor selection, researcher uses PIR sensor instead of IR sensor and ultrasonic sensor because PIR has broad range of object detection. In actuators, the researcher uses camera for the image capturing. Researcher uses 5 V DC power adapters for power supply with UPS as a backup with jumper wires as shown in Table 1. Researcher required a LCD or screen for the implementation process.
- Software Requirement: Integrated development environment (IDE) for coding, Python language, Wireless Internet.

Table 1 Showing the components used to design the surveillance system

Microcontroller	Raspberry Pi 3
Sensor	PIR sensor
Actuator	Camera
Jumper wire	Male and female cable
Power supply	Adapter of DC 5 V/2.4 A

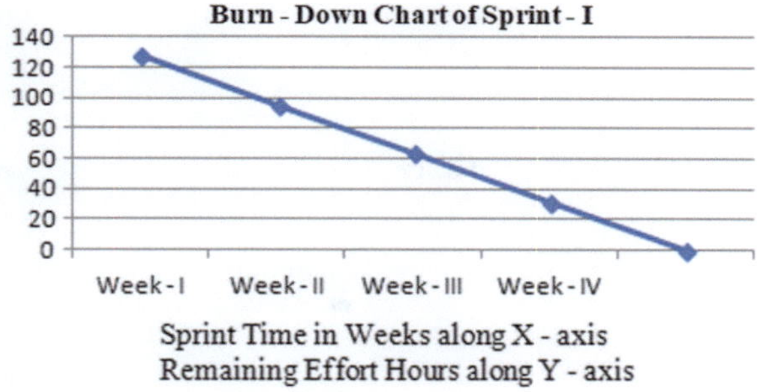

Fig. 1 Showing burn down chart of sprint—I

- Design the system.
- Explain the working of the system.

Third Phase: In this phase, the testing and release of the developed product have been encountered. In testing, the unit testing, integration testing and regression testing are being done and in release the product has been released.

Sprint—I: In this sprint, the hardware integration will be done by the team which includes the following functionalities:

- Check the working of UPS.
- Check the working of microcontroller and adaptor cable.
- Check the working sensor and testing sensor value.
- Calibration of sensor value and check the working of camera.
- Integrate microcontroller, sensor and camera, and check the working of the integrated system.
- QAE checking report of the sprint—I.
- Create the burn down chart of the Sprint—I.

The burn down chart for sprint—I is shown in Fig. 1 where the remaining efforts in hours are along *Y*-axis and sprint time in weeks is along *X*-axis. This chart shows that the effort goes down as the progress goes up.

- Create the Dashboard of the Sprint—I.

The dashboard for the sprint—I gives the work progress in the parts, i.e., not done, in progress and done. Here in Fig. 2, the progress of the work has been displayed in four-week sprint.

Sprint—II: In this sprint, the following activities will be done.

- Design data flow diagram of the system.

Figure 3 shows the data flow diagram (DFD) of level 0 in which the smart surveillance system received the motion detected by the PIR sensor and gives the outputs which are camera activation, alerts and database update.

Figure 4 shows the data flow diagram of level 1 in which the PIR sensor activates when the condition fulfills, i.e., 6:30 PM–9:30 AM. The PIR sensor detects the motion, and value is sent to the microcontroller. Raspberry Pi 3, the microcontroller,

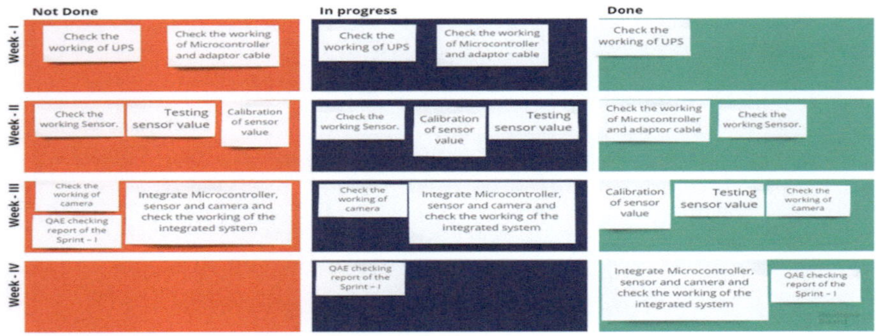

Fig. 2 Showing dashboard for sprint—I

Fig. 3 DFD 0th level

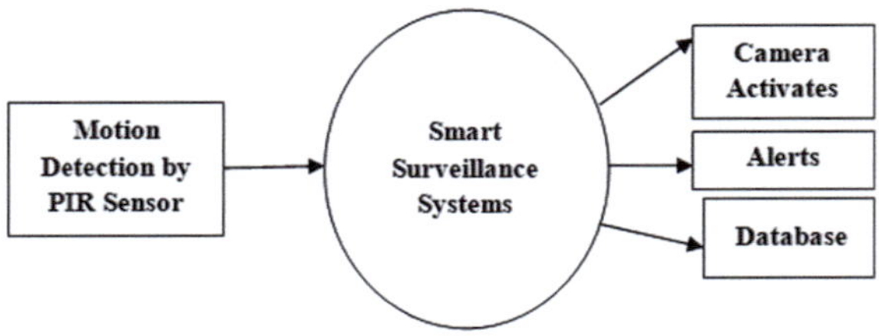

Fig. 4 DFD 1st level

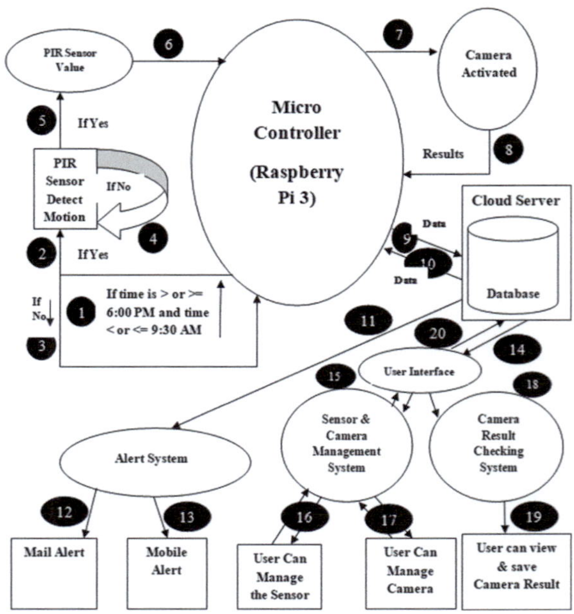

activates the camera to capture, and results reflect in the microcontroller. After that, the data of every type is sent to the cloud server where the images are stored in the database. Cloud server provides the facility like user interface and alert system module. In alert system, the user can manage sensor and camera both, and in the camera result checking system user can view and save the image. In alert system module if a motion is detected in the given period of time, then the system sends the alert to the user on e-mail and on the mobile number which is given. Cloud server has a major impact that user can view the system activities from any remote location.

- Design flowchart of the system.

In Fig. 5, the data flowchart of smart surveillance system has been shown in which the flow of data is lined. When the smart surveillance system starts, then

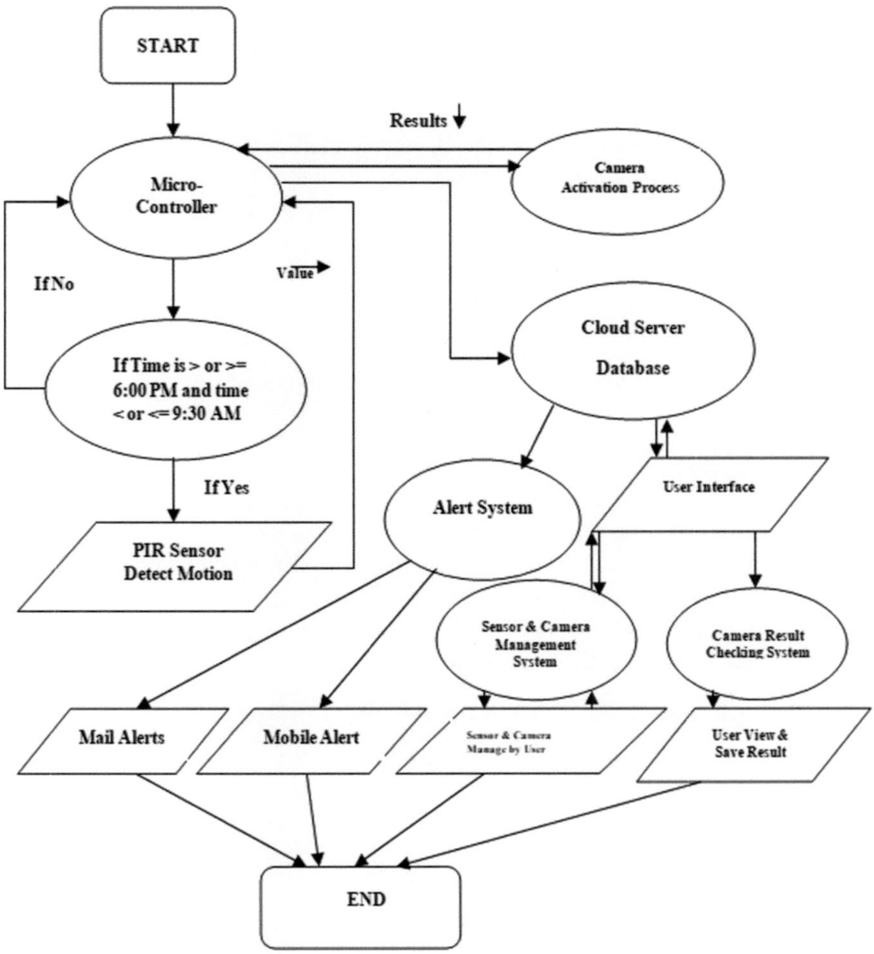

Fig. 5 Data flowchart of surveillance system

the microcontroller which is Raspberry Pi 3 here inspects the time period which is given. If the motion is detected by the PIR sensor then the value is send to the microcontroller. After that the instructions are transferred to the camera module and camera captures the images. Captured images are stored in the microcontroller and send to the cloud server with the time stamp. The cloud server provides the facility of alert system and user interface. In the alert system, the alerts are sent to the designated e-mail and mobile number of the user while in user interface the sensor and camera management system and camera result checking system provide the facility to manage sensor and camera inside the sensor and camera management system and facility to view and save the results inside the camera checking system.

- QAE checking report of sprint—II.

Sprint—III: In this sprint, the software part will be completed which includes the following

- Setting up the working environment for front end Raspbian operating system including downloading libraries and Python and at back end the cloud server.
- Unit testing the working environment.
- Setting up logical connection (coding) between microcontroller and sensor.
- Unit testing about the working of the logically connected module.
- Setting up logical connection (coding) between microcontroller and camera.
- Unit testing of the logically connected module.
- Creating the burn down chart of the sprint—III.
- Creating the dashboard of sprint—III.
- QAE checking report of the sprint—III.

Sprint—IV: In this sprint, the following activities will be done.

- Creating a database on the server.
- Creating the database connectivity.
- Unit testing of the table created.
- Setting up logical connection (coding) with wireless technology.
- Unit testing of the logical connection (coding) with wireless technology.
- Creating the burn down chart of the sprint—IV.
- Creating the dashboard of sprint—IV.
- QAE checking report of the sprint—IV.

Sprint—V: In this sprint, the following activities will be done.

- Integrated testing.
- System testing.
- Behavioral testing of the system.
- Creating the burn down chart of the sprint—V.
- Creating the dashboard of sprint—V.
- QAE checking report of the sprint—V.

Sprint—VI: In this sprint, the following activities will be done.

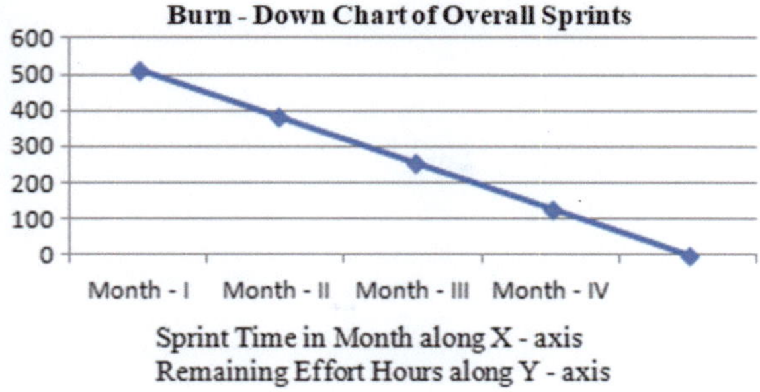

Fig. 6 Showing burn down chart of overall sprints

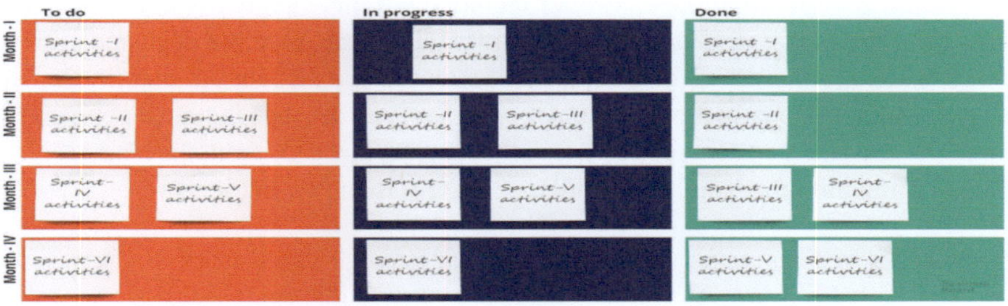

Fig. 7 Showing sprint dashboard

- QAE checking report of overall system.
- Deployment of the system.
- Maintenance of the system.
- Creating burn down chart of overall sprints.

Figure 6 shows the burn down chart of all the sprints. All the sprints go end in four months. This sprint is showing the detail of the all sprints where the sprint time in month is along *X*-axis and the remaining efforts in hours are along *Y*-axis.

- Create dashboard of overall sprints.

Figure 7 shows the dashboard of overall sprint which was designed to complete the project. Here, the work progress is displayed in three parts, i.e., to do, in progress and done, along with the month detail.

4 Working

When there is a movement before the PIR sensor at that time the time period between 6:00 PM and 9:30 AM has to be checked if the condition is true, then the PIR sensor value has been sent to the microcontroller and then according to the instruction by the controller the camera has been activated and image is received by the controller and sent to the cloud server where these images are stored in the designed database. Some

Fig. 8 Showing the working module of surveillance system

images are also stored in the controller memory, but this memory is not enough for storage. If the movements are in this given frame of time, then the designed system will send an alert to the given e-mail ID of the user and also on the mobile also by means of alert system. In the user interface, the user can manage the sensor or camera for better use and user can view and save the result at any time from the cloud server. The working module of smart surveillance system is shown in Fig. 8.

5 Results

Novel software development technique-based surveillance system works when the motion is detected by the PIR sensor in the given time frame and then the results are sent to the cloud server. The alerts are also sent to the specified e-mail ID and mobile of the user. The images are saved along with the current time stamp as shown in Fig. 9.

Fig. 9 Showing the results of surveillance system

6 Conclusion

Novel software development technique-based and IoT-enabled surveillance system is very helpful for the security point of view. Researcher after successful implementation of the system and fruitful results concluded that this system is helpful for the security at any place. The system's main advantage is that it sends the alerts to the user by means of e-mail and mobile. Another advantage is the energy-efficient nature of the system. The camera does not capture at all time; it captures only when motion is detected in the defined time period so the energy is saved. This energy-saved nature of the system will refine in the future along with the enhancement of the more features in the system.

References

1. Saini MK, Atrey PK, Saddik AE (2014) From smart camera to smarthub: embracing cloud for video surveillance. Hindawi Publishing Corporation. Int J Distrib Sens Netw 2014:1–10
2. Talari S (2017) A review of smart cities based on the internet of things concept. Energies 10(421):1–23
3. Qureshi F (2008) Smart camera networks in virtual reality. Proc IEEE 96(10):1–17
4. Shi Y, Lichman S Smart cameras: a review, pp 1–35
5. Nichal SN, Prof. Singh JK (2015) Raspberry pi based smart supervisor using internet of things (IoT). Int J Adv Res Electron Commun Eng (IJARECE) 4(7):1922–1926
6. Sugumaran N, Vijay GV, Annadevi E (2017) Smart surveillance monitoring system using Raspberry pi and pir sensor. Int J Innov Res Adv Eng (IJIRAE) 4(4):11–15
7. Kaundanya C (2017) Smart surveillance system using Raspberry pi and face recognition. Int J Adv Res Comput Commun Eng (IJARCCE) 6(4):621–624
8. Ovseník L (2010) Video surveillance systems. Acta Electrotechnica et Informatica 10(4):46–53

An Investigation on Distributed Real-Time Embedded System

Manjima De Sarkar, Atrayee Dutta and Sahadev Roy

Abstract This paper deals with the working of multiple processors systems. Embedded system consists of a large number of processing elements which make the system become more complex. Real-time embedded systems are widely being connected with the wired as well as wireless network. To design any distributed system, the performance analysis is an important consideration. Performance depends on the scheduling methods, i.e. process allocation, which is also discussed in this paper.

Keywords Distributed systems · Embedded systems · Performance analysis · Real-time systems · Scheduling

1 Introduction

The demand of high performance in embedded system [1] is increasing in our day-to-day life as well as in various applications like medical unit, security, industrial and automation. To meet these criteria, system designed by using single processor will not satisfy the basic timing criteria. Therefore, the system development trend changes from centralized to distributed systems. Distributed system [2] has higher degree of heterogeneity with the similarity of other real-time embedded systems. Any distributed real-time embedded system is heterogeneous in different aspects like hardware structure, software components, communication protocols and memory allocations. These complexities introduce new challenges in coordinating and maintaining the system.

The demand of high performance in embedded system is increasing in our day-to-day life as well in such as various applications like medical unit, security, industrial

M. De Sarkar · A. Dutta (✉) · S. Roy (✉)
Department of Electronics and Communication Engineering, National Institute of Technology Arunachal Pradesh, Papum Pare, Yupia, Arunachal Pradesh 791112, India
e-mail: atrayee2013@gmail.com

S. Roy
e-mail: sahadevroy@nitap.ac.in

M. De Sarkar
e-mail: manjima19@gmail.com

R. K. Shukla et al. (eds.), *Social Networking and Computational Intelligence*, Lecture Notes in Networks and Systems 100, https://doi.org/10.1007/978-981-15-2071-6_44

automation, etc. To meet these criteria, system designed by using single processor will not satisfy the basic timing criteria. Modern embedded system consists of more numbers of programmable CPUs and integrated circuit. Software platform executes the programmable CPUs, combine and split into several concurrent processes. High-timing accuracy is required for distributed real-time embedded system. Timing analysis plays an important role which must be taken care of during the hardware–software co-synthesis process.

This paper will describe about:

- The industrial platform of the computational embedded systems in actuator, different kinds of sensors, industrial-automated vehicles, robots, etc.
- Technology updated on real-time distributed and embedded systems and how this changes shall implement on the manufacturer to control any industrial-automated systems.
- Development of technologies which target to expand distributed real-time embedded systems for industrial purpose.
- Different architecture and protocols are studied in distributed embedded system.

Structure of this paper is as follows: Sect. 2 describes about proposed work of the distributed real-time embedded systems and Sect. 3 tells about the task scheduling concepts. Finally, Sect. 4 gives concluding remarks.

2 Generic Architecture of Real-Time Embedded System

Distributed system requires more than one control units for a single system. Each control unit is directly connected with the sensors and actuators may work independently and complete any subprocess or multiple subprocess under the control of the main processor. Generally different actuator and sensors are directly connected with the processors to perform dedicated tasks. This kind of sub network is called 'node' [3]. In the distributed embedded system, the whole system is constructed in a manner that all the nodes are connected with another node or only with the main host depending on the application of each node, one is directly connected with the actuators and the sensors, the operation performance improved, higher degree of transparency compared to centralized and decentralized system. One of the most important factors in real-time embedded system is meeting the timing requirement. It is essential for normal operation as well. If a single deadline is missed, it gives cataclysmic result. Multiple processors work together to meet the deadline criteria of the different tasks.

Generic model of distributed embedded system is presented in Fig. 1. Four processors are performed in a single task. For an example, if three robot manipulators are controlled by a central processor. Robot [1] manipulators will work by three independent processors. Their task is to pick and place and is initiated by central processor. Let central processor is equipped with colour sensors. When a particular colour is detected, a particular robot will start work. The first one is main processor

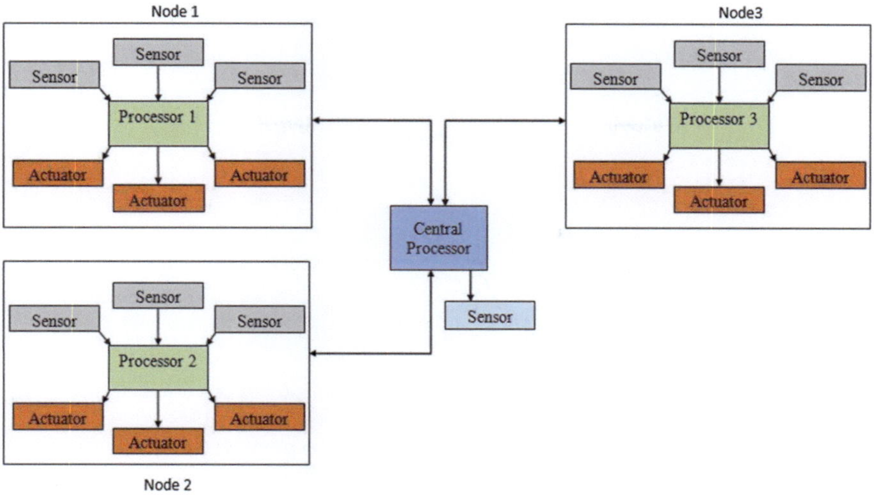

Fig. 1 Generic model of distributed embedded system

(P_{Host}) and rest three (P_1, P_2, P_3) are independent processors in the sense that they are not working with the same clock pulses but their activity will be controlled by the central processor. The timing analysis is important as they do not run on the same clock pulse. Not only the clock pulse but their processes are also different and they work independently. The main processor (P_{Host}) triggers the other processors. When the individual task is completed by the other nodes, they send a signal to the main processor that its task is over and it is free. Therefore, main processor (P_{Host}) exhibits time management. Basically, the work is to involve in the processor to know how it scheduled the task dynamically. Thus, the concept of dynamic scheduling arrives in the main processor (P_{Host}).

The task of the main processor (P_{Host}) is to take the decision that which processor will activate next task. For this purpose, it uses the colour sensor for detecting three colours (red, green and blue). These three colours are detected by the main processor (P_{Host}) and it detects the colour particularly in three band areas that one segment for red colour, and other two segments for green and blue colours, respectively. It identifies the colours based on the coloured three robotic arms which are driven by other processors (P_1, P_2 and P_3). When red colour is detected, one robotic arm will be in action, similarly for detection of green and blue colour, other two robotic [4] arms will go on action. When the process of the robot 1 is over, the main processor will send a signal to processor 1 for the next cycle. According to the instruction, robot 1 will do the process and the process will be already in the program for processor 1 (P_1). So it is not dependent on the main processor (P_{Host}) except the interrupt which will come from the main processor (P_{Host}). When green and blue colours are detected, same things will happen. The other processors (P_1, P_2 and P_3) will work in the same manner. When the processor P_1 is working, the other two processors (P_2, P_3) can also work in parallel manner because all the processors are independent of

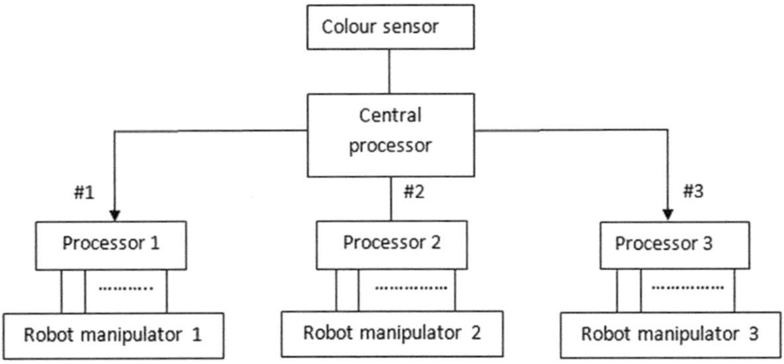

#1: Task initiated when red colour detected and processor1 is free.
#2: Task initiated when green colour detected and processor2 is free.
#3: Task initiated when blue colour detected and processor3 is free.

Fig. 2 Basic block diagram of multiprocessor embedded system

each other. However, the main processor (P_{Host}) sends the triggered signal for the other processors (P_1, P_2 and P_3) as to how much time is required to complete it is undetermined and a specific task is loaded on it. We do not need to calculate this because when the task is completed, it gives the ready signal to the main processor (P_{Host}) (Fig. 2).

3 Task Scheduling and Resource Sharing in Distributed Embedded Systems

A real-time system generally consists of a set of periodic tasks and each task has a deadline, they must be complete within the deadline. A futuristic performance analysis is important for software and hardware co-design. Previously, in most embedded systems, various processors can run simultaneously on a single CPU and get involved with one another accordingly to the scheduling [5]. The total execution time of any processor consists of the request and finish of the process. Task scheduling in cloud computing area is recent and burning topic of research and it is also known as NP-complete problem. The process of allocating the task in the available virtual machine is known task scheduling. The concept behind the task scheduling is that one task run until it fulfils its task and then terminates. Then next task will start running in the same process. The simplicity of task scheduling is the drawback also. Each task allocation of the total process is affected by others. Scheduling algorithm depends on task goal. If there is a lot of work and all work has to complete within the deadline, then scheduling is the best approach. Task scheduling having three-stage process such as find priority of the task of given DAG using priority attribute method. Sort the task as per priority value of the tasks and finally allocate these tasks in the available virtual machine. There are number preassumed which are number of tasks in DAG known in advance, allocation of tasks statistically, no deadline for task, task priority known in advance and allocation in batch mode as per priority. Various types of different task scheduling such as preemptive, non preemptive, static, dynamic, distributed and centralized which has owned merits and demerits.

A most common scheduling [6] which is widely used is 'Rate Monotonic Scheduling' (RMS) [3] which is also used in distributed real-time embedded system as all processors are independent. In this case of fixed priority, rate monotonic scheduling is applicable.

Let the total period is P_i and each process is characterized by its execution time C_i to complete any operation. Deadline of the process is at the end of its period. The processes may have different periods. There is a set of n independent periodic task which are scheduled by the rate monotonic scheduling algorithm. Here, 'n' number of independent task must meet their deadline [7]. Total period P_i divided into n independent task period. Let for task 1, the time period is T_1, for task 2, time period is T_2 and so on. Utilization of the resources is defined by C_n, where C is T_n n the execution time for 'n' tasks and T_n is the time period for 'n' tasks.

If we summarize the whole tasks, it becomes

$$\frac{C_1}{T_1} + \frac{C_2}{T_2} + \cdots\cdots + \frac{C_n}{T_n} \leq n\left(2^{\frac{1}{n}} - 1\right)$$

For a set of n independent numbers of periodic tasks, if the task meets its first deadline, then $D_i < T_i$,that is deadline must be smaller than the time period. When the entire higher priority tasks are started at any instant of time, then the task must meet its entire future deadline which is the start time of another task.

$$W_n(t) = C_1\left(\frac{t}{T_1}\right) + C_2\left(\frac{t}{T_2}\right) + \cdots\cdots + C_n\left(\frac{t}{T_n}\right) = \sum_{j=0}^{n} C_j\left(\frac{t}{T_j}\right)$$

Each task has two inbuilt nodes—start node and end node. These two nodes are used to explain any algorithm. Each task has a period and a deadline. Period is the time between two consecutive initiations and deadline [5] is the maximum time allowed from initiation to termination of the task. A tasks deadline is satisfied if it is greater than or equal to the worse case delay.

To analyse the timing behaviour of distributed embedded system, it is required to develop a system in real time where input of the system is considered in each and every moment and based on the inputs, the system computes some specific tasks.

A computational resource plays a vital part to form a bigger system. According to the Moore law, at every 18 months, the number of the transistors on a single die becomes double. As the number of the transistor increases, complexity of the system also increases. In the fabrication processes, the semiconductor companies use more number of the processors and memories on a single chip.

The convergence has helped exterminate the silos between operational methodology and information technology, enabling unstructured machine-generated data to be analyzed for insights to drive improvements [8]. As RF devices have low cost, it give new access for plant observation and control the system also. A new trend observed in real-time [9] embedded system is rearrangement of the processor themselves either by imposing extra hardware which is connected with the instruction set or by changing the instruction set. As system complexity increased, cost of the whole system also increased. From the manufacturing and controlling point of view, the number of control point and observation point might increase without using extra wire for connection and large investment. We must say that high processing power in the embedded system on chip gives software upgradation at the higher level with minimum developing cost.

Hardware components in system on chip consist of one or multiple processors, memories, different components and interfaces [10]. All the components are directly connected by communicating network which lies from simple to hierarchical buses.

As given in Fig. 3, the embedded system on chip design starts with definition and authorization of a functional specification which is not depended on the architecture. The next part of chip design is architectural designing which is followed by authorization of the functional specification. Nowadays, architectural designing part works in manually. The term hardware/software partitioning [11] comes from mapping implementation where some functions are worked as software tasks, and

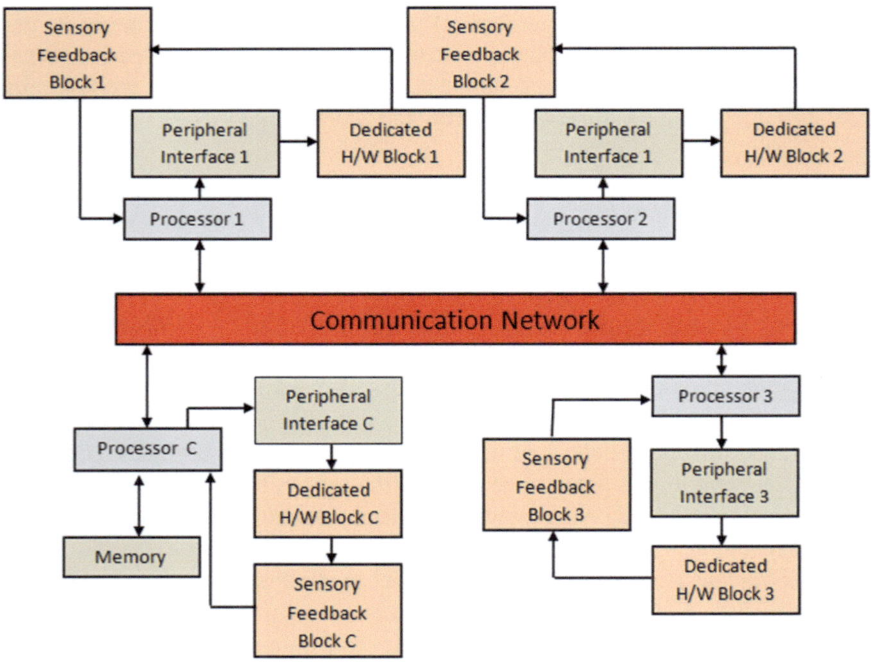

Fig. 3 Communication network

some worked as hardware task. Partitioning concept works automatically to support multiprocessor system, interface and share with hardware [12]. Comparing with both software and hardware parts, software part is easily modified. But the hardware part gives better performance. Combining different individual applications makes a complete distributed system. Individual application must be connected to each other for better communication. An application must know its own current status. There must be a middleware technology which can improve the distributed system for better performance. Because of the presence of the middleware, we can separate application from the operating system also. Middleware is a kind of software that gives basic functional units and helps to perform the task accurately. It must be designed transparently, so that the user will not consider it in the hardware structure. The quality of the middleware is depended upon the transparency level. There are different approaches to build up a middleware. If the middleware took care of the operation for individual system, then it will be more complicated. That's why the term interface comes. Interfaces must be applied for different individual operating system having the same mechanism and it can also hide the complexity of the system. Middleware has other advantages, like if we want, we can integrate any new technology after development. There are several kinds of middleware are available in worldwide. Microelectronics and software are allowing embedded system to merge a set of processing elements. As environment is dynamic, the total system must be operated challengingly. If we want to change any environment, there must be capability of extra adaptive power from the software and hardware resources. Middleware consists of different individual layers. Each of the layers concentrates on different aspects but they have a common idea to execute the whole distributed system successfully. Any distributed systems can come under the region of real-time system, embedded system and distributed domain also. Real-time system not only required the right performance but also required to execute within the accurate time. The system has the ability to handle the running task within the dead line. Initially, execution time, period and deadline must be considered for the task purpose. As there is no shared memory, so communication happens between two nodes by passing the messages. Communication and connections between nodes may be complicated due to its hierarchical network architecture. In any distributed system, several subsystems are used and each has different individual protocol for communication along with individual operating system. Hence, heterogeneity comes in the system. To overcome this heterogeneity, a middleware must be implemented to achieve the accurate communication between each nodes of the system. Middleware is a kind of software that gives basic functional units and helps to perform the task accurately. It must be designed transparently, so that the user will not consider it in the hardware structure. The quality of the middleware is depending on the transparency level. There are different approaches to build up a middleware. If the middleware takes care of the operation for individual system, then it will be more complicated. That's why the term interface comes. Interfaces must be applied for different individual operating systems having the same mechanism and it can also hide the complexity of the system. Presence of the interface gives good flexibility and scalability of the middleware system. Middleware has such kind of mechanism which can allow the system to run as well as hide the failure also if any component failing happens. The advantage of using the middleware is that the system should not compromise breakdown for a single component failure. A successful middleware should survive with the system and avoid single node failure also at the same time. Generally, middleware operates and describes the protocols that are used in the system for the application development, as each node in the distributed real-time system consumes more than enough processing capabilities. Communication protocols are established that give a node to manage a communication from one to another node. Network protocol in distributed real-time embedded system based on the concept that one node acts like monitor and it is responsible for monitoring and also regenerating if the node has been lost. The protocol gives the facilities to transmit any message in the whole system.

In distributed real-time embedded system, real-time communication protocols are the important components to make the system punctual and safe. Communication protocol is applied in both software and hardware part [13]. There is no existence of computer network without communication protocol. It is normally described in layered architecture model. In industrial plant or in embedded system Ethernet is also an important consideration.

There are some challenging approaches comes in distributed real-time embedded systems. Different Quality of Service (QoS) properties play an important role in the real-time applications. QoS defines the understanding of overall performance of any service. This service may be cloud computing or may be computer network or any kind of networking under this category. QoS is especially important in few particular applications. Services are categorized in different levels. Levels may be environmental condition, may be physical or cost also. One level of service must be coordinated with other levels. In the distributed embedded system, various micro controllers are connected with inputs individually. As all microcontrollers work independently, so there is no connection between one input to the other input. As various resources are used to build a distributed real-time embedded system, there may be chance for reduction of the competition.

At the initial condition when we establish the system, it becomes quite hard to compare the analysis of the system. All the nodes in the distributed systems are connected by one wire cabling in a simplified manner. So if any failure happens, it will be easily identified and recovered due to simple connectivity.

4 Conclusion

Nowadays, to design any embedded system, 90% of the system is on software. A real-time operating system must satisfy the deadline of the task and the frequency of activation. Few restrictions have an effect on scheduling and on response time. Most important guideline for system on chip designing is the clear understanding between function and architecture and between communication and computation. Our work is to investigate the multiprocessor scenario. Sometimes, tasks are so complicated that our processor is not able to handle the whole task where multiple processors are essentially required. The performance of various regions in the system is examined and corresponding results also evaluated. It can be easily concluded that the system is functioning in distributed system manner.

References

1. Gajski DD, Vahid F (1995) Specification and design of embedded systems. IEEE Des Test Comput 12(1):53–67
2. Pereira CE, Carro L (2006) Distributed real-time embedded systems: recent advances, future trends and their impact on manufacturing plant control. IFAC Proc 39(3):21–32
3. Sha L, Rajkumar R, Sathaye SS (1994) Generalized rate-monotonic scheduling theory: a framework for developing real-time systems. Proc IEEE 82(1):68–82
4. Ha S, Coros S, Alspach A, Bern JM, Kim J, Yamane K, (2018) Computational design of robotic devices from high-level motion specifications. IEEE Trans Robotics 34(5):1240–1251
5. Yen TY, Wolf W (1998) Performance estimation for real-time distributed embedded systems. IEEE Trans Parallel Distrib Syst 9(11):1125–1136
6. Bakshi S, Gajski DD(1999) Partitioning and pipelining for performance-constrained hardware/software systems. IEEE Trans Very Large Scale Integr VLSI Syst 7(4):419–432
7. Zhao Y, Gala V, Zeng H (2018) A unified framework for period and priority optimization in distributed hard real-time systems. IEEE Trans Comput-Aided Des Integr Circuits Syst 37(11):2188–2199
8. Palencia JC, Harbour MG, Gutiérrez JJ, Rivas JM (2017) Response-time analysis in hierarchically-scheduled time-partitioned distributed systems. IEEE Trans Parallel Distrib Syst 28(7):2017–2030
9. Durairaj G, Koren I, Krishna, CM (2001) Importance sampling to evaluate real-time system reliability: a case study. Simulation 76(3):172–182
10. Bolsens I, De Man HJ, Lin B, Van Rompaey K, Vercauteren S, Verkest D (1997) Hardware/software co-design of digital telecommunication systems. Proc IEEE 85(3):391–418
11. Xie Y, Li L, Kandemir M, Vijaykrishnan N, Irwin MJ (2007) Reliability-aware co-synthesis for embedded systems. J VLSI Sig Proc Syst SIViP 49(1):87–99.
12. Smith RG (1980) The contract net protocol: high-level communication and control in a distributed problem solver. IEEE Trans Comput 12:1104–1113
13. Singhal G, Roy S (2019) A novel method to detect program malfunctioning on embedded devices using run-time trace. In Advances in signal processing and communication, Springer, Singapore, pp 491–500.

Real-Time Robust and Cost-Efficient Hand Tracking in Colored Video Using Simple Camera

Richa Golash and Yogendra Kumar Jain

Abstract Dynamic hand gesture recognition field has high potential to change the interaction mechanism between human and machine. But user interfaces (UIs) working on hand movements are still a challenge because of the lack of cost-effective and robust hand tracking techniques. To avoid the challenges encountered in tracking, non-rigid and subtle object-Hand, researchers use advance cameras which overall increases the cost and complexity of any technique. In this paper, we have focused on two important stages of dynamic hand tracking, first is hand modeling and second is robust hand tracking. We have developed a prototype of hand tracking using graphical user interface (GUI) of MATLAB software, working on live videos captured using a normal camera. The proposed system is tested on videos in intelligent biometric group hand tracking (IBGHT) database.

Keywords Human gesture recognition · Computer vision · Feature extraction · SIFT · Normal camera · Depth images · Hand tracking

1 Introduction

Computer vision and pattern recognition have made possible to design and develop natural user interfaces (NUIs). Researchers have acknowledged the fact that field of dynamic hand gesture recognition can bring revolutionary change in the interaction mechanism between human and machine. But NUI's working on hand movements are still a challenge because of the lack of cost-effective and robust hand tracking techniques [1–3].

One of the most common and severe problem encountered by researchers is freedom factor in hand skeleton, that create occlusion while recording hand movement [3–6]. Another problem is effect of illumination and non-uniform pattern of hand movement which produces blurredness in images. Trajectory of hand motion for any movement also has less correlation with its previous movement These are some of the most prominent reasons that scholars are forced to apply static background condition.

R. Golash (✉) · Y. K. Jain
E&I Department, Samrat Ashok Technological Institute, Vidisha, Madhya Pradesh 464001, India
e-mail: golash.richa@gmail.com

R. K. Shukla et al. (eds.), *Social Networking and Computational Intelligence*, Lecture Notes in Networks and Systems 100, https://doi.org/10.1007/978-981-15-2071-6_45

The current state of art indicates that hand tracking avoids the use of normal camera and relies more on advance camera which works on depth images [5–9], as colored images obtained from normal camera are sensitive to variation in illumination.

This paper proposes a novel methodology to track hand motion, it is unique in two ways, first the video is captured using normal camera and we have worked completely on colored images. The second distinguishing factor is the methodology to create hand model by applying connected component on CIE color space, without doing any background segmentation. This hand model is independent of hand structure. The use of local feature for hand detection in subsequent frames and tracking by the use of second-level scale-invariant feature transform (SIFT) matched feature give robustness against variation in illumination. The proposed algorithm is cost efficient and makes the real-time hand tracking free from complex computation.

As per the literature survey, the two important aspects in dynamic HGR are hand initialization and hand tracking. Initialization of hand, this stage is very important as it gives foundation for detecting correct ROI in the remaining frame. One of the common approaches for detecting hand is motion history images produced by accumulating difference of two consecutive images. Shan et al. [10] have detected moving hand region by applying logical AND operator on accumulated difference images and color probability distribution. Weights are given according to the distance between target and sample object and then tracked using mean shift embedded particle filter (MSEPF). [11, 12] have used similar technique for hand detection and after creation of hand ROI [11] have used adaptive Kalman filter (AKF) to estimate hand location. Kalman filter considers object as a set of moving pixel and thus gives error if the number of frames increases, to overcome this inconsistency, Asaari have used Eigen values of hand as features. In [12], Joo et al. features of depth image difference are used as a classifier for seed point. This classifier is then tracked using depth adaptive mean shift algorithm. In [9], Park et al. have accumulated, depth difference images using 3D depth Prime sensor and ToF camera. They have discussed that depth images include lots of noise at the edges of the object; hence require spatial filter and morphological operation. Initial hand region is detected using clustering method. After initial hand localization, it is further tracked using Kalman algorithm, the overall process of tracking is complex in this method.

In [13], an indirect approach is used to find initial ROI, they have first detected face by Viola Jones algorithm and then by using skin histogram, hand region is determined. Target is tracked by iteration process using continuously adaptive mean shift algorithm (CAMShift). Our work is influenced by [14] where Bao et al. have proposed tracking of hand movement direction without any segmentation, using SURF feature. In this algorithm, it is assumed that hand covers large area in a frame and it has substantial displacement between two adjacent frames. This work was improved further by Yao et al. in [15] by considering some range in displacement parameter for matching key-points of current frame with previous frame. But the pruning process used to eliminate outlier SURF point increases ROI every time, which make overall process difficult.

2 Proposed Algorithm Description

The complete system is divided into two stages, first stage deals with detection of hand and defining the region of interest (ROI). The second stage is hand localization in each frame and tracking using centroid of the matched features. The centroids of each frame are plotted to track the path of hand movement. The pictorial description of the methodology is described in Fig. 1.

2.1 Defining ROI

In our case, we have combined color clue and connected component to find hand region in the initial frame. Hand is a biological area, one of the best way to locate hand is by using skin segmentation. In vision-based image processing, RGB, YCbCr, HSV, and CIE are commonly used color spaces. The ideal color space for detection of skin region is the one, which separates luminance component from the color component. In RGB color space, R, G, and B components are highly correlated to each other, i.e., if the intensity changes, all the three components will change accordingly. In case of using YCbCr color space, segmentation results in blurred image and thus object detection is difficult. In contrast to RGB and YCbCr, CIE $L^*a^*b^*$ color space gives clearer image for object detection, after skin color segmentation [16].

In this work, we have used '$L^*a^*b^*$' channel of CIE color space, it is basically designed to approximate human vision. 'L' channel stands for luminance (Lightness) and 'a' and 'b' are two color channels. This color space is perceptually uniform, and its 'L' component has very close match with human perception of lightness. In this

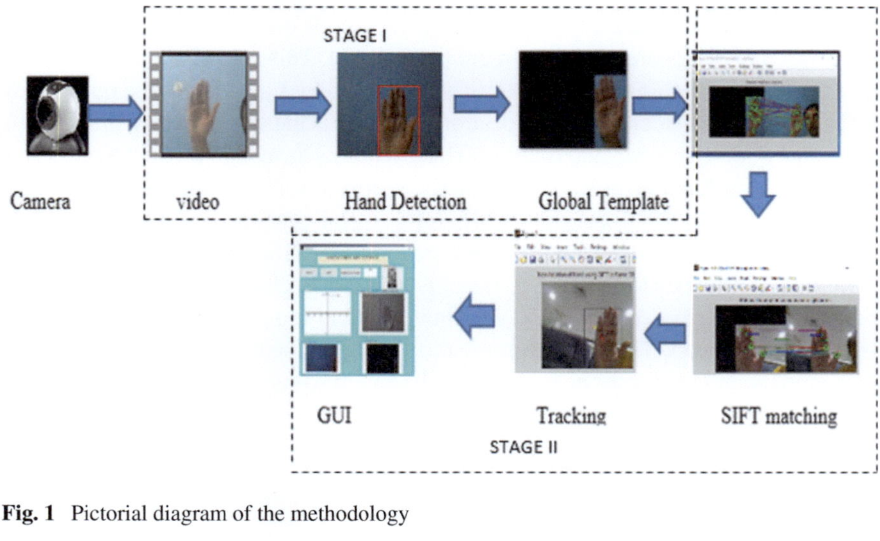

Fig. 1 Pictorial diagram of the methodology

color space, the value of the channels is determined as follows:

$$L_{xy} = 116.h\left(\frac{Y}{Y_w}\right) - 16 \tag{1}$$

$$a_{xy} = 500.\left[h\left(\frac{X}{X_w}\right) - h\left(\frac{Y}{Y_w}\right)\right] \tag{2}$$

$$b_{xy} = 116.\left[h\left(\frac{Y}{Y_w}\right) - h\left(\frac{Z}{Z_w}\right)\right] \tag{3}$$

where

$$h(q) = \begin{cases} \sqrt[3]{q} & \text{if} \quad q > 0.008856 \\ 7.787q + 16/116 & \text{otherwise} \end{cases} \tag{4}$$

And X, Y, Z are tristimulus and X_w, Y_w, Z_w are white tristimulus.

In real-time system, color information is not sufficient to extract hand template; therefore, to make the ROI detection robust, we have combined color thresholding and connected component method to determine ROI. The overall advantage of using this algorithm is (1) The technique is simple to design even in presence of natural noise occurred when video is captured. (2) Template based on the color image are more informative in extracting SIFT features in the system design.

2.2 *Hand Localization and Tracking*

Tracking of an object means finding object in every frame and then locating its center of movement in each frame. Efficient tracking is one of the crucial factors in HGR which operates on hand movement [5, 6, 9, 11]. This part of algorithm is built on staged filter-matching approach of SIFT proposed by Lowe [17]. The purpose of SIFT algorithm is to identify those locations in image scale space that are invariant to image translation, scaling, and rotation. These features have high distinctiveness and better detection accuracy toward local image distortions and viewpoint change. The rich information in SIFT descriptors is helpful in real-time fast and exact matching of the object even in the presence of background noise [18–20].

Region of interest (ROI) created in stage I is used as global template throughout the matching process. The SIFT key-points of global template are saved as global key point descriptors. The key-points are the extremum location at all pyramid level. Low contrast and responses at an edge are discarded using predetermined value:

$$D\left(\widehat{X}\right) = D + \frac{1}{2}\frac{\partial D^T}{\partial X}\widehat{X} \tag{5}$$

$\widehat{X}$ is calculated by setting $D(x, y, \sigma)$ to zero

The most important step in real-time tracking is accurate detection of hand in each frame. Hence, robust matching is crucial step in tracking. For this stage, we have used modified *k-d* tree algorithm known as the best-bin-first search method proposed by Lowe. This process identifies the nearest neighbors with high probability using only a limited amount of computation. The nearest neighbors are defined as the key-points with minimum Euclidean distance from the given descriptor vector.

$$\text{distance } (a_G, b_T) = \sqrt{\sum_{i=1}^{128} (a_i - b_i)} \tag{6}$$

where a_G and b_T are descriptor vector in global template and the current frame. To remove ambiguous matches and to identify closest match for each descriptor a_G in the global template, second-level match is performed. In this, two nearest neighbors in the current frame are determined. Let they are b_T and c_T, and their distances from a_G are distance (a_G,b_T), and distance ($a_{G,}c_T$). The ratio distance$(a_G, b_T)\big/$ distance $\left(a_{G,}c_T\right)$ is calculated and a threshold of 0.8 is applied in our method. If the obtained ratio is less than 0.8, then b_T is selected as close match to a_G as compared to c_T and vice versa. Out of all the unique and distinctive features obtained in the previous step, the feature with minimum distance is selected as the dominant feature. A bounding box equal to ROI is created around the dominant feature.

3 Experimental Results

We have used normal camera enabled system to capture hand motion. Each image sequence consisting of 100–120 frames. Technique is implemented on MATLAB-8 software and results of tracking are demonstrated with the help of graphical user interface (GUI) of MATLAB. The proposed methodology is also tested on intelligent biometric group hand tracking (IBGHT) database [21]. This database is generated for purpose of research and development in the area of visual hand tracking. The algorithm of the proposed method is given as:

Algorithm 1

1. Input: real time recording of video. Convert it into frames.
2. Choose initial search window size and resize all frames according to that size.
3. Initial frame is selected and ROI enclosing hand region is detected using CIE color space and connected component method.
4. Tracking of hand region using SIFT.

   a. Determine the SIFT feature from global hand detected in step: 2. Save the feature descriptor in temporary folder.

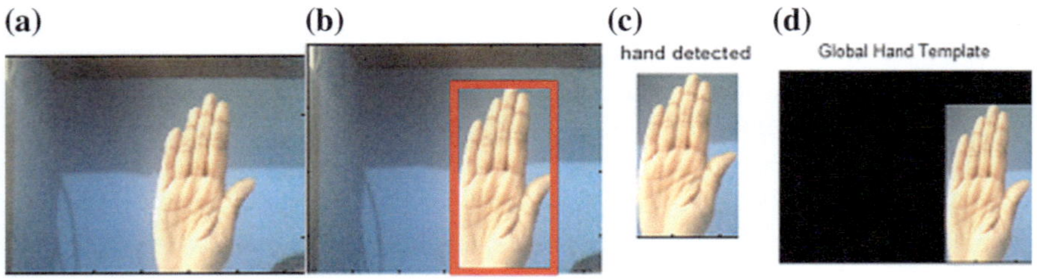

Fig. 2 Steps to create hand model for a particular input video. **a** Initial frame selection. **b** Hand detection. **c** Extraction of hand template. **d** Creation of global hand template with frame size for SIFT feature matching

   b. Determine SIFT feature in every current frame.
   c. Match descriptor of global template with the current frame descriptors.
   d. Determine the 2nd nearest neighbors of the matched descriptor.
   e. The key-centroid is the closest match from all 2nd nearest neighbor.
   f. Based on the key-centroid determine the new position of bounding box.

5. Repeat step 5–step 8 till the frames assigned are finished.
6. Plot of all new centroid will be the tracking path of hand.
7. Demonstrate using Graphical User Interface (GUI).

Figure 2 shows the steps to create hand model in stage I. Result demonstrates that technique is robust with an average efficiency of 88.75%. Figure 3 shows SIFT feature matching of global template with every current frame coming. Global template of hand for that particular video is concatenated with every current frame. Figure 3a is the result of first nearest neighbor matching and Fig. 3b is the result of second-level nearest matching of features. Out of all second nearest pair features, feature with minimum distance is selected as dominant feature or the centroid of hand in that frame. The result, reflect the robustness and efficiency of SIFT algorithm in finding the spatial and temporal properties of hand in every current frame. Figure 4 and shows tracking of hand movement of different shape and background conditions. The proposed methodology is also tested on IBGHT database [21] videos and gives an average efficiency of 93.1%.

4 Conclusion

The various results show that the proposed methodology is capable of tracking non-rigid object-Hand, efficiently and in cost-effective manner. The proposed methodology is also tested on IBGHT database [21] videos and gives an average efficiency of 93.1%. The work demonstrates that technique is user independent and use of local feature SIFT gives better result for tracking non-rigid object-Hand in real time and can track efficiently even if normal camera is used. In the future, the technique shall

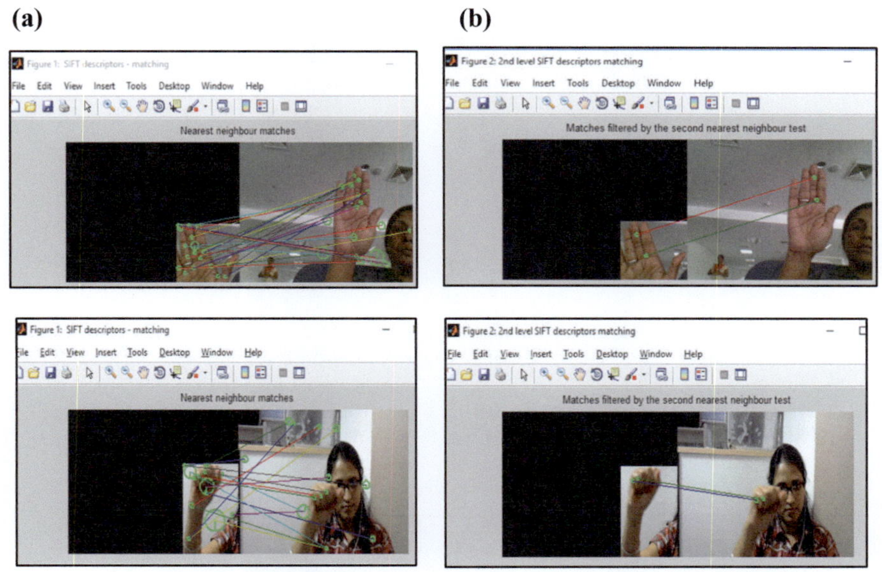

Fig. 3 Two-level SIFT feature matching of global template with every current frame. **a** Euclidean distance matching. **b** Threshold ratio match

be extended to complex real-time background and will track multiple hand movement in a video. The technique can be further developed to create complete contactless user interface for any appliance based on hand movement.

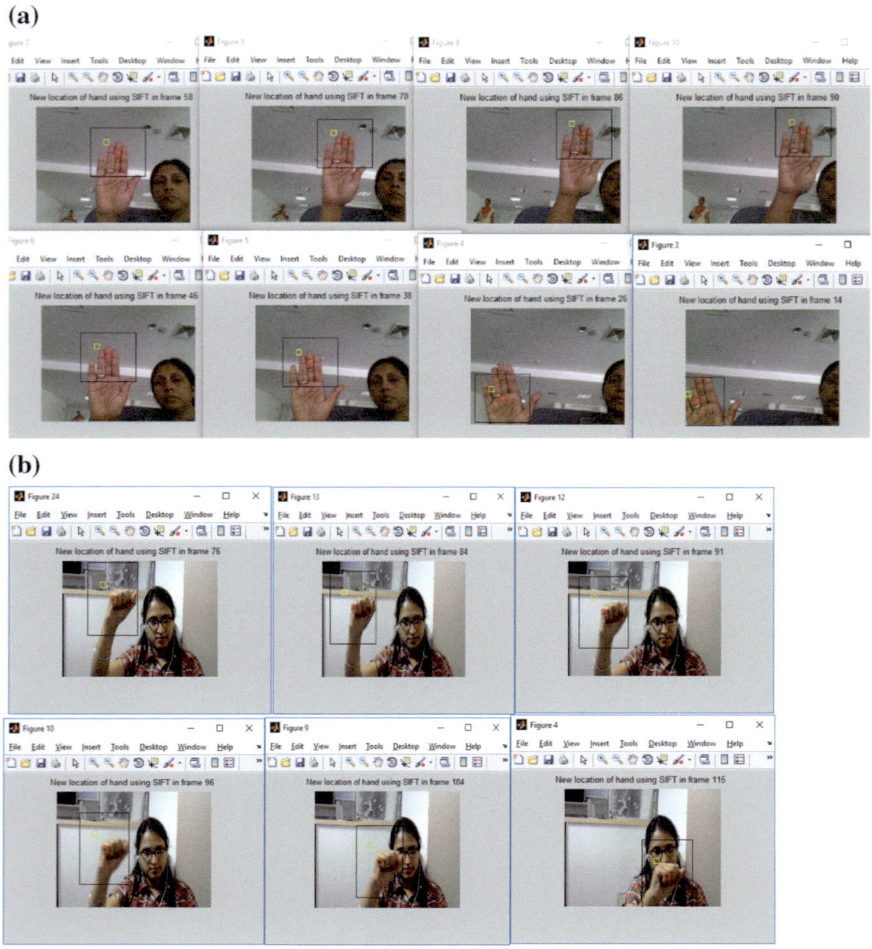

Fig. 4 Tracking result of hand movement. **a** Frame no. 14, 26, 38, 46, 58, 70, 86, 90 in real-time background and, **b** Frame no. 76, 84, 91, 96, 104, 115 from video of IBGHT data set

References

1. Wachs JP, Kölsch M, Stern H, Edan Y (2011) Vision-based hand-gesture applications. Commun ACM 54:60–71
2. Yang S, Premaratne P, Vial P (2013) Hand gesture recognition: an overview. In: Proceedings 5th IEEE, IC-BNMT, Guilin, China, pp 63–69
3. Pisharady PK, Saerbeck M (2015) Recent methods and databases in vision-based hand gesture recognition: a review. Comput Vis Image Und 141:152–165
4. Stergiopoulou E, Sgouropoulos K, Nikolaou N, Papamarkos N, Mitianoudis N (2014) Real time hand detection in a complex background. Eng App Artif Intel 35:54–70
5. De Smedt Q, Wannous H, Vandeborre JP (2016) Skeleton-based dynamic hand gesture recognition. In: Proceedings IEEE-CVPRW, Las Vegas, USA, pp 1–9
6. Chong Y, Huang J, Pan S (2016) Hand gesture recognition using appearance features based on 3D point cloud. J Softw Eng Appl 9:103–111
7. Suarez J, Murphy RR (2012) Hand gesture recognition with depth images: a review. In: Proceedings 2012 IEEE RO-MAN, Paris, France, pp 411–417
8. Fu Q, Santello M (2010) Tracking whole hand kinematics using extended Kalman filter. In: Annual international conference of the IEEE engineering in medicine and biology, pp 4606–4609
9. Park S, Yu S, Kim J, Kim S, Lee S (2012) 3D hand tracking using Kalman filter in depth space. EURASIP J Adv Signal Process 2012(1):36
10. Shan C, Wei Y, Tan T, Ojardias F (2004) Real time hand tracking by combining particle filtering and mean shift. In: IEEE international conference on automatic face and gesture recognition, pp 669–674
11. Asaari MS, Rosdi BA, Suandi SA (2015) Adaptive Kalman filter incorporated eigen hand (AKFIE) for real-time hand tracking system. Multimed Tools Appl 74:9231–9257
12. Joo SI, Weon SH, Choi HI (2014) Real-time depth-based hand detection and tracking. Sci World, Article ID 284827, http://dx.doi.org/10.1155/2014/284827, vol 17
13. Kovalenko M, Antoshchuk S, Sieck J (2014) Real-time hand tracking and gesture recognition using semantic-probabilistic network. In: Computer modelling and simulation (UKSim), IEEE, pp 269–274
14. Bao J, Song A, Guo Y, Tang H (2011) Dynamic hand gesture recognition based on SURF tracking. In: Electric information and control engineering (ICEICE), pp 338–341
15. Yao Y, Li CT (2013) Real-time hand gesture recognition for uncontrolled environments using adaptive SURF tracking and hidden conditional random fields. In: Proceedings ISVC 2013, Crete, Greece, pp 29–31
16. Wang X, Hänsch R, Ma L, Hellwich O (2014) Comparison of different color spaces for image segmentation using graph-cut. In: Proceedings 2014 International Conference on VISAPP, Lisbon, Portugal, pp 301–308
17. Lowe DG (2004) Distinctive image features from scale-invariant key points. Int J Comput Vision 60:91–110
18. Lin W, Wu Y, Hung W, Tang V (2013) A study of real-time hand gesture recognition using SIFT on binary images. Adv Intell Syst Appl, pp 235–246
19. Tuytelaars T, Mikolajczyk K (2008) Local invariant feature detectors: a survey. Found Trends® Comput Graph Vis 3:177–280
20. Sykora P, Kamencay P, Hudec R (2014) Comparison of SIFT and SURF methods for use on hand gesture recognition based on depth map. AASRI Procedia 9:19–24
21. Asaari MS, Rosdi BA, Suandi SA (2014) Intelligent biometric group hand tracking (IBGHT) database for visual hand tracking research and development. Multimed Tools Appl 70:1869–1898

Communication and Networks

A State of the Art on Network Security

Vinay Kumar, Sairaj Nemmaniwar, Harshit Saini and Mohan Rao Mamidkar

Abstract The rapid enhancement in computer network technology field brings great conveniences to all users, but at the same time, it also brings security threats. Users need to be aware of the importance of network security. Network security is a critical issue because many types of attacks are increasing nowadays, and the protection of the user's system is a high priority. After gathering, evaluating and analyzing the information on network security, we have done an extensive literature survey. This paper deals the various techniques based on system security and data security, viz. firewall technology and antivirus technology.

Keywords Security · Active attack · Passive attack · Firewall technology

1 Introduction

Network security problem is increasing prominently because intellectual properties (e.g., private data) are no longer safe and can be easily accessed through Internet. Network security is a set of rules and regulations adopted by the administration of network to monitor and restrict uncertified access of system and resources. The whirlwind development of computer network makes every information popular and broadly spread. However, all kinds of information are transmitted via public communication methods, which can be illegally hacked or tampered by enemies of network security, thus resulting in an immeasurable loss. The network security problem is everywhere, i.e., among personal computer users and in public and private organizations too. There are different ways to overcome network security problem. Firewall and antivirus software help in detecting viruses and malwares. Encryption–decryption technique used in communication between two networks also reduces the chances of a breach in system and maintains a privacy policy.

The rest of the paper is structured as follows. The literature survey has been illustrated in Sect. 2. Network security, issues and its need have been explained in Sect. 3. Various types of attack are explained in Sect. 4. Section 5 explains various

V. Kumar (✉) · S. Nemmaniwar · H. Saini · M. R. Mamidkar
National Institute of Technology Jamshedpur, Jamshedpur, India
e-mail: vkumar.cse@nitjsr.ac.in

R. K. Shukla et al. (eds.), *Social Networking and Computational Intelligence*, Lecture Notes in Networks and Systems 100, https://doi.org/10.1007/978-981-15-2071-6_46

existing techniques for system security and data security. The conclusion of this paper is explicated in Sect. 6.

2 A State of the Art: Network Security

Li et al. [1] implemented a security visual prototype based on various factors such as pie charts, line charts, etc. which solves the technical problems that visualization is not intuitive.

de los Reyes et al. [2] proposed a new security architecture for mobile enterprise which uses network-based security and cloud computing. It is helpful in detecting and handling advanced persistent threats (APT) and attacks.

Agrawal et al. [3] proposed novel solutions to the existing schema, of cloud computing and robustness of in cloud computing.

Chen et al. [4] have proposed vCNSMS to address the various network securities in multiple-tenant data center and have demonstrated a vCNSMS with a very centralized and collaborative scheme.

Shaikh et al. [5] have proposed a firewall model that protects against malicious insiders and also solves the problem of general laziness in applying patches to the software.

Zhao et al. [6] proposed a new intrusion detection method which is based on improved K-means and designed to meet the security requirements in cloud computing.

Shin et al. [7] introduced a concept of network security virtualization (NSV) that can virtualize security resources/functions and provide proper security response functions from various network devices.

Khyavi et al. [8] created a model to address the security concerns at the application level for the next-generation network which could be used for cloud computing.

Sun [9] proposed a network security situation prediction and also proposed a Markov prediction method which is based upon the complex network and also aims at achieving an effective prediction of security status.

Alabdulatif et al. [10] introduced a framework to provide a lightweight and scalable privacy-preserving anomaly detection service for sensor data. They developed a model in a cloud-based anomaly detection model that is both lightweight and scalable but also ensures data privacy.

Hurley-Smith et al. [11] proposed a VCN approach that protects routing and forms a secure network environment prior to routing operations.

Table 1 summarizes all the above existing approaches on the basis of reference number, year of publication, name of the author(s) and technique used.

Table 1 Summary for the existing methodologies of network security in the paper

Ref. No.	Year	Author	Techniques used
[1]	2011	Li et al.	Network security situation awareness model, security situation visualization prototype system
[2]	2012	de los Reyes et al.	Network-based security, cloud computing and a transformed enterprise
[3]	2016	Agrawal et al.	Antivirus with detection techniques such as signature-based, heuristics-based
[4]	2014	Chen et al.	VCNSMS
[5]	2010	Shaikh et al.	Disarming firewall components used
[6]	2016	Zhao et al.	Improved *K*-means and simulation analysis
[7]	2015	Shin et al.	Various routing algorithms used like multipath-naive, shortest-inline, etc
[8]	2016	Khyavi et al.	Level-wise security plan for the next-generation network: infrastructure security plan; services security plan; applications security plan
[9]	2015	Sun	Coarse graining processing of the data Markov prediction based on complex network
[10]	2017	Alabdulatif et al.	Data clustering and homomorphic encryption
[11]	2015	Hurley-Smith et al.	Virtual closed network techniques are used

3 Network Security, Issues and Its Need

There are various common nouns we associate with computer security in general; some of this are as follows: data security, information system security, system security among others. Two inferences can be drawn from these nomenclatures, firstly to guarantee an unaltered functioning of data system in an inter-connected environment. And secondly to prevent the breach of data that is stored, manipulated and transferred in the information system. In a nutshell, in this paper, "network security" is in reference to the credibility of network environment, non-disclosure, easy-access and integrity of information in transmission, which is one of the major ethos of network security.

Network security can be logically defined under the following aspects [12]: **Security Attack**: means all those actions which seek to damage information pertaining to message change, data denial offerings, traffic control among others.

Security Mechanism: comprises of all the functionalities orchestrated to sense and prevent the attack on security the subsequent recovery (if it comes to that). Data encryption, exchange of authentication, routing control are some of the few actions under it.

Security Service: includes the services that tend to combat against the security attacks while improving the data processing security with varied security mechanisms. Peer entity authentication, confidentiality of data, information integration are a part of it.

3.1 Security Issues

Security issues arising in different layers are discussed below [12]:

Security in Physical Layer: Prevention from breach or hacking.
Security in Data Link Layer: Technical methods such as encryption communication are used here for secure information flow.
Security in Transport Layer: Makes sure safe data flow.
Security in Operating System: Securing the OS with controlled access inclusive of mailing server system and Web server.

3.2 Need of Network Security

The major aims of securing a network are mentioned below:

(i) **Auditability**: such that all the activities can be reviewed.
(ii) **Integrity**: maintaining the consistency of the information.
(iii) **Dependability**: making sure the sender and receiver acknowledge the message sent and received.
(iv) **Non-Disclosure**: information shared only between the authorized users.
(v) **Availability**: authorized users should be able to retrieve the required information from the system.

4 Types of Attacks

4.1 Active Attacks [11, 13]

(i) **Modification**: unauthorized manipulation and modification made to the data in transit.
(ii) **Spoofing**: when a node fakes its identity so as to make changes in the information.
(iii) **Wormholes**: These attacks occur when an attacker transfers the packet that it receives to some other point in the network and then replays them from the same point in the network.
(iv) **Denial of service**: it occurs when a valid user is stopped or not allowed to access the devices, computer systems and other IT resources. The attackers achieve this by increasing the traffic in the network.
(v) **Fabrication**: Fabrication is an attack in which an unauthorized user tries to insert a fake message as if it was a valid user in that network which leads to the loss in originality and confidentiality of the information.

(vi) **Sinkhole**: In sinkhole, the root station is prevented to receive the correct and complete data. The attacker achieves this by sucking and transferring all the information to the adjacent neighboring node.

(vii) **Sybil**: Duplicate copies of the malicious nodes are created and they share the secret key among themselves, thereby increasing the probability of attack.

4.2 Passive Attack [11]

(i) **Traffic attack**: The attacker tries to figure out the amount of data that is transferred from the sender to the receiver via that route. The best part about traffic attack is that data is not changed or modified.

(ii) **Eavesdropping**: The main motive of eavesdropping is to find the secret data that the sender or receiver may have. The data can be anything from the secret password to any other confidential information.

(iii) **Monitoring**: Here, the malicious node can read the private data but modification or updating is not possible.

5 Techniques for System Security and Data Security

5.1 Firewall Technology

Firewall [5] is a network security system. It comes between internal and external network. It keeps the networks safe from unauthorized access. Thus, it protects the data of network.

The basic functions are to filter data passing through a network, control the accessing behavior, recording of activities, etc. It also has the responsibility of detecting and alarming the network administrators. The main technologies implemented in firewall are:

(i) Packet filtering means to block packets at a point between sender and receiver.

(ii) It can be considered as a checkpoint. It is totally based on network layer, and it acts on IP layer (router). On the arrival of a packet, it checks and determines whether it can pass it or not.

(iii) Application gateway is an application program present between two different networks. Whenever a connection is established between two sides, client and destination, then a connection is established with the application gateway. After this connection, the client has to deal with this application gateway.

5.2 Data Encryption Technology

The primary idea behind data encryption technology is to improve the secrecy of information by encryption (encryption key) and keep it safe from threats of decoding. The main methods applied here are of two categories: Encryption with symmetric key and Encryption with asymmetric key [10].

(i) **Encryption with Symmetric Key**: It refers to a system in which both sender and receiver of data share among them a single, common key used in the process of encryption and decryption. This is a simple process and considered fast. Therefore, it has a widespread use. Some popular algorithms of this category: RC2, CAST-128, etc.
(ii) **Encryption with Asymmetric Key**: It makes use of two keys to encrypt data. And, these keys are sent over the network. It makes sure that wrong persons cannot use the keys. Here, we have a lock mechanism in which a particular private key is needed to decrypt an encrypted data (using public key). This is more secure than the former. A popular example of it is RSA.

5.3 Intrusion Detection Technology

It is an active network security technology [6]. It is a very reliable add on to the firewall. This gathers information from the system's internals and other network resources. It takes precautions against the possible network attacks.

It encompasses the areas of attack monitoring, security activities log maintenance, etc. Misuse detection takes the assumption that every attack is of same trend/type. And therefore, all the popular seen attacks can be found out by process of comparison. Here, a significant part is to define the intrusion trends to differentiate the normal behaviors and attacks. Some famous algorithms used here are trend–comparison, state–transition–inspection, etc. Intrusion detection mechanism has the following mentioned functionalities:

(i) Statistical analysis of the activity trends.
(ii) Maintenance of the system's configuration architecture and weak points/vulnerabilities.
(iii) Identification of any attack trend.

5.4 Antivirus Technology [3]

Computer virus spreads mostly by using unverified storage devices on the systems or through some other means in a network. A significant role is played by prevention of virus. Data from removable storage media (e.g., pen drives) must be checked thoroughly. Many softwares are available in the market that claim to identify such malicious programs. Such softwares are made on the basis of the viruses seen day to day, and updates too are available to identify newly introduced viruses.

5.5 *Virtual Private Network (VPN)*

Virtual private network (VPN) [11] deals with creation of secure connection over a network, e.g., Internet. This can also be thought as having a private form of network inside the architecture of a public network. This is mostly used by remotely located users who want to access some data. In terms of cost and reliability, it is highly recommended over the typical private networks.

6 Conclusion

In today's world, network security is among one of the biggest challenges that the computer network field is facing. One of the major tasks in network security is to maintain a certain amount of network security level. Another important aspect that is important in achieving a global network security is to keep track of all the real threats or attacks and then deploy an advanced mechanism that can counter the threats. Taking into account all the study done, there is still a sense a threat among the multinational industries. This paper focuses on the latest technologies in network security technologies. A lot of research challenges have been thoroughly identified which we expect to become major research trends in the coming years. Summing up with the result of this survey are expected to be useful for researchers, students, health, banking sectors and policy makers working in the field of the network technology.

References

1. Li X, Wang Q, Yang L, Luo X (2011) Network security situation awareness method based on visualization. In: 2011 third international conference on multimedia information networking and security (MINES), IEEE, pp 411–415
2. de los Reyes G, Macwan S, Chawla D, Serban C (2012) Securing the mobile enterprise with network-based security and cloud computing. In: 2012 35th IEEE Sarnoff symposium (SARNOFF), IEEE, pp 1–5
3. Agrawal A, Wahie K (2016) Analyzing and optimizing cloud-based antivirus paradigm. In: 2016 international conference on innovation and challenges in cyber security (ICICCS-INBUSH), IEEE, pp 203–207
4. Chen Z, Dong W, Li H, Zhang P, Chen X, Cao J (2014) Collaborative network security in multi-tenant data center for cloud computing. Tsinghua Sci Technol 19(1):82–94

5. Shaikh ZA, Ahmed F (2010) Disarming firewall. In: 2010 international conference on information and emerging technologies (ICIET), IEEE, pp 1–6
6. Zhao X, Zhang W (2016) An anomaly intrusion detection method based on improved *K*-means of cloud computing. In: 2016 sixth international conference on instrumentation & measurement, computer, communication and control (IMCCC), IEEE, pp 284–288
7. Shin S, Wang H, Gu G (2015) A first step toward network security virtualization: from concept to prototype. IEEE Trans Inf Forensics Secur 10(10):2236–2249
8. Khyavi MH, Rahimi M (2016) Conceptual model for security in next generation network. In: 2016 30th international conference on advanced information networking and applications workshops (WAINA), IEEE, pp 591–595
9. Sun S (2015) The research of the network security situation prediction mechanism based on the complex network. In: 2015 international conference on computational intelligence and communication networks (CICN), IEEE, pp 1183–1187
10. Alabdulatif A, Kumarage H, Khalil I, Yi X (2017) Privacy-preserving anomaly detection in cloud with lightweight homomorphic encryption. J Comput Syst Sci 90:28–45
11. Hurley-Smith D, Wetherall J, Adekunle A (2015) Virtual closed networks: a secure approach to autonomous mobile ad hoc networks. In: 2015 10th international conference for internet technology and secured transactions (ICITST), IEEE, pp 391–398
12. Simmonds A, Sandilands P, Van Ekert L (2004) An ontology for network security attacks. In: Asian applied computing conference, Springer, Berlin, Heidelberg, pp 317–323
13. Yue X, Chen W, Wang Y (2009) The research of firewall technology in computer network security. In: Asia-Pacific conference on computational intelligence and industrial applications PACIIA 2009, vol 2, IEEE, pp 421–424

A Survey on Wireless Network

Vinay Kumar, Aditi Biswas Purba, Shailja Kumari, Amisha, Kanishka and Sanjay Kumar

Abstract The society is progressing toward the centralization of information, and hence, there is an urgent need to have the information available at any dimension, anywhere, and anytime. To fulfill the need, the radio signals that have high frequency are used to communicate among the PCs or computers including other network devices from last three decades. In this paper, a survey on secure transmission is done to identify the problems faced while transmitting data over wireless network and ensure safe transmission. Further, various types of protocols and issues related to wireless networks are identified.

Keywords Wi-Fi · WLANS · Security · Protocols

1 Introduction

The network that is established by the radio signals that have high frequency to communicate among the PCs or computers including other network devices by taking advantage of Ethernet protocol is called wireless network. Occasionally, this network can also be referred to as Wi-Fi network or WLAN.

This network has enabled multimedia communication between people and numerous devices from any distant or nearby location, and thus, wireless network has become an integral part of our life in the last few years. It initiates primary differences in data networking, developing integrated networks and telecommunication. A portable network has been created by digital modulation, compression of information, adaptive modulation, multiplexing, and wireless access. Several technologies have been designed that help in reduction of time and various types of hindrance put forward by cables and more suitable than networking that uses wired communication. It helps the stimulating apps like sensor networks, automated highways, smart homes, telemedicine, etc. The ones who were first users of wireless technology are the military, emergency services, and law enforcement organizations. The society is progressing toward the centralization of information, and hence, there is an urgent

V. Kumar (✉) · A. B. Purba · S. Kumari · Amisha · Kanishka · S. Kumar
National Institute of Technology Jamshedpur, Jamshedpur, India
e-mail: vkumar.cse@nitjsr.ac.in

R. K. Shukla et al. (eds.), *Social Networking and Computational Intelligence*, Lecture Notes in Networks and Systems 100, https://doi.org/10.1007/978-981-15-2071-6_47

Table 1 Different types of networks and its applications

Type	Applications	Standards	Range
Personal area network (PAN)	Cable replacement for peripherals	Bluetooth, ZigBee, NFC	In the reach of a person
Local area network (LAN)	Wired network extension(wireless)	IEEE 802.11 (Wi-Fi)	Building
Metropolitan area network (MAN)	Inter-network connectivity	IEEE 802.15 (WiMAX)	City
Wide area network (WAN)	Access network	Cellular (UMTS, LTE, etc.)	Worldwide

need to have the information available at any dimension, anywhere, and anytime. We know that any communication network makes use of transmitter (i.e., wireless router or hotspot) as well as a receiver which may be any device that can be run on Wi-Fi like laptop, mobile, tablet etc. This method also uses the same technologies. It has been found that it is convenient, flexible, and easy-to-use wireless communications. Wireless communications can be instrumental in helping organizations to reduce their wiring costs. Various types of networks are given in Table 1.

The rest of paper is structured as follows: The literature survey has been illustrated in Sect. 2. Wireless protocols have been explained in Sect. 3. Various issues and challenges are discussed in Sect. 4. The conclusion of this paper is explicated in Sect. 5.

2 A Survey on Wireless Network

Kumar and Hancke [1] proposed a ZigBee-based animal health monitoring system for real-time monitoring of physiological parameters such as body temperature, rumination, and heart rate. They also represented a model of an animal health monitoring system.

Vibhuti [2] discussed about Wired Equivalent Privacy (WEP) protocol which prescribes a group of instructions and rules by which, over the airwaves, wireless data is transmitted where security is applied. It is discussed about conceptual weakness of WEP protocol and listed available solution for WEP venerability.

Kosek-Szott et al. [3] proposed various open future experiment fields related to coexistence of Wi-Fi such as effective transmission in multi-hopping

environments, short- and long-term adaptability and many more, and coexistence of many Wi-Fi devices in dense environment. They identified Wi-Fi future use case and analyzed the required functionalities of this use case.

Ali et al. [4] proposed OLSR as the best protocol for large wireless networks because of the large number of mobile stations. In terms of bandwidth utilization and media access delay, performance of various routing protocols had been analyzed with a developed model of MANET by them.

Wen et al. [5] proposed the adaptation of hopping progress to anisotropic wireless sensor networks where hop counting computation and anchor selection both are took into account. They proposed a localization algorithm which is free ranged, depended on EHP and PSO.

Wu et al. [6] presented a method to design a Bluetooth-based coal mine gas concentration monitoring system design. This technology can measure the gas concentration in coal mines by installing front-end Bluetooth module, and the hat or devices of the workers can be placed. Then, the data can be sent to the control room such that staff may remain updated about the underground gas concentration and take preventive measures.

Kumar and Gupta [7] presented an idea of preventing passive eavesdropping in Bluetooth active devices with the help of BlipTrack Bluetooth Detector (BBD). BBD is a sensor device which aims to detect the victim device and the intruder who has attacked the device. It stores the MAC addresses of victim and intruder both and sends it to the server from which server could find the attacker.

Kurt et al. [8] have discussed optimization in bundle volume in wireless sensor networks because of grid applications which is smart.

Chung and Yang [9] proposed a innovative structural idea of the near-field communication (NFC) antenna design for a tablet PC. The goal was to achieve a miniaturized loop antenna design by attaching ferrite sheets on both sides of the loop antenna. There is a high chance that these ferrite sheets minimize the eddy currents produced on the adjacent metallic back cover of the tablet PC by the loop antenna. As a result, it improves the range of communication of the NFC.

Gardill et al. [10] proposed a technique for bringing up an inter-radio access technicality cell reselection in UMTS user materials. In this technique, every mobile device is forced to leave the UMTS cells that are in the circumference of transmitter such that it is putting out a noise-like RF signal on the UMTS downlink frequency bands, and then, they camp on non-UMTS cells, e.g., GSM. As a result even in areas with UMTS coverage, GSM-based limiting technique can be used.

Markevicius et al. [11] claimed that two thermocouples network is used for experiments and technologies in wireless network and showed that the energy consumption takes less time at the time of establishing network.

Finogeev and Finogeev [12] have presented a routing protocol for WSNs by use of path centrality which is dependent on the operator calculus approach. They

have claimed that from base station to main path each node is sensored in the network.

Table 2 summarizes all the above existing approaches on the basis of reference number, year of publication, name of author(s), technique used, and other parameter.

3 Short-Range Wireless Protocols

For low limit wireless communions with less energy expense, four protocols are there—(1) Bluetooth (over IEEE 802.15.1), (2) Ultra-wideband (UWB, over

Table 2 Summary for the existing methodologies of wireless network in the paper

References	Year	Author's name	Techniques	Current work and future works
[1]	2015	Kumar and Hancke	ZigBee	The exploration of UWB can target animal location and tracking applications
[2]	2005	Vibhuti	Wired Equivalent Privacy protocol	Small IV space and imperfect sorting of data integrity verification can be improved by WEP protocols for better security
[3]	2017	Kosek-Szott et al.	Multi-hop ad hoc network devices	It is providing high-throughput WLAN, traffic offloading, and supporting IOT devices
[4]	2018	Ali et al.	Hurst index	Mobile ad hoc network model can be developed in future according to the factors affecting the QoS
[5]	2018	Wen et al.	Cumulative distribution function, particle swarm optimization	Study on heterogeneous wireless sensor network with nodes of different ranges and the impact of the informal communication range in AWSN
[6]	2014	Wu et al.	Controller Area Network bus technology	By CAN bus technology, debugging and transmission of data point can be realized by block diagram
[7]	2015	Kumar and Gupta	BlipTrack Bluetooth Detector	The data collected by cloud server can be analyzed and stored for future references. BBD can also be enhanced for IOT

(continued)

Table 2 (continued)

References	Year	Author's name	Techniques	Current work and future works
[8]	2017	Kurt et al.	Tmote sky platform model, MIP framework	The influence of conjoint optimization of transmission and data packet size power level on WSN lifetime can be investigated
[9]	2015	Chung and Yang	Near-field communication (NFC)	Good for the NFC antenna of the tablet PC. If the antenna is set over the center of the tablet PC, then a metaphor design is needed
[10]	2011	Gardill et al.	Universal Mobile Telephony System (UMTS), GSM, inter-radio access technology (RAT) cell reselection	A high-power noise jamming system may be required to generate a sufficient amount of additional interference at the UE antenna. In a next step, field tests should examine the usability of the noise jamming approach under more realistic conditions
[11]	2018	Markevicius et al.	Bluetooth v4.0, low energy, ZigBee, ANT	Creating a wireless network which consists of different types of sensors (temperature, humidity, pressure, etc.) by ensuring minimum energy consumption
[12]	2017	Finogeev and Finogeev	Operator calculus approach	In order to raise the fault tolerance power and reliability, it should be focused on balancing load, stability etc.

802.15.3), (3) ZigBee (over IEEE 802.15.4), and (4) Wi-Fi (over 802.11). It is defined by IEEE that the physical and MAC layers are established with 10–100 m for wireless networking. In terms of different prosody, together with astringency, network model, power consumption, safety, characteristic of worship assistance, the important properties, and behaviors of Ferro and Potorti are compared by themselves for Bluetooth and Wi-Fi.

Bluetooth over IEEE 802.15.1 Bluetooth is called wireless term for the short-range mutual connection of electronic gadgets. It works as an open wireless technology criterion for sending fixed and digital data over short distances. Bluetooth, familiar as IEEE 802.15.1 criterion, does use radio wave which only works within short-range distance. It allows to the users passing connections of voice and message between several devices in real time. Bluetooth was introduced in 1994 as a wireless alternative for RS-232 cables. It is that personal area network which assures security against interferences and safety in sending the messages.

UWB over IEEE 802.15.3 Ultra-wideband is a term related to radio technology which needs high-bandwidth communications over the main parts of the radio spectrum and immensely less energy level for short range. For non-cooperative radar imaging, UWB uses traditional applications. It has been recently attached with indoor short-range high-speed wireless communication. Digital pulses are broadcast by UWB. These are timed exactly on a carrier signal across a highly mass spectrum (number of frequency channels) at the same time. There must be coordination between transmitter and receiver for the accuracy of trillionths of a second in sending and receiving pulses.

ZigBee Overpowers IEEE 802.15.4 ZigBee protocol is mainly intended for those types of networks which are wirelessly controlled and monitored. Those devices that utilize less energy and work within reach of person are supported for small rate WPAN (LR_WPAN) by defining specifications by ZigBee over IEEE 802.15.4. ZigBee caters networking which are organized by it, and hops are of multiform and are mainly meshed, consisting of battery of long lifetime. There are two types of devices in an LR-WPAN network. They are full functioning device (FFD) and reduced-functioning device (RFD).

Wireless LAN over IEEE 802.11 a, b and g Standards The b and g standards of IEEE 802.11 are meant for wireless LAN connectivity. Whether devices in network are connected to an AP (access point) (i.e., infrastructure mode) or not (i.e., ad hoc mode), the users are permitted to access the Internet at broadband speeds. The various components of IEEE 802.11 collaborate to give WLAN which help dynamism of snations to the higher strata transparently. A computer must be equipped with a wireless network interface controller in order to establish a connection to a Wi-Fi LAN. As a result, anyone with a wireless network interface controller and within the range of a network can access it. This is the reason why wireless network is more prone to attacks (known as eavesdropping) than networks which are in wired form.

4 Issues and Challenges

Issues are the problems and challenges are difficulties, we face while implementing the real life scenario in to end product.

Issues are as follows:-

1. **Slow Connection**: Wireless network can get halted despite of faster speed of network in the regions. It is just because of far connection of router from homes, and due to this, Wi-Fi speed slows down.
2. One of the issues related to wireless network is interference. It is mainly in those areas where people get the first time Internet setup there and just use the default frequency channel like 1, 6, and 11. This results in increased crowding and slowing down of Internet, also during rush hour.
3. **No Internet Connection**: Sometimes, it just occurs that routers and modem stops working, there is no clarification for it.
4. **Security**: Wireless network brings a lot of security issues such that wireless network security is itself a vast topic to discuss. This includes the network administrator must contain the information about the person accessing the network, how they were accessing it, and from where they were retrieving it. And then afterward, policies must be applied by them to check that level of access.
5. **Physical Connectivity**: Issues of wireless network related to physical association is that ground floors, basement areas and some buildings which have glazed windows change the direction of cellular signals set limit to the wireless networks. Mountains and unfavorable territory were also seen as an issue due to coverage gaps.

Challenges are as follows:

1. **Capacity**—Capacity addresses the number of devices connected to the network, types of devices present in the network, potential of those devices, types of applications used, and locations on your campus where most activities are being done.
2. **Coverage**—This is the most obvious challenge that colleges and universities have to face while dealing with wireless networks.
   (i) The critical areas on campus must be understood.
   (ii) Though coverage might seem simple, it is difficult to cover the entire campus just in first attempt.
   (iii) We have to plan in such a way that total campus is covered.
3. **Price**—Because of the rapid evolvement of wireless network devices and security threats, the physical threats to network are increasing as well, causing the decrement in the lifespan of wireless environment. Today, the lifespan is 3–4 years.
   (i) After 3–4 years, it will become hard to keep up the required Wi-Fi completion levels because the numbers of users and their dependency are incrementing day by day.

(ii) As the wireless network is being improved and advanced in technology and speed, so is the cost associated with it. It is costly even with monthly payments and still a challenge to figure out how to afford everything.

4. **Meeting User Demands**—Wireless technology improves efficiency ensuring that users need not be experts in technology.

   (i) It was observed that when doctors meet new problems, they themselves are able to find solutions by using certain technology IT department fails to support.
   (ii) According to the participants, detailed policy statements must be defined specifying devices IT department can support.

5 Conclusion

This survey provides us with an overall view of technologies, various issues, and upcoming challenges and works related to wireless networks. This survey tells us about the importance of wireless network in people's life, generating new opportunities for today's generation, and leading to evolve various technologies. In wireless networks, fifth generation is to a large extent in the research stage and the upcoming 5G networking architectures will support a large number of connected devices with various bandwidth requirements. New technologies like automation in transport, self-driving cars, and warrior drones will request much more dependency on wireless infrastructure and will need much greater dependency and guaranteed connectivity. Research is being done for reprocessing the existing infrastructure in new ways. Wireless has not finished changing the world. The survey results are expected to be helpful for the students, researchers, and professionals working in the field of the wireless networks.

References

1. Kumar A, Hancke G (2015) A ZigBee-based animal health monitoring system. IEEE Sens J 15(1):610–617
2. Vibhuti S (2005) IEEE 802.11 WEP (wired equivalent privacy) concepts and vulnerability. San Jose State University, CA, USA, CS265 Spring
3. Kosek-Szott K, Gozdecki J, Loziak K, Natkaniec M, Prasnal L, Szott S, Wagrowski M (2017) Coexistence issues in future WiFi networks. IEEE Netw 31(4):86–95
4. Ali D, Yohanna M, Silikwa WN (2017) Routing protocols source of self-similarity on a wireless network. Alex Eng J
5. Wen W, Wen X, Yuan L, Haixia X (2018) Range-free localization using expected hop progress in anisotropic wireless sensor networks. EURASIP J Wirel Commun Netw 2018(1):299
6. Wu Y, Feng G, Meng Z (2014) The study on coal mine using the Bluetooth wireless transmission. In: 2014 IEEE workshop on electronics, computer and applications, pp 1016–1018. IEEE

7. Kumar M, Gupta BK (2015) Security for Bluetooth enabled devices using BlipTrack Bluetooth detector. In: 2015 international conference on advances in computer engineering and applications (ICACEA), pp 155–158. IEEE
8. Kurt S, Yildiz HU, Yigit M, Tavli B, Gungor VC (2017) Packet size optimization in wireless sensor networks for smart grid applications. IEEE Trans Ind Electron 64(3):2392–2401
9. Chung M-A, Yang C-F (2016) Miniaturized NFC antenna design for a tablet PC with a narrow border and metal back-cover. IEEE Antennas Wirel Propag Lett 15:1470–1474
10. Gardill M, Zorn S, Weigel R, Kölpin A (2011) Triggering UMTS user equipment inter-rat cell reselection using noise jammers. In: Microwave conference (GeMIC). 2011 German, pp 1–4. IEEE
11. Markevicius V, Navikas D, Andriukaitis D, Cepenas M, Valinevicius A, Zilys M, Malekian R, Janeliauskas A, Walendziuk W, Idzkowski A (2018) Two thermocouples low power wireless sensors network. AEU-Int J Electron Commun 84:242–250
12. Finogeev AG, Finogeev AA (2017) Information attacks and security in wireless sensor networks of industrial SCADA systems. J Ind Inf Integr 5:6–16

Jaya Algorithm Based Optimal Design of LQR Controller for Load Frequency Control of Single Area Power System

Nikhil Paliwal, Laxmi Srivastava and Manjaree Pandit

Abstract In this paper, Jaya algorithm has been proposed for optimizing the parameters of the linear quadratic regulator (LQR) for load frequency control (LFC). In feedback control system, LQR is an advanced and modern control technique. The LQR technique is based on minimizing the quadratic performance index. In the LQR controller, the main problem is to shape the weighting matrices Q and R. The problem to shape the weighting matrices Q and R in LQR controller can be solved using various evolutionary computing techniques. In this paper, the weighting matrices for load frequency control of the electrical power network are designed by using genetic algorithm and Jaya algorithm. It is shown that Jaya algorithm is the most powerful method, as it provides improved system performances by optimal design of the matrices Q and R to minimize settling time.

Keywords LFC · LQR · Single area power network · Genetic algorithm · Jaya algorithm

1 Introduction

These days, the demand of electrical power is rising and the main concern for the electrical power network operator is to provide better quality of electrical power at the end of consumers under various condition of changing load. The frequency of the electrical power network should be kept constant as far as possible for the satisfactory operation of the electrical power network. Many techniques have been proposed to overcome any deviations and to maintain constant value of frequency.

N. Paliwal · L. Srivastava (✉) · M. Pandit
Madhav Institute of Technology and Science, Gwalior, Madhya Pradesh 474005, India
e-mail: laxmigwl@gmail.com

N. Paliwal
e-mail: pnikhil02@gmail.com

M. Pandit
e-mail: drmanjareep@gmail.com

R. K. Shukla et al. (eds.), *Social Networking and Computational Intelligence*, Lecture Notes in Networks and Systems 100, https://doi.org/10.1007/978-981-15-2071-6_48

The continuous control of generated electric power to meet the changes in load is a vital problem and is a main challenge for power system operator. Unpredictable variation in load always causes mismatching in power generation and power demand and also adversely affects the quality of electrical power which is generated. The deviation in frequency from its nominal or rated value should be reduced and kept within the limits for the satisfactory operation of frequency-dependent equipment. For restoring the balance between generation and load, load frequency control is used by maintaining the synchronous speed of alternator. The fundamental goal of LFC is to reduce the deviation of frequency to zero. This can be obtained by using a typical controller, e.g., PID but this type of controller is quite moderate in the operation. Modern controller is much faster and also gives good output response as compared with the conventional and traditional controllers.

Day-by-day, the electric power system is becoming more and more complex due to large interconnections. The electric power system requires load frequency control in order to keep the system reliable and safe [1–4]. The change of reactive power affects the voltage magnitude, whereas the change of real power affects the frequency of a power system.

To deal with the effects of load variation and maintain frequency and constant voltage level, there is a requirement of control system. Frequency deviation can affect the overall stability of the system; therefore, in order to maintain system stability, imbalances between generation and load must be corrected in as minimum time as possible to avoid frequency deviation. The problem of controlling the frequency in large electric power systems is handled by regulating the generating unit production in responses to changes in the load which is popularly known as load frequency control. LFC is a very challenging problem in power system control and operation for supplying reliable and sufficient electrical power of good quality. The electric power system becomes more and more complex with day to day increasing power demand. The electric power network is concerned to local variations of load. As the load varies, the frequency related to that area is affected. The transients in frequency should be removed as fast as possible. Generators working in that control area always alter their speeds (accelerate or decelerate) to maintain the relative power and frequency angle to the predefined values with tolerance boundary in static and dynamic conditions. [5].

2 Load Frequency Control

Frequency must remain almost persistent for the satisfactory and reliable operation of the electrical power system. The deviation in frequency can directly impact or affect the power system reliability, performance of system, and also efficiency [6]. If the frequency is large enough it can damage equipment and can also degrade the load performance. There are so many varieties of control strategies which has been proposed and probed by many researchers for the design of LFC in electrical power system.

Many traditional as well as classical approaches have been already used to provide a supplementary control that will drive to the normal operating value of the frequency within a very short span of time [7]. This substantial research is due to the fact that load frequency control is a very crucial parameter of the electric power system, where the primary focus is to maintain frequency fluctuations within preset limits. The load frequency control incorporates a suitable control system that has the ability to re-adjust the power system frequency to original preset value or very close to the preset value effectively after the sudden change in load [8].

In this paper, a very recently developed evolutionary computing technique, Jaya algorithm [9, 10], has been proposed for optimal design of LQR controller for LFC in single area power network. The performance of Jaya algorithm based LQR Controller has been compared with that of GA [11] based LQR controller.

The fundamental task of LFC is to sustain a balance in real power in the electric power network via adjusting the frequency of system. A change in frequency occurs whenever the demand in real power change [12].

Basically, the components of electric power network are not linear, so around a nominal operating point, in general, a linearization is typically performed to get a model which is linear and can be used further in the controller design process. Generally, electrical power generating unit comprises of:

(i) Turbine
(ii) Generator
(iii) Hydraulic valve actuator
(iv) Rotating mass and load
(v) Governor.

2.1 *Load Frequency Control Model of Single Area Power System*

Figure 1 represents the important components of the load frequency control for single area electric power network which can also be presented in the matrix form, as given in Eq. (1).

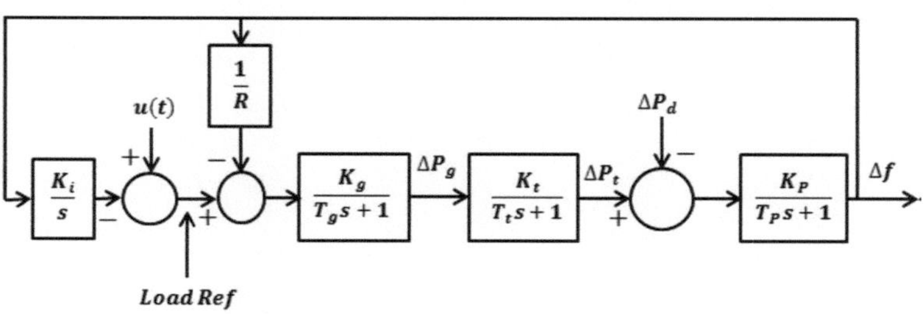

Fig. 1 LFC for single area power system

$$\begin{bmatrix} \Delta \dot{P}_{\mathrm{g}} \\ \Delta \dot{P}_{\mathrm{t}} \\ \Delta \dot{\omega} \end{bmatrix} = \begin{bmatrix} \frac{1}{T_{\mathrm{g}}} & 0 & -\frac{1}{RT_{\mathrm{g}}} \\ \frac{1}{T_{\mathrm{t}}} & -\frac{1}{T_{\mathrm{t}}} & 0 \\ 0 & \frac{1}{2H} & -\frac{D}{2H} \end{bmatrix} \begin{bmatrix} \Delta P_{\mathrm{g}} \\ \Delta P_{\mathrm{t}} \\ \Delta \omega \end{bmatrix} + \begin{bmatrix} 0 \\ 0 \\ -\frac{1}{2H} \end{bmatrix} \Delta P_{\mathrm{l}} + \begin{bmatrix} \frac{1}{T_{\mathrm{g}}} \\ 0 \\ 0 \end{bmatrix} \tag{1}$$

2.2 The Optimal Control Problem

The fundamental task of the optimal control problem is to search an optimal controller which can work for linear as well as for nonlinear system [8]. This optimal controller, within the physical constraints of the existing system, minimizes the cost function.

Mathematically, the problem of optimal control can be considered as to find the optimal controller $u^*(t) = -Kx(t)$ that minimize a performance index:

$$\mathrm{J} = \int_0^{\mathrm{t_f}} \left(\mathrm{X^T Q x} + \mathrm{U^T R u}\right)\mathrm{dt} \tag{2}$$

which is subject to the initial condition and linear constraints as follows:

$$x(0) = x_0;\ \dot{x} = Ax(t) + Bu(t) \tag{3}$$

where $t \in \left[0, t_f\right], x, x_0 \in R^n, u \in R^m, A\, and\, B\, are\, n \times n\, and\, n \times m$ const. matrices, respectively, and $Q\, is\, n \times n\ positive\, semi-definite\, matrix\, and\, R\, is\, a\, m \times m\ positive\, definite\, matrix.$

3 Evolutionary Computing Techniques

In this section, the utilized evolutionary computing based optimization methods are briefly reviewed.

3.1 Genetic Algorithm

Genetic algorithm (GA) is a probabilistic heuristic search algorithm which is similar to the natural selection mechanics and the survival of fittest [13].

3.2 *Jaya Algorithm*

All the swarm intelligence and evolutionary based algorithms, apart from tuning of common controlling parameters, also require the proper tuning of some algorithm-specific parameters. The proposed Jaya algorithm [9, 10] does not require any algorithm-specific parameters tuning and requires the tuning of only common controlling parameters.

Computation steps to apply Jaya Algorithm are as follows:

Step 1. Initialize the population for weighting matrices, number of design variables, and set maximum iteration count $k^{\max.}$
Step 2. Set iteration to $k = 0$.
Step 3. Identify the solutions which is worst and best in the population by observing the deviation in the frequency.
Step 4. Update the solutions based on worst and best solutions.
Step 5. If updated solution is better than previous solution, then move to step 6, else move to step 7.
Step 6. Replace the previous solution by the updated solution. Move to step 8.
Step 7. Keep the previous solution.
Step 8. Increase the iteration count by 1, $k = k + 1$.
Step 9. If $k < k^{\max}$, then move to step 3, otherwise stop.

4 Results and Discussions

A comparison analysis between an optimal LQR controller based on GA and Jaya algorithm, and LQR controller is carried out. The performance comparison here is done in the terms of frequency deviation, maximum overshoot, settling time, and also steady-state error. The variables of the example [8], which is solved here using MATLAB, are shown in Table 1.

The performance index and state space system of above example are as follow [8]:

Table 1 Parameter of the single area electric power station

Parameter	Value
Turbine time const. (T_t)	0.5 s
Governor time const. (T_g)	0.2 s
Inertia const. (H)	5
Speed regulation (R)	0.05
Frequency sensitive load coeff. (D)	0.8
Nominal frequency	50 Hz
Base power (S_{base})	1000 MVA

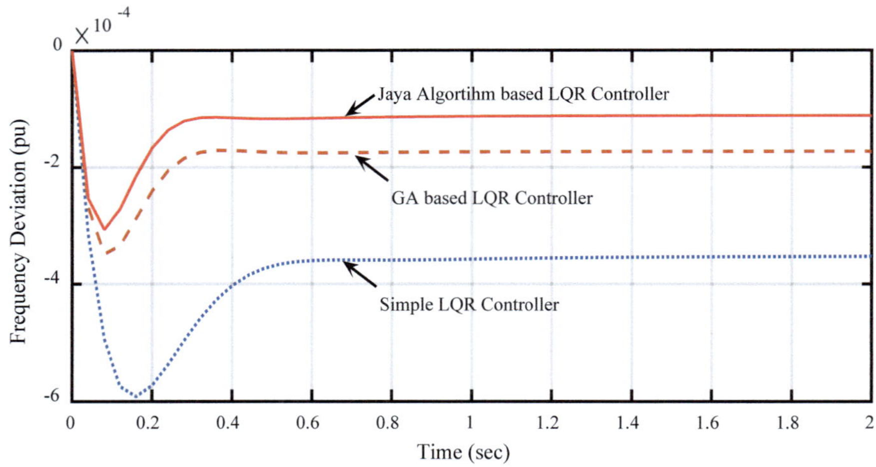

Fig. 2 System response when load change is 10%

$$J = \int_0^{\infty} \left(20x_1^2 + 10x_2^2 + 5x_3^2 + 0.15u^2\right) \tag{4}$$

$$A = \begin{bmatrix} -5 & 0 & -100 \\ 2 & -2 & 0 \\ 0 & 0.1 & -0.08 \end{bmatrix} \tag{5}$$

$$B = \begin{bmatrix} 0 \\ 0 \\ -0.1 \end{bmatrix} \tag{6}$$

$$Q = \begin{bmatrix} 20 & 0 & 0 \\ 0 & 10 & 0 \\ 0 & 0 & 5 \end{bmatrix}, R = 0.15 \tag{7}$$

The system response of simple LQR, GA based LQR, and Jaya algorithm based LQR controllers for the LFC of a single area for load change of 10, 20, and 30% are shown in Figs. 2, 3, and 4, respectively. The corresponding values of the three vital parameters, frequency deviation, maximum overshoot, and settling time are shown in Tables 2, 3, and 4 for load change of 10%, 20%, and 30%, respectively.

5 Conclusion

In this paper, GA and Jaya algorithm are implemented for optimal design of LQR controller for the LFC in the single area power network and the results are compared. From the results obtained, this can be clearly concluded that Jaya algorithm based

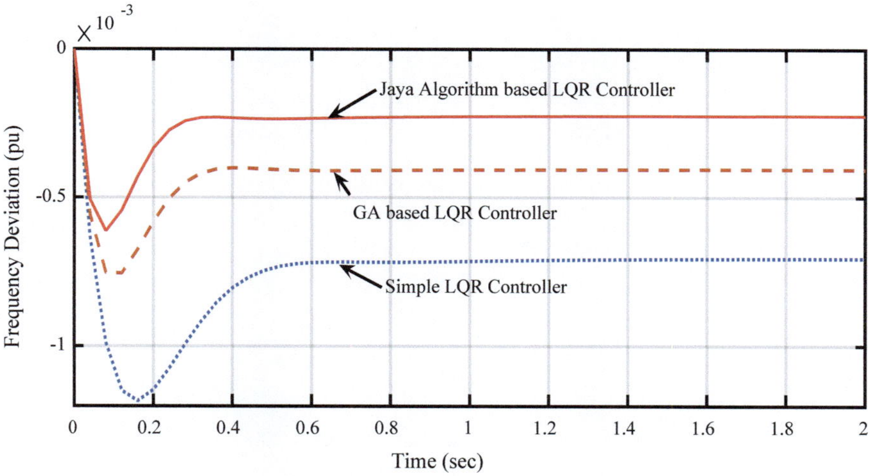

Fig. 3 System response when load change is 20%

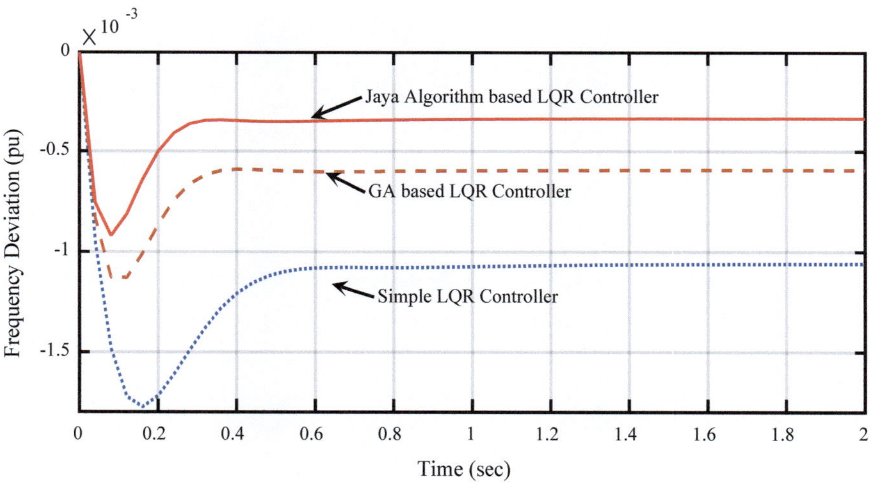

Fig. 4 System response when load change is 30%

LQR controller gives the best results and performance considering the three vital parameters, namely frequency deviation, settling time, and maximum overshoot as compared with that of GA based and simple LQR controllers.

Table 2 Comparison of various LQR controllers (load change = 10%)

Criteria↓	Simple LQR controller	GA based LQR controller	Jaya algorithm based LQR controller
Frequency deviation (pu)	−0.00035295	−0.00017118	−0.00011185
Frequency deviation (Hz)	−0.0176475	−0.008559	−0.0055925
Maximum frequency deviation (pu)	−0.00059147	−0.00034897	−0.00030716
Settling time (s)	0.6	0. 5	0.4
Steady-state error	-3.52×10^{-4}	-1.71×10^{-4}	-1.11×10^{-4}

Table 3 Comparison of various LQR controllers (load change = 20%)

Criteria↓	Simple LQR controller	GA based LQR controller	Jaya algorithm based LQR controller
Frequency deviation (pu)	−0.00070590	−0.00040099	−0.00022371
Frequency deviation (Hz)	−0.035295	−0.0200495	−0.0111855
Maximum frequency deviation (pu)	−0.0012	−0.00075633	−0.00061433
Settling time (s)	0.6	0. 5	0.4
Steady-state error	-7.05×10^{-4}	-4.00×10^{-4}	-2.23×10^{-4}

Table 4 Comparison of various LQR controllers (load change = 30%)

Criteria↓	Simple LQR controller	GA based LQR controller	Jaya algorithm based LQR controller
Frequency deviation (pu)	−0.0011	−0.00059321	−0.00033556
Frequency deviation (Hz)	−0.055	−0.0296605	−0.016778
Maximum frequency deviation (pu)	−0.0018	−0.0011	−0.00092149
Settling time (s)	0.6	0. 5	0.4
Steady-state error	-1.10×10^{-3}	-5.93×10^{-4}	-3.35×10^{-4}

Acknowledgements The authors sincerely acknowledge and are thankful to MHRD, New Delhi for providing financial assistance under TEQIP-III and to AICTE, New Delhi for providing financial assistance under RPS Project File No.8/36/RIFD/RPS/POLICY-I/2016-17, dated August 2, 2018. The authors are also thankful to the Director, MITS for providing the facilities for carrying out this research work.

References

1. Benjamin NN, Chan WC (1982) Variable structure control of electric power generation. IEEE Trans Power Apparus Syst PAS-101(2):376–380
2. Kothari ML, Nanda J, Kothari DP, Das D (1989) Discrete-mode automatic generation control of a two-area reheat thermal system with new area control error. IEEE Trans Power Syst 4(2):730–738
3. Pan CT, Liaw CM (1989) An adaptive controller for power system load-frequency control. IEEE Trans Power Syst 4(1):122–128
4. Wang Y, Zhou YR, Wen C (1994) New robust adaptive load-frequency control with system parametric uncertainties. IEEE Trans Gener, Transm Distrib 141(3):184–190
5. Yousef AM, Zahran M, Moustafa G (2015) Improved power system stabilizer by applying LQG controller. WSEAS Trans Syst Control 10:278–288
6. Thakur GS (2014) Load frequency control (LFC) in single area with traditional Ziegler-Nichols PID tuning controller. Int J Res Advent Technol 2:49–53
7. Haupt RL, Haupt SE (2004) Practical genetic algorithms (2nd edn.), Wiley, Inc.
8. Saadat H (2002) Power system analysis, 3rd edn. McGraw-Hill, New York, NY, USA
9. Rao VG (2016) Jaya: a simple and new optimization algorithm for solving constrained and unconstrained optimization problems. Inter Journ Ind Engg Comput 7:19–34
10. Adel A, Sajjad A, Hossein S (2017) Using Jaya algorithm to optimal tuning of LQR based power system stabilizers. In: 2nd International conference on computational intelligence and application, pp 482–486
11. Poodeh MB, Eshtehardiha S, Kiyoumarsi A, Ataei M (2007) Optimizing LQR and pole placement to control buck converter by genetic algorithm. In: International conference on control, automation and system, pp 2195–2200
12. Kundur P (1994) Power system stability and control. McGraw-Hill Inc, The EPRI Power Engineering Series
13. Bhatt P, Roy R, Ghoshal SP (2010) GA/particle swarm intelligence based optimization of two specific varieties of controller devices applied to two-area multi-units automatic generation control 32(4):299–310

A Review on Performance of Distributed Embedded System

Atrayee Dutta, Manjima De Sarkar and Sahadev Roy

Abstract A distributed embedded system is said to be a system which consists both hardware and software parts interacting all via some channels or interconnection network. For analyzing the performance of a distributed embedded system, it is required to know many basic terms which are associated with it along with its problems and ways to solve it. The approaches to solve these problems include simulation technique and also scheduling technique such as holistic scheduling technique. Other testing techniques are also applied such as fault injection technique to detect and solve the faults in a particular location where the fault has been occurred. Such technique reduces our time in correcting error or faults. For designing a distributed embedded system or any system which involves both hardware and software, safety measures are needed to be taken which has been discussed in this paper.

Keywords Compositional · Distributed · Holistic scheduling technique · Fault injection technique · Simulation · Stuck–at fault

1 Introduction

An embedded system is a special function processor that processes the information of the system which is closely integrated to its environment. The properties of embedded system comprises of small size, consume less power, rugged operating ranges and low per unit cost because of that it has less processing units that make it difficult to program and interact [1]. However, an external hardware can interface with

A. Dutta · M. De Sarkar (✉)
Department of Electronics and Communication Engineering, National Institute of Technology Arunachal Pradesh, Yupia, Arunachal Pradesh 791112, India
e-mail: manjima19@gmail.com

A. Dutta
e-mail: atrayee2013@gmail.com

S. Roy (✉)
Department of Electronics and Communication Engineering, National Institute of Technology Arunachal Pradesh, Yupia, Arunachal Pradesh 791112, India
e-mail: sahadevroy@nitap.ac.in

R. K. Shukla et al. (eds.), *Social Networking and Computational Intelligence*, Lecture Notes in Networks and Systems 100, https://doi.org/10.1007/978-981-15-2071-6_49

making them an intelligent system [2]. The modern embedded systems are based on microcontroller and microprocessor. Both microcontroller and microprocessor range from a simple to a complex class of computation and even for the application at hand. Embedded system ranges from devices such as digital watch to factory controller, hybrid vehicles and MRI. The complexity also ranges from single chip to multiple units [3].

A distributed embedded system comprises both hardware and software components interacting each other via some channels or interconnection network [4]. It differs from parallel computing or multiprocessor with the reason that its individual nodes have much higher independence for passing the messages. There is a problem with the distributed embedded system. It is very difficult to maintain their global states as every node of the system has the independence of its own scheduling.

The same problem may be encountered in the interconnection networks to [5]. The interconnection network may also consist of smaller subnetworks and which will also have its own scheduling and network protocol [6].

This paper discusses the following points:

- An example for analyzing the performance of the distributed embedded system.
- Basic terms which are required to understand before analyzing further on distributed embedded system.
- A list of required properties for distributed embedded system.
- Different approaches to perform the analysis on distributed embedded system. This method includes methods such as simulation-based methods, holistic scheduling analysis and compositional method.
- Testing technique to test the relationship between the hardware and software. As the distributed component consists both hardware and software components, the testing technique applied for this purpose is to selection of data technique to test the embedded system.
- At last point, the paper describes different safety measures are discussed for the interaction of hardware and software.

This paper divides as follows: Sect. 2 describes the analysis of performance. It describes basic performance difficulties faced by distributed system. It also gives brief description of basic terms which are required to know before performing any analysis on distributed system. It also gives a brief knowledge about the requirements should is needed to kept in mind before and after performing any analysis on distributed system.

2 Analysis of Performance

2.1 Distributed Embedded System

Let A1 is a sensor node which sends data to CPU1. The CPU1 stores this data in a memory with a task P1. The CPU2 processed the data with task P2 which have worst-case execution time (WCET) and a best-case execution time (BCET). This processes CPU data and transmitted it through allocated bus to the hardware input and output devices using task P#. Suppose the CPU1 is using preemptive fixed-priority scheduling. The preemptive fixed-priority scheduling executes the highest priority task from the entire task which is currently ready to execute [7]. In this case, P1 is taken as highest priority task. Also, the maximum load of CPU1 is obtained when the task P2 is continuously using WCET and all the sensors are continuously submitting the data. Now, sensor node A2 receives real-time data through the input interface using task P2. This data is sent to another CPU2 for the processing of task P3. The processed packets from the CPU2 are then sending to the play out buffer. The task P4 regularly removes buffer from every packet (Fig. 1).

Let's assume the bus is using the FCFS (first come first serve) pattern. Since A1 and A2 interface on a similar bus, it leads to jitter in the packet stream which causes an undesirable buffer underflow or overflow.

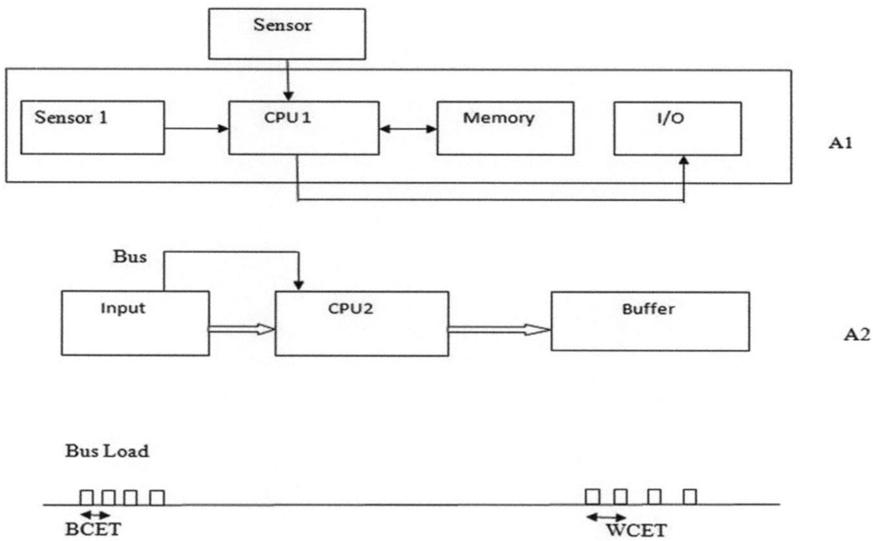

Fig. 1 Block diagram showing an example of a network of distributed embedded system

2.2 Basic Terms

Before starting the analysis there are some basic terms that are needed to be clarified. These are as follows:

1. Arrival events: The beginning of task, eventuality of the sensor input, or arrival of a packet which begins the task. Such events are said to be arrival events [8].
2. Finishing events: The finishing of a task or part of it again is model as an event.

In distributed embedded system, physical location of the arrival events and finishing events are different. The function embedded system is to process the data with arrival events. The computational time needs to change, and communication of an embedded system depends upon the input data and on the arrival events [9]. Considering the situation of conservative resource sharing strategy like the time-triggered architecture (TTA), interaction of different processes can be eliminated with application of static sharing policy [10]. Suppose the shared policy is dynamically controlled policies, every activity will communicate with each other effecting the timing properties.

3. Worst case and Best case: It is defined to be maximum interval of time between arrival and finishing events under every environmental state.
4. Upper and Lower bound: It restrains the worst-case and best-case behavior. These quantities cannot be computed at the same time as run-time of the system. The upper bound and lower bound focus more on upper bound and lower bounds. They are used to verify whether the system is meeting the desired timing requirements, e.g., deadlines
5. Statistical Measures: In place of computing bounds on worst case and best case, we can conclude the statistical characteristics of run-time, e.g., expected values, variance and quantiles.

Terms used for performance analysis:

1. Non-Determinism and Interference: This case arises when we have limited information of the system environment, for example, we do not know when the external input will arrive at the system. In addition to this, there are interferences present in the CPU, memory, bus or network. The system is said as non-deterministic, and there will enormous difference of behavior in worst case and best case.
2. Limited Analyzability: In this case, there is no appropriate way to determine the upper and lower bound with worst and best case.

2.3 Requirements

The requirements of distributed embedded system for performance analysis are listed as follows:

1. Correctness: The result must be correct. There should not be any reachable system states and feasible reaction in the system environment which violated calculated bounds.
2. Accuracy: Both upper bound and lower bound should be closer to worst case and best-case timing.
3. Embedding into the design process: All method must integrate into a one functional requirement and design methodology.
4. Short analysis time: There should be short analysis time and the design should also have the property of configurability.

3 Approaches to Performance Analysis

3.1 Method Based on Simulation

The process of, estimating of performance is achieved with simulation. The softwares which are used for this process are System C, Synopsys and Keil.

For determining the properties of timing of embedded system, the process of simulation is only sufficient. It also requires concept of time, functional computation, account and communication process and elements which underlying the hardware, resource sharing policies [11]. This leads to high computation time, and as a result, performance estimation quickly becomes bottleneck.

There are some additional problems with simulation such as it has insufficient corner case coverage [12]. There are some appropriate simulation stimuli which are used to cover this corner.

3.2 Holistic Scheduling Analysis

This scheduling first investigate the pattern of the system that whether it is sporadic, periodic, jitter and bursts [13]. The scheduling has 'one model approach' which allows all different approaches into one model. Since this approach has 'one model approach', it is used in distributed system.

3.3 Compositional Method

There are few problems that arise in case of complex distributed embedded system which are as follows:

1. The architecture of the complex distributed system is highly heterogeneous.

2. Multiple control threads complicate the timing analysis.

To solve the above problem, a method based on the classical real-time scheduling can be used. This method combines abstraction of event task, and it also provides extra interface between them [14]. This approach is based upon the following principle:

1. The main goal is to achieve results in real-time scheduling, particularly for the case sharing in a single processor.
2. The application model is simple and is combination of several tasks.

Thus, the compositional can be said to be more preferable method for analyzing the performance any distributed embedded system.

But there is limitation with the compositional method also. It is that the compositional method and the other two approaches discussed above are appropriate only with the incoming patterns. To overcome the following limitation, two types of interfaces are used along with the above methods, those are as follows:

1. EMIF: This stands for event model interfaces which is required only for performance analysis. They perform specific conversion of the arrival events which changes the mathematical representation of the event streams.
2. EAF: When EMIF is present in a system, it uses event adaptation functions (EAF). Here, the hardware or software implementation is changed to make the system analyzable.

4 Testing Technique Between Hardware and Software Interaction in Embedded System

Distributed embedded system consists of both hardware and software components, so it is expected to contain some failures while interaction [15]. To detect this failure a test selection technique is needed to be developed. The test selection technique is used with fault injection technique so as to detect the failure in interaction. The fault injection technique check for the fault in particular location of the hardware or software, and it also examines how the system is behaving. It does so by injecting faults to the target system and checks the communication faults between hardware and software. The process can be classified into following steps:

1. Simulate the behavior of the embedded system into software program.
2. Hardware faults or errors are transformed to software faults or errors and are given into the program.
3. The faults or errors in data are selected and are used to detect the faults or errors which happen because of hardware–software interaction.

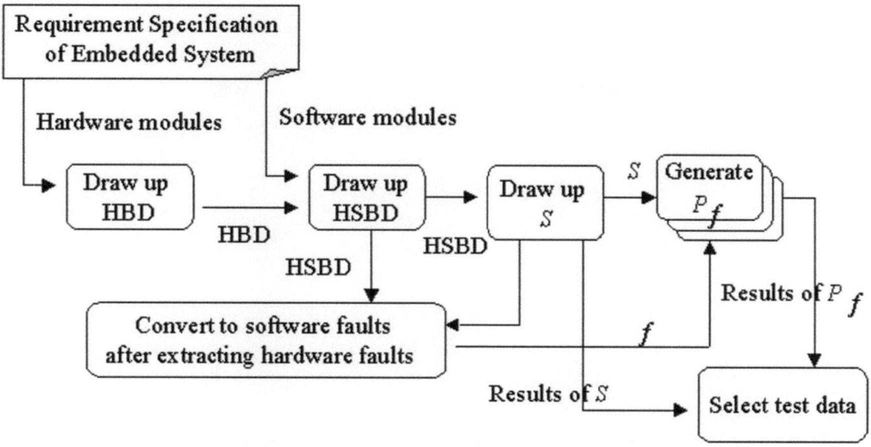

Fig. 2 Block diagram showing selection of data technique

4.1 Selection of Data Technique for Testing Embedded Systems

1. From the Fig. 2, the hardware block diagram (HBD) describes the module of the hardware and its input and output information.
2. For simulating the behavior of the HSBD, the hardware errors are converted to software errors and are given to the program S to form program Pf. A test data is formed using program Pf.

4.2 Analyzing Hardware–Software Modules in Embedded System

4.2.1 Hardware Block Diagram (HBD)

The hardware block diagram represents the hardware module and its respective input-output signals from the required objective of the targeted system. In Fig. 3, the rounded rectangular represents the hardware module and the lines connected to them represent their input-output signals.

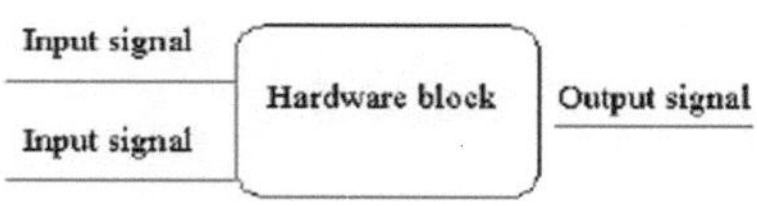

Fig. 3 Composition of HSBD

Fig. 4 Composition of HSBD

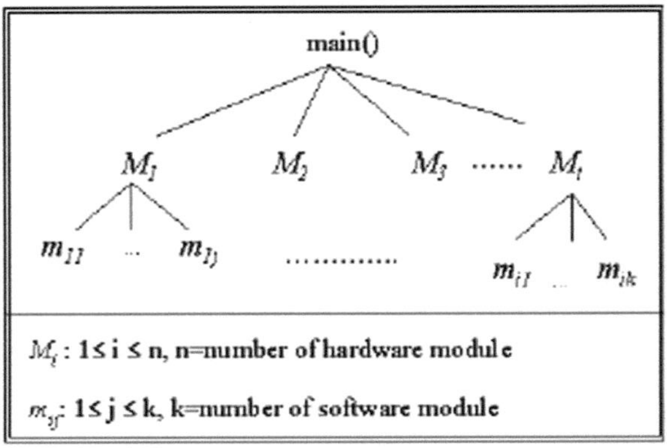

Fig. 5 Composition of S

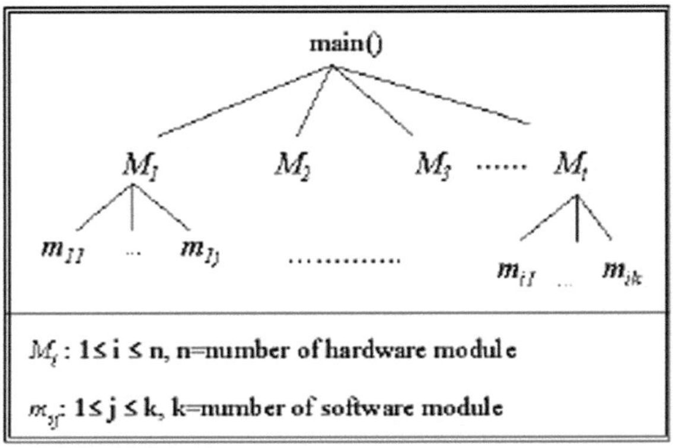

4.2.2 Hardware–Software Block Diagram (HSBD)

The HSBD represents the software module. In Fig. 4, the software module is represented in the form of an ellipse and their respective directions are represented by arrows.

4.2.3 Combing up the Program S

In Fig. 5, the software (mik) in the hardware module (Mi) of the HSBD is made for a function which is written in C language.

4.2.4 Formulating the Program Pf

To detect the errors in the interaction of the hardware and software module, a program Pf is created by injecting software errors into program S. The software errors in the program S are hardware errors which are converted to software errors. To perform this program, it is required to study the different types of hardware errors.

The hardware errors of the embedded system are divided into physical fault and logical fault. The physical fault includes bad connection and inconstant power supply. Logical faults are caused by environmental factors, for example, exposure to heavy iron. Hardware faults are divided into stuck-at faults, bridging faults, open faults, power disturbance faults and spurious current faults. The faults which are mentioned above are explained as follows:

1. Stuck-at faults represent the flaws in lines or rounded rectangular in HSBD.
2. Bridging faults caused by switching of values when two or more lines crosses.
3. Open faults are due to the resistance of lines or bad connection.
4. Power disturbance faults are due to variance in power supplies.

5 Safety Requirements of Hardware–Software Interactions in Complex System

One approach to do the safety analysis is to considering all the hardware and software interaction in terms of physical resources and time.

For any hardware and software combination system, the resources are divided into the following classes depending on the criticality of the resource usage.

a. Intrinsically critical

These contain safety critical data at every point of execution. Examples are I/O and RAM used for safety critical functions, processor registers, etc.

b. Primary control

The resource controls the function of an intrinsically critical resource. Examples are memory management unit (MMU) registers and I/O control registers.

c. Secondary control

These act as a backup for the primary resources and also control access of primary resources. Example: Key registers should be set to particular values before MMU registers are changed.

d. Non-Critical

These resources are never used by the critical software and do not affect any part of the hardware which is used by the critical function.

e. Used.

This resource locates the memory map which does not relate to any physical device. The requirement of this location is that any part of the software cannot access this memory.

Thus, a safety system is one which powers up in a safe state and also respects minimal safety requirements throughout each stage of the initialization.

References

1. Yen TY, Wolf W (1995) Communication synthesis for distributed embedded systems. In: Proceedings of the 1995 IEEE/ACM international conference on computer-aided design, pp 288–294
2. Xie G, Zeng G, Liu L, Li R, Li K (2016) High performance real-time scheduling of multiple mixed-criticality functions in heterogeneous distributed embedded systems. J Syst Arch 70:3–14
3. Mallak A, Weber C, Fathi M, Holland A (2017, October) Active diagnosis automotive ontology for distributed embedded systems: In: 2017 IEEE European technology and engineering management summit (E-TEMS), pp 1–6
4. Ballesteros A, Proenza J, Barranco M, Almeida L (2018, June) Reconfiguration strategies for critical adaptive distributed embedded systems. In: 2018 48th annual IEEE/IFIP international conference on dependable systems and networks workshops (DSN-W), pp 57–58
5. Ebeid E, Fummi F, Quaglia D (2015) Model-driven design of network aspects of distributed embedded systems. IEEE Trans Comput-Aided Des Integr Circuits Syst 34(4):603–614
6. Zhang X, Mohan N, Torngren M, Axelsson J, Chen DJ (2017) Architecture exploration for distributed embedded systems: a gap analysis in automotive domain. In: 2017 12th IEEE international symposium on industrial embedded systems (SIES), pp 1–10
7. Xie Guoqi, Chen Yuekun, Xiao Xiongren, Cheng Xu, Li Renfa, Li Keqin (2018) Energy-efficient fault-tolerant scheduling of reliable parallel applications on heterogeneous distributed embedded systems. IEEE Trans Sustain Comput 3(3):167–181
8. Mubeen S, Sjödin M, Nolte T, Lundbäck J, Gålnander M, Lundbäck KL (2015) End-to-end timing analysis of black-box models in legacy vehicular distributed embedded systems. In: 2015 IEEE 21st international conference on embedded and real-time computing systems and applications (RTCSA), pp 149–158
9. Gu Zonghua, Han Gang, Zeng Haibo, Zhao Qingling (2016) Security-aware mapping and scheduling with hardware co-processors for flexray-based distributed embedded systems. IEEE Trans Parallel Distrib Syst 27(10):3044–3057
10. Xie Guoqi, Chen Yuekun, Li Renfa, Li Keqin (2018) Hardware cost design optimization for functional safety-critical parallel applications on heterogeneous distributed embedded systems. IEEE Trans Industr Inf 14(6):2418–2431
11. Krzywicki K, Adamski M, Andrzejewski G (2015) EmbedCloud–design and implementation method of distributed embedded systems. In: Doctoral conference on computing, electrical and industrial systems, pp 157–164
12. Honig WL, Läufer K, Thiruvathukal GK (2015) A framework architecture or student learning in distributed embedded systems. In: 2015 10th IEEE international symposium on industrial embedded systems (SIES), pp 1–4
13. Wandeler E, Thiele L, Verhoef M, Lieverse P (2006) System architecture evaluation using modular performance analysis: a case study. Int J Softw Tools Technol Transfer 8(6):649–667
14. Roy S, Saha R, Bhunia CT (2016) On efficient minimization techniques of logical constituents and sequential data transmission for digital IC. Indian J Sci Technol 9(9):1–9
15. Singhal G, Roy S (2019) A novel method to detect program malfunctioning on embedded devices using run-time trace. In: Advances in signal processing and communication. Springer, Singapore, pp 491–500

A Comparative Study of DoS Attack Detection and Mitigation Techniques in MANET

Divya Gautam and Vrinda Tokekar

Abstract Mobile ad hoc network is a self-configured, decentralized, constellation of machines that together form architecture less movable network. Because of dynamic changing nature of the network, it is more prone to various attacks. DDoS attacks are the major security risk on mobile ad hoc networks (MANET). DDoS attacks have the tendency to make large volume of unauthorized traffic, due to which the legitimate users cannot use the resources. In this work, various DDoS detection and mitigation techniques have been analyzed. This work has abridged various types of DDoS techniques and attack detection methods. It has also identified advantages and disadvantages of various DDoS defense mechanisms. Volumes of academic research have been discussed that depicts a diverse array of methodologies in detecting, preventing, and mitigating the impact of DDoS attacks.

Keywords MANET · DDoS attack · DDoS algorithms · Zero-day attack · Resource depletion

1 Introduction

Mobile ad hoc network is a type of network having no specific architecture suitable for mobile devices. In MANET, nodes have the freedom to move in any direction. MANET security has been burning issues for the research and analysis community. Because of dynamic changing nature of MANET, it is vulnerable to various attacks. Besides, different parameters also contribute to its vulnerability, like shared radio channels, the open design, and restricted resources, etc. Presently, MANETs are vulnerable to numerous attacks together like masquerading, distortion of messages, eavesdropping, and denial of service (DoS) attacks [1].

D. Gautam (✉)
Amity University Madhya Pradesh, Gwalior, India
e-mail: divyagautam06@gmail.com

V. Tokekar
IET, DAVV, Indore, India
e-mail: vrindratokekar@yahoo.com

R. K. Shukla et al. (eds.), *Social Networking and Computational Intelligence*, Lecture Notes in Networks and Systems 100, https://doi.org/10.1007/978-981-15-2071-6_50

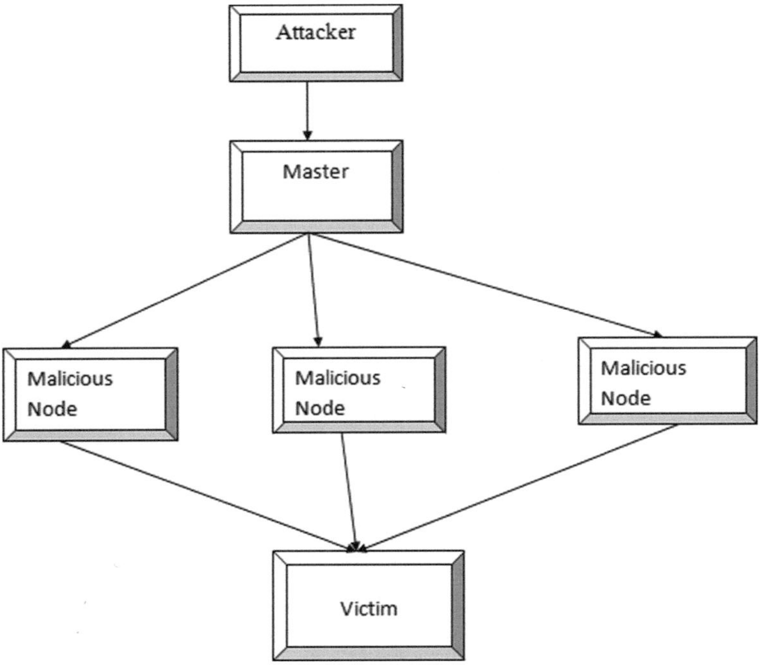

Fig. 1 Distributed denial of service attacks

DoS attacks are well-known for restricting the authorized user from gaining access to the network and using other resources of network. One of the most vulnerable attacks related to DoS is distributed denial of service (DDoS) attack which actually drains victim resources such as bandwidth, processing capacity, or memory of a victim machine. In this attack, the victim machine is flooded with incoming messages which makes machine shutdown forcefully, by simply denying for the services to authenticated users (Fig. 1). In this, services of the victim networks are also compromised. Numerous efforts had been made in order to combat DDoS attacks in the past decades. In today's scenario, there are many important services are dependent on Internet for communication which actually is a part on their infra, and ramification of DoS attacks can be exceptionally dangerous. So the biggest issue with wireless mobile ad hoc networks combat today is security, since there is no control from center.

2 Phases for Launching DoS/DDoS Attacks

The DoS/DDoS attacks occur in mainly three phases, i.e., (a) acquiring botnets, (b) propagation, and finally, (c) attack.

2.1 Acquiring Botnets

This is the initial phase of the attack when the attack army or botnet is generated. To achieve this, malicious programs called worms are used which are highly self-propagating in nature. The main idea behind the success of the creation of botnets is to detect and exploit the vulnerabilities in the system to acquire control.

There are the different ways in which botnets are acquired like random scanning; in this method, the already infected machine infects new machines by randomly probing IP addresses from the available address space. This random nature of the scanning makes this distributed throughout various networks and also generates huge network traffic which can also be used as a red flag for identifying these kinds of malicious activity. Random scanning also generates duplicate probes due to its random nature.

Another way to acquire the bots is hit-list scanning; this type of determination of targets involves preparing a hit list as the name suggests including all the vulnerable machines which is determined by the attacker. A worm is then sent to affect these machines. The time taken to infect machines is reduced to a significant extent. Here, also the large size of the list which is sent along with the worm can be used as a red flag to detect this kind of attacks.

Third method is Permutation Scanning which is an efficient method in terms of lower duplicates of the probes of the same IP addresses. This is achieved by self-coordination between the infected nodes. This also provides a mechanism to stop and to assess the potential benefits in terms of spreading the infection.

Fourth way to get botnets is Topological Scanning where once a node is infected, information is extracted from the node about other potential nodes which can be affected. This is an alternate for the hit-list scanning method. For MANETs which are peer-to-peer systems, if the worm hits a node, then it can extract information about the other nodes in the network and thus can infect all. Thus, in topological scanning, there is no pre-existing list of the target nodes. This method though is slower but advantageous for the attackers as large volume of traffic is not produced, and thus, these attacks can go undetected.

2.2 Propagation

After completion of the initial phase, i.e., acquiring new botnets, the propagation stage aims at spreading the malicious code to the infected nodes to launch the attack. This malicious code includes details like the target, date, and duration of the attack, etc. The propagation of the malicious code can be achieved through methods. First is central source propagation method, in this propagation method, the malicious code is propagated from a central server. This exploit is carried out by the attack node which attacks the target, and after successfully disrupting the normal operations of the target, the infected node uses the target to propagate the code to the next target node and this process goes on until all nodes are successfully disrupted. Second method is called as back-chain propagation; in this method of propagation, the target machine downloads a copy of code from the compromised node. This happens during the acquiring phase, and a connection is set up between the infected node and the target node which is used later to propagate the malicious piece of code. Third method is autonomous propagation; in this method of propagation, the malicious code is transferred to the target nodes from the infected nodes at the time of exploitation automatically. This method generates lower volume of network traffic and thus has a higher chance of going undetected.

2.3 *Attack*

This is the final and the main phase in which the attack is launched. There are several categorizations of these attacks. Bandwidth depletion attack method targets to consume entire network's bandwidth to deny legitimate users of the target. This is achieved by using the extensive network of botnets. This can be done by exploiting the protocols or via amplification and thus is further classified as Protocol Exploit Attack and Amplification Attack. Resource depletion attacks aim at exhausting the target system's important resources like processor, sockets, and memory which are essential for the proper running of the system. This is achieved by two forms of attacks, namely Protocol Exploit Attacks and Malformed Packet Attacks. Infrastructure attacks are the most extreme form of DDoS attacks which tend to disrupt the backbone of networks by damaging significant, important elements of the Internet thereby making these attacks a hybrid type of attacks combining both resource depletion attacks as well as Bandwidth depletion attacks. Example of this type of attacks can be DNS flooding technique in which the root DNSs are targeted as they are the top level service point for the whole world's Internet connection and thus have a potential to disrupt the world's Internet service, thereby causing losses in billions. Zero-day attacks are one of classic forms of DoS attacks as the vulnerabilities of the system are not revealed until one day later after the attack is complete. Though they have a huge potential for disrupting the functioning of a network, they do reveal vulnerabilities present in the system which can be fixed and thus any later events like that can be avoided. Sometimes organizations provide bounty for discovering such vulnerabilities.

3 Literature Review

Around 40% of the participants in a survey were under DoS attack between 1999 and 2005 [2] as per the survey report of "Computer Security Institute-computer crime and security." About 29% of the participants were targeted by a DoS attack in between year 2006 and year 2009 [3]. In 2010, Worldwide Infrastructure Security Report [3] identified that denial of service attacks came in mainstream, and network administrators were handling bigger and more frequent DDoS attacks. Hence, it is difficult to detect or minimize the effect of DDoS. The 2012 Annual DDoS Attack and Impact Survey reports [4] found that around 35% companies have faced this attack.

The study [5] found that the DDoS attacks were coming up very rapidly in the first six months of 2013 as larger scale attacks and causing more problems to the Internet. DDoS attacks slow down or collapse servers by overloading the servers with extremely large traffic.

In DDoS attack, the attacker packet is under the control of the master node and agent node. Once attack array is completed, the attacker breaks the networks. Now each node (attacker bot) shoots many numbers of compromised packets to the target node, and it causes depletion of resources of the target node and ultimately it crashes the target machine. Hence, it is difficult to generate an accurate defense mechanism to break down the DDoS attack by early detection [6].

Douligeris and Mitrokotsa [7] have proposed four ways for combating with DDoS attacks. Attack detection, characterization, prevention, trace back, tolerance, and mitigation are possible by their proposed approach. The exact idea of attack prevention is to identify the security loopholes like weak authentication methods, insecure protocols, and defenseless computer machines which are more susceptible to become a bot (attacker node) for the centralized attacker node.

Chauhan and Nandi [8] proposed a protocol based on quality of service aware on-demanding routing protocol. They have used signal stability as one of the routing criteria with quality of service matrices.

Jun et al. [9] have given a detection technique for DDoS flooding attack by using step-by-step scheme for investigations, where they have used entropy for detecting the DDoS attacks in order to provide reliable transmission and stopping the abnormal traffic flow of data [10]. A variety of DoS types have been documented, including SYN flooding, teardrop, ICMP ping flooding, Smurf attacks, and Fraggle attacks [11–13]. Detecting DoS and DDoS attacks often depends on the type of attack that is occurring. Carl et al. mentioned that anomaly detection, measuring network flow rates of packets, and sequential change-point detection are few detection methods that offer different statistical evaluations of network traffic [13]. Each approach assumes anomalies occur within standard network signals and that allowed user activity could be differentiated from abnormal DoS events through real-time software attack detectors.

Carnegie Mellon University's site [14] lists several common DoS prevention techniques, such as implementing router filters to lessen exposure, installation of patches that guard against various flooding techniques, disabling unused or nonessential network services, and a variety of other methods. The real-time DoS detection software, using pattern detection, can be implemented [14] to filter out nefarious activity, as well. By having smart hardware and software services, preventing known attacks becomes easier. However, new DoS attacks could be unleashed that may not fit patterned behavior and this might allow defensive systems to be circumvented.

4 Comparative Study

Many studies have been carried out on the detection mechanism as the extent of damage caused by DDoS attacks has increased. However, the present or existing security mechanisms have failed in providing an effective defense against DDoS attacks or can only provide defense against specific types of these attacks. Few DDoS attack detection methods are based on trace back, while some others are on feature monitoring of a server. A comparative study has been done between techniques used for detecting and mitigating the DoS Attacks in MANET (Table 1).

5 Conclusion

There are various issues that can be concluded after analysis of various DDoS attack detection techniques. First, it is a tough task to keep a trade-off [32] in between the performances of the real-time defense methods and utilization of the resources of the target machines. These (DDoS) attacks exhaust the host resources like their BW, power of processing, or memory, etc. In order to achieve better results of the DDoS defense techniques, the consumption of resources of victim resources machine should be minimized.

Second, scalability is one more challenging aspect in DDoS attack concepts. In, DDoS attacks multitude of signatures and attack scenarios are involved. Therefore, the researchers should always keep in mind about these different aspects of attacks while designing the best possible defense mechanism for attacks. The noteworthy point from this study is that the researchers should carry out the research in real-time environment also tests the result on the real-life performance of those researches rather than with fixed data set and fixed attack signatures. In most cases, the real-time attack performance is quite dissimilar to the virtual environment. Hence, it is required to consider scalability while developing the defense mechanism for real-time networks.

Third, it is quite required to ensure the defense mechanism against one open-research problem such as the zero-day attack, which is also an emerging issue. In this, the DDoS attackers attack with newer version of attack with all its capacity in terms of power and complexity. Therefore, the researchers are striving hard to provide defense against zero-day attack. Besides, the technical skills, it is prerequisite for the researchers to understand attacker's psychology and skills in order to defend against new types of DDoS attacks.

Fourth, DDoS attack can take place even in the absence of software vulnerabilities in a system. Researchers are striving hard to distinguish between attackers' requests from other legitimate requests. Thus, defense mechanisms that based on detecting and filtering attacker's requests have limited effectiveness. Traceback techniques aim

Table 1 Comparison table for various solutions available for DDoS attack in MANET

Ref. no.	Author name	Techniques used	Conclusion	Benefit/limitation
1	Lai et al. [15]	DDoS attacks are being identified by computing the harmonic mean (average arriving rate of packets). Also, they have calculated the difference in harmonic mean	The harmonic mean is not varying; instead, it is quite fixed even after network is flooded with packets at allowable bandwidth. This is a different activity on the BW; hence, there is an abnormal traffic in the network	It is an effective technique for examining the attack traffic. But continuous watch is required on the bandwidth
2	Mehfuz and Doja [16]	SPA-ARA (secure power-aware ant routing algorithm) is introduced for detection	This study adopts a trade-off between selection of fast paths and a better use of network resources. They also introduced a protocol to incorporate a trust model in order to detect unauthorized or compromised node in the network	Time complexity is high due to the presence of routing pheromone table and trust pheromone table based on ant routing algorithm
3	Arunmozhi and Venkataramani [17]	Flow monitoring table has been introduced for each node contains flow, source, packet sending rate, and destination. Table is to be updated after every flow of data between source and destination	On monitoring MAC layer signals, explicit congestion notification is sent by destination for reducing the sending rate. If still sender does not reduce the rate, then rates are compared with the table to detect the attacker node. Packets can discard once node has been detected as malicious	Accuracy to identified DDoS attack flows is high in this study due to the application of rate limiting to the malicious network flows
4	Wu et al. [18]	This work deals with replay attack, and each node in MANET has to store the MAC and verify it with upcoming nodes MAC. If both are same, then drop rather than forwarding the packet	This is costlier method, so let nodes to decide whether to forward the packet or not. By giving incentives, nodes will be ready to forward the packet, as nodes are selfish	In this method, authors have mentioned the key game parameters, such as the penalty for forwarding a bad packet without verification; this can affect the probability that a node will verify a received packet
5	Rao [19]	DDoS mitigation techniques such as using router-based access control lists (ACL), rate limiting, combining both ACL, and rate-limiting techniques	Specific rule sets are created to detect and ignore anomalous patterned IP activity	This study offers better result for IPv4 addresses, but serious issues arise when IPv6 needs to be established globally and transition from IP version 4 to IPv6 has to be done

(continued)

Table 1 (continued)

Ref. no.	Author name	Techniques used	Conclusion	Benefit/limitation
6	Xu et al. [20]	Detection technique based on KPCA and PSO-SVM	This technique helps in improving the detection time and provides more accurate results of finding DDoS in networks	Method can enhance the speed of DDoS detection effectively based on some characteristics extracted
7	Mishra et al. [21]	They have defined various attacking tools are described like Trinoo, TFN, and Knight also have divided the intrusion tolerance and mitigation into two parts (a) fault tolerance (b) quality of service	DDoS attack effect can be minimized by increasing the tolerance limit and QoS	Lot of work is required to improve the QoS by increasing the tolerance limit
8	Michalas et al. [22]	The method is having two parts as game theory and cryptographic puzzles. First part is a client puzzle which is able to provide security from DoS attacks in such networks. The second part is based on fundamental principles of game theory, where a multiplayer takes place between various nodes of the ad hoc network	The combination of puzzles and computational problems gave an improvement in efficiency and latency of involved machines (nodes) and is capable of handling DDoS attacks. The simulation results show the approach is quite effective for the devices having lesser resources under mobile ad hoc environment, where the quick information exchange is the requirement	Improved preventive technique for DoS and DDoS attacks
9	Kim and Helmy [23, 24]	Abnormality is defined in terms of increment in packets at network layer, increment in frames, and collisions and busy time at MAC layer. If there is any abnormality, then the information is gathered and saved. Later, it is characterized as time series	Traffic pattern matching and Kolmogorov–Smirnov (KS)-fitness test both are used for searching the abnormality. Later the countermeasures and rate limiting factor are also discussed in the work	It is a better method for computation having lesser memory overhead

(continued)

Table 1 (continued)

Ref. no.	Author name	Techniques used	Conclusion	Benefit/limitation
10	Ye et al. [25]	They have extracted the six-tuple characteristic values related to DDoS attack and then use the support vector machine algorithm to judge the traffic and carry out DDoS attack detection. We focus on the analysis of the changes in the characteristic values of traffic and verify the feasibility of this method by deploying the SDN experimental environment	By analyzing the ICMP traffic, we have come to the conclusion that the ICMP flow has no source port and destination port, so SSP and RPF are zero, which makes the six-tuple characteristic value matrix change into four-tuple characteristic value matrix, whether attacked or not	As false alarm rate is very slow, so it is required to simulate the normal data flow more comprehensively
11	Noh et al. [26]	They have applied inductive learning approach to detect DDoS attack. The ratio of no. of TCP flags to the total no. of packets is calculated based on proposed network traffic analysis methods	The presence of DDoS attacks is detected based on TCP flag rates and by using machine learning algorithms	Machine learning algorithms seem to be effective for layer 4 attacks. This applies only on the specific flags depending on the rate of such flags in the traffic. This is not proven a method for application layer attacks
12	Mankins et al. [27]	The proposed architecture takes into consideration of different pricing and purchase functions. Service quality differentiation is provided by architecture. This will help in selecting clients based on good behavior and differentiate bad behavior clients	The pricing of resources of DDoS is done by using distributed gateway-based architecture	Lesser probability of predicting attacks
13	Mukhopadhyay et al. [28]	DDoS attack detection implementation is discussed. Many approaches are used here	Various approaches of DDoS attacks are discussed and analyzed in this paper. Methodologies used by all these approaches are highlighted	There are no good methods to detect slowloris and RUDY attacks discussed here

(continued)

Table 1 (continued)

Ref. no.	Author name	Techniques used	Conclusion	Benefit/limitation
14	Lipson [29]	They have developed enhanced ICMP traceback-cumulative path to detect DDoS attack	Victim uses I trace-CP messages to track source and path used. These I trace-CP messages are generated by intermediate routers and entire attack path will be constructed in very short time	It is difficult to adapt to changing topology as changes need to be done on every router, and more space is needed to process packets
15	Thatte et al. [30]	Dynamic detection method is applied for DDoS attack detection	This is deployed near to victim site, normally on the nearest router of the victim site. It is effective for non-distributed low DoS attacks	No extra memory is needed if any modification is needed to existing infrastructure
16	Ansari and Waheed [31]	DDoS attack-flooding attack is detected and prevented based on assessment of signal properties	MAC layer of different nodes is analyzed by understanding the signal properties and updating the routing tables	Better approach for detecting flooding attack more accurately then the current approach used for detection of attack at network layer

to locate the attack sources irrespective of the spoofed source IP addresses either in the process of attack (active) or in the postattack (passive).

Various efforts have been made to combat DDoS attack. Though, none of these approaches independently accomplishes avoidance or provides enough countermeasures to conquer and determine DDoS threats over wireless networks. Various schemes have been adopted based on these approaches, but still researchers are struggling to provide comprehensive solution to tackle changing nature of DDoS attacks. One of the main reasons behind it is lack of comprehensive knowledge about DDoS incidents.

References

1. Sandoval G, Wolverton T (2000) Leading web sites under attack. Tech Report. CNET News Tech Rep. TR 2100–1017, 9 Feb 2000
2. Gordon LA, Loeb MP, Lucyshyn W, Richardson R (2004) 2004 CSI/FBI computer crime and security survey. Computer Security Institute, San Francisco, CA
3. Dobbins R, Morales C (2010) Worldwide infrastructure security report. Arbor Networks Annual Survey
4. http://www.neustar.biz/enterprise/resources/ddos-protection/2012ddosattackreport#.UlT_otdR5DI
5. http://www.itproportal.com/2013/07/31/ddos-attacks-rise-dramatically-first-half-2013

6. Kumar S, Varalakshmi G (2011) Detection of application layer DDoS attack for a popular website using delay of transmission. Int J Adv Eng Sci Technol 10(2):181–184
7. Douligeris C, Mitrokotsa A (2004) DDoS attacks and defense mechanisms: classification and state of the art. Comput J Netw 44(5):643–666
8. Chauhan G, Nandi S (2008) QoS aware stable path routing (QASR) protocol for MANETs. In: First international conference on emerging trends in engineering and technology, pp 202–207
9. Jun JH, Oh H, Kim SH (2011) DDoS flooding attack detection through a step-by-step investigation. In: IEEE 2nd international conference on networked embedded systems for enterprise applications. ISBN: 978-1-4673-0495-5
10. Erickson J (2008) Hacking—the art of exploitation (2nd edn.). No Starch Press Inc., San Fancisco, CA, p 50, p 250–258
11. Singh A Demystifying denial-of-service attacks. Part one. A report on symantec site
12. Goodrich M, Tamassia R (2011) Introduction to computer security. Morgan Kaufmann Publishers, Boston, MA, pp 256–260
13. Carl C, Kesidis G, Brooks RR, Rai S (2010) Denial-of-service attack-detection techniques. IEEE Internet Comput 82–89
14. www.cert.org
15. Lai WS, Lin CH, Liu JC, Huang HC, Yang TC (2008) Using adaptive bandwidth allocation approach to defend DDoS attacks. Int J Softw Eng Its Appl 2(4):61–72
16. Mehfuz S, Doja MN (2008) Swarm intelligent power-aware detection of unauthorized and compromised nodes in MANETs. J Artif Evol Appl
17. Arunmozhi SA, Venkataramani Y (2011) DDoS attack and defense scheme in wireless ad hoc networks. Int J Netw Secur Its Appl (IJNSA) 3(3). https://doi.org/10.5121/ijnsa.2011.3312
18. Wu X, Yau DKY (2006) Mitigating denial-of-service attacks in MANET by distributed packet filtering: a game theoretic approach. In: Proceedings of the 2nd ACM symposium on information, computer and communication security, pp 365–367
19. Rao SRS (2011) Denial of service attacks and mitigation techniques: real time implementation with detailed analysis. SANS Institute
20. Xu X, Wei D, Zhang Y (2011) Improved detection approach for distributed denial of service attack based on SVM. IEEE 978-1-4577-0856
21. Mishra A, Gupta BB, Joshi RC (2011) A comparative study of distributed denial of service attacks, intrusion tolerance and mitigation techniques. In: EISIC'11 Proceedings-European intelligence and security informatics conference pages pp 286–289. IEEE Computer Society Washington, DC, USA ISBN: 978-0-7695-4406-9
22. Michalas A, Komninos N, Prasad NR (2011) Mitigate DoS and DDoS attack in mobile ad hoc networks. Int J Digit Crime Forensics 3(1):14–36
23. Kim Y, Helmy A (2009) CATCH: a protocol framework for cross-layer attacker traceback in mobile multi-hop networks. Elsevier
24. Kim Y, Helmy A (2006) Attacker traceback with cross-layer monitoring in wireless multi-hop networks. SASN'06, 30 Oct 2006
25. Ye J, Cheng X, Zhu J, Feng L, Song L (2018) A DDoS attack detection method based on SVM in software defined network. Hindawi, Secur Commun Netw 2018, Article ID 9804061, p 8
26. Noh S, et al (2003) Detecting distributed denial of service (DDoS) attacks through inductive learning. LNCS 2690, pp 286–295
27. Mankins D, Krishnan R, Boyd C, Zao J, Frantz M (2001) Mitigating distributed denial of service attacks with dynamic resource pricing. IEEE
28. Mukhopadhyay D, Oh BJ, Shim SH, Kim YC (2010) A study on recent approaches in handling DDoS attacks. Cornell University Library
29. Lipson HF (2002) Tracking and tracing cyber-attacks: technical challenges and global policy issues. CERT Coordination Center, Special Report: CMU/SEI-2002-SR-009
30. Thatte G, Mitra U, Heidemann J (2005) Detection of low-rate attacks in computer networks. University of Southern California IEEE

31. Ansari A, Waheed MA (2017) Flooding attack detection and prevention in MANET based on cross layer link quality assessment. In: 2017 international conference on intelligent computing and control systems (ICICCS). Electronic ISBN: 978-1-5386-2745-7, IEEE
32. Fakieh KA (2016) An overview of DDoS attacks detection and prevention in the cloud. Int J Appl Inf Syst (IJAIS)—ISSN: 2249–0868, Foundation of Computer Science FCS, New York, USA, 11(7), December 2016. www.ijais.org

Prediction of Software Effort Using Design Metrics: An Empirical Investigation

Prerana Rai, Shishir Kumar and Dinesh Kumar Verma

Abstract These days, prediction of effort in software project is the shove area for the researchers. The estimation of effort in software process is as essential as software product. Primarily, estimation models consist of relation between dependant and independent variable(s). The effectiveness of these models is to bring more accuracy to the work plan and reduce financial cost. The variables in these models may be considered as complexity, size, person per month, and other different software metrics. Most of these models only considered the static behaviour of the software product, in which the fixed value of the effort predicted at the starting of project. Hence, there is a need to formulate a methodology which considered the future changes in the software project for effort estimation. In this paper, a model has been formulated which can be use to make the prediction of software efforts with the help of software metrics, primarily design metrics, such as Depth of Inheritance Tree, Line of Code, Weighted Method per Class. The correlation between the metrics and effort is been shown with the help of regression model formulated in this paper. The model has been validated by the data set collected from the PROMISE repository.

Keywords Software quality · Software effort estimation · Regression · Software metrics

1 Introduction

Software engineering is a combination of design, development, and maintenance which gives a vast area to explore. Prediction plays a vital role in developing software in this process [1]. Prediction of software effort is the most recent research area in

P. Rai (✉) · S. Kumar · D. K. Verma
Computer Science and Engineering, Jaypee University of Engineering and Technology, Guna, India
e-mail: prerana.rai99@gmail.com

S. Kumar
e-mail: dr.shishir@yahoo.com

D. K. Verma
e-mail: dinesh.hpp@gmail.com

R. K. Shukla et al. (eds.), *Social Networking and Computational Intelligence*, Lecture Notes in Networks and Systems 100, https://doi.org/10.1007/978-981-15-2071-6_51

software assessment. There were many different methods which helped in predicting effort among them expert judgement was in trend [2]. Initially, software estimation was based on reasoning and intuitive [3]. There are many researchers who are trying to predict an optimal way to known the effort at the initial of software development life cycle (SDLC) used to make software. After the effort has been evaluated than effective planning can be done with the help of these evaluated values.

Hence, an effective planning is done which helps to complete the project within given budget and time. Size and complexity were the factors which brought a major change in effort prediction. During the development process of software the measurement of the software metrics is done, which can be used as the variables in developing model to predict the development process quality [4]. The measured software metrics are used to develop estimation model which are useful to predict other software metrics, i.e., effort, re-usability, cost etc. [5]. There is a lot of work going on to predict effort at initial level of software development. The need is to estimate the accurate cost and time to avoid consequences that may direct to poor quality of software [6]. The key factor is the quality estimation process which helps developers to complete the project within the given cost and time. Due to the advancement in the technology and growth of software development, huge amount of project data are available. In respect to accuracy and efficiency, non-algorithmic estimation models are giving best results compared to traditional algorithmic models [7].

In this paper, a model has been designed, by exploring different software design metrics with regression analysis, to predict effort estimation. In Sect. 2 the work related to effort has been described. In Sect. 3, software metrics have been explained in detail. Further, in Sect. 4, research problems have been formulated in 7 different hypotheses. A brief description about the data used in this paper has been described in Sect. 5. In Sect. 6, detailed explanation about the research approach has been presented. Results from the experimentation have been presented in Sect. 7. Finally, performance evaluation and conclusion of this research have been presented in Sects. 8 and 9, respectively.

2 Related Work

Many efforts are done to overcome or minimize this rate. The key process by which project manager's identify the approaches for improvement in software product is software estimation. The improvement in the software product leads to reduce the failure rate of product when in use. A large numbers of estimation models are proposed since the last 30 years, those are categorizing as algorithmic and non-algorithmic [8]. The methods used in these models are regression, expert judgment, and analogy etc. The models based on regression provide good accuracy in their estimates when compared to other models [9]. Agrawal et al. used neural network for estimating the lines of codes of software projects [10]. The dataset was extracted from International Software Benchmarking Standards Group (ISBSG) repository which consists of projects having the record of lines of code which is used in this approach. Some techniques of machine learning, i.e., genetic algorithm, decision tree, artificial neural network, and case-based reasoning are also applied in software estimation [11].

A non-linear relationship was examined by Nan et al. between budget and schedule of the software that helps in minimizing the negative effect on budget and schedule of software. In this, the researcher has formulated four hypotheses and cross-sectional data was collected from international technology firm to validate these hypotheses [12]. There are various different parameters such as time, man-power which can help in reducing the effort of developing the software [13]. An effort estimation method based on function point was proposed by Zheng et al. which help in estimating software effort accurately [14]. Function point also helps in predicting effort of software a way to accurate [15]. A model has been proposed to estimate effort using use case point, which gives the number of workers and time required in developing software project [16].

3 Software Metrics

A suite of software metrics has been used for measurement of object-oriented software. There are 6 different software metrics which together forms software metrics suite. The conceptual definitions of each software metrics are as follows:

Weighted Methods per Class (WMC): The complexity found in the methods of program helps in predicting the effort which tells the development and maintenance of class to be done. WMC is related to polymorphism, as high WMC makes the class complex whereas low WMC makes class simpler [17].

$$\mathrm{WMC} = \sum_{i=1}^{n} C_{\mathrm{i}} \tag{1}$$

where,

C_{i} complexity of the methods,
n number of methods.

Depth of Inheritance Tree (DIT): The depth of the class specifies method number. If the DIT is deeper than the complexity of the program increases which gives an option of reuse the method [14].

Number of Children (NOC): The breadth is reduced and depth is increased due to increase in inheritance level. When the number of child classes increases the NOC of program is high, which conversely increase the possibility to reuse the base class. High NOC indicates fewer faults which are due to high reuse [18].

Bugs: A bug in a program is a problem which causes to produce invalid output. There can be many different reasons for the cause of a bug. The functionality of a program is affected by bug which leads to the failure of the program. The insertion of bug is due to mistakes and errors made in source code or component or design. In some cases, the compiler can also generate bug [19].

Response For a Class (RFC): The classes with high RFC complicate the testing and debugging which makes the classes more complex and harder to understand. The value of the response class is low than it predicts inappropriate testing time [4].

$$\mathrm{RFC} = \sum_{i=1}^{n} \mathrm{M}C_{\mathrm{i}} \quad (2)$$

where,

MC_{i} number of methods called in response to a message that invokes method [4].

Line of Code (LOC): LOC tells the number of line of code written to develop software [20]. It shows the complexity and effort used to design and develop software.

Effort: It defined as the amount of effort which can be in person-hours or money required for developing and maintaining the software system. Hence, it is critical job of managers to estimate accurate effort using this kind of information. Further, the planning, budget, time to develop the system, etc. are based on this estimated effort [21].

4 Research Hypothesis

Research hypothesis is designed to test an essential procedure in statistics. Hypothesis is a well explained phenomenon for a proposed model which can be tested accordingly. The objective to plan and design hypothesis is to evaluate the proposed model on different input parameter. It helps to know the exact parameter which is suitable for the model. Hence, the research hypothesis helps in evaluating the results in a better way with the help of different parameters.

Hypothesis Formulation
The following seven hypotheses have been formulated to analyze the impact of software metrics on effort:
Hypothesis 1 (H1): The higher WMC found in the structure of software has a negative relationship with effort.
The complexity of the software is increased due to the strong coupling between the classes in the object-oriented design which introduces multiple kinds of interdependencies among the classes. As the number of method in a class increases which gives a large impact on child class, it is due to inheritance of the parent from the child. The high value of WMC indicates higher complex classes which can increase the effort of the software [21]. In this paper, a relationship of WMC with development effort has been analyzed which is used to identify the impact on the effort.

Hypothesis 2 (H2): The low value of DIT has a positive effect on the effort.

The complexity, potential reuse, and behaviour of a class are represented by DIT. The higher value of DIT indicates the reuse of methods which makes it more complex to predict the behaviour of class, whereas, low DIT values indicate low complexity which seems easy to develop software. Hence, development time and effort can be minimized [14]. In this paper, a relationship of DIT with development effort has been analyzed which is used to identify the impact on the effort.

Hypothesis 3 (H3): The greater NOC value has a positive effect on the effort.

The greater number of children in the program increases the reuse level of children. As the increasing number of children improve the re-usability in the program, which helps to reduce the complexity of the code as well as KLOC of the software [18]. In this paper, a relationship of NOC with development effort has been analyzed which is used to identify the impact on the effort.

Hypothesis 4 (H4): The minimum number of Bugs found in software has a positive impact on effort.

Presence of bugs in the program highlighted the flaws in the program. A program having maximum number of bugs increases the failure rate of the system when it is in use. Hence, substantial efforts are incorporated by the project manager to minimize this number before delivering the system to the user [19]. In this paper, a relationship of number of bugs with development effort has been analyzed which used to identify the impact on the effort.

Hypothesis 5 (H5): Low RFC in a program has a positive relationship with effort.

The testing and debugging become complex due to higher value of RFC which requires the tester to have deep knowledge of its functionality. Hence, it increases the effort of tester for making prediction about the time for testing [14]. A relationship of RFC with development effort has been analyzed in this paper which is used to identify the impact on the effort.

Hypothesis 6 (H6): Less KLOC (Kilo Line of Code) has a positive relationship with the effort.

The program with larger KLOC value takes more time to develop the program which means that KLOC is key metric which makes the more accurate effort estimates comparison to other metrics [21]. In this paper, a relationship of KLOC with development effort has been analyzed which is used to identify the impact on the effort.

Hypothesis 7 (H7): A combined effects of WMC, DIT, NOC, BUG, RFC, and KLOC on effort.

In the above six hypotheses, an impact of each software metrics on effort has been hypothesized individually. The impact varies as positive or negative according to the software metric. The combined effects of all the six metrics need to be analyzed. This combined analysis helps to predict effort by all the metrics jointly. So the model has been formulated that uses all the six metrics as predictors of the effort. The analysis will give the result of finding the suitable result with multiple regressions. A relationship of all the six metrics with effort has been analyzed in this paper which is used to identify the impact on the effort.

5 Data Collection

In this paper, the formulated hypotheses have been analyzed on real dataset. The data used has been collected from the Promise repository [22] which has object-oriented metrics with 92 versions of 28 proprietary, open-source, and academic projects with 718 tuple [23]. The metric suite consists of six object-oriented based design metrics. These metrics help in calculating software effort. The extraction of these types of metrics information from the source code is done through the automated tools which reduce flaws from data. The Promise Data Repository is a platform which consists of different datasets which helps in building a predictive model for software. This repository helps in making the availability of software engineering dataset public. To extract, the data hierarchical and *k*-means clustering is used from the software project which is available in the repository. Kohonen's neural network is used to identify groups of similar projects [23].

Effort has been considered as a dependent metric on other independent metrics in this paper. The data set used does not include the values of effort metric for software projects. So, to calculate effort from the given metrics, Constructive Cost Model (COCOMO) model [24] is begin used, where KLOC metrics are used for calculation for effort. The data set collected from the repository categories as organic projects data [22]. Following equation of COCOMO model has been used to estimate the values for effort of each entity of data:

$$\text{Effort} = a(\text{KLOC})^b \tag{3}$$

where, KLOC is Kilo Line Of Code, $a = 2.4$ and $b = 1.05$ are constant values.

The summary of descriptive statistics of the dataset is given in Table 1.

Table 1 Descriptive statistics of dataset

Metric	Mean	Median	Standard deviation	Maximum	Minimum
WMC	11.40	7.00	0.449	120	0
DIT	36.31	27.00	32.669	267	1
NOC	25.15	8.00	31.025	201	1
BUG	0.47	0.00	0.043	10	0
RFC	35.51	25.00	1.349	288	0
KLOC	0.289	0.149	0.155	4.541	0.001
Effort	0.677	0.326	0.038	11.75	0.001

6 Research Methodology

6.1 Simple Linear Regression

It deals with single explanatory variable, which is concerned with one independent variable and one dependent variable [25].

- Variable denoted as x is predictor, explanatory, or independent variable.
- Variable denoted as y is response, outcome, or dependent variable.

After this hypothesis is begin created to minimize the residuals in which a line through a sample point is measured by the sum of squared residuals and the motive is to make this sum as small as possible [26].

6.2 Multiple Regressions

It helps in finding the relationship between several independent or predictor variables and a dependent or criterion variable. It is used when you have more than one measurement variables, where one measurement variables are dependent variable and rest variable are independent variables which has an effect on dependent variable [27].

7 Experimental Result

In this section, the results obtain from methodology are discussed. Initially, the data set extracted from the promise repository has been pre-processed to remove the inconsistency and redundancy. The normality graph of the dependent and independent variables has been shown in Fig.1. After the pre-processing of the dataset analysis has been performed with the cleaned data for testing the hypotheses formulated in Sect. 4. The objective of this testing is to explain the dependency of software metrics on effort, estimate the predication percentage of effort with the help of software metrics, and formulation of composite model which establish the relationship of effort with all the software metrics considered in this research. In this paper, an approach has been proposed which uses different design metrics to establish the relationship with effort and with this relation, accurate estimates of the effort can be predicted using these metrics.

According to the result linear regression statistical test can be performed on the value of each metrics data set. In the normality tests, the p-value of individual metrics data set is seen and the p-value for all metrics data set is equal or less than 0.05. The result is demonstrated in Table 2. The rejection and acceptance is done on the basis of p-value, where the p-value is less than or equal to 0.05 than the hypothesis will

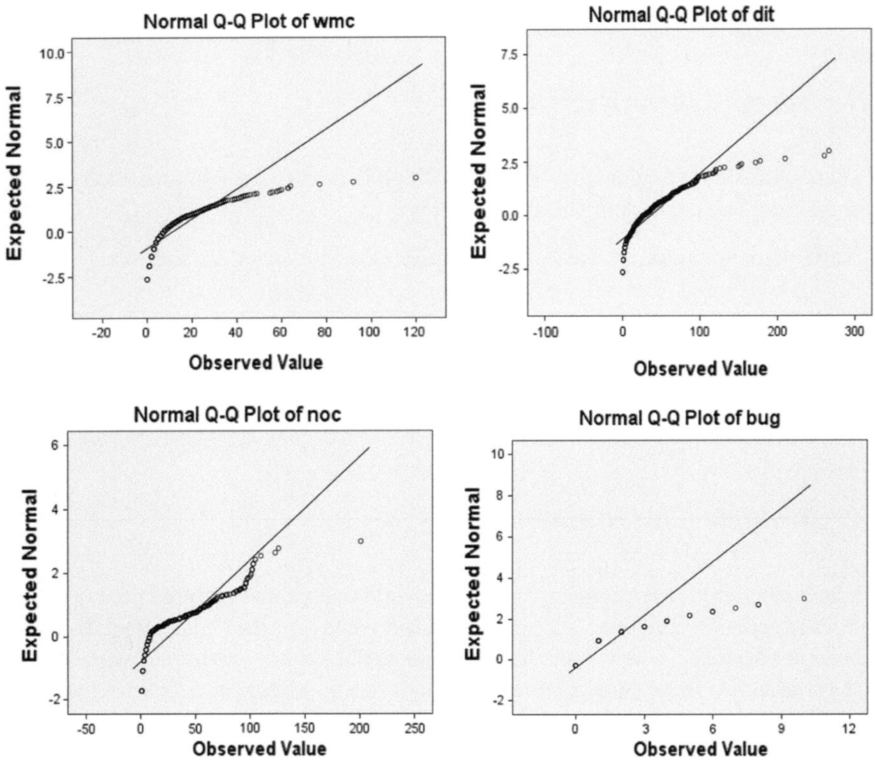

Fig. 1 Normality graph for all metrics

be accepted otherwise it will be rejected. With the help of the above technique each hypothesis has been analyzed using the linear regression statistical method.

8 Performance Evaluation

On the basis of experimental results shown in previous section, a summary of acceptance or rejection of each hypothesis has been tabulated in Table 2.

If the value of beta coefficient is positive, than there would be increase of 1 unit in every predictor variable and the outcome variable will increase by the beta value. If the value of beta coefficient is negative, than there predictor variable will increase by 1 unit and outcome variable decrease by beta value. As per the outcome for H1, the hypothesis can be accepted on the basis of the *p*-value; where the *p*-value is below the threshold value. The analysis of the hypothesis H1 indicates that WMC is oppositely related to the effort of software. The percentage of predicting the effort by weighted method per class of a code is low which shows the declining trend with effort. The hypothesis H2 is not accepted as the *p*-value is above the threshold value; it shows a

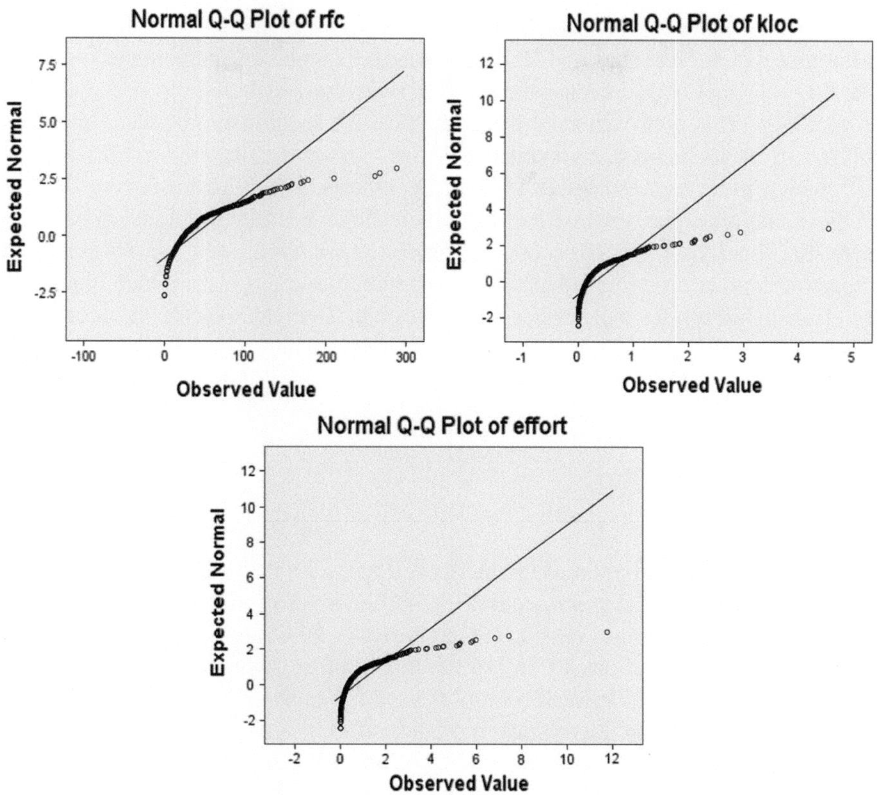

Fig. 1 (continued)

Table 2 Summary of hypothesis results

Hypothesis id	R^2 value	P-value	Result
H1	0.586	0.000	Accepted
H2	0.002	0.483	Rejected
H3	0.001	0.460	Rejected
H4	0.417	0.000	Accepted
H5	0.680	0.000	Accepted
H6	0.699	0.000	Accepted
H7	0.799	0.000	Accepted

very weak relationship between effort and DIT. The value of R^2 also indicates a very small percentage of prediction for effort. Hence, it shows that the hypothesis H2 has to be rejected. The H3 result shows that the p-value calculated is greater than the threshold value which is not accepted. This shows that NOC does not affect effort at a significance level. It is seen that the R^2 value of NOC is also very low and it does not make any major effect to calculate the effort of software. So H3 is rejected.

The analysis result of H4 shows that if the p-value is smaller than the threshold value that can be accepted and BUG predicts the significant percentage of effort. The low values of bugs decrease the effort of software which helps in predicting the effort easily. This approach establishes the positive relationship between the bugs and effort. In the result of hypothesis H5, RFC shows that the p-value is less than 0.05 which indicate the positive level of hypothesis. The hypothesis result of H6 predicts a significance level with effort which shows a strong relationship between them. So, this hypothesis is accepted. The above six hypothesis metrics show the impact of metrics on effort independently. In hypothesis H7, a combined impact on the effort by all the six metrics has been analyzed. The result shows the acceptable value and more percentage of effort by all six metrics jointly. So, hypothesis H7 is accepted.

9 Conclusion

In this paper, the formulation of the model is done, with the help of design metrics. The model is been used to make accurate prediction of effort needed for development or maintenance of software. WMC, DIT, NOC, BUG, RFC, and KLOC are the design metrics used for prediction. To design the model effort value is considered as the dependent variable [27]. IBM SPSS tool is used for analysis of the dataset and find the relationship between the metrics to predict effort.

The result shows that the p-value of DIT and NOC is greater than 0.05, which cannot be accepted. In case of WMC and bugs the p-value is less than the threshold value (0.05), it is accepted. The p-value of RFC is less than 0.05 which indicates that it can be accepted. The p-value of KLOC is also less than 0.05 that is also accepted. It is detected that out of six metrics four metrics, i.e., WMC, Bugs, RFC, and KLOC helps in predicting the effort of software. Further, some other metrics can be used to evaluate the prediction on some different parameters. Static metrics can also be used with design metrics for prediction of effort.

References

1. Albrecht AJ, Gaffney JE (1983) Software function, source lines of code, and development effort prediction: a software science validation. IEEE Trans Softw Eng SE-9(6):639–648
2. Attarzadeh I, Ow SH (2009) Proposing a new high performance model for software cost estimation. In: 2009 international conference on computer and electrical engineering. ICCEE 2009, vol 2, pp 112–116
3. Balaji N, Shivakumar N, Ananth VV (2013) Software cost estimation using function point with non algorithmic approach. Glob J Comput Sci Technol 13(8)
4. Benton NE, Neil M (1999) A critique of software defect prediction models. IEEE Trans Softw Eng 25(5):675–689
5. Boehm B, Clark B, Horowitz E, Westland C, Madachy R, Selby R (1995) Cost models for future software life cycle processes: COCOMO 2.0. Ann Softw Eng 1(1), 57–94

6. Ebrahimpour N, Gharehchopogh FS, Khalifehlou ZA (2016) A new approach with hybrid of artificial neural network and ant colony optimization in software cost estimation. J Adv Comput Res 7(4):1–12
7. Gray AR, MacDonell SG (1997) A comparison of techniques for developing predictive models of software metrics. Inf Softw Technol 39(6):425–437
8. Jorgensen M, Shepperd MA (2007) Systematic review of software development cost estimation studies. IEEE Trans Softw Eng 33(1):33–53
9. Jureczko M, Madeyski L (2010) Towards identifying software project clusters with regard to defect prediction. In: Proceedings of the 6th international conference on predictive models in software engineering—PROMISE'10, vol 1
10. Kamei Y, Matsumoto S, Monden A, Matsumoto KI, Adams B, Hassan AE (2010) Revisiting common bug prediction findings using effort-aware models. In: IEEE international conference on software maintenance (ICSM). https://doi.org/10.1109/ICSM.2010.5609530
11. Karunanithi N, Whitley D, Malaiya YK (1991) Using neural networks in reliability prediction. IEEE Softw 9(4):53–59
12. Kulkarni UL, Kalshetty YR, Arde VG (2010) Validation of CK metrics for object oriented design measurement. In: Proceedings—3rd international conference on emerging trends in engineering and technology, ICETET 2010, pp 646–651
13. Nan N, Harter DE (2009) Impact of budget and schedule pressure on software development cycle time and effort. IEEE Trans Softw Eng 35(5):624–637
14. Olague HM, Etzkorn LH, Gholston S, Quattlebaum S (2007) Empirical validation of three software metrics suites to predict fault-proneness of object-oriented classes developed using highly Iterative or agile software development processes. IEEE Trans Softw Eng 33(6):402–419
15. Primandaria PL, Sholiq (2015) Effort distribution to estimate cost in small to medium software development project with use case points. Procedia Comput Sci 72:78–85
16. Rijwani P, Jain S (2016) Enhanced software effort estimation using multi layered feed forward artificial neural network technique. Procedia Comput Sci 89:307–312
17. Tang MH, Kao MH, Chen MH (1999) An empirical study on object-oriented metrics. In: Software metrics symposium IEEE, pp 242–249
18. Kulkarni UL, Kalshetty YR, Arde VG (2010) Validation of ck metrics for object oriented design measurement. In: International conference on emerging trends in engineering and technology, vol 159, pp 646–651
19. Kamei Y, Matsumoto S, Monden A, Matsumoto KI, Adams B, Hassan AE (2010) Revisiting common bug prediction findings using effort-aware models. In: International conference on software maintenance IEEE, pp 1–10
20. Albrecht AJ, Gaffney JE (1983) Software function, source lines of code, and development effort prediction: a software science validation. IEEE Trans Softw Eng 9(6):639–648
21. Subramanyam R, Krishnan MS (2003) Empirical analysis of ck metrics for object-oriented design complexity: implications for software defects. IEEE Trans Softw Eng 29(4):297–310
22. http://openscience.us/repo/defect/ck/ant.html
23. Jureczko M, Madeyski L (2010) Towards identifying software project clusters with regard to defect prediction. In: Proceedings of the 6th international conference on predictive models in software engineering, pp 9–18. https://doi.org/10.1145/1868328.1868342
24. Boehm B, Clark B, Horowitz E, Westland C, Madachy R, Selby R (1995) Cost models for future software life cycle processes: COCOMO 2.0. Ann Softw Eng 1(1):57–94
25. Chatterjee S, Hadi AS (2015) Regression analysis by example (4th edn.). Wiley, pp 25–26
26. Elliott AC, Woodward WA (2007) Statistical analysis quick reference guidebook with SPSS examples. Sage Publications (1st edn.), pp 155–157
27. Verma DK, Kumar S (2017) Prediction of defect density for open source software using repository metrics. J Web Eng 16(3–4):293–310

Recent Advancements in Chaos-Based Image Encryption Techniques: A Review

Snehlata Yadav and Namita Tiwari

Abstract The transmission rate of multimedia data becomes exponential due to the tremendous development in network and communication facilities in last decades. Legitimate and illegitimate both kind of users can access these facilities easily 24 × 7. During transmission of data, protecting the authenticity, confidentiality and integrity of data is paramount because it affects the privacy as well as reputation of the person. Data encryption method is used to provide secure transmission of data among end devices. With the revolution in digital photography, image authenticity and integrity problems are hot topics. This paper reviews the resources offered within scope of research community so far to cover recent developments in chaotic image encryption field.

Keywords Image encryption · Chaotic system · Cat map

1 Introduction

Security aspect of information has become more essential in data during transmission. Images have been used in various application areas such as medical field, engineering, e-commerce and social networking. With the development of digital photography and smart mobile phones, image generation and sharing rate become very high in last decades. The information shared through communication channel has to be in secure manner to maintain privacy and integrity of data. The integrity and authenticity of image become paramount while transmitting using insecure channel. Image has some inherent features such as bulkiness and high data redundancy. Providing secure communication, impenetrable and plausible image encryption constructions are essential.

An American scientist Lorenz developed Chaos theory [1]. Unpredictability, ergodicity and sensitivity three essential properties of chaotic cryptosystem that

S. Yadav (✉) · N. Tiwari
Maulana Azad National Institute of Technology, Bhopal, India
e-mail: yadavsnehlata@gmail.com

N. Tiwari
e-mail: namita_tiwari21@rediffmail.com

R. K. Shukla et al. (eds.), *Social Networking and Computational Intelligence*, Lecture Notes in Networks and Systems 100, https://doi.org/10.1007/978-981-15-2071-6_52

makes it appropriate candidate for cryptographic applications. These properties are linked to confusion and diffusion of plausible ciphertext [2]. Confusion is the correlation between the key and the encrypted plaintext while diffusion is associated with output on input bits. Chaotic map possesses all these properties. Chaotic maps are categorized into one-dimensional map (1D) as well as high dimensional map. 1D chaotic map is easy to implement as they have basic structures [3, 4], while the HD chaotic system is complex structured and has more number of variables [5]. With the recent revolution in digital photography, image authenticity and truthfulness problems are hot topics. One solution to these problems is image encryption. Classical cryptography provides various different seminal methods, data encryption standards and advanced encryption standards which fulfil the requirement of confusion and diffusion. Unlike text, image has large volume and high correlation among pixel values. Theses conventional methods lead to bad encryption results and long execution time [6]. Various image encryption constructions like SCAN [7] and ECC Elgamal [8]. Visual cryptography [9] and algorithms constructed on chaos have been developed. Chaos-based cryptosystems are the most famous and accepted methods among these methods as it ensures satisfactory encryption performance.

2 Background

In this section, we have brief discussion of chaotic system and concept applied in cryptography and Literature review.

2.1 Chaotic Image System

Chaos behaviour is associated with a nonlinear property of physical system. It occurs for a particular value of parameters. The chaotic behaviour is constructed and developed using one dimensional or high dimensional systems, i.e. chaotic maps. Chaotic maps are constructed on discrete time or continuous time parameter. The potency of cryptography lies in appropriate selection of keys for encryption of data. This decryption key remains confidential and secure from adversary. The chaotic cryptosystems in cryptography can be constructed as Pseudorandom bit stream generator. Figure 1 depicts a conventional image cryptosystem constructed built on chaos hardness assumption [10].

Brief evolution of image cryptosystem is depicted as

1963—Lorenz proposed Chaos theory [1].
1968—Seminal work on Cat map proposed by Arnold.
1990—Securing communication is potential application of Chaos theory [11].
1995—Cryptosystems built on chaos were considered as substitute to classical/conventional cryptography [12].

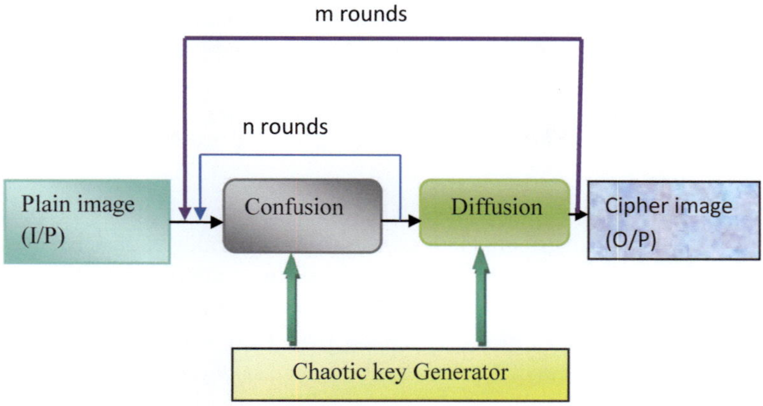

Fig. 1 Chaos-based image system

1998—Ergodic nature of chaotic trajectory was recommended and logistic maps were applied for chaotic cryptosystem [13] and also high dimensional chaotic maps were proposed.
1999—Implemented chaos functions in symmetric key cryptographic setting [14].
2010—NPCR and UACI randomness test for image cryptosystem were proposed in [15].
2010—Image cryptosystem with DNA addition with chaotic maps was proposed [16].
2011—Chaotic image cryptosystem constructed using magic cube transformation that creates chaotic behaviour [17].
2016—Chaotic maps with SHA-3 [18] and Beta maps for chaotic image cryptosystem were proposed [19].
2017—Chaotic systems along with elliptic curve Elgamal system were proposed [20].
2018—SRM based chaos system for compressed sensing has been deployed for simultaneous compression and enciphering of image.

2.2 *Literature Review*

This section represents the research work of some prominent authors in last decade the relevant area of image cryptosystems and short description of various variants of chaos-based techniques used for image cryptosystem.

Chaotic algorithms use discrete systems that has received much consideration to generate chaotic keys [13, 21, 22]. Colour images are directly encrypted with the help of chaotically coupled chaotic maps; high security was provided by chaotic amalgamation of pixels' colours. It was very much suitable for numerous applications in real-time communication systems, internet of things and cloud storage [23].

For image cryptosystem, a chaotic block cipher method was proposed by [24] that was based on essential primitive operations, nonlinear transformation function and chaotic tent map. This method uses 256 bits for session keys and is applicable for real-time applications.

By encrypting the input plain image, a DNA sequence matrix has been applied by author [16] that is then distributed into same block sizes, followed by DNA sequence complement operation, is executed to the output of sum of matrix with 2 logistic maps and decodes matrix of DNA sequence to obtain cipher image. This cryptosystem is resistive to statistical and differential attack [25]. Comparative analysis of work done in the area of image cryptosystem built on Chaos assumption is depicted below (Table 1).

Enhancement in encryption speed is achieved by author [26]. He proposed improved diffusion scheme using chaotic orbit turbulence technique for message before encryption that considerably increases speed of spreading process. This construction is found suitable for image as well as video communication.

The encryption method constructed by [17] uses the principal of magic cube transformation to shuffle pixel positions of image and transforms the values of pixels along with pseudorandom sequences that are outputs of chaotic maps.

In year 2013 [27], pixels contained in the image are shuffled by permuting specific blocks of predetermined size of input colour image followed by the mixing, masking and scrambling of shuffled pixels, are performed with the help of 3D cat map. The three rules that were used in this construction lead to confusion and/or diffusion association between unencrypted image and cipher image.

In [18], the author efficiently integrates double chaotic maps, hash algorithm SHA-3 and auto-updating system. Plain image's pixel positions are first shuffled, and then, its hash value is evaluated to obtain control parameter value and primary conditions of logistic map. After this, for row and column, permutation is executed to exchange pixels of image. To enlarge key space 3D chaotic map is applied in diffusion process.

The author [19] integrates two beta chaotic maps to implement image cryptosystem. This is inspired from beta function of mathematics. Mourad Zaied presented mathematical proofs of chaotic behaviour from nonlinear dynamic equations. Predetermined key length was of 512 bits. Two beta chaotic maps generate pseudorandom sequence in substitution stage as well as permutation stage of encryption process. Correlation coefficient value was not calculated while other parameters such as histogram analysis, NPCR, UPCI and PSNR values were evaluated and compared.

SHA-1 and chua attractor integration produces chaotic behaviour that is applied by author [28] for image cryptosystem. Plain image size gets increased after encryption process, [20] overcomes this problem by breaking compressed plain image into independent equal-sized RGB blocks and then uses 4D Cat map and 3D Lorenz map for confusion and EC Elgamal for diffusion purpose. This scheme was efficiently implementable and provides secure transmission of image.

Table 1 Comparative study of cryptographic schemes constructed on chaos

Author and year	Types of chaotic map	Type of input image	Key space		Correlation coefficient		Differential attack measures	
					Plain image	Cipher image	NPCR (in %)	UACI (in %)
[24]	Chaotic tent map	Colour image	2^{256}	H-	0.9954	−0.0209	99.61	33.41
2010				V-	0.9903	−0.0144		
				D-	0.9846	−0.035		
[16]	2D and 1D logistic map	Grey image	10^{72}	H-	0.9468	0.0036	99.61	38
2010				V-	0.9697	0.0023		
				D-	0.9153	0.0039		
[17]	Arnold cat map	Colour image	2^{148}	H-	0.9156	0.001	99.62	33.19
2011				V-	0.8808	0.006		
				D-	0.8603	0.091		
[26]	Chirikov and Chebyshev map	Grey image	2^{167}	H-	0.9404	0.0088	99.609	33.464
2012				V-	0.9299	−0.0087		
				D-	0.9257	−0.0060		
[6]	Magic cube transformation	Grey image	10^{42}	H-	0.9861	−0.0014	99.61	33.44
2011				V-	0.9735	−0.0278		
				D-	0.9568	−0.0098		
[27]	3D cat map	Colour and grey image	NA	H-	0.97646	0.002909	99	33
2013				V-	0.98077	−0.01503		
				D-	0.96627	0.012901		
[18]	1D logistic map, 3D chaotic cat map	Colour and grey image	10^{30}	H-	NA		99.6	33.3
2016				V-				
				D-				
[19]	Beta map	Colour and grey image	NA	H-	0.9187	0.006687	99.6227	33.01
2016				V-	0.9557	0.006687		
				D-	0.8877	0.007019		
[28]	Nested map	Colour image	$2^{80} \times 10^{84}$	H-	NA		99.6	32.01
2016				V-				
				D-				
[20]	4D Arnold cat map and 3D Lorenz Chaotic map	Colour image	NA	H-	0.9727	-0.00774	1	33.47
2017				V-	0.9448	0.000718		
				D-	0.000718	0.005889		
[29]	3D Cat map	Grey image	2^{149}	H-	0.9849	0.0018	NA	NA
2018				V-	0.9693	0.0014		
				D-	0.9562	0.0034		

To compress and encrypt image almost simultaneously [29], SRM construction on chaos assumption was used while PACT is applied for shuffling operation.3D cat map generates key stream. Statistical test is performed to ensure that the algorithm is found appropriate for secure image transmission.

3 Statistical Test

3.1 Pixel Correlation Coefficients

The adjacent value of pixels of an unencrypted image is very close in vertical, horizontal and diagonal directions. A high correlation value denotes the best similarity between input unencrypted image and its corresponding output encrypted image. The correlation coefficient (ρ) will always be one in above said directions albeit for encrypted image, and the correlation coefficient tends to zero. Correlation coefficient is evaluated as [30]

$$\rho = \frac{c(x, y)}{\sqrt{D(x)}\sqrt{D(y)}}$$

where

$$c(x, y) = \frac{1}{n}\sum_{i=1}^{n}\left(x_i - \frac{1}{n}\sum_{j=1}^{n} x_j\right)\left(y_i - \frac{1}{n}\sum_{j=1}^{n} y_j\right)$$

$$D(x) = \frac{1}{n}\sum_{i=1}^{n}\left(x_i - \frac{1}{n}\sum_{j=1}^{n} x_j\right)$$

3.2 Histogram Analysis

An image histogram is representative of the pixel distribution of any given image that is acquired by plotting the total number of pixels at each intensity level. It represents the statistical characteristics of an image. Histogram of encryption method reveals that the random numbers that are generated from the chaotic map are uniformly spread like white noise. For a computationally strong encrypted image, the distribution must be uniform.

3.3 Entropy

It is the essential measure of randomness of pixels of image. Randomness is introduced to avoid predictability. Suppose a binary source generates 2^n symbols having equal probabilities, where n is the length of symbol, the entropy measure of binary source is evaluated as [31]

$$\text{Entropy} = -\sum_{i=1}^{2^n} P_i \log_2 P_i$$

4 Sensitivity Tests

4.1 Differential Attack Measures

If there is a minor alteration in input image has taken place then image cryptosystem should be sensitive enough to generate dissimilar encrypted image. Numerous methods have been proposed and constructed that provides protection levels against differentials attacks [15]. Let consider P be the encrypted image relating to the source unencrypted image without changes and Q be the corresponding encrypted image comparing to the unencrypted source image with pixel value change then **mean absolute error**(MAE) measures change among E_i and P_i, the ciphered image and source image, respectively. Let W and H are taken as the width and height of the unencrypted source image, MAE is evaluated as

$$\text{MAE} = \frac{1}{W * H} \sum_{i=1}^{H} \sum_{j=1}^{W} |P_i(i, j) - E_i(i, j)|$$

The **number of pixel change rate** (NPCR) measures the change in per cent of unlike pixels between P and Q. It is evaluated as

$$D(i, j) = \begin{cases} 0 & P(i, j) = Q(i, j) \\ 1 & P(i, j) \neq Q(i, j) \end{cases}$$

$$\text{NPCR} = \frac{1}{W * H} \sum_{i=1}^{H} \sum_{j=1}^{W} D(i, j) * 100\%$$

The unified average changing intensity (UACI) evaluates the average intensity variation between P and Q. It is evaluated as

$$\text{UACI} = \frac{1}{W * H} \sum_{i=1}^{H} \sum_{j=1}^{W} \frac{|P(i, j) - Q(i, j)|}{255} * 100\%$$

Sensitivity to single-bit change in the encryption key: Encryption methods ought to be profound to change in parameter also, single change in bit of public key prompt to huge change in conduct of encryption process [31]. This sensitivity is measured by the mean squared error (MSE) that indicates, how far the encrypted (encoded) image is from the genuine image. The larger value is an emblematic of good quality encryption algorithm. MSE is evaluated as

$$\text{MSE} = \frac{1}{W * H} \sum_{i=1}^{H} \sum_{j=1}^{W} (P_i(i, j) - E_i(i, j))^2$$

5 Conclusion

Development and advancement of cryptographic primitives and methods for image security, privacy and authenticity have been a functioning exploration subject up until now. Exponential image transfer rate is due to social media such as Instagram, Whatsapp, Facebook and e-commerce websites like Alibaba, Flipcart, Amazon and so forth. Security and privacy of image data are vital. This paper represents recent improvements in the area image cryptosystem constructed chaos assumption in summarized manner so that researchers can have a quick look and appreciative scenario of the same. As chaos-based systems are burgeoning techniques, many authors develop image cryptosystem using various chaotic map along with other techniques like EC Elgamal, DNA sequence, etc. All methods resist to attacks and satisfy the required test criteria. These strategies and constructions can be utilized progressively in real time in public network and are open for cryptanalysis for the scientist's community. Numerous authors recommended extending their algorithms to video cryptosystems.

References

1. Lorenz EN (1963) Deterministic non periodic flow. J Atmos Sci
2. Shannon CE (1949) Communication theory of secrecy systems. Bell Syst Tech J 28(4):656–715
3. Wang X, Teng L, Qin X (2012) A novel color image encryption algorithm based on chaos. Signal Process 1101–1108
4. Bhatnagar QG (2012) Selective image encryption based on pixels of interest and singular value decomposition. Digit Signal Process 648–663
5. Jakimoski G, Subbalakshmi KP (2007) Discrete lyapunov exponent and differential cryptanalysis. IEEE Trans Circuits Syst II, pp 499–501

6. Zhu ZL, Wang C, Chai H, Yu H (2011) A chaotic image encryption scheme based on magic cube transformation. In: IEEE fourth international workshop on chaos-fractals theories and applications. China
7. Chen RJ, Horng SJ (2010) Novel SCAN-CA-based image security system using scan and 2-D von Neumann cellular automata. Signal Process Image Commun 413–426
8. Li L, El-Latif AAA, Niu X (2012) Elliptic curve Elgamal based homomorphic image encryption scheme for sharing secret images. Signal Process 1069–1078
9. Chen TH, Tsao KH, Lee YS (2012) Yet another multiple-image encryption by rotating random grid. Signal Process 2229–2237
10. Yang S, Sun S (2008) A video encryption method based on chaotic maps in DCT domain. Sci Direct Prog Nat Sci 18
11. Pecora LM, Carroll TL (1990) Synchronization in chaotic systems. Phys Rev Lett 64:821
12. Bradley E (1995) Causes and effects of chaos. Comput Graph Elsevier Sci Ltd. 19(5):755–778
13. Baptista MS (1998) Cryptography with chaos. Phys Lett A 50
14. Bose R, Banerjee A (1999) Implementing symmetric cryptography using chaos function. In: 7th international conference on advanced computing and communications. Roorkee
15. Wu Y, Noonan JP, Agaian S (2011) NPCR and UACI randomness test for image encryption. J Sel Areas Telecommun (JSAT) 31(8)
16. Zhang Q, Guo L, Wei X (2010) Image encryption using DNA addition combining with chaotic maps. Math Comput Model (Elsevier) 52:2028–2035
17. Zhu ZL, Wang C, Chai H, Yu H (2011) A chaotic image encryption scheme based on magic cube transformation. In: Fourth international workshop on chaos-fractals theories and applications. China
18. Ye G, Huang X (2016) A secure image encryption algorithm based on chaotic maps and SHA-3. Secur Commun Netw, Wiley online library, pp 2015–2023
19. Zahmoul R, Zaied M (2016) Toward new family beta maps for chaotic image encryption. In: IEEE international conference on Systems, man and cybernatics. Hungary
20. Wu J, Liao X, Yang B (2017) Color image encryption based on chaotic systems and elliptic curve ElGamal scheme. Signal Process 141:109–124
21. Pareek NK, Patidar V, Sud KK (2003) Discrete chaotic cryptography using external key. Phys Lett A 309:75–82
22. Huang F, Guan ZH (2005) Cryptosystem using chaotic key. Chaos Solut Fractals 23:851–855
23. Pisarchik AN, Zanin M (2008) Image encryption with chaotically coupled chaotic maps. Physica D (Elsevier), 237:2364–2648
24. Amin M, Faragallah OS, El-Latif AAA (2010) A chaotic block cipher algorithm for image cryptosystems. Commun Nonlinear Sci Number Simulat 15:3484–3497
25. Gupta K, Silakari S (2011) A new approach for fast color image encryption using chaotic map. J Inf Secur 2:139–150
26. Fu C, Chen JJ, Zou H, Meng WH, Zhan YF, Yu YW (2012) A chaos-based digital image encryption scheme with an improved diffusion strategy. Opt Soc Am 20(3):2363–2378
27. Kanso A, Ghebleh M (2012) A novel image encryption algorithm based on a 3D chaotic map. Commun Nonlinear Sci Numer Simulat (Elsevier) 17:2943–2959
28. Slimane NB, Bouallegue K, Machhout M (2016) Nested chaotic image encryption scheme using two-diffusion process and the secure hash algorithm SHA-1. In: 4th IEEE international conference on control engineering and information technology. Tunisia
29. Chen J, Zhang Y, Qi L, Fu C, Xu L (2018) Exploiting chaos-based compressed sensing and cryptographic algorithm for image encryption and compression. Opt Laser Technol 99:238–248
30. Mao Y, Chen G (2005) Chaos based image encryption. In: Handbook of geometric computing. Berlin Heidelberg, Springer
31. Zhang NA (2013) Colour image encryption algorithm combining compressive sensing with Arnold transform. J Comp 8(11):2857–2863

Image Fusion Survey: A Comprehensive and Detailed Analysis of Image Fusion Techniques

Monica Manviya and Jyoti Bharti

Abstract Image fusion is very trending area of research in the field of image processing. It has vast range of applications in miscellaneous fields like surveillance, diagnosis, and photography. In this survey paper, all the methodologies have been discussed and various approaches are given to overcome the shortcomings of sundry problems. Detailed analysis is carried out in this paper. Many efforts have been made to find out the challenges and also the achievements done so far. We undergo various research papers and tried to cover every aspect relating image fusion. Multi-scale transform image fusion methods have been discussed on the basis of their domain, i.e., in spatial and frequency domain. First, image fusion is explained in details with its major applications as well as techniques entangled. Second, comparison of various fusion methods is done in tabulation. Finally, concluded with the current insightful discussions of image fusion and future aspects. This survey will be used as a reference for image fusion related domain.

Keywords Fusion · Multi-sensor · Pixel level · Feature · Pyramid · DWT · PCA

1 Introduction

Image fusion is the most trending area of research in the field of image processing. Image fusion basically combines two different types of images taken from various sensors taken at different positions and time which gives a new image called fused image [1]. It is more suitable for machines as well as humans interpretations for performing different image processing tasks such as object detection, target recognition, and in various other fields of medical research, remote sensing etc. [2]. Figure 1 explains the fusion process by the help of a fusion technique where two input images of different kind here we have taken visible and infrared image and with simple

M. Manviya (✉) · J. Bharti
Maulana Azad National Institute of Technology, Bhopal, India
e-mail: monicamanviya26@gmail.com

J. Bharti
e-mail: jyoti2202@gmail.com

R. K. Shukla et al. (eds.), *Social Networking and Computational Intelligence*, Lecture Notes in Networks and Systems 100, https://doi.org/10.1007/978-981-15-2071-6_53

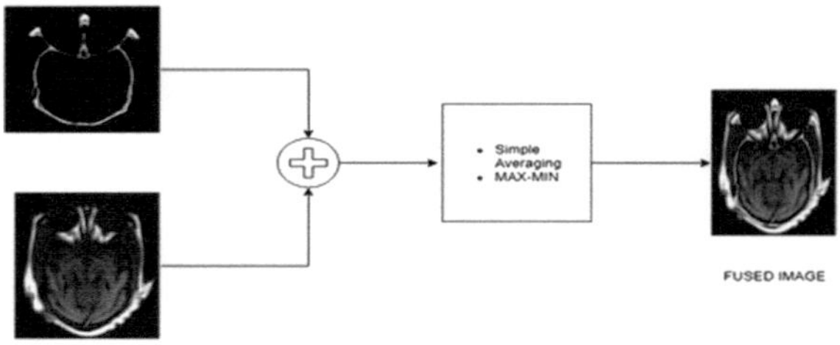

Fig. 1 Block diagram showing image fusion

averaging or Min-Max method they are fused and we get fused image as shown in Fig. 1.

The purpose of image fusion is to uplift the recognition accuracy, and also it works against the drawbacks of sensitivity of images having irregular illumination and different temperature changes [3]. Several algorithms based on multi-scale analysis were developed for the use of merits like source image into final fused image [4]. The fused image must meet the following requirements: (i) Fused image must contain the most significant information, i.e., all relevant information must be present, (ii) the fused image must not contain any artifacts and inconsistencies, and (iii) It must be free from noise and mis-registration they must be suppressed [5–7]. Before fusion process takes place, there is point to point correspondence and it can be done using two sensors and this is as (i) single sensor—In this sensor, series of images are fused together. (ii) multi sensor: In this sensor, composite image is formed and it also overcomes the limitation of single sensor [8]. Figure 2a, b shows the two sensors used for image fusion.

There are different ways of image-registration like scaling, translation, rotation, or some non-rigid transformation with the help of which local wrapping can be done [9]. Multi-sensor images have many advantages over single sensor, and these are robustness in fused image. Multi sensors also reduce uncertainty [10]. Image Fusion has various applications in Medical Diagnostics [11, 12]: satellite imaging, remote sensing, and robotics, distributed and parallel processing multi-model plant recognition, Biomedical Imaging [6, 7, 12, 13]. Fusing different images also help

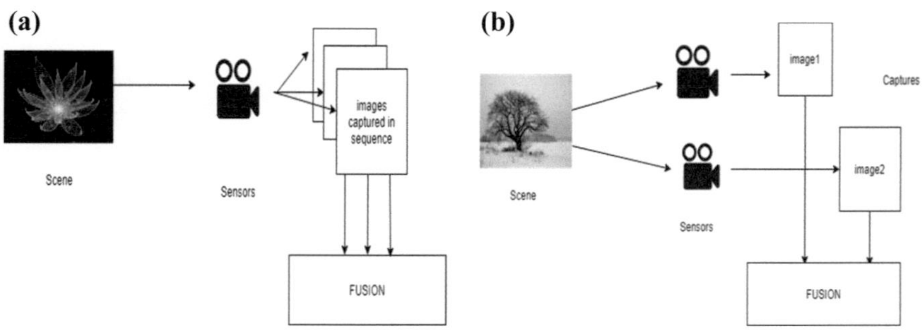

Fig. 2 **a** Single sensor based image fusion **b** multi-sensors based image fusion

to differentiate anatomy of patient and also his metabolism, optical remote sensing, and in image processing [14].

Any image fusion process have three stages (i) image accession/acquisition which is nothing but acquiring several images from different sensors. (ii) image registration which is having proper alignment like pixel to pixel correspondence and (iii) final fusion of image which is combining the set of all input images into single image [14, 15].

In this paper, key ideas of fusion methods are included with the overview of all image fusion methods; however, results of particular algorithm and details of particular one algorithm are not described here. Many efforts have made for pointing out the interesting side and only important methods. This literature survey is a investigation of recent year research papers, comprehensive evaluation and comparison of various image fusion methods have been done [1, 7, 14, 15]. For instance, Bikash Mehar have also discussed about the region level also called decision level in a different manner [1]. Ma [7] have conducted a comparative survey on visible and infrared image fusion(including the three levels of fusion) based on TNO dataset, they have used and also they have discussed applications in a very vast manner. All the classic methods have been discussed in this paper [7–9]. Pixel level has been discussed in detail. These papers have explored the pyramid methods mainly the Laplacian Pyramid is used with the combination of new methods as well as filters [16–22]. Improved methods more efficient than existing were explored also the implemented results also, and all types of survey papers regarding image fusion with their solution are analysed. However, the results of comparative experiments have not been described. Efforts have been made for the summarization and gathering the interesting ideas of existing methods of fusion and its applications [3, 7, 8, 23–29]. In Fig. 3, flow chart is also given for our survey overview.

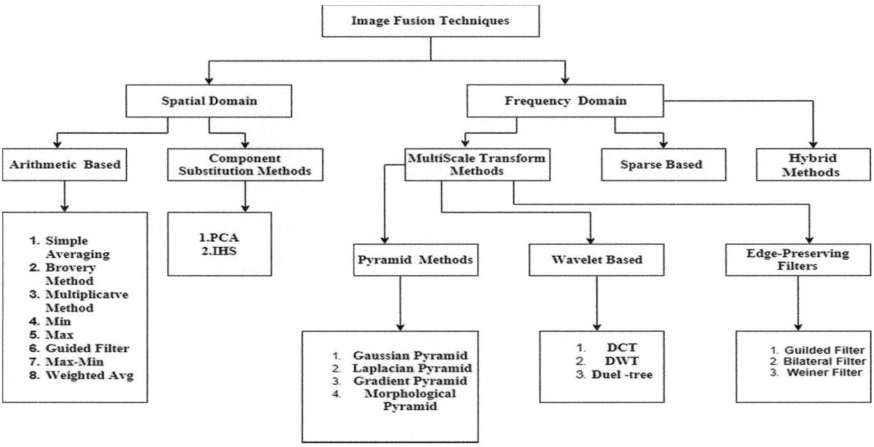

Fig. 3 Comparative flow diagram showing different levels of fusion based

In this paper, survey is done on different image fusion and comparative approach is taken. This paper is divided in sections and First, image fusion is explained in details with its major applications as well as techniques entangled. Second, comparison of various fusion methods is done in tabulation. Finally, concluded with the current insightful discussions of image fusion and future aspects.

2 Overview of Literature Survey

Figure 3 shows an overview of image fusion techniques, and after this a level-based comparison is shown in Table 1 [24, 27–32] as image fused is classified into 3 levels hence we compared each level and it can be concluded that among the three pixel-level is better than the other two.

2.1 *Level-Based Comparison of Image Fusion*

See Table 1.

Table 1 Subjective comparison table of fusion levels

Sr. no.	Pixel-level	Feature–level	Decision-level
1.	In pixel-level image fusion, the fusion process is carried out at pixel to pixel basis. It is the lowest level in context of integrating the raw source/input images into a single fused image [28]	Feature level basically operate on extracted objects on the features like shape, size, pixel, edge, texture, or intensities. The similar features from source image are fused to produce an output image [28]	Decision level work on higher level of abstraction as it integrates information of more than one algorithm and fused image. It depends on probabilistic variables or feature descriptors [28]
2	It is the most popular among the three because the original source image measured quantities are directly involved in fusion moreover it is the efficient in computation and have easy implementation	This level identifies the distorted features from image. This levels in basically extracted similar features from input images and then fusion rule is applied to get more efficient fused image	It is the highest level. It is multi-scale representation and different decisions are taken at levels. It is least popular among three
3	It gives best performance	Performance in detection is medium	Performance in detection is worst
4	Fused information have minimum data loss	It contains medium data loss in fused image	Information loss is maximum in fused image
5	They are totally sensor dependent	There is moderate dependence on sensors	Minimum dependence on sensors

2.2 Spatial and Frequency Domain

Spatial Domain The Spatial Domain is also called simple techniques of fusion. They work directly on pixels [3]. The input image pixel is added to fused pixel. Although it is the most simple technique of fusion still it faces distortions in output fused image [27].

Transform Based In Transform domain, the input image is firstly transformed to a transform domain and the fusion rules are applied to the transformed image and then it undergo an inverse transform so that to acquire the resultant fused image [33].

2.3 Arithmetic Based Methods

1. **Simple Averaging Method:**

   This is the simplest method in which corresponding pixel values are taken from input image is taken average is calculated. So the fused image will be having the averaged pixel intensity [27].

2. **Brovery Method:**

   In this method, each band of multispectral image is multiplied with panchromatic image and then normalize with multi spectral image and this is done repeatedly for each individual band. This gives high-resolution fused image, and it is based on modulation of intensity [33].

Multiplicative Method
This method combines MS and PAN image for giving color preservation. The arithmetic methods like (multiplication, division, subtraction, and addition) are used to produce chromatic image to intensity image. This multiplicative method is simply give color preservation. This method produce bands called spectral bands having high correlation having characteristics of source image [25, 32].

MIN
In this method, input image is of low contrast in which pixel having minimum value is selected from corresponding pixels of input images [25, 26].

MAX
This method is just opposite as compared to minimum selection. This method takes the maximum pixel value and thus produced a high contrast fused image but it ignored the pixel value having low pixel intensity highly sensitive to noise and artifacts [25, 26].

2.4 *Component Substitution Methods*

IHS Method

IHS stands for Intensity Hue Saturation. Intensity stands for total reflected light that reaches our eyes. Hue is the wavelength of colors used and basically shows saturation. It gives visual representation of by replacing one of the component (I, H, S) of input image by another component of input image. Thus, it contains the spectral information so it needs to be meticulously controlled. It adds high resolution to it [25, 26]. It is one of the oldest method and frequently used for sharpening.

PCA

PCA is a technique principal components are formed using correlated variables. PCA basically depends on data sets. It does not have fixed vectors like DCT, FFT etc., and it depends on data sets. PCA algorithm goes like

(i) column vectors are produced from input image.
(ii) co-variance matrix is calculated.
(iii) Eigen vectors and Eigen values are computed using the matrix in step ii.
(iv) normalization is done using mean of Eigen vectors.
(v) step (iv) Eigen value is multiplied with each pixel of source image.
(vi) final summation of both matrices is done to get the fused image [7, 25, 26].

2.5 *Multi-scale Transform Methods*

Pyramid Methods

This method form pyramid structure of an original image at certain levels. A selective approach is applied on source image. This method firstly decompose the pyramid structure, after this step all the decomposed images are combined, and then lastly the inverse PT is applied to get the fused image. The commonly used pyramid structures are Laplacian Pyramid, Gaussian Pyramid etc. [7, 16, 17].

Laplacian Pyramid (LP)

Laplacian pyramid is derived from Gaussian Pyramid the generation of Gaussian pyramid followed by the formation of the Laplacian pyramid. The Gaussian pyramid is created by convolving the image with a Gaussian low-pass filter and then down sampled by a factor of 2. Laplacian Pyramid is one of the method of Multi-Scale image fusion and basically it is a method by which low resolution images are converted to high-resolution images [34]. LP method undergo these four phases (i) blurring (low-pass filtering), (ii) sub-sampling, (iii) interpolating, and (iv) differencing are used at every pyramid level. Original image is taken at the first level, and after that they are divided into further levels [18–22].

2.6 *Wavelet Methods*

DWT

DWT stands for discrete wavelet transform and has advantage over the fourier transform by giving results in frequency and time domain. In this method, the source images are converted into informative and approximate coefficients at specific level then these coefficients are combined using fusion rules and then inverse wavelet transform is applied to get final fused image result [3].

But this method suffers from problems like lack of directionality since it uses frequency and time domain and have the shortcomings like aliasing and shift variance. To overcome these problems, a new method was introduced and this was duel-tree complex wavelet transform as it reduces the shortcomings of shift invariance and selection of directions but since this method also use wavelets hence it has shortcomings like having improper edges and curve. This method minimize the spectral distortion problem of an image but with this has a demerit of low spatial resolution. Duel-tree wavelet transform was introduced to overcome the demerits of DWT [5–7, 26–33].

DCT (Discrete Cosine Transform)

This methods break the input image into blocks of size n × n after this discrete cosine transform are calculated and fusion rules are on each block to get these fused coefficients and then at last inverse discrete cosine transform is applied to get the final fused result. It has excellent compactness properties [26–33].

2.7 *Sparse Based*

In this method, a high quality over-complete dictionary is used and the input images are shown sparsely in it and thus making it better in representation. This method uses sliding window to divide the input images into different patches that overlap for this they use sliding window and then sparse coding is performed using over-complete dictionary thus in this way making the method robust. Methods like online dictionary learning, multi-scale dictionary learning, adaptive sparse representation, PCA, and clustering are applying this method in mage fusion domain [3, 9].

2.8 *Hybrid Methods*

This method integrates different kind of methods like hybrid of sparse representation and multi-scale transform, hybrid of sparse methods with sparse methods with neural networks. The main aim of this method is combine the features of multi-scale transform and region detection which preserve the details and also give more efficient results [7, 26].

3 Metrics for Performance Evaluation

To check the efficiency of fused image certain performance measures are there such as Entropy, PSNR, Mean, Standard Deviation, Mutual Information, SNR, RMSE, MSE, co-relation coefficient, and image quality index (Q_{w}), Average gradient, spatial frequency etc. helps in evaluating the fused image efficiency [1, 6, 7, 25–29, 33].

3.1 Comparison Table

See Table 2.

4 Conclusion

In this paper, we have demonstrated the various image techniques in a comprehensive and comparative manner. Various different types of techniques as discussed here are very useful in creating fused image more suitable for recognition, detection, and visual perception. Table 2 is the summarised form of all the methods surveyed, and their merits and demerits are listed and from the table we can conclude that no specific one method us the best result and for getting better result we have to use methods like DWT with the combination of filters or we can use hybrid methods in which two or more methods can be combined. In future work, we will work on these hybrid fusion methods for getting best results.

Table 2 Comparison table of fusion methods

Method name	Merits	Demerits
Spatial domain		
Simple averaging technique [7]	Simplest method of fusion having easy implementation	Reduce contrast
Brovery method and multiplicative method [7, 26]	It has very fast processing which takes very less time and produce image having good contest	Distortion in color
Max method [7, 26, 33]		Produce blurred image of low contest
MIN		
MAX–MIN		
Weighted avg	Gives detection reliability	SNR is high
HIS [33]	Having high sharpening ability and fast processing power	It only process three RGB bands and thus results will have color distortion
Laplacian, Gaussian, Gradient pyramid, Morphological pyramid [7, 26–33]	Edge preservation, aliasing free, self inverting, rotational invariant. High visual effect	Levels of decomposition levels that affects the fused image result
DCT [26–33]	It converts the image in series of waveform thus widely used for real applications	Do not produce good quality fused image
DWT [7, 26–33, 35–37]	It provide better SNR ratio as compared to pixel based methods	Lack of directionality, self-aliasing, oscillations, shift variance
Hybrid methods [7, 33]	Give more enhance quality of fused image	It is very slow and takes lots of time
Hybrid of PCA and DWT [7, 26–33]	Efficient and high spatial resolution	Complex in nature

References

1. Meher B, Agrawal S, Panda R, Abraham A (2019) A survey on region based image fusion methods. Inf Fusion 48
2. Ying H, Rong X (2018) A block image fusion algorithm based on algebraic multi-grid method. Procedia Comput Sci 131:273–281. ISSN 1877-0509
3. Ashalatha B, Reddy MB (2017) Image fusion at pixel and feature levels based on pyramid imaging. In: 2017 IEEE international conference on smart technologies and management for computing, communication, controls, energy and materials (ICSTM). Chennai, pp 258–263
4. Yin W, Zhao W, You D, Wang D (2019) Local binary pattern metric-based multi-focus image fusion. Opt Laser Technol 110:62–68. ISSN 0030-3992
5. Yang Y (2011) A novel DWT based multi-focus image fusion method. Procedia Eng 24:177–181. ISSN 1877-7058
6. Li S, Kang X, Fang L, Hu J, Yin H (2017) Pixel-level image fusion: a survey of the state of the art. Inf Fusion 33:100–112. ISSN 1566-2535
7. Ma J, Ma Y, Li C (2019) Infrared and visible image fusion methods and applications: a survey. Inf Fusion 45
8. Ma J, Zhang D (2018) An image fusion method based on content cognition. Procedia Comput Sci 131:177–181. ISSN 1877-0509. pp 119-132, ISSN 1566-2535
9. Wu R, Yu D, Liu J, Wu H, Chen W, Gu Q (2017) An improved fusion method for infrared and low-light level visible image. In: 2017 14th international computer conference on wavelet active media technology and information processing (ICCWAMTIP). Chengdu, pp 147–151
10. He K, Zhou D, Zhang X, Nie R (2018) Multi-focus: focused region finding and multi-scale transform for image fusion. Neurocomputing 320:157–170. ISSN 0925-2312
11. Wei C, Zhou B, Guo W (2017) A three scale image transformation for infrared and visible image fusion. In: 2017 20th international conference on information fusion (fusion), Xi'an, pp 1–6
12. Ashalatha B, Reddy B (2017) Enhanced pyramid image fusion on visible and infrared images at pixel and feature levels. In: 2017 international conference on energy, communication, data analytics and soft computing (ICECDS). Chennai, pp 613–618
13. Li M, Dong Y, Wang X (2013) Pixel level image fusion based the wavelet transform. In: 2013 6th international congress on image and signal processing (CISP). Hangzhou, pp 995–999
14. Rajini KC, Roopa S (2017) A review on recent improved image fusion techniques. In: 2017 international conference on wireless communications, signal processing and networking (WiSPNET). Chennai, pp 149–153
15. Mao R, Fu X, Niu P, Wang H, Pan J, Li S, Liu L (2018) Multi-directional Laplacian pyramid image fusion algorithm 568–572. https://doi.org/10.1109/icmcce.2018.00125
16. Meng W, Huisheng Z, He H (2008) A pseudo cross bilateral filter for image denoising based on Laplacian pyramid. In: 2008 IEEE international symposium on knowledge acquisition and modeling workshop. Wuhan, pp 235–238
17. Burt P, Adelson E (1983) The Laplacian pyramid as a compact image code. IEEE Trans Commun 31(4):532–540
18. Xingmei L, Liang C, Jin W (2010) The application of Laplacian pyramid in image super-resolution reconstruction. In: 2010 2nd international conference on signal processing systems. Dalian, pp V3-157–V3-159
19. Pei L, Xie Z, Dai J (2010) Joint edge detector based on Laplacian pyramid. In: 2010 3rd international congress on image and signal processing. Yantai, pp 978–982
20. Teng Y, Liu F, Wu R (2013) The research of image detail enhancement algorithm with Laplacian pyramid. IEEE Int Conf Green Comput 2013:2205–2209
21. Pradeep M (2013) Implementation of image fusion algorithm using MATLAB (Laplacian pyramid). In: 2013 international mutli-conference on automation, computing, communication, control and compressed sensing (IMac4s). Kottayam, pp 165–168

22. Hong S, Yu X, Chen Q, Wang L (2016) Improved nonlinear resolution enhancement based on Laplacian pyramid. In: 2016 16th international symposium on communications and information technologies (ISCIT). Qingdao, pp 146–150
23. Lakshmi A, Rakshit S (2010) Gaussian restoration pyramid: application of image restoration to Laplacian pyramid compression. In: 2010 IEEE 2nd international advance computing conference (IACC). Patiala, pp 66–71
24. Aslantas V, Bendes E, Toprak AN, Kurban R (2011) A comparison of image fusion methods on visible, thermal and multi-focus images for surveillance applications, (ICDP). London 2011:1–6
25. Li M, Dong Y (2013) Review on technology of pixel-level image fusion. In: Proceedings of 2013 2nd international conference on measurement, information and control. Harbin, pp 341–344
26. Anita SJN, Moses CJ (2013) Survey on pixel level image fusion techniques. 2013 IEEE (ICECCN). Tirunelveli, pp 141–145
27. Sumathi M, Barani R (2012) Qualitative evaluation of pixel level image fusion algorithms (PRIME-2012). Tamilnadu, Salem, pp 312–317
28. Wu KX, Wang CH, Li HL (2010) Image fusion at pixel level algorithm is introduced and the evaluation criteria. In: 2010 international conference on educational and network technology. Qinhuangdao, pp 585–588
29. Mishra D, Palkar B (2015) Image fusion techniques: a review IJCA. 130:7–13. https://doi.org/10.5120/ijca.2015.907084
30. Zhang XD (2012) Study on feature layer fusion classification model on text/image information. Phys Procedia 33:1050–1053. ISSN 1875-3892
31. Kumar K (2010) Total variation regularization-based adaptive pixel level image fusion. In: 2010 IEEE workshop on signal processing systems. San Francisco, CA, pp 25–30
32. Liu J, Wang Q, Shen Y (2005) Comparisons of several pixel-level image fusion schemes for infrared and visible light images. In: 2005 IEEE instrumentation and measurement technology conference proceedings. Ottawa, Ont., pp 2024–2027
33. Kekre HB, Mishra D, Saboo R (2013) Review on image fusion techniques and performance evaluation parameters. Int J Eng Sci Technol 5(4)
34. Aishwarya N, Abirami S and Amutha R (2016) Multifocus image fusion using discrete wavelet transform and sparse representation. In: 2016 international conference on wireless communications, signal processing and networking (WiSPNET). Chennai, pp 2377–2382
35. Li X, Zhou F, Li J (2018) Multi-focus image fusion based on the filtering techniques and block consistency verification, pp 453–457. https://doi.org/10.1109/icivc.2018.8492825
36. Naidu V (2011) Multi-resolution image fusion by FFT. In: 2011 international conference on image information processing. Shimla, pp 1–6
37. Ma Y, Chen J, Chen C, Fan F, Ma J (2016) Infrared and visible image fusion using total variation model. Neurocomputing 202. https://doi.org/10.1016/j.neucom.2016.03

Some New Methods for Ready Queue Processing Time Estimation Problem in Multiprocessing Environment

Sarla More and Diwakar Shukla

Abstract Ready queue processing time estimation problem deals with many constraints. Because the processes which reside in the ready queue of computer memory come in varieties such as process size, process requirement indifferences and process types. To match up all these differences is a difficult task to solve so that the processes can be used to perform its task efficiently at any platform. A prior estimation of ready queue processing time helps to meet the system reliability and robustness. A pre-calculated time will ensure the system from failure; also, the backup of task performed can be maintained. In this paper, the existing methods on this approach are described, and how some new methods can be used for the better performance is demonstrated. For this purpose, some sampling techniques are used, and the lottery scheduling procedure is explained which very efficiently performs this task of scheduling on the basis of probabilistic approach and randomness property. The estimation is performed by using sampling methods; with the help of some mathematical calculations, the results are obtained, and finally, confidence interval will ensure the accuracy of the result. So that some new methods can be generated; this provides the result as more efficient than the previous ones. Although various scheduling schemes are available, the lottery scheduling scheme provides the fairness and also removes starvation. Rather than working on the complete data set, some samples can be generated to modularize the work which will be efficient too. So, this paper proposed some new methods in ready queue processing time estimation in multiprocessor environment.

Keywords Ready queue · Multiprocessing · Process management · Lottery scheduling · Estimation · Sampling

---

S. More (✉) · D. Shukla
Dr. Harisingh Gour University, Sagar, Madhya Pradesh 470003, India
e-mail: sarlamore@gmail.com

D. Shukla
e-mail: diwakarshukla@rediffmail.com

R. K. Shukla et al. (eds.), *Social Networking and Computational Intelligence*, Lecture Notes in Networks and Systems 100, https://doi.org/10.1007/978-981-15-2071-6_54

1 Introduction

A process entering in ready queue waits for the CPU to be allocated for completing their intended tasks. Processes have their own properties according to which the CPU allocation can be decided, and processes are having many variations according to size measure, type measure and requirement requests, so these types of heterogeneity can be handled by using various approaches to create some productive function. Each process comes with its process control block where all its properties and to do functions can be decided. Processes come in various states where the requirement request will decide at what time a process should be in which state, and this task is performed by a scheduler which comes in three varieties. The short-term scheduler brings the processes from job queue to ready queue, midterm scheduler provides proper context switch and long-term scheduler is responsible to assign the CPU to each individual process at a particular time. If we talk about multiprocessor environment where more than one processor is associated to perform the scheduling task and CPU allocation and execution of process, there is a lot of heterogeneity associated with processes, but the task here is to process the request in timely and efficient manner so that power management can be maintained, and the system efficiency can be achieved in terms of any type of system failure. For performing this task, lottery scheduling is used here with some sampling techniques and estimation measures to demonstrate that how we can perform the ready queue processing time estimation in multiprocessor environment so that the estimated time will ensure the system to be safe with any type of system failure. In this paper, the procedure is explained that how the ready queue processing time estimation be performed. The concept of lottery scheduling is opt here from Waldspurger and Weihl [1] who described lottery scheduling as randomized proportional resource share scheduling which has the fairness of the resource allotment to each and every process so that not a single process becomes idle. It is a randomized approach so that for performing this whole concept some sampling concepts have been adopted by Shukla et al. [2–9]. This gives variations in lottery scheduling schemes and provides a systematic way that how these techniques can be used effectively and how the efficient result can be obtained. The basic concepts of operating system have been described in [10–12] from which one can understand the *a* to *z* of these concepts. Cochran [13] and Thompson [14] give the concepts of various sampling methods which can deal on a large population, and out of this huge population, the best sampling method which is compatible to the sampled data can be acquired for the intended operations. The basic terminology and concepts of this research have been proposed in More and Shukla [15]. This proposed the working phenomena of the existing methods.

2 Concept of Ready Queue

Ready queue or run queue has the purpose to be executed by the CPU by using appropriate scheduling approach. At the same time, different processes came, but it is the responsibility of the CPU to run only one process at a time. The processor keeps all the processes in a queue which are ready to be processed. Processes which do other function rather than execution such as waiting for an I/O event to occur or interrupted with other events do not reside in the ready queue. The threading concept is also supported by the ready queue. Multithreading concept demonstrates that while the processing of a running thread, the upcoming thread is kept by the virtual processor with its associated priority in the ready queue [10] (Fig. 1).

2.1 Ready Queue Estimation Problem

- In case when a system shuts down suddenly, this may be threatening to the processes. In this case, it is a problem of ready queue for estimating the total processing time of all processes [15]. In distributed manner, if some processes are running on other machines, it is required to calculate the total processing time of all the processes so that the process can be finished, and the system properly shuts down in secure and systematic way.
- Heterogeneity is another issue in ready queue processing time estimation. Size measure, type variant and requirement indifferences are some heterogeneous properties by which it is problematic to communicate and estimate the total processing time in ready queue.
- If random breakdown occurs, the running process requires estimating the remaining process time estimation in ready queue; for this reason, a backup manager is a must requirement to estimate how many jobs remain un-allotted in CPU and their time estimation in ready queue.
- A common belief that more related input information can predict the better results. In case of large number of ready queue processes, the CPU utilization time can be predicted with some auxiliary information which may reduce the length of computed time interval.

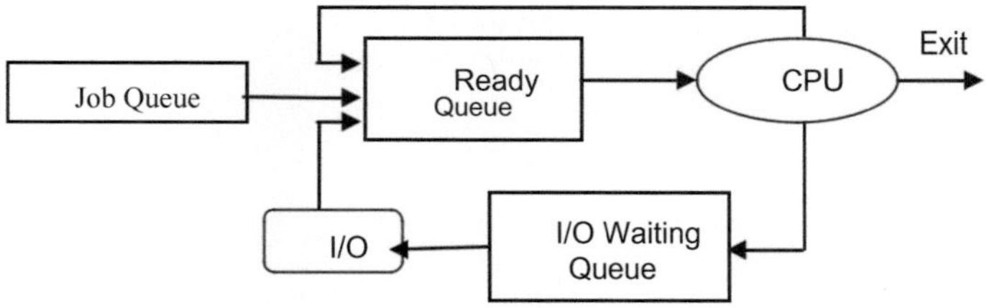

Fig. 1 Queuing diagram

3 Need of Estimation

Estimation is necessary to obtain the accuracy of results. For performing estimation, various approaches can be applied in order to gain effective results. Sampling is one of the methods for estimation which is further divided into many forms. Probabilistic and non-probabilistic samplings are the two major variants of sampling.

3.1 Sampling

Sampling is one of the techniques by which a fine decision can be made from the abundant of data. The technique works in a modularized way in which samples are made, and by applying the operation on each sample, the result is obtained, but collectively, the main goal is the comprehensive collection of all sample surveys. Sampling is a term used in statistics that describes the methods of selecting a predefined representative number of data from a larger data population. Cochran [13] provides clear concepts of sampling techniques and various other methods associated with sampling. The advantages of sampling methods are reduced cost, greater speed, greater scope and greater accuracy. In sampling schemes [14], variability from sample to sample can be estimated using the selected single sample by which we can collect and construct estimates for the parameter of the population of interestingness. There are many ways to construct estimates associated with the guidelines of desirability. Some desirable properties for estimators are: unbiased or nearly unbiased. When the estimator is unbiased we have a low MSE (mean square error) or a low variance. MSE measures how far the estimate is from the parameter of interest whereas variance measures how far the estimate is from the mean of that estimate. Thus, when an estimator is unbiased, its MSE is the same as its variance). Robust means answer does not fluctuate too much with respect to extreme values.

3.2 Confidence Intervals

The sample mean by itself is a single point. In population mean, it does not give any idea about how good the estimation is. If we want to assess the accuracy of this estimate, we have to use confidence intervals which provide us with information as to how good our estimation is. A confidence interval, viewed before the sample is selected, is the interval which has a prespecified probability of containing the parameter. To obtain confidence interval, we need to know sampling distribution of the estimate.

$$\text{C.I} = \text{point estimate} \pm \text{margin of error}$$

In the population the sample size n, variability V and the desired level of confidence are some factors affecting C.I estimates. The terminology associated with C.I determined by various properties such as $\bar{t}$ which denotes to sample mean, z denoted the z value for a particular confidence level. $\bar{b}$ denotes the population standard deviation and n is the number of observation in sample. The confidence interval for population mean with $\bar{b}$ known is

$$\pm Z\bar{b}/$$

100 (1-α) % confidence interval for μ can be derived as

$$\frac{\bar{y}-\mu}{\sqrt{\mathrm{Var}(\bar{y})}} \sim N(0,1) \quad \text{whereas,} \quad \frac{\bar{y}-\mu}{\sqrt{\hat{\mathrm{V}}\mathrm{ar}(\bar{y})}} \sim t_{n-1}$$

and an approximate 95% CI for μ is

$$\bar{y} \pm t_{\alpha/2}\sqrt{\left(\frac{N-n}{N}\right)\left(\frac{s^2}{n}\right)}$$

The z values associated with C.I are as follows: for 90%, it is 1.645; for 95%, it is 1.960; for 99%, it is 2.576, and for 99.9%, the z value is 3.291.

3.3 Auxiliary Data and Ratio Estimation

The auxiliary information about the population includes a known variable to which the variable of interest is approximately related. The auxiliary information is easy to measure, but the variable of interest may be expensive to measure.

- Population units: 1, 2, …, N
- Variable of interest: $y1, y2, \ldots, yN$ (expensive or costly to measure)
- Auxiliary variable: $x1, x2, \ldots, xN$ (known).

Ratio estimator is

$$\text{If } \tau_y = \sum_{i=1}^{N} y_i \text{ and } \tau_x = \sum_{i=1}^{N} x_i \text{ then, } \frac{\tau_y}{\tau_x} = \frac{\mu_y}{\mu_x} \text{ and } \tau_y = \frac{\mu_y}{\mu_x} \cdot \tau_x$$

$$\hat{\tau}_r = \frac{\bar{y}}{\bar{x}} \cdot \tau_x$$

The estimator is useful in situation when X and Y are highly linearly correlated through the origin and the Var ($\hat{\tau}_r$) is less than Var (N). In case when N is unknown, it provides a way to estimate τ_y because when N is unknown, one cannot use N.

4 Contribution of Methods

The research is based on the lottery scheduling scheme [1]. In this scheme of scheduling, the framework of ticket or currency associates scheduling mechanism and also describes proportional share resource management technique other than the probabilistic-based lottery algorithm. Some features are adopted from the systematic lottery (SL) scheduling (2). Group lottery scheduling (GLS) [3] model is able to estimate the total ready queue processing time in multiprocessing environment. Type I and type II allocations are used, and the variations in both are compared; by using some numerical analysis, results are represented with efficiency. The algorithm works in k processor environment ($k > 1$) [4], and the sample data can be collected in random selection without replacement method to calculate the ready queue processing time. The PPS-LS [5] is a fruitful idea to be considered for the size measurement of processes. In multiprocessor environment the concept of auxiliary variables adopted [7] where the ready queue mean time is estimated by using the Lottery Scheduling algorithm. Here, the auxiliary data source is considered for the strengthening of the proposed methodology. Process size, process priority and process expected time are the three additional data sources used in this work. Factor-Type (F-T) Estimator [8] in Multiprocessor environment is presented to estimate ready queue processing time where T_A and T_B are used as two new estimators and for total processing time they are compared with each other by using ratio estimation technique. In the setup of lottery scheduling, two estimators' bias and m.s.e have been processed to obtain some calculated values under large sample approximation. The transformed factor-type (T-F-T) estimator [9] in multiprocessor environment is used to estimate ready queue processing time and tries to find out the solution of the problem of ready queue time estimation which occurs when there is a sudden failure in system, and many processes remain in the ready queue. An immediate action has to be taken for the remaining jobs in the ready queue that how much time is required to be estimated before shutting down the system so that if we restart the system again, it will be held in a secure and safe manner. For this prediction, a sampling technique is used in lottery scheduling. So, the task here is to make a hybrid type of method which contains all these characteristics and performed the estimation in efficient manner. Systematic lottery scheduling and group lottery scheduling schemes are described here which have some merits and demerits from which an idea is derived, and on the basis of process heterogeneity, the challenge is sorted out, and with some advance process characteristics prediction, the ready queue process time estimation is performed.

5 Method of Estimation of Ready Queue in Multiprocessing Environment

See Fig. 2.

Fig. 2 Ready queue length estimation method

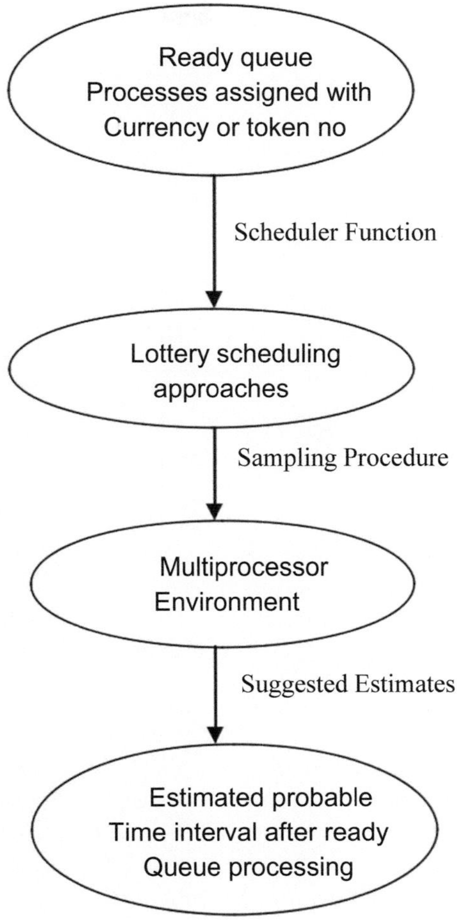

6 Scheduling Approaches

Systematic lottery scheduling is a randomized systematic approach where the samples are generated, and the first process selected at random with a predefined logic after which other processes follow the same pattern. The total n processes are computed, and their processing time is estimated at the session end. Now, for the sample mean calculation, the following formulas can be

$$\text{used } t_i = 1/k \sum t_{ij}$$

$$\mathrm{E}(t_i) = 1/k \sum t = t$$

$$\mathrm{var}(t_{\mathrm{sys}}) = 1/n \sum = (t_i - t_i)^2$$

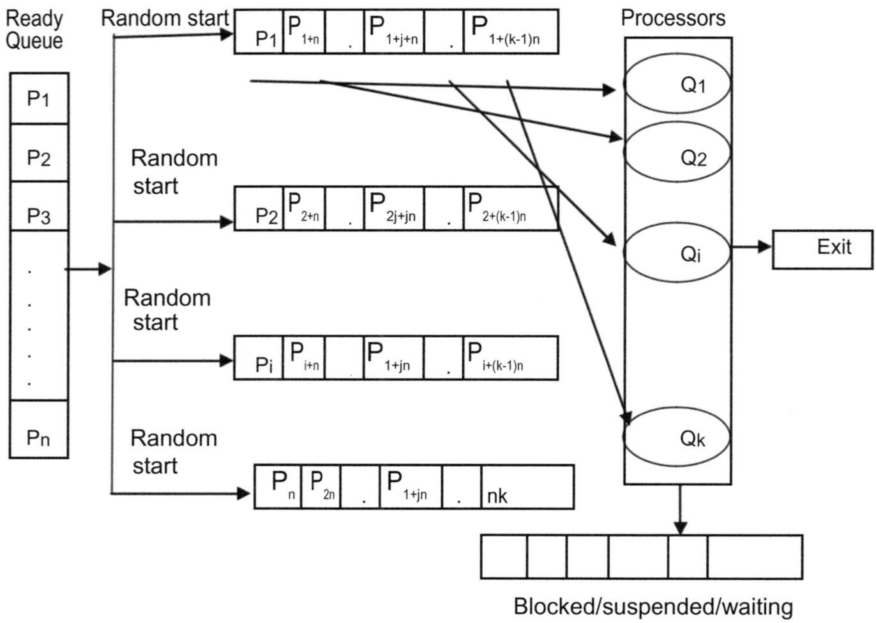

Fig. 3 Processing of ready queue under SLS

For the given data set of processes, some random samples are formed, and their sample mean time is calculated. Some computational values are formed like mean time, square of mean time, total sum of square, mean square and variance of SLS. At last, confidence interval is calculated which shows that there is 99% confidence limit of true values. So here, we obtained that this procedure provides the better sample representation, and the queue time estimation is sharper. But, it has one drawback too. It creates very high differences for the predicted estimates observed, and also, the size measure of processes is not handled in a proper way (Fig. 3).

The Group lottery scheduling approach has two variants type I allocation and type II Allocation. We can consider that there are some processors r such that $Q1$, $Q2$, $Q3$, …, Qr. Every one processor is having a random sample of jobs from the respected ready queue. In the ready queue, all processes are in homogeneous manner. All ready queue processes are divided into r groups on the basis of size measure. The process selection for processing is performed on the basis of matching of token number of the random process to that of the processor of a particular group. At session end, CPU gives average of time as

$$t* = 1/k \sum i \sum j t_{ij}$$

For the given data set, the groups are formed according to size measure and weight index, mean time, square of mean time, total sum of square, mean square, and finally, the variance of GLS is calculated followed by the confidence interval. The problem encounter in this scheme is the random selection of processes so that the allocation is formed and found that type II allocation is better over type I allocation. In this scheme, we can also consider the type measure and the requirement measure so that better prediction of ready queue processing time has been performed.

The estimation of ready queue time processing is done with the help of factor type estimation in multiprocessing environment. In lottery scheduling environment, the auxiliary information produces better results so that two estimators T_A and T_B are introduced and compared with the ratio estimator. The BIAS, M.S.E and confidence interval are calculated to obtain the efficiency of the proposed algorithm. Class of FT estimator is

$$T_d = \overline{y}\left[\frac{(A+C)\overline{X} + fB\overline{x}}{(A+fB)\overline{X} + C\overline{x}}\right]$$

BIAS and M.S.E of T_A and T_B are

$$B(T_d) = \frac{(C - fB)\overline{Y}}{(A + fB + C)}\left\{\frac{C}{A + fB + C}V_{02} - V_{11}\right\}$$

$$M(T_d) = E\left(T_d - \overline{Y}\right)^2$$

$$= \bar{Y}^2\left[V_{20} + P^2V_{02} + 2PV_{11}\right]$$

and the confidence intervals is as follows

$$P\left[T_A \pm 1.96\sqrt{V(T_A)} \leq \overline{Y} \leq T_A \pm 1.96\sqrt{V(T_A)}\right] = 0.95$$

$$P\left[T_B \pm 1.96\sqrt{V(T_B)} \leq \overline{Y} \leq T_B \pm 1.96\sqrt{V(T_B)}\right] = 0.95$$

7 Summary/Conclusion

The problem undertaken herewith is related to the estimation of ready queue processing time. There are some existing methods, but each one has merit and demerit. It is difficult to find a uniformly better method. After computation of the proposed topic, it is expected to come up with new methods of estimation more efficient than earlier in view of multiprocessor environment. The new technique takes care of size variant of processes, the type differences of processes and the different requirements of the processes. Finally it will perform the scheduling task efficiently to serve the purpose of time estimation of ready queue in a precise manner so that if any threaten occurs in the system, it will estimate the time before the crash, and we can safe our system. So, a combine method can be adopted by analyzing these approaches which can better serve the purpose of time estimation in multiprocessing environment.

References

1. Waldspurger CA, Weihl WE (1994) Lottery scheduling: flexible proportional share resource management. In: Proceedings of the 1994 operating system design and implementation conference (OSDI'94). Monterey, California, pp 1–11
2. Shukla D, Jain A, Choudhary A (2010) Estimation of ready queue processing time under SL scheduling scheme in multiprocessor environment. Int J Comput Sci Secur 4:74–81
3. Shukla D, Jain A, Choudhary A (2010) Estimation of ready queue processing time under usual group lottery scheduling in multiprocessor environment. Int J Comput Appl 8:39–45
4. Shukla D, Jain A, Choudhary A (2010) Prediction of ready queue processing time in multiprocessor environment using lottery scheduling. Int J Comput Internet Manag 18:58–65
5. Shukla D, Jain A (2012) Analysis of ready queue processing time under PPS-LS and SRS-LS scheme in multiprocessing environment. GESJ: Comput Sci Telecommun 33:54–65
6. Shukla D, Jain A (2012) Estimation of ready queue processing time using efficient factor type estimator (E-F-T) in multiprocessor environment. Int J Comput Appl 48:20–27
7. Shukla D, Jain A (2012) Ready queue mean time estimation in lottery scheduling using auxiliary variables in multiprocessor environment. Int J Comput Appl 55:13–19
8. Shukla D, Jain A (2013) Estimation of ready queue processing time using Factor type (F-T) estimator in multiprocessor environment. Compusoft, Int J Adv Comput Technol 2:256–260
9. Shukla D, Jain A, Verma K (2013) Estimation of ready queue processing time using transformed factor type (T-F-T) estimator in multiprocessor environment. Int J Comput Appl 79:40–48
10. Silberschatz A, Galvin P (1999) Operating system concepts, 5th edn. Wiley (Asia)
11. Stalling W (2000) Operating system. Pearson education, Singapore, Indian edition, 5th edn. New Delhi
12. Tanenbaum A, Woodhull (2000) Operating system, 8th edn. Prentice Hall of India, New Delhi
13. Cochran WG (2005) Sampling techniques. Wiley Eastern Publication, New Delhi
14. Thompson S Sampling, 3rd edn. Wiley Eastern Publication
15. More S, Shukla D (2018) A Review on ready queue processing time estimation problem and methodologies used in multiprocessor environment. Int J Comput Sci Eng 6:1186–1191

Review of Various Two-Phase Authentication Mechanisms on Ease of Use and Security Enhancement Parameters

Himani Thakur and Anand Rajavat

Abstract We are living in a digital era, where most of the critical information whether it is confidential, academic documents or chats and even money transaction is being done in digital format. Digital technology is progressing at its large because of its ease and speed. With this digital data advancement, responsibility and core need arises for its security arises. For protecting data, authentication mechanism plays a major role. Authentication mechanism assures security of data by allowing legitimate users to access it. Using single-phase authentication mechanism (UID and password) is the easiest and convenient mechanism to implement and use in past days. For preserving data from false access, there is a need to develop more protected authentication mechanism which cannot be easily penetrated. This gave birth to two-phase authentication mechanisms that are commonly known as two-factor authentication mechanisms. Two-phase authentication mechanisms provide additional security by adding one more factor for authentication in traditional single-phase authentication mechanisms. There are lots of two-phase authentication mechanisms that developed with technological advancement. In this paper, we will review various two-phase authentication mechanisms by considering security enhancement they provide and ease of use in their implementation.

Keywords Authentication · Digital data · Security · Single-phase authentication · Two-phase authentication

1 Introduction

Authentication mechanism allows managing access of structures by means of checking to verify if a user's credentials match the credentials stored in the database of authorized users existing in data authentication server.

H. Thakur (✉) · A. Rajavat
Department of Computer Science and Engineering, SVIIT, SVVV, Indore, Madhya Pradesh 453111, India
e-mail: himanihts2017@gmail.com

A. Rajavat
e-mail: anandrajavat@yahoo.co.in

R. K. Shukla et al. (eds.), *Social Networking and Computational Intelligence*, Lecture Notes in Networks and Systems 100, https://doi.org/10.1007/978-981-15-2071-6_55

Clients are commonly known having a client ID and password. Verification is practiced once the client gives credentials, for instance a secret phrase (password) and client ID that matches therewith client ID and password in the stored database of authentication server. Most clients are most mindful of utilizing a password, which, as a piece of data that should be notable exclusively to the client, is named as data authentication factor.

Authentication is imperative because of its license associations to keep their systems secure by permitting exclusively genuine clients (or procedures) to get to its ensured resources, which can incorporate pc frameworks, networks, databases, sites, and diverse network-based applications or administrations.

Once genuine, a method or user is sometimes subjected to an authorization method as well, to see whether or not the genuine individual ought to be permissible right to use to a sheltered system or resource. A resource user may be genuine however not succeed to be resource access if that user of it was not approved authorization to access it.

1.1 Types of Authentication Mechanisms [1]

Two-factor Authentication This authentication mechanism adds an extra layer of security in the authentication process. This system needs that second authentication offers by user considering an addition to the password. This method usually needs that a verification code enters by user that received in the registered mobile phone via text message or enter a code enter that generated by application.

Multifactor Authentication (MFA) This method needs users to evidence with over one factor of authentication, together with face recognition or fingerprint biometric factor, security key sort of possession factor, or a sort of token created by an application.

One-Time Password This method is a mechanically generated character set or numeric string of short characters that provide a user authentication. This character password is merely legal for one group action or login session, and is used by the users who forgot their passwords and are agreed to a one-time password for log in, sometimes used for brand new users and is used for password forgot process.

Three-factor Authentication This authentication (3FA) process could be a sort of MFA that uses three factors of authentication, typically a data factor (password) combined with a possession factor (security token) and immanency factor (biometric).

Biometrics Whereas some authentication mechanism will rely exclusively on recognition, biometrics are typically used as a second or third authentication factor. The additional common kinds of identification out there embody facial, fingerprint scans, or voice recognition and retina scan.

Mobile Authentication It is that the method of confirming user via their mobile devices or confirming the devices itself. It provides secure locations login to user and resources from anyplace. This method includes multifactor authentication that may embody one-time passwords, identification or QR code validation.

Continuous Authentication In this method user being either logged out or in, a company's application frequently calculate an "authentication score" that provide whether the account owner has permission to access application or not.

API Authentication The quality ways of managing API authentication are HTTP basic authentication; OAuth, and API keys.

In HTTP authentication, the server requests authentication info, i.e., a username and password, from a client. The client then passes the authentication data to the server in an authorization header.

In the API key methodology, a user that comes first time is appointed a singular generated value that indicates that the user is well-known. Then, every time the user tries to access the system again; his distinctive keys accustomed verify that he is an identical user who entered the system antecedently.

Open authorization is an open standard for token-based authentication and authorization on the Web. OAuth permits a user's account data to be utilized by third-party services, like LinkedIn, while not revealing the user's password. OAuth acts as an intermediary on behalf of the user, given that service with an access token that authorizes specific account data to be shared.

2 Literature Review

In this paper [2], authors provided a novel and efficient authentication system which uses the cloud data centers and mobile phone to discover the uniqueness of the item. The system uses fast response codes to spot the item details. The project when enforced within the reality can offer a simpler authentication system that the people will use to search out the originality of the item before shopping for it. They will conjointly ensure that the item that these obtain is originally manufactured by the various manufacturers and not from any cheat. The system conjointly reduces the value of the authentication method as there is no want for adding up any expensive tags to every product. Printing the QR codes is a lot of economical than different authentication systems. The foremost necessary advantage of the system is that the authentication is completed by the user itself, and there is middleman within the method that will increase the trustiness and security of the system.

In this paper [3], authors used principal curves approach for fingerprint trivialities' extraction; then, these keep them in an exceedingly database on a cloud; then, authors used the bio-hash operate to secure the biometrics templates. Additionally, they compared their method with the method given in previous researches and intended the error rates for his or her approach and verified that these increase the system performance by twenty-fifth.

In this paper [4], authors present some advances system on offline signature identification system. From the analysis of the recent literature within the field, a number of the foremost valuable approaches is given and also the most fascinating directions for more analysis are highlighted.

In this paper [5], authors propose the implementation of a voice-based fuzzy vault authentication mechanism, for secure access and encryption support at intervals cloud platforms and cloud shared storage. The experimental outcome, stress on assessing the performances of the biometric intermediary, has shown FRR variable from one-third to thirty-second and much rates variable from 2.5 to 11.3%.

In this paper [6], authors propose a brand new image integrity authentication method supported fixed point theory. Within the proposed scheme, the subsequent three criterions square measure considered for selecting an acceptable transform fk (•) whose fixed points are used for image integrity authentication. (1) Fragility: the fixed points of fk (•) should be sparse; (2) straightforward calculation: a fixed point is simply found by few iterations; (3) transparence: a fixed point is found in a very little neighborhood of a given image function. They construct an acceptable transform fk (•) satisfying these criterions, based on the Gaussian convolution and de-convolution, known as GCD transform. Once establishing a theorem for the existence of fixed points of the GCD transform fk (•), these provide algorithms for a fast calculation of a fixed point image that is extremely on the point of the given image and for the complete image integrity authentication scheme mistreatment the obtained fixed point image. The semi-fragility drawback is additionally mathematically thought about via the commutatively of transforms. Experimental results show that the proposed method has superb performance.

In this paper [7], authors developed an iris recognition algorithm supported Fisher algorithm which may be run in a very lighter computing platform. Experiments conducted with CASIA information show exciting results, wherever the system achieves an awfully high accuracy. Iris recognition may be a biometrics authentication system mistreatment iris image. It is one in all the foremost reliable biometrics systems. The systems need substantial computing power. Therefore, it is not been able to penetrate the market however.

This paper [8] discusses the various ICA-based mostly techniques that are utilized in the last decade. This paper reviews the comparative study of various face recognition techniques that is predicated on ICA. The vital part of this survey is that the discussion of previous work of face recognition associated with ICA. There are totally different strategies obtainable associated with ICA. Also, compared the various strategies in tabular form. During this survey, the paper offers the transient summary of "How to recognize face using image processing".

In this paper [9], the human behavior is recognized from a collection of video samples, and therefore, the features are extracted victimization HOG transform. KNN classifiers are accustomed to classify the features extracted from the videos. The HOG feature primarily based on analysis has achieved higher recognition and accuracy of 93% compared to the prevailing ways. There are many factors affects this

Gait authentication which may be classified into two classes. They are (i) external factors: angles, lighting atmosphere, garments that have the same color as background and alternative external objects and (ii) internal factors: changes in gait because of natural effects like illness, aging, pregnancy, gaining, or losing weight.

In this paper [10], authors projected a sturdy face recognition technique by victimization native binary pattern and bar graph of adjusted gradient feature extractor and descriptors. During this study, author has found that LBP feature extractors have virtually simple fraction higher accuracy result than the HOG feature extractors that does not have that a lot of distinction. They need to test it for authentication purpose to register to their device by taking one label as an administrator, and it gave important results.

In this paper [11], authors present a completely unique security framework for NFC Secure Element-based Mutual Authentication and Attestation for IoT access with a user device like a mobile device using NFC based mostly. Host card emulation (HCE) is a mode for the primary time. The recent framework for NFC Secure Element-based Mutual Authentication and Attestation for IoT access provides a completely unique on-demand communication and management of IoT devices with security, privacy, trust, and proof of locality using the NFC-based HCE mode and secures tamper-resistant SE and TPM. This method cannot verify the dynamic device state like control-flow integrity.

Author proposes a noisy vibration method for cloaking vibration sounds throughout pairing against such attacks. The method only needs a speaker for emitting the masking sound throughout key transmission [12]. They conjointly study motion sensor exploits against this scheme and compliment it with extra measures to mask vibration effects on motion sensors. Their analysis shows that whereas vibration pairing could seem to be a beautiful mechanism for guaranteeing the protection and trust in an IoT network, it must be protected against acoustic aspect channel attacks by defensive measures like masking signals that are low price and straightforward to implement.

In this paper [13], author studied the ensemble performance of biometric authentication system that is supported secret key generation. Bearing on an ensemble of codes supported Slepian-Wolf binning, we have provided elaborate, sharp analyses of the false–reject and false–accept chances, in terms of error exponents, for a large category of stochastic decoders that covers the optimum MAP decoder, in addition as many extra decoders, as special cases. Converse bounds are derived in addition.

Author propose a physical-layer challenge-response authentication approach, this paper [14] supported combined shared secret key and channel state information (CSI) between 2 legitimate nodes in an orthogonal frequency division multiplexing (OFDM) system. The projected approach used although the correlation of channel coefficients exists, which might be exploited to extract the shared secret key in standard approaches. Moreover, channel coding is utilized to mitigate the distinction between the two calculable channels in addition to channel fading and background noise. Thus, they ascertained that within the projected approach as a physical-layer authentication approach, the decoder's output is often used for authentication and provided a reliable decision below active attack.

3 Conclusion

In this paper, we review various privies described authentication techniques. In our review, we find out that all techniques that used in authentication use ID and password scheme.

The server verifies user password related to user id from the verification table store in database. If the user password matches in verification table, then server authenticates this user. But this process have a risk, a hacker will capture the message from the network and intercept it then it login to server and access user data. Although the password is encrypted throughout communication from user to server, such an attack continues to be possible.

Here arises a necessity for a mechanism that is most dynamic that if an attacker intercepts the message, then additionally he should not be ready to build successful authentication try.

Acknowledgements I am highly thankful to my guide Dr. Anand Rajavat (Head of Department of Computer Science). His suggestion and interest have helped me in integrating the work. His accommodating nature tolerates my persistent queries and provided the best solution to my problem. I am also thankful to Department of Computer Science for providing all facilities and resources needed for this research paper.

References

1. https://searchsecurity.techtarget.com/definition/authentication
2. Umanandhini D, TamilSelvan T, Udhayakumar S, Vijayasingam T (2012) Dynamic authentication for consumer supplies in mobile cloud environment. In: IEEE conference at ICCCNT'12. Coimbatore, India
3. Sabri HM, Ghany KKA, Hefny HA, Elkhameesy N (2014) Biometrics template security on cloud computing. At 978-1-4799-3080-7/14/$31.00 @ 2014. IEEE
4. Impedovo D, Pirlo G, Russo M (2014) Recent advances in offline signature identification. In: IEEE 2014 14th international conference on frontiers in handwriting recognition
5. Velciu MA, Pătraşcu A, Patriciu VV (2014) Bio-cryptographic authentication in cloud storage sharing. In: 9th IEEE international symposium on applied computational intelligence and informatics. Timişoara, Romania, pp 15–17
6. Li X, Sun X, Liu Q (2015) Image integrity authentication scheme based on fixed point theory. IEEE Trans Image Process 24(2)
7. Nugroho H, Al-Absi HRH, Shan LP (2018) Iris recognition for authentication: development on a lighter computing platform. In: IEEE 978-1-5386-8369-9/18/$31.00 ©2018
8. Naik R, Singh DP, Choudhary J (2018) A survey on comparative analysis of different ICA based face recognition technologies. In: Proceedings of the 2nd international conference on electronics, communication and aerospace technology (ICECA 2018). IEEE Conference Record #42487; IEEE Xplore ISBN: 978-1-5386-0965-1
9. Monisha SJ, Sheeba GM (2018) Gait based authentication with hog feature extraction. IEEE 978-1-5386-1974-2/18/$31.00 ©2018
10. Tsigie MW, Thakare R, Joshi R (2018) Face recognition techniques based on 2D local binary pattern, histogram of oriented gradient and multiclass support vector machines for secure document authentication. In: Proceedings of the 2nd international conference on inventive

communication and computational technologies (ICICCT 2018). IEEE Xplore compliant—part number: CFP18BAC-ART; ISBN: 978-1-5386-1974-2
11. Sethia D, Gupta D, Saran H (2018) NFC secure element-based mutual authentication and attestation for IoT access. J Trans Consum Electron 14(8). https://doi.org/10.1109/tce.2018.2873181. IEEE, transactions on consumer electronics
12. Anand SA, Saxena N (2018) Noisy vibrational pairing of IoT devices. https://doi.org/10.1109/tdsc.2018.2873372, IEEE
13. Merhav N (2018) Ensemble performance of biometric authentication systems based on secret key generation. https://doi.org/10.1109/tit.2018.2873132, IEEE
14. Choi J (2018) A coding approach with key-channel randomization for physical-layer authentication. https://doi.org/10.1109/tifs.2018.2847659, IEEE

An Efficient Network Coded Routing Protocol for Delay Tolerant Network

Mukesh Sakle and Sonam Singh

Abstract Delay Tolerant Networks are the networks designed especially for the environment where there is no continuous connectivity. Network coding concept is not only an evolution but also a resolution in the area of wireless network because with the use of it, we cannot only utilize the network bandwidth effectively; it also decreases the power consumption and increases the throughput taking care to some aspect of security of packets in the network. In this paper, we propose an efficient routing protocol namely EDIF and EPIF which is better than the DIF and PIF proposed in [1] in terms of delay, throughput and packet delivery ratio. Through simulation results, it is shown that our protocol is better than the existing system.

Keywords DTN · PROPHET · DIF · PIF · EDIF · EPIF

1 Introduction

In delay tolerant network (DTN), packets from source to destination are send using store and forward paradigm, i.e., every node in DTN can besides routing the packets, can also store the packets. Hence, when link fails in DTN, it need not to be retransmitted from the source and when the link is available, again packet is transmitted from the node where the link has broken. While in the existing TCP/IP protocol, link failure results in retransmission of packets from the source node. The main characteristics of delay-tolerant network are large delays, opportunistic contacts and high disconnectivity or no end-to-end connectivity. There are various routing protocols [2] designed for DTN such as Epidemic routing, Spray and wait routing, PROPHET (Probabilistic Routing Protocol Using History of Encounters and Transitivity) and MaxProp. Epidemic routing floods the network with the replicas of packet or bundle to be transmitted to the destination giving high delivery ratio, but epidemic routing

M. Sakle (✉)
Shri Govindram Seksaria Institute of Technology and Science, Indore, India
e-mail: mukeshsakle@gmail.com

S. Singh
Parul Institute of Engineering and Technology, Vadodara, India
e-mail: sonam.6191@gmail.com

R. K. Shukla et al. (eds.), *Social Networking and Computational Intelligence*, Lecture Notes in Networks and Systems 100, https://doi.org/10.1007/978-981-15-2071-6_56

uses more bandwidth and energy. Spray and Wait limits the number of replicas of the packet to be transmitted to L and wait for the destination to come in range to deliver the packet and hence overcomes the disadvantage of epidemic routing. PROPHET uses delivery predictability value (which is based on previous encounters of nodes with each other) for forwarding of bundles in the network.

There are many issues in DTN such as buffer management, security, reliability, routing and energy. Network coding can be used to utilize buffer and bandwidth efficiently and reduces the transmission rate increasing the throughput.

2 Related Work

Basically, network coding involves coding or mixing of different packets we want to transmit in network into a resultant packet and route it in the network by broadcasting it. There are many ways (techniques) in which network coding is done. In linear network coding, the packets are multiplied with coefficient chosen from finite set of Galois field to form linear combination and then transmitted in network [3]. In random linear network coding, nodes transmit random linear combinations of the packets they receive, with coefficients chosen from a Galois field [4]. In XOR-coding, packets are X-ored with each other and then the intermediate node can broadcast the coded packet. COPE [5] uses opportunity coding in which the node has knowledge of which and how many packets are with its neighbours and hence XOR multiple packets into one packets in such a way that the neighbour can decode easily. Besides this, we can classify network coding into two categories: inter-session network coding [4] where coding is performed in the different session and intra-session network coding where coding can be done between packets belonging to the same sessions.

3 EDIF and EPIF

DIF (Delay Inferred Forwarding) and PIF (Probability Inferred Forwarding) are routing protocol designed in [1] which uses expected delay and expected probability as parameters to find the multihop path to the destination. It is shown by the author in [1] that these protocols are better than single-hop delegation forwarding in [6] and multihop opportunistic forwarding in terms of packet delivery ratio and delay. Now, our aim is to improve the performance of this algorithm using network coding.

3.1 EDIF

```
N ← the number of nodes
Ia,b ← the mean intermeeting time of node a & b
Initialize Dmin=Da,b,k-1
for b in 1……….N do
    if b≠a and b≠d then
        if Ia,b/2 + Db,d,k-1 < Dmin then
            Dmin=Ia,b/2 + Db,d,k-1
            end if
    end if
end for
Da,d,k=Dmin
```

→Decide the routing path according to expected delay value.

The steps of forwarding in the DIF algorithm can be listed as follows.

1. If $D\,b, d, k - 1 < D\,a, d, k$, forwarding the message to *b* will decrease the expected delay *D*; therefore, *a* will forward the message.
2. If $D\,b, d, k - 1 \geq D\,b, d, k$, forwarding the message to *b* will either increase the expected delay *D* or keep it the same; therefore, *a* will not forward the message.

→ Encode the packets routed to eventually reach the destination.
→ Decode the encoded packet at the destination when it reaches there.

The encoding and decoding procedure of EDIF and EPIF are same, and hence, it is described in Sect. 3.3.

3.2 EPIF

EPIF (Efficient Probability Inferred Forwarding) works as follows:

→ Calculate the expected probability of nodes

Let the remaining hop count be *k* and the expected probability between the node *a* and the destination node *d* be $P\,a, d, k$.

Suppose that with the remaining hop count $k - 1$, the expected probability between the node *b* (that a meets with) and the destination node *d* is $P\,b, d, k - 1$.

We will calculate expected probability $P\,a, d, k$ for PIF using the algorithm given in [1] as follows:

```
N ← the number of nodes
Ia,b ← the mean intermeeting time of node a and b
Ma,b ← the meeting probability of  node a and b
T ← the time slot width
 Initialize Pmax = Pa,d,k-1
for j in 1………N do
    if b≠a  and  b≠d  then
        Ma,b = 1- exp(-T/Ia,b)
        if Ma,b × Pb.d,k-1 > Pmax  then
          Pmax = Mi,j × Pb,d,k-1
            end if
     end if
end for
Pa,d,k =  Pmax
```

→ Decide the routing path according to expected probability value.

The steps of forwarding in the PIF algorithm can be listed as follows.

1. If $P\,b, d, k\ - 1 > P\,a, d, k$, forwarding the message to b will increase the expected probability; therefore, a will forward the message.
2. If $P\,b, d, k - 1 \leq P\,a, d, k$, forwarding the message to b will decrease the expected probability or keep it the same; therefore, a will not forward the message.

→ Encode the packets routed to eventually reach the destination.
→ Decode the encoded packet at the destination.

Encoding and decoding process are described below.

3.3 Encoding and Decoding Process

Network coding is used to improve the throughput and use the bandwidth efficiently further improving the performance of DIF and PIF algorithm. Suppose, we have to transfer L number of packets from node a to b, we can transfer coded packet α as a linear combination of packets S_1, S_2, S_3, …, S_L such that

$$\alpha = \sum_{n=0}^{L} S_n \beta_n$$

where $\beta_1, \beta_2, \ldots, \beta_n$ are coding coefficients which are taken randomly from the Galois field (2^8). This is the encoding process performed at the relay node. Now, when this encoded packet is sent to the next node c which already holds some k coded packets in its buffer in this case, c will generate a coded packet.

$$c = \sum_{n=0}^{k} \alpha_n \gamma_n$$

where $\gamma_1, \gamma_2, \gamma_3, \ldots, \gamma_n$ are coding coefficients taken randomly from the Galois field. Now, when c transfers this encoded packets to next node *d*, it stores c in its buffer if space is available or it can again encode *c* with each packet in its buffer.

When the encoded packets along with the coding coefficient reach the destination, the destination can get the packets $S_1, S_2, S_3, \ldots, S_L$ from the encoded packets using the coding coefficient. This is similar to solving the linear equation. For example to get α, we have decoded packets c_n and coding coefficient γ_n with us and hence

$$\alpha = \sum_{n=0}^{k} c_n \gamma_n$$

Similarly, we can get *S* using coding coefficient β_n and α_n (which we got by the above equation) in following way:

$$S = \sum_{n=0}^{k} \alpha_n \beta_n$$

But here, we have to take care that the destination cannot decode to get the source packets $S_1, S_2, S_3, \ldots, S_L$ till it receives all the coding coefficient γ_n and β_n along with the all the encoded packets α_n, c_n and so on. Hence, the destination has to wait for the packets.

4 Performance Evaluation

We have implemented our algorithms EDIF and EPIF in NS2 using DTN attachments in NS2 to make it work for DTN environment. We compare the existing work DIF and PIF in [1] with our proposed work EDIF and EPIF and show the results (Table 1).

We have taken two sources and two destinations and perform network coding using linear coding technique. The various parameters such as delay, packet delivery ratio and throughput are taken with respect to simulation time (Fig 1).

Delay is time interval from sending the message from the source till the time when it reaches the destination. Lower the end-to-end delay, better the performance is. Through simulation, we found that the end-to-end delay of EDIF and EPIF is lower than that of DIF and PIF (Fig. 2).

Packet delivery ratio is the ratio between total number of messages sent to total number of message received. The packet delivery ratio of EDIF and EPIF is greater than that of DIF and PIF (Fig. 3).

Table 1 Simulation parameters

Parameters	Values
Simulation used	NS-2.35
Number of nodes	90
Dimensions of simulation area	1407 × 732
Routing protocol	DIF, PIF, EDIF, EPIF
Simulation time	50
Antenna type	Omni antenna
MAC protocol	802.11
Queue	DropTail
Channel type	Wireless channel

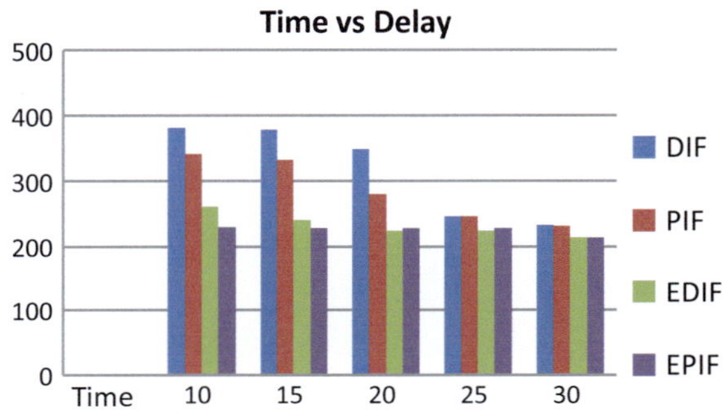

Fig. 1 Time versus delay

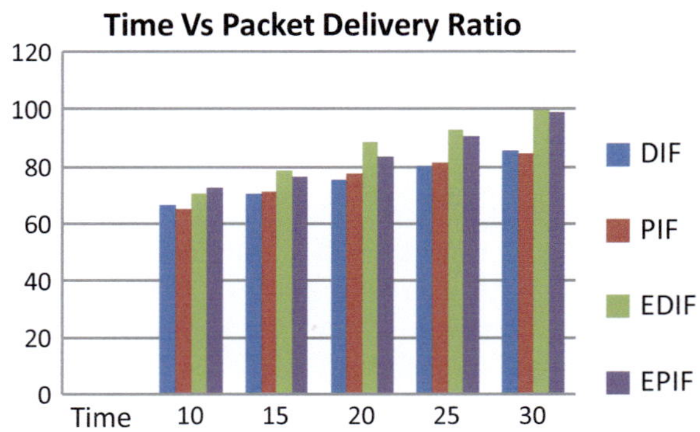

Fig. 2 Time versus packet delivery ratio

Throughput is the rate at which message is sent over the network channel. The throughput of EDIF and EPIF is greater than that of DIF and PIF.

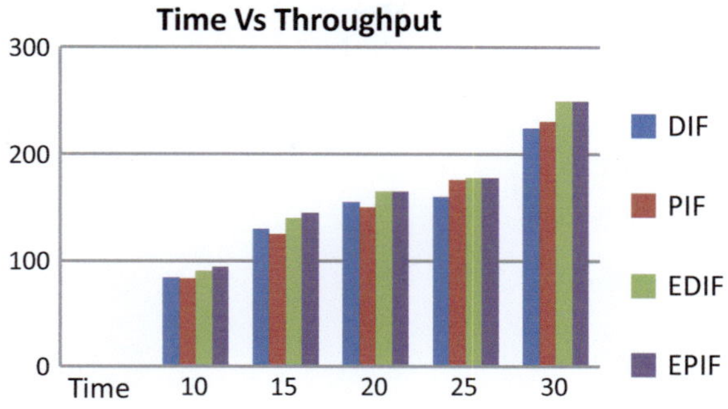

Fig. 3 Time versus throughput

5 Conclusion

Linear network coding is used to improve the performance of the existing routing protocol which uses multihop delivery quality for forwarding in delay-tolerant network. By comparison, we have conclude that as compare to the existing protocols used in [1], our protocol performs better in terms of delay, throughput and packet delivery ratio and hence is more efficient.

6 Future Scope

We can further improve our algorithm by using the inter-session concept using COPE. In this, if two sources wanted to send packets to two different destinations and have some common nodes in their respective path towards destination, packets can be intelligently coded analysing the packets of the nexthop and thus utilizing less energy and bandwidth.

References

1. Liang M, Zhang Z, Liu C, Chen L (2015) Multihop-delivery-quality-based routing in DTNs. IEEE Trans Veh Technol 64(3)
2. Puri P, Singh MP (2013) A survey paper on routing in delay tolerant networks. In: IEEE international conference on information systems and computer networks, pp 400–405
3. Pertz A, Fok CL, et.al (2011) Network coded routing in delay tolerant networks: an experience report. ACM
4. Wang Y, Jain S, Martonosi M, Fall K (2005) Erasure-coding based routing for opportunistic networks. ACM, pp 229–236
5. Widmer J, Le Boudec JY (2005) Network coding for efficient communication in extreme networks. ACM
6. Erramilli V, Crovella M, Chaintreauand A, Diot C (2008) Delegation forwarding. In: Proceedings of ACM MobiHoc, pp 251–260

Hybrid Text Illusion CAPTCHA Dealing with Partial Vision Certainty

Arun Pratap Singh, Sanjay Sharma and Vaishali Singh

Abstract The term CAPTCHA was introduced as the Turing test that claims to classify human and robot which may intend to intervene in the security traits of a database. CAPTCHA stands for "Completely Automated Public Turing test to tell Computers and Humans Apart." CAPTCHA has been represented in various customs such as distorted texts, 3D texts, audio, graphical, gaming and many more. The recent approach is gaming CAPTCHA which is a bit heavier for the server to load over a browser. The logic behind the gaming CAPTCHA is dealing with dragging and dropping object to the target position which does not belong to the hard AI problems. Gaming CAPTCHA is a time-consuming and complicated approach which confuses users to deal with the games. Turing test or CAPTCHA should be as easier as possible for human and almost impossible for bots. Here the proposed system is able to provide a new challenge of CAPTCHA in front of users that forms a hybrid text illusion which only deals with human's partial vision certainty. Hybrid text illusion signifies that a word observed by normal eyes actually gets differ from partially opened eyes. A user is required to recognize the partial one which is not possible with normal eyesight even a robot is not able to recognize by using any kind of text scanning approaches. It is a new generation CAPTCHA which creates illusion where only human can deal with.

Keywords CAPTCHA · Hybrid text illusion · Gaming CAPTCHA · Partial vision · Text recognition

A. P. Singh (✉) · V. Singh
The Right Click Services Pvt. Ltd., Bhopal, India
e-mail: singhprataparun@gmail.com

V. Singh
e-mail: vaishali.ec0709@gmail.com

S. Sharma
Oriental Institute of Science & Technology, Bhopal, India
e-mail: sanjaysharmaemail@gmail.com

R. K. Shukla et al. (eds.), *Social Networking and Computational Intelligence*, Lecture Notes in Networks and Systems 100, https://doi.org/10.1007/978-981-15-2071-6_57

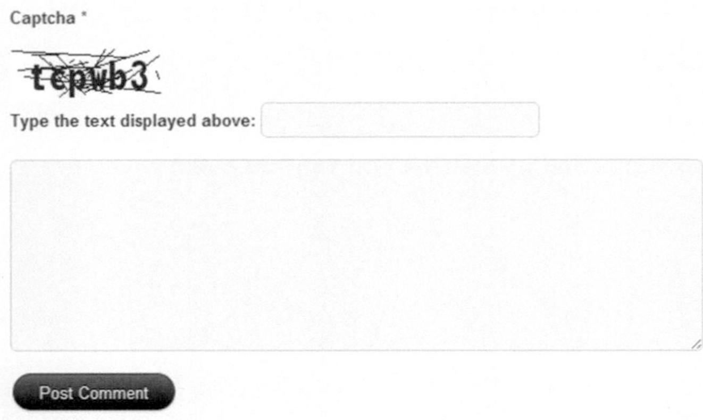

Fig. 1 Conventional CAPTCHA [2]

1 Introduction

A CAPTCHA is a type of challenge-response test designed to detach humans from robot software programs. CAPTCHAs are used as security checks for spammers and hackers to use the form on web pages to insert malicious or trivial code [1]. There are various types of CAPTCHA, and the conventional way is text recognition where user is required to recognize the distorted string belongs to alphanumeric characters in the form of image. Rather than conventional, there are so many methods have been introduced such as picture recognition CAPTCHA where user requires finding similar images from various distinct one. The most interesting and trending one is gaming CAPTCHA where user plays a game for authenticating his human tendency (Figs. 1, 2 and 3).

2 Related Works

JingSong et al. proposed a method of moving letters instead of steady one that claimed to provide better level of security comparing to the conventional one in the field of CAPTCHA. Video is nothing in the field of image processing because video is a sequence of frames, so attack is intended to apply over a frame which is ideally still [5]. Author again proposed another method which was based on 3D animation where letters are perplexed with complicated background along with continuous motion. But it becomes more complicated for human to understand the actual string, and as per the ethics; CAPTCHA should be as easy as possible for human and almost impossible for robot [6].

Aadhirai et al. proposed a new idea of CAPTCHA where object is required to identify with respect to the distance from a subject. It is considered as the hardest

Fig. 2 Picture recognition CAPTCHA [3]

Fig. 3 Gaming CAPTCHA [4]

AI problem for bots where security level is at the peak. But it is often complicated for human because of horde objects over an image [7] (Fig. 4). Song Gao et al. proposed a technique through which gaming CAPTCHA can be solved without human intervention. Approach is based on Auto-attack with Offline human Learning (AOffL) that helps to understand the game for offline knowledge. There are various steps through which an attack can be applicable such as learning phase which involves scanning frames, recognizing background and foreground, moving objects, target

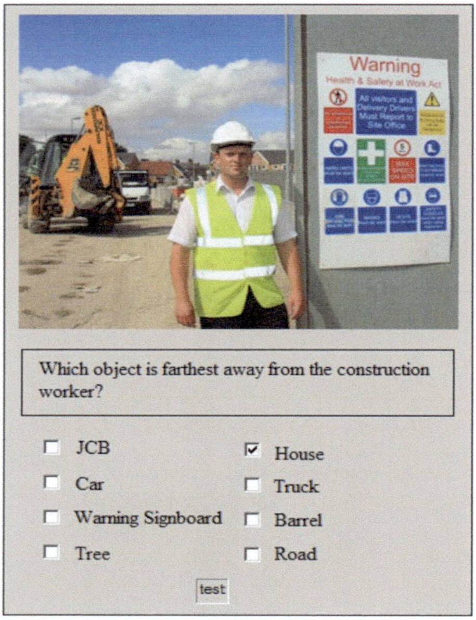

Fig. 4 Distance-based CAPTCHA [7]

and many more. The most important step is to design attack operation, for example, drawing lines for drag and drop based games. The final step is to mimic the human solver based on relay attacks [8] (Fig. 5).

Cao Lei et al. proposed a CAPTCHA which is based on gesture recognition that comes in the category of graphical CAPTCHA. User is required to recognize particular gesture from different kind of images. CAPTCHA is designed as finger guessing games which intend to confuse the bots to understand the problem by putting irrelevant gestures [9] (Fig. 6).

S. Ashok Kumar et al. proposed a CAPTCHA which is based on color recognition and clicks. User is required to recognize the objects based on color and perform mouse click events. But recognizing object with respect to color is not a big deal in the field of image processing [10] (Fig. 7).

3 Problem Identification

As per the researches in the past few years, the latest one is Bird Shooting CAPTCHA that works with color recognition. Color recognition is often easy from an image in the field of image processing. The method involves color thresholding technique where a particular color can be detected by the help of color threshold values (Fig. 8).

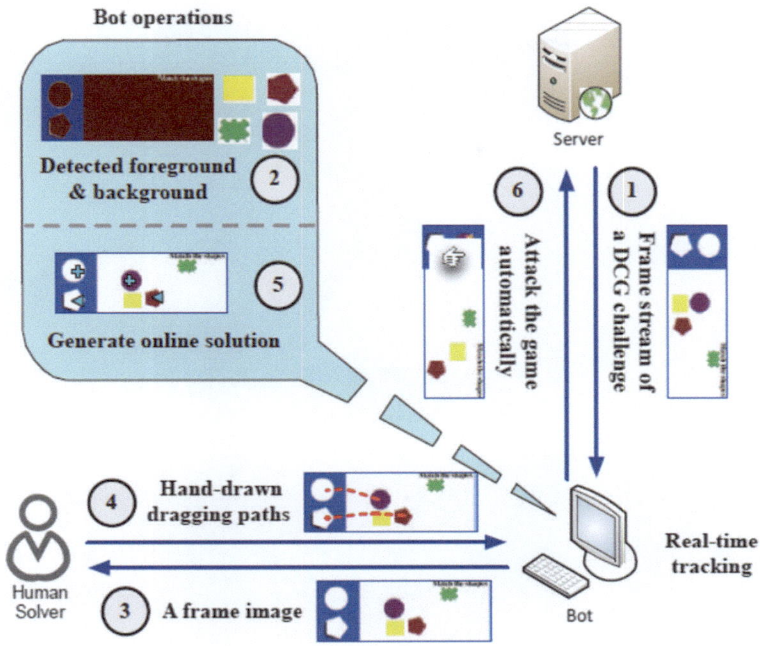

Fig. 5 Relay attack [8]

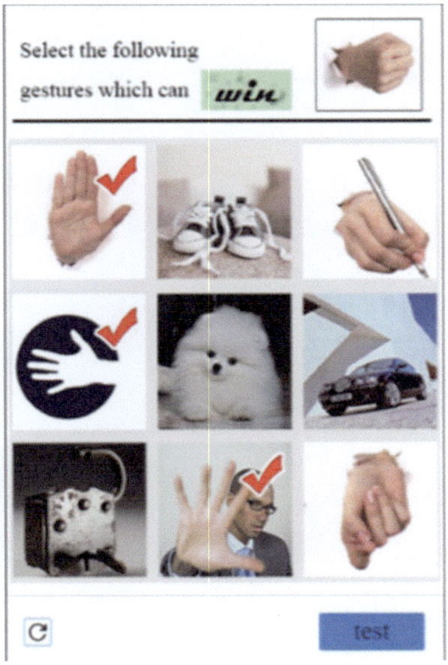

Fig. 6 Finger guessing game [9]

Fig. 7 Duck hunting game [10]

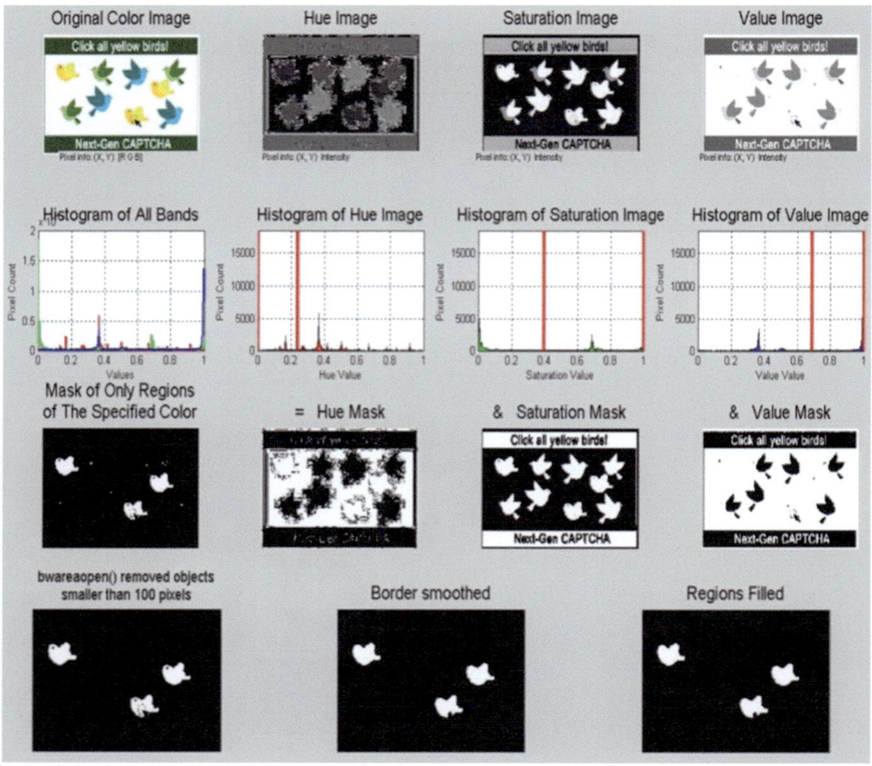

Fig. 8 Color detection

4 Proposed Work

The proposed CAPTCHA is a bit different from the earlier ones because it is neither based on clicks nor drag and drop, and it is based on hybrid text illusion (HTI) that can only be solved using partial certainty. A word which seems actually contains another word behind the same that can only visible when a human partially opens his eyes. No image processing or optical character reader (OCRS) can recognize the faded one from an image except human. Only human tendency can solve this HTI-based CAPTCHA, and no intervention can be made through machine-based programs. For example, an image which seems to look like a word "**CHANGE**," but actually there is another word behind CHANGE, i.e., "**UR LIFE**," and user is required to recognize the second word and submit with exact string (Figs. 9 and 10).

This is the simplest problem that a human can solve within few seconds but a big challenge for machine-based programs to recognize it correctly. As per the human perception, the optical graphic illusion allows a system to differentiate human and machines which may form the Turing test that can become an optimal CAPTCHA. It is the renowned hybrid text optical illusion (or bi-focal text illusion). There are two different words associated with each other, one of which can read from the nearest and a different from a far away. The optical illusion was originally developed by Philippe G. Schyns and Aude Oliva. It can be achieved through Gaussian smoothing or blur that allows reducing details from an image. Gaussian is an image blurring filter that expresses through a normal distribution for calculating the transformations. The formula of a Gaussian function in one dimension is

$$G(x) = \frac{1}{\sqrt{2\pi\sigma^2}} \mathrm{e}^{-\frac{x^2}{2\sigma^2}}$$

Gaussian function in two dimensions is

$$G(x, y) = \frac{1}{2\pi\sigma^2} \mathrm{e}^{-\frac{x^2+y^2}{2\sigma^2}}$$

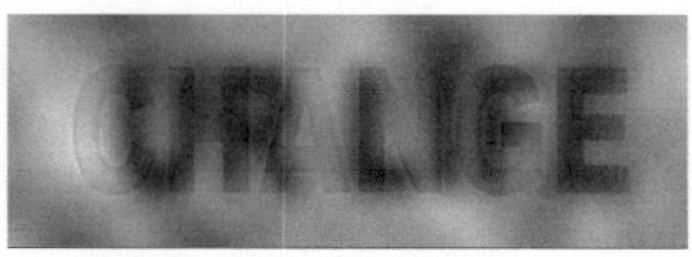

Fig. 9 Hybrid text illusion CAPTCHA-I [11]

Fig. 10 Hybrid text illusion CAPTCHA-II [11]

where x is the distance from the origin of horizontal axis, and y is from vertical axis. σ is the standard deviation. When applied in two dimensions, then this formula creates a surface, with focal circle with Gaussian distribution from its center point. The values of this distribution are used to create a convolution matrix that is applied to the original image. This determination process is visually depicted in the shape of the right side. The new value of each pixel is set to the weighted average of that pixel neighborhood. The value of the original pixel receives the heaviest weight (with the highest Gaussian value) and the smaller pixels in the form of the distance to increase the original pixel. It is a result of blur attribute that protects the boundaries and edges; better than the others, it is a more uniform blurred filter. The reduced standard deviation can be computed as (Figs. 11 and 12)

$$\sigma_r \approx \frac{\sigma_x}{\sigma_f 2\sqrt{\pi}}$$

Fig. 11 Gaussian blur on text [11]

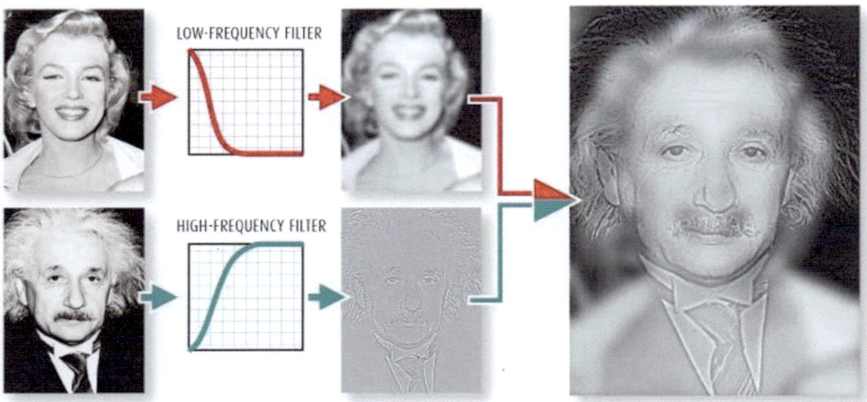

Fig. 12 Gaussian blur on hybrid images [12]

5 Result Analysis

The result has been analyzed with different candidates, and the average mean time for successful completion is less than the others. Color recognition affects the base paper's technique which perceives the unsuccessful and insecure CAPTCHA model. The proposed CAPTCHA has been tested with various attacks such as OCR, TensorFlow-based system; for example, Google Lens, background subtraction and many more but no tactics could work so far. The completion success rate and time consumption chart are given below along with comparison (Graphs 1 and 2).

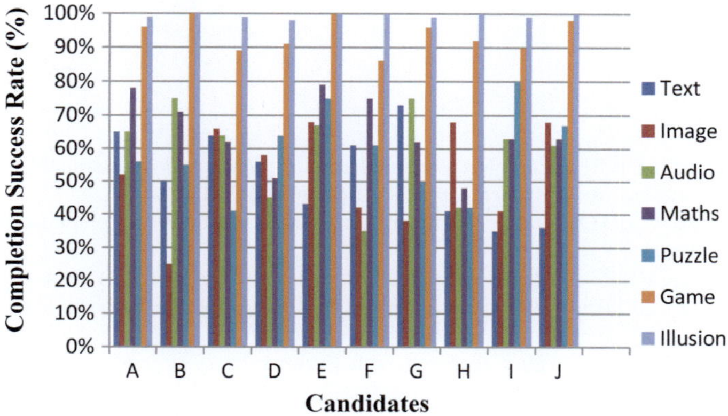

Graph 1 Completion success rate [10]

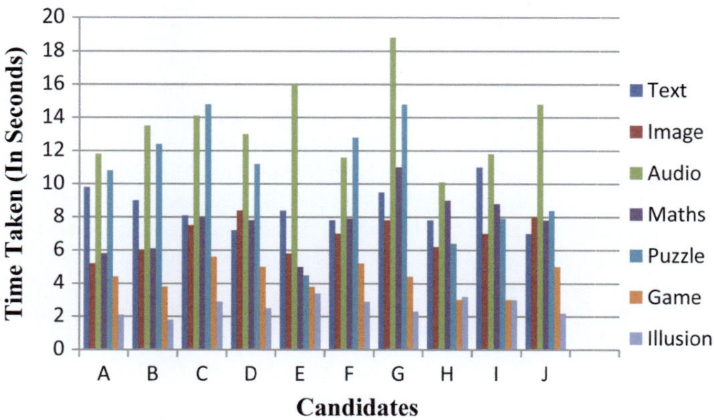

Graph 2 Completion time comparison graph [10]

6 Conclusion and Future Scope

The current proposed concept of hybrid text illusion CAPTCHA represents a new challenging problem that can only be solved by human eyes and no image processing tactics can recognize the desired string. This HTI CAPTCHA can be solved within few seconds as compared to all other CAPTCHAs. As per the optimal web security, a CAPTCHA should be as easy as possible along with less time consumption and almost impossible for robots. So, this kind of CAPTCHA can replace the existing conventional method and some interesting challenges can be proposed in front of human as well as machines in the future.

References

1. TechTarget: CAPTCHA (Completely automated public turing test to tell computers and humans apart), https://searchsecurity.techtarget.com/definition/CAPTCHA
2. Word Press: CAPTCHA code authentication, https://wordpress.org/plugins/captcha-code-authentication
3. The Register: Facebook serves up shaved tattooed 'butterfly' as CAPTCHA, https://www.theregister.co.uk/2015/04/30/facebook_serves_up_shaved_pierced_tattooed_butterfly_as_captcha
4. Geek: Gamified CAPTCHAs dispense with distorted text, https://www.geek.com/games/gamified-captchas-dispense-with-distorted-text-1573300
5. Cui J et al (2010) CAPTCHA design based on moving object recognition problem. In: The 3rd international conference on information sciences and interaction sciences, Chengdu, pp 158–162
6. Cui J, Mei J, Wang X, Zhang D, Zhang W (2009) A CAPTCHA implementation based on 3D animation. In: 2009 international conference on multimedia information networking and security, Hubei, pp 179–182. https://doi.org/10.1109/mines.2009.298
7. Aadhirai R, Kumar PJS, Vishnupriya S (2012) Image CAPTCHA: based on human understanding of real world distances. In: 2012 4th international conference on intelligent human computer interaction (IHCI), Kharagpur, pp 1–6. https://doi.org/10.1109/ihci.2012.6481832
8. Gao S, Mohamed M, Saxena N, Zhang C (2014) Gaming the game: defeating a game captcha with efficient and robust hybrid attacks. In: 2014 IEEE international conference on multimedia and expo (ICME), Chengdu, pp 1–6. https://doi.org/10.1109/icme.2014.6890287
9. Lei C (2015) Image CAPTCHA technology research based on the mechanism of finger-guessing game. In: Third international conference on cyberspace technology (CCT 2015), Beijing, pp 1–4. https://doi.org/10.1049/cp.2015.0843
10. Kumar SA, Kumar NR, Prakash S, Sangeetha K (2017) Gamification of internet security by next generation CAPTCHAs. In: 2017 international conference on computer communication and informatics (ICCCI), Coimbatore, pp 1–5. https://doi.org/10.1109/iccci.2017.8117754
11. Fiverr: hybrid text illusion, https://www.fiverr.com/opillusionist/create-hybrid-text-illusion
12. Air Freshener: far Marilyn look Einstein, https://airfreshener.club/quotes/far-marilyn-look-einstein.html

"By Using Image Inpainting Technique Restoring Occluded Images for Face Recognition"

Usha D. Tikale and S. D. Zade

Abstract Facial recognition is also known as the face recognition it describes to predict and to establish individual identity by using face. Face recognition system is used to identify people in real time system. In image processing, hair, moustache, beards, sunglasses on face by scarf and other clothing accessories, etc are some reasons for occlusion. This is refers to as facade of the facial image. For past years, face recognition has been researched in the picture in controlled environment. Typically, partial occlusion is concerned with the identifying under uncontrolled condition. In this research paper, main aim to solve the challenges for recognition face occurs due to the presence of partial occlusion on the basis of classification made literature review and analysis are done. The image inpainting technique-based method such as exemplar-based inpainting technique, extraction of features, fast-weighted principal component analysis, etc., is used in face recognition. In this project express the restores occluded part of image or the removed occlusion of images using exemplar-based image inpainting technique, combination of fast-weighted principal component analysis (FW-PCA) and extraction of feature.

Keywords Detection of face · Extract the feature: restore image · FW-PCA occlusion · Inpainting technique · Exemplar-based inpainting technique

1 Introduction

Manipulation and analysis of digitized image are done in order to improve the class in image processing. Various technical fields such as medical field, remote sensing, video processing and robot vision are impacted by digital image processing (DIP). In today's era, analysis of human biometric such as hand, fingerprint, ears, eye face recognition is required lot of attention in image processing in the research area.

U. D. Tikale (✉) · S. D. Zade (✉)
PIET, Nagpur, India
e-mail: ushatikalevru@gmail.com

S. D. Zade
e-mail: cdzshrikant@gmail.com

R. K. Shukla et al. (eds.), *Social Networking and Computational Intelligence*, Lecture Notes in Networks and Systems 100, https://doi.org/10.1007/978-981-15-2071-6_58

1.1 Challenges for Face Recognition

- **Lights**—The light is the parameter due to which recognition process may be affected. Some of the parameters are—the lightening conditions, sources of light, environmental conditions, focus of light, illuminations, blurring levels, etc., may affect the face recognition systems.
- **Image Capturing devices**—Vary in image capturing devices varies in image quality which will lead to image parameters variations. Such Variations in image capturing devices lead to variation in pixels and dimensions as well, which further may lead to mismatch.
- **Different Angles/Poses in image**—Images may vary along with variations in capturing angles and poses in images. So, there is mismatch in images.

1.2 Overview of Occlusion

Any unneeded object or obstacle in image disturbing the recognizing process is known as Occlusion. The view of an object is to refer an hindrance in the occlusion, and it can be synthetic as well as natural.

In real time, occlusion occurs via accessories like:

- Hand on face
- Sunglasses, hat, scarf, beards, mustache
- Face behind any obstacle
- Face images texture, etc.

Various kinds of facial occlusions have shown in Fig. 1.

1.3 Causes of Occlusion

Several problems may occur due to the presence of occlusion. Some of them likely to be stated as

- Developing system which tracks an object such as cars, people, etc., then by another object occlusion occurs if the object are hidden image is tracking.
- Where you do not have any information of the occlusion areas people are using a range of camera.
- Most of the critical issue in this research papers in many face recognition system is the facial occlusion such as in video surveillance.
- The problems of any image are corrupted by noise, scanning old photo resting dust on the scratching images or scanner of a scanning glass or others have stamps or logos.

Fig. 1 Kinds of facial occlusion

2 Objective

Important goal of research paper is to deal with face recognition problem of one class where occlusion of some appears badly in a given face image deforms the image in the presence of facial occlusions due to hair, mustache, sunglasses, etc. In uncontrolled or in intentional situations, some variations in facial images are commonly encountered, and big trouble to the FR like face recognition is cause based on the face investigation system but studied are less in this research paper.

2.1 Objective

- To build a system capable of taking input as occluded images, process it and provide relevant output which seems like dis-occluded.
- To design a module capable of processing on minor facial occlusions and provide faster as well as accurate results.
- Another module capable of processing on major occlusions, restore such regions and provide recognizable face images as proper results.
- To find face image dataset feasible in performing facial recognition activities.

- Normalize the dataset and perform clustering operation to get relevant clusters.
- In order to restore occluded image, recreation of images is to be done after getting matching expected image.

3 Literature Survey

A lot of research has been done on recognition of faces to get more accurate result in the given facial images under different environmental conditions and on various approaches available in occlusion are to solve the many problem in face recognition. The methodology used in our project is taken from various researches and sources. Several techniques like image inpainting with exemplar-based techniques, dataset clustering, face detection, FW-PCA (fast-weighted principal component analysis) and extraction of feature are referred from various research papers and from research organizations as mentioned [1–4].

For the formation of this project, main focus has been given in the area of

- Image inpainting and its techniques
- Face recognition under uncontrolled conditions
- Feature extraction
- Principal component analysis (PCA)
- Fast-weighted PCA
- Face detection.

The algorithms have been defined recognition under a large variety of conditions in the face image. In recent years, partially, the problem is received considerable attention is that of hidden faces. Since the work a variety of methods has proposed by Martinez, partially, matching images is not occluded training sample, but test images are occluded. Goal of this algorithm is to define a matching process of the non occluded faces. In research article, "In the Training and Testing Sets Recognition of Face with Occlusions," by Martinez and Jia [5] redefine as a reconstruction process of the face recognition problem.

Christine Guillemot and Olivier Le Meur in January 2014 introduced two new technique in Image inpainting [6]. The first method of this category is known as diffusion-based inpainting technique. These methods are to complete curves, straight lines and for small regions inpainting face. However, these methods are not suited for reconstructing the large areas of texture but are to blur image. In other technique, the texture to be synthesized is learned from the known part of the image or similar regions in a sample texture. Exemplar is taken known part of the image and stitching together patches by copying and sampling that learning is done. This method is also called as exemplar-based inpainting techniques. Diffusion-based techniques are better than the exemplar-based methods and sparse-based methods for large texture areas. Diffusion-based inpainting technique or exemplar-based inpainting technique has a major application in dis-occlusion of an image. This technique used to improve the accuracy and result of occlusion.

4 Analysis of Existing Systems

- **Image Inpainting**: Image inpainting is a technique used of recovering the image lost or detected parts. Basically, image inpainting is a process of recovering missing or damaged region in an image. Exemplar is taken known part of the image and stitching together patches by copying and sampling that learning is done.
- **Fast Digital Inpainting Method**: All the inpainting techniques depend on the size of the gap needs to be filled and to require minutes and hours. In inpainting algorithm is to complete fully in-paint the missing areas, they make it in-efficient applications for interactive user. Thus, the conventional inpainting techniques are very speed up of the new class of fast inpainting techniques.
- **Exemplar-based Image Inpainting**: Based on similar work on texture synthesis having a goal of better recovering texture of the damaged area, the other category of image inpainting method has appeared in last decade. The image inpainting technique is slightly different from texture synthesis problem.
- **Face Recognition by Using Principal Component Analysis (PCA)**: The recognized images can be normalized by using the principal component analysis. PCA transformed the various possibly correlated variables into a smaller number of uncorrelated variables by using a mathematical method.
- **Comparative Analysis**: The accuracy of the proposed approach is compared with some existing approaches is recognition of the face. To measure the biometric recognition, accuracy is to compute false rejection rate (FRR), and the false acceptance rate (FAR) is the common way.

5 Proposed Research Methodology

The proposal is to design a robust occlusion removal system so that it would be convenient to restore occluded facial images and retrieve occlusion-free image so that face recognition system would recognize the required face. The recognition system challenge will be reduced to certain extent and accuracy to be improved. Presenting system would provide input image to face recognition system after the removal of occlusion.

In this work, an approach for restoring facially occluded images is proposed. A task is to obtain occlusion-free image from partially occluded facial images for face recognition purpose.

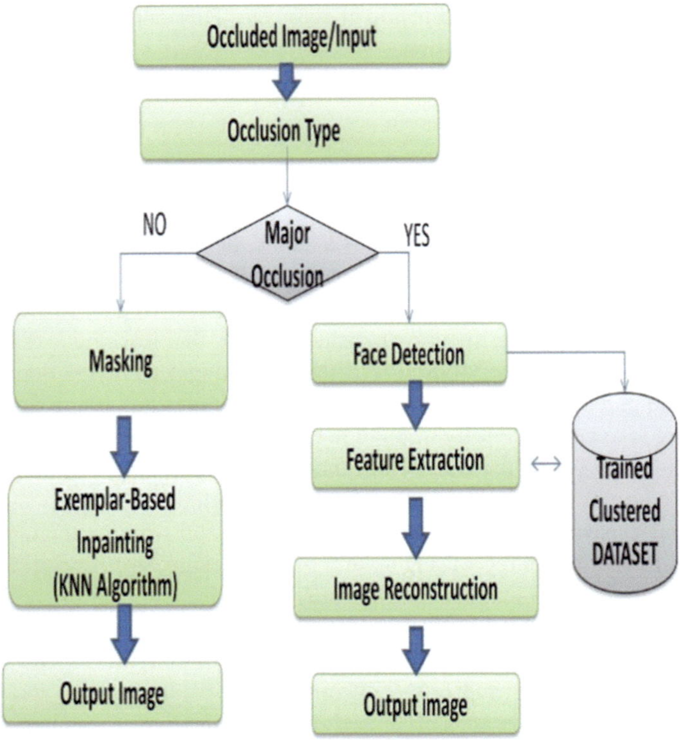

Fig. 2 Flow diagram of Inpainting Technique Restoring Occluded Image for Face Recognition

5.1 Flow of Project

The model of the proposed system follows: (Fig. 2)

1. **Input Image**

The first stage in input image process is to select the image which finds occluded and browsing the origin in the organization and classify image in the category minor and major occlusion depending on the basis of occlusion.

2. **Occlusion type**

Occlusion may be categorized either as minor occlusion or as major occlusion. The differentiation done is on the basis of background texture present in image.

3. **Masking**

After selecting inpainting technique, the first step is to find the occluded area is to be masked. Wherever occlusion present in image is to be selected which is known as masking of occluded region. As the occluded area in image is masked, the algorithm gets X_1, Y_1 coordinates by on click event of mouse and respective X_2, Y_2 coordinated by click release event of mouse. On the basis of coordinating points, KNN algorithm gets input those points to find nearest neighbors and fills the unknown region by nearest known pixels [7].

4. **Exemplar-Based Image Inpainting Technique**

The technique uses KNN algorithm which could remove the occluded pixels using nearest background pixels. Here, the approach is to use exemplar-based image inpainting technique since it is a generalized algorithm mostly used for classifying purpose where k-nearest neighbor algorithm works on the known pixels to replace occluded pixel with reference to surrounding pixels. Thus, the technique is to be selected here is inpainting. The need of mentioning these techniques is to reduce complexity on the basis of occlusion.

Three processes of major occlusion are as follows: (A) Detection of Face, (B) Extraction of Face and (C) Reconstruction of Face

(A) **Detection of Face**

Before begins the further process the input image is to organized. Here, only facial image is detecting and extracting from entire image excluding surrounding region. The need for extracting facial image from entire image is because to perform further processes with better accuracy and efficiency for replacement tasks. The output image is found to be organized and relevant for further process.

(B) **Extraction of Face and Clustering**

It gets the obtained image by face detection which is to be organized. In this, extracted feature in the form of vectors and reproduce of code like color segmentation, texture and edges. The dataset was initially trained and clustered. K-medoids is a widely used algorithm which uses Euclidean distance among feature vectors of each image. K- medoid clustering applied on raw data set based on feature vector and Euclidean distance calculated and compared.

By obtaining matching vectors, particular cluster is retrieved on the label. Again, the matching is to be done only with available images in retrieved cluster. Output will be matched with input image may called relevant images.

(C) **Reconstruction of Face**

Every image remained relevant, it is possible to simply detect occluded part of input image, thereafter, those pixels are substituted with the pixels of relevant images, and output will be recreated or reconstructed image which seems to be dis-occluded.

Thus it is stated that Occlusion can be removed virtually and produce occlusion free image. There may be more than one image obtained after restoration of occluded image, but specifically or optically or visually, maximum matched image is to be selected. Get the image is occlusion-free image [1].

6 Proposed Plan of Work

Module 1: Image Inpainting Technique

- Exemplar-based image inpainting technique (KNN algorithm)
- Masking Process
- To find the occluded area to be masked. Wherever occlusion present in the image and is to be select which is known as masking of occluded region.

Module 2: (1) Dataset Normalization

(2) Clustering of dataset
(3) Occluded dataset made for testing purpose.

Module 3: Face Detection

- To detect the face across the input image.

Module 4: Face Extraction

- Matching Process
- Feature Vectors Comparison.

Module 5: Image Reconstruction

- It is the intermediate process where occluded parts of image are replaced referring to selected image.
- To calculate the difference between occluded input image and selected image.

7 Conclusion

- The challenges to reduced the face recognition and Reconstructing the Occluded Facial Image with quick relevant to output image.

References

1. Hosoi T, Nagashima S, Ito K (2012) Restoration of occluded regions using FW-PCA for face recognition. In: 2012-IEEE
2. Banday M, Sharma R (2014) Image inpainting—an inclusive review of the underlying algorithm and comparative study of associated technique
3. Min R, Dugelay J-L (2012) Inpainting of sparse occlusion in face recognition. IEEE Trans
4. Kotsia I, Buciu I, Pitas I (2008) An analysis of facial expression recognition under partial facial image occlusion
5. Jia H, Martinez AM (2008) Face recognition with occlusions in the training and testing sets. In: 2008-IEEE
6. Guillemot C, Le Meur O (2014) Image inpainting: overview and recent advances. IEEE Sig Process Mag 31(1):127–144
7. Cunningham P, Delany SJ (2007) k-nearest neighbour classifiers. Technical report UCD-CSI march-2007

Social Networking

Personality Prediction and Classification Using Twitter Data

Navanshu Agarwal, Lokesh Chouhan, Ishita Parmar, Sheirsh Saxena, Ridam Arora, Shikhin Gupta and Himanshu Dhiman

Abstract Twitter is a prevailing online social media platform which is available for a large number of users. On Twitter, users communicate through messages and posts. On it, the messages are termed as tweets. This paper is intended to predict the user's personality from an analysis of the tweets which the user has shared. There are many techniques for personality prediction but there are some drawbacks of these techniques which are addressed in this paper. The objective of this paper is to provide a general outlook of the measures which are taken to predict the user's personality and draw comparisons between the results obtained from passing the available data through different classifiers.

Keywords Classification · Machine learning algorithms · Personality prediction · Regression · Twitter

N. Agarwal (✉) · L. Chouhan · I. Parmar · S. Saxena · R. Arora · S. Gupta · H. Dhiman
Department of Computer Science and Engineering, National Institute of Technology Hamirpur, Hamirpur, Himachal Pradesh 177005, India
e-mail: cs14mi515@nith.ac.in

L. Chouhan
e-mail: lokesh@nith.ac.in

I. Parmar
e-mail: cs14mi505@nith.ac.in

S. Saxena
e-mail: cs14mi523@nith.ac.in

R. Arora
e-mail: cs14mi529@nith.ac.in

S. Gupta
e-mail: cs14mi537@nith.ac.in

H. Dhiman
e-mail: cs14mi553@nith.ac.in

R. K. Shukla et al. (eds.), *Social Networking and Computational Intelligence*, Lecture Notes in Networks and Systems 100, https://doi.org/10.1007/978-981-15-2071-6_59

1 Introduction

Personality traits are distinctive characteristics which describe an individual. Social media platforms such as Twitter, Facebook have proliferated over the last few years, which have provided opportunities to researchers to analyze the data that users share on these social media platforms. The rate at which data is produced on these platforms is changing continuously. A large number of users create accounts and use these services to interact with other users. Views shared by users on this platform can be used to predict personality traits. This is a rapidly growing research area and has attracted much attention recently [1].

The dataset which has been used in the models has been taken from Kaggle which was put up by the Technical University of Munich. The university conducted a survey, displaying tweets taken from random Twitter profiles and asked people to choose a personality type which would best describe the person. Based on the input provided by the users, a personality type was chosen for the person. Twitter has been preferred in this paper, due to the 280 character limit of tweets, imposed by the company which makes it convenient to analyze.

Personality traits which have been used in this paper for predicting personality are: conscientious (being organized and guided by principles conforming to one's conscience), extrovert (being sociable and outgoing), agreeable (being kind, warm and cooperative), novelty seeking (being open to new experiences), rigid (being unable to change habits), impulsive (being able to do things without thinking much about consequences), psychopath (being excessively aggressive) and obsessive (becoming obsessed with a particular thought). Many machine learning-based classifiers have been implemented for this purpose which include: random forest, K-nearest neighbor, multiclass support vector machine, logistic regression, AdaBoost classifier, etc.

The rest of the paper consists of the following: Sect. 2 discusses the related work. Section 3 puts forward the classifiers used and discusses them in detail. Section 4 consists of the system model and the system design flow. Section 5 discusses the results produced with different plots for all the classifiers in detail. Section 6 consists of the conclusion and the last section involves the references.

2 Related Work

From the literature survey, it was observed that the usage of social media data in order to determine the personality of a human being is being explored very rapidly. Machine learning techniques are proving themselves to be very handy in providing accurate results in a much quicker way as compared to contemporary prediction and classification techniques.

Significant work is being done throughout the world in this field. The work of Ngatirin et al. [2], which involves a survey and comparison of different techniques for personality prediction, is worthy of note. The research is not just limited to Twitter data, with data on other social media sites like Facebook also being utilized. Laleh and Shahram [3] came up with a novel way of analyzing a person's Facebook activities in order to determine their personality.

Xue [4] applied the machine learning paradigm of label distribution learning to personality recognition and came up with quite promising results. Aydin et al. [5] performed personality prediction through application of random forest regression to audiovisual data which served as inspiration to apply the same technique to textual twitter data used in our research work.

Pratama and Sarnos [6] work on classifying personalities using Naive Bayes, KNN, and SVM that offered useful insights on navigating around the difficulties faced while using machine learning algorithms for personality prediction. Ong et al. [7] conducted similar work and designed their own model for using Twitter data for personality prediction and classification with lots of scope for improvements to prediction accuracy.

3 Methodology

This research paper involves the usage of various classifiers to get the score for personality prediction and comparison purpose. Although machine leaning is known to have many models, the paper mainly focuses on five main classifiers and makes predictions using them. The classifiers used are listed below.

3.1 Random Forest

Random forest is one of the multipurpose algorithms which is capable of performing both regression as well as classification with almost same accuracy and efficiency. It is an ensemble classifier which uses multiple machine algorithms to obtain predictive performance. It uses multiple decision trees for prediction to come up to the final outcome. It compiles the results from all the decision tress and leads to final output [5]. Random trees select only a fraction of rows that too at random and are trained upon a particular number of features. Random forest is the most accurate learning algorithm because single decision trees are prone to high variance whereas random tree averages all the variance across the decision trees.

3.2 K-Nearest Neighbor

KNN is a classification method for differentiating objects based upon the training examples provided in the feature space. KNN is a type of instance based or easy learning, where the function is only approximated locally and all computation is delayed until the classification. It is one of the most easily comprehendible techniques, and it can work even when one has scarce or no prior knowledge about the data distribution [6]. K informs us about the number of nearest neighbors that are utilized for prediction. Proximity metrics for KNN are: Euclidean distance, Hamming distance, Manhattan distance, Minkowsky distance, and Chebychev distance.

3.3 Multiclass Support Vector Machine

In the case of multiclass SVM, response variables should have multiple categories, whereas in the case of binary classification, response variables have two categories. Algorithms for implementing multiclass SVM are as follows: Weston Walkins multiclass SVM and Crammer singer multiclass SVM [6]. These are not math-oriented algorithm. It should have response variables which are more and more categorical. The most appropriate dataset is selected and model based on that data is used for prediction. Dataset is categorized into response variables and attributes. These attributes are the drivers for the value associated with response variables. Attributes are also called variates which may be categorical and numerical in nature.

3.4 Logistic Regression

Regression analysis is one of the methods to identify relationships between complex and random data. Logistic regression is the variation of above used for classification which can handle linear as well as nonlinear relationship among the dichotomous-dependent variable and the dichotomous-independent variable. The following equation represents logistic regression:

$$Y = \ln \frac{p}{1-p} = b_0 + b_1 x_1 + b_2 x_2 + \cdots + b_k x_k$$

where $0 < y < 1$ and $p =$ probability of presence of required attribute.

3.5 AdaBoost

Adaptive boosting or AdaBoost is an ensemble classifier which forges a strong classifier from a subset of weak ones, thereby enhancing the performance of classifiers. A model is created from the training data and then a second model is built upon it to correct the errors from the previous model. In the same way, models are added until the training set prediction accuracy is up to the requirements [8]. This is used for binary classification. At the end, the prediction is made from the set of weak classifiers by calculating their weighted average.

4 System Model

This research paper involves the usage of various classifiers to get the score for personality prediction and comparison purpose. Although machine leaning is known to have many models, the paper mainly focuses on five main classifiers and makes predictions using them. The classifiers used are listed below (Fig. 1):

1. **Data Collection**: Data is collected from a survey. In this survey, tweets from random Twitter profiles were taken and people taking the survey were asked to choose from the given ten personality types which would best describe the person: conscientiousness, obsessive, perfectionist, psychopath, impulsive, rigid, novelty seeking, extrovert, empathetic, and agreeable.
2. **Preprocessing**: Tweets are preprocessed by converting it to lowercase, transforming www.* or https?//* to URL, @username to AT USER, eliminating white spaces, trimming the tweet and non-ASCII characters. The steps followed are demonstrated in Fig. 2.
3. Feature vectors are defined by splitting tweets into words and removing stop words to ensure full word coverage in training and evaluation dataset. New training set is created based on personality labels predicted from Survey results.
4. Class labels are mapped to numbers, like conscientiousness is mapped to zero and extrovert is mapped to one, and so on.
5. Dataset is split and trained.
6. Bag of words, TF-IDF and N grams are implemented as features and are concatenated together. Predicted output labels are written to the file.
7. Different classifiers comprising of random forest, multiclass SVM, KNN, logistic regression, and AdaBoost are applied to the model and bar graph on the different models is plotted displaying categorization accuracy scores.

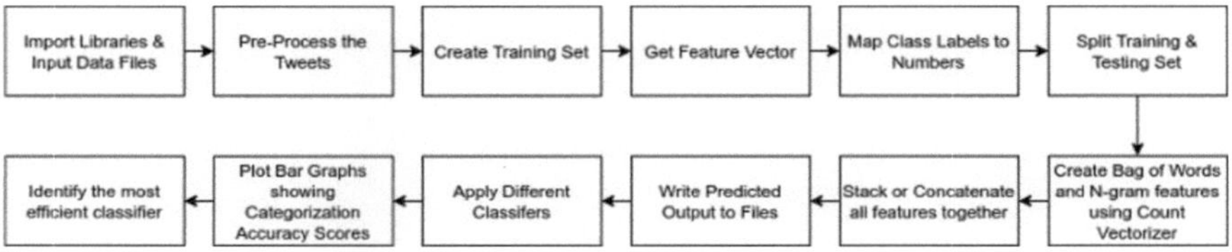

Fig. 1 System design flow

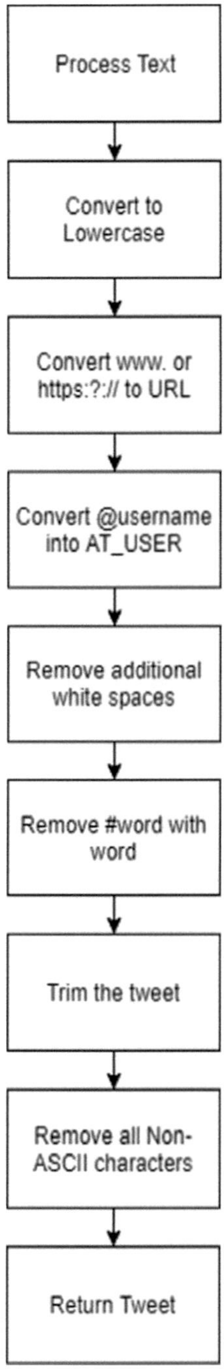

Fig. 2 Flowchart for preprocessing

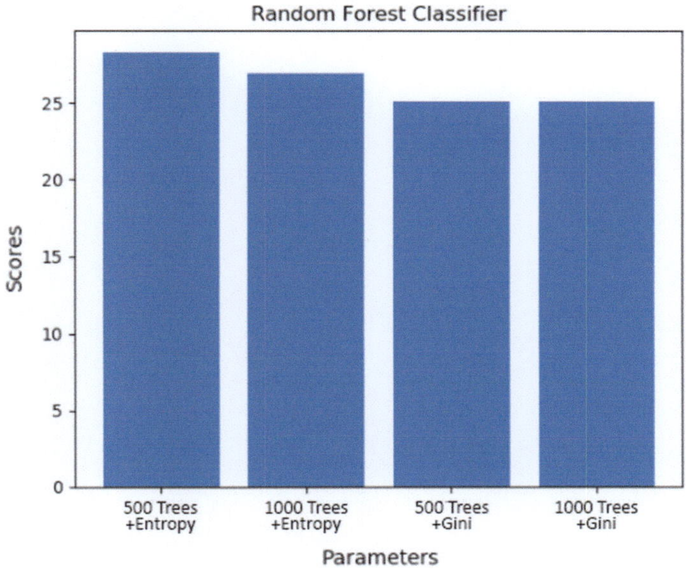

Fig. 3 Bar graph for random forest classifier

5 Experimental Results

The proposed system is trained and tested over the provided dataset using the classifiers of random forest, K-nearest neighbor, multiclass support vector machine, logistic regression, and AdaBoost. Parameters were varied in order to adjudge which configuration could come up with the best outcomes. Categorization accuracy scores were then calculated for each configuration corresponding to each classifier and then plotted in the form of bar graphs given in Figs. 3, 4, 5, 6, and 7.

It is found that among the chosen configurations for simulation, $k = 25$ for KNN classifier, 1000 Trees + Gini index for random forest classifier, rbf kernel for multiclass SVM classifier, multi = true for regression classifier and max depth = 10 and n estimators = 500 for AdaBoost classifier led to the most accurate categorization of personalities.

6 Conclusion

The paper was aimed at exploring the novel technique of determining a human's personality on the basis of the tweets made by them on Twitter. Various machine learning models have been utilized for this purpose. The primary contribution of the researchers being investigating the best possible classifier can be used to predict personality traits using Twitter data. The insights made were very promising and helped

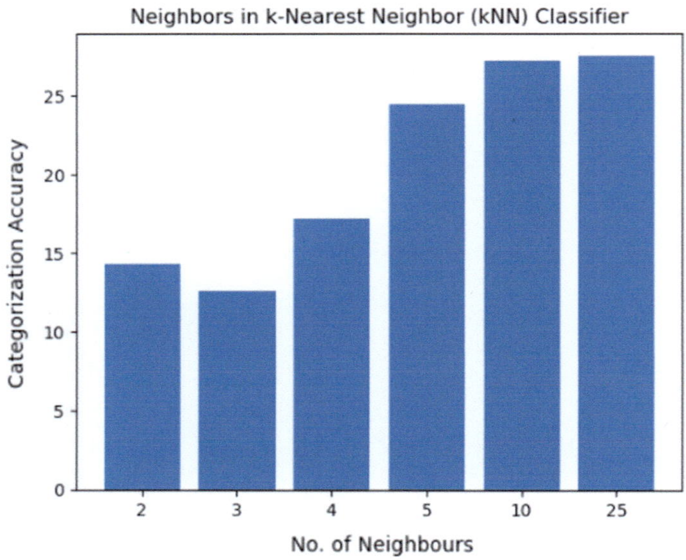

Fig. 4 Bar graph for KNN classifier

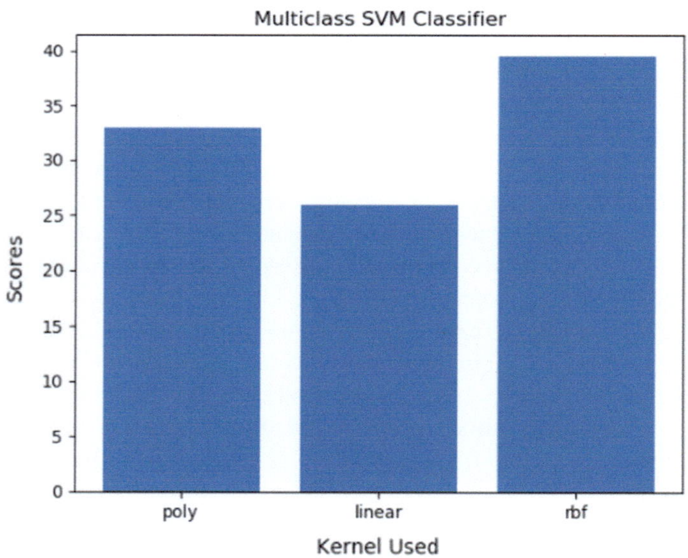

Fig. 5 Bar graph for multiclass SVM classifier

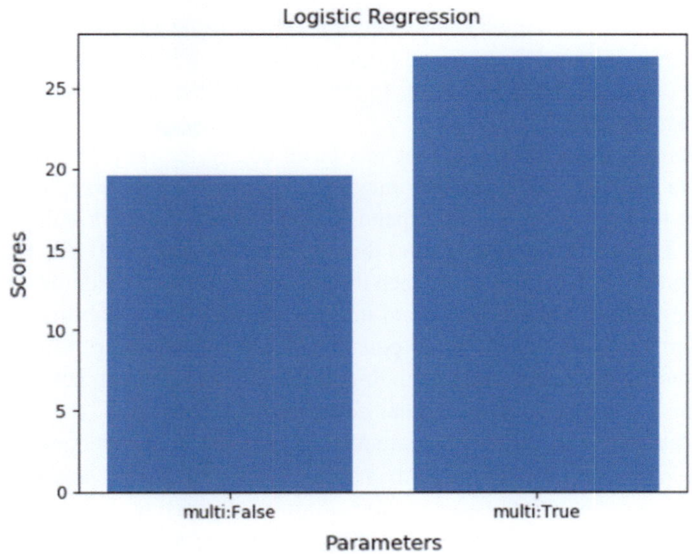

Fig. 6 Bar graph for logistic regression classifier

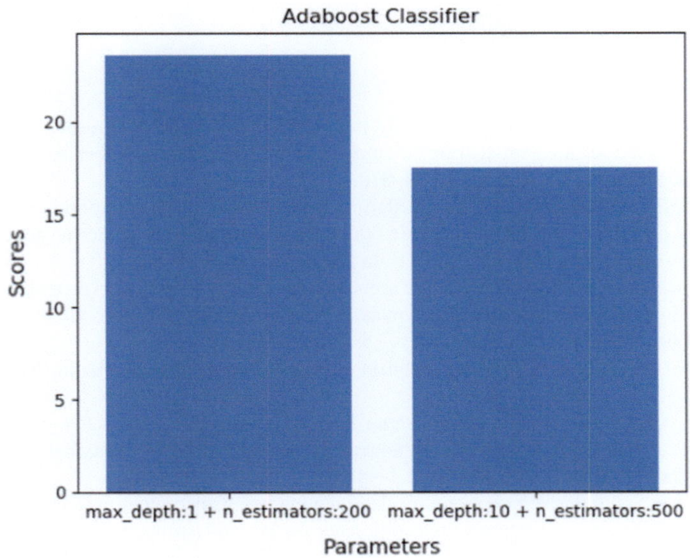

Fig. 7 Bar graph for AdaBoost classifier

to conclude that the best model is K-nearest neighbor using TF-IDF as features. The use of machine learning for personality prediction has yielded promising results by helping to target audience according to their personalities and making the system more market oriented.

The most significant problem that was faced during the project was the lack of sample data as input. For better accuracy of categorization, the size of the training set and the testing set need to be expanded vastly. Since there are millions of fake profiles in Twitter, they might impact the accuracy of analysis. If there is a doubt about the integrity of the input data, then the results cannot be considered reliable and proper investigation need to be carried into the integrity and reliability of the data. Another limitation that needs to be considered while developing future iterations of the project is the dynamically changing user behavior on social networking sites which needs to be answered for in later developments.

In the future, work can also be extended to consider the emoticons included in the tweets instead of just the text. It has been observed that more and more people are using emoticons in their tweets as a medium of expressing their emotions. This provides one with a new avenue for research work to be carried out and may lead to significant results and improvements in personality prediction.

References

1. Li A, Wan J, Wang B (2017) Personality prediction of social network users. In: 2017 16th international symposium on distributed computing and applications to business, engineering and science (DCABES), Anyang, pp 84–87
2. Ngatirin NR, Zainol Z, Yoong TLC (2016) A comparative study of different classifiers for automatic personality prediction. In: 2016 6th IEEE international conference on control system, computing and engineering (ICCSCE), Batu Ferringhi, pp 435–440
3. Laleh A, Shahram R (2017) Analyzing Facebook activities for personality recognition. In: 2017 16th IEEE international conference on machine learning and applications (ICMLA), Cancun, pp 960–964
4. Xue D et al (2017) Personality recognition on social media with label distribution learning. IEEE Access 5:13478–13488
5. Aydin B, Kindiroglu AA, Aran O, Akarun L (2016) Automatic personality prediction from audio-visual data using random forest regression. In: 2016 23rd international conference on pattern recognition (ICPR), Cancun, pp 37–42
6. Pratama BY, Sarno R (2015) Personality classification based on Twitter text using Naive Bayes, KNN and SVM. In: 2015 international conference on data and software engineering (ICoDSE), Yogyakarta, pp 170–174
7. Ong et al (2017) Personality prediction based on Twitter information in Bahasa Indonesia. In: 2017 federated conference on computer science and information systems (FedCSIS), Prague, pp 367–37
8. Shu X, Wang P (2015) An improved adaboost algorithm based on uncertain functions. In: 2015 international conference on industrial informatics—computing technology, intelligent technology, industrial information integration, Wuhan, pp 136–139

A Novel Adaptive Approach for Sentiment Analysis on Social Media Data

Yashasvee Amrutphale, Nishant Vijayvargiya and Vijay Malviya

Abstract Sentiment analysis (SA) is the approach of determining polarity of any content whether the given sentence contains positive, negative, or neutral sentiments. In many real-world situations, it is required to know the public emotions about happening in surrounding environment. Thus, this analysis helps in decision making on a particular task. There is a huge area where sentiment analysis can be utilized to improve the decision making like while launching a new product, adding additional features in existing products, announcing of a new government policy, etc. This paper shows sentiment analysis system which is based on machine learning algorithm used by TextBlob API using python. Proposed system uses natural language tool kit (NLTK) dataset for training the algorithm. This newly implemented application is used to do sentiment analysis on "twitter" (a social networking application) real-time data. Its experimental results are also presented in this paper. The results/analysis can help big brands, companies, and governments in planning future activities.

Keywords Machine learning · NLP · Sentiment analysis · TextBlob · NLTK and twitter

1 Introduction

Sentiment analysis is a task to classify whether a short paragraph of text is being positive or negative. Sentiment analysis algorithm calculates the text content polarity value. If polarity is positive, it means sentience contains the positive emotions. If sentience polarity is negative, it means sentience contains the negative emotions.

Y. Amrutphale (✉) · N. Vijayvargiya · V. Malviya
Malwa Institute of Technology, Indore, India
e-mail: yamrutphale@gmail.com

N. Vijayvargiya
e-mail: nishant.vijay23@gmail.com

V. Malviya
e-mail: vijaymalviya@gmail.com

Rajiv Gandhi Proudyogiki Vishwavidyalaya, Bhopal, India

R. K. Shukla et al. (eds.), *Social Networking and Computational Intelligence*, Lecture Notes in Networks and Systems 100, https://doi.org/10.1007/978-981-15-2071-6_60

Sometime polarity may be zero which means sentience has no emotions; it just contains some information. Many tools and techniques are used to calculate the sentiments, but the most common technique is machine learning. In machine learning, we train the algorithm from some predefine dataset to identify the word polarity.

Recently, sentiment analysis has become important for many companies. Merchant subscribes on social networking sites like Instagram, twitter, Facebook, and other to get reviews for their various products. If the company wants to track tweets about their brand to command over the impact on time and also many Web site analyze the comments on their articles. This will help them to track comments and impact. So the sentiment analysis is an automated system that collects and analyzes the content and generates the desired results. Most of the public maters are directly related to the public emotions, like government policies, politics, marketing, advertisements, share market, etc.

Two different approaches are used to calculate sentiments from any text. First one is lexicon-based sentiment analysis, and second is machine learning-based sentiment analysis. In lexicon-based sentiment analysis, text is divided into tokens, and then polarity of words is calculated. According to number of words, positive, negative, or neutral sentiment polarity is calculated. This approache also popular as "bag of word model." In machine learning model, a classifier is trained by some existing training set to identify the subjectivity and polarities of new sentiments. There are many classifiers which may be used in supervised learning like probabilistic classifiers (Naïve Bayes classifier [1], Bayesian network [2], and maximum entropy classifier [3]), linear classifiers (support vector machine classifiers [4] and neural network [5]), decision tree classifiers, and rule-based classifiers [6].

There are many applications which are sentiment analysis approach. Sentiment analysis is used for quick review and rating of movies [7]. There are many stock market analysis systems which are using sentiment analysis to predict the future prices of stock [8, 9]. Political parties can use it to know the public opinion about their candidates. Government can also use it to review their policies through sentiment analysis. Many online recommendation systems are developed on the basis of sentiment analysis. Advertisements and marketing field are frequently using sentiment analysis to develop their systems.

Objective of this research work is to implement a general-purpose sentiment analysis system. This implementation takes twitter posts for SA. Targeted data is real-time twitter data to know the current sentiments about given product, issue, or person.

We are including the details in our paper which are as follows. Section 2 contains the literature review. In this section, some research paper summary related to sentiment analysis is presented. Section 3 has presented the sentiment analysis through social media. Methodology and implementation details are included in Sect. 4. All the experimental results and their analysis are shown in Sect. 5. The entire research work summary is concluded in Sect. 6.

2 Literature Review

Moreo et al. [10] have proposed a lexicon-based comment-oriented news sentiment analyzer, i.e., (LCN-SA). This analyzer works for the common tendency of web users, to express their views in un-standard language, and analyzer also works in multi-domain scenario by targeting users. Authors proposed one automatic focus detection module along with sentiment analysis module, and these modules are capable of finding and using public opinions of topics in news items. Taxonomy-lexicon is used by these modules that are specifically designed for news analysis. Analyzer experiment results show that the results obtained are extremely promising. Linguistic modularized knowledge model with low-cost adaptability and hierarchical lexicon specifically designed to analyze news comments are also the features of LCN-SA.

Nguyen and Shirai [8] built a model for perdition of stock prices. They included sentiment on social media impacts on stock prices in their implementation. Some new features are added to calculate stock prices in the traditional stock price analysis system. As a result, they found that including sentiment information which is captured from social media can improve the prediction accuracy. In this research, a model called topic sentiment latent directed allocation (TSLDA) was proposed. This method can show the topic and sentiment at the same time.

Olivera et al. [9] proposed a methodology to examine the impact of micro-blogging data on stock market values. Their methodology used sentiments and attention indicates from social media. They used a Kalman filter to merge survey sources and micro-blogging. Their experimental results clearly show that micro-blogging data put impact on stock market prices.

Sun et al. [11] presented a study based on natural language processing (NLP) techniques for sentiment analysis. They discussed general NLP technique which is used for text preprocessing like tokenization, word segmentation, part of speech (PoS), tagging, parsing, etc. According to authors, sentiment analysis can be divided into three different levels: first is document level, second is sentence level, and third is fine-grained level. Authors also included different sentiment analysis techniques in their research paper.

Piryani et al. [12] presented a study of opinion mining and SA on research done from the year 2000 to 2016. They included many Web of Science (WoS) indexed papers. In their study, they found many approaches of sentiment analysis like machine learning and lexicon-based sentiment analysis. They also identify different levels of analysis (document, sentiment, or aspect-level). Author presented a detailed and wide analysis mapping of opinion mining and sentiment analysis.

Salas et al. [13] proposed a sentiment analysis method for classification of features and news polarity using opinion mining. Proposed method is based on ontology-driven approach. Their research is focused on financial news sentiment analysis. They address the difficulty of feature-based opinion mining.

3 Sentiment Analysis on Social Networking Platforms

Social sentiment analysis is the use of social media like twitter and Facebook to understand the wisdom of the crowd [14]. So analyst takes twitter firehouse and thus puts certain keywords about a certain subject matter, and they try and understand from NLP that what the people are saying about that specific topic. A lot of work analyst tries to do is to understand that how to filter out sarcasm, what is positive, and what is negative and also understand hundreds of mode of icons used by people on twitter. This is very useful for marketer to understand what the pulses about the product. It also helps to understand where company putting their marketing amount. What is the reaction to advertising campaign? It is ability to look at millions of tweets in a couple of time about a given subject. It is like having millions of people focus group. So that sentiment analysis of social media is much more representative and useful.

Twitter may be a treasured tour of sentiments. Folks round the word place thousands of reaction and opinion on each topic beneath the sun daily. It is like one huge platform of psychological information and is endlessly being updated. We are able to use it to research legion text in seconds with the facility of machine learning.

Sentiment instrument receives some input text like twitter tweets. First the text needs to split into many words or sentences. This method is termed as tokenization, and result of this method creates tiny token type huge text. The method simply counts every word shows up once the text is tokenized. This can be known as bag of word model. Then we tend to loom up the sentiment values for every word from the sentiment lexicon that is all prerecorded. The classifier told the sentiment values of tweets. This method will be occurring in three main steps.

i. Register for twitter API,
ii. Install dependencies,
iii. Write script for sentiment analysis.

Twitter API is associated with nursing application programming interface. It is the entryway that allows user to access some server's internal practicality. One will scan or write tweets from own application victimization twitter API. In second step, user has to be compelled to install dependencies that needed reading the text type authentic account and calculate the sentiment values, and then scriptwriting is needed. Presently, python artificial language is usually utilized in scriptwriting for machine learning construct, and through scriptwriting, sentiment analysis results will be calculated and given in desired format.

4 Methodology and Implementation

4.1 Data Extraction (Datasets)

Proposed sentiment analysis system works on real-time twitter posts (tweets). So that in this implementation, there are no pre-stored datasets. This proposed solution is taking real-time tweets directly from the twitter.

We need a twitter account, which works as a gateway for application to access twitter data. To connect proposed developed application to the twitter, we have to create access token and access token secret on twitter using https://apps.twitter.com/. These keys and tokens are also known as API keys and tokens. Twitter has provided the facility to create a customized twitter application for developers. For that, twitter has provided these API keys and tokens. By the use of these keys and tokens, developers can develop their own customized application with many features of twitter.

4.2 API Keys and Tokens Generation

API stands for application program interface. API keys are series of code produced by twitter to allow twitter profile to poll twitter feeds. For that, a twitter application needs to be created by going to the http://apps.twitter.com and for signing in, we have to create an application and need five information as requested by the dialog boxes. Once after giving all the basic data, twitter application is created that allows user to move social profile to function. Before this entire process, it is mandatory to have a valid mobile number associated with profile, which is going to be used. To check this, we can visit mobile tab in settings of twitter account. Twitter requires mobile number to validate the twitter profile.

After creating of application twitter, application manager area will be there. Here, tab "Key and Access Token" gives all credential required for twitter application development. Four main information are required to develop twitter app development, and they are user key (API key), user secret (API secret), access token, and access token secret. We can also mange access level through permission tab whether app required read-only permission or read/write or read/write and access direct message permission.

4.3 API Used

Python provides supports for implementation of machine learning concepts. To do this, many APIs are implemented and provided by python or by individuals. Some of them have used in this implementation. APIs used for this implementation are

a. tweepy- provides streaming classes to extract real-time twitter data form twitter account.
b. nltk- module provides download access to the natural language tool kit dataset. It is basically a training dataset which is used to training any machine learning algorithm.
c. tkinter- is used to design the GUI of application. It provides classes like button, label, etc.
d. numpy- provides functions and data structures for numeric operations.
e. text blob- model is used to perform text analysis operations.

4.4 GUI Design

Application GUI is divided into four parts as shown in Fig. 1. First part, top-left section is used as an input section. Mainly two inputs are required for sentiment analysis: number of tweets and search text. They do filtration tasks. The number of tweets is the count of tweets that is used for analysis. Search text filters the tweets, and

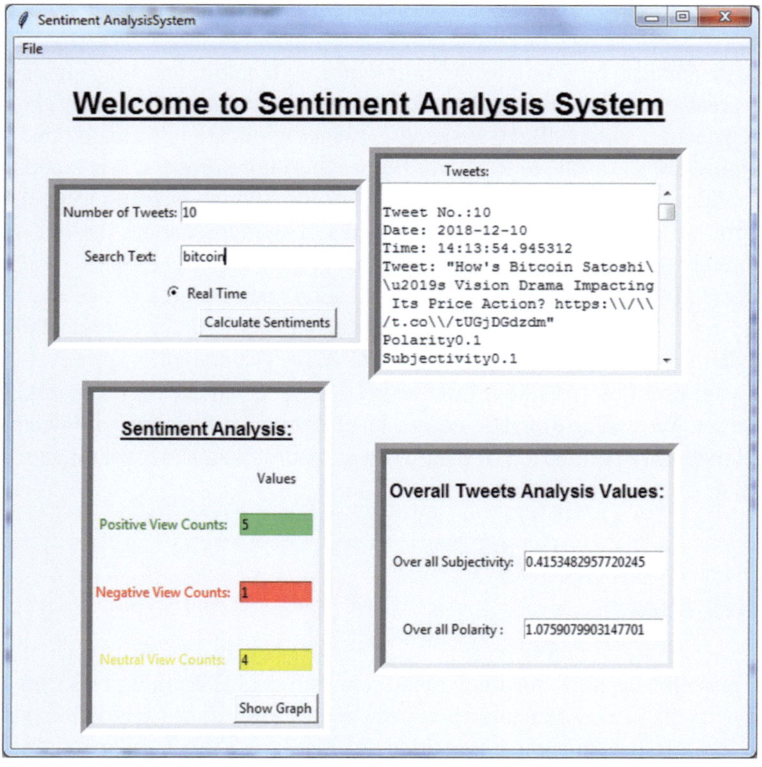

Fig. 1 Sentiment analysis system GUI

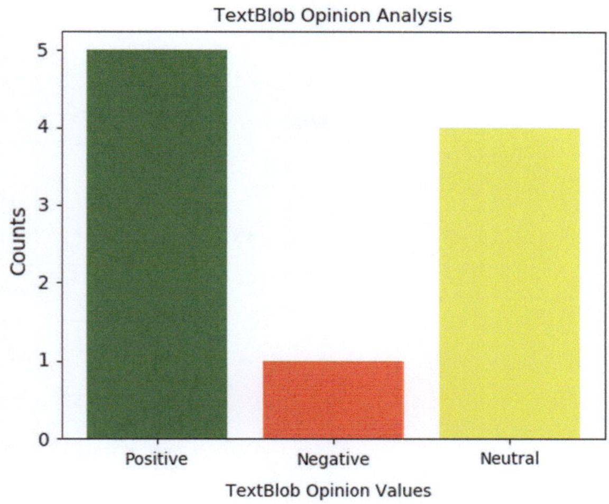

Fig. 2 Bar chart of sentiment analysis result

algorithm searches the tweets related to the text entered in search text. After entering both values, we need to click on calculate sentiment button. It starts extracting the tweets and calculates the sentiments. All extracted tweets are visible in tweet section (top-right). This section contains tweet number, data of tweet, time of tweet, tweet text, and polarity and subjectivity of individual tweets. Positive, negative, and neutral tweet count summary is given in the bottom-left section of GUI. In this section, "show graph" gives the facility to show the graphical bar chart view of results as represented in Fig. 2. Bottom-right section represents the overall tweets analysis, this section has two values overall subjectivity that is average of all tweet subjectivity and second value is overall polarity, and it is the sum of all polarity values.

4.5 Algorithm Process

Figure 3 represents the flow chart of proposed system execution process. Two main inputs are essential, i.e., total tweet count and search text. "Number of tweets" input represents the sample size of analysis. "Search text" works as filter for twitter. Algorithm tries to search and extract tweets only related to search text given. Initially, each tweet is analyzed. Sentiment polarity and subjectivity is calculated of each tweet individually. Finally, when tweet count reaches the limit of sample size, then overall subjectivity and polarity is counted. In the last step, all the results are displayed in GUI. Proposed application also has the facility to display the results in graphical (bar chart) form.

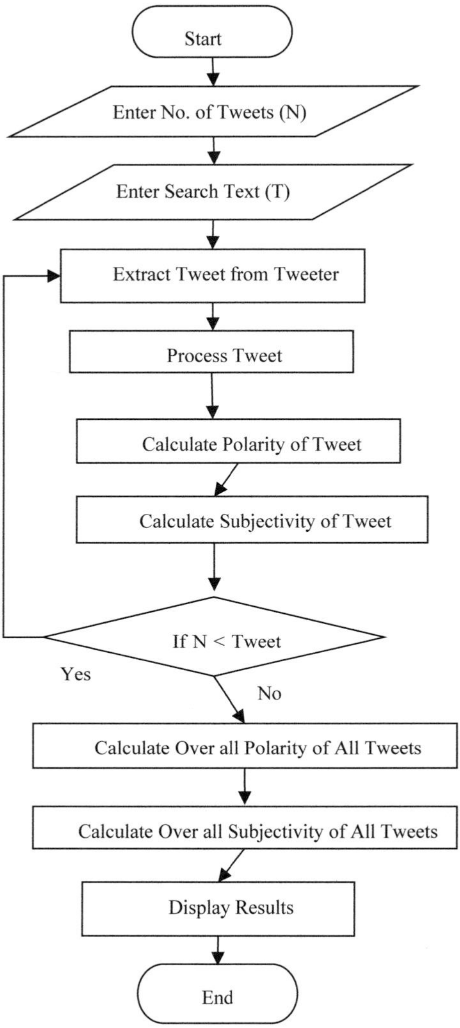

Fig. 3 Flow chart of algorithm

5 Result Analysis

Table 1 shows the experimental results of sentiment analysis done by proposed implementation. This analysis is done in the month of December 2018. Table represents the following results:

Number of tweets: This column represents the analysis sample size.

Topic: This column shows the filter text, i.e., this analysis is about the given topic.
Positive: This column shows the number of tweets representing positive sentiments.
Negative: This column shows the number of tweets representing negative sentiments.
Neutral: This column shows the number of neutral sentiments. These tweets do not have any sentiments.

Table 1 Experimental results of sentiment analysis system

No. of tweets	Topic	Positive	Negative	Neutral	Subjectivity	Polarity
1372	Bitcoin	390	356	626	0.346816	10.62543
221	BJP	55	61	105	0.3256	5.15011
402	Congress	100	89	213	0.323289	0.346816
270	Cricket	85	67	118	0.368665	11.10077
200	MP election	2	35	113	0.268147	3.810519
250	Rohingya	37	81	132	0.282281	10.2568
85	Viratkohli	25	5	55	0.44602	8.30645

Subjectivity: This column represents the overall subjectivity of all tweets. Subjectivity is in the range of 0–1.
Zero number represents tweet does not contain emotion. Tweet contains only information, and one represents tweet is full of emotions.
Polarity: This field contains the overall polarity of all tweets. This number can be positive, negative, or zero. Positive polarity means discussion about issue, topic, or person which is positive. Negative value represents the negative emotions, and zero shows the neutral emotions.

While doing analysis of experiments, following points are observed.

a. It is a real-time sentiment analysis. Proposed application extracts real-time tweets from twitter. Speed of data extraction depends on topic. If topic is from current issues or from popular things, then we get tweets frequently. But if topic is not from current issues, then extraction takes time.
b. This application takes data online from Internet, so Internet connectivity is also a major factor. If Internet connectivity gets break during run application will stop immediately and not return complete analysis.
c. One important observation we found that some time algorithm is not able to calculate the sentiments of any tweet then it will consider that as neutral sentiments.
d. By result analysis, it has been found that proposed application is working for different types of issues, person, etc.

6 Conclusion

Sentiment analysis can play an important role in many real-word applications. It can help in increasing accuracy of any prediction system. Nowadays many recommendation systems can also use sentiment analysis to improve the efficiency of system and increase their product sells. Thus, we are using advanced supervised machine learning algorithm for generating results on sentiment analysis. We have used most popular NLTK dataset to train the algorithm and to give the results. This system works on twitter real-time data to give current sentiment analysis. In our implementation, we proposed a generalized, real-time, and machine learning-based sentiment analysis system. This system is a new and advanced approach for generating sentiment analysis results as compared to older methods. This system can give emotion values of any issue, product, or person using the semantic method. As compared to "bag of word" method, the semantic method can assume as more appropriate approach as it contains the analysis of whole semantics of the word formation. It is algorithm based and utilizes machine learning, so it does produce the results in lesser time and give the output quickly on bigger datasets. This way this sentiment analysis tool can be used in several environments to get desired reviews and sentiment analysis in various fields.

References

1. Kang H, Yoo SJ, Han D (2012) Senti-lexicon and improved Naïve Bayes algorithms for sentiment analysis of restaurant reviews. Exp Syst Appl 39:6000–6010
2. Inza I, Lozano JA (2012) Approaching sentiment analysis by using semi-supervised learning of multi-dimensional classifiers. Neurocomput 92:98–115
3. Kaufmann M (2012) JMaxAlign, a maximum entropy parallel sentence alignment tool. In: Proceedings of COLING'12: demonstration papers, Mumbai, pp 277–288
4. Chien CC, You-De T (2011) Quality evaluation of product reviews using an information quality framework. Decis Supp Syst 50:755–768
5. Ruiz M, Srinivasan P (1999) Hierarchical neural networks for text categorization. In: Presented at the ACM SIGIR conference
6. Medhat W, Hassan A, Korashy H (2014) Sentiment analysis algorithms and applications: a survey. Ain Shams Eng J 5:1093–1113
7. Singh VK, Piryani R, Uddin A, Waila P (2013) Sentiment analysis of movie reviews: a new feature-based heuristic for aspect-level sentiment classification. In: International multi-conference on automation, computing, communication, control and compressed sensing (iMac4s), 22–23 Mar 2013
8. Nguyen TH, Shirai K (2015) Topic modeling based sentiment analysis on social media for stock market prediction. In: Proceedings of the 7th international joint conference on natural language processing. Beijing, China, July 26–31, pp 1354–1364
9. Oliveira N, Cortez P, Areal N (2017) The impact of microblogging data for stock market prediction: using twitter to predict returns, volatility, trading volume and survey sentiment indices. Exp Syst Appl 73:125–144
10. Moreo A, Romero M, Castro JL, Zurita JM (2012) Lexicon-based comments-oriented news sentiment analyzer system. Exp Syst Appl 39:9166–9180
11. Sun S, Luo C, Chen J (2016) A review of natural language processing techniques for opinion mining systems. Inf Fusion. https://doi.org/10.1016/j.inffus.2016.10.004
12. Piryani R, Madhavi D, Singh VK (2017) Analytical mapping of opinion mining and sentiment analysis research during 2000–2015. Inf Process Manag 53:122–150
13. Salas MP, Valencia R, Ruiz A, Colomo R (2016) Feature-based opinion mining in financial news: an ontology-driven approach. J Inf Sci, 1–20. https://doi.org/10.1177/0165551516645528
14. Araque O, Corcuera-Platas I, Fernando Sánchez-Rada J, Iglesias CA (2017) Enhancing deep learning sentiment analysis with ensemble techniques in social applications. Exp Syst Appl 77:236–246

Sentiment Analysis and Prediction of Election Results 2018

Urvashi Sharma, Rattan K. Datta and Kavita Pabreja

Abstract Social media is becoming very popular throughout the planet. People feel very comfortable in expressing their views freely. It was decided to take the help from this fact for sentiment analysis of the discussion and their free comments with regard to the forthcoming assembly elections in India. The social media data is analyzed by applying different mining techniques to predict the possible outcomes of the assembly elections and watching the positions of various political parties in India. An attempt has been made to analyze the user behavior on social media in order to find out the prediction of political parties' position in election.

Keywords Sentiment analysis · Social media · Big data analytics · R language · Data mining · Election prediction

1 Introduction

There are various definitions of big data as per different sources. According to Gartner's definition, "Big Data are information assets with Volumes, Velocities and/or Variety requiring innovative forms of information processing for enhanced insight discovery, decision—making and process automation". Big data is a collection of datasets so large and complex that it becomes difficult to process using on-hand database management tools, as per Wikipedia, 2014. Big data refers to datasets whose size is beyond the ability of typical database software tools to capture, store, manage, and analyze, as mentioned by The Mc Kinsey Global Institute, 2012. It is a

U. Sharma (✉)
IPS Academy, Indore, India
e-mail: tamnnas143@gmail.com

R. K. Datta
Mohyal Educational and Research Institute of Technology, New Delhi, India
e-mail: rkdatta_in@yahoo.com

K. Pabreja
Maharaja Surajmal Institute, GGSIPU, New Delhi, India
e-mail: kavita_pabreja@rediffmail.com

R. K. Shukla et al. (eds.), *Social Networking and Computational Intelligence*, Lecture Notes in Networks and Systems 100, https://doi.org/10.1007/978-981-15-2071-6_61

term that describes the large volume of data—both structured and unstructured—that inundates a business on a day-to-day basis, as per definition given by SAS Company. "Big Data" exceeds the reach of commonly used hardware environments and software tools to capture, manage, and process it with in a tolerable elapsed time for its user populations", according to Tera data Magazine article, 2011.

2 Big Data History and Current Considerations

The presence of "Big Data" or this massive amounts of increasing data, gathering and storing large amounts of information for eventual analysis is ages old and it offers both an opportunity and a challenge to researchers. The term big data is used to describe the growth and availability of huge amount of data. Big data may appear as a new discipline; it has been developing for years. Amazing, 90% of the data in the world today has been created only in the last two years. The concept became popular in the early 2000s when industry, analyst Doug Laney expressed the definition of big data as the three Versus. Now a day's big data is defined using five Vs as shown in Fig. 1.

Volume: Many factors contribute to the increase in data volume. High data volume imposes distinct data storage and processing demands, as well as additional data preparation, creation, and management process. Variety of sources includes a business tractions, social media such as Facebook, Twitter, and LinkedIn. Stock exchange and information from sensors like GPS, smart meter and telemetric scale, or machine to machine data online transactions such as point-of-scale and banking, scientific and research experiments, such as large Hadrons collider the past, storing it would have been a problem. But now a new technologies (such as Hadoop) have eased the burden.
Velocity: In big data environments, data is streaming in at unprecedented speed and must be dealt with in a timely manner. Radio-frequency identification tags (RFID), sensor, and smart metering are driving the need to deal with torrents of data in near

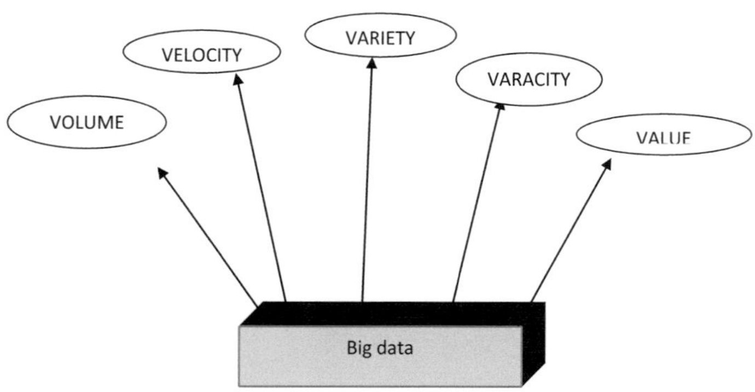

Fig. 1 Figure of big data as five verses

real time. Data velocity is put into perspective when considering that the following data volume can easily be generated in a given minute 350,000 tweets 300 h may be a more of video footage uploaded to YouTube.
Variety: Nowadays data has come in all types of variety and in multiple formats that need to be supported by big data solutions. Data variety brings challenges for enterprises in terms of data integration, transformation, processing, and storage, which includes structured data in the form of video, audio, stock ticker, data and financial transactions, unstructured data, semi-structured data, e-mail, semi-structured data, from structured numeric data, unstructured data text documents, images, video, audio, managing, merging and governing different varieties of data is something many organizations still grapple.
Veracity: In addition to the increasing velocities and varieties of data, veracity refers to the quality or fidelity of data. A big data environment needs to be assessed for quality that can lead to data processing activities to resolve invalid data and remove noise. In relation to veracity, data can be part of the signal or noise of dataset.
Value: Value is defined as the usefulness of data for an enterprise. It includes how we can use this big data for spanning and enhancing the business and living style. Value is also dependent on how long data processing takes. Value and time are inversely related. The longer it takes for data to be turned into meaningful information, the lesser is the value.

The data being created and stored on global scale is almost in-conceivable and it keeps on growing but only a very small percent of data is actually analyzed and used that show the big data has tremendous potential but is being used at a very small label.

Big data is important because better the data, better analysis, and better prediction when we use big data with high-power analytical tools you can achieve a lot. For example, big data term itself is evolving from terabytes to zettabytes.

Big data is very huge heterogeneous and complex data. There are heterogeneity, scale, timeliness, complexity, and privacy problems with big data. There are various resources for big data, viz. audio, videos, and post in social media, various database tables, e-mail attachments, etc. People use Twitter in diverse form and store 250 Million tweets per day. Big data has many opportunities like financial services, healthcare, retail, Web/social, manufacturing and government.

3 Big Data Analytics and Overview

Big Data Analytics is the process of collecting, organizing, analyzing large datasets to discover different patterns and other useful information. Big data analytics is a set of technologies and techniques that require new forms of integration to disclose large hidden values from large datasets that are different from the usual ones, more complex, and of a large enormous scale.

Big data analytics has the following types: **Prescriptive**: This type of analytics helps to decide what actions should be taken. It is very valuable but not used largely. It focuses on the answer to specific questions like, hospital management, diagnosis of cancer patients, diabetes patients that determine where to focus treatment. **Predictive**: This type of analytics helps to predict future or what might happen. For example, some companies use predictive analytics to take decision for sales, marketing, production, etc. **Diagnostic**: In this type, one looks at past and analyzes the situation what happened in past and why it happened and how we can overcome this situation, for example, weather prediction, customer behavioral analysis, etc. **Descriptive**: It describes what is happening currently and predicts near future, for example, market analysis, complaints behavioral analysis, etc.

Social media is an Internet-based communication tool that empowers people to share information. To understand better the term social media, social indicates associating with people and spending time in order to develop their relationships whereas media indicates tool for communication such as Internet, TV, radio, newspaper, and so on. Social media is stated as an electronic platform for socializing people. Social media sites provide data which are vast, noisy, distributed, and dynamic. Hence, data mining techniques provide researchers the tools needed to analyze such large, complex, and frequently changing social media data. Understanding how users behave when they connect to these sites is important for a number of reasons.

4 Sentiment Analysis

Sentiment analysis is the process of determining whether a piece of writing is positive, negative, or neutral, as shown in Fig. 2. It is also known as opinion mining, deriving the opinion or attitude of a speaker. The sentiment analysis is a complex process that involves five different steps to analyze sentiment data. These steps are shown in Fig. 3.

5 Analysis of Data Mining Techniques for Social Network—A Review

Data mining that involves pattern recognition, mathematical, and statistical techniques to search data warehouses and help the analyst in recognizing significant trends, facts, relationships, and anomalies [1, 2]. The techniques (specially related to machine learning) in order to gather, store, process, and analyze this vast amount of data are very important to understand [3].

Data mining of social networks can be done using the graph mining methods such as classification/topologies, prediction, efficiency, pattern detection, measurement and metrics, modeling, evolution and structure, data processing, and communities. To

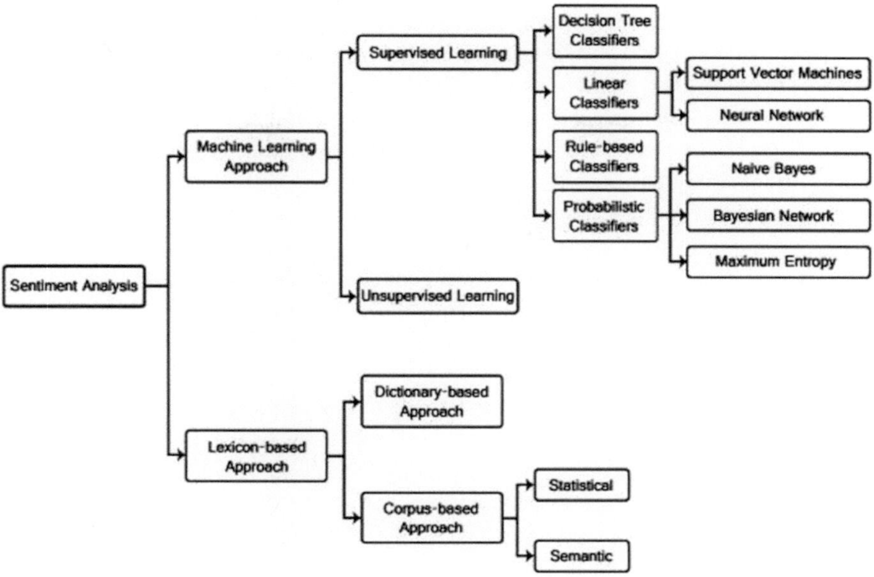

Fig. 2 Process of sentiment analysis

Fig. 3 Steps of sentiment analysis

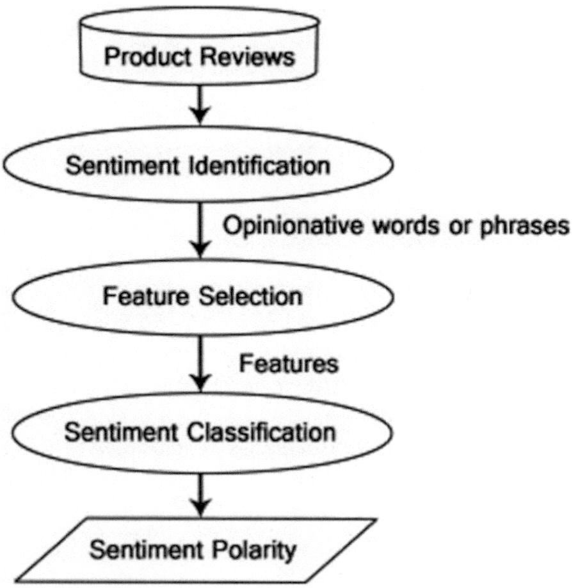

extract the information represented in graphs, we need to define metrics that describe the global structure of graphs, find the community structure of the network, and define metrics that describe the patterns of local interaction in the graphs, develop efficient algorithms for mining data on networks, and understand the model of generation of graphs [4]. The social networks contain millions of unprocessed raw data. By analyzing this data, new knowledge can be gained. Since this data is dynamic and unstructured, traditional data mining techniques will not be appropriate. Web data mining is an interesting field with vast amount of applications. The growth of online social networks has significantly increased data content availability because profile holders become more active producers and distributors of such data [4].

Web mining is the process of analyzing and discovering patterns on Web data. It can be defined as searching data automatically from various online resources. Web mining can be categorized into three aspects, Web usage mining, Web content mining, and Web structure mining [5]. Social media sites provide data which are vast, noisy, distributed, and dynamic. Hence, data mining techniques provide researchers the tools needed to analyze such large, complex, and frequently changing social media data. Some representative research issues [6] in mining social networking sites using data mining techniques are as follows:

1. Influence Propagation
2. Community or Group Detection
3. Expert Finding
4. Link Prediction
5. Recommender Systems
6. Predicting Trust and Distrust among Individuals
7. Behavior and Mood Analysis
8. Opinion Mining.

Some other important data mining applications related to OSNs include information and activity diffusion, topic detection and monitoring, marketing research for businesses, data management and criminal detection. Online social networks (OSNs) have fetched the interest of researchers for their analysis of usage as well as detection of abnormal activities. Anomalous activities in social networks represent unusual and illegal activities exhibiting different behaviors than others present in the same structure. Data mining approaches are used to detect anomalies. A special reference is made to the analysis of social network-centric anomaly detection techniques which are broadly classified as behavior based, structure based, and spectral based [7].

The social media has become main communication way to connect with our family, friend, and colleagues. People are sharing all information in many forms such as text, audio as well as video on the social networking site to share their feeling. The average global Internet user spends two and a half hours daily on social media. This way social media users produce huge amount of data which cannot be handled with traditional data management techniques [8]. There are many issues and challenge in big data and social media such as: analyses of social media data, technology challenge in big data, management challenges, and security issues.

In recent years, social media has become ubiquitous and important for social networking and content sharing. The authors showed that a simple model built from the rate at which tweets are created about particular topics can outperform market-based predictors [9]. The identification of different classes of user behavior has the potential to improve, for instance, recommendation systems for advertisements in online social networks [10].

Social media allows customers and prospects to communicate directly to your brand representative or about your brand with their friends. Study regarding the online activities of 236 social media users, by identifying different types of users, a segmentation of these users and a linear model to examine how different predictors related to social networking sites have a positive impact on the respondents' perception of online advertisements, has been done to discover how to engage with different types of audiences in order to maximize the effect of the online marketing strategy [11]. The success of a social networking site is directly associated with the quality of content users share. Given that users participate in multiple social networks, it is expected that a user may share the same content across multiple sites. If content is easily replicated across sites, then one can detect rising content from one social networking site and implant it into another site [12]. Using profile browsing events, latent interaction graphs have been constructed as a more accurate representation of meaningful peer interactions [13]. The mined information from social platforms can significantly impact business strategy of any business enterprise [14]. In addition, installing and running software often costs less than hiring and training personnel. Computers are also less prone to errors than human investigators, especially those who work long hours [15]. Social media is a social platform that is made up of people who are connected by several interdependencies. Social media has changed the nature of information in terms of availability, importance, and volume. Through social media like Twitter and Facebook, participants reveal personal information that has real values, as they can be extracted and mined to improve decision making. A small-sized data was extracted from Twitter and analyzes using a data mining classification algorithm known as sentiment analysis (also referred to as opinion extraction, opinion mining, sentiment mining, and subjective analysis). Additionally, the result of the analysis was used to predict the outcome of the aforementioned election [16]. Sentiment analysis is a new area in text analytics where it focuses on the analysis and understanding of the emotions from the text patterns. This new form of analysis has been widely adopted in customer relation management especially in the context of complaint management. With increasing level of interest in this technology, more and more companies are adopting it and using it to champion their marketing efforts. However, sentiment analysis using Twitter has remained extremely difficult to manage due to the sampling bias. Various aspects using sentiment analysis to predict outcomes as well as the potential pitfalls in the estimation due to the anonymous nature of the Internet [17] has been analyzed. Social media has been used profoundly all over the world for analysis of political campaigns prior to election process and post-election as well. Many researchers have been analyzing the tweets by citizens of a nation on Twitter which is a micro-blogging Web site where users read and write millions of tweets on a variety of topics on daily basis [18].

6 Case Study: Sentiment Analysis of Assembly Election Results

6.1 Methodology

To predict the assembly elections in India, we have done data collection, preprocessing, and data analysis.

6.2 Data Collection and Preprocessing

Social media has been used for watch user behavior about the assembly elections in India. The sentiments of user changing based on position of political parties and their candidates. The data collection step is the initial phase in the research, where data is collected from Facebook. We have collected and analyzed more than 5000 Facebook comments of user on political parties and their candidates. The user's comment on Facebook page few days before to the assembly election in India has been collected. We have collected all comments that contained names of popular political parties and candidates contesting for elections. These are BJP, Congress, Rahul Gandhi, and Narendra Modi.

The Facebook comments have useful information related to political parties besides special characters, punctuation marks, and emojis. The data collected hence has been cleaned so as to remove punctuation symbols, special characters. All comments have been converted to lowercase and finally a word corpus has been generated.

6.3 Data Analysis

We have used the R language which is a powerful language used widely for data analysis and statistical computing. R language was developed in early 90 s. R has enough provisions to implement machine learning algorithms in a fast and simple manner. The collected data was loaded into the R package. This way the algorithm was able to identify and classify the sentiments of each of the 5000 comments into the positive, negative, or neutral sentiment per candidates. We have generated word cloud using R corresponding to four different sets of comments focusing on BJP, Congress, Modi, and Rahul.

The analysis of frequent terms and their associations for all four sets of comments has been accomplished. This has been carried out to understand which terms are more frequently used while discussing a particular political party.

The approach followed here is to count the positive and negative words in each Facebook comment and assign a sentiment score. This way, we can ascertain how positive or negative a FB comment is. Nevertheless, there are multiple ways to calculate such scores; here is one formula to perform such calculations.

1. Score = Number of positive words − Number of negative words
2. If Score > 0, means that the FB comment has 'positive sentiment'
3. If Score < 0, means that FB comment has 'negative sentiment'
4. If Score = 0, means that the FB comment has 'neutral sentiment'.

To find out the list of positive and negative words, an opinion lexicon (English language) can be utilized.

The code below showcases how sentiment analysis is written and executed.

```
score. sentiment = function(sentences, passwords, neg.words, .progress = 'none')
{
# Parameters
# sentences: vector of text to score
# pos.words: vector of words of positive sentiment
# neg.words: vector of words of negative sentiment
# .progress: passed to lapply() to control of progress bar
# create a simple array of scores with laply
scores = laply(sentences,
function(sentence, pos.words, neg.words)
{
# remove punctuation
sentence = gsub("[[:punct:]]", "", sentence)
# remove control characters
sentence = gsub("[[:cntrl:]]", "", sentence)
# remove digits?
sentence = gsub('\\d + ', '', sentence)
# define error handling function when trying tolower
tryTolower = function(x)
{
# create missing value
y = NA
# tryCatch error
try_error = tryCatch(tolower(x), error = function(e) e)
# if not an error
if (!inherits(try_error, "error"))
y = tolower(x)

# result
return(y)
}
# use tryTolower with sapply
sentence = sapply(sentence, tryTolower)
```

```
# split sentence into words with str_split (stringr package)
word.list = str_split(sentence, "\\s + ")
words = unlist(word.list)
# compare words to the dictionaries of positive & negative terms
pos.matches = match(words, pos.words)
neg.matches = match(words, neg.words)
# get the position of the matched term or NA
# we just want a TRUE/FALSE
pos.matches = !is.na(pos.matches)
neg.matches = !is.na(neg.matches)
# final score
score = sum(pos.matches) - sum(neg.matches)
return(score)
}, pos.words, neg.words, .progress = .progress)
# data frame with scores for each sentence
scores.df = data.frame(text = sentences, score = scores)
return(scores.df)
}
```

Sentiment analysis is performed on 5000 comments to understand the emotions of public toward all four different sets of comments and tried to relate the extracted emotions prior to elections date with the polling results.

7 Experimental Result

All comments in Facebook mentioning "**Rahul**", "**Modi**", "**BJP**", and "**Congress**" have been analyzed from the following perspective:

- Word cloud generation;
- Document matrix of frequent terms;
- Sentiment analysis.

The word cloud has been generated corresponding to Facebook data mentioning "Rahul" and it has been found from the size of word that "Rahul" has been discussed less in comparison with the keyword "MP elections" as shown in Fig. 4.

The generation of bar plot showing frequent terms is shown in Fig. 5. The most talked words have been plotted as shown for "Rahul", "Modi", "BJP", and "congress" comments, respectively.

Finally, the sentiment analysis of all four sets of comments has been done to understand the emotions of public prior to day of assembly elections. Emotions of anger, anticipation, disgust, fear, joy, sadness, and surprise have been extracted using get_nrc_sentiment() function of "R" software and other preprocessing of comments. From the sentiment analysis of "BJP" tweets, it is very clear that there is enough anger and anticipation for "BJP" party and the same was visible in the results of elections

Fig. 4 Word cloud

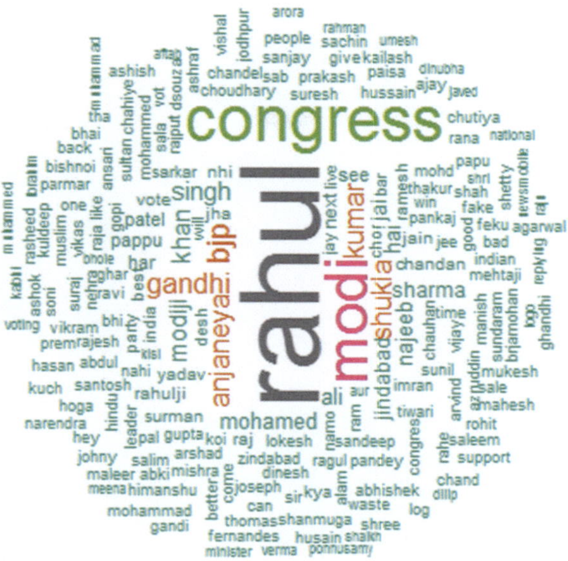

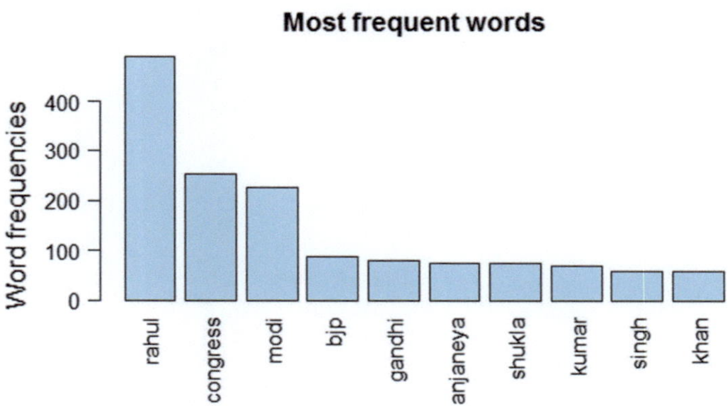

Fig. 5 Frequently talked terms

where “BJP” suffered a loss. As was also apparent from the results that “Congress” had won and made the government, same is clearly observed in sentiments of people. There is too much of joy and anticipation for “Congress” in the comments. These are shown in Figs. 6 and 7, respectively.

The results of MP election are given in Table 1 that verifies the sentiments of public.

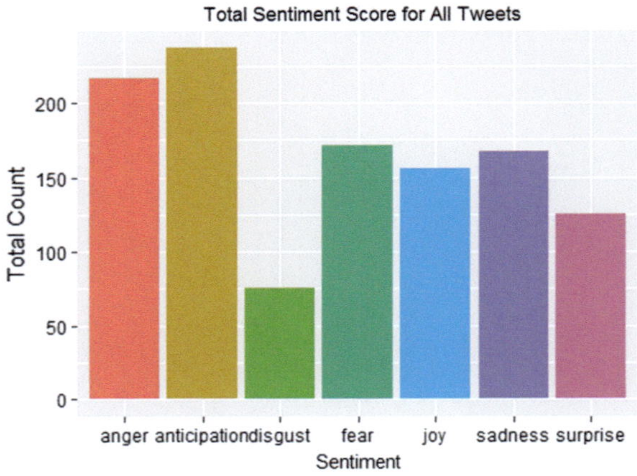

Fig. 6 Sentiment analysis of tweets mentioning "BJP"

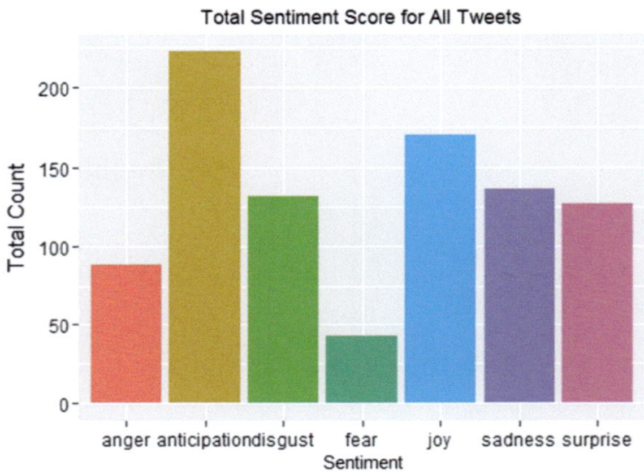

Fig. 7 Sentiment analysis of tweets mentioning "Congress"

Table 1 Election results

Party name	No. of seats won
Others and BSP	7
BJP	109
Congress	114

8 Conclusion

This paper has described a refinement of the original research aims and describes the achievements of the research. The above analysis clearly indicates that the rising uses of social media will certainly give indication of a completely new era in the field of research on human behavior. We want to analyze the user behavior on social media and find out the outcome of our study on prediction of political parties' position in election. Social media is a powerful source of public opinion about elections. To derive the public opinion from Facebook by using R language software on political parties' position in election, we have done sentiment analysis of emotions of people which shows there is consistent correlation between social media results and the traditional declared results. Social media has accomplished to capture emotions of the people prior to elections and has ability to forecast the electoral results on behalf of public behavior.

References

1. Chamatkar AJ, Butey PK (2014) Importance of data mining with different types of data applications and challenging areas. Int J Eng Res Appl 4(5) (Version 3):38–41. ISSN: 2248-9622, www.ijera.com
2. Verma JP, Agrawal S, Patel B, Patel A (2016) Big data analytics: challenges and applications for text, audio, video, and social media data. Int J Soft Comput Artif Intell Appl (IJSCAI) 5(1)
3. Ali A, Qadir J, Rasool R, Sathiaseelan A, Zwitter A, Crowcroft J (2016) Big data for development: applications and techniques. Big Data Anal 1:2. https://doi.org/10.1186/s41044-016-0002-4
4. Vedanayaki M (2014) A study of data mining and social network analysis. Ind J Sci Technol 7(S7):185–187
5. Fernando SGS, Perera SN (2014) Empirical analysis of data mining techniques for social network websites. Compusoft 3(2):582
6. Nandi G, Das A (2013) A survey on using data mining techniques for online social network analysis. Int J Comput Sci (IJCSI) 10(6):162. ISSN (Print): 1694-0814, ISSN (Online): 1694-0784, www.IJCSI.org
7. Kaur R, Singh S (2016) A survey of data mining and social network analysis based anomaly detection techniques. Egypt Inf J 17(2):199–216
8. Kumari S (2016) Impact of big data and social media on society. Glob J Res Anal 5(3). ISSN No. 2277-8160
9. Asur S, Huberman BA (2010) Predicting the future with social media. arXiv:1003.5699v1 [cs.CY], 29 Mar 2010
10. Maia M, Almeida J, Almeida V (2008) Identifying user behavior in online social networks. Proceedings of the 1st workshop on social network systems. ACM, pp 1–6
11. Vinerean S, Cetina I, Dumitrescu L, Tichindelean M (2013) The effects of social media marketing on online consumer behavior. Int J Bus Manage 8(14). ISSN 1833-3850. E-ISSN 1833-8119. Published by Canadian Center of Science and Education
12. Benevenutoy F, Rodriguesy T, Cha M, Almeiday V (2009) Characterizing user behavior in online social networks. Proceedings of the 9th ACM SIGCOMM conference on Internet measurement. ACM, 2009
13. Jiang J, Wilson C, Wang X, Huang P, Sha W, Dai Y, Zhao BY (2010) Understanding latent interactions in online social networks. In: IMC'10, Melbourne, Australia, 1–3 Nov 2010
14. Pippal S, Batra L, Krishna A, Gupta H, Arora K (2014) Data mining in social networking sites: a social media mining approach to generate effective business strategies. Int J Innov Adv Comput Sci (IJIACS) 3(2):22–27. ISSN 2347-8616
15. Chen H, Chung W, Xu JJ, Wang G, Qin Y, Chau M (2004) Crime data mining: a general framework and some examples. Computer 4:50–56
16. Umar KI, Chiroma F (2016) Data mining for social media analysis: using twitter to predict the 2016 US presidential election. Int J Sci Eng Res 7(10):1972–1980. ISSN 2229-5518
17. Choy M, Cheong MLF, Laik MN, Shung KPChoy M, Cheong MLF, Laik MN, Shung KP (2011) A sentiment analysis of Singapore presidential election 2011 using Twitter data with census correction. IJRET Int J Res Eng Technol, eISSN: 2319-1163, pISSN: 2321-7308
18. Narwal N, Pabreja K (2018) Social media analytics. CSI Commun 42(5&6):10–13

Toward the Semantic Data Inter-processing: A Semantic Web Approach and Its Services

Anand Kumar and B. P. Singh

Abstract Semantic data and information make the applications and its services intelligent by providing reasoning mechanism and inter-processing capability of data and information. Semantic Web research communities have developed standard and more complex use-cases applications that need semantic data inter-processing for better utilization of data and information. Currently, Semantic Web is frequently used to develop the innovative applications in various areas with unbelievable benefits and services, and people quickly agree to adopt the upgraded technology. Hence, this paper will broadly explain the Semantic Web and its supporting technology with their applications and benefits. We will also introduce here about the Web services that will be utilized for the development of innovative applications in various areas which will relay the innovative services to applications developers and users.

Keywords Semantic Web · Web services · Knowledge construction · Semantic data inter-processing · Web-based applications

1 Introduction

The researches of Semantic Web are spread at the convergence of the number of various streams. These research streams are artificial intelligence, information theory, distributed systems, databases, philosophy and many other areas of researches. Many ancestries have done the researches in various streams with different domains for distinct problems. Most of the researchers have contributed to software agents and Semantic Web technology to construct the automated production system [1] and resolve the problems of various domains. Some exciting features of the Semantic Web

A. Kumar (✉)
Department of Computer and Information Sciences, J. R. Handicapped University, Chitrakoot, Uttar Pradesh, India
e-mail: anand_smsvns@yahoo.co.in

B. P. Singh
Dayalbagh Educational Institute, Agra, Uttar Pradesh, India
e-mail: bp.sing76@gmail.com

R. K. Shukla et al. (eds.), *Social Networking and Computational Intelligence*, Lecture Notes in Networks and Systems 100, https://doi.org/10.1007/978-981-15-2071-6_62

are capable to increase the inter-process capabilities of information, enhance the services and also resolve the limitations and issues of present information management systems. Semantic Web technology improves the elementary facilities and services of information processing and allows the information management systems to make the interoperability with other Semantic Web applications to enhance the facility and functionality of information management systems. Therefore, Semantic Web has built up a link between the existing contents of Web and semantic descriptions (i.e., metadata) of knowledge. These semantic descriptions can be utilized to make the Semantic Web technologies [2] based new information management systems.

This paper is organized and elaborated as following sections and points. We are describing about the Semantic Web and its elementary technologies in Sect. 2 and also depicting the layered architecture of Semantic Web in this section. We demonstrate the vision and benefits of Semantic Web in Sects. 3 and 4, respectively. Sections 5 and 6 are focusing on the applications and services of Semantic Web orderly. Finally, we briefly describe the conclusions of this manuscript in Sect. 7.

2 Semantic Web

The Semantic Web term and idea were invented by Tim Berners-Lee with his colleagues [3] in year 2001, which is raising the semantical representation of Web contents that are machines interpretable and understandable as well as a Web content readable by human being. Tim Berners-Lee and others mentioned that "The Semantic Web is not a separate Web but it is an extension of the current one, in which information is given in a well-defined meaning, enabling computers and people to work in better cooperation" [3–5]. The World Wide Web Consortium (W3C) has depicted a well-defined layered architecture of Semantic Web [6] as shown in Fig. 1.

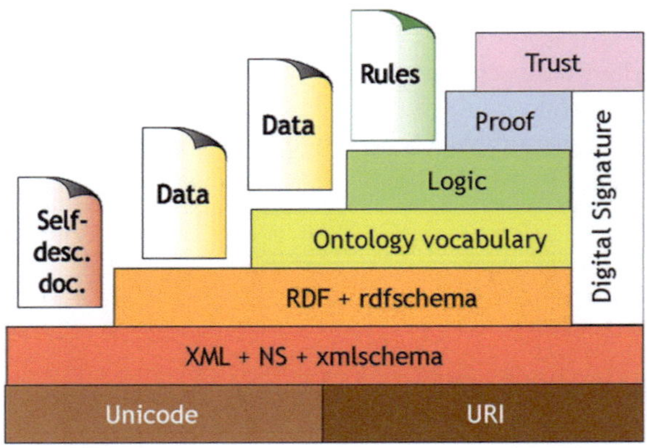

Fig. 1 Layered architecture of Semantic Web

In Fig. 1, the lowest layer that includes Unicode and Uniform Resource Identifier (URI) provides standardized international characters set for identifying and defining the semantical meanings of resources in Semantic Web. The next layer from the lowest layer which includes eXtensible Markup Language (XML) with namespace and schema definitions provides the facilities to integrate the semantic definitions of resources with other XML-based standards. The Resource Description Framework (RDF) and Resource Description Framework Schema (RDFS) are used to construct the statements about resources, i.e., objects with significant URI's and also define with URI's referred vocabularies. The ontology layer played the key role in Semantic Web that enables evolution of vocabularies with defining the relations between the different concepts that make enrich for reasoning mechanism and inferencing of information. Digital signature layer as shown in figure vertically will be capable for detecting the alterations of documents. Remaining three top layers logic, proof and trust are under in research, and currently, these things are used for simple application demonstrations. The writing of rules is enabled by logic layer, while the execution of these rules and its evaluation is done together with the trust layer mechanism for applications.

At present, World Wide Web (WWW) is one of the major sources that has huge amount of information and also provides uncounted services. It is fully decentralized, dynamic and too much rich about the information and services. It is also growing so fast and accelerating the pace of knowledge. But the current Web becomes untouched about full potential of utilizing the information because information and its meaning are articulated to utilize by human being only. Therefore, the perception of the Semantic Web technology as mentioned with its invention is that this technology will use to define and represent the web contents in such a manner which will be machine-interpretable and enable the semantics of information for automation, integration and reuse of data within the applications and among the different applications.

3 Semantic Web Vision

As mentioned, vision of Semantic Web by its inventor, contents of the Web should be semantically available for processible and understandable by human users and as well as by machines. This vision makes the user's life easy with exciting knowledge world and capable to create the new and trustful information over the World Wide Web. One of the major visions is to make the available Web content for intelligent knowledge processing and for utilizing it with full potential. Computer will theoretically be able to solve the problems that are out of reach today. But it will come possible in future. The representation of Web content is expressed in semantic languages to make accessible and processible by computer. The computer can provide the facility of automated task based on millions of small specialized reasoning services and accessible information.

4 Benefits of Semantic Web

At present, people are focusing the intention to develop the Web applications based on Semantic Web technology and will try to inherit the benefits of this technology in these applications. This emerging technology owns various benefits on current Web and other applications that are based on Semantic Web. We are looking these benefits when we are going to develop Semantic Web-based applications that are discussed as follows.

4.1 Knowledge Integration

Currently, a huge amount of information and knowledge resides on the Web, and it is also increasing day by day in which some information are redundant. People cannot utilize this information potentially because this information is not easy to integrate without its semantical representation. This information and knowledge stand and fall on knowledge integration. The wresting of valuable information from overcrowded and growing data world begins and that will be ended with knowledge integration. Semantic Web is used to express the information and knowledge of various resources in integrated phenomenon.

4.2 Knowledge Construction

The construction of knowledge will manage the information in knowledge about the resources in systematic manner and also enable the semantics of these information and knowledge. The semantical construction will make to utilize the information and knowledge potentially and also capable to rectify the redundant information. Semantic Web provides the facility to construct the higher forms of information with highly machine interpretability and operability. There are six ways, which is used to construct the hierarchy of knowledge.

4.2.1 Commonality

In Semantic Web, the knowledge is constructed through the use of synonyms. Various data items declared as equivalent in structural ontology to remove to complexity and reduce the redundancy of knowledge.

4.2.2 Inheritance

Inheritance of knowledge is declaring that a data instance is a form of another data instance, but they are not exactly equivalent. One data instance is a subset of other data instances in the ontology declaration by using the term 'subClassOf.'

4.2.3 Restrictions

Restrictions are used to define a term relatively to other terms of limits. This adds the useful vocabulary by using a new term that is a condition of an existing term. For example, the access of file of university must have the access constraint such as 'canRead' with user type.

4.2.4 Properties

The property of properties describes the wealth of knowledge regarding the data element with additional context. It is used to define the metadata on the resources, which make rich and wealthy knowledge.

4.2.5 Collections

It forms the concepts by combining data terms together. The concept is referred as a particular data term that may be related to several other data terms simultaneously.

4.2.6 Rules

Rules are deployed to extract the knowledge from the given conditions. It authorized the user to access the knowledge from different resources with various authorization levels such as insert, delete, read and write.

4.3 Knowledge Searching

The major goal of any knowledgebase or database is getting the useful answer on query. Presently, there are two methods: keyword matching and relational database queries, which are widely used to search the answer on query. Keyword matching method provides some useful answer when the data pattern of query is matching otherwise unusual. The relational database query follows the standard language to get the answer on query. Semantic Web supports wide range of query languages that enable powerful and flexible searching of knowledge with inference and resolution.

4.4 Knowledge Inferencing

The Semantic Web induces the meaningful content to computers, which enables the computer to infer new knowledge from known knowledge. For example, a given knowledgebase contains organizational structure of university and their relationship. If School for Information Science and Technology is affiliated to Babasaheb Bhimrao Ambedkar University and Department of Computer Science is affiliated to School for Information Science and Technology, then the system could infer that Department of Computer Science is affiliated to Babasaheb Bhimrao Ambedkar University.

4.5 Knowledge Perspective

The Semantic Web exposes the users to represent the knowledge in their perspective. The knowledge in Semantic Web is aligned as per users need. This benefit of Semantic Web is accomplished by using various technologies such as RDF, RDFS and OWL, which enable user to construct the knowledge for their specific needs. The users can start to construct the knowledge from existing ontology and knowledgebase or only one of them.

5 Applications of Semantic Web

The effort of Semantic Web research makes the available contents on Web for intelligent knowledge processing by computer and also uses for Semantic Web-enabled applications. It is not the separate from the WWW. This technology provides the extension of current Web in which information is constructed in such a manner that will infer the well-defined meanings. These meanings are enabling the computers and people for better working with coordination and cooperation [3]. It focuses on typical and narrative procedures of various significant disciplines of Computer Science stream which are including knowledge management, artificial intelligence, databases, Internet technology, software agents, e-Commerce, etc. Number of methods and tools are developed and available for the construction and integration of knowledge for Semantic Web-based applications, which comes under the Semantic Technologies. The Semantic Technologies are currently being in research for the deployment and integration in various disciplines like Computer Science and including with ambient intelligence, software engineering, cognitive systems, corporate intranets, knowledge management, E-governance, [7] bioinformatics, etc.

6 Semantic Web Services

Service-oriented approach is used to relay the services in application. Hence, this approach is especially used to develop such kind of software that is capable to combine the reusable and distributed components [8]. In the way to fulfill the vision of Semantic Web, researchers are proposed Semantic Web services (SWS) and continuously added day by day. The intelligent software agents can exploit the semantic descriptions to resolve the complex tasks by machines rather than humans [9]. SWS is well connected among Semantic Web, agents and Web-concerned service technologies. Therefore, combinations of Web services and technologies available for Semantic Web are potentially used to discovery the innovative services. These services can employ for configuration and implementation of web-based services and agent-based systems. In the following subsection, we are focusing some major Web services and also describing about SWS.

6.1 OWL-S

A group of DAML Services were developed OWL-based Web service ontology (OWL-S) [10], and these services are used for the Semantic Web services description, which are articulated in Web Ontology Language (OWL). The OWL-S group published the OWL-S version 1.0 in year 2004 and designates an upper ontology for describing Web services semantically along with three main significant features: Service Profile, Service Model and Service Grounding.

6.1.1 Service Profile

The Service Profile of OWL-S offers the essential functional description of services that are used for advertising purpose. This functional description stipulates the high-level representation of services in such a manner that is appropriate for software agents and also checked about services that are fit for required purposes or not. The services are defined in its functional parameters such as inputs, outputs, preconditions and effects as shown in Fig. 2.

Generally, inputs and outputs stipulate the different parameters semantically, and services handled these parameters. Preconditions provide the logical expressions with conditions that are essential for the service to be executed successfully. Finally, effects denoted to results in OWL-S, which specify the changes after successful execution of the services. The Service Profile also includes the capability to capture the information from various resources. This information is related to classifications with respect to reference taxonomies, service name and textual descriptions.

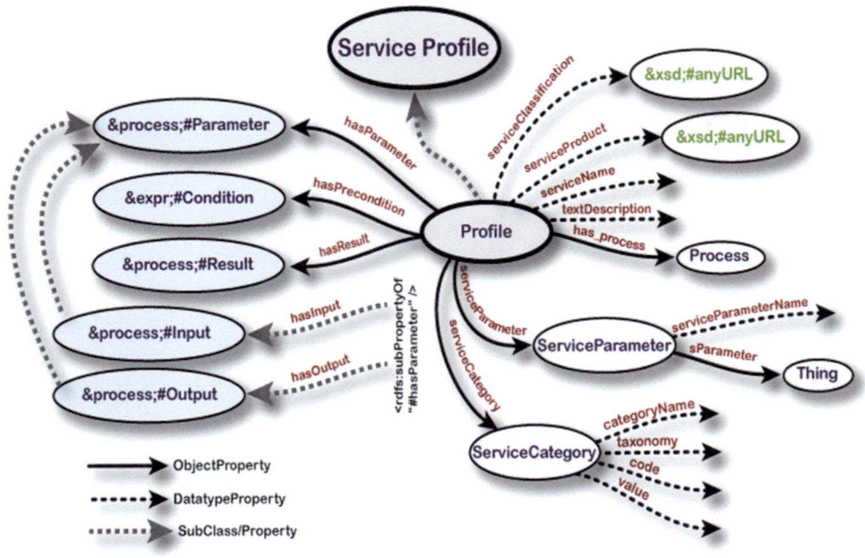

Fig. 2 The OWL-S ontology (figure adapted from Martin et al. [10])

6.1.2 Service Model

The Service Model specifies the semantic about content of requests, replies, conditions under which certain results hold with the description of how service invoked by clients. In order to specify that how the client can establish the interaction with a service, OWL-S provides services that process a collection of inputs, outputs, preconditions and effects, and a process model is specifying the efficient way under which a client can interact with services. OWL-S supports initial level of reasoning as OWL-DL reasoners, which is not adequately expressive or does not appropriate for expressing the definitions of preconditions and effects. Therefore, OWL-S provides the facility for description of preconditions and effects by the literals of XML to accept arbitrary languages, i.e., SWRL [11].

6.1.3 Service Grounding

The Service Grounding offers necessary facts that are required for initiating the services. It used the protocol, a message format, their serialization, transport protocol and address of the end point [12]. There is no language that is predefined and to be used for grounding. But, for its broad implementation, the reference grounding enactment is facilitated for WSDL.

6.2 WSMO

WSMO [13, 14] is an ontology which intends to describe entirely applicable features for the fractional or comprehensive automation of discovery, configuration, selection, intervention, implementation and monitoring of Web services. WSMO is describing those Web services which can be built based on Web Service Modeling Framework (WSMF) [15]. There are four top-level foundations of WSMF which are ontologies, Web services, goals and mediators that can be defined by WSMO. Ontologies deliver the prescribed semantics about the terminology which is utilized within all other components of WSMO. It should be necessary that all descriptions of resources and all about data, which are interchanged throughout the services, must be semantically defined based on ontologies. Web services are computational units which represent some valuable description for a specified domain. Goals are the representation of user's need for certain functionality in perspective of clients. At last, mediators handle the interoperability complications among pair of WSMO elements. A principle about WSMO is the significance of intervention to decrease the coupling and pact with the heterogeneity to characterize the Web.

6.3 WSDL-S

WSDL-S [16] has been invented in year 2005 in LSDIS Laboratory at University of Georgia and IBM [17]. It is a trivial method to correlate semantic annotations to the services of Web. The ability to extension of elements and characteristics is approached by WSDL specification [18]. These elements and characteristics are the key innovation of WSDL-S. The annotations based on semantics are the conceptual models in form of URIs, which can be added for interfacing, operation and construction of message by using the extensibility of WSDL. WSDL services are used to define the semantic models that are independent form of language. It anticipates the opportunity of using WSML, OWL and UML as latent languages [16]. There are two new children elements such as precondition and effect for the WSDL operation element, which is defined by WSDL-S. These elements provide the facility for definition of the conditions. These conditions should be held before execution of an operation and effects the execution. Generally, such information is utilized to discover the appropriate Web services.

6.4 SAWSDL

The W3C working group was introduced the Semantic Annotations for WSDL (SAWSDL) and specification of XML Schema [19]. It was accepted as a W3C Recommendation in year 2007. Basically, SAWSDL is a controlled and standardized version of WSDL-S that is including a small number of changes. Such changes are trying to facilitate a good level of annotations and disregard some existing problems. These problems have not mentioned any agreement among different community at the time of creating the specification. Mainly, there are three essential alterations between SAWSDL and WSDL-S. Firstly, there was no agreement about model to define the precondition and effect within the Semantic Web and its service communities. This precondition and effect are not considered directly. SAWSDL is unable to prevent such kinds of annotations which are demonstrated in the guide of usage that is created by the SAWSDL working group [20]. Furthermore, the modelReference extension attribute exchanged the category annotation in SAWSDL that can be used to annotate XML Schema, complex-type definitions, simple-type definitions, declarations of elements and declarations of attributes as WSDL interfaces, operations and faults. Lastly, WSDL-S schemaMapping annotation was disintegrated into two distinct extension attributes as liftingSchemaMapping and loweringSchemaMapping.

6.5 WSMO-Lite

WSMO-Lite supported three categories of annotations such as modelReference, liftingSchemaMapping and loweringSchemaMapping. These annotations can introduce the representation of elements semantically which are defined on Web Wide World. Particular representation language about documents and explicit vocabulary for the adaptation of users are not provided by SAWSDL. This characteristic provides the extensibility and forces the users to prefer their particular ontologies on choice for defining the semantics of services. WSMO-Lite is an incremental building of a heap of technologies for the implementation of Semantic Web services by exactly locating the lack of services [21] of SAWSDL. WSMO-Lite introduces four major kinds of semantic annotations for Web services [22] which are elaborated in following subsections:

6.5.1 Functional Semantics

Functional semantics are used to describe the functional aspect of services. Functions based on semantics will be built to relay the services. These functions offer the services to its clients when the function is invoked.

6.5.2 Nonfunctional Semantics

Nonfunctional semantics express the explicit details that are regarding the execution and capable to provide the environment to run the services on its price or quality. Nonfunctional semantics deliver service-oriented supplementary information. This supplementary information can support to rank and select the furthermost suitable services.

6.5.3 Behavioral Semantics

Behavioral semantics identify such protocol which is ordering the operations that are required to client to follow them when a service is invoking. Behavioral semantics provide the ordering of operations in the applications. The users must require to act on application according to the ordering of operations.

6.5.4 Information Model

Information model specifies the semantics of input, output and fault messages that are used in services. The semantics of input, output and fault messages provide the innovative operability of services.

6.6 MicroWSMO

MicroWSMO is a micromechanism that provides the semantic annotation of Web APIs and RESTful services [23] for the better support of discovery, configuration and invocation. Microformats provide the facility for annotating human concerned Web pages to construct the information for machine readable and processible [24]. MicroWSMO may be adopted the WSMO-Lite ontology as reference for semantical annotations of RESTful services. Hence, WSDL services and RESTful services both can be treated as homogeneously, and they are annotated with WSMO-Lite and MicroWSMO separately. MicroWSMO also introduces four main kinds of semantic annotations like annotations of WSMO-Lite for services. These are functional semantics, nonfunctional semantics, behavioral semantics and information model semantics. Model references on the service are facilitating the functional and nonfunctional semantics. Behavioral semantics are incorporated as model references on the operations. At the last, semantics of information model can be confined on input and output messages of operations.

6.7 SA-Rest

SA-REST [22, 25] is a standardized mechanism to provide semantic annotations to RESTful services and the APIs of Web. Essentially, SA-REST tackles the descriptions of grounding service to semantic meta-models by applying model reference type of annotations from SAWSDL. Though, SA-REST is unable to initiate the machine-processable description of a RESTful service or Web APIs comparable to WSDL files. Web APIs are only defined in human-readable HTML Web pages for various purpose and situation. Such kind of HTML Web pages does not contain those elements that a machine can utilize to recognize the services. The microformat of SA-REST embedded semantic annotations among Web pages, which is elaborating RESTful services. These formats can use the GRDDL [26] and RDF [27] that are recommended by W3C Recommendations. SA-REST provides the facility to embed the RDF triples with the text. Hence, it can be extended among both the document and clustered.

7 Conclusion

After exploring the vital literatures related to Semantic Web technology, its services and various agent-based applications, we observed that the Semantic Web is competent to handle the agent-based and web-based applications. Ontology is the powerful technology of the Semantic Web that has vital potential for the organization, management and understandability about information for machine. It is capable to inherit the reasoning mechanism through conceptualization of information and relationship among these concepts. Applications, benefits and vision of Semantic Web have been elaborated in this manuscript which will inherit in applications development by using appropriate Semantic Web technology and its services. Various kind of Semantic Web services is discussed in this paper that can provide the facility to develop Web services embedded application of intelligent knowledge processing. These services support the semantic annotations in Web applications based on the semantic description of knowledge resources.

References

1. Feldmann S, Herzig SJI, Kernschmidt K, Wolfenstetter T, Kammerl D, Qamar A, Lindemann U, Krcmar H, Paredis CJJ, Vogel-Heuser B (2015) Towards effective management of inconsistencies in model-based engineering of automated production systems. In: IFAC symposium on information control in manufacturing
2. Fensel D, Bussler C, Ding Y, Kartseva A Klein V, Korotkiy M, Omelayenko M, Siebes R (2002) Semantic Web application areas. In: Proceedings of the 7th international workshop on applications of natural language to information systems. Stockholm, Sweden, June 2002, pp 27–28
3. Berners-Lee T, Hendler J, Lassila O (2001) The Semantic Web. Scientific American, May 2001
4. Klein M, Visser U (2004) Semantic Web challenge 2003. In: IEEE intelligent system. IEEE Computer Society, pp 31–33
5. Berners-Lee T (2000) Semantic Web-XML2000. http://www.w3.org/2000/Talks/1206-xml2ktbl/slide10-0.html
6. Daconta M, Obrst L, Smith K (2003) The Semantic Web: a guide to the future of XML, web services, and knowledge management. Wiley, Indianapolis

7. Misra DC (2006) Ten emerging E-government challenges today: the future may be sober and not hype, keynote address. In: 4th international conference on E-governance. Indian Institute of Technology, New Delhi. In: Sahu GP (ed) (2006) Delivering E-government, New Delhi, Gift Publishing, Global Institute of Flexile Systems Management, 15 Dec 2006, pp 6–14 (Chapter 2)
8. Erl T (2007) SOA principles of service design. Prentice Hall
9. McIlraith S, Son T, Zeng H (2001) Semantic Web services. IEEE Intell Syst 16:46–53
10. Martin D, Burstein M, Hobbs J, Lassila O, Mcdermott D, Mcilraith S, Paolucci M, Parsia B, Payne T, Sirin E, Srinivasan N, Sycara K (2004) OWL-S: Semantic markup for web services. W3C Member Submission. http://www.w3.org/Submission/OWL-S
11. Horrocks I, Patel-Schneider PF, Boley H, Tabet S, Grosof B, Dean M (2004) SWRL: a Semantic Web rule language combining OWL and RuleML. W3C Member Submission. http://www.w3.org/Submission/2004/SUBM-SWRL-20040521/
12. Kopecky J, Roman D, Moran M, Fensel D (2006) Semantic Web services grounding. In: Advanced international conference on telecommunications and international conference on internet and web applications and services (AICT-ICIW'06), Guadelope, French Caribbean, p 127
13. Fensel D, Lausen H, Polleres A, De Bruijn J, Stollberg M, Roman D, Domingue J (2007) Enabling Semantic Web services: the web service modeling ontology. Springer
14. Roman D, Lausen H, Keller U (2006) Web service modeling ontology (WSMO). WSMO working draft. http://www.wsmo.org/TR/d2/v1.3/
15. Fensel D, Bussler C (2002) The web service modeling framework WSMF. Electron Commer Res Appl 1:113–137
16. Akkiraju R, Farrell J, Miller J, Nagarajan M, Schmidt MT, Sheth A, Verma K (2005) Web service semantics—WSDL-S. W3C Member Submission. http://www.w3.org/Submission/WSDL-S/
17. WSDL-S (2006) https://www.ibm.com/developerworks/webservices/tutorials/ws-understand-web-services2/ws-understand-web-services2.html, July 2006
18. Sivashanmugam K, Verma K, Sheth A, Miller J (2003) Adding semantics to web services standards. In: Proceedings of the 2003 international conference on web services (ICWS)
19. Farrell J, Lausen H (2007) Semantic annotations for WSDL and XML schema. W3C recommendation. http://www.w3.org/TR/sawsdl/
20. Akkiraju R, Sapkota B, Semantic annotations for WSDL and XML schema-usage guide. Working group note. http://www.w3.org/TR/sawsdl-guide/
21. Vitvar T, Kopecky J, Viskova J, Fensel D (2008) WSMO-lite annotations for web services. In: Proceedings of the 5th European Semantic Web conference
22. Sheth A (2003) Semantic Web process lifecycle: role of semantics in annotation, discovery, composition and orchestration
23. Richardson L, Ruby S (2007) RESTful web services. O'reilly Media, Inc
24. Maleshkova M, Kopecký J, Pedrinaci C (2009) Adapting SAWSDL for semantic annotations of RESTful services. In: Workshop: beyond SAWSDL at on the move federated conferences & workshops
25. Lathem J, Gomadam K, Sheth A (2007) SA-REST and (S)mashups: adding semantics to RESTful services. ICSC'07: proceedings of the international conference on semantic computing
26. Connolly D (ed) (2007) Gleaning resource descriptions from dialects of languages (GRDDL). W3C recommendation. http://www.w3.org/TR/grddl/
27. Adida B, Birbeck M, Mccarron S, Pemberton S (eds) (2008) RDFa in XHTML: syntax and processing. W3C recommendation. http://www.w3.org/TR/rdfa-syntax/

A Big Data Parameter Estimation Approach to Develop Big Social Data Analytics Framework for Sentiment Analysis

Abdul Alim and Diwakar Shukla

Abstract We are living in modern society, and mobile phones, tablets, laptops and personal computers have become part of our daily lives. These technologies are providing facility for social interaction and running all over the world through the expansion of Internet. The social media are producing huge amount of behavioral data through using different social media applications which such activities (likes, comments, share, etc.) have performed by users. This large volume of data can be generated in the form of structured, unstructured or semi-structured, and the 80 percent of social big data has been produced unstructured. By analyzing these data, we can predict people's sentiments and opinions. This opinion we can use for business perspective such as ranking of any particular products, rating of online courses, rating of online shopping sites and human behavior of buying a car. In this paper, we have proposed a big social data analytics framework to analyze human behavior using sampling technique.

Keywords Big data · Big social data · Sampling · Big data analytics · Human behavior · Sensor data

1 Introduction

The social media have become important medium between people and spreading the information in different domains like business, entertainment, politics, health care and crisis management. The social media have been providing new opportunities to receive, create and share the public messages at low cost ubiquitously. The numerous growth of social media offers many possibilities, categorization of data formats

A. Alim (✉)
Department of Computer Science and Applications, Dr. Harisingh Gour Vishwavidyalaya, Sagar, Madhya Pradesh, India
e-mail: abdulaleem1990@gmail.com

D. Shukla
Department of Mathematics and Statistics, Dr. Harisingh Gour Vishwavidyalaya, Sagar, Madhya Pradesh 470003, India
e-mail: diwakarshukla@rediffmail.com

R. K. Shukla et al. (eds.), *Social Networking and Computational Intelligence*, Lecture Notes in Networks and Systems 100, https://doi.org/10.1007/978-981-15-2071-6_63

such as text, images, videos, audios and log files. Such types of data have led to an increase, day to day, in high speed because the huge volume of datasets is produced daily and *n* number of active users on social media platform. Those data volumes are produced in different data formats like structured, unstructured and semi-structured that are known as big social data. The social media analytics plays a very important role in various sectors and generates visible benefits. One of the benefited areas is business which is made for several purposes such as the prediction result useful for detecting new trends in the business communications or issues. At present, the social media becomes an important medium to spread emergency information management and aware to people for the current status [1]. The United Nation Institute has estimated that 2.5 quintillion bytes of data are produced every day on Global Geospatial Information Management, and according to Google, 25 petabytes of data volume is being generated per day. The traditional big data has three Vs: volume, velocity and variety [2], but after 3 Vs IBM has extended one another V known as veracity [3]. Big data typically includes large datasets of unstructured data that need to process real-time analytics and discover new values from the hidden data. Figure 1 explores the phenomenon of big data.

According to Chen et al. [4], the Internet companies are growing rapidly; for example, Google is processing hundred petabytes data per day. Facebook generates 10 petabytes of large datasets per month. Figure 1 explores how different sectors are

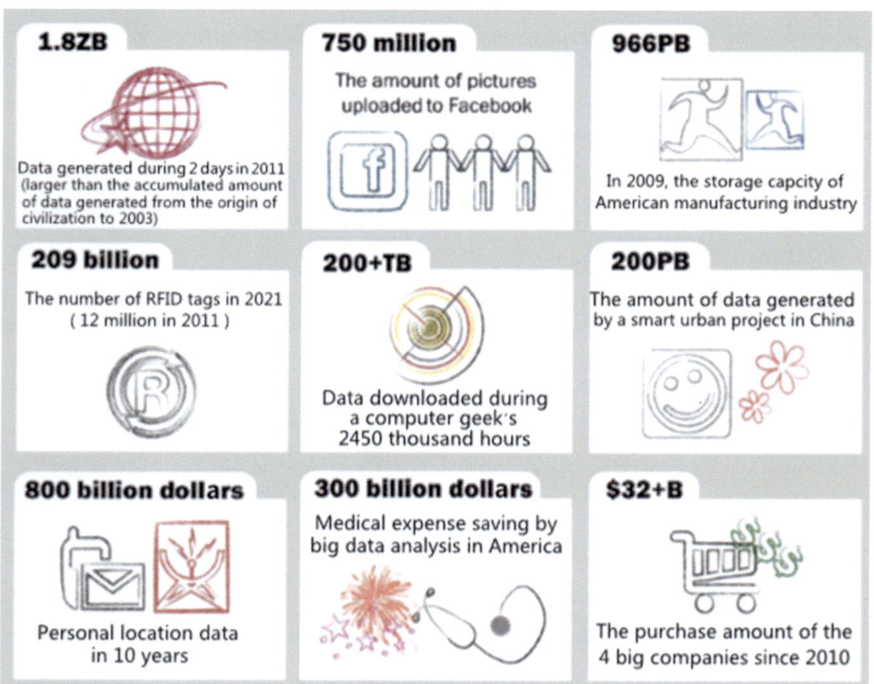

Fig. 1 Exploration of big data

producing the huge amount of data. YouTube also generates huge data by uploading 72 h of videos. But in present time the Internet of Things (IoT) devices are generating largest data because the IoT concept is that the sensor devices are connected to each other and communicate to all over the world. By 2030, the number of IoT devices will increase on trillion [4]; thus, the data will be heterogeneous, unstructured, missing data and high redundancy. Figure 2 shows and clarifies the architecture of IoT with big data analytics.

In Fig. 2, the IoT conceptual architecture is divided into layers, the IoT devices have number of sensors which are connected to each other for communication, and network devices will provide Internet facility for connecting the devices and storing data on cloud by using IoT gateways. The big data analytics layer will process and predict the business value using different technologies such as Hadoop, MapReduce

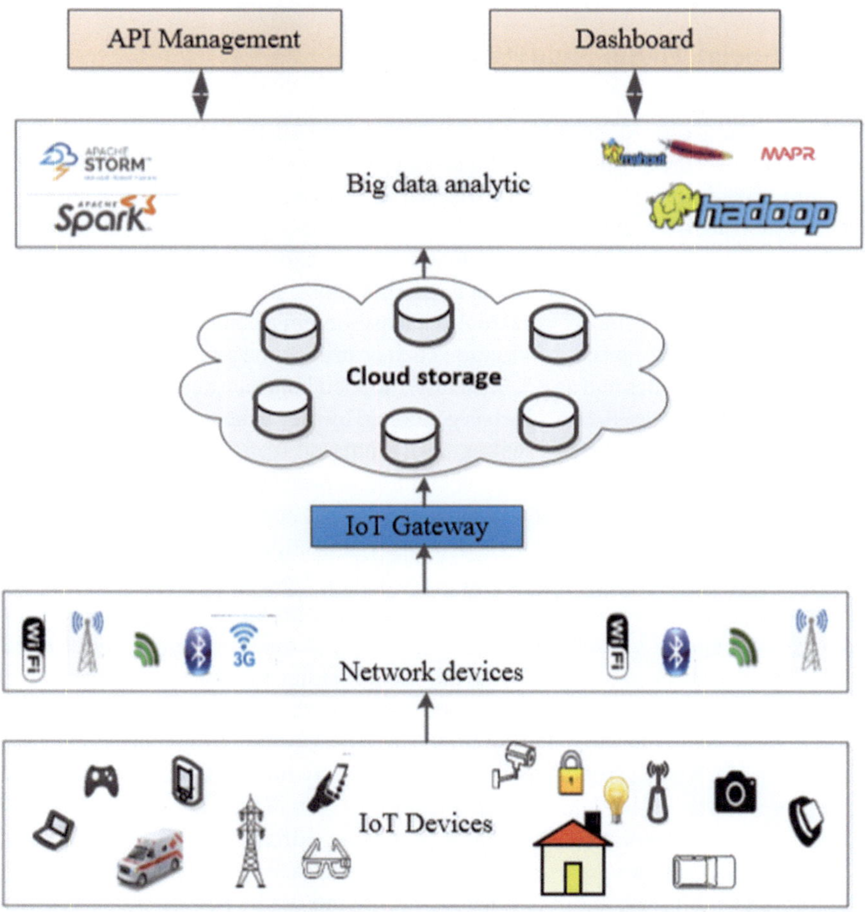

Fig. 2 Architecture of Internet of Things

and Spark and provide to user through the application program interface [5]. Currently, social media are producing large datasets with high speed. The social big data has stored a large amount of raw material data; thus, high and appropriate computational techniques are required to extract useful patterns from those raw data. Social media have become a unique information source to deal with information and discover more valuable opportunities for social and economic exchange [6]. The social media analytics traditionally has to focus on rich description of network data, but recently, the development of statistical methods is playing an important role in big social data analytics. The parameter estimation method improves the existing method of model selection through using appropriate estimation method [7]. In this paper, we have proposed a model for social big data analytics framework by using parameter estimation.

2 Big Social Media Analytics

The big data analytics to provide evidence-based decision making, the every organizations need an efficient process to return high volumes of fast-moving and diverse data into meaningful insights. The overall process can be broken down into five stages, which are acquisition and recording, extraction cleaning annotation, integration, aggregation and representation, modeling and analysis and interpretation. The big data analytics techniques represent a relevant subset of tools which are text analytics (extract information from textual database on email and blogs), information extraction (extract structured information such as drug name and dosage), sentiment analysis, audio analytics and so on. Basically, the social big data analytics has focused on to process unstructured data which is generated by the social media like Facebook, LinkedIn, blogger, Wordpress, Twitter, Instagram and YouTube, and mobile apps such as Find My Friends. Through the social interaction, the user generates different contents—for example, sentiment, images, videos and bookmarks. The relationship and their interactions happen between people, organization and products. The social media data analytics can be applied to drive insight from such data. Furthermore, the social media can be categorized into two different groups: One is content-based analysis and the other is structure-based analysis. The content-based analytics refers to the data that is posted by the user on social media; the data can be customer feedback, product review, text, audio, video and noisy data. The structure-based analytics refers to model through a set of node and edges, representing participants and relationship [8]. Many researchers have done the study on social media and focused on to process data, clean data and visualization of data. There are many literatures available on data exploration of social media using different sampling techniques. The sampling techniques are used to analyze large datasets, and also, there are many sampling algorithms available that select data points with specific properties that can be used in classification-related problem. Some sampling techniques are query by committee (maintaining a set of classifiers that all trained on a labeled dataset), uncertain sampling (select data points for which there is a low posterior probability) and density sampling (does not select data points in low probability regions, resulting in a low number of selected outlier) [9].

3 Big Data Parameter Estimation

The social media provide an unparalleled platform for relationship between consumer posting behavior and marketing. For the knowledge discovery from big data, the researchers have proposed various models, algorithms, software, hardware and technologies, and try to have more accurate and reliable results. However, in big data analytics environment, many big social data parameters should be considered such as compatibility, deployment complexity, cost, efficiency, performance, reliability, support and security risk. Parameter estimation has the ability to find more accuracy in result because we are estimating the nearby value of unknown parameter. For example, a company wants to know the customer interest for a particular car which will be launched in next some days. The parameter estimation will help to predict the customer interest using sampling techniques. Some other examples like prediction of symptoms, disease evaluation and cost in the optimize hospital operations with expenditure through considering different parameters [10]. In the digital era, people can use web-based applications such as social media through the digital devices like laptop, mobile and tablets for creation, access and exchange of user-generated content. The social media have led to numerous data services, tools and analytics platform for the estimation of various quantitative and quantitative attributes of unstructured data in huge dataset. Basically, the social media data analytics refers to extract user's query by retrieval, lexical analysis, pattern recognition, information extraction, data mining techniques like association analysis, visualization and predictive analysis. But such analysis is not an easy task; in this regard, there are many challenges over the unstructured data also. The challenges include cleaning the unstructured textual data specially processing to high frequency streamed real-time data, data protection and data visualization. The main aim of social big data analytics is providing information to the customer, and it should be error-free like missing data, incorrect data and inconsistent data [11]. This paper has focused on to estimate big social data parameters and predict reliable with useful information such as volume.

4 The Concept of Social Big Data

Today big data has become a challengeable problem for many research areas such as data mining, machine learning, computational intelligence, the semantic web, information fusion and social networks. The different big data frameworks are available for processing large datasets such as Apache Hadoop, MapReduce, Spark, RStudio and Python with *n* number of libraries. Mostly, the big data comes from sensor devices which are used to gather climate, traffic and flight information, post to social

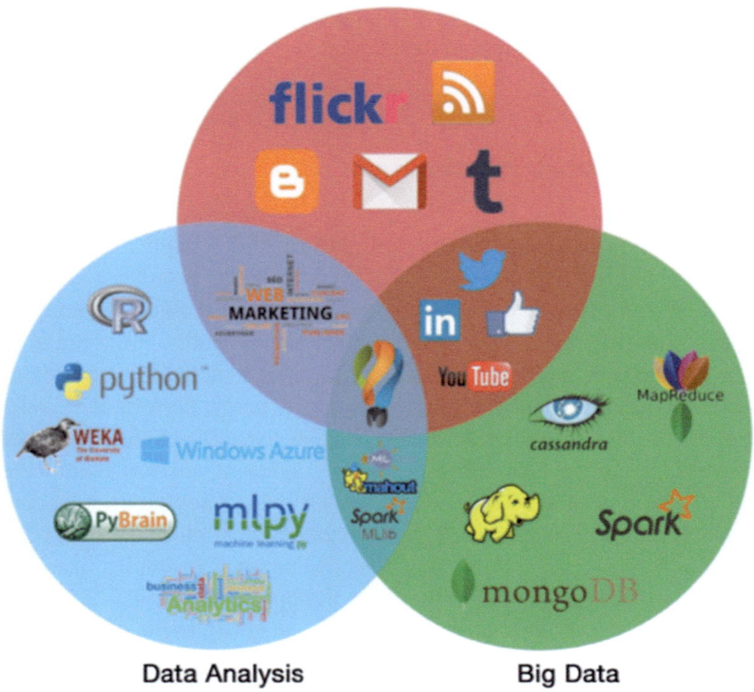

Fig. 3 Conceptual map of big social data

site, digital picture and videos (YouTube users upload 72 h of new video contents per minute), transactional records, etc. Figure 3 shows the conceptual map of big social data.

The big data refers to datasets that are terabyte to petabytes in size, and the big data has been defined through the 3 Vs model which was defined in 2011 by Laney. After the basic 3Vs of big data, some authors also have extended some another big data Vs. The following Table 1 has described to basic and extended big data Vs.

Table 1 Brief summary of big data parameters

Volume	The collection of large amount of data with different formats such as structured, unstructured and semi-structured
Velocity	The speed of transferring data; in other words the online streaming data which is constantly changing through the absorption complementary data collection
Variety	Different types of data which are collected via sensors, smartphone, social networks, text, images, video, audio, log file, etc.
Value	The process of extracting valuable information from large set of social data through the big data analytics
Veracity	Accuracy and correctness of information when the analytics techniques have performed on large datasets

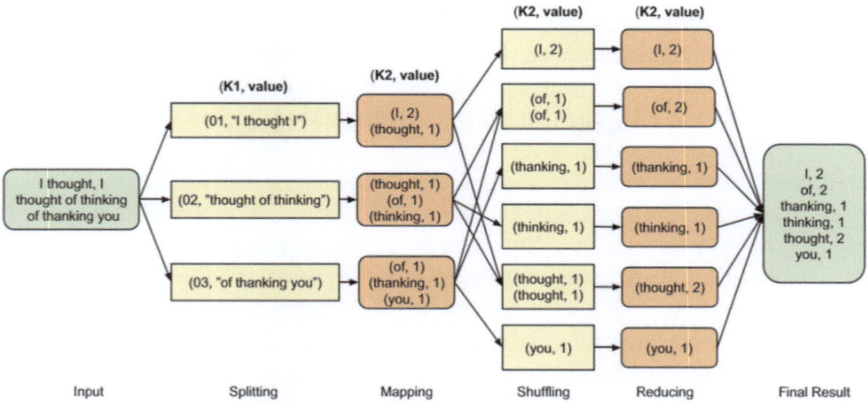

Fig. 4 MapReduce process for word count in text

When the data is large, then the processing of data also becomes a challenging task. The main challenging problems are data processing, data storage, data representation and data pattern mining, analyzing the user behaviors from the social media. The MapReduce is one of the most efficient big data solutions for big social data processing. The MapReduce framework has provided significant improvements in large-scale data-intensive applications in cluster and has basic elements for processing the large datasets mapper and reducers. Figure 4 has explored the working of mapper and reducer together.

Figure 4 has explained the classical word count problem. The MapReduce is an efficient paradigm which can be processed on textual data analysis problem in social media environment. As shown in Fig. 4, the input dataset split into a subset of chunks that will later processed by the mappers. The shuffling process is used to combine those key values which are used to count the words that generate output as a result [12].

5 The Proposed Framework

The social media are one of the most powerful media which allow millions of users to express and spread their opinion using corresponding social media applications about a particular topic and showing their behavior by liking and disliking content. The big social datasets continuously generate the datasets in high-volume, high-variability, high-variety and high-value. Already we have discussed that almost 80% of social data is in text format; therefore, the social big data analytics has become key element for public sentiment and their opinion. The sentiment analysis aims to determine people's sentiment and mining people's opinion related to particular topic by analyzing their post such as positive and negative. The sentiment analysis can be divided into two types: one is lexicon (mining the popularity of text datasets from the phrases in the document), and second is machine learning. It provides facility to build a model based on labeled training datasets for determining the attitude of document [13]. This research article has proposed a social big data analytics framework and focused on text analysis. The text analysis determines people's attitude liking, disliking, positive comment and negative comment. In this framework, we have applied sampling technique to estimate the unknown parameters such as the possibility of a particular product selling, opinion of people's about liking- and disliking-based comment on particular product. By estimating these parameters, we can predict the useful value of a specific product. The result will be best and bad which depends on the confidence interval. Figure 5 explores the overall framework of proposed system. The proposed framework will measure the average size of like (positive response) and dislike (negative response) on a fixed time (t). In the proposed method, we have considered three social media applications Facebook, Twitter and Instagram. The main aim is to calculate the average size of hits per fixed interval time (t) on a product when every second people like and dislike; thus, we can predict people's behavior for the future use. Here, we have assumed that any one manufacturer company wants to know the attitude of people for their particular product, and then it has sent the product advertise on three social media platforms: Facebook, Twitter and Instagram.

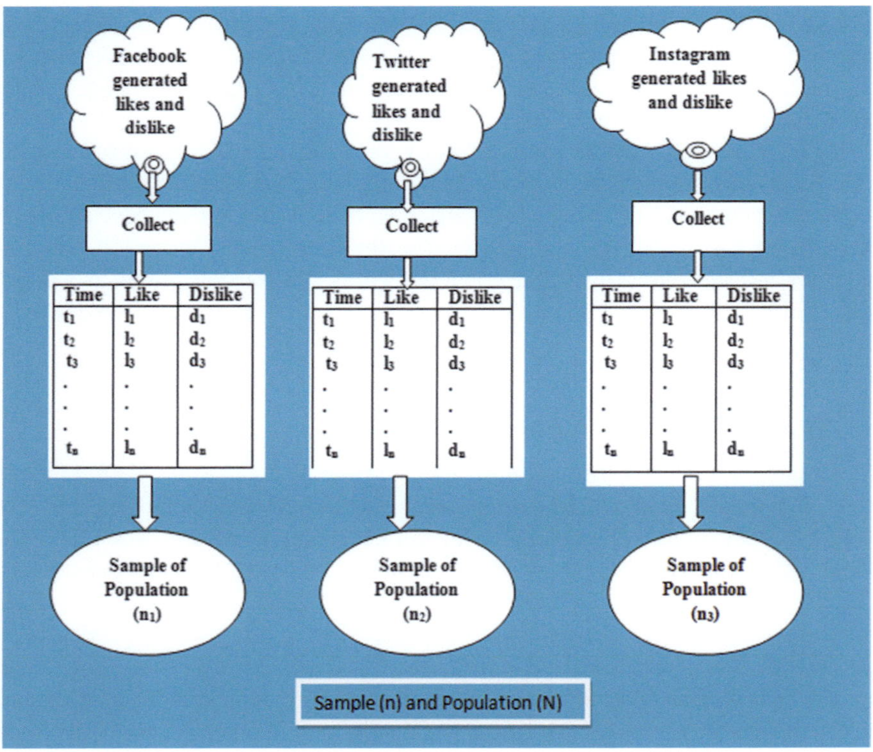

Fig. 5 Proposed framework for sentiment analysis

In Fig. 5, there are K social media servers available from different places. Let $Y_{ij}(t)$ be the size of jth positive hits in ith server has $(i = 1, 2, 3, \ldots K,\ j = 1, 2, 3, \ldots N_i(t))$, the ith server has total $N_i(t)$ hits at time t. The mean size of ith serve is $\overline{Y}_i(t) = \frac{1}{N_i(t)} \sum_{i=1}^{K} Y_{ij}(t)$ which is unknown parameter. Here we have considered K social media servers and one product advertisement; let there are N servers, then

$$N(t) = N_1(t_1) + N_2(t_2) + N_3(t_3) + \cdots N_K(t) \tag{1}$$

A total n sample has to be studied for impact evolution, which has been decided through random sample technique. The size of the sample in different strata considering the phased operation as stratification criteria will be

$$n(t) = n_1(t) + n_2(t) + n_3(t) + \cdots n_K(t) \tag{2}$$

Due to the dynamic nature of the population from time to time general weight is

$$W_i = \frac{N_i}{N} \tag{3}$$

Estimation of mean and variance for the proposed method, to estimate the population (hits) mean $\overline{Y} = \sum_{i=0}^{n} W_i \overline{Y}_{\mathrm{i}}$ of the study parameter Y, the estimator used is given by

$$\bar{y}_{\mathrm{st}} = \sum_{i=1}^{L} W_i \bar{y}_{\mathrm{i}} \tag{4}$$

which is an unbiased estimator of the population mean and the variance of $\bar{y}_{\mathrm{st}}$ under proportional allocation is given by

$$V(\bar{y}_{\mathrm{st}}) = \sum_{i=1}^{L} \left(\frac{1}{n_i} - \frac{1}{N_i} \right) W_i^2 S_i^2 \tag{5}$$

In stratified sampling, before taking a sample, the sample has to decide about the allocation of the sample size to the strata (equal, proportional and optimum) [14].

6 Conclusion

Emerging mobile phone technologies are playing a key role in human interaction via social media applications. The big social media have collected lots of human behavior data like ideas, political affiliation, opinions, human's stress, positive and negative attitude. The big social data analytics has the capability to analyze the attitude of human being using social media datasets. This research paper has proposed a social big data analytics framework which will help product manufacturer companies for launching their products based on customer demand. In this framework, we have used stratified sampling technique for calculating the user response in the form of liking and disliking to any particular product. The proposed methodology assumed that the whole population has the three different social media servers in different time intervals with dynamic nature of population. After that, the sample size has selected from each stratum and the average size of pooling hits is calculated which is performed by the user. The future suggestion is that if anyone wants to calculate average mean based on sentimental tokens, then this model will also help by using word count algorithm.

References

1. Stieglitz S, Mirbabaie M, Ross B, Neuberger C (2018) Social media analytics-challenges in topic discovery, data collection, and data preparation. Int J Inf Manage 39:156–168
2. Lee J-G, Kang M (2015) Geospatial big data: challenges and opportunities. J Big Data Res 2:74–81
3. Jagadish HV (2015) Big data and science: myths and reality. J Big Data Res 2:49–52
4. Chen M, Mao S, Liu Y (2014) Big data: a survey. J Mob New Appl 19:171–209
5. Marjani M, Nasaruddin F, Gani A, Karim A, Hashem IAT, Siddiqa A, Yaqoob I (2017) Big IoT data analytics: architecture, opportunities, and open research challenges. IEEE Access 5:5247–5261
6. Zeng D, Chen H, Lusch R, Li S (2010) Social media analytics and intelligence. IEEE Intell Syst 6:13–16
7. Snijders TAB, Koshkinen J, Schweinberget M (2010) Maximum likelihood estimation for social network dynamics. J Ann Appl Stat 4:567–588
8. Gandomi A, Haider M (2015) Beyond the hype: big data concepts, methods, and analytics. Int J Inf Manage 35:137–144
9. Rojas JAR, Kery MB, Rosenthal S, Dey A (2017) Sampling techniques to improve big data exploration. In: 2017 IEEE 7th symposium on large data analysis and visualization. Phoenix, AZ, pp 26–35
10. Qussous A, Benjelloun F-Z, Lahcen AA, Belfkih S (2018) Big data technologies: a survey. J King Saud Univ Comput Inf Sci 30:431–448
11. Batrinca B, Treleaven PC (2015) Socil media analytics a survey of techniques tools and platforms. J Knowl Cult Commun 30:89–116
12. Bello-Orgaz G, Jung JJ, Camacho D (2016) Social big data: recent achievements and new challenges. Inf Fusion 28:45–59
13. Alaoui IEI, Gahi Y, Messoussi R, Chaabi Y, Todoskoff A, Kobi A (2018) A novel adoptable approach for sentiment analysis on big social data. J Big Data 5:1–8
14. Pandey R, Verma MR (2018) Sample allocation in different strata for impact evaluation of development programme. Rev Mat Estat 26:103–112

A Novel Approach of Vertex Coloring Algorithm to Solve the *K*-Colorability Problem

Shruti Mahajani, Pratyush Sharma and Vijay Malviya

Abstract The graph theory is one among the most studied fields for research. There are several algorithms and concepts of graph theory, which is used to solve mathematical and real-world problems. Graph coloring is one of them. The graph coloring problem is a kind of NP-hard problem. Graph coloring has the capability of solving many optimization problems like air traffic management, train routes management, time table scheduling, register allocation in operating system, and many more. The vertex coloring problem can be defined if given a graph of vertices and edges; every vertex has to color in such a manner that two adjacent vertices should not have the same color, using the minimum number of colors. This paper introduced the implementation prospective of a vertex coloring algorithm based on k-colorability using recursive computation. Implemented algorithm is tested on DIMACS graph benchmarks. Result of novel method shows it has reduced execution time and improved chromatic number.

Keywords Vertex coloring problem · Time complexity · Chromatic number · Graph coloring problem · DIMACS · Recursion

1 Introduction

The need for the graph coloring is arisen while coloring the map of the country so that no two adjacent cities have the same color. There are many real-world applications in which the concept of graph coloring is used such as scheduling, exam timetable, sudoku game, communication network, and air traffic management.

S. Mahajani · P. Sharma · V. Malviya (✉)
Malwa Institute of Technology, Indore, India
e-mail: vijaymalviya@gmail.com

S. Mahajani
e-mail: shrutimahajani@yahoo.com

P. Sharma
e-mail: pratyushsharma2109@gmail.com

Rajiv Gandhi Proudyogiki Vishwavidyalaya, Bhopal, India

R. K. Shukla et al. (eds.), *Social Networking and Computational Intelligence*, Lecture Notes in Networks and Systems 100, https://doi.org/10.1007/978-981-15-2071-6_64

We can define the graph coloring problem with the help of a graph. Suppose $G = (V, E)$ is a large undirected (when directions are not defined) graph where V is our set of vertices and E is our set of edges. In our graph, edges can be denoted as (i, j) where i and j are edges. Graph coloring problem is when we have to give a color to each vertex i which belongs to the set of vertices V such that i and j should have different colors.

Chromatic number can be defined as the minimum number of colors required to color G. It is denoted as X (G). Suppose a graph is colored using maximum k colors, then it is termed as real-world colorability problem and can be called as the GCP decision version. The graph coloring problem, i.e., GCP is a way to assign one color to every vertex of the graph G in such a way that no two adjacent vertices should have the same color. Also the number of colors used to color the graph is minimized. Most real-world problems can resemble with GCP which can be solved using this algorithm. Graph coloring problem is used in many applications such as scheduling [1, 2], timetable [3, 4], register allocation [5], frequency assignment [6], and communication networks [7].

Edge coloring is a technique to assign a specific color to all edges so that all two adjacent edges get a different color, and a face coloring of a planar graph is a method to assign one color to each face so that no two faces which share the same boundary receive the same color.

2 Related Work

Graph coloring problem is one of the NP-hard problems [8] for finding an optimum solution for a graph coloring problem. This problem gets generally divided into two classes: exact algorithms [9, 10] and approximate algorithms. The exact algorithm is an algorithm which solves a problem in a polynomial time; they give an optimal solution to problem. The approximate algorithm is algorithms that do not solve the problem in a polynomial time; they do not give guaranty to give an optimal solution of the problem.

Generally, two types of implementation approaches are preferred for implementing graph coloring algorithm. One is sequential, and the next one is parallel. Both the approaches have their own importance. Sequential approaches are reliable for small size problem and provide minimum number of color for a graph; whereas, parallel approaches provide good result for a big problem and provide fast computation for any graph problem. There are many sequential coloring algorithms such as Sequential Greedy Algorithm (SGA), First Fit (FF), Largest-Degree-First-Ordering, Incidence-Degree-Ordering [4], Smallest-Degree-Last-Ordering, and Saturation-Degree-Ordering. There are many parallel graph coloring approaches such as Parallel Maximal Independent Set (PMIS), Jones and Plassmann, i.e., JP, the Largest-Degree-First, i.e., LDF, Smallest-Degree-Last, i.e., SDL, Graph Partitioning Approach, i.e., GPA and Block Partitioning Approach, i.e., BPA.

3 Proposed Implementation

This section of paper is all about the implementation details of algorithm. Proposed implementation of graph coloring algorithm is based on recursion. Recursion is a well-known problem-solving strategy in computing. There are many mathematical problems like The Knight's tour problem, *N* Queen Problem, Subset Sum, Hamiltonian Cycle, Sudoku and Solving Crypt-arithmetic Puzzles which can be solved using concepts of recursion. Here, this paper is solving graph coloring problem using same recursion. Hence, we can name the algorithm as VCAURA approach.

3.1 Implementation Platform

Java programming language is used to the implementation of algorithm. For compilation and execution, jdk1.8.0_144 platform is used. Algorithm is developed on Window operating system platform using Eclipse editor. Pentium Dual Core CPU E5200 @ 2.50 GHz processor is used for execution of the application. Primary memory is 2.00 GB used.

3.2 Algorithm Flowchart

Figure 1 shows the flowchart of implementation. As this algorithm is using graph input in the form of DIMACS format, so that algorithm read graph data forms DIMACS graph instances, and then this data is converted into adjacency matrix. Flowchart clearly shows the recursive behavior of algorithm. Each recursive call picked up a vertex and assigns a suitable color to this vertex.

3.3 Algorithm Methodology

There are basically two methods, which are the core of proposed implementation. One is VertexColor method, which is used to assign colors to vertices of graph. Second is isSafe method, which checks the assignment of color to vertex is safe or not.

a. **VertexColor**

Figure 2a shows the source code of VertexColor method. This method takes a vertex *k* as an input. Here, *k* is the vertex on which algorithm is trying to color. Inside of this method, there is a loop which will be executed m times. Here, m is the number of colors which are needed to color the graph vertices. There is a method called isSafe.

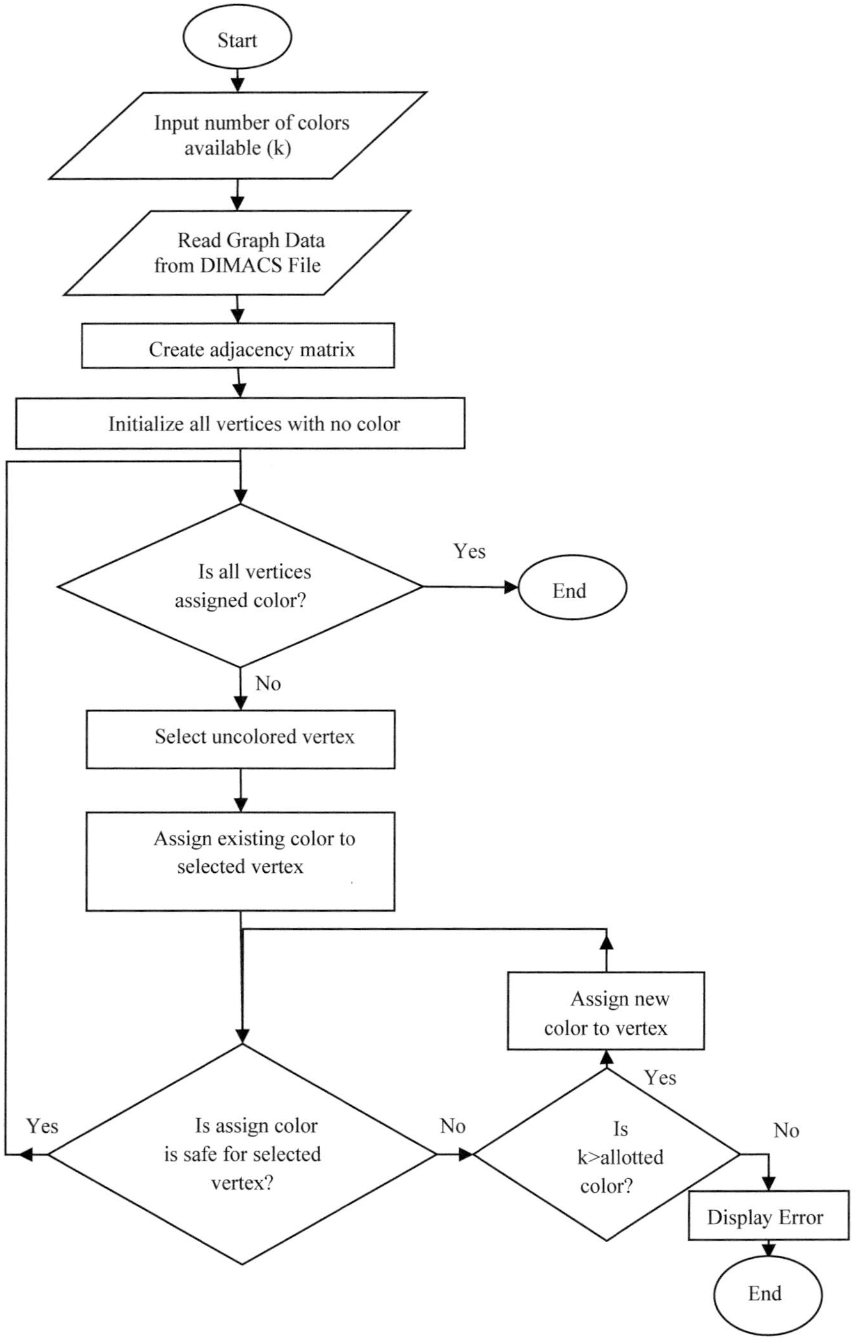

Fig. 1 Flowchart for the process of proposed algorithm (VCAURA)

Fig. 2 VertexColor and isSafe method

(a)

```
vertexColor(int k)
{
    for(int c=1; c<=m; c++)
    {
        If(isSafe(k,c))
        {
            x[k]=c;
            if((k+1)<n)
                vertexColor(k+1);
            else
            {
                print x[];
                return;
            }
        }
    }
}
```

(b)

```
isSafe(int k, int c)
{
    for(int i=0;i<n;i++)
    {
        If(G[k][i]==1 && c==x[i])
        {
            return false;
        }
    }
    return true;
}
```

The isSafe method checks that k vertex is safe for coloring with c color or not. If it is safe to color, then method returns true otherwise method will return false. If coloring is safe, then algorithm assigns the color, and VertexColor method is called recursively by increasing value of k by one. This recursive call will be continued till value of k is less than or equal number of vertices (n). After completion of all recursions, this method prints the ssolution.

Here, $k =$ the vertex that is going to color in current level of the recursion.

$x[k]$ is an array that holds the current color at each vertices.

b. **IsSafe Method**

Figure 2b shows the source code of isSafe method. This method takes two parameters: one is current vertex, which is going to be colored by algorithm in current recursion. Parameter c is the color, by which this current vertex will be colored. IsSafe performs a loop operation, and this loop is iterating through all the graphs and checks the vertex k's all adjacent vertices and their colors. If vertices are found colored and if any color matches with current color c, then this method will return false otherwise method will return true. If isSafe method returns false, it means algorithm can place color c to vertex k, because there are no adjacent edge with same color.

3.4 Data Set Used

Discrete Mathematics and Theoretical Computer Science (DIMACS) graph instances are used as a data set to analyze the performance of proposed implementation. It sponsors implementation challenges to establish practical algorithm performance on problems. Graph coloring is one of them. DIMACS challenge was sponsored in 1992 by DIMACS. DIMACS defined a format for undirected graph, which has been used as a standard format for problems in undirected graphs. This format was used for several DIMACS computational challenges.

4 Result

On DIMACS provided graph instances, we have tested the proposed implementation. Graph instance details are already mentioned in the previous section. Here, this section represents the comparisons of few well-known graph coloring algorithms with proposed implementation. Two main parameters number of colors (Chromatic number) and execution time (in seconds) are taken for comparison. Graphical comparison of execution time is also included in this section. We executed the VCAURA algorithm using the data sets and compared it with standard available graphs.

The below graph shows the difference between the execution of proposed algorithm and existing edge cover-based graph coloring algorithm (ECGCA). In the given graph, we have taken the Y-axis as execution time in milliseconds; whereas, the X-axis is used to demonstrate the various standard graphs used for testing purpose. We can observe that in the below graph, the results are fluctuating to the baseline. Proposed method is showing improvements in chromatic number and execution time. For few sets, the results are remarkably noticeable. For instance, C200.5 where ECGCA is generating time complexity of 19,091.7 ms with a chromatic number value 239, our proposed method shows the effective change with a time complexity value 5.949, and chromatic number value is 226 (Figs. 3 and 4).

Now, we compared VCAURA for another sets of test graphs. The below depicts the difference. This time we have taken the hybrid parallel genetic algorithm for GCP (HPGAGCP). Change can be observed from miles 250 up to miles 1000. For miles 1000, the best of the performance for proposed method comes out to take the 0.093 ms; whereas, the traditional method is taking 48.559 ms.

Similarly, now taking the next approach to compare with is modified cuckoo optimization algorithm for GCP (MCOACOL). Then, this is compared with our proposed VCAURA which shows improvements in various data sets (Figs. 5 and 6).

Now taking the forth comparison in light is with MPGCA which is multipoint guided mutation GCA. For few sets, it has provided the same execution time, and for

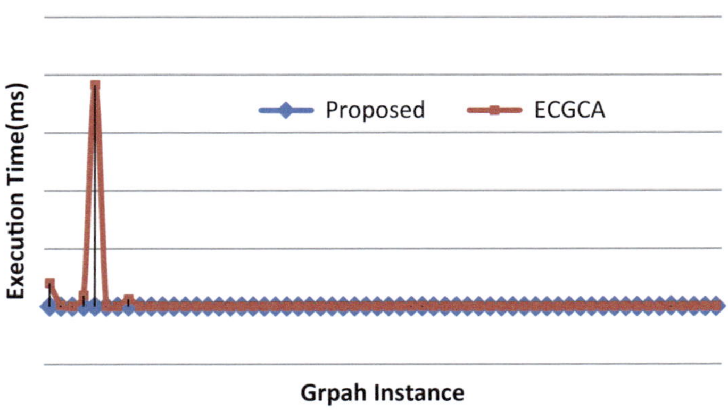

Fig. 3 Execution time comparison in proposed (VCAURA) versus EEECGCA

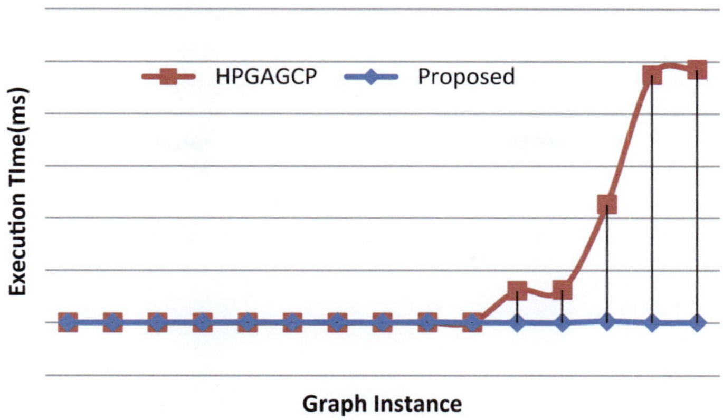

Fig. 4 Execution time comparison in proposed versus HPGAGCP

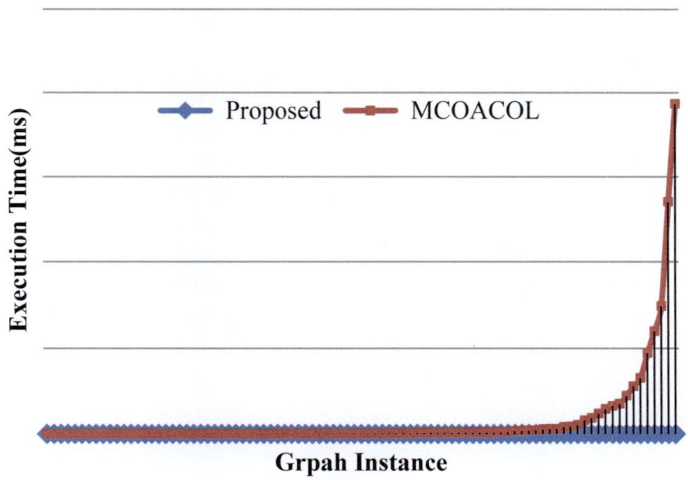

Fig. 5 Execution time comparison in proposed versus MCOACOL

few it has improved the execution time. In these cases, chromatic number can also be observed for choosing the suitable approach for implementing the algorithm. In this test, we have found that 4-Insertions_4 type graphs provide best performance with proposed solution. Chromatic numbers are same, but time has improved from 1071 ms to just 0.092 ms.

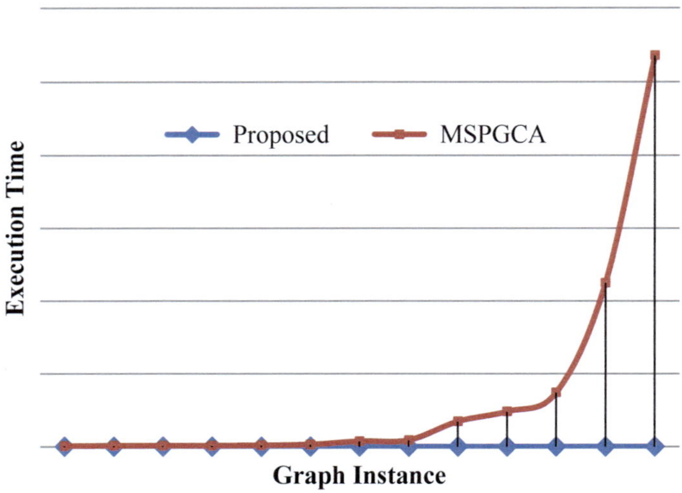

Fig. 6 Execution time comparison in proposed versus MSPGCA

5 Conclusion and Future Enhancement

This paper is focused on implementation and experiment of a graph coloring algorithm based on recursion. We have implemented algorithm with the help of Java programming language and experimented on DIMACS graph instances. We have also compared the experimental results with five other algorithms. It has been found that proposed algorithm solves graph coloring problem for undirected graph in optimum time with improved chromatic number at various places. It has been also observed that for some graph instances, it gives optimum solution. But for many graph instances, solutions (chromatic number) are not optimum.

In this research and implementation, certain areas of improvement are observed. To get better results in the context of time complexity, parallel computing concept can be applied to implement algorithm. In this research, algorithm is designed only according DIMACS graphs, and one can implement the same concept for any real-word problem. By the result, it has been found that the algorithm is not generating optimum chromatic number. So, there is also a possibility of improvement in algorithm to get better chromatic number.

References

1. Gamache M, Hertz A, Ouellet JO (2007) A graph coloring model for a feasibility problem in monthly crew scheduling with preferential bidding. Comput Oper Res 34(8):2384–2395
2. Zufferey N, Amstutz P, Giaccari P (2008) Graph colouring approaches for a satellite range scheduling problem. J Sched 11(4):151–162
3. de Werra D (1985) An introduction to timetabling. European J Oper Res 19:151–162

4. Burke EK, McCollum B, Meisels A, Petrovic S, Qu R (2007) A graph-based hyper heuristic for timetabling problems. Eur J Oper Res 176:177–192
5. de Werra D, Eisenbeis C, Lelait S, Marmol B (1999) On a graph-theoretical model for cyclic register allocation. Discr Appl Math 93(2):191–203
6. Smith DH, Hurley S, Thiel SU (1998) Improving heuristics for the frequency assignment problem. Eur J Oper Res 107(1):76–86
7. Woo TK, Su SYW, Wolfe RN (2002) Resource allocation in a dynamically partitionable bus network using a graph coloring algorithm. IEEE Trans Commun 39:1794–1801
8. Garey MR, Johnson DS (1979) Computers and intractability. Freeman
9. Malaguti E, Monaci M, Toth P (2011) An exact approach for the vertex coloring problem. Discr Optim 8(2):174–190
10. Segundo PS (2012) A new DSATUR-based algorithm for exact vertex coloring. Comput Oper Res 39:1724–1733

Predicting the Popularity of Rumors in Social Media Using Machine Learning

Pardeep Singh and Satish Chand

Abstract With the spread of digital technologies, social media sites like Twitter, Instagram, Facebook, etc. become a major attraction for people of all ages. Being a rich source of information at an unprecedented speed, it is very prone to misinformation also. This misinformation in the form of rumors (unverified information whose veracity is uncertain) observed to be propagated in events like political campaigning, natural disasters or any event of an emergency through social media. The underlying motivation of this work is to show that the machine learning approaches are successful for rumor popularity prediction also. By referencing three rumor datasets mostly related to unrest and riots, we use machine learning models to predict rumor popularity. A comparative study has been made by taking five machine learning methods such as logistic regression, gradient boosting regression, support vector regression, extreme gradient boosting regression and Bayesian ridge. We have extracted tweet features using TF-IDF approach and two user features like follower count (number of followers of the user who posted the tweet) and status count (history of the number of tweets posted by a user). These features have been used to train various machine learning models which have shown the potential in rumor popularity prediction task. Our empirical results show that on an average extreme gradient boosting regression (XGBoost) outperformed other methods in terms of predicting rumor popularity.

Keywords Social network analysis · Rumor propagation · Machine learning · Twitter

---

P. Singh (✉) · S. Chand
School of Computer and Systems Sciences, Jawaharlal Nehru University,
New Delhi 110067, India
e-mail: pardeepsinghinfo@gmail.com

S. Chand
e-mail: schand20@gmail.com

R. K. Shukla et al. (eds.), *Social Networking and Computational Intelligence*, Lecture Notes in Networks and Systems 100, https://doi.org/10.1007/978-981-15-2071-6_65

1 Introduction

Rumors in Social Media: According to the Oxford English Dictionary, a rumor is "a currently circulating story or report of uncertain or doubtful truth". It is basically a piece of unverified circulating information whose veracity may be partially true, true or an entirely false. In the absence of a reliable or authentic source, individuals use social networks to share information quickly. In such situations, rumors may be the best medium. Examples of such rumors for various events on social networks are:

- *Car chase under way in Paris as a gunfight breaks out. Reports of hostages. Latest from Paris.*
- *Police in Ferguson plans to release the name of the officer who shot unarmed teen Michael Brown.*

Figure 1 shows a rumor example from the Charlie Hebdo rumor dataset. It starts with a source tweet stating "Hostage taker in supermarket siege killed", reports say and is followed by retweets around that statement. As per the timelines, in one hour of time, many retweets came for the source tweet which shows its popularity. The major concern with retweets is their veracity, i.e., to classify whether new replies may again be a rumor or not.

There are a variety of processes behind this diffusion of information like the influence of users, tweet text and other features which are responsible for propagating this information rapidly. In these types of situations, when there is a high probability of content popularity, persons with many motives try to spread rumors among the influential users so that it becomes viral. Regarding the dynamics of rumors, there have been several descriptive studies. Hence research on rumor retweet prediction follows various directions. One is through biological virus propagation models like SIS and SIR. These are two-compartment models in epidemiology which divide the population into three components susceptible, infected and recovered. Susceptible means population leads to be infected and as per the infection rate items in the susceptible region move toward infection region. With recovery rate, the items move

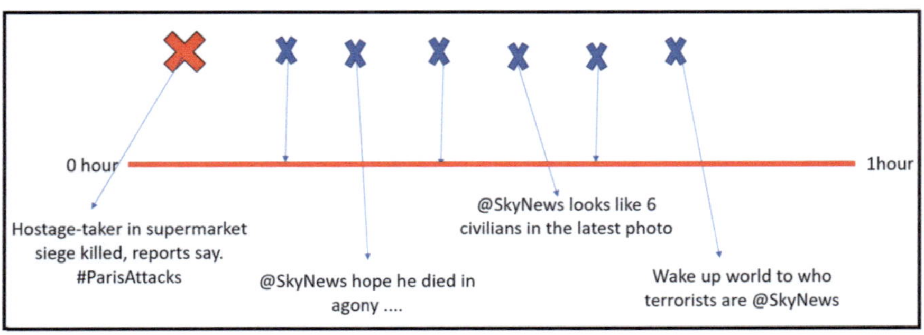

Fig. 1 An illustrative rumor example about Charlie Hebdo in Paris killing eleven people

toward recovery. Leskovec et al. [1] extended the SIS model to apply compartment models to the online social environment. Another approach is point process models in which we focus on modeling social media tweets. These models use history of events along with the magnitude of influence for retweet prediction [2]. The other way to build retweet prediction model is to use content features, structural properties of users and temporal information of tweets [3]. These models are feature-based which used various features of tweet body to predict how much popular the tweet will be.

We experiment and justify our approach with an openly available dataset called PHEME which contains tweets about various social events which triggered rumors.

The novelty of this paper lies in the comparative study of various popular machine learning regression models on rumor datasets to check for the better performance in terms of predicting popularity.

2 Related Work

This section discusses some existing work related to rumors propagation. In [4], the authors studied fake content on Twitter by applying regression model on real-world data which calculates the number of nodes that are affected.

Prediction of tweet popularity related to breaking news studied in [5]. Similar to other information propagation models, it was discussed in this paper that the retweet count is predicted from super-nodes on Twitter and its features, i.e., follower count of super-node receive the update and retweet probability. Similarly, retweet behavior also depends on the probability of a re-tweeter follower. Another feature in this approach is the time when the tweet is posted, hop distance from super-node and saturation point.

Another exploratory study [6] proposes multi-order Markov model to predict the cascade size of Twitter with graph features such as degree distribution, edge growth rate, clustering and diameter.

In [7], researchers use deep learning models with tweet content, user profile and similarity between tweet text and user interest for retweet prediction.

It has been observed in [8] that information diffusion in Twitter relies on linguistic as well as the influence of initial creator of the tweets and represents it as a tree pattern, i.e., tweet trees are created based on tweets data and user activity matrix. Authors also find the group of influential people in Twitter and tree pattern they belong to.

Another study [9] use information cascade modeling to predict how many retweets a source tweet gain during a fixed time period. Authors focused on important features such as the flow of cascade and page rank which shows user influence on prediction task.

The paper [10] summarizing situational tweets during natural disaster scenario while extracting important tweets from whole information by integer linear programming optimization technique and use content and graph-based summarization techniques for predictions.

The study [11] showed research into social media rumors with four major components, i.e., detection of rumors, tracking rumors, stance classification of rumors and veracity classification of rumors. Authors discussed scientific approaches to analyzing and developing models for these four components.

In the paper [12], authors revealed that most of the reactions for social media content are received within one hour of its posting. They conducted their analysis on Facebook pages and showed that 34% reactions are obtained in the first hour of its posting. They further divided one hour time period into 15-minute buckets as the bucket can capture essential reactions.

Different studies have focused on the applications of micro-blogging services in various fields. Due to the importance of predicting popular content in social networks, there have been quite a number of studies on how and why content gets popular in social networks. We explored different studies which are based on rumors propagation and popularity predictions in a social network like Twitter.

3 Dataset Used

This section discusses rumor dataset used for experiments. We used standard rumor dataset PHEME part of the FP7 research project described in detail in [13]. It contains seven rumor datasets which contain tweets generated mostly related to riots and unrest. But we use only three datasets in our case study. All the three datasets contain rumors in the form of tweets and their attributes. The three datasets are:

- Charlie Hebdo Shooting
- Sydney Siege
- Ottawa Shooting.

Charlie Hebdo shooting: This dataset contains rumors related to an incident in which two brothers entered into French magazine office in Paris and killed eleven people in 2015. Various rumors are spread in social media like Twitter and one of the rumor is “French terror suspects want to be martyrs”.

Sydney Siege: This dataset contains rumors for Sydney siege also known as Sydney siege hostage crisis. It was a hostage situation when a lone gunman kept people of Sydney as hostages. One example of rumor during this hostage event is “Hostage situation erupts in Sydney cafe, Australian prime minister says it may be politically motivated”.

Ottawa Shooting: This dataset contains rumors related to shooting at parliament hill in Ottawa Canada results in the death of a Canadian soldier in 2014. One example of rumor spread during this event is “FBI assisting in the case of the Ottawa shooting, sources have confirmed to CTV News”.

	created_at	favorite_count	followers_count	statuses_count	text	user_id	ret_count
0	Wed Oct 22 13:58:25 +0000 2014	10	1209	22152	Only photo I will tweet. CPR being performed o...	44685373	8
1	Wed Oct 22 15:38:19 +0000 2014	250	20040953	35650	Ottawa Police Service: There were "numerous gu...	428333	801
2	Wed Oct 22 14:25:19 +0000 2014	45	7000118	87771	Canada's Parliament Hill is on lockdown as pol...	6017542	170
3	Wed Oct 22 14:10:21 +0000 2014	66	365930	25593	Shooter still on loose after uniformed soldier...	19038934	409
4	Wed Oct 22 16:11:09 +0000 2014	134	11834319	21529	Police say shots fired at 3 #Ottawa sites - N...	5402612	567
5	Wed Oct 22 15:20:55 +0000 2014	129	66735	38429	CONFIRMED | 1 shooter shot dead at Parliament ...	18999969	468
6	Wed Oct 22 23:05:35 +0000 2014	232	20038478	35649	The Canadian soldier killed in today's Ottawa ...	428333	338
7	Wed Oct 22 22:59:49 +0000 2014	351	1361	4	Our thoughts and prayers go out to Nathan Ciri...	2326482060	230

Fig. 2 Attributes of the rumor dataset

	id	retweet_time	Dates	Time	Rtime
0	5.249227e+17	Wed Oct 22 16:14:00 +0000 2014	2014-10-22	16:14:00	58440
1	5.249227e+17	Wed Oct 22 16:30:31 +0000 2014	2014-10-22	16:30:31	59431
2	5.249227e+17	Wed Oct 22 16:55:41 +0000 2014	2014-10-22	16:55:41	60941
3	5.249227e+17	Wed Oct 22 16:57:51 +0000 2014	2014-10-22	16:57:51	61071
4	5.249227e+17	Wed Oct 22 17:10:51 +0000 2014	2014-10-22	17:10:51	61851
5	5.249227e+17	Wed Oct 22 17:17:22 +0000 2014	2014-10-22	17:17:22	62242
6	5.249227e+17	Wed Oct 22 18:27:59 +0000 2014	2014-10-22	18:27:59	66479
7	5.249227e+17	Wed Oct 22 18:56:08 +0000 2014	2014-10-22	18:56:08	68168
8	5.249227e+17	Wed Oct 22 18:59:36 +0000 2014	2014-10-22	18:59:36	68376

Fig. 3 Attributes of the rumor dataset

3.1 Data Preprocessing

Each dataset contains metadata tweets in JSON format. First, we convert it into CSV format and perform data cleansing while removing non-English tweets, zero values in the attributes, etc. We also select important attributes of the tweet like Text, Retweet_time, Retweet_count, Follower_count, Statuses_count, etc. which are important for analysis (Figs. 2 and 3).

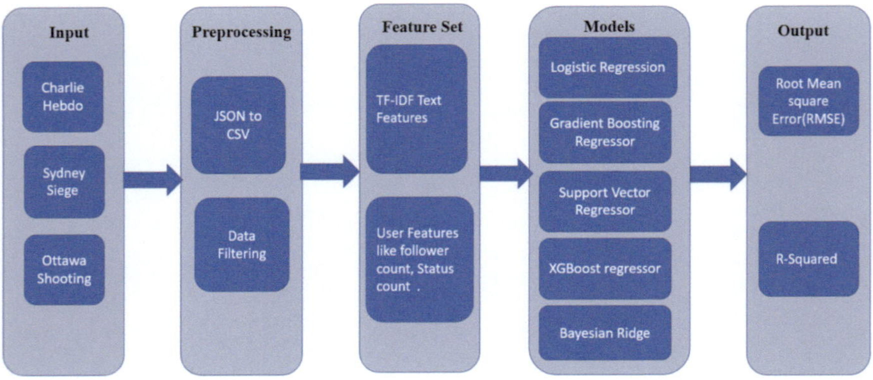

Fig. 4 Schematic representation of proposed work

4 Methodology

In this paper, we present rumor retweet prediction as a regression analysis problem. We focused on applying various five machine learning models like logistic regression, support vector regression, gradient boosting regression, extreme gradient boosting regression (XGBoost) and Bayesian ridge to predicting the scale of misinformation diffusion. We are interested to see up to what extent we can predict the popularity of rumors.

Figure 4 presents an architectural view of our proposed approach in the form of layers. First layer illustrates three rumor datasets used for prediction tasks. Data preprocessing is done in second layer while removing irrelevant data and converting the tweets from JSON to CSV. Third layer is feature set layer comprises of text features, user features and temporal features used for training machine learning models. Fourth layer comprised of machine learning models. Finally, the output of machine learning models is calculated in fifth layer with RMSE and R-squared as the evaluation metrics.

5 Feature Representation

This section discusses several features that are used for training the machine learning models.

5.1 Content Features

A problem with machine learning models is that they cannot work with raw text directly. To process the text data, it must be converted into numbers especially numeric vectors. In our experiments, we use TF-IDF vectorizer for converting tweet text into numerical feature vectors.

Term Frequency(TF): It measures the frequency of words, i.e., a term repeats how many times in a document. As the length of every document is different and it may be the case that in long documents terms have more frequency than shorter documents. To normalize it, we divide it by document length. TF(w) = (Number of times term w appears in a document) / (Total number of terms in the document).

Inverse Document Frequency (IDF): Terms like 'of', 'for', 'is' appears frequently in the document and have less importance. Hence IDF measures the term importance by taking the log of the total number of documents by the total number of documents with term w in it. IDF(w) = $\log_e$ (Total number of documents / Number of documents with term w in it).

5.2 User Features

User features capture the social influence and helps for propagating the information rapidly in the social network like Twitter. We experiment with following most intuitive user features.

- **Statuses_count**: Statuses_count in Twitter shows the activity rate of the user. Higher statuses count shows that a user is more active.
- **Follower_count**: Follower_count shows the popularity of a user. Generally, more influential users have a higher follower count.

5.3 Temporal Features

Temporal features helps us to analyze the tweets from a time series point of view. We use temporal features like Retweet_time and Creation_time.

- **Retweet_time**: The exact time when a particular user makes a retweet for a source tweet.
- **Creation_time**: Time of creation of source tweet.

6 Machine Learning Models for Rumor Popularity

The task of rumor retweet prediction can be carried out with the help of different machine learning models. Each model has some set of parameters used for the training the models. The performance of these models is checked on testing data. We tuned the hyper-parameters of these models with a various set of values and select the optimum out of these for better results.

Logistic Regression (LR): It is a statistical model used for solving classification as well as regression problems. The task of logistic regression is to model a real-valued function over events. The basic equation for logistic regression with multiple independent variables is

$$\text{logit}(r_j) = \left(\frac{r_j}{1 - r_j}\right) = a_j + b_1 x_1 + b_2 x_2 + ...b_q x_q \tag{1}$$

where $x_1, x_2, \ldots, x_p$ are the independent variables, i.e., tweet characteristics and a_j is the intercept, p_j is the dependent variable, i.e., retweet time. We train a logistic regression on the training data with L2 regularization along with parameter C to 1.0.

Gradient boosting Regression (GBR): It is an ensemble-based machine learning technique that uses an ensemble of week models at each step of decision trees and producing better error rate than random guessing. The basic intuition behind gradient boosting is to repetitively leverage the patterns in residuals and strengthen a model with week predictions and make it better. This model forms the additive model forward stage wise. We train a gradient boosting regression on the training data with least absolute deviation as loss function, learning_rate = 0.4, n_estimators = 1000, max_depth = 5.

Support Vector Regression (SVR): It is one of the robust methods of support vector machine used for regression tasks. It is a non-parametric approach relies on Kernel function. We use Radial Basis Kernel in this model along with SVM Meta parameters C and ϵ. The parameter C tells the model that how much you avoid misclassification of training samples whereas ϵ provide a level of accuracy of the approximated function.

Extreme Gradient Boosting Regression (XGBR): It is an optimized distributed gradient boosting model which use gradient boosting framework for both classification and regression tasks. It is very popular for solving machine learning challenges posted by various sites like kaggle. This model also known as regularized gradient boosting because it uses regularized model formulation to control overfitting. We train XGBR on the training data with parameters (max_depth = 7, learning_rate = 0.03, n_estimators = 500, booster = 'gblinear', reg_lambda = 0.05).

Bayesian Ridge (BR): This model is a ridge regression from a Bayesian perspective. This model estimates a probabilistic regression model with a zero prior mean using spherical Gaussian. The basic approach is the introduction of uninformative priors over the hyper-parameters. We train a Bayesian ridge model on the training data with parameters ($\alpha 1 = 0.0001, \alpha 2 = 0.001, \lambda 1 = 0.0001, \lambda 2 = 0.5$).

7 Data Splitting and Validation

Data splitting plays an important role for the performance of models. In our experiments, we partition the dataset into 70:30 ratio, i.e., seventy percent of data used for training purpose and thirty percent of data used for testing purpose. we also performed 10-fold cross-validation on training sets to tune the hyper-parameters of models and compare their performance.

8 Evaluation

We evaluate the models with standard machine learning metrics like root-mean-square error (RMSE), R-squared to compare the accuracy of prediction. These metrics are explained below:

Root-Mean-Square Error (RMSE): RMSE is negatively scored-oriented evaluation metric (i.e., the lower value is better) often used to judge the quality of prediction.

Mathematically, it can be calculated by taking the square root of average of squared distance between predicted values and observed values.

$$\text{RMSE} = \sqrt{\frac{1}{n}\sum_{j=1}^{n}\left(y_j - \hat{y_j}\right)^2} \quad (2)$$

R-squared: R-squared is an intuitive measure of how close our linear model fits a set of observations. It is an estimation of the strength of the relationship between the given model and response variable. It is positively score evaluation metric means higher the R-squared value, the better the model fits your data. Mathematically, it can be calculated as:

$$R = 1 - \frac{\text{Error Sum of Square}}{\text{Total Sum of Square}} = 1 - \frac{\sum\left(y_j - \hat{y_j}\right)^2}{\sum\left(y_j - y\right)^2} \quad (3)$$

Once we instantiate TF-IDF vectorizer, we get to each text feature extracted and fit this TF-IDF transformed data along with user features into a single feature set for training the models for prediction task.

9 Analysis of Results

Frequency of Popular words from the three rumor datasets: In Figs. 5, 6 and 7 we show word clouds, a visualization method displays frequency of words appears

Fig. 5 Word cloud of the tweets related to Charlie Hebdo shooting data

Fig. 6 Word cloud of the tweets related to Sydney siege data

in tweet text of rumor datasets while making the size of each word proportional to its frequency with the purpose of summarizing its content. Figures 8, 9 and 10 shows mean TF-IDF score of text features.

Tables 1 and 2 show the results in the form of root-mean-square error and R-squared as a performance metric. From the simulation results, it is revealed that logistic regression performed better in Charlie Hebdo dataset while extreme gradient boosting regression (XGBR) provide better results on other two rumor datasets. It is concluded from the results that no model performs better on all configuration and datasets and hence on an average extreme gradient boosting regression performs better on rumor datasets (Figs. 11 and 12).

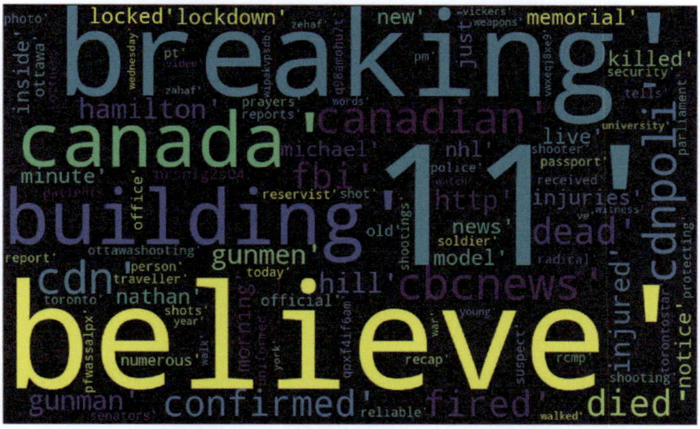

Fig. 7 Word cloud of the tweets related to Ottawa shooting data

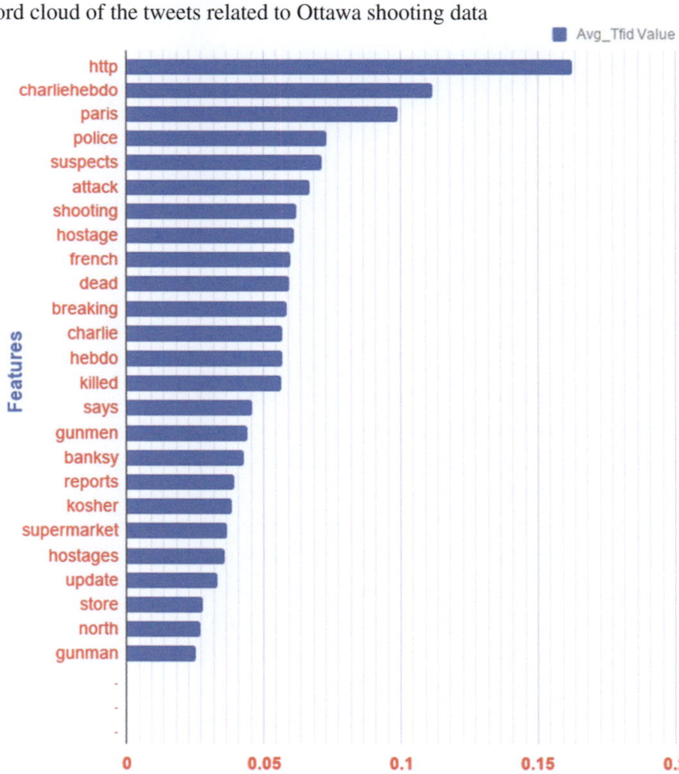

Fig. 8 Mean TF-IDF score of text features for Charlie Hebdo dataset which shows variable importance

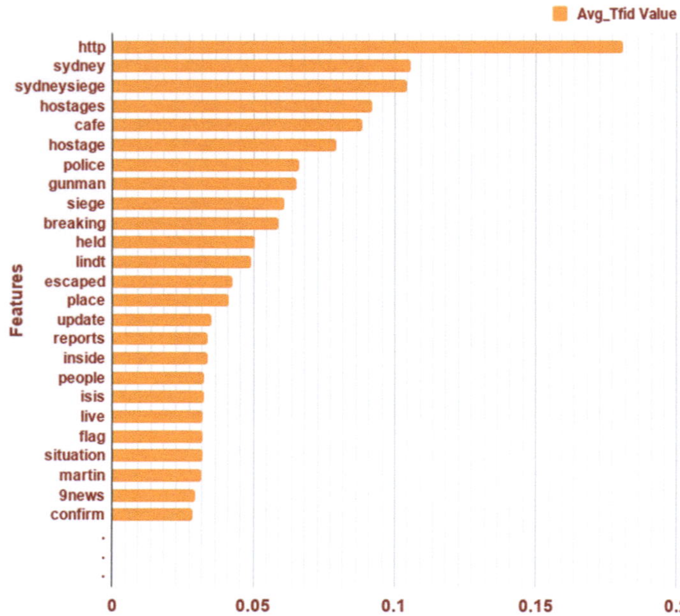

Fig. 9 Mean TF-IDF score of text features for Sydney siege dataset which shows variable importance

10 Conclusions

In the case of a social network like Twitter, predicting the popularity of rumor tweets is quite important for several applications such as viral marketing, disastrous events, election process, and riots. In this work, we analyzed the machine learning models on standard rumor datasets that use tweet text and user features and tries to predict the number of retweets for that rumored tweet. We experimentally tested our approach using three rumor datasets that we collected from PHEME dataset. We extract 100 text-based features from each rumor dataset with the help of TF-ID vectorizer. Thereafter, we take user features like Follower_count, Statuses_count, Retweet_time, Creation_time for training the models.

We took one hour of retweets for a source tweet and apply machine learning models like logistic regression, gradient boosting regression, support vector regression, extreme gradient boosting regression and Bayesian ridge and performed a comprehensive set of experiments to find out the optimal machine learning model. We tuned the hyper-parameters of these models with a various set of values and select the optimum out of these for better results.

Finally, the performance of models is checked by RMSE and R-squared. Hence, we empirically found that on an average XGBoost provide better results for rumor popularity prediction task.

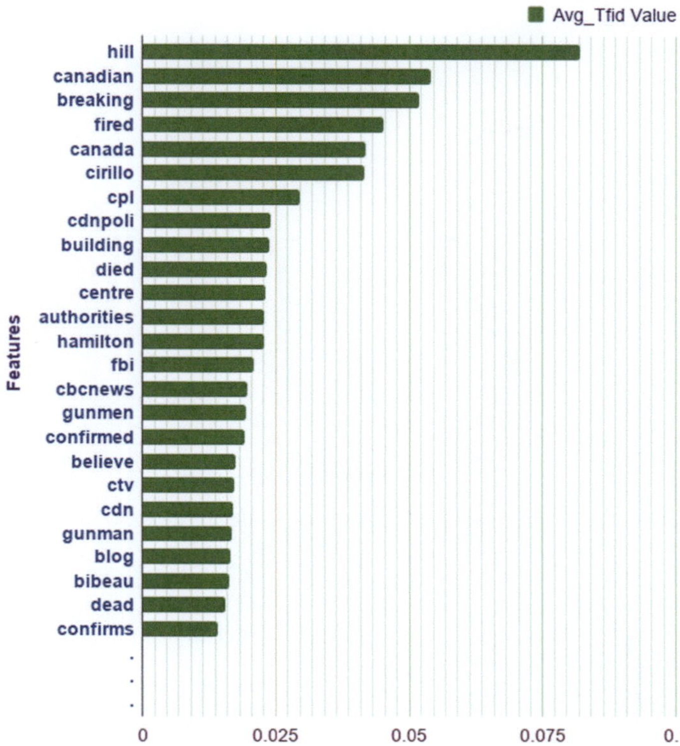

Fig. 10 Mean TF-IDF score of text features for Ottawa shooting dataset which shows variable importance

Table 1 Value of RMSE obtained by various machine learning models on rumor datasets

Rumor datasets	LR	GBR	SVR	XGBR	BR
Charlie Hebdo	**147**	165	177	156	166
Sydney siege	120	**118**	123	**118**	119
Ottawa shooting	228	191	227	**183**	194
Mean	165.0	158.0	175.6	**152.3**	159.6

Bold indicates better results

Table 2 Value of *R*-squared obtained by various machine learning models on rumor datasets

Rumor datasets	LR	GBR	SVR	XGBR	BR
Charlie Hebdo	**0.34**	0.17	0.05	0.26	0.16
Sydney siege	0.04	0.06	−0.006	**0.07**	0.05
Ottawa shooting	−0.05	0.26	−0.05	**0.31**	0.23
Mean	0.11	0.16	−0.002	**0.21**	0.14

Bold indicates better results

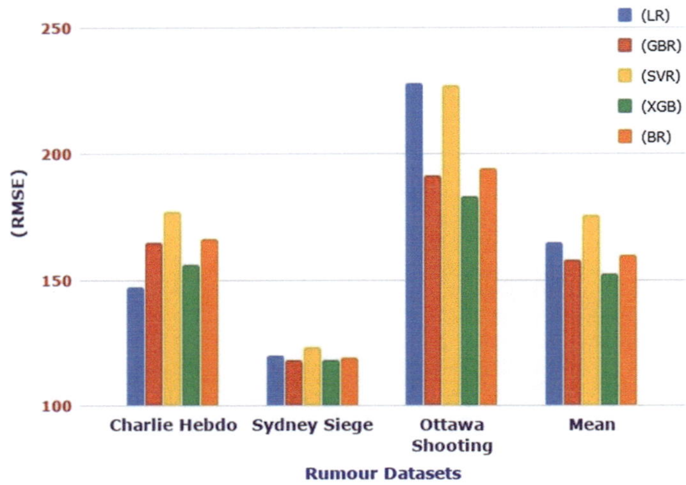

Fig. 11 RMSE value obtained after applying machine learning models

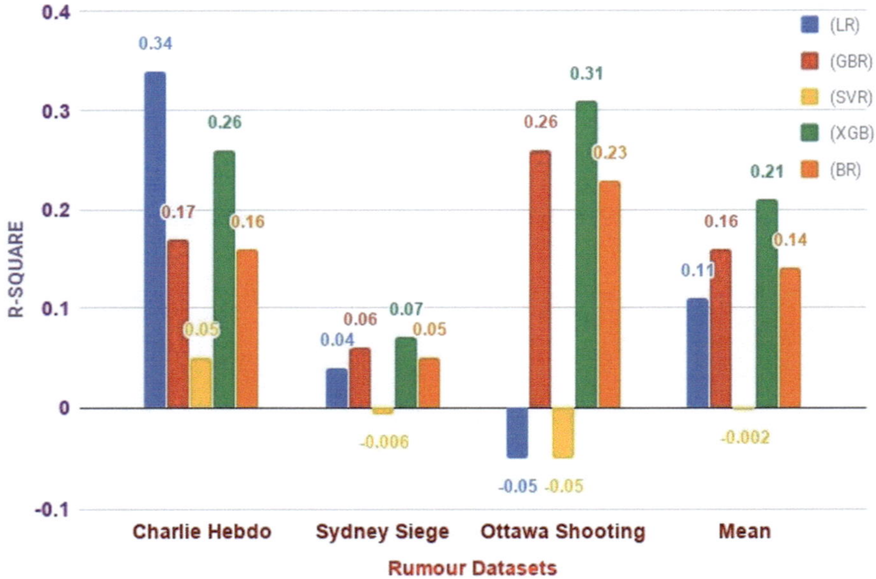

Fig. 12 *R*-squared value obtained after applying machine learning models

11 Future Scope

We make an attempt to predict the popularity of rumors with machine learning approaches but this work can be extended further with advanced models like deep neural networks and point process models. In our experiments, we took one hour of retweets for a rumored tweet. However, it may possible that some rumors spread after hours. Hence an approach would be desirable for handling such a scenario.

Acknowledgements The author thanks Department of Computer Science IIT Hyderabad for a two-month research internship and especially Dr. Srijith P. K. and Uddipta Bhattacharjee for their valuable suggestions.

References

1. Leskovec J, Adamic LA, Huberman BA (2007) The dynamics of viral marketing. ACM Trans Web (TWEB) 1(1):5
2. Mishra S, Rizoiu MA, Xie L (2016) Feature driven and point process approaches for popularity prediction. In: Proceedings of the 25th ACM international on conference on information and knowledge management. ACM, pp 1069–1078
3. Hong L, Dan O, Davison BD (2011) Predicting popular messages in twitter. In: Proceedings of the 20th international conference companion on world wide web. ACM, pp 57–58
4. Gupta A, Kumaraguru P, Castillo C, Meier P (2014) Tweetcred: real-time credibility assessment of content on twitter. In: International conference on social informatics. Springer, Cham, pp 228–243
5. Wu B, Shen H (2015) Analyzing and predicting news popularity on twitter. Int J Inf Manage 35(6):702–711
6. Shafiq Z, Liu A (2017) Cascade size prediction in online social networks. In: IFIP networking conference (IFIP networking) and workshops. IEEE, pp 1–9
7. Zhang Q, Gong Y, Wu J, Huang H, Huang X (2016) Retweet prediction with attention-based deep neural network. In: Proceedings of the 25th ACM international on conference on information and knowledge management. ACM, pp 75–84
8. Kafeza E, Kanavos A, Makris C, Vikatos P (2014) Predicting information diffusion patterns in twitter. In: IFIP international conference on artificial intelligence applications and innovations. Springer, Berlin, pp 79–89
9. Kupavskii A, Ostroumova L, Umnov A, Usachev S, Serdyukov P, Gusev G, Kustarev A (2012) Prediction of retweet cascade size over time. In: Proceedings of the 21st ACM international conference on information and knowledge management. ACM, pp 2335–2338
10. Rudra K, Banerjee S, Ganguly N, Goyal P, Imran M, Mitra P (2016) Summarizing situational tweets in crisis scenario. In: Proceedings of the 27th ACM conference on hypertext and social media. ACM, pp 137–147
11. Zubiaga A, Aker A, Bontcheva K, Liakata M, Procter R (2018) Detection and resolution of rumours in social media: a survey. ACM Comput Surv (CSUR) 51(2):32
12. Kumar N, Ande G, Kumar JS, Singh M (2018) Toward maximizing the visibility of content in social media brand pages: a temporal analysis. Socl Netw Anal Min 8(1):11
13. Zubiaga A, Liakata M, Procter R, Hoi GWS, Tolmie P (2016) Analysing how people orient to and spread rumours in social media by looking at conversational threads. PLoS ONE 11(3):e0150989

Optimizing Memory Space by Removing Duplicate Files Using Similarity Digest Technique

Vedant Sharma, Priyamwada Sharma and Santosh Sahu

Abstract In this paper, we proposed a data cleaning technique, for memory space optimization. We are using sdhash techniques for effective, fast and efficient duplicate files detection and removing in memory. The correct identification of duplicate files is the first critical step in data cleaning process. The fast growth of the data targets demands new automated methods for removing data duplication quickly, accurately, and reliably. Sdhash tool is used for calculation of similarity score of a data files, store, and compare its similarity hashes referred to as similarity digests (sdhash). In contrast, compare whole file, to brute force method, our method compares only the finger prints of all files and is able to efficiently distinguish among duplicate files. In addition, our evaluation data which contains hundreds of files, provides insights into the typical levels of content similarity across related Files. The proposed method is excellent in metric of time and space complexity.

Keywords Data cleaning · Fingerprinting · Ssdeep · Sdhash

1 Introduction

Day by day the usage of internet [1] will get increase because of advent of recent technology and varied online application, technological advancement, social network, on-line promoting, IT environment [2] growth information shared through the network become will increase. This may create tremendous volume of information such as text file, images, audio and video. Because of such type of duplicate files stored on HDD, SSD or in the Cloud, we may run out of memory space which create a great problem for us and may your device [3] gets slowdown in its performance. Hence, it is necessary to find and delete all those files which are having the same

V. Sharma
University Institute of Technology, Rajiv Gandhi Proudyogiki Vishwavidyalaya, Bhopal, Madhya Pradesh, India

P. Sharma (✉) · S. Sahu
School of Information Technology, Rajiv Gandhi Proudyogiki Vishwavidyalaya, Bhopal, Madhya Pradesh, India
e-mail: priyamwada14@gmail.com

R. K. Shukla et al. (eds.), *Social Networking and Computational Intelligence*, Lecture Notes in Networks and Systems 100, https://doi.org/10.1007/978-981-15-2071-6_66

contents. Finding duplicate files manually in your hard drive is not an easy task as it takes a lot of time and effort, and also it could be risky as you may delete an important file by mistake.

Finding duplicate files is a crucial importance for several industries over a good sort of applications [4]. planning to sight the duplicate or about duplicate records that seek advice from the identical real-life entity, duplicate record elimination may be a vital knowledge cleansing task making an attempt to form the info additional concrete and reach higher data quality. The foremost naive technique is to pair-wisely compare all record pairs within the info so as to sight the duplicate records. Obviously, this technique is much impracticable thanks to its intolerable complexness of O (N2), wherever N is that the variety of records within the info.

To lower the time complexity, various techniques have been proposed. We can broadly classify the duplication methods into two major categories, direct content-based searching and indirect signature-based (derived from content) searching. Direct-based method compared every file to other file is called brute force techniques. Something is driven from the content file is called signature-based techniques or approximate matching [5].

2 Related Work

Brute force method [6, 7] of duplication search process is by all-against-all comparison. Every object in the target system is compared to all objects in the reference list. In the operational phase, the examiner compares it to all reference lists, using the comparison function of the approximate matching tool [8]. The best match is the one sharing the higher similarity with the queried object, and if it is above a threshold, the corresponding object is separated. The major drawback of this approach is the time complexity, which is O (n^2), wherever N is that the variety of records within the system.

Hash-based duplication method [8] uses a hashing algorithm to identify "chucks" of data. Commonly used algorithms are SHA-1 and MD5. When file is processed by a hashing algorithm, a hash is created that represent the data. A hash is a bit string that represents the file processed. If you find the same data through the hashing algorithm multiple times, the same hash value is created each time. Compare the new [9] hash value with the existing values; If it exists earlier in the system, then system says, the content of the file is duplicate and it can be removed.

Ssdeep: context triggered [6] piecewise hashing (CTPH) is another methodology going to sight content similarity within the computer memory unit-wise level. The most plan of this tool is to make variable size blocks employing a rolling hash formula to see once blocks begin and stop (set boundaries). The rolling hashing produces a random worth supported a window that moves through the input byte-by-byte. When the primary worth is generated, the following ones are created terribly quickly given the recent hash, the removed a part of the window, and also the new additional one. The formula adopted by ssdeep was impressed by the Adler-32 verification.

Sdhash: One of the most known methods of approximate matching is sdhash [10]. This tool is based on the idea of identifying and picking from an object features that are least likely to occur in other ones by chance and use them to create the digest. It defines feature as a 64 bytes sequence extracted from the input by a fixed-size window that moves byte-by-byte. Technologist entropy is calculated for all extracted options and a collection of rock bottom ones in predefined intervals are selected [11, 12]. These options are hashed victimization SHA-1 and, therefore, the result split into five sub hashes, will not insert the feature into a Bloom filter. Every filter [13, 14] features a limit on the quantity of components (features) which will [15] be inserted thereon, and once it reaches its capability, a brand new one is formed. The ultimate digest could be a sequence of all Bloom filters [16], and its size corresponds to concerning a 2.6% of the input.

3 Proposed Methodology

For removing duplicate files, we tend to used sdhash tool. It produces the similarity digests of every one files and prints them to plain output. Every digest is totally self-contained and is strictly one line of printable American Standard Code for Information Interchange characters. It consists of many header fields separated by semicolons, followed by base 64-encoding of the (binary) digest data. To the purpose of experiments, initially we take dataset of 20 images and out of them, five are duplicate images. To generate the fingerprints of images, we use command > sdhash *.jpg -o testset. The command will save the fingerprints of whole images in a single file testset.sdbf as shown in Fig. 1.

We generate the similarity digest of whole jpg file present in folder in single command. This command produces a file testset.sdbf which contains the similarity digest of all jpg file present in the testset folder.

The procedure we use consists of the following steps:

1. Obtain a set of fingerprints of interested files.
2. Compare the fingerprints by using command > sdhash -c testset.sdbf | sort, shown in Fig. 2.

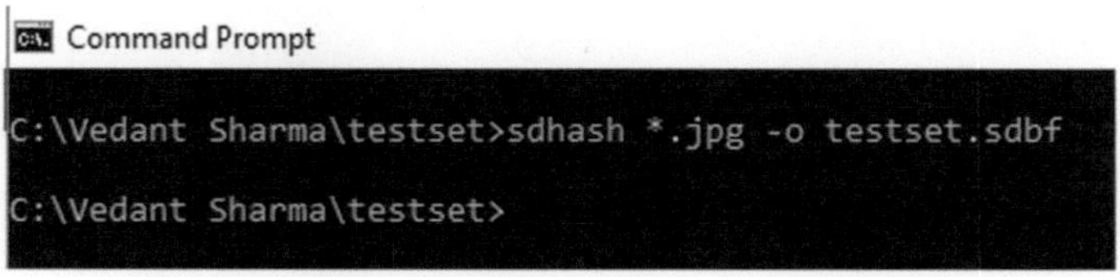

Fig. 1 Generation of fingerprints of image folder testset

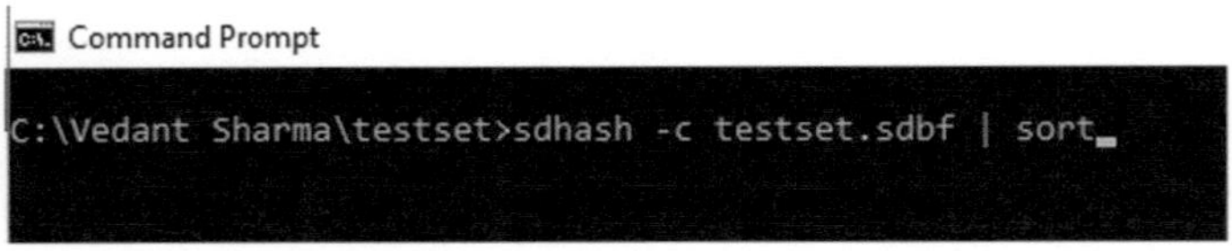

Fig. 2 Comparision command of fingerprints of images

This command compares each image fingerprint with other image fingerprints and gives the similarity score in sorted form. It should compare all unique hash pairs $n \times (n - 1)/2$ where n number of hashes.

Similarity score is shown in Fig. 3. The result of the similarity score is a number between 0 and 100. If *S*-score is 100, then it is a duplicate files. For data cleaning process, we can use a threshold value; if *S*-score becomes more than threshold value then, we can remove such a files. In brute force method of finding duplicate files, we required to compare whole file to the other files rather than comparing the fingerprints.

4 Result Analysis

We have analyzed the experiment with two metric time and space complexity. Detail analyses are given below.

4.1 Time Complexity

The times required for brute force method and similarity digest method is shown in Fig. 4. As shown in the figure, time required for brute force method is much higher than the similarity digest method. Minimum time required for text files and maximum time required for video files as the size of video file is more. With the help of *S*-score, we can find a duplicate file in a less time than brute force methods.

4.2 Space Complexity

The space required for brute force method and similarity digest method is shown in Fig. 5. As shown in the figure, space require for brute force method is much higher than the similarity digest method. Minimum space required for text files and maximum space required for video files as the size of video file is more than text file. With the help of *S*-score, we can find a duplicate files in a less space than brute force methods.

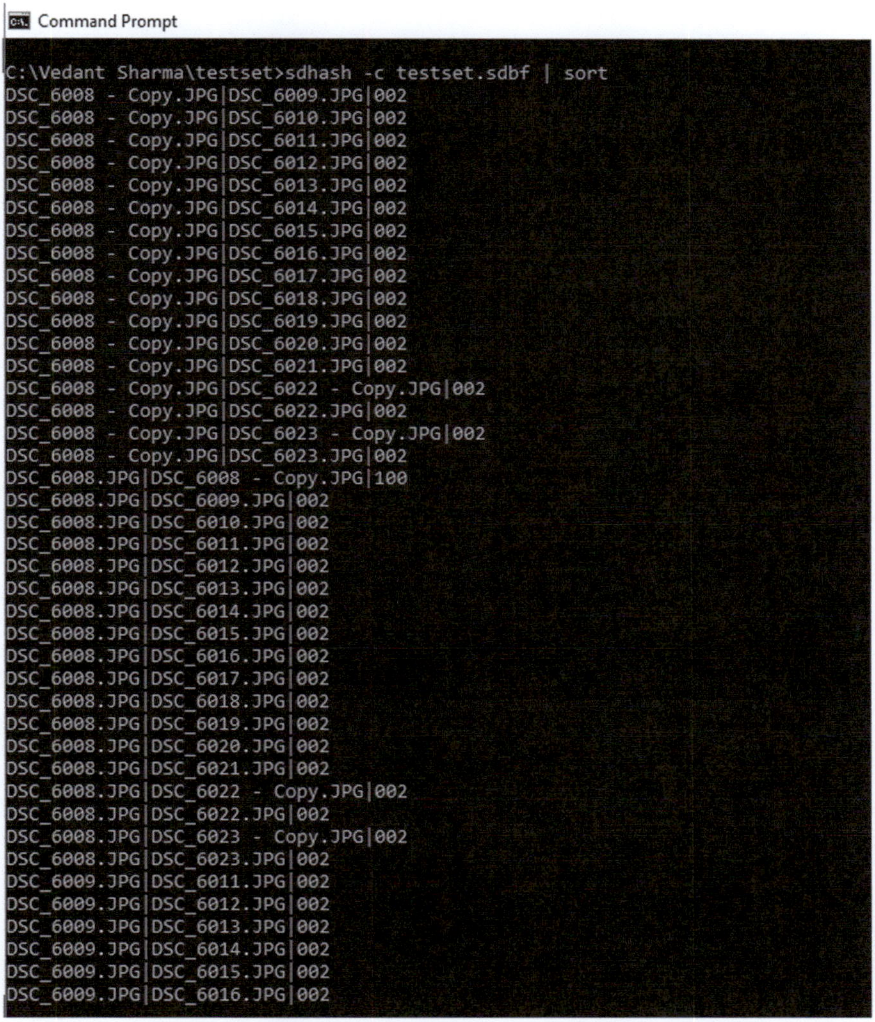

Fig. 3 Result of comparison of fingerprints of images

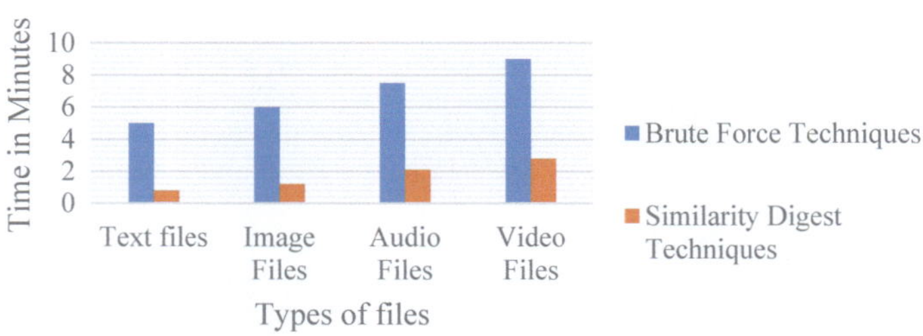

Fig. 4 Time required for removing duplicate files

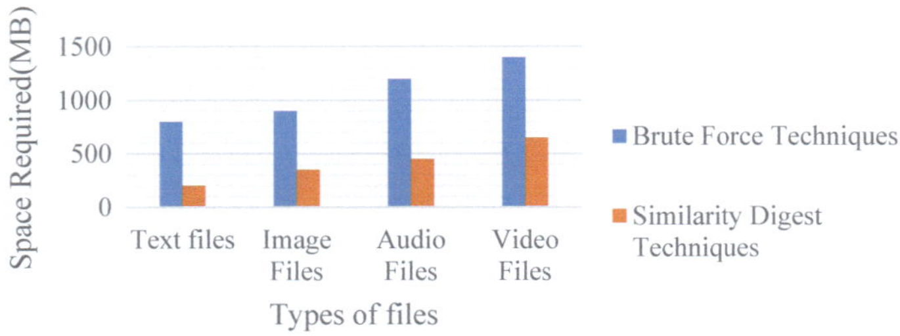

Fig. 5 Space required for removing duplicate files

5 Conclusion

In this paper, we propose a new duplication detection method called similarity digest method. Compared with its previous version, brute force methods have two major new features. It has two steps for removing data duplication methods. We perform experimental evaluations in the performance of time and space complexity. We performed a similarity digest of several types of files result show that our method is more efficient, fast and reliable compare to brute force method. We developed a practical approach to building and optimizing different file signatures by utilizing similarity digests. The signature is derived from the whole set of files. The method affords excellent time and space complexity. Unlike previous brute force method, which relies heavily number of comparison on all possible combination of files. In all cases, sdhash demonstrated accuracy and scalability with respect to target sizes. Signature-based sdhash demonstrated the better performance for all four types of files.

References

1. Hernandez MA, Stolfo SJ (1995) The merge/purge problem for large databases. In: Proceedings of the ACM SIGMOD'95. San Jose, CA, pp 127–138
2. Weis M, Naumann F (2004) Detecting duplicate objects in XML documents. In: Proceedings of IQIS'04. Paris, France, pp 10–19
3. Zhang J, Ling TW, Bruckner RM, Liu H (2004) PC-filter: a robust filtering technique for duplicate record detection in large databases. In: DEXA'04, Zaragoza, Spain
4. Zhang J (2010) An efficient and effective duplication detection method in large database applications. In: 2010 fourth international conference on network and system security. IEEE, pp 494–501
5. Breitinger F, Guttman B, McCarrin M, Roussev V, Approximate matching: definition and terminology. http://csrc.nist.gov/publications/drafts/800–168/sp800_168_draft.pdf
6. Roussev V, Ahmed I, Sires T (2014) Image-based kernel fingerprinting. In: Digital forensics research workshop. Elsevier Ltd
7. Lin Z, Rhee J, Zhang X, Xu D, Jiang X (2011) Siggraph: brute force scanning of kernel data structure instances using graph-based signatures. NDSS. http://dx.doi.org/10.1.1.188.7659

8. Bjelland PC, Franke K, Arnes A (2014) Practical use of approximate hash based matching in digital investigations. Digit Invest 11(1):s18–s26
9. Ranjithaa S, Sudhakara P, Seetharamanb KS (2016) A novel and efficient de-duplication system for HDFS. In: 2nd international conference on intelligent computing, communication & convergence. ICCC-2016
10. Moia VHG, Henriques MAA (2017) Similarity digest search: a survey and comparative analysis of strategies to perform known file filtering using approximate matching. In: Security and communication networks
11. Roussev V, Quates C (2012). Content triage with similarity digests: the m57 case study. In: Proceedings of the 12th annual digital forensics research conference, S60e8. http://dx.doi.org/10.1016/j.diin.2012.05.012
12. Roussev V (2010) Data fingerprinting with similarity digests. In: Advances in digital forensics VI. Springer, pp 207–226
13. Roussev V (2009) Building a better similarity trap with statistically improbable features. In: Proceedings of the 42nd Hawaii international conference on system sciences. Waikoloa Village Resort. IEEE, Hawaii, HI
14. Roussev V (2012) Managing terabyte-scale investigations with similarity digests. In: Advances in digital forensics VIII. Springer, pp 19–34
15. Roussev V (2010) Data fingerprinting with similarity digests. In: Chow K-P, Shenoi S (eds) Advances in digital forensics VI, IFIP AICT, vol 337, pp 207–225
16. Roussev V (2011) An evaluation of forensic similarity hashes. In: The proceedings of the digital forensic research conference DFRWS 2011, USA

Sentiment Analysis to Recognize Emotional Distress Through Facebook Status Updates

Swarnangini Sinha, Kanak Saxena and Nisheeth Joshi

Abstract Youngsters during their adolescent experience various changes in their life. Some youngsters could not handle pressure created due to these changes and tend to feel isolated and distressed. Social media is a popular medium of communication amongst youngsters to remain connected with their friends. Facebook is one of the most preferred social media sites which stores the gigantic amount of data which can be explored for sentiment analysis. We have studied Facebook status updates in English and Hindi to determine the degree of emotional distress using bag-of-words model enhanced with weighted affect words, negation, discourse relation, emoticons, and punctuation marks. SVM classifier has given a promising result with an accuracy of 90.49%.

Keywords Sentiment analysis · Affect · Social networking sites · Bag of words

1 Introduction

Sentiment analysis is the process of recognizing emotions specified in textual information [1]. These emotions can be extracted at different levels ranging from word, sentence to document. Text mining is predominantly used in sentiment analysis. Nowadays, enormous information is available on social networking sites due to the penetration of Internet amongst 3.5 billion users' worldwide.[1] People use social networking sites as the most popular and fast communication media to remain in touch

[1] https://www.brandwatch.com/blog/96-amazing-social-media-statistics-and-facts-for-2016/.

S. Sinha (✉) · N. Joshi
Department of Computer Science and Engineering, Banasthali Vidyapith, Vanasthali, Rajasthan, India
e-mail: swarnangini@gmail.com

N. Joshi
e-mail: jnisheeth@banasthali.in

K. Saxena
Department of Computer Application, Samrat Ashok Technological Institute, Vidisha, India
e-mail: ks.pub.2011@gmail.com

R. K. Shukla et al. (eds.), *Social Networking and Computational Intelligence*, Lecture Notes in Networks and Systems 100, https://doi.org/10.1007/978-981-15-2071-6_67

with their friends, family, peer, or the world as a whole. These sites play a vital role in spreading mass opinion; hence, they can be used to build the positive public opinion about different societal issues [2]. There are 3.03 billion active social media users (see Footnote 1), most of which range in the age group of 18–49 years who share their emotions, photographs, daily activities, chats, opinions about products; political parties; social issues, movies, and many more. Sentiments spread through social media are contagious which can be used as a tool for the well-being of mankind [3–5].

Facebook is the most preferred social networking sites with a share of 2.072 billion users (see Footnote 1) across the globe. Eighty-eight per cent users of Facebook belong to 18–29 years of age.[2] It is generally considered as the network of friends and family members. Thus, people use this medium to share their thoughts, emotions, achievements, and events taking place in their day-to-day life in an informal manner. Youngsters are the main users of social media; hence, we can use these posts to detect their emotional distress.

1.1 Facebook Status Updates

Facebook status updates exhibit peculiar characteristic features as stated under.

1. The term 'status update' suggests giving information about oneself. So it is best used as the medium for self-expression [6].
2. The language is informal without strictly following grammatical rules of any language [6].
3. Messages are conveyed using keywords, abbreviations, and emoticons.
4. The status updates are short and constitute maximum 420 words.
5. Status updates are retained on the timeline for a longer period.
6. Status updates shared by the users are accessible to only their friends unless profiles are shared publically. Hence, others cannot comment or like their posts.
7. Mostly there is no hesitation amongst friends to share their emotions.
8. There is no parental control.
9. The emotions shared are contagious whether good or bad.
10. Youngsters feel more connected with friends while conversing through Facebook [7].
11. Psychological state of a youngster and his peer can be predicted through these posts [8].

[2]https://sproutsocial.com/insights/new-social-media-demographics/.

1.2 Objectives

Adolescence is the most sensitive phase of life when youngsters experience biological changes and become ready to face socio-economic pressure. Most of them cope up with this pressure and lead a normal life. But there are some youngsters who cannot deal with this pressure and become either anti-social or tend to get involved in activities to harm themselves [9, 10]. This adolescent pressure, if not dealt properly, could lead to loneliness, sadness, depression, and sometimes suicide. To overcome this grave problem of the society and to prevent youngsters from taking such drastic steps, we have decided to give our contribution in the form of the following.

1. Identify basic emotions of Facebook status updates written in English and Hindi.
2. Enhance bag-of-words model by incorporating fine-gained affect analysis taking into account discourse relation combined with weighted affect word evaluation.
3. Identify sentiment polarity.
4. Calculate the degree of emotional distress.

The paper is organized as introduction stating the importance of the problem statement followed by Sect. 2 depicting the state- of- the- art of sentiment analysis using bag-of-words (BOW) model. Section 3 describes the methodology used, and Sect. 4 gives information about experimental setup. Section 5 represents results obtained followed by a conclusion.

2 Present Research

Most of the studies for sentiment analysis have been carried out using BOW model [11]. BOW model treats the entire document as a collection of words without considering a word order or any grammatical rule. There is no relation between words, and they are considered as independent of each other. The main objective of this model is to identify the positive or negative polarity of the document based on either word presence or its frequency. The analysis is carried out at different levels like document level, sentence level, or word level [12]. The analysis performed using BOW model can be further enhanced by taking into account the fine-grained aspects of emotions expressed through these informal textual posts. This includes handling the following features of text data carefully.

1. Identifying domain-specific keywords: The accuracy of classifier is greatly influenced by the context in which words are used in the sentence. Sentiment words change with respect to domain. Hence taking this into consideration, only domain-specific words are selected.
2. Negation handling: When negation occurs in a sentence, it mainly influences the original meaning of positive or negative emotion words by inverting their polarities [13–15].

3. Double negation handling: It is observed that if negation is used more than once in a sentence, then it invalidates the effect of negation on emotion words. The emotion words in such sentences are generally adjectives or adverbs [16].
4. Intensifiers and diminishers: They increase or decrease the polarities of negative or positive affect words. They do not have their own sentiment orientation, but their presence strongly conveys the sentiments which they are associated with. They never invert polarities of the affect words [17, 18].
5. Effects of conjunctives: Conjunctives are used to connect words, clauses, or sentences. They provide meaningful information about the sentence. The presence of conjunction in a sentence makes the calculation of polarity difficult. When it appears in a sentence, we need to find which part of the sentence contributes more to the final emotional polarity of the sentence [19, 20].
6. Punctuation marks: The punctuations like exclamation mark and question mark are used to further increase or decrease the strength of the emotion expressed. An exclamation mark used in a sentence conveys strong emotions such as surprise, astonishment, and any other such emotions. It adds additional emphasis to the emotion expressed. In contrast, question mark indicates confusion.
7. Use of emoticons: The emoticons are the most popular and strong way of expressing sentiments on social media platform. Youngsters prefer to express their emotions with the help of emoticons than to follow traditional way of text messaging.
8. Inclusion of slangs and abbreviations: They represent the language of Internet which is mainly used in social media.

3 Proposed Methodology

Before performing sentiment analysis on these posts, the key features of these status updates are studied thoroughly.

3.1 Key Features of Status Updates

Post-category: Status updates are self-expressive and mainly give information about a youngster himself. Youngsters who are low in self-esteem are vocal about their love life by expressing emotions of love in the form of a poem called Shayari (short love poetry written in Urdu), pictures depicting love, love quotes, etc. [21]. Some are self-obsessed who update their photographs 3–4 times a day or even more.
Sentiment Category: Sentiments expressed include both positive and negative emotions. On social networking sites, positive sentiments are prevailing than negative sentiments [22]. Negative emotions are articulated in the form of anger, sadness, disgust, loneliness, and stress.

Contents: Youngsters experience biological, physiological, and socio-economical changes in this significant and sensitive phase of their life. They mainly post status updates in the form of photographs, achievements, love life, crush, girlfriends, boyfriends, breakups, friends, and activities happening in their daily life (listening to music, watching movies, visiting different places, picnic with friends, etc.) which focus on them as individuals. A few youngsters post about school or college life.

Length of Content: Youngsters with negative sentiments post status updates which are lengthy as compared to others. The average length of posts is 48 words.

Frequency of Posts: Generally, frequency of posts depends upon the personal interest and the amount of time spent by the user on Facebook. On an average, they update status 2–3 times a day. But it is observed that youngsters going through emotional distress tend to post more as compared to others. More posts are found around festivals in India like Eid, Raksha Bandhan, Ganesh Chaturthi, and Janmashtami, special days like Friendship Day, Independence Day, etc., indicating sharing of the positive emotions amongst friends. Similarly, there were more posts on the death of former President Dr. APJ Abdul Kalam expressing grief.

Time of posts: It is observed that young people who are experiencing some psychological problem in their life prefer to update their status at night until midnight. This may be because they want to be aloof while uploading their posts. Others, on the contrary, upload their posts at any time of the day mainly when they are either not in college, school, or at the workplace.

Use of Language: They generally use slag language with grammatically incorrect and informal texts. Lots of abbreviations and emoticons are used to become more expressive in conveying their sentiments.

Gender: Males update status more frequently than females. Males are more expressive about their love life than females. Females post mainly about their friends, family, fashion trends, positive quotations, etc. They usually receive or write more positive posts than negative posts.

Working Status: Youngsters, who are working, spend less time on Facebook than the one who is studying. Status is updated repeatedly by students because they are eager to inform about their life's happenings to their friends.

3.2 Architecture

Sentiment analysis of these status updates is carried out in five phases which are depicted in the architecture given below in Fig. 1.

3.2.1 Preprocessing

Status updates which are in the form of noisy data are subjected to preprocessing which include converting text into lowercase, spelling corrections, removal of

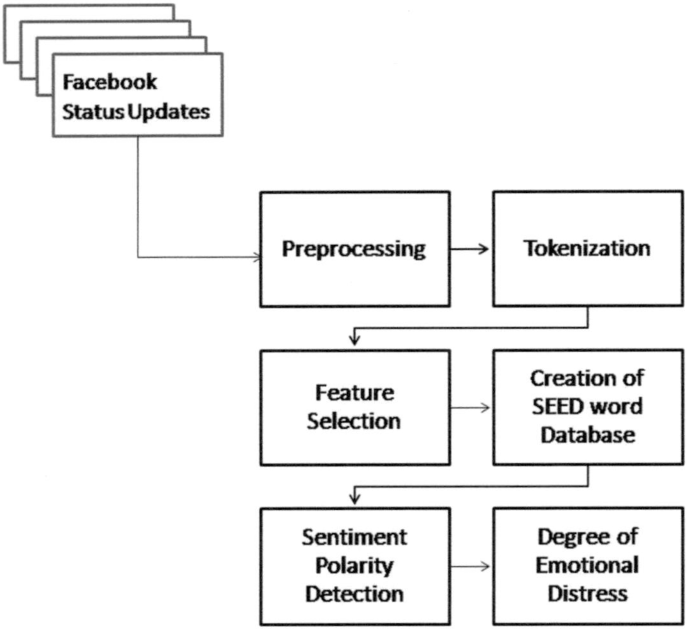

Fig. 1 Architecture of proposed method

special symbols except exclamation and question mark which convey strong emotions as studied by Xue et al. [21], removal of unnecessary spaces, removing URL links and removing stop words. Preprocessed data gives better results of sentiment identification than the noisy data.

3.2.2 Tokenization

The entire textual data is converted into a matrix of words. The white spaces are used as the separators. This helps in separating and identifying stop words, affect words, punctuation marks, emoticons, and any other special symbol present in a sentence.

3.2.3 Feature Selection

The features depicting sentiments are selected from the tokenized data. These features are basically categorized as positive and negative sentiments. Sentiment polarity and weight associated with each feature word is assigned using SentiWordNet[3] and Hindi SentiWordnet.[4] The core of accurate sentiment polarity depends on the extraction of

[3]sentiwordnet.isti.cnr.it/.

[4]http://www.cfilt.iitb.ac.in/wordnet/webhwn/downloaderInfo.php.

sentiment-oriented words called features from this informal textual data. Their value can be stored as a binary number representing the presence or absence of sentiment in the feature vector, or it can be stored as an integer or decimal value used to depict intensity of sentiment in the text. Better the selection of sentiment word accurate would be the results of the sentiment polarity.

3.3 Seed Word Creation

According to [22] model, human sentiments are basically categorized into six broad categories like anger, sadness, fear, love, joy, and surprise. Based on these basic emotions, synonyms and antonyms are collected along with their polarity value. Apart from those words, adverbs, adjectives, intensifiers, and diminishers depicting the emotions are also incorporated in the database. A list of emoticons (see Footnote 4) with their corresponding score of negative or positive sentiments is also created.

3.4 Sentiment Polarity Detection

The proposed methodology detects polarity value of each sentiment word by locating it in the seed word database and its corresponding weight is assigned to it. While deciding the sentiment polarity of a sentence, various criteria are taken into consideration.

1. Negation handling: The weights of emotion words and the part of speech tags are taken from SentiWordNet. We are not using POS tagger. The sentiment score of a sentence, I *love* you is 0.625 and I *don't love* you is calculated as −0.625.
2. Double negation handling: The sentiment polarity of the sentence, I *don't* say I *don't* love you is positive because of double negation which restores the original sentiment of a word 'love'. Otherwise, the sentence would have been interpreted as negative.
3. Intensifiers and diminishers: In a sentence, jyaada pyaar (more love), the word 'jyaada' increases the intensity of the word 'pyaar'. Thus, it conveys a strong positive emotion of love. On the contrary, thoda pyaar (less love) decreases the strength of positive emotion of love, thereby affecting the intensity of sentiment expressed.
4. Effects of conjunctives: In a sentence, I like you *but* I don't love you. The first sentence depicts the positive strength of emotion unlike the second sentence. But the sentiment polarity of this sentence is finalized on the basis of second sentence because of the presence of conjunctive 'but'.
5. Punctuation marks: An exclamation mark used in a sentence 'ham jaan de dete hain magar jaane nahin dete !!' (We give life but do not let go!!) indicates strong negative emotion. In contrast, question mark used in a sentence 'maine apni

zindgi mein bahut dhoke khaye h ???????' (I have a lot of deception in my life) indicates strong negative emotion.
6. Use of emoticons: The emoticons are the most popular and strong way of expressing sentiments on social media platform. They prefer to express their emotions with the help of emoticons than to follow traditional way of text messaging. The weights associated with these emoticons are taken from emoji sentiment ranking.[5]
7. Inclusion of slangs and abbreviations: They represent language of Internet which is mainly used in social media. The original word for the slang used is searched in the database and finally taken into account.

3.4.1 Degree of Emotional Distress

The frequency of negative and positive affect words in a sentence is used to decide the sentiment polarity of a sentence. The aggregate score of all affect words in a sentence and consecutively in a document is calculated using their weights which help to find emotional distress.

4 Experimental Setup

4731 Facebook status updates of 100 youngsters are collected for the period of three months from July 2015 to September 2015. There are total 4731 sentences and 13,013 words. The standard bag-of-words model with enhanced word-level polarity detection method is applied on these status updates.

4.1 Calculation of Degree of Emotional Distress

Negation: Sentiment score of affect word is inverted before adding to final score.
Double negation: Sentiment score of affect word is taken into account as given in the database.
Intensifiers and diminishers: Their respective scores are added to aggregate score to increase or decrease the score.
Conjunctives: Depending upon the conjunctive used, corresponding weights of affect words are used.
Punctuation marks: For each exclamation or question mark, 0.1 is added to aggregate score.
Emoticons: The weight of an emoticon used in a sentence is added to final score.

[5]kt.ijs.si/data/Emoji_sentiment_ranking/.

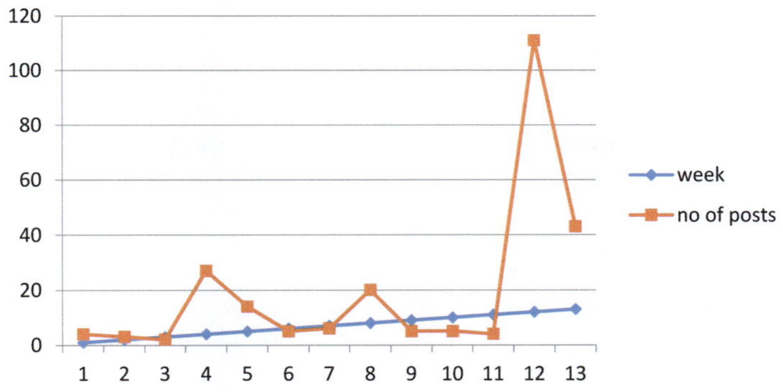

Fig. 2 Weekly frequency of posts

Slang: The weight of original word for the corresponding slag word used is taken into consideration.

Hence, the aggregate score of emotional distress depends on the sum of weights of all affect words found in a sentence coupled with discourse relation.

4.2 Impact of Frequency

The frequency of posts has crucial impact on the emotional distress of a youngster. Generally, a pattern is followed by an individual when updating his Facebook status. A change in frequency is sometimes affected by the external events taking place in his surroundings in the form of festivals, birthdays, and special days like friendship day, valentine day, etc., which are eagerly awaited and celebrated by youngsters. This change in frequency is treated as normal when it is governed by the external events.

Because it is reflected in the status updates of other people also. But sudden high or low post frequency without being affected by the external events definitely demands our attention. The graph shown in Fig. 2 is the representation of number of posts updated every week by one of the youngsters. Total 13 weeks were taken into consideration from July 2015 to September 2015. The graph shows sudden growth in frequency of posts which indicates that he might have been undergoing some psychological pressure.

4.3 Impact of Time

There is no specific time to post status update, but youngsters prefer to update their status during their free time. The time duration of 24 h is divided into four quarters of 6 h each, and posts were grouped in these quarters. The graph shown in Fig. 3 depicts

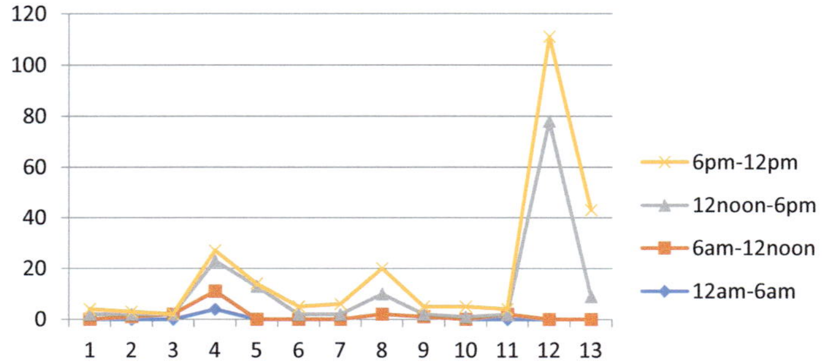

Fig. 3 Time of posts

time usually followed by a college going student who used to update his status after his college was over.

4.4 Impact of Affect Expressed

In Fig. 4, the graph represents how different emotions were expressed through status updates from week 1 to week 13. Generally, emotions of love are associated with joy and they are rarely co-associated with emotions of anger, sadness, and fear. But from fourth week onwards, we can see emotions of sadness were also present with love and joy which might have been due to some problem in his love life or some other circumstances which might have made him sad. Towards 12th and 13th week, there was sudden eruption of all the different emotions simultaneously which indicates that something was not good in his life and needs to be tackled carefully.

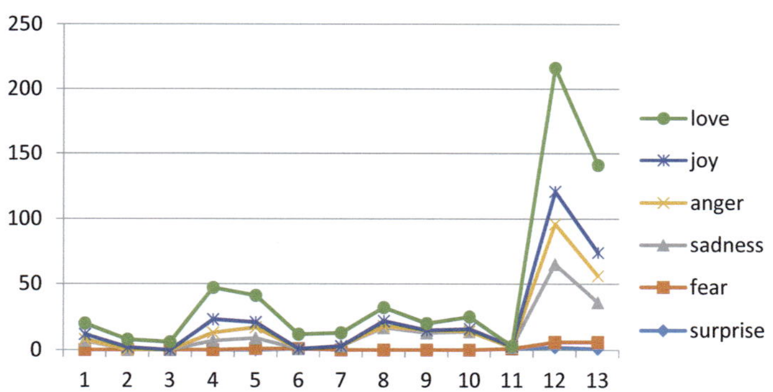

Fig. 4 Expression of affect

5 Classification Algorithms

In order to perform our experiment, we used Weka machine learning toolkit, version 3.8. The dataset is divided into training dataset with 60% data and test dataset with 40% data. The training dataset is trained using different classifiers, and results are tested on test dataset for predicting the outputs. Naïve Bayes, random tree, and support vector machine (SMO) classification algorithms with two parameters were used to check the polarity of the status updates. The classifiers were trained with parameters like the presence or absence of affect words depicting six different emotions and their frequencies.

Naïve Bayes is a simple probabilistic algorithm which is based on Bayes theorem. It treats each attribute as independent value. It is simple to implement and can be applied on small training dataset. The algorithm envisages the probability that a given record can be classified into a particular class with the highest probability.

Random forest algorithm can be used for both classification and regression. It randomly creates decision tree using the test attributes by applying some rules. The tree predicts the outcome and retains the predicted outcome. The votes for each predicted outcome are then calculated. And the final prediction from the forest of trees is selected depending on the high vote of the predicted outcome.

Support vector machine is a dominant classification algorithm. It treats the dataset as the points plotted in space. They are expected to be separated by sufficient space. It calculates a maximum margin hyperplane which divides the data points into two classes.

The weights assigned to affect words and their discourse relations in the sentence were used to find the level of emotional distress. The degree of emotional distress was further divided as low, medium, and high.

6 Results

The results obtained after applying classification algorithms with the presence or absence of affect words or frequencies are depicted in Table 1.

Naïve Bayes classifier when used to interpret the polarity of a sentence based on the presence or absence of affect words, we could get only 60% accuracy and random forest gave 57.26% accurate result while SVM has shown the lowest performance. Because mere presence or absence does not give the actual strength of the sentiment

Table 1 Comparative analysis of classifiers

Classifier	Presence or absence of affect words	Frequency of affect words
Naïve Bayes	60.17	73.96
Random forest	57.26	87.19
SVM (SMO)	51.86	88.42

expressed in the sentence. These results were improved when used with frequency of affect words in a sentence.

The results are evaluated based on precision, recall, and accuracy. The precision represents the fraction of correctly classified instances with the total instances of a class. While recall represents the percentage of predictions for positive instances, the accuracy gives the information about percentage of results calculated correctly.

SVM algorithm with 91.49% accuracy, 0.903 precision rate, and 0.915 recall performed well in determining the level of emotional distress.

7 Conclusion

Facebook as popularly known as is a network of friends. It is widely used by youngsters to remain connected with their friends. We have used Facebook status updates to identify affect expressed by young people. We found that simple presence or absence of affect words or their frequencies do not give the desired results. Thus, we have used weights of affect words coupled with their discourse relation in the sentence to get more accurate results. It was observed that SVM algorithm gave the most effective outcome in determining the level of emotional distress.

References

1. Liu B (2015) Sentiment analysis and subjectivity. Handbook of NLP, 2nd edn. Cambridge University Press
2. Cheong F, Cheong C (2011) Social media data mining: a social network analysis of tweets during the 2010–2011 Australian floods. In: 15th Pacific Asia conference on information systems (PACIS), Brisbane, Australia, 7–11 July 2011, pp 1–16
3. Kramer ADI, Chung K (2011) Dimensions of self-expression in Facebook status updates. In: Proceedings of the fifth international association for the advancement of artificial intelligence conference on weblogs and social media, pp 169–176
4. Ortigosa A, Martin JM, Carro RM (2014) Sentiment analysis in Facebook and its application to e-learning. Comput Hum Behav 31:527–541
5. Bollen J, Goncalves B, Ruan G, Mao H (2011) Happiness is assortative in online social networks. Artifi Life 17:237–251
6. Spear LP (2000) The adolescent brain and age-related behavioral manifestations. Neurosci Biobehav Rev 24:417–463
7. Yessenov K, Misailoovic S (2009) Sentiment analysis of movie review comments. 6.863 Spring 2009 final report
8. El-Din DM (2016) Enhancement bag-of words model for solving the challenges of sentiment analysis. Int J Adv Comput Sci Appl 7:244–252
9. Whitelaw C, Garg N, Argamon S (2005) Using appraisal groups for sentiment analysis. In: CIKM'05 proceedings of the 14th ACM international conference on information and knowledge management, Bremen, Germany, 31 Oct–05 Nov 2005, pp 625–631
10. Asmi A, Ishaya T (2012) Negation identification and calculation in sentiment analysis. In: IMMM 2012: the second international conference on advances in information mining and management, pp 1–7

11. Farooq U, Mansoor H, Nongaillard A, Ouzrout Y, Muhammad AQ (2017) Negation handling in sentiment analysis at sentence level. J Comput 12(5):470–478
12. Hogenboom A, Iterson PV, Heerschop B, Frasincar F, Kaymak U (2011) Determining negation scope and strength in sentiment analysis. In: IEEE international conference on systems, man, and cybernetics (SMC),Anchorage, AK, USA, 9–12 Oct. 2011
13. Rzepka R, Takizawa M, Araki K (2016) Emotion prediction system for japanese language considering compound sentences, double negatives and adverbs. In: Conference proceedings of language on computers IJCAI 2016 workshop, New York City, New York, Jul–2016
14. Dragut E, Fellbaum C (2014) The role of adverbs in sentiment analysis. In: Proceedings of France semantics in NLP: a workshop in honor of Chuck Fillmore (1929–2014), Baltimore, Maryland, USA, 27 June 2014, pp 38–41
15. Dadvar M, Hauff C, de Jong F (2011) Scope of negation detection in sentiment analysis. In: Proceedings of the Dutch-Belgian information retrieval workshop, DIR 2011, University of Amsterdam, Amsterdam, pp 16–20
16. Farooq U, Muhammad AQ (2013) Product reputation evaluation: the impact of conjunction on sentiment analysis. In: Proceedings of the 7th international conference on software, knowledge, information management and applications (SKIMA'2013), Chiang-Mai, China, Dec 2013, pp 590–602
17. Mukherjee S, Bhattacharyya P (2012) Sentiment analysis in twitter with lightweight discourse analysis. In: Proceedings of the 24th international conference on computational linguistics, technical papers, pp 1847–1864
18. Thelwall M (2015) Data mining emotion in social network communication: gender differences in MySpace. J Am Soc Inform Sci Technol 61:190–199
19. Kramer ADI (2010) An unobtrusive behavioral model of "gross national happiness". In: CHI' 10 proceedings of the SIGCHI conference on human factors in computing systems, Atlanta, Georgia, USA, 10–15 Apr 2010, pp 287–290
20. Marshall TC, Lefringhausen K, Ferenczi N (2015) The big five, self-esteem, and narcissism as predictors of the topics people write about in Facebook status updates. Pers Individ Differ 85:35–40
21. Xue Y, Li Q, Jin L, Feng L, Clifton DA, Clifford GD (2014) Detecting adolescent psychological pressures from micro-blog. In: International conference on health information science. HIS, pp 83–94
22. Shaver P, Schwartz J, Kirson D, O'Connor C (2001) Emotional knowledge: further exploration of a prototype approach. In: Parrott G (ed) Emotions in social psychology: essential readings. Psychology Press, Philadelphia, pp 26–56

MIX
Papier aus verantwortungsvollen Quellen
Paper from responsible sources
FSC® C105338
FSC
www.fsc.org

If you have any concerns about our products,
you can contact us on
ProductSafety@springernature.com

In case Publisher is established outside the EU,
the EU authorized representative is:
Springer Nature Customer Service Center GmbH
Europaplatz 3, 69115 Heidelberg, Germany

Printed by Libri Plureos GmbH
in Hamburg, Germany